# NUMERICAL ANALYSIS

# NUMERICAL ANALYSIS

## VITHAL A. PATEL
Humboldt State University

**Saunders College Publishing**
*Harcourt Brace College Publishers*

Fort Worth     Philadelphia     San Diego     New York

Orlando     San Antonio     Austin     Toronto

Montreal     London     Sidney     Tokyo

Text Typeface: Times Roman
Compositor: Monotype Composition, Inc.
Executive Editor: Jay Ricci
Managing Editor: Carol Field
Project Editors: Martha Brown, Sarah Fitz-Hugh
Copy Editor: Colleen Cranney
Manager of Art and Design: Carol Bleistine
Art Director: Jennifer Dunn
Cover Designer: Jennifer Dunn
Text Artwork: Grafacon
Director of EDP: Tim Frelick
Production Manager: Carol Florence
Marketing Manager: Monica Wilson

Cover Credit: © Joe Cornish/Tony Stone Images (clock);
© Tony Stone Images (gravel background)

Printed in the United States of America

ISBN 0-03-098330-4

Library of Congress Catalog Card Number: 93-086061

3 4 5 6   0 9 0   9 8 7 6 5 4 3 2 1

This book was printed on paper made from waste paper, containing 10% post-consumer waste and 40% pre-consumer waste, measured as a percentage of total fiber weight content.

To the Loving Memory of
My Parents and Raman

 # PREFACE

This text is written for students at the upper-undergraduate to the beginning-graduate level in mathematics, the physical sciences, and engineering. The reader is assumed to have studied calculus and linear algebra, and to have had an introduction to differential equations, and a structured programming language such as BASIC, C, FORTRAN, or PASCAL. Some basic results in these areas are given in appendices A, B, C, D, and E for quick reference or easy review.

The material in this text is developed from a collection of lecture notes amassed over a period of years of teaching numerical analysis, primarily to undergraduates. I have taught this course many times since 1969. It introduces students to methods and algorithms that are required for scientific computing.

This text emphasizes the mathematical underpinnings of these methods as well as their algorithmic aspects, since numerical analysis requires both a theoretical knowledge of the subject and computational experience with it. The object of the text is to introduce students to numerical analysis as a branch of applied mathematics.

In order to solve a problem using a numerical method, one has to know the derivation of the method and its error analysis. Even if one is using a software package, one must know the program's limitations. The majority of real problems cannot be solved by a standard program. For such problems it is necessary to adapt a standard numerical method to the new situation or to develop one's own method. The ability to do so requires a good theoretical foundation in numerical analysis. My hope in writing this text is to provide students with a good, readable text that not only will enable them to solve problems successfully, but also will stimulate their interest in developing new numerical methods.

It is very important for students to have computational experience. Mainframe computers are accessible to students, and minicomputers and micros are abundant. Even calculators can perform computations that were once reserved for large computers. Students will learn much more by programming an algorithm and experimenting with it. In order to help students learn and understand an algorithm, there are problems in the exercises where the reader is asked to go through at least three initial steps of the algorithm using a calculator. This approach has helped my students to write their own programs smoothly (and with a lot less effort).

Algorithms are presented in pseudocodes so that students can write computer programs based on these pseudocodes in any standard computer language. These pseudocodes can be easily modified. Students are required to write computer programs in the computer exercises. The purpose of these exercises is to provide a deeper understanding of algorithms. The development of accurate, efficient, general-purpose software is not the specific purpose of this text. If the instructor prefers, students may use popular softwares such as Mathcad, Mathematica, IMSL, and NG for the computer exercises.

Instructors can request a Solutions Manual providing program listings of the pseudocodes and detailed solutions to all of the exercises presented throughout the text. A 3.5-inch disk within the manual contains the solutions to the computer exercises and the listed programs. For your convenience, the solutions and programs contained on the disk are provided in three languages—PASCAL, FORTRAN, and C—and are formatted for a DOS platform.

## ORGANIZATION

This book contains more than enough material for a standard one-year course. Instructors can teach a standard course of one semester using the unstarred sections (see the table of contents) of chapters 1–8. (I have taught the material in chapters 1–8 in sequence for one semester, and most of the material in chapters 1–15 in sequence for two semesters.) Instructors may not have time to cover all of the topics. The following diagram shows the prerequisites of various chapters. Many other sequences are possible, depending on the instructor's preference and the time available.

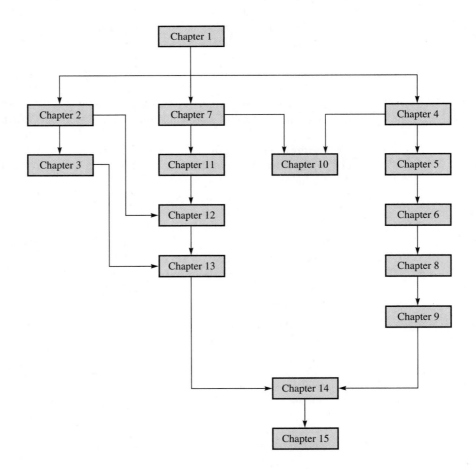

# ACKNOWLEDGMENTS

Two close friends, Roy Tucker and Charles Biles, have been of considerable help over the years and I extend special thanks to them. Also, I am thankful to my students at Humboldt State University for their comments and suggestions at various stages of manuscript preparation. I express my gratitude to Samsuhadi for technical typing. I also want to express my gratitude to the following reviewers who made numerous valuable suggestions that led to the improvements in the manuscript:

Neil Berger
University of Illinois, Chicago

Homer Brown
Middle Tennessee State University

H. W. Brown
Middle Tennessee State University

James Daniel
The University of Texas at Austin

Bruce Edwards
University of Florida

Richard Falk
Rutgers University, New Brunswick

G. S. Gill
Brigham Young University

Ian Gladwell
Southern Methodist University

Ronald B. Guenther
Oregon State University

Fabio Guerinoni
Florida State University, Tallahassee

Richard Hansen
Brigham Young University

Israel Koltracht
The University of Connecticut

Jian-Tong Lin
University of Texas at Arlington

Leonard Lipkin
University of North Florida

Giles Wilson Maloof
Boise State University

Hedley C. Morris
San Jose State University

Larry Riddle
Agnes Scott College

Everett Riggle
California State University, Chico

Lawrence Shampine
Southern Methodist University

David Sibley
The Pennsylvania State University

Then there is my family who waited patiently over the repeated promises that "it will be done soon." My special thanks go to my wife, children, and brothers without whose encouraging support the book would probably have not been completed. Thus, in a way, the book is theirs, too. Finally, my thanks go to the staff of Saunders College Publishing, especially to Jay Ricci, executive editor, and Martha B. Brown and Sarah Fitz-Hugh, project editors, for their encouragement and efficient collaboration. Also, I am thankful to Beth Sweet, assistant editor, for her help.

I would welcome suggestions for improvements and additions to the book from instructors and students who have used the book. You may correspond directly with me via EMAIL: PATELV@AXE. HUMBOLDT.EDU or Department of Mathematics, Arcata, California 95521-8299.

# BRIEF CONTENTS

# CONTENTS

## APPENDICES  *573*

## ANSWERS TO SELECTED ODD-NUMBERED PROBLEMS  *621*

## INDEX  *647*

*Chapter*

# 1 NUMERICAL COMPUTATIONS

---

## 1.1 INTRODUCTION

In this text we are concerned with numerical methods used to solve the most common mathematical problems that arise in the physical, biological, and social sciences, and many other disciplines. The problem is stated in mathematical terms by using various assumptions. The next step is to solve the stated mathematical problem. Unfortunately, many practical problems do not have an analytical solution; consequently, we look for an approximation or numerical solution. Also, an analytical solution may not be convenient for numerical evaluation. Therefore, we look for methods that give an approximate solution to our formulated problem. Because these methods work with numbers and produce numbers, they are called numerical methods. Numerical analysis provides a means of proposing and analyzing numerical methods for the study and solution of mathematically stated problems. To facilitate computation, numerical methods are programmed for execution on a computer. A poorly written computer program can spoil a good numerical method both with inaccurate answers and by using excessive computer time for computation. Therefore, it is very important to take into account the programming aspects of a numerical method. A computer output must also be analyzed for its correctness.

A complete and unambiguous set of directions to solve a mathematical problem to the desired accuracy in a finite number of steps is called an **algorithm**. Thus a numerical method can be considered an algorithm.

We imagine a program library containing subroutines, written by experts, for every conceivable situation. In fact, there exists a large number of computer packages like IMSL (international mathematical and statistical library), NAG (numerical algorithm group), and more through which many subroutines are available on mainframe computers. Also, IMSL and NAG have special subsets of their full libraries for microcomputers. For microcomputers, MATHCAD, MATLAB, and MATHEMATICA also provide programs. Other packages are being developed continually. While it is easy to call these subroutines, there are many pitfalls in numerical computation. One should be able to recognize symptoms of numerical ill health and diagnose a problem. It is important to have a clear understanding of the numerical methods used by these subroutines.

In the next section, we develop some fundamental notions about digital computers since they are the principal means for our calculations.

1

## 1.2  NUMBER REPRESENTATION

As is well known, our usual number system is the decimal system. The number 796.85 is expressed as

$$796.85 = 7 \times 10^2 + 9 \times 10^1 + 6 \times 10^0 + 8 \times 10^{-1} + 5 \times 10^{-2}$$

The number 10 is the base of the decimal system.

Electrical impulses are either on or off and computers read pulses sent by their electrical components. If "off" state represents 0 and "on" state represents 1, then computers can use a system that needs only 0 and 1 as digits to represent a real number. This system is called the **binary system** and has base 2. Consider

$$\begin{aligned}(1001.101)_2 &= 1 \times 2^3 + 0 \times 2^2 + 0 \times 2^1 + 1 \times 2^0 \\ &\quad + 1 \times 2^{-1} + 0 \times 2^{-2} + 1 \times 2^{-3} \\ &= 8 + 0 + 0 + 1 + \frac{1}{2} + 0 + \frac{1}{8} \\ &= 9 + \frac{1}{2} + \frac{1}{8} = \frac{77}{8}\end{aligned}$$

Further, consider

$$\begin{aligned}(1101011111)_2 &= 1 \times 2^9 + 1 \times 2^8 + 0 \times 2^7 + 1 \times 2^6 + 0 \times 2^5 \\ &\quad + 1 \times 2^4 + 1 \times 2^3 + 1 \times 2^2 + 1 \times 2^1 + 1 \times 2^0 \\ &= 512 + 256 + 0 + 64 + 0 + 16 + 8 + 4 + 2 + 1 = 863\end{aligned}$$

In order to represent 863, we need 10 binary digits. This is a major drawback of the binary system. The octal or hexadecimal number system (with base system 8) presents a compromise between the binary and decimal system when we discuss how numbers are stored in the computer. IBM 3033 uses the base 16 and the numbers 10, 11, 12, 13, 14, and 15 are usually denoted by A, B, C, D, E, and F, respectively. Most computers have an integer mode to represent integers and a floating-point mode to represent real numbers within given limits. The floating-point representation is closely connected to scientific notation. Letting $x$ be any real number, $x$ can be represented in floating-point form as

$$x = (\text{sign } x)(.d_1 d_2 \ldots d_t d_{t+1} \ldots)_\beta \times \beta^e \tag{1.2.1}$$

where the characters $d_i$ are digits in the base $\beta$ system. In other words, $0 \le d_i \le \beta - 1$ for $i = 1, 2, \ldots$ with $d_1 \ne 0$ and $e$ an integer. The number $(.d_1 d_2 \ldots d_t d_{t+1} \ldots)$ is called the **mantissa** and $e$ is called the **exponent** or **characteristic** of $x$. Usually $e$ is restricted by

$$-N \le e \le M \tag{1.2.2}$$

for some large positive integers $N$ and $M$. If during the calculations some computed quantity has an exponent $n > M$ then usually the result is meaningless and is called **overflow**. If an exponent $n < -N$, then usually the result is returned as zero without any warning message by many computers and is called **underflow**. Similarly an infinite representation of a mantissa cannot be used and so the mantissa of $x$ has to be terminated at $t$ digits. Let us denote this terminated number by fl($x$). This termination is done in two ways. The first way is to delete the digits $d_{t+1}$, $d_{t+2}$, ... to get the following **chopped** representation:

$$\text{fl}(x) = (\text{sign } x)(.d_1 d_2 \ldots d_t)_\beta \times \beta^e \qquad (1.2.3)$$

The second way is to add $\beta/2$ to $d_{t+1}$ and then chop off the resulting digits $\delta_{t+1}$, $\delta_{t+2}$, ... to get the **rounded** representation

$$\text{fl}(x) = (\text{sign } x)(.\delta_1 \delta_2 \ldots \delta_t)_\beta \times \beta^{e_1} \qquad (1.2.4)$$

where $\delta_i$ may or may not be $d_i$ and $e_1$ may or may not be $e$.

**EXAMPLE  1.1.1**

The number 13/6 has an infinite decimal representation given by $13/6 = 2.166666\ldots = 0.2166666\ldots \times 10^1$. Letting $t = 5$, the chopping representation of 13/6 is

$$\text{fl}\left(\frac{13}{6}\right) = 0.21666 \times 10^1 = 2.1666$$

For the rounding representation, add 5 to the sixth digit, $6 + 5$, and then chop off the digits after the fifth digit. Thus

$$\text{fl}\left(\frac{13}{6}\right) = 0.21667 \times 10^1 = 2.1667 \qquad ■ ■ ■$$

The error that results from replacing a number by either its chopped representation or rounded representation is called **round-off error**.

In Table 1.2.1 the floating-point characteristics are given for commonly used digital computers (Atkinson 1989) for single precision.

**Table 1.2.1**

| Computer | $\beta$ | $t$ | $N$ | $M$ | $\beta^{1-t}$ |
|---|---|---|---|---|---|
| CDC 6600 & Cyber Series 170 | 2 | 48 | 975 | 1070 | $7.11 \times 10^{-15}$ |
| Cray-1 | 2 | 48 | 16384 | 8191 | $7.11 \times 10^{-15}$ |
| Hewlett Packard HP-45, 11C, 15C | 10 | 10 | 98 | 100 | $1.00 \times 10^{-9}$ |
| IBM 3033 | 16 | 6 | 64 | 63 | $9.54 \times 10^{-7}$ |
| DEC VAX 11/780 | 2 | 24 | 128 | 127 | $1.19 \times 10^{-7}$ |
| PDP-11 | 2 | 24 | 128 | 127 | $1.19 \times 10^{-7}$ |
| Prime 850 | 2 | 23 | 128 | 127 | $2.38 \times 10^{-7}$ |

The question of rounding or chopping in Table 1.2.1 often depends on the installation or the compiler.

Let us denote the set of all numbers represented by Equation (1.2.3) or Equation (1.2.4) and zero by

$$F = F(\beta, t, N, M)$$

The real numbers, program instructions, integers, alphabetic symbols, and so on, are stored in words in digital computers. These words have a fixed number of digits.

The number of digits (bits) in a word is called the **word length**. For scientific calculations a long length is desirable; a short length, on the other hand, is significantly less expensive and perhaps more useful for business calculations and data processing. Word lengths range from 12 bits to 60 bits. In some computers a longer word is broken into smaller pieces called **bytes** (each consisting of 8 bits) for ease in handling.

Consider a hypothetical computer that uses 32 bits in a word. Of the 32 bits, the first bit is used to hold the sign of the number, 0 for $+$ and 1 for $-$. The remaining 31 bits hold a binary number 0 to 1111111111 1111111111 1111111111 1. This applies only to integers. For a floating-point number, the first bit holds the sign, the next 7 bits hold the exponent (including one for the sign of the exponent), and the last 24 bits hold the mantissa.

Consider for example

$$1 \ 1111001 \ 1111111111 \ 1111111111 \ 1111$$

The first bit indicates that the number is negative. The next bit indicates that the exponent is negative. The next 6 bits, 111001, are equivalent to

$$1 \times 2^5 + 1 \times 2^4 + 1 \times 2^3 + 0 \times 2^2 + 0 \times 2^1 + 1 \times 2^0 = 32 + 16 + 8 + 1$$
$$= 57$$

The last 24 bits indicate that the mantissa is

$$1 \times 2^{-1} + 1 \times 2^{-2} + \cdots + 1 \times 2^{-24} = \frac{1}{2}\left(1 + \frac{1}{2} + \cdots + \frac{1}{2^{23}}\right)$$

$$= \frac{1}{2}\frac{\left(1 - \left(\frac{1}{2}\right)^{24}\right)}{\left(1 - \frac{1}{2}\right)} = 1 - \left(\frac{1}{2}\right)^{24}$$

$$\approx 1 - (0.596046) \times 10^{-7}$$
$$\approx 0.999999 \text{ to seven places}$$

The closest 7 digit decimal number is $-0.6938893 \times 10^{-17}$ for $-(1 - (\frac{1}{2})^{24}) \times 2^{-57}$. Thus the machine number is used to represent any real number in the interval $(-0.69388935 \times 10^{-17}, -0.69388925 \times 10^{-17})$.

Since we are representing real numbers with approximate real numbers, our interest is to find the maximum error involved. Let $x^*$ be an approximation of $x$ in the decimal system for the following cases:

| (1) | (2) | (3) |
|---|---|---|
| $x = 2.1666$ | $x = 0.0004$ | $x = 10000.0001$ |
| $x^* = 2.1667$ | $x^* = 0.0003$ | $x^* = 10000.0000$ |
| $\|x - x^*\| = 0.0001$ | $\|x - x^*\| = 0.0001$ | $\|x - x^*\| = 0.0001$ |

In all cases, the absolute error $|x - x^*|$ is $10^{-4}$. In case (1) $x^*$ is a good approximation for $x$, while in case (2) $x$ is so small that $|x - x^*|$ represents a significant change. In case (3) $x^*$ seems to be an excellent approximation. Thus it is clear that the ratio of $|x - x^*|$ to $|x|$ is important.

Let $x^*$ be an approximation to $x$. The **relative error** in $x^*$ is given by

$$\left|\frac{x - x^*}{x}\right| \quad \text{provided } x \neq 0$$

**EXAMPLE 1.2.2**

In case (1) the relative error $|(x - x^*)/x| = |(2.1666 - 2.1667)/2.1666| = 0.00005$; in case (2) the relative error $|(x - x^*)/x| = |(0.0004 - 0.0003)/0.0004| = 0.25$; and in case (3) the relative error $|(x - x^*)/x| = |(10000.0001 - 10000)/10000.0001| = 10^{-8}$.

In case (2) the relative error indicates that $x^*$ is a poor approximation to $x$.

■ ■ ■

Since we do not know the true real number in a practical situation, we do not know what the error is. We will be happy with some bounds on the error. Let us find the relative error when we chop or round a given real number $x$ in our decimal system. Let $x$ be represented by

$$x = (\text{sign } x)(.d_1 d_2 \ldots d_t d_{t+1} \ldots) \times 10^e \tag{1.2.5}$$

We approximate $x$ by simply chopping off the digits $d_{t+1}$, $d_{t+2}$, $\ldots$, to get the chopped representation

$$\text{fl}(x) = x^* = (\text{sign } x)(.d_1 d_2 \ldots d_t) \times 10^e \tag{1.2.6}$$

Thus the relative chopping error in $x^*$ is given by

$$\left|\frac{x - x^*}{x}\right| = \frac{(.00 \ldots d_{t+1} d_{t+2} \ldots)}{(.d_1 d_2 \ldots d_t d_{t+1} \ldots)}$$
$$= \frac{(.d_{t+1} d_{t+2} \ldots) \times 10^{-t}}{(.d_1 d_2 \ldots d_t d_{t+1} \ldots)}$$

Since $1 \leq d_1 \leq 9$, the minimum value of the denominator is 0.1 and $.d_{t+1} d_{t+2} \ldots < 1$, the relative chopping error is

$$\left|\frac{x - x^*}{x}\right| \leq \frac{10^{-t}}{0.1} = 10^{1-t} \tag{1.2.7}$$

For rounding, if $d_{t+1} < 5$, then $.d_{t+1} d_{t+2} \ldots \leq \frac{1}{2}$. Therefore,

$$\left|\frac{x - x^*}{x}\right| = \frac{(.d_{t+1} d_{t+2} \ldots) \times 10^{-t}}{(.d_1 d_2 \ldots d_t d_{t+1} \ldots)} \leq \frac{1}{2} 10^{1-t}$$

If $5 \leq d_{t+1} < 10$, then

$$\text{fl}(x) = x^* = (\text{sign } x) \times (.\delta_1 \delta_2 \ldots \delta_t) \times 10^{e_1}$$

where $e_1 = e$ or $e + 1$. If $e_1 = e$, then the relative rounding error in $x^*$ is given by

$$\left|\frac{x - x^*}{x}\right| = \frac{|.d_1 d_2 \ldots d_t d_{t+1} \ldots - .\delta_1 \delta_2 \ldots \delta_t|}{(.d_1 d_2 \ldots d_t d_{t+1} \ldots)}$$
$$\leq \frac{0.00 \ldots 05}{(0.1)} = \frac{1}{2} 10^{1-t}$$

If $e_1 = e + 1$, it can be proved that the relative rounding error in $x^*$ is also given by

$$\left| \frac{x - x^*}{x} \right| \leq \frac{1}{2} 10^{1-t} \qquad \textbf{(1.2.8)}$$

It can be proved for the base system $\beta$ that the relative chopping error (Exercise 9) is

$$\left| \frac{x - x^*}{x} \right| \leq \beta^{1-t} \qquad \textbf{(1.2.9)}$$

and the relative rounding error is

$$\left| \frac{x - x^*}{x} \right| \leq \frac{1}{2} \beta^{1-t} \qquad \textbf{(1.2.10)}$$

In Table 1.2.1 the maximum relative chopping error in $\mathrm{fl}(x)$ is given by the quantity $\beta^{1-t}$. Equations (1.2.9) and (1.2.10) depend only on the floating-point number system and the value of $t$. Therefore, they are independent of the size of the number. We may increase $t$ to reduce round-off error. For instance, use of double precision instead of single precision reduces round-off error. The value of $t$ is said to be the number of **significant digits** of a computer.

The smallest positive floating-point number $\epsilon$, when added to the floating-point number 1.0 to produce a floating-point number different than 1.00, is known as the **machine epsilon**.

**EXAMPLE 1.2.3**    Let $x = 13/6$. For $t = 5$ and $\beta = 10$, the chopping representation of 13/6 is $\mathrm{fl}(13/6) = 0.21666 \times 10^1$. From Equation (1.2.7)

$$\left| \frac{13/6 - \mathrm{fl}(13/6)}{13/6} \right| \leq 10^{-4}$$

while $\left| \dfrac{13/6 - \mathrm{fl}(13/6)}{13/6} \right| \approx 0.3 \times 10^{-4}$.

For $t = 5$, the rounding representation of 13/6 is $\mathrm{fl}(13/6) = 0.21667 \times 10^1$. From Equation (1.2.8)

$$\left| \frac{13/6 - \mathrm{fl}(13/6)}{13/6} \right| \leq \frac{1}{2} 10^{-4}$$

while $\left| \dfrac{13/6 - \mathrm{fl}(13/6)}{13/6} \right| \approx 0.15 \times 10^{-4}$.    ■ ■ ■

---

## 1.3  FLOATING-POINT ARITHMETIC

In order to understand the kind of arithmetic done by computers, let us analyze computer arithmetic on a five-digit hypothetical machine that is similar to the actual arithmetic carried out by several common computers.

Let $x = 0.24689 \times 10^3$ and $y = 0.13579 \times 10^2$. These numbers are loaded in arithmetic registers. The arithmetic registers are capable of shifting numbers right and left in order to align the decimal point or to adjust the exponents during arithmetic operations. The length of an arithmetic register relative to a word length is an important characteristic of a computer. If the single precision number has $t$ digits for its mantissa, then the arithmetic register should contain $t + p$ digits where $p$ is comparable to $t$. This decision involves not just mathematical but economical factors. A register is called a single length register if $p = 0$, and a double length register if $p = t$. Several computer manufacturers use $p = 1$ because of economic considerations. This single extra digit is called a guard digit. The guard digit makes a noticeable effect on the accuracy of computed results. We will assume that our hypothetical machine has a double length register and uses chopping. In order to determine what values will be produced by a particular machine, one has to study the manuals supplied by a manufacturer.

Let us add $x$ and $y$ on our machine. The exponent of the smaller number is to be adjusted so that the exponents of both numbers are the same. Thus

$$x = 0.24689\ 00000 \times 10^3$$
$$y = 0.01357\ 90000 \times 10^3$$

These numbers are added in the accumulator and we get

$$x + y = 0.26046\ 90000 \times 10^3$$

Since this number is to be stored, it is converted to the following five-digit floating-point number. Hence

$$\text{fl}(x + y) = 0.26046 \times 10^3$$

The result $0.26046 \times 10^3$ is stored as a computer word. Since the exact sum $= 0.260469 \times 10^3$, the relative error $|(x + y - \text{fl}(x + y))/(x + y)| = 0.345530 \times 10^{-4}$. Let us subtract $y$ from $x$. In the accumulator

$$x - y = 0.23331\ 10000 \times 10^3$$
$$\text{fl}(x - y) = 0.23331 \times 10^3$$

The result $0.23331 \times 10^3$ is stored.

Multiplication is simple because the exponents do not have to be aligned. Since

$$xy = 10^3 \times 10^2 \times 0.24689 \times 0.13579$$
$$= 0.335251931 \times 10^4$$

and $\text{fl}(xy) = 0.33525 \times 10^4$, the relative error $|(xy - \text{fl}(xy))/(xy)| = 0.57598 \times 10^{-5}$.

Division is not allowed when $y = 0$. If the mantissa of the numerator is greater than the mantissa of the denominator, we shift the mantissa of the numerator one place to the right. Thus

$$\frac{x}{y} = 10 \times \frac{0.24689}{0.13579}$$
$$= 10^2 \times \frac{0.0246890000}{0.1357900000} = 0.1818175000 \times 10^2$$

therefore $\text{fl}(x/y) = 0.18181 \times 10^2$ and the relative error $|((x/y) - \text{fl}(x/y))/(x/y)| = 0.41250 \times 10^{-4}$.

Normally $fl(x) \neq x$. Using Equations (1.2.9) and (1.2.10),

$$\frac{x - fl(x)}{x} = -\epsilon \tag{1.3.1}$$

where $-\beta^{1-t} \leq \epsilon \leq 0$ if chopped and $-\frac{1}{2}\beta^{1-t} \leq \epsilon \leq \frac{1}{2}\beta^{1-t}$ if rounded. Equation (1.3.1) can be expressed in a more useful form:

$$fl(x) = x(1 + \epsilon) \tag{1.3.2}$$

Denote machine addition, subtraction, multiplication, and division by the symbols $\oplus$, $\ominus$, $\otimes$, and $\oslash$ respectively. For any floating-point numbers $x$ and $y$, we have from Equation (1.3.2)

$$fl(x + y) = x \oplus y = (x + y)(1 + \epsilon_1)$$
$$fl(x - y) = x \ominus y = (x - y)(1 + \epsilon_2)$$
$$fl(xy) = x \otimes y = xy(1 + \epsilon_3)$$

and

$$fl\left(\frac{x}{y}\right) = x \oslash y = \frac{x}{y}(1 + \epsilon_4) \tag{1.3.3}$$

where $\epsilon_1$, $\epsilon_2$, $\epsilon_3$, and $\epsilon_4$ may be different. It can be seen from Equations (1.3.2) and (1.3.3) that

$$\begin{aligned}
fl(x + (y + z)) = x \oplus (y \oplus z) &= x \oplus (y + z)(1 + \epsilon_5) \\
&= (x + [(y + z)(1 + \epsilon_5)])(1 + \epsilon_6) \\
&= x(1 + \epsilon_6) + (y + z)(1 + \epsilon_5)(1 + \epsilon_6)
\end{aligned}$$

$$\begin{aligned}
fl((x + y) + z) = (x \oplus y) \oplus z &= (x + y)(1 + \epsilon_7) \oplus z \\
&= [(x + y)(1 + \epsilon_7) + z](1 + \epsilon_8) \\
&= (x + y)(1 + \epsilon_7)(1 + \epsilon_8) + z(1 + \epsilon_8)
\end{aligned}$$

Hence, often

$$x \oplus (y \oplus z) \neq (x \oplus y) \oplus z$$

In other words, the associative law breaks down. Similarly (Exercise 11) the distributive law often fails:

$$x \otimes (y \oplus z) \neq (x \otimes y) \oplus (x \otimes z)$$

**EXAMPLE 1.3.1**

Illustrate that the associative law breaks down. Let $x = 0.52867 \times 10^4$, $y = 0.38234 \times 10^2$, and $z = 0.25678 \times 10^1$. Find $x \oplus (y \oplus z)$ and $(x \oplus y) \oplus z$.

Since $y$ and $z$ are first added, the exponent of $z$ is adjusted so that the exponents of $y$ and $z$ are the same. Hence

$$y = 0.38234\,00000 \times 10^2$$
$$z = 0.02567\,80000 \times 10^2$$
$$y + z = 0.40801\,80000 \times 10^2$$

and

$$fl(y + z) = y \oplus z = 0.40801 \times 10^2$$

Now to add $(y \oplus z)$ to $x$, the exponents of $(y \oplus z)$ and $x$ are adjusted. Therefore,

$$x = 0.52867\ 00000 \times 10^4$$
$$y \oplus z = 0.00408\ 01000 \times 10^4$$
$$x + (y \oplus z) = 0.53275\ 01000 \times 10^4$$

and

$$x \oplus (y \oplus z) = 0.53275 \times 10^4 \tag{1.3.4}$$

One can verify that

$$x \oplus y = 0.53249\ 00000 \times 10^4$$

and

$$(x \oplus y) \oplus z = 0.53274 \times 10^4 \tag{1.3.5}$$

Comparing Equations (1.3.4) and (1.3.5), $x \oplus (y \oplus z) \neq (x \oplus y) \oplus z$. ∎∎∎

One must not implicitly assume the validity of the associative law. Although the associative law for addition is not valid in floating-point arithmetic, it is comforting to know that the commutative law $x \oplus y = y \oplus x$ still holds and should be valuable in our programming.

Another important source of error is the subtraction of a number from a nearly equal number. Consider $x = \sqrt{457} \approx 0.2137755 \times 10^2$ and $y = \sqrt{456} \approx 0.2135415 \times 10^2$. Subtract $y$ from $x$ on the five-digit machine. First $x$ and $y$ would be stored as $\mathrm{fl}(x) = 0.21377 \times 10^2$ and $\mathrm{fl}(y) = 0.21354 \times 10^2$. Since our hypothetical machine has double length register,

$$\mathrm{fl}(x) = 0.21377\ 00000 \times 10^2$$
$$\mathrm{fl}(y) = 0.21354\ 00000 \times 10^2$$
$$\mathrm{fl}(x) - \mathrm{fl}(y) = 0.00023\ 00000 \times 10^2$$

and

$$\mathrm{fl}(\mathrm{fl}(x) - \mathrm{fl}(y)) = \mathrm{fl}(x) \ominus \mathrm{fl}(y) = 0.23000 \times 10^{-1} = 0.02300$$

The last three zeros at the end of the mantissa are of no use. Since the exact value of $x - y \approx 0.000234 \times 10^2 = 0.0234$, we have the relative error

$$\left| \frac{x - y - \mathrm{fl}(\mathrm{fl}(x) - \mathrm{fl}(y))}{x - y} \right| = 0.17170 \times 10^{-1} \tag{1.3.6}$$

This relative error is quite large when compared to the relative errors of $\mathrm{fl}(x)$ and $\mathrm{fl}(y)$. How can a more accurate result be obtained? Sometimes the problem can be reformulated to avoid the subtraction. In this example,

$$\sqrt{457} - \sqrt{456} = \frac{(\sqrt{457} - \sqrt{456})(\sqrt{456} + \sqrt{457})}{\sqrt{457} + \sqrt{456}}$$

$$= \frac{1}{\sqrt{457} + \sqrt{456}} \approx 1 \oslash (\mathrm{fl}(x) + \mathrm{fl}(y))$$

$$= 1 \oslash (0.42731 \times 10^2) = 0.23402 \times 10^{-1} = 0.023402$$

The relative error $0.72643 \times 10^{-5}$ is very small compared to Equation (1.3.6).

**EXERCISES**

1.  Express the following base $\beta$ numbers in floating-point form.
    (a) $(123.456)_{10}$   (b) $(27.653)_8$   (c) $(10101.1101)_2$   (d) $(AB.168)_{16}$

2.  Express the following base $\beta$ numbers in decimal form.
    (a) $(10.001101)_2$   (b) $(0.775)_8$   (c) $(1.89ABC)_{16}$   (d) $(67.015)_8$

3.  Write the elements of the set $F(2, 3, 1, 2)$. Convert the elements in the decimal system and then represent the numbers as points on a straight line.

4.  Convert $(0.1)_{10}$ to binary form. Are ten steps of length $(0.1)_{10}$ in binary form the same as one step of length 1.0? (*Hint:* $\frac{1}{10} = \frac{0}{2} + \frac{0}{2^2} + \frac{0}{2^3} + \frac{1}{2^4} \cdots.$)

5.  Let $fl(x)$ be given by chopping. Prove that (a) $fl(-x) = -fl(x)$ and (b) $fl(\beta^r x) = \beta^r fl(x)$ by assuming that underflow or overflow does not occur.

6.  If the following approximations are used, find the absolute error and relative error.
    (a) $\frac{1}{7} \approx 0.14$   (b) $\frac{1}{7} \approx 0.1428$

7.  How many significant digits are there if (a) 852.045 is approximated by 852.01, (b) 0.000452 is approximated by 0.00041, and (c) 1.2345 is approximated by 1.234?

8.  Find the absolute error and the relative error for Exercise 7.

9.  Prove Equations (1.2.9) and (1.2.10).

10. Let $x = 0.5678 \times 10^1$, $y = 0.3456 \times 10^2$, and $z = 0.1234 \times 10^3$. Perform the following operations on a hypothetical machine that uses four-digit floating-point arithmetic, double precision accumulator, and chops the resulting number to four digits before any subsequent operation is performed.
    (a) $x \oplus y$   (b) $y \oplus x$   (c) $(x \oplus y) \oplus z$   (d) $x \oplus (y \oplus z)$
    (e) $x \otimes (y \oplus z)$   (f) $(x \otimes y) \oplus (x \otimes z)$

11. Prove that $x \otimes (y \oplus z) \neq (x \otimes y) \oplus (x \otimes z)$ in some cases.

12. Consider a machine that uses four-digit floating-point arithmetic. Let $x = \ln 2.10 = 0.74194$, and $y = \ln 2.11 = 0.74669$. Compute $fl(x) \ominus fl(y)$ and determine its relative error.

13. Indicate how the following formulas should be written to avoid the loss of significant digits due to subtraction:
    (a) $\ln x - \ln y$   (b) $e^x - x - 1$ if $x$ is close to 0   (c) $1 - \cos x$ if $x$ is close to 0   (d) $e^{x-y}$
    (e) $\dfrac{-b + \sqrt{b^2 - 4ac}}{2a}$ if $b > 0$ and $\dfrac{-b - \sqrt{b^2 - 4ac}}{2a}$ if $b < 0$

14. Consult your computer manuals and indicate whether your computer has a single length register, a double length register, a register with a guard digit, or none of the above. Write a brief description of the addition operation.

15. Find out the base and number of digits in the mantissa of your computer. What is the largest number and the smallest number represented by your computer?

**COMPUTER EXERCISE**

C.1.  Write a program to find the machine epsilon of your computer.

## 1.4   NUMERICAL COMPUTING

In order to solve a problem on a digital computer, we perform the following steps:
**Step 1**   Construct a mathematical model for a given problem by using a variety of simplifying assumptions. As a consequence of these assumptions, the resulting mathematical model has inherent limitations. Be aware that a numerical solution of a mathematical model cannot improve the accuracy of a mathematical model except by coincidence. Since we cannot judge a priori the error due to either the mathematical model or the numerical method, we need to find as accurate a numerical solution of a mathematical model as possible. This mathematical model is a set of formulas, inequalities, equations and so on.
**Step 2**   Decide what mathematical or numerical method to use to solve a mathematical problem. Our concern is to find the best numerical method for solving a given mathematical problem. How should a numerical method be evaluated? Two important properties come to mind—accuracy and efficiency. The amount of computer time used in the execution of a particular method can be used to measure the efficiency. The time it takes to complete floating-point arithmetic is usually much longer than the time it takes to store, test, perform integer arithmetic, fetch, and so on. Also, there are comparatively few nonarithmetic steps in most numerical algorithms, so a reasonable estimate of the required time can be obtained by counting the number of required additions, subtractions, multiplications, and divisions. Because addition and subtraction take approximately the same amount of time on nearly all computers, these two operations will be counted together. Division time is longer than multiplication time for one operation. We can use an expression

$$E = pA + qM + rD \qquad (1.4.1)$$

where $p$, $q$, and $r$ are the total number of floating-point additions, multiplications, and divisions respectively in a given algorithm and $A$, $M$, and $D$ are the amounts of time it takes a computer to perform one addition, one multiplication, and one division respectively. If many algorithms with equal accuracy are available, then the algorithm with the smallest value of $E$ is preferred.

The computation times in seconds for 1000 repetitions of common arithmetic operations on an IBM PC are given in Table 1.4.1. The installation of a mathematical coprocessor on an IBM PC or a compatible machine increases the speed of floating-point arithmetic considerably. For comparison, the computation times with and without the mathematical coprocessor Intel 8087 using Turbo Pascal on an IBM PC (Lastman and Sinha 1988) are given in Table 1.4.1.

**Table 1.4.1**

| Operation | Time without Intel 8087 | Time with Intel 8087 |
|---|---|---|
| Addition | 0.28 | 0.17 |
| Subtraction | 0.28 | 0.17 |
| Multiplication | 1.04 | 0.17 |
| Division | 1.71 | 0.22 |

The closeness of a computed answer to the exact answer is the accuracy of the computed answer. By their very nature, in most cases numerical methods do not give exact answers. There are many sources that contribute to the inaccuracy of the answers produced by a numerical method.

1. **Input error:** Data from measurements of practical problems contain errors that affect the accuracy of calculations based on the data.

2. **Round-off error:** A computer has a fixed word length and therefore most numbers, including those obtained by arithmetic operations, cannot be expressed exactly. Each number is represented by its nearest machine number. This type of error is called round-off error and is a characteristic of the computer or the computer language. The subtraction of two nearly equal numbers can be avoided by reformulating the problem. Thus the resulting vast increase in round-off error can be avoided. The generated round-off error contaminates subsequent calculations and the error propagation becomes an important source of error.

3. **Propagated error:** In order to see how errors accumulate in a complicated algorithm, consider the sum of two positive numbers $x_1$ and $x_2$. The first step is to convert these numbers into floating-point numbers by chopping or rounding. Errors are thus introduced. Denoting the converted numbers with an overbar and using Equation (1.3.2), we get

$$\bar{x}_1 = \text{fl}(x_1) = x_1(1 + \epsilon_1) \quad \text{and} \quad \bar{x}_2 = \text{fl}(x_2) = x_2(1 + \epsilon_2) \qquad \textbf{(1.4.2)}$$

where $|\epsilon_i| \leq \mu$ for $i = 1$ and 2.

When we add these two numbers, we actually compute $\bar{x}_1 \oplus \bar{x}_2$ (recall $\oplus$ denotes machine addition). The actual error is given by

$$\begin{aligned}(x_1 + x_2) - (\bar{x}_1 \oplus \bar{x}_2) &= [(x_1 + x_2) - (\bar{x}_1 + \bar{x}_2)] + [(\bar{x}_1 + \bar{x}_2) - (\bar{x}_1 \oplus \bar{x}_2)] \\ &= [(x_1 - \bar{x}_1) + (x_2 - \bar{x}_2)] + [(\bar{x}_1 + \bar{x}_2) - (\bar{x}_1 \oplus \bar{x}_2)]\end{aligned}$$
$$\textbf{(1.4.3)}$$

The term in the first brackets on the right-hand side of Equation (1.4.3) is the error propagated because of the initial conversion to floating-point numbers. Consequently, this error is called the propagated error. The term in the second brackets on the right-hand side of Equation (1.4.3) is the error generated because of the machine's arithmetic by rounding or chopping. It is called the round-off error. Using Equation (1.3.3), we have

$$\bar{x}_1 \oplus \bar{x}_2 = (\bar{x}_1 + \bar{x}_2)(1 + \epsilon_3) \qquad \textbf{(1.4.4)}$$

where $|\epsilon_3| \leq \mu$. Using Equations (1.4.2) and (1.4.4) in Equation (1.4.3), we get

$$|(x_1 + x_2) - (\bar{x}_1 \oplus \bar{x}_2)| \leq |-\epsilon_1 x_1 - \epsilon_2 x_2| + |-\epsilon_3(\bar{x}_1 + \bar{x}_2)| \qquad \textbf{(1.4.5)}$$

Further, using Equation (1.4.2), the last term of Equation (1.4.5) simplifies and is given by

$$\begin{aligned}\epsilon_3(\bar{x}_1 + \bar{x}_2) &= \epsilon_3(x_1 + \epsilon_1 x_1 + x_2 + \epsilon_2 x_2) \\ &= \epsilon_3(x_1 + x_2) + \epsilon_3(\epsilon_1 x_1 + \epsilon_2 x_2)\end{aligned}$$

Using this and $|\epsilon_i| \leq \mu$ for $i = 1$, 2, and 3 in Equation (1.4.5), we get

$$|(x_1 + x_2) - (\bar{x}_1 \oplus \bar{x}_2)| \leq (2\mu + \mu^2)(x_1 + x_2) \qquad \textbf{(1.4.6)}$$

Since $|\epsilon_i| \le \mu$, we replaced $|\epsilon_i|$ by $\mu$ in Equation (1.4.5). Thus Equation (1.4.6) gives the error bound for the worst case that may be encountered. When large numbers of operations are involved, the bounds may be several times larger than the actual error involved and therefore are too pessimistic. Often, the round-off errors tend to cancel each other, but, under the right circumstances, the round-off errors can grow like a rolling snowball. This phenomenon is referred to as instability and will be treated later.

**EXAMPLE 1.4.1**

Add $x_1 = 0.36789$ nad $x_2 = 2.5678$ using four-digit floating-point arithmetic with chopping.

Our machine transforms these numbers as $\bar{x}_1 = 0.3678 \times 10^0$ and $\bar{x}_2 = 0.2567 \times 10^1$. For adding, we have

$$\bar{x}_1 = 0.0367\,8000 \times 10^1 \quad \text{and} \quad \bar{x}_2 = 0.2567\,0000 \times 10^1$$

Then

$$\bar{x}_1 + \bar{x}_2 = 0.2934\,8000 \times 10^1 \quad \text{and so} \quad \bar{x}_1 \oplus \bar{x}_2 = 0.2934 \times 10^1$$

Since the exact sum $x_1 + x_2 = 2.93569$, the exact absolute error $|x_1 + x_2 - (\bar{x}_1 \oplus \bar{x}_2)| = 0.00169$. The propagated error is $|x_1 + x_2 - (\bar{x}_1 + \bar{x}_2)| = |(x_1 - \bar{x}_1)| + (x_2 - \bar{x}_2)| = |2.93569 - 2.93480| = 0.00089$, and the round-off error is $|\bar{x}_1 + \bar{x}_2 - (\bar{x}_1 \oplus \bar{x}_2)| = |2.9348 - 2.934| = 0.0008$.

Since we are chopping, $\mu = \beta^{1-t} = 10^{-3} = 0.001$. Hence, from Equation (1.4.6)

$$|(x_1 + x_2) - (\bar{x}_1 \oplus \bar{x}_2)| \le (2\mu + \mu^2)(x_1 + x_2) = 0.00587$$

Thus

$$|(x_1 + x_2) - (\bar{x}_1 \oplus \bar{x}_2)| = 0.00169 < 0.00587$$

is true but not very accurate.    ■ ■ ■

4. **Truncation error:** This error occurs when we truncate the process after a certain number of steps because, for an exact result, an infinite sequence of steps or too many steps was required. For example, Maclaurin's series for $e^x$ is given by

$$e^x = 1 + x + \frac{x^2}{2!} + \cdots + \frac{x^n}{n!} + \cdots \tag{1.4.7}$$

Equation (1.4.7) can be written as the iteration

$$r_i = r_{i-1} + \frac{x^i}{i!} \quad \text{for } i = 1, 2, \ldots$$

with

$$r_0 = 1 \tag{1.4.8}$$

This formulation becomes more accurate as we use more and more iterations. However, because of computational resources, a decision must be made to stop the iterative process. Truncation error is the error introduced when we truncate the process after a certain finite number of steps on a hypothetical perfect computer that had no round-off errors (all digits were retained). Since we do not have a perfect computer, more iterations mean more arithmetic operations, and more arithmetic operations mean more round-off errors and more propagated errors. In many cases, reducing truncation error means increasing round-off error.

Let us find $e^{1/2}$ using Equation (1.4.8). Since we cannot go on for ever, we stop somewhere. For simplicity, let us stop at $i = 3$. Then

$$r_3 = r_2 + \frac{1}{3!}\left(\frac{1}{2}\right)^3 = r_1 + \frac{1}{2!}\left(\frac{1}{2}\right)^2 + \frac{1}{3!}\left(\frac{1}{2}\right)^3$$

$$= r_0 + \frac{1}{1!}\left(\frac{1}{2}\right) + \frac{1}{2!}\left(\frac{1}{2}\right)^2 + \frac{1}{3!}\left(\frac{1}{2}\right)^3$$

We approximate $e^{1/2}$ by $r_3$. Thus

$$e^{1/2} \approx r_3 = 1 + \frac{1}{1!}\left(\frac{1}{2}\right) + \frac{1}{2!}\left(\frac{1}{2}\right)^2 + \frac{1}{3!}\left(\frac{1}{2}\right)^3 = \frac{79}{48}$$

An infinite sequence of steps is required for the exact result. By stopping at $i = 3$ we introduce a truncation error given by (Theorem A.8)

$$tr = \left(\frac{1}{2}\right)^4 \frac{1}{4!} e^{\xi}$$

where $\xi$ is between 0 and 1/2.

Since $\xi$ is not known, we estimate the maximum value of $e^{\xi}$ on the closed interval [0, 1/2]. Thus

$$|tr| \leq \frac{e^{1/2}}{2^4\,4!} \approx 0.00429$$

The actual truncation error $= e^{1/2} - (79/48) \approx 0.00289 < 0.00429$.

**Step 3**    After selecting a numerical method, we must carry out the programming of our numerical method. It is expected that the student is familiar with the rudiments of programming. The programming language could be Basic, Fortran, Pascal, C, or any other language.

A program written for a particular set of numbers must be rewritten for another set of numbers. In most cases, few additional statements are required to write a program to handle a general case. It will be worth the extra effort. Write out the mathematical algorithm in complete detail and write the code in a style that is easy to read and understand. Check your code thoroughly for omissions and errors before heading for a terminal. Do not rush. It pays to trace through the code with pencil and paper on a typical and simple example.

Always print the input data and initially print intermediate results to understand the program's operations. It helps to have labels for the output. Write a long program in steps by writing and testing a series of subroutines and function subprograms.

Use comments so that another person can understand what the code does. Insert blank comment lines in order to improve the readability of the code. For each subroutine, explain through comments the parameters used.

**Step 4** The chance of a human being making an arithmetic error is very high, while the chance of a machine making an arithmetic error is very low. The main concern is programming error. Fortunately, some programming errors are repeated many times during the execution of a program and therefore the existence of those errors are identified by absurd numerical output. For a complex and lengthy computer program, it becomes difficult to detect and correct small but important logical programming errors. This makes debugging a very important component of the computer program-ming process and sometimes it makes a crucial difference in the numerical results. Therefore, it is extremely important to test a computer program for known results, since a poorly written computer program can spoil an excellent method both by provid-ing inaccurate results and by using excessive computer time.

In this text we will discuss numerical methods and analyze them. It is assumed that the reader will write computer programs for these methods.

## EXERCISES

1. Prove that $|x_1 x_2 - \bar{x}_1 \otimes \bar{x}_2| \le |x_1 x_2|(3\mu + 3\mu^2 + \mu^3)$ where $x_1$ and $x_2$ are real numbers and

$$\mu = \begin{cases} \beta^{1-t} & \text{if chopped} \\ \frac{1}{2}\beta^{1-t} & \text{if rounded} \end{cases}$$

2. Let $x_1 = 0.12345 \times 10^2$ and $x_2 = 0.23456 \times 10^1$. Use four-digit chopped floating-point arithmetic to find $\bar{x}_1 \otimes \bar{x}_2$. Find the absolute propagated and round-off errors. Compare the exact error with the upper error bound given in Exercise 1.

3. Add $x_1 = 0.12345 \times 10^2$ and $x_2 = 0.23456 \times 10^1$ using four-digit chopped floating-point arithmetic. Find the absolute propagated and round-off errors. Compare the exact error with the upper error bound given by Equation (1.4.6).

4. Prove that

$$|(((x_1 + x_2) + x_3) + x_4) - (((\bar{x}_1 \oplus \bar{x}_2) \oplus \bar{x}_3) \oplus \bar{x}_4)| \le \mu(4x_1 + 4x_2 + 3x_3 + 2x_4)$$
$$+ \mu^2(6x_1 + 6x_2 + 3x_3 + x_4) + \mu^3(4x_1 + 4x_2 + x_3) + \mu^4(x_1 + x_2)$$

where $x_1, x_2, x_3,$ and $x_4$ are positive real numbers and

$$\mu = \begin{cases} \beta^{1-t} & \text{if chopped} \\ \frac{1}{2}\beta^{1-t} & \text{if rounded} \end{cases}$$

The first term on the right-hand side of the inequality contributes significantly. In that term $x_1$ is multiplied by 4 while $x_4$ is multiplied by 2. This and other terms on the right-hand side of the inequality suggest that the best strategy for addition is to add from the smallest to the largest.

5. Add $x_1 = 0.36789$, $x_2 = 2.5678$, $x_3 = 0.12345$, and $x_4 = 0.034567$ using four-digit chopped floating-point arithmetic. Rearrange these numbers from the smallest to the largest and then add them using four-digit chopped floating-point arithmetic. Compare these sums with the exact sum.

## SUGGESTED READINGS

K. E. Atkinson, *An Introduction to Numerical Analysis,* 2nd ed., John Wiley & Sons, New York, 1989.

G. J. Lastman and N. K. Sinha, *Microcomputer-Based Numerical Methods for Science and Engineering,* Saunders College Publishing, Philadelphia, 1989.

# Chapter

# 2 NONLINEAR EQUATIONS

To sketch the graph of $f(x) = x \cos x$ on the interval $[0, \pi]$, numbers must be found in $[0,\pi]$ at which $f$ is maximum and minimum. The critical numbers satisfy the equation $f'(x) = \cos x - x \sin x = 0$. To sketch the graph of $f(x) = x \cos x$ on $[0, \pi]$, find $x$ in the interval $[0, \pi]$ such that

$$\cos x - x \sin x = 0$$

This chapter deals with methods that approximate the roots of equations of this type.

## 2.1  INTRODUCTION

A basic problem encountered in practice is finding the roots of an equation or a system of equations. For example,

1. $\tan x - x = 0$
2. $x^3 - x^2 - x + 1 = 0$
3. $e^{x^2} - \cos x = 0$
4. $3x^2 + y = 4, \qquad x + 2y^2 = 3$

Example 4 is a system of two nonlinear equations in two unknowns while 1, 2, and 3 are examples of nonlinear equations in one unknown. Consider a single nonlinear equation in one unknown of the form

$$f(x) = 0 \tag{2.1.1}$$

When there exists a number $r$ such that $f(r) = 0$, then $r$ is called a **root** of the equation $f(x) = 0$; also, $r$ is known as a **zero** of $f(x)$. The roots of the equation $x^3 - x^2 - x + 1 = 0$ in example 2 are 1 and $-1$, and the zeros of $f(x) = x^3 + x^2 - x + 1$ are 1 and $-1$.

This chapter is concerned with numerical solutions of equations of the form $f(x) = 0$, where $f$ is a real valued function and differentiable as many times as the derivatives are required. In numerical applications it must be understood that the equation usually cannot be exactly satisfied due to round-off errors. Therefore, the mathematical definition of a root is modified by thinking that $r$ is a root of $f(x) = 0$ if $|f(r)| < \epsilon$, where $\epsilon > 0$ is a given tolerance. The inequality $|f(r)| < \epsilon$ defines intervals instead of points, but this can hardly be avoided.

The zero $r$ is of **multiplicity** $m$ if $(x - r)^m$ is a factor of $f(x)$ and $f(x) = (x - r)^m q(x)$ where $\lim_{x \to r} q(x) \neq 0$. If $m = 1$, then often $r$ is called a simple zero. If $m$ is a positive integer, it can be shown that (Exercise 8) $f(r) = f'(r) = \cdots = f^{(m-1)}(r) = 0$ and $f^{(m)}(r) \neq 0$. For

$$f(x) = x^3 - x^2 - x + 1 = (x - 1)^2(x + 1)$$

the zero 1 has multiplicity 2 and $f'(1) = 0$.

## 2.2    THE BISECTION METHOD

One of the oldest methods for finding a zero of $f(x)$ is to trap that zero in an interval and then make that interval smaller and smaller. For this technique, find two numbers $a$ and $b$ such that $f(a)$ and $f(b)$ have opposite signs, in other words $f(a)f(b) < 0$. Since $f$ is assumed to be continuous on $[a, b]$, there is at least one number $r$ in $(a, b)$ by the Intermediate Value Theorem (Theorem A.2) for which $f(r) = 0$. For simplicity, assume that $(a, b)$ contains only one zero of $f$ (Figure 2.2.1).

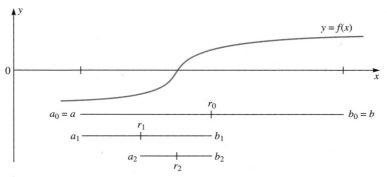

**Fig. 2.2.1**

We seek to find $r$ by determining a sequence of intervals, each with length that is half of the previous interval and each containing the zero of $f$.

Set $a_0 = a$ and $b_0 = b$. Let $r_0 = (a_0 + b_0)/2$ be the midpoint of the closed interval $[a_0, b_0]$. Then exactly one of three possible cases exists: (1) $f(a_0)$ and $f(r_0)$ have opposite signs; (2) $f(a_0)$ and $f(r_0)$ have same signs; (3) $f(r_0) = 0$. If $f(r_0) = 0$, then $r_0$ is the zero of $f(x)$. Assume that $f(r_0) \neq 0$ and $f(a_0)$ and $f(r_0)$ have opposite signs, so that, $f(a_0)f(r_0) < 0$. Then the zero is between $a_0$ and $r_0$ by the Intermediate Value Theorem.

Set $a_1 = a_0$ and $b_1 = r_0$. Let $r_1 = (a_1 + b_1)/2$ be the midpoint of $[a_1, b_1]$. Then, again, one of three cases exists: (1) $f(a_1)$ and $f(r_1)$ have opposite signs; (2) $f(a_1)$ and $f(r_1)$ have same signs; (3) $f(r_1) = 0$. Assume that $f(r_1) \neq 0$ and $f(a_1)$ and $f(r_1)$ have the same signs, so that $f(a_1)f(r_1) > 0$. Then the zero is between $r_1$ and $b_1$ $(= r_0)$.

Set $a_2 = r_1$ and $b_2 = b_1$. Let $r_2 = (a_2 + b_2)/2$ be the midpoint of $[a_2, b_2]$. Continue this procedure to generate $r_0, r_1, \ldots r_n$ until $|f(r_n)|$ or $|r_n - r_{n-1}|$ is as small as desired.

**EXAMPLE 2.2.1**

Find the root of $e^x - 2 \cos x = 0$ in the closed interval $[0, 2]$.

We are looking for a zero of the function $f(x) = e^x - 2 \cos x$. Starting with $a_0 = 0$ and $b_0 = 2$, $f(0) = -1$ and $f(2) = 8.22135$ and so there is a root of $e^x - 2 \cos x = 0$ in the interval $[0, 2]$.

Then $r_0 = (a_0 + b_0)/2 = 1$ and $f(1) = 1.63767$. Since $f(0)$ and $f(1)$ have opposite signs, the root lies between $[0, 1]$. Setting $a_1 = 0$ and $b_1 = r_0 = 1$ results in $r_1 = (a_1 + b_1)/2 = 0.5$ and $f(0.5) = -0.10644$. Since $f(0.5)f(1.0) < 0$, the root lies between $[0.5, 1.0]$. Set $a_2 = 0.5$ and $b_2 = 1.0$ so $r_2 = (a_2 + b_2)/2 = 0.75$. In this manner the sequence $\{r_i\}$ is obtained and is given in Table 2.2.1.

**Table 2.2.1**

| $n$ | $a_n$ | $b_n$ | $r_n$ | $f(r_n)$ |
|---|---|---|---|---|
| 0 | 0.00000 | 2.00000 | 1.00000 | 1.63767 |
| 1 | 0.00000 | 1.00000 | 0.50000 | -0.10644 |
| 2 | 0.50000 | 1.00000 | 0.75000 | 0.65362 |
| 3 | 0.50000 | 0.75000 | 0.62500 | 0.24632 |
| 4 | 0.50000 | 0.62500 | 0.56250 | 0.06321 |
| 9 | 0.53906 | 0.54297 | 0.54102 | 0.00338 |
| 12 | 0.53955 | 0.54004 | 0.53979 | 0.00003 |
| 13 | 0.53955 | 0.53979 | 0.53967 | -0.00031 |
| 14 | 0.53967 | 0.53979 | 0.53973 | -0.00014 |
| 15 | 0.53973 | 0.53979 | 0.53976 | -0.00006 |
| 16 | 0.53976 | 0.53979 | 0.53978 | -0.00002 |
| 17 | 0.53978 | 0.53979 | 0.53979 | 0.00001 |

**Algorithm**

**Subroutine** bisect($f$, $a$, $b$, root, maxit, eps)
Comment: $[a, b]$ contains the root; maxit is the maximum number of iterations; eps is tolerance.
1 **Print** $a$, $b$, maxit, eps
2 $n = 0$
3 $an = a$
4 $bn = b$
5 **Repeat** steps 6–10 **while** $n \leq$ maxit
    6 root $= (an + bn)/2$
    7 Print root, $f(\text{root})$, and $n$
    8 **If** $|f(\text{root})| \leq$ eps or $|bn - an| \leq$ eps, **then print** 'finished' and exit.
    9 **If** sign($f(\text{root})$) · sign($f(bn)$) $< 0$, **then** $an = $ root; **otherwise** $bn = $ root.
    10 $n = n + 1$
11 **Print** 'algorithm fails: no convergence' and exit.

How far is $r_n$ from the exact root $r$? Since we know that $r$ is in the closed interval $[a_n, b_n]$ and $r_n = (a_n + b_n)/2$, then

$$|r_n - r| = \left|\frac{a_n + b_n}{2} - r\right| \le \left|\frac{a_n + b_n}{2} - a_n\right| = \left|\frac{b_n - a_n}{2}\right| \tag{2.2.1}$$

From Figure 2.2.1,

$$b_1 - a_1 = \frac{b_0 - a_0}{2} = \frac{b - a}{2} \qquad b_2 - a_2 = \frac{b_1 - a_1}{2} = \frac{b - a}{2^2}$$

Continuing

$$b_n - a_n = \frac{b - a}{2^n} \qquad \text{for } n = 0, 1, \dots \tag{2.2.2}$$

Substituting Equation (2.2.2) in Equation (2.2.1) yields

$$|r_n - r| \le \frac{|b - a|}{2^{n+1}} \tag{2.2.3}$$

Equation (2.2.3) provides an upper bound for the error of the computed root $r_n$.

**EXAMPLE 2.2.2**

Determine the accuracy of the computed $r_9$ in Example 2.2.1

The true solution is $r \approx 0.53978517$. From Equation (2.2.3), $|r_9 - r| \le (2 - 0)/2^{10} = 2^{-9} \approx 0.00195$. The actual error $|r_9 - r| = 0.00123$ is less than the estimated error bound. ∎∎∎

Equation (2.2.3) can also be used to determine the number of iterations for the given accuracy. In order to have $|r_n - r| < \epsilon$, choose $n$ from Equation (2.2.3) such that

$$\frac{b - a}{2^{n+1}} < \epsilon \tag{2.2.4}$$

Equation (2.2.4) can be written as

$$2^{n+1} > \frac{b - a}{\epsilon} \tag{2.2.5}$$

Taking the log on both sides of Equation (2.2.5) and then solving for $n$ yields

$$n > \frac{\log\left(\dfrac{b - a}{\epsilon}\right)}{\log 2} - 1 \tag{2.2.6}$$

This gives only an upper bound for the number of iterations necessary. In this analysis, the effects of the round-off errors were not considered although they do affect the design of the algorithm.

In Example 2.2.1, in order to obtain the zero to within $\epsilon = 10^{-5}$,

$$n \ge \frac{\log(2. \times 10^5)}{\log 2} - 1 \approx 16.61$$

It would require 17 iterations.

The convergence of the bisection method is slow. On average one binary digit in accuracy is gained at each step. Since $1/10 \sim 1/2^{3.3}$, we gain one decimal point every 3.3 steps or about three decimal points in every ten steps. The convergence of the bisection method is completely independent of $f$. The major advantage of this method is that it is foolproof if an interval $[a, b]$ that contains a zero of $f$ can be found. Sometimes two numbers $a$ and $b$ cannot be found such that $f(a)f(b) < 0$. For example, $f(x) = x^2 - 4x + 4 = (x - 2)^2 \geq 0$ for all $x$. The bisection method cannot be used to find the zero 2 of $f(x) = (x - 2)^2$.

## 2.3    THE REGULA FALSI METHOD

The bisection method finds the successive values of $r_n$ using only the signs of $f(a_n)$ and $f(b_n)$ and not their relative magnitudes. For instance, if $f(a_n) = -0.5$ and $f(b_n) = 50$, then $r_n$ is selected as the midpoint of the closed interval $[a_n, b_n]$. It makes sense to select $r_n$ closer to $a_n$ than to $b_n$. The regula falsi method provides a way to select $r_n$ closer to $a_n$ than to $b_n$.

Assume that $f$ is continuous on a closed interval $[a_0, b_0]$ and $f(a_0)f(b_0) < 0$. First, we approximate the graph of $f$ by the straight line $L$ joining $A_0(a_0, f(a_0))$ and $B_0(b_0, f(b_0))$. Then we approximate the root $r$ of $f$ by the root $r_0$ of the straight line $L$ as shown in Figure 2.3.1.

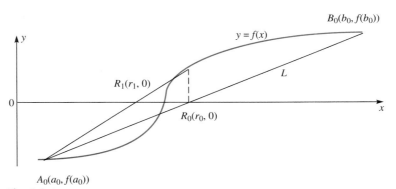

Fig. 2.3.1

The straight line $L$ intersects the $x$-axis at $R_0(r_0, 0)$. To find $r_0$, equate the slope of $R_0B_0$ to the slope of $B_0A_0$ and get

$$\frac{0 - f(b_0)}{r_0 - b_0} = \frac{f(b_0) - f(a_0)}{b_0 - a_0} \tag{2.3.1}$$

Solving Equation (2.3.1) for $r_0$

$$r_0 = b_0 - \frac{f(b_0)(b_0 - a_0)}{f(b_0) - f(a_0)} \tag{2.3.2}$$

In the denominator, $f(b_0)$ and $f(a_0)$ have opposite signs and therefore do not cause large errors. As in the bisection method, if $f(r_0) \neq 0$, then $f(a_0)f(r_0) > 0$ or $f(a_0)f(r_0) < 0$. Assuming $f(a_0)f(r_0) < 0$, the root $r$ lies in $[a_0, r_0]$. Set $a_1 = a_0$, $b_1 = r_0$, and

$$r_1 = b_1 - \frac{f(b_1)(b_1 - a_1)}{f(b_1) - f(a_1)} \tag{2.3.3}$$

If $f(r_1) = 0$, then $r_1$ is the root. If $f(r_1) \neq 0$, then $f(a_1)f(r_1) > 0$ or $f(a_1)f(r_1) < 0$. Assuming $f(a_1)f(r_1) > 0$, the root $r$ lies in $[r_1, b_1]$. Set $a_2 = r_1$ and $b_2 = b_1$ and determine $r_2$. The process is repeated using

$$r_n = b_n - \frac{f(b_n)(b_n - a_n)}{f(b_n) - f(a_n)} \tag{2.3.4}$$

until $|f(r_n)|$ or $|r_n - r_{n-1}|$ is as small as desired.

**EXAMPLE 2.3.1**

Find the root of $e^x - 2 \cos x = 0$ in the closed interval $[0, 2]$ using the regula falsi method.

Since $a_0 = 0$ and $b_0 = 2$, $f(0) = -1$ and $f(2) = 8.22135$. Using Equation (2.3.2),

$$r_0 = 2 - \frac{8.22135(2 - 0)}{8.22135 + 1} = 0.21689 \quad \text{and} \quad f(0.21689) = -0.71094$$

Since $f(b_0)f(r_0) = f(2)f(0.21689) < 0$, the root is in the closed interval $[0.21689, 2.00000]$.

Set $a_1 = r_0 = 0.21689$ and $b_1 = b_0 = 2$. Using Equation (2.3.3),

$$r_1 = 2 - \frac{8.22135(2 - 0.21689)}{8.22135 + 0.71094} = 0.35881 \quad \text{and} \quad f(0.35881) = -0.44101$$

Since $f(b_1)f(r_1) < 0$, the root is in the closed interval $[0.35881, 2.00000]$. Continuing the computed results are given in Table 2.3.1.

**Table 2.3.1**

| $n$ | $a_n$ | $b_n$ | $r_n$ | $f(r_n)$ |
|---|---|---|---|---|
| 0 | 0.00000 | 2.00000 | 0.21689 | $-0.71094$ |
| 1 | 0.21689 | 2.00000 | 0.35881 | $-0.44101$ |
| 2 | 0.35881 | 2.00000 | 0.44236 | $-0.25110$ |
| 3 | 0.44236 | 2.00000 | 0.48853 | $-0.13613$ |
| 4 | 0.48853 | 2.00000 | 0.51315 | $-0.07186$ |
| 5 | 0.51315 | 2.00000 | 0.52603 | $-0.03741$ |
| 6 | 0.52603 | 2.00000 | 0.53271 | $-0.01933$ |
| 7 | 0.53271 | 2.00000 | 0.53615 | $-0.00995$ |
| 8 | 0.53615 | 2.00000 | 0.53792 | $-0.00511$ |
| 9 | 0.53792 | 2.00000 | 0.53883 | $-0.00262$ |

In this example, $r_n$ replaces $a_n$ rather than $b_n$ because the graph of $f$ as shown in Figure 2.3.2 is concave upward ($f''(x) > 0$) in the interval $[0, 2]$.

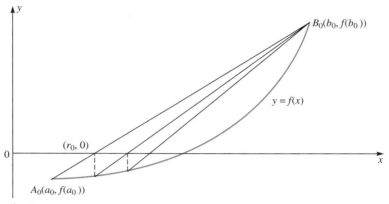

**Fig. 2.3.2**

The convergence is particularly slow when the interval endpoint on one side of the root becomes frozen far from the root, producing approximations that do not improve quickly. Although, as the iteration converges, it will almost always converge from one side. The method of regula falsi is designed to produce faster convergence than the bisection method. For some functions it does, but for others it does not. Therefore it is essential to have a procedure that will switch over to the bisection method if the regula falsi method fails to converge with fixed $N$ iterations.

## 2.4    THE SECANT METHOD

The regula falsi method can be modified in several ways. A popular modification, called the secant method, retains the use of the secant but gives up the bracketing of the roots. It is not necessary that $f(a_0)f(b_0) < 0$. The two most recent values of $r_i$ are used to determine the secant line.

Assume that $f$ is continuous on a closed interval $[a_0, b_0]$. Then the secant line joining $(a_0, f(a_0))$ and $(b_0, f(b_0))$ crosses the $x$-axis when $x = r_0$, which is given by Equation (2.3.2) as

$$r_0 = b_0 - \frac{f(b_0)(b_0 - a_0)}{f(b_0) - f(a_0)}$$
(2.4.1)

It is not necessary to check whether $f(a_0)f(r_0) > 0$ or $f(a_0)f(r_0) < 0$ as it was for the bisection and regula falsi methods. We use the secant line joining $(b_0, f(b_0))$ and $(r_0, f(r_0))$ which crosses the $x$-axis when $x = r_1$. This is given by

$$r_1 = r_0 - \frac{f(r_0)(r_0 - b_0)}{f(r_0) - f(b_0)} \quad \text{if } f(r_0) \neq f(b_0)$$
(2.4.2)

Continuing the process generates a sequence given by

$$r_{n+1} = r_n - \frac{f(r_n)(r_n - r_{n-1})}{f(r_n) - f(r_{n-1})} \quad \text{if } f(r_n) \neq f(r_{n-1}) \tag{2.4.3}$$

for $n = 0, 1, \ldots$ with $r_{-1} = a_0$ and $r_0 = b_0$.

Since we do not have $f(r_n)f(r_{n-1}) < 0$ for all $n$, the convergence of the sequence to the root is not guaranteed by the secant method. However, it can be shown that if the sequence generated by the secant method converges to a root, then it generally converges noticeably faster than the sequence generated by the bisection method. If the secant method does not converge and if $f(a)f(b) < 0$, then the bisection method or the regula falsi method can be used.

EXAMPLE 2.4.1    Find the root of $e^x - 2 \cos x = 0$ in the closed interval $[0, 2]$ using the secant method.
From the last example, $r_0 = 0.21689$ and $f(0.21689) = -0.71094$. Also $b_0 = 2$ and $f(2) = 8.22135$. Then from Equation (2.4.2), $r_1$ is given by

$$r_1 = r_0 - \frac{f(r_0)(r_0 - b_0)}{f(r_0) - f(b_0)} = 0.21689 - \frac{(-0.71094)(0.21699 - 2)}{(-0.71094 - 8.22135)} = 0.35881$$

The computed values of $r_i$ are given in Table 2.4.1.

**Table 2.4.1**

| $n$ | $r_n$ | $f(r_n)$ |
|---|---|---|
| 0 | 0.21689 | −0.71094 |
| 1 | 0.35881 | −0.44101 |
| 2 | 0.59068 | 0.14408 |
| 3 | 0.53358 | −0.01696 |
| 4 | 0.53959 | −0.00053 |
| 5 | 0.53979 | 0.00001 |

It is clear from Table 2.4.1 that it takes only six iterations for the secant method to have $|f(r_N)| < \frac{1}{2} \times 10^{-5}$, while the bisection and regula falsi methods give $|f(r_N)| > \frac{1}{2} \times 10^{-3}$ even after ten iterations.

EXERCISES

1. Carry out four steps of the bisection method to find the roots of the following equations in the given intervals:
   (a) $xe^x - 2 = 0$    $[0, 1]$        (c) $e^x - 5 \sin x = 0$    $[0, \pi/2]$
   (b) $x - 2 \cos x = 0$    $[0, \pi/2]$        (d) $x^3 - x^2 - x - 1 = 0$    $[1, 2]$
2. Carry out four steps of the regula falsi method to find the roots of the equations in Exercise 1.

3. Carry out four steps of the secant method to find the roots of the equations in Exercise 1.
4. Find $\sqrt[3]{9}$ using the bisection method, regula falsi method, or secant method. (*Hint*: Use $x^3 - 9 = 0$.)
5. Show that if the regula falsi method is applied to the equation $x^2 - 8 = 0$ with $a = 0$ and $b = 4$, then $b$ stays fixed and $a$ gradually approaches $\sqrt{8}$.
6. Denote the intervals that arise in the bisection method by $[a_0, b_0]$, $[a_1, b_1]$, ..., $[a_n, b_n]$, ... and $r_n = (a_n + b_n)/2$. Show that $\lim_{n \to \infty} a_n = \lim_{n \to \infty} b_n = \lim_{n \to \infty} r_n = r$, where $f(r) = 0$.
7. Show that if (1) $f''(x) \neq 0$ in $[a, b]$ and (2) $f(b)f''(b) > 0$, then $b$ remains one of the points of the regula falsi method.
8. Let $f$ be $(m + 1)$ times differentiable. Show that if $f$ has a zero of multiplicity $m$, then $f^{(m-1)}(r) = f^{(m-2)}(r) = \cdots = f'(r) = f(r) = 0$, but $f^{(m)}(r) \neq 0$.
9. Determine an upper bound for the number of iterations necessary to find the root of $e^x - 5 \sin x = 0$ with accuracy $10^{-5}$ in the interval $[0, \pi/2]$ when the bisection method is used.
10. Determine an upper bound for the number of iterations necessary to find the root of $x^3 - x^2 - x - 1 = 0$ with accuracy $10^{-5}$ in the interval $[1, 2]$ when the bisection method is used.

## COMPUTER EXERCISES

C.1. Write a subroutine that accepts $f$, $a$, $b$, maxit, and eps and carries out the bisection procedure.
C.2. Write a subroutine that accepts $f$, $a$, $b$, maxit, and eps and generates a sequence $r_n$ given by Equation (2.3.4).
C.3. Write a subroutine that accepts $f$, $a$, $b$, maxit, and eps and generates a sequence $r_n$ given by Equation (2.4.3).
C.4. Write a computer program to find a root $r$ of $f(x) = 0$ on a given interval $[a, b]$ such that $|f(r)| \leq 10^{-5}$ using Exercises C.1 and C.2. Print $n$, $a_n$, $b_n$, $r_n$, and $f(r_n)$.
C.5. Write a computer program to find a root $r$ of $f(x) = 0$ on a given interval $[a, b]$ such that $|f(r)| \leq 10^{-5}$ using Exercises C.1 and C.3. Print $n$, $r_n$, and $f(r_n)$.
C.6. Using Exercise C.4 or C.5 find the roots of Exercises 1(a) and 1(c) or Exercises 1(b) and 1(d). Compare the number of required iterations for 1(c) or 1(d) to an upper bound in Exercise 9 or 10.
C.7. Archimedes' principle states that when a solid of density $\rho_s$ is placed in a liquid of density $\rho_l$ where $\rho_s < \rho_l$, it displaces an amount of liquid whose weight equals that of the solid. A sphere of radius $r$ sinks to a height $h$ in liquid as shown in the figure. Because of Archimedes' principle

$$\frac{4}{3} \pi r^3 \rho_s = \frac{1}{3} \pi h^2 (3r - h)\rho_l$$

Find $h$ if $r = 1$, $\rho_l = 62.5$ lb/ft$^3$ (density of water), and $\rho_s = 0.6$ lb/ft$^3$.

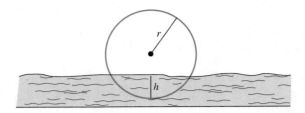

C.8.    The relation between the pipe friction $f$ and the Reynolds number Re for turbulent flow of a fluid in a smooth pipe known as the Kàrmàn-Prandtl universal relation is given by (Pao 1966)

$$\frac{1}{\sqrt{f}} = -0.8 + 2.0 \log(\text{Re}\sqrt{f}\,)$$

Find $f$ for Re $= 10^5$ and $10^6$.

C.9.    A mass moving in simple harmonic motion is described by

$$y(t) = 2e^{-t} \sin\left(2t - \frac{\pi}{6}\right), \quad t \geq 0$$

Find $t$ for which $|y(t)| = 0.5$.

## 2.5   RATES OF CONVERGENCE

We tried to improve the bisection method by the regula falsi and secant methods. In Example 2.4.1 we saw that the sequence generated by the secant method converges with a greater speed than the sequence generated by the regula falsi method. How fast does the sequence generated by a particular method converge to the solution? This question cannot be answered precisely because the speed with which the sequence converges may depend on the starting values as well as the round-off errors. However, there is a way to compare the rates of convergence of various methods.

Consider

$$\left\{\frac{11}{2}, \frac{21}{4}, \frac{41}{8}, \dots, 5 + \frac{1}{2^n}, \dots\right\} \quad \text{and} \quad \left\{\frac{11}{2}, \frac{41}{8}, \frac{641}{128}, \dots, 5 + \frac{1}{2^{2^n-1}}, \dots\right\}$$

The second sequence seems to converge faster to 5 than the first sequence. We would like to define this character by defining the order of convergence.

Let the sequence $\{r_n\}$ converge to $r$. Denote the difference between $r_n$ and $r$ by $e_n$; in other words, $e_n = r_n - r$. If there exists a positive number $p \geq 1$ and a constant $c \neq 0$ such that

$$\lim_{n \to \infty} \frac{|r_{n+1} - r|}{|r_n - r|^p} = \lim_{n \to \infty} \frac{|e_{n+1}|}{|e_n|^p} = c \tag{2.5.1}$$

then $p$ is called the **order of convergence** of the sequence. The constant $c$ is called the **asymptotic error** constant. If $p$ is large, then the sequence $\{r_n\}$ converges rapidly to $r$. If $p = 1$, then the convergence is said to be **linear** and we also require $c < 1$; if $p = 2$, then it is **quadratic**. It is not necessary that $p$ be an integer.

**EXAMPLE 2.5.1**    Find the order of convergence of the sequences

$$\left\{\frac{11}{2}, \frac{21}{4}, \frac{41}{8}, \ldots, 5 + \frac{1}{2^n}, \ldots\right\} \quad \text{and} \quad \left\{\frac{11}{2}, \frac{41}{8}, \frac{641}{128}, \ldots, 5 + \frac{1}{2^{2^n - 1}}, \ldots\right\}$$

Since $\lim_{n\to\infty} (5 + (1/2^n)) = 5$, the sequence converges to 5. Hence $e_n = r_n - r = 1/2^n$. Therefore

$$\lim_{n\to\infty} \frac{|e_{n+1}|}{e_n} = \lim_{n\to\infty} \frac{1}{2^{n+1}} 2^n = \frac{1}{2}$$

Thus the sequence

$$\left\{\frac{11}{2}, \frac{21}{4}, \frac{41}{8}, \ldots, 5 + \frac{1}{2^n}, \ldots\right\}$$

converges linearly to 5.

For

$$\left\{\frac{11}{2}, \frac{41}{8}, \frac{641}{128}, \ldots, 5 + \frac{1}{2^{2^n - 1}}, \ldots\right\}$$

since $\lim_{n\to\infty} (5 + (1/2^{2^n - 1})) = 5$, $e_n = r_n - r = 1/2^{2^n - 1}$. Hence

$$\lim_{n\to\infty} \left|\frac{e_{n+1}}{e_n^2}\right| = \lim_{n\to\infty} \frac{(2^{2^n - 1})^2}{2^{2^{n+1} - 1}} = \lim_{n\to\infty} \frac{2^{2^{n+1} - 2}}{2^{2^{n+1} - 1}} = \frac{2^{-2}}{2^{-1}} = \frac{1}{2}$$

Thus the sequence

$$\left\{\frac{11}{2}, \frac{41}{8}, \frac{641}{128}, \ldots, 5 + \frac{1}{2^{2^n - 1}}, \ldots\right\} \text{ converges quadratically to 5.}$$

■ ■ ■

Another form is sometimes very convenient for linear convergence; therefore, we derive it. Since $\lim_{n\to\infty} |e_{n+1}/e_n| = c$, there exists $k > 0$ such that $|e_{n+1}/e_n| \leq k$ for $n \geq 0$. Therefore

$$|e_{n+1}| \leq k\,|e_n| \quad \text{for } n = 0, 1, \ldots. \tag{2.5.2}$$

Hence $|e_{n+1}| \leq k\,|e_n| \leq k\,(k|e_{n-1}|) = k^2|e_{n-1}| \leq k^2\,(k|e_{n-2}|) = k^3|e_{n-2}| \ldots \leq k^{n+1}|e_0|$. This gives

$$|e_{n+1}| \leq k^{n+1}|e_0| \quad \text{for } n = 0, 1, \ldots. \tag{2.5.3}$$

We can use $k$ as a measure of convergence. If $k < 1$, then iterations converge; therefore, we require $c < 1$ in Equation (2.5.1) for linear convergence. In Equation (2.5.1) for $p = 1$, $c$ is called the rate of convergence.

In the case of the bisection method, Equation (2.5.1) may not exist for $p = 1$, yet comparison of Equation (2.2.3) with Equation (2.5.3) shows that the bisection method converges linearly with a rate of 1/2.

**EXAMPLE 2.5.2**

Discuss the convergence of Example 2.2.1.

The true solution $r \approx 0.53978517$. Since $|e_{12}| = |r_{12} - r| \approx 0.00000483$, $|e_{13}| = |r_{13} - r| \approx 0.00011517$, $|e_{14}| = |r_{14} - r| \approx 0.00005517$, $|e_{15}| = |r_{15} - r| \approx 0.00002517$, and $|e_{16}| = |r_{16} - r| \approx 0.00000517$. Then

$$\frac{|e_{13}|}{|e_{12}|} \approx 23.84472 \qquad \frac{|e_{14}|}{|e_{13}|} \approx 0.47904 \qquad \frac{|e_{15}|}{|e_{14}|} \approx 0.45623 \qquad \frac{|e_{16}|}{|e_{15}|} \approx 0.20540$$

We do not have a limit, but it can be verified that $|e_n| \leq (\frac{1}{2})^n |e_0|$ for $n = 0, 1, \dots$.

■ ■ ■

It can be shown that (Ralston 1965) the regula falsi method converges linearly if the function is concave upward or downward in a given interval $[a_0, b_0]$ which contains a zero of $f$.

It can be verified (Exercises 13 and 14) that the order of convergence of the secant method is $(1 + \sqrt{5})/2 \approx 1.6$ which is faster than linear but not quite quadratic. Therefore, if the sequence generated by the secant method converges to a root, then it converges faster than the sequence generated by the bisection method or the regula falsi method.

**EXAMPLE 2.5.3**

Discuss the convergence of Example 2.4.1

We have $|e_2| = |r_2 - r| \approx 0.05089$, $|e_3| = |r_3 - r| \approx 0.00621$, and $|e_4| = |r_4 - r| \approx 0.0002$ and $|e_3|/|e_2|^{(1+\sqrt{5})/2} \approx 0.768$, and $|e_4|/|e_3|^{(1+\sqrt{5})/2} \approx 0.72742$, while we have (Exercises 13 and 14)

$$\lim_{n \to \infty} \frac{|e_{n+1}|}{|e_n|^{(1+\sqrt{5})/2}} = M^{-(1-\sqrt{5})/2} = \left[ \frac{f''(r)}{2 f'(r)} \right]^{-(1-\sqrt{5})/2}$$

$$= \left[ \frac{e^r + 2 \cos r}{2(e^r + 2 \sin r)} \right]^{-(1-\sqrt{5})/2} \approx 0.74815$$

■ ■ ■

To investigate the meaning of order of convergence, let us consider linear and $p$th order methods to find a root $r$ of $f(x) = 0$. Let $\{r_0, r_1, \dots\}$ be a sequence generated by a linear method. Assume that the sequence $\{r_0, r_1, \dots\}$ converges to $r$ for which $r_0$ is very close to $r$ such that

$$|r_{n+1} - r| \approx c |r_n - r| \quad \text{for } n = 0, 1, \dots$$

Then

$$|e_{n+1}| \approx c |e_n| \approx c^2 |e_{n-1}| \cdots \approx c^{n+1} |e_0| \tag{2.5.4}$$

Suppose that another sequence $\{\hat{r}_0, \hat{r}_1, \ldots\}$ generated by a $p$th order method converges to $r$ for which $\hat{r}_0$ is very close to $r$ such that

$$|\hat{r}_{n+1} - r| \approx k\,|\hat{r}_n - r|^p \quad \text{for } n = 0, 1, \ldots$$

Then

$$\begin{aligned}|\hat{e}_{n+1}| &\approx k\,|\hat{e}_n|^p \approx k\,(k|\hat{e}_{n-1}|^p)^p = k^{1+p}\,|\hat{e}_{n-1}|^{p^2} \\ &\approx \cdots \approx k^{1+p+p^2+\cdots+p^n}\,|\hat{e}_0|^{p^{n+1}}\end{aligned} \quad (2.5.5)$$

Since $(1 + p + p^2 + \cdots + p^n)(1 - p) = 1 - p^{n+1}$, $1 + p + p^2 + \cdots + p^n = (1 - p^{n+1})/(1 - p)$. So Equation (2.5.5) reduces to

$$|\hat{e}_{n+1}| \approx k^{(1-p^{n+1})/(1-p)}\,|\hat{e}_0|^{p^{n+1}} \quad (2.5.6)$$

To compare the speed of convergence, consider Example 2.2.1. For the bisection method $c = 1/2$, $r_0 = 1$, and $|e_0| = |r_0 - r| = |1 - 0.53797| = 0.46021$. Determine the minimum value of $n$ such that $|e_n| \leq 10^{-5}$ for the bisection method which is linear. From Equation (2.5.4), we have $|e_n| \approx c^n\,|e_0| \leq 10^{-5}$. Substituting $c = 1/2$ and $|e_0| = 0.46021$, we get

$$\left(\frac{1}{2}\right)^n (0.46021) \leq 10^{-5}$$

Taking the log on both sides and solving for $n$, we get

$$n \geq \frac{5 + \log(0.46021)}{\log(2.0)} \approx 15.49$$

The secant method has $(1 + \sqrt{5})/2\ (= p)$ order of convergence (Exercise 14). When the secant method is used to find the root of $f(x) = e^x - 2\cos x = 0$ starting with $r_0 = 0.21689$ (Example 2.4.1), then $|\hat{e}_0| = |r_0 - r| = |0.21689 - 0.53979| = 0.32290$. From Example 2.5.3, $k = M^{-(1-\sqrt{5})/2} \approx 0.74815$. To determine the minimum value of $n$ such that $|\hat{e}_n| \leq 10^{-5}$ for the secant method, from Equation (2.5.6)

$$|\hat{e}_n| \approx k^{(1-p^n)/(1-p)}\,|\hat{e}_0|^{p^n} \leq 10^{-5}$$

Substituting $k = 0.74815$ and $|\hat{e}_0| = 0.32290$ yields

$$(0.74815)^{(1-p^n)/(1-p)}\,(0.32290)^{p^n} \leq 10^{-5}$$

Taking the log on both sides and solving for $p^n$ yields

$$p^n \geq \frac{-5 + (\log(0.74815)/(p-1))}{(\log(0.74815)/(p-1)) + \log(0.32290)}$$

Substituting $p = (1 + \sqrt{5})/2 \approx 1.61803$ and solving for $n$, we get $n \geq 4.18$. The bisection method requires 16 iterations while the secant method requires only 5 iterations. Thus the higher order method is preferable to the linear method.

Even by knowing the order of convergence and the asymptotic error constant, the required number of iterations to achieve a desired accuracy cannot be known since the limiting behavior holds when the values of iterates are very close to the root.

## 2.6   AITKEN'S $\Delta^2$ METHOD

Let a sequence $\{r_n\}$ converge linearly to $r$. We want to construct a sequence $\{\hat{r}_n\}$ that converges to $r$ faster than the sequence $\{r_n\}$. Since $\{r_n\}$ converges linearly to $r$, for large $n$ we have

$$\frac{r_{n+1} - r}{r_n - r} \approx c \tag{2.6.1}$$

As the iterates $r_{n+1}$ approach $r$, $c$ will remain nearly constant at each stage. Thus at the next stage

$$\frac{r_{n+2} - r}{r_{n+1} - r} \approx c \tag{2.6.2}$$

From Equations (2.6.1) and (2.6.2), for large $n$ we have

$$\frac{r_{n+1} - r}{r_n - r} \approx \frac{r_{n+2} - r}{r_{n+1} - r}$$

Thus

$$(r_{n+1} - r)^2 \approx (r_n - r)(r_{n+2} - r) \tag{2.6.3}$$

Solving Equation (2.6.3) for $r$ (Exercise 6), we get

$$r \approx \frac{r_n r_{n+2} - r_{n+1}^2}{r_n - 2 r_{n+1} + r_{n+2}} \tag{2.6.4}$$

Adding $r_n$ and $-r_n$ on the right-hand side of Equation (2.6.4) and simplifying (Exercise 7), results in

$$r \approx r_n - \frac{(r_{n+1} - r_n)^2}{r_{n+2} - 2 r_{n+1} + r_n} \tag{2.6.5}$$

The **forward difference** of the sequence $\{r_n\}$ is defined by

$$\Delta r_n = r_{n+1} - r_n \quad \text{for } n = 0, 1, \ldots$$

Then

$$\Delta^2 r_n = \Delta(\Delta r_n) = \Delta(r_{n+1} - r_n) = r_{n+2} - 2 r_{n+1} + r_n$$

Thus Equation (2.6.5) becomes

$$r \approx r_n - \frac{(\Delta r_n)^2}{\Delta^2 r_n} \tag{2.6.6}$$

By using Equation (2.6.6), a new sequence $\{\hat{r}_n\}$ is formed and is given by

$$\hat{r}_n = r_n - \frac{(\Delta r_n)^2}{\Delta^2 r_n} \tag{2.6.7}$$

Thus it is possible to construct a sequence $\{\hat{r}_n\}$ that converges to $r$. By using Equation (2.6.7), it can be shown that the sequence $\{\hat{r}_n\}$ converges faster to $r$ than $\{r_n\}$ by proving

$$\lim_{n \to \infty} \frac{\hat{r}_n - r}{r_n - r} = 0 \tag{2.6.8}$$

**EXAMPLE 2.6.1**

Use Aitken's $\Delta^2$ method to generate the sequence $\{\hat{r}_n\}$ for Example 2.3.1 (the regula falsi method).

For $n = 4$, we have

$$\hat{r}_4 = r_4 - \frac{(r_5 - r_4)^2}{(r_6 - r_5) - (r_5 - r_4)}$$
$$= 0.51315 - \frac{(0.01288)^2}{(-0.00620)} = 0.53991$$

Similarly, the computed values for $\hat{r}_5$, $\hat{r}_6$, and $\hat{r}_7$ are given in Table 2.6.1.

**Table 2.6.1**

| $n$ | $r_n$ | $\hat{r}_n$ | $f(\hat{r}_n) = e^{\hat{r}_n} - 2\cos\hat{r}_n$ |
|---|---|---|---|
| 4 | 0.51315 | 0.53991 | 0.00034 |
| 5 | 0.52603 | 0.53980 | 0.00004 |
| 6 | 0.53271 | 0.53980 | 0.00004 |
| 7 | 0.53615 | 0.53979 | 0.00001 |
| 8 | 0.53792 | | |
| 9 | 0.53883 | | |

The values for $\hat{r}_5$, $\hat{r}_6$, and $\hat{r}_7$ are significantly more accurate than $r_5$, $r_6$, and $r_7$. ■ ■ ■

**Algorithm**

**Subroutine** Aitken (**r**, M, **rh**, eps)
Comment: **r** $= [r_0, r_1, \ldots, r_M]$; **rh** $= [\hat{r}_0, \hat{r}_1, \ldots, \hat{r}_{M-2}]$; eps is tolerance.
   1 **Print r**, M, eps
   2 $n = 0$
   3 **Repeat** steps 4–10 **while** $n \leqslant M - 2$
      4 $p = r_{n+1} - r_n$
      5 $q = r_{n+2} - r_{n+1} - p$
      6 $rh_n = r_n - (p^2/q)$
      7 **If** $n = 0$, **then** go to 9
      8 **If** $|rh_n - rh_{n-1}| <$ eps, **then print** 'finished' and exit.
      9 **Print** $n$, $r_n$, $rh_n$
      10 $n = n + 1$
   11 **Print** 'algorithm fails; no convergence' and exit. ◀

**Comments**

1. It is of no use to use Equation (2.6.7) for a sequence that converges quadratically.
2. If the convergence factor $c$ in Equation (2.6.1) is very close to one, then the denominator in Equation (2.6.7) is very small. Be very cautious of the acceleration process.

## 2.7 STOPPING CRITERIA

Imagine designing a black box that inspects the output of an algorithm and then decides the best stopping point. For finding a root, find a sequence $\{r_n\}$ such that $\lim_{n\to\infty} r_n = r$ where $f(r) = 0$. We want $r_N$ such that

$$|r_N - r| \leq \epsilon \qquad (2.7.1)$$

where $\epsilon$ is the required tolerance. The tolerance $\epsilon$ should be greater than $k \cdot eps$ where eps is a machine epsilon and $k = \max\{f(x)|x \in [a, b]\}$. Since $r$ is not known, we cannot find $r_N$ such that Equation (2.7.1) is satisfied. How do we find $r_N$? In other words, when do we stop? Unfortunately, there is no definite answer. However, one of the following simple tests can be used.

$$(1)\ |f(r_N)| \leq \delta \qquad (2)\ |r_N - r_{N-1}| \leq \delta \qquad (3)\ \frac{|r_N - r_{N-1}|}{|r_N|} \leq \delta \qquad (2.7.2)$$

where $\delta$ is a given tolerance. Often Equation (2.7.2) is used with $\delta = \epsilon$ to find $r_N$ and we hope Equation (2.7.1) is satisfied. When Equation (2.7.2) is used, be aware of the problems involved in using it.

Consider the first test. If $|f(r_N)| \leq \delta$, then stop and use this $r_N$ with the hope that it satisfies Equation (2.7.1). Geometrically, $|f(r_N)| \leq \delta$ means that calculations are terminated once $P_N(r_N, f(r_N))$ is in the horizontal band bounded by the lines $y = \delta$ and $y = -\delta$ as shown in Figure 2.7.1. Sometimes, as shown in Figure 2.7.1, $\delta \neq \epsilon$. In Case 2, $|r_N - r|$ is many times larger than $\delta$. In Case 3, $|r_N - r|$ is smaller than $\delta$.

We have

$$f(r_N) = f(r_N) - 0 = f(r_N) - f(r) = f'(\xi)\,(r_N - r)$$

where $\xi \in (r_N, r)$ by the Mean Value Theorem. Therefore,

$$|r_N - r| = \frac{|f(r_N)|}{|f'(\xi)|} \leq \frac{\delta}{|f'(\xi)|} \qquad (2.7.3)$$

If $|f'(\xi)|$ is very small, then $|r_N - r|$ is likely to be much larger than $\delta$, our required tolerance. In other words. we stopped too early (Figure 2.7.1, Case 2). On the other hand, if $|f'(\xi)|$ is very large, then $|r_N - r|$ is much smaller than $\delta$, our required tolerance. In other words, we stopped too late (Figure 2.7.1, Case 3). This leads to unnecessary computations. If $|f'(\xi)| \approx 1$, then $|r_N - r| \leq \epsilon$ (Figure 2.7.1, Case 1).

**EXAMPLE 2.7.1** Let $f(x) = (x - \sqrt{2})^5$. Start with the bisection method on the interval $[1, 2]$. Verify that the root lies in the interval $[1.0, 1.5]$.

Our stopping criterion is to have $|f(r_N)| < \frac{1}{2} \times 10^{-5}$. Since $f(1.5) = (1.5 - \sqrt{2})^5 < 4.6 \times 10^{-6}$, we stop and conclude that 1.5 is the root. But if we look at $|r_N - r| = |1.5 - \sqrt{2}| = |-.0858| = 8.5 \times 10^{-2}$, then $|f(r_N)| \leq \delta$ is not a very good stopping criteria in this case because $|f'(\xi)| = 5|\xi - \sqrt{2}|^4 \leq 2.7 \times 10^{-4}$.

■ ■ ■

Similarly one can depict the case of $|f'(\xi)|$ being very large.

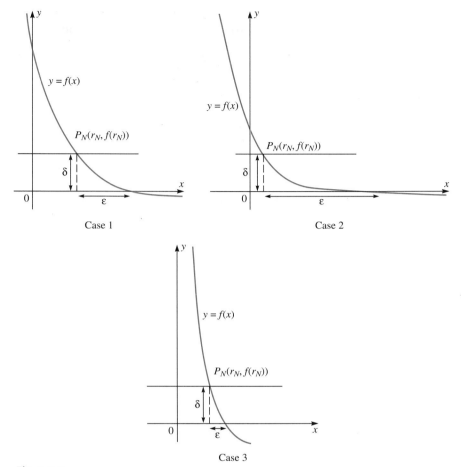

**Fig. 2.7.1**

Look at the second stopping criteria. It is possible that $|r_N - r_{N-1}| \to 0$ as $N \to \infty$ but $\lim_{N \to \infty} r_N \neq r$. Consider the harmonic series $1 + \frac{1}{2} + \cdots + \frac{1}{N} + \cdots$. Define $r_1 = 1$, $r_2 = 1 + \frac{1}{2}$, $r_3 = 1 + \frac{1}{2} + \frac{1}{3}, \ldots, r_N = 1 + \frac{1}{2} + \frac{1}{3} + \cdots + \frac{1}{N}$. Then $r_N - r_{N-1} = 1/N$ where $r_N = \sum_{i=1}^{N} \frac{1}{i}$. We know that $\lim_{N \to \infty} r_N$ does not exist, however $\lim_{N \to \infty} (r_N - r_{N-1}) = 0$. The sequence $\{r_1, r_2, \ldots, r_N, \ldots\}$ does not converge and yet $\lim_{N \to \infty} (r_N - r_{N-1}) = 0$.

Thus it seems hopeless to design a black box that works in every case; however, calculations are terminated in each of the following cases:

1.  When $|r_N - r_{N-1}| < \epsilon$ and $|r_N - r_{N-1}|/|r_N| < \epsilon$ provided $r_N \neq 0$. Also, it is convenient to use $|r_N - r_{N-1}| < \epsilon_1 |r_N| + \epsilon_2$ which is a combination of $|r_N - r_{N-1}| < \epsilon$ and $|r_N - r_{N-1}|/|r_N| < \epsilon$.

2.  When an upper bound on the number of iterations is reached. The stopping criteria will not be satisfied if the sequence of iterates diverges or the computer program is incorrect. Therefore, in writing a general purpose program, it is a good practice to set an upper bound on the number of iterations to be performed.

Check whether sequences 1–4 converge linearly to 0:

1. $\{1, 1/3, 1/3^2, \ldots, 1/3^{n-1}, \ldots\}$
2. $\{1, 1/2^2, 1/3^2, \ldots, 1/n^2, \ldots\}$
3. $\{1, 1/2^k, 1/3^k, \ldots, 1/n^k, \ldots\}$ where $k$ is any positive integer
4. $\{1, 1/10, 1/10^2, \ldots, 1/10^{n-1}, \ldots\}$
5. Show that $\{1/10, 1/10^2, 1/10^4, 1/10^8, \ldots, 1/10^{2^{n-1}}, \ldots\}$ converges quadratically to zero.
6. Verify Equation (2.6.4).
7. Verify Equation (2.6.5).
8. By adding $r_{n+2}$ and $-r_{n+2}$ on the right-hand side of Equation (2.6.4) and simplifying, verify that

$$r \approx r_{n+2} - \frac{(r_{n+2} + r_{n+1})^2}{r_{n+2} - 2r_{n+1} + r_n}$$

   Is there any preference between this formula and Equation (2.6.5)?
9. Prove Equation (2.6.8).
10. Let $r_n = 1/4^n$. find $\hat{r}_{16}, \hat{r}_{17}, \hat{r}_{18}$, and $\hat{r}_{19}$ using Equation (2.6.7).
11. Let $r_n = \sum_{i=1}^n (0.9)^i$. Find $\hat{r}_{16}, \hat{r}_{17}, \hat{r}_{18}$, and $\hat{r}_{19}$.
12. Another method to find a zero of $f$ can be developed by starting with initial approximations $r_0$ and $r_1$. The value of $r_0$ is kept fixed and successive iterations are given by

$$r_{n+1} = r_n - \frac{(r_n - r_0)}{f(r_n) - f(r_0)} f(r_n)$$

   If the process converges to a zero of $f$, then prove that the rate of convergence to a simple zero is linear.

In Exercises 13 and 14, we determine the order of convergence of the secant method.

13. (a) By subtracting $r$ from both sides of Equation (2.4.3) and denoting $e_n = r_n - r$, verify that Equation (2.4.3) reduces to

$$e_{n+1} = \frac{e_{n-1} f(r_n) - e_n f(r_{n-1})}{f(r_n) - f(r_{n-1})} \quad \text{if} \quad f(r_n) \neq f(r_{n-1})$$

   (b) Verify that (a) can be written as

$$e_{n+1} = e_n e_{n-1} \frac{\left[\dfrac{f(r_n)}{e_n} - \dfrac{f(r_{n-1})}{e_{n-1}}\right]}{r_n - r_{n-1}} \frac{r_n - r_{n-1}}{f(r_n) - f(r_{n-1})}$$

   (c) Use $f(r) = 0$, $g(x) = (f(x) - f(r))/(x - r)$ and the Mean Value Theorem to reduce (b) to

$$e_{n+1} = e_n e_{n-1} \frac{g'(\xi_n)}{f'(\zeta_n)}$$

   where $\xi_n$ and $\zeta_n$ are in $(r_{n-1}, r_n)$ assuming $f'(x)$ exists and $f'(x) \neq 0$ for all $x \in [a, b]$.

(d) Show that $\lim_{n\to\infty} e_{n+1}/(e_n e_{n-1}) = f''(r)/(2f'(r)) = M \neq 0$ assuming $f''(r) \neq 0$.

(e) Take the ln, and using $q_n = \ln(Me_n)$, reduce the difference equation $e_{n+1} = Me_n e_{n-1}$ to $q_{n+1} - q_n - q_{n-1} = 0$.

**14.** (a) Solve the difference equation $q_{n+1} - q_n - q_{n-1} = 0$. (*Hint*: Use Appendix D.)

(b) Since $q_n = \ln(Me_n)$, verify that the solution of (a) reduces to $e_n = e^{c_1((1+\sqrt{5})/2)^n}/M$ for large $n$ where $c_1$ is a constant.

(c) Using (b), show that $\lim_{n\to\infty} |e_{n+1}|/|e_n|^{(1+\sqrt{5})/2} = M^{-((1-\sqrt{5})/2)}$. Thus the order of convergence of the secant method is $(1 + \sqrt{5})/2 \approx 1.618$.

---

**COMPUTER EXERCISES**

**C.1.** Find the zero of $f(x) = (x - \sqrt{3})^{1/5}$ using the bisection method subroutine (Exercise C.1, p. 25) such that (a) $|f(r_n)| < 10^{-5}$, and (b) $|r_n - r_{n-1}| < 10^{-5}$. Print $n$, $r_n$, $f(r_n)$, and $r_n - r$ in both cases. (c) Compare the number of required iterations in (a) and (b). Analyze the stopping criteria in (a) and (b).

**C.2.** Find the zero of $f(x) = (x - \sqrt{3})^5$ using the bisection method subroutine (Exercise C.1, p. 25) such that (a) $|f(r_n)| < 10^{-5}$ and (b) $|r_n - r_{n-1}| < 10^{-5}$. Print $n$, $r_n$, $f(r_n)$, and $r_n - r$ in both cases. (c) Compare the number of required iterations in (a) and (b). Analyze the stopping criteria in (a) and (b).

**C.3.** Write a subroutine that accepts elements $r_0, r_1, \ldots, r_M$ and generates a sequence $\hat{r}_n$ given by Equation (2.6.7).

**C.4.** Find the zero of $f(x) = \cos x - x \sin x$ in the closed interval $[0, \pi]$ using the regula falsi method subroutine (Exercise C.2, p. 25). Use Exercise C.3 to generate the sequence $\hat{r}_n$. Compare $\hat{r}_n$ and $\hat{\hat{r}}_n$.

---

## 2.8   THE NEWTON–RAPHSON AND CAUCHY METHODS

In Section 2.4, the secant line was used to approximate the root of $f(x) = 0$. The tangent line may also be used to approximate the root of $f(x) = 0$. Let $f$ take a zero in $[a, b]$. To approximate $r$, begin with an initial approximation $r_0$ to $r$. Then construct the tangent line to the graph of $f$ at $(r_0, f(r_0))$ and call the $x$-intercept of this tangent line $r_1$ as shown in Figure 2.8.1.

The equation of the tangent line through $P_0(r_0, f(r_0))$ with slope $f'(r_0)$ is given by

$$y - f(r_0) = f'(r_0)(x - r_0) \tag{2.8.1}$$

Since the tangent line crosses the $x$-axis at $Q_1(r_1, 0)$, the coordinates $(r_1, 0)$ of $Q_1$ satisfy Equation (2.8.1). Therefore

$$0 - f(r_0) = f'(r_0)(r_1 - r_0) \tag{2.8.2}$$

Solving Equation (2.8.2) for $r_1$, results in

$$r_1 = r_0 - \frac{f(r_0)}{f'(r_0)} \tag{2.8.3}$$

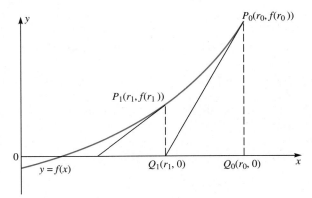

**Fig. 2.8.1**

provided $f'(r_0) \neq 0$. Similarly a tangent line is drawn at $P_1(r_1, f(r_1))$ and its $x$-intercept $r_2$ is given by

$$r_2 = r_1 - \frac{f(r_1)}{f'(r_1)}$$

Continuing, we get

$$r_{n+1} = r_n - \frac{f(r_n)}{f'(r_n)} \quad \text{for } n = 0, 1, \ldots \tag{2.8.4}$$

provided $f'(r_n) \neq 0$. The algorithm defined by Equation (2.8.4) is known as the **Newton–Raphson method** or sometimes just the **Newton method**. Isaac Newton solved a cubic polynomial using this method and the first systematic account of this method was given by Joseph Raphson in 1690.

We can also derive Newton–Raphson's formula using Taylor's Theorem (Theorem A.8). Let $r_n$ be close to the root $r$ of $f(x) = 0$. Then

$$\begin{aligned} 0 = f(r) &= f(r_n + (r - r_n)) \\ &= f(r_n) + (r - r_n)f'(r_n) + \frac{(r - r_n)^2}{2!}f''(\xi_n) \end{aligned} \tag{2.8.5}$$

where $\xi_n \in (r_n, r)$. For small values of $(r - r_n)$ in well-behaved functions, the $(r - r_n)^2$ term is negligible and therefore yields

$$f(r) \approx f(r_n) + (r - r_n)f'(r_n)$$

Since $f(r) = 0$,

$$f(r_n) + (r - r_n)f'(r_n) \approx 0 \tag{2.8.6}$$

Solving Equation (2.8.6) for $r$, produces

$$r \approx r_n - \frac{f(r_n)}{f'(r_n)}$$

Since $r_n$ was assumed to be close to $r$,

$$r_{n+1} = r_n - \frac{f(r_n)}{f'(r_n)} \tag{2.8.4}$$

is presumably closer to $r$.

Keep $(r - r_n)^2$ and assume that $(r - r_n)^3$ is neligible. Then

$$f(r) \approx f(r_n) + (r - r_n) f'(r_n) + \frac{(r - r_n)^2}{2!} f''(r_n) \tag{2.8.7}$$

Since $f(r) = 0$,

$$(r - r_n)^2 f''(r_n) + 2(r - r_n) f'(r_n) + 2f(r_n) \approx 0 \tag{2.8.8}$$

This is a quadratic equation in $(r - r_n)$. Solving Equation (2.8.8) for $(r - r_n)$, yields

$$r - r_n \approx \frac{-f'(r_n) \pm \sqrt{(f'(r_n))^2 - 2f(r_n)f''(r_n)}}{f''(r_n)}$$

$$\approx -\frac{\left[ f'(r_n) \mp \sqrt{(f'(r_n))^2 - 2f(r_n)f''(r_n)} \right]}{f''(r_n)} \tag{2.8.9}$$

In order to avoid subtraction of almost equal terms, multiplying and dividing the right-hand side of Equation (2.8.9) by

$$f'(r_n) \pm \sqrt{(f'(r_n))^2 - 2f(r_n)f''(r_n)}$$

and selecting the positive sign produces (Exercise 17) **Cauchy's formula**:

$$r_{n+1} = r_n - \frac{2f(r_n)}{f'(r_n) \left[ 1 + \sqrt{1 - \dfrac{2f(r_n)f''(r_n)}{(f'(r_n))^2}} \right]} \tag{2.8.10}$$

provided $f'(r_n) \neq 0$. This formula requires two derivatives, which is not a very serious problem when working with polynomial equations.

Cauchy's formula can be simplified by approximating

$$\sqrt{1 - \frac{2f(r_n)f''(r_n)}{(f'(r_n))^2}} \qquad \text{by} \qquad 1 - \frac{f(r_n)f''(r_n)}{(f('(r_n))^2}$$

using the binomial expansion yielding **Halley's formula**:

$$r_{n+1} = r_n - \frac{f(r_n)}{f'(r_n) \left[ 1 - \dfrac{f(r_n)f''(r_n)}{2(f'(r_n))^2} \right]} \tag{2.8.10a}$$

Further, approximating

$$\left[ 1 - \frac{f(r_n)f''(r_n)}{2(f'(r_n))^2} \right]^{-1} \qquad \text{by} \qquad 1 + \frac{f(r_n)f''(r_n)}{2(f'(r_n))^2}$$

using the binomial expansion yielding **Chebyshev's formula**:

$$r_{n+1} = r_n - \frac{f(r_n)}{f'(r_n)} \left[ 1 + \frac{f(r_n)f''(r_n)}{2(f'(r_n))^2} \right] \tag{2.8.10b}$$

**EXAMPLE 2.8.1**

Find the root of $e^x - 2 \cos x = 0$ in the closed interval $[0, 2]$ using the Newton–Raphson and Chebyshev methods.

Let $f(x) = e^x - 2 \cos x$. Then $f'(x) = e^x + 2 \sin x$ and Equation (2.8.4) reduces to

$$r_{n+1} = r_n - \frac{f(r_n)}{f'(r_n)} = r_n - \frac{e^{r_n} - 2 \cos r_n}{(e^{r_n} + 2 \sin r_n)} \quad \text{for } n = 0, 1, \ldots$$

The bisection and secant methods required two values to start computations, while the Newton–Raphson and Cauchy methods only require one value.

For $r_0 = 2$ and $n = 0$,

$$r_1 = 2 - \frac{(e^2 - 2 \cos 2)}{(e^2 + 2 \sin 2)} = 2 - \frac{8.22135}{9.20765} = 2 - 0.89288 = 1.10712$$

Continue until $|f(r_n)| < 10^{-5}$ and $|r_{n-1} - r_n|/|r_n| < 10^{-5}$. The computed values are given in Table 2.8.1

**Table 2.8.1**

| $n$ | $r_n$ | $f(r_n)$ | $\dfrac{\lvert r_n - r_{n-1} \rvert}{\lvert r_n \rvert}$ |
|---|---|---|---|
| 0 | 2.0 | | |
| 1 | 1.10712 | 2.13114 | 0.80649 |
| 2 | 0.66000 | 0.36895 | 0.66619 |
| 3 | 0.54832 | 0.02354 | 0.21181 |
| 4 | 0.53983 | 0.00012 | 0.01573 |
| 5 | 0.53979 | 0.00000 | 0.00008 |
| 6 | 0.53979 | 0.00000 | 0.00000 |

Since $f''(x) = e^x + 2 \cos x$, Equation (2.8.10b) reduces to

$$r_{n+1} = r_n - \frac{(e^{r_n} - 2 \cos r_n)}{(e^{r_n} + 2 \sin r_n)} - \frac{(e^{r_n} - 2 \cos r_n)^2(e^{r_n} + 2 \cos r_n)}{2(e^{r_n} + 2 \sin r_n)^3} \quad \text{for } n = 0, 1, \ldots$$

For $r_0 = 2$ and $n = 0$,

$$r_1 = r_0 - \frac{(e^2 - 2 \cos 2)}{(e^2 + 2 \sin 2)} - \frac{(e^2 - 2 \cos 2)^2(e^2 + 2 \cos 2)}{2(e^2 + 2 \sin 2)^3}$$

$$= 2 - 0.89288 - \frac{(67.59059)(6.55676)}{2(9.20765)^3}$$

$$= 2 - 0.89288 - 0.28386 = 0.82326$$

The computation is continued until $|f(r_n)| < 10^{-5}$ and $|r_{n-1} - r_n|/|r_n| < 10^{-5}$. The computed values are given in Table 2.8.2.

**Table 2.8.2**

| $n$ | $r_n$ | $f(r_n)$ | $\dfrac{\|r_{n-1} - r_n\|}{\|r_n\|}$ |
|---|---|---|---|
| 0 | 2 | | |
| 1 | 0.82326 | 0.91825 | 1.42936 |
| 2 | 0.54884 | 0.02498 | 0.50001 |
| 3 | 0.53979 | 0.00000 | 0.16770 |
| 4 | 0.53979 | 0.00000 | 0.00000 |

■ ■ ■

**Algorithm**

**Subroutine** Newton ( $f$, $df$, rn, root, maxit, eps)
Comment: $df = f'$; $rn = r_0$; maxit is the maximum number of iterations to be computed; eps is tolerance.
  1 **Print** rn, maxit, eps
  2 $n = 0$
  3 **Repeat** steps 4–10 **while** $n \leq$ maxit
    4 deno $= df(rn)$
    5 **If** deno $= 0$, **then print** 'algorithm fails: deno $= 0$' and exit.
    6 root $= rn - (f(rn)/\text{deno})$
    7 **Print** root, $f$(root), and $n$
    8 **If** $|\text{root} - rn| <$ eps and $|rn - \text{root}| <$ eps $|\text{root}|$, **then** print 'finished' and exit.
    9 $rn =$ root
  10 $n = n + 1$
11 **Print** 'algorithm fails: no convergence' and exit.

◀

## Rate of Convergence

Subtracting $r$ from both sides of Equation (2.8.4) yields

$$r_{n+1} - r = r_n - r - \frac{f(r_n)}{f'(r_n)}$$

$$= \frac{(r_n - r)f'(r_n) - f(r_n)}{f'(r_n)} \tag{2.8.11}$$

Using Equation (2.8.5), Equation (2.8.11) reduces to

$$r_{n+1} - r = \frac{(r_n - r)^2}{2} \frac{f''(\xi_n)}{f'(r_n)} \tag{2.8.12}$$

Denoting $r_n - r = e_n$ in Equation (2.8.12), rewriting, and taking the limit, we obtain

$$\lim_{n \to \infty} \left| \frac{e_{n+1}}{e_n^2} \right| = \lim_{n \to \infty} \left| \frac{f''(\xi_n)}{2f'(r_n)} \right| = \frac{1}{2} \left| \frac{f''(r)}{f'(r)} \right| \tag{2.8.13}$$

Thus **Newton–Raphson's method** converges **quadratically** if $f'(r) \neq 0$. The quadratic convergence means that near the root the error at the end of one step is proportional to the square of the error at the beginning of the step. The **Cauchy method** has a **third order** convergence (Exercise 16); however, it requires the second derivative so more computations are needed at each step. Therefore, the Newton–Raphson method has become the method of choice for functions with simple zeros.

## Multiple Roots

Let $f(x) = (x - r)^m q(x)$ where $m$ is a positive integer and $\lim_{x \to r} q(x) \neq 0$. We can verify (Exercise 11) that Equation (2.8.11) reduces to

$$r_{n+1} - r = (r_n - r) \frac{[(m - 1) q(r_n) + (r_n - r) q'(r_n)]}{m q(r_n) + (r_n - r) q'(r_n)} \qquad (2.8.14)$$

Using $e_n = r_n - r$ in Equation (2.8.14), we can verify that

$$\lim_{n \to \infty} \left| \frac{e_{n+1}}{e_n} \right| = \frac{m - 1}{m} \qquad (2.8.15)$$

Thus, the convergence is linear in the case of multiple roots. If it is known that the root has multiplicity $m$, then the Newton–Raphson method could be accelerated using the formula

$$r_{n+1} = r_n - m \frac{f(r_n)}{f'(r_n)} \quad \text{for } n = 0, 1, \ldots \qquad (2.8.16)$$

For $r_0$ sufficiently close to $r$, Equation (2.8.16) is quadratically convergent (Exercise 12). Rarely we know in advance the multiplicity of a root, so Equation (2.8.16) is of little practical use. Another method to find a root of unknown multiplicity is to define

$$\mu(x) = \frac{f(x)}{f'(x)}$$

which has a simple zero at $r$ (Exercise 13). Use

$$r_{n+1} = r_n - \frac{\mu(r_n)}{\mu'(r_n)} \quad \text{for } n = 0, 1, \ldots \qquad (2.8.17)$$

which converges quadratically. This is sometimes called the **modified Newton–Raphson method**. We have to calculate one higher derivative of $f(x)$ than in the Newton–Raphson method. Since $\mu'(r_n) = \{[f'(r_n)]^2 - f(r_n) f''(r_n)\}/[f'(r_n)]^2$ consists of the differences of two small numbers, the denominator of Equation (2.8.17) can cause serious round-off errors.

**EXAMPLE 2.8.2**

Find the roots of $x^3 - x^2 - x + 1 = 0$ by the Newton–Raphson and modified Newton–Raphson methods and Equation (2.8.16).

Let $f(x) = x^3 - x^2 - x + 1 = (x - 1)^2(x + 1) = 0$. This has the repeated root 1 and simple root $-1$. Equation (2.8.4) reduces to

$$r_{n+1} = r_n - \frac{f(r_n)}{f'(r_n)} = r_n - \frac{r_n^3 - r_n^2 - r_n + 1}{3r_n^2 - 2r_n - 1} \quad \text{for } n = 0, 1, \ldots$$

**Table 2.8.3**

| $n$ | $r_n$ | $f(r_n)$ | $\left|\dfrac{e_{n+1}}{e_n}\right|$ |
|---|---|---|---|
| 1 | 6.80645 | 263.19358 | 0.64516 |
| 2 | 4.69024 | 77.48923 | 0.63554 |
| 3 | 3.26692 | 22.66992 | 0.62243 |
| 4 | 2.39068 | 6.55752 | 0.60545 |
| 5 | 1.81367 | 1.86281 | 0.58509 |
| 6 | 1.45823 | 0.51616 | 0.56316 |
| 7 | 1.24865 | 0.13902 | 0.54263 |
| 8 | 1.13084 | 0.03648 | 0.52620 |
| 9 | 1.06737 | 0.00938 | 0.51489 |
| 10 | 1.03422 | 0.00238 | 0.50802 |
| 11 | 1.01725 | 0.00060 | 0.50417 |
| 12 | 1.00866 | 0.00015 | 0.50213 |
| 13 | 1.00434 | 0.00004 | 0.50108 |
| 14 | 1.00217 | 0.00001 | 0.50054 |

For $r_0 = 10$ and $n = 0$,

$$r_1 = 10 - \frac{1000 - 100 - 10 + 1}{300 - 20 - 1} = 10 - \frac{891}{279} = 6.80645$$

Continue until $|f(r_n)| < 10^{-5}$. The computed values are given in Table 2.8.3. From the table, $|r_{14} - r| = 0.00217 = 2.17 \times 10^{-3}$. Thus $|f(r_n)| < \epsilon$ is not a good stopping criterion in this case. Equation (2.8.4) starting with 10 leads to 1. The ratio $|e_{n+1}/e_n|$ converges to $\frac{1}{2} (= (m - 1)/m)$.

If we start with $-10$, then Equation (2.8.4) leads to $-1$ and the ratio $|e_{n+1}/e_n^2|$ converges to $1(= \frac{1}{2}|f''(r)/f'(r)|)$ as shown in Table 2.8.4. Continue until $|f(r_n)| < 10^{-5}$ and $|(r_{n+1} - r_n)/r_n| < 10^{-5}$. In this case $|r_9 - r| = 0$.

**Table 2.8.4**

| $n$ | $r_n$ | $f(r_n)$ | $\left|\dfrac{r_{n+1} - r_n}{r_n}\right|$ | $\left|\dfrac{e_{n+1}}{e_n^2}\right|$ |
|---|---|---|---|---|
| 1 | $-6.58621$ | $-321.48920$ | 0.51832 | 0.06897 |
| 2 | $-4.32708$ | $-94.41509$ | 0.52209 | 0.10662 |
| 3 | $-2.84780$ | $-27.35768$ | 0.51945 | 0.16693 |
| 4 | $-1.90526$ | $-7.64085$ | 0.49470 | 0.26513 |
| 5 | $-1.34755$ | $-1.91537$ | 0.41386 | 0.42411 |
| 6 | $-1.07940$ | $-0.34332$ | 0.24843 | 0.65732 |
| 7 | $-1.00563$ | $-0.02266$ | 0.07335 | 0.89358 |
| 8 | $-1.00003$ | $-0.00013$ | 0.00560 | 0.99162 |
| 9 | $-1.00000$ | 0.00000 | 0.00003 | 0.99995 |
| 10 | $-1.00000$ | 0.00000 | 0.00000 | |

Since 1 is repeated root, the modified formula given by Equation (2.8.17) gives a sequence that converges quadratically to 1. In this case,

$$\mu(x) = \frac{f(x)}{f'(x)} = \frac{x^3 - x^2 - x + 1}{3x^2 - 2x - 1}$$

and

$$\mu'(x) = \frac{3x^4 - 4x^3 + 2x^2 - 4x + 3}{(3x^2 - 2x - 1)^2}$$

so Equation (2.8.17) becomes

$$r_{n+1} = r_n - \frac{\mu(r_n)}{\mu'(r_n)} = r_n - \frac{(r_n^3 - r_n^2 - r_n + 1)(3r_n^2 - 2r_n - 1)}{3r_n^4 - 4r_n^3 + 2r_n^2 - 4r_n + 3} \quad \text{for } n = 0, 1, \ldots$$

For $r_0 = 10$ and $n = 0$,

$$r_1 = 10 - \frac{(891)(279)}{(26163)} = 0.49845$$

Similarly we find $r_2, r_3, \ldots$. The computed results are given in Table 2.8.5. The ratio $|e_{n+1}/e_n^2|$ converges to $|\mu''(r)/(2\,\mu'(r))| = \frac{1}{4}$ as seen in Table 2.8.5.

**Table 2.8.5**

| $n$ | $r_n$ | $f(r_n)$ | $\left|\dfrac{r_{n+1} - r_n}{r_n}\right|$ | $\left|\dfrac{e_{n+1}}{e_n^2}\right|$ |
|---|---|---|---|---|
| 1 | 0.49845 | 0.37694 | 19.06211 | 0.00619 |
| 2 | 0.89391 | 0.02132 | 0.44239 | 0.42174 |
| 3 | 0.99687 | 0.00002 | 0.10328 | 0.27836 |
| 4 | 1.00000 | 0.00000 | 0.00313 | 0.25078 |
| 5 | 1.00000 | 0.00000 | 0.00000 | |

Since 1 is a repeated root of multiplicity 2, Equation (2.8.16) reduces to

$$r_{n+1} = r_n - 2\frac{f(r_n)}{f'(r_n)} = r_n - 2\frac{(r_n^3 - r_n^2 - r_n + 1)}{(3r_n^2 - 2r_n - 1)} \quad \text{for } n = 0, 1, \ldots$$

For $r_0 = 10$ and $n = 0$,

$$r_1 = r_0 - 2\frac{(1000 - 100 - 10 + 1)}{(300 - 20 - 1)} = 10 - 2\frac{891}{279} = 3.61290$$

The computed values are given in Table 2.8.6.

**Table 2.8.6**

| $n$ | $r_n$ | $f(r_n)$ | $\left|\dfrac{r_{n+1} - r_n}{r_n}\right|$ | $\left|\dfrac{e_{n+1}}{e_n^2}\right|$ |
|---|---|---|---|---|
| 1 | 3.61290 | 31.49350 | 1.76786 | 0.03226 |
| 2 | 1.57669 | 0.85693 | 1.29145 | 0.08447 |
| 3 | 1.05804 | 0.00693 | 0.49020 | 0.17452 |
| 4 | 1.00081 | 0.00000 | 0.05719 | 0.23957 |
| 5 | 1.00000 | 0.00000 | 0.00081 | 0.24985 |
| 6 | 1.00000 | 0.00000 | 0.00000 | |

**Comment**

1.  Consider $f(x) = \sqrt[5]{x}$. Since $f'(x) = 1/(5x^{4/5})$, Equation (2.8.4) reduces to

$$r_{n+1} = r_n - r_n^{1/5}\, 5r_n^{4/5} = r_n - 5r_n = -4r_n$$

This leads to $r_{n+1} = -4r_n = -4(-4r_{n-1}) = \cdots = (-4)^{(n+1)} r_0$ and $r_{n+1}$ grow proportional to $(-4)^{(n+1)}$ for any starting point $r_0$ as shown in Figure 2.8.2. We cannot reach $r = 0$ unless we start with $r_0 = 0$.

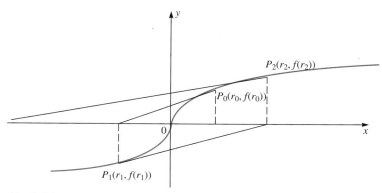

Fig. 2.8.2

Further consider $f(x) = \arctan x$. Then Equation (2.8.4) reduces to $r_{n+1} = r_n - (1 + r_n^2)\arctan(r_n)$. If $r_0$ is chosen so that $\arctan(r_0) > 2|r_0|/(1 + r_0^2)$, then the sequence $\{r_n\}$ diverges. Under what conditions can the sequence given by Equation (2.8.4) be expected to converge? These conditions are given by the following theorem.

**Theorem 2.8.1**

Let $f''(x)$ be continuous for all $x$ in some open interval containing $r$ such that $f(r) = 0$ and $f'(r) \neq 0$. Then there exists an $\epsilon > 0$ such that the iterates $r_n$ given by Newton's method [Equation (2.8.4)] converge quadratically to $r$ whenever $|r_0 - r| < \epsilon$.

*Proof*   Consider an interval $I = [r - \delta, r + \delta]$ on which $f'(x) \neq 0$ and $f''(x)$ is continuous. Define

$$M = \frac{1}{2}\frac{\max_{x \in I}|f''(x)|}{\min_{x \in I}|f'(x)|}$$

Select $\epsilon > 0$ such that $M\epsilon < 1$. This is possible because as $\epsilon \to 0$, $M \to \frac{1}{2}|f''(r)|/|f'(r)|$ and $M\epsilon \to 0$. Select $r_0$ such that $|r_0 - r| = |e_0| \leq \epsilon$.
From Equation (2.8.12)

$$|e_1| = \frac{1}{2}\frac{|f''(\xi_1)|}{|f'(r_0)|}|e_0|^2 \leq M|e_0|^2 = M|e_0||e_0| \leq M\epsilon|e_0| < |e_0| \leq \epsilon$$

and $M|e_1| \leq M(M|e_0|^2) = (M|e_0|)^2 < 1$. Consequently, $|r_1|$ is also within distance $\epsilon$ of $r$ and $M|e_1| < 1$. Applying the same argument to $r_2, r_3, \ldots, r_n$ inductively, $|r_n - r| = |e_n| < \epsilon$ and $M|e_n| < 1$ for all $n \geq 0$. Hence

$$M|e_n| \leq (M|e_{n-1}|)^2 \leq (M|e_{n-2}|)^{2^2} \cdots \leq (M|e_0|)^{2^n} \leq (M\epsilon)^{2^n}$$

Therefore $|e_n| \leq (M\epsilon)^{2^n}/M$. Since $M\epsilon < 1$, this shows that $r_n \to r$ as $n \to \infty$.

We proved that if the Newton–Raphson method converges, then it converges quadratically if $f'(r) \neq 0$. This proves the theorem. ∎

From Theorem 2.8.1, we conclude that the Newton iterates converge to a root $r$ provided that $r_0$ is sufficiently close to $r$. This theorem does not help us decide if the iterates will converge from a given $r_0$ since $r$ is unknown. Since in practice divergence is quickly detected in the iterations, there is nothing to be gained by investigating convergence in advance.

2.  For computational purposes, care must be taken when applying Newton–Raphson's method to a function that has zero or near-zero derivatives between the root and any point that may be reached by the iterates. In such cases the value $r_{n+1}$ may be in the wrong direction or an oscillation may take place about the zero derivative as shown in Figures 2.8.3 and 2.8.4.

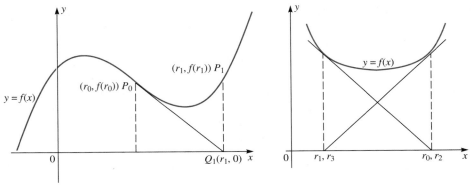

Fig. 2.8.3, 2.8.4

In order to account for this in a general purpose program, include a test whether $|f(r_{n+1})| > |f(r_n)|$ at each step. If this happens for a couple of iterations, then the bisection method can be executed before Newton's method is tried again.

3.  Sometimes $f'(x)$ can be complicated. For example, $f(x) = x^3 e^x \cos x$. Then $f'(x) = 3x^2 e^x \cos x + x^3 e^x \cos x - x^3 e^x \sin x$. When using a calculator it is complicated to compute $f'(x)$.

    Since

$$f'(r_n) = \lim_{r_{n-1} \to r_n} \frac{f(r_n) - f(r_{n-1})}{r_n - r_{n-1}}$$

replace $f'(r_n)$ by $(f(r_n) - f(r_{n-1}))/(r_n - r_{n-1})$ in Equation (2.8.4). Then Equation (2.8.4) becomes

$$r_{n+1} = r_n - \frac{f(r_n)(r_n - r_{n-1})}{f(r_n) - f(r_{n-1})} \qquad \textbf{(2.8.18)}$$

which is the secant formula. To use the Newton–Raphson formula, $f'(r_n)$ has to be provided. Though Newton–Raphson's formula will normally require fewer steps than the secant method, it will require more computations per step if $f'(x)$ is complicated. It has been shown (Atkinson 1989) that if the time to evaluate $f'(x)$ is more than 44 percent of that necessary to compute $f(x)$, then the secant method is more efficient.

## EXERCISES

1.  Let $f(x) = x^2 - A$ where $A > 0$. Derive the Newton–Raphson formula $r_{n+1} = \frac{1}{2}(r_n + (A/r_n))$. This algorithm can be used to find the square root of $A$ since $f(x) = 0$ gives $x = \pm\sqrt{A}$. To find $\sqrt{2}$, compute four iterates $r_1, r_2, r_3,$ and $r_4$ starting with $r_0 = 1$.
2.  Find $\lim_{n\to\infty} |e_{n+1}|/e_n^2$ for Exercise 1.
3.  Let $f(x) = 1 - (A/x^2)$ where $A > 0$. Derive the Newton–Raphson formula $r_{n+1} = \frac{1}{2}r_n(3 - (r_n^2/A))$. This algorithm can also be used to find the square root of $A$ since $f(x) = 0$ gives $x = \pm\sqrt{A}$. To find $\sqrt{2}$, compute four iterates $r_1, r_2, r_3,$ and $r_4$ starting with $r_0 = 1$.
4.  Find $\lim_{n\to\infty} |e_{n+1}|/e_n^2$ for Exercise 3.
5.  The $m$th root of $A$ can be obtained by using the Newton–Raphson formula for $f(x) = x^m - A$ or $f(x) = 1 - (A/x^m)$. Verify that the iterations are given by

$$r_{n+1} = \frac{1}{m}\left[(m-1)r_n + \frac{A}{r_n^{m-1}}\right] \quad \text{or} \quad r_{n+1} = \frac{1}{m}\left[(m+1)r_n - \frac{r_n^{m+1}}{A}\right]$$

6.  Find $\sqrt[8]{2}$ using Exercise 5.
7.  Find $\lim_{n\to\infty} |e_{n+1}|/e_n^2$ for Exercise 5.
8.  Let $f(x) = R - (1/x)$. Then the solution of $f(x) = 0$ can be used to find the reciprocal of $R$. Verify that the Newton–Raphson formula gives $r_{n+1} = r_n(2 - Rr_n)$ for $f(x) = R - (1/x)$. This formula can be used to find the reciprocal of a number $R$ without division. Find the reciprocal of 7 by computing the first four iterates starting with $r_0 = 0.1$.
9.  Compute the first four iterates of Newton–Raphson's method starting with $r_0 = 3$ for each of the following equations:
    (a) $xe^x - 2 = 0$   (c) $e^x - 5\sin x = 0$
    (b) $x - 2\cos x = 0$   (d) $x^3 - x^2 - x - 1 = 0$
10. The Newton–Raphson method can be used to find complex roots. The formula is

$$z_{n+1} = z_n - \frac{f(z_n)}{f'(z_n)} \quad \text{for } n = 0, 1, \ldots, \tag{2.8.19}$$

where $f(z)$ is a complex valued function of a complex variable $z = x + Iy$, $x$ and $y$ are reals and $I = \sqrt{-1}$. Decomposing $z_n = x_n + Iy_n$, $f(z_n) = g(x_n, y_n) + Ih(x_n, y_n)$ and

$$f'(z_n) = h_y(x_n, y_n) - Ig_y(x_n, y_n) + g_x(x_n, y_n) + Ih_x(x_n, y_n)$$

and then substituting in Equation (2.8.19) and decomposing it into real and imaginary parts, show that

$$x_{n+1} = x_n - \frac{g g_x + h h_x}{g_x^2 + g_y^2}, \qquad y_{n+1} = y_n + \frac{h g_x - g h_x}{g_x^2 + g_y^2} \quad \text{for } n = 0, 1, \ldots$$

$$(2.8.20)$$

Thus complex arithmetic can be avoided by using Equation (2.8.20). Carry out four iterations given by Equation (2.8.20) for solving $z^2 + 1 = 0$ starting with value $z_0 = 1 + I$.

**11.** Verify Equation (2.8.14).

**12.** In the case of a zero of multiplicity $m$, the Newton–Raphson method converges linearly. If $m$ is known, then the Newton–Raphson formula can be modified by

$$r_{n+1} = r_n - m \frac{f(r_n)}{f'(r_n)}$$

Show that the modified procedure converges quadratically.

**13.** Suppose $f(x)$ has a multiple zero at $r$. Verify that the new function defined by $\mu(x) = \lim_{t \to x} f(t)/f'(t)$ has a simple zero at $r$.

**14.** Let $P(x)$ be a cubic polynomial with three real roots $x_1$, $x_2$, and $x_3$. Show that if the Newton–Raphson method is started with $r_0 = (x_1 + x_2)/2$, then we get $x_3$ in one step.

**15.** Let $f(x) = x^2 - A$. Derive Chebyshev's formula $r_{n+1} = \frac{1}{8}[3r_n + (6A/r_n) - (A^2/r_n^3)]$. This algorithm can also be used to find the square root of $A$. To find $\sqrt{2}$, compute four iterates $r_1$, $r_2$, $r_3$, and $r_4$ starting with $r_0 = 1$.

**16.** Verify that $\lim_{n \to \infty} |e_{n+1}|/|e_n|^3 = \frac{1}{6}|f'''(r)/f'(r)|$ for Equation (2.8.10). In other words, Cauchy's method has a third-order convergence if $f'(r) \neq 0$.

**17.** Derive Equation (2.8.10) from Equation (2.8.9).

## COMPUTER EXERCISES

**C.1.** Write a subroutine that accepts as parameters a function $f$ and its first derivative $f'$, the initial guess $r_0$, the maximum number of iterations maxit, tolerance eps and carries out the Newton–Raphson procedure.

**C.2.** Using Exercise C.1, write a computer program to find a root of $f(x) = 0$ accurate to within $10^{-5}$. Print $n$, $r_n$, and $f(r_n)$.

**C.3.** Using Exercise C.2, find the roots of the equations in Exercises 9(a) and (c). Compare the required number of iterations with the bisection and secant methods.

**C.4.** To study the rate of convergence of the Newton–Raphson method, write a computer program to compute $|e_{n+1}/e_n|$ and $|e_{n+1}|/e_n^2$. Use Exercise (C.1). Using your program, compute $|e_{n+1}|/e_n^2$ for $x^3 - 8 = 0$ and verify that $|e_{n+1}|/e_n^2$ converges to $|f''(r)/(2f'(r))|$.

**C.5.** Using Exercise C.1, find the roots of $f(x) = x^3 - 0.5x^2 - 0.25x - 0.125 = 0$ accurate to within $10^{-5}$ with $r_0 = 0, 0.1, -0.1, 1.0$, and $-1.0$. Also, use Equation (2.8.17) to find the multiple roots of $f(x) = 0$. Using Exercise C.1 with $\mu(x) = f(x)/f'(x)$, for the multiple root verify that $|e_{n+1}|/e_n^2$ converges to $|\mu''(r)/(2\mu'(r))|$.

**C.6.** It can be shown that the shape of a cable suspended from its two ends is a catenary. An equation of a catenary is given by $y = \cosh(x/a)$ where $a$ is a parameter to be determined. The $y$-axis passes through the midpoint of the two endpoints. A telephone wire is suspended between two poles 200 ft apart. The cable is allowed to sag 10 ft in the middle. Using Exercise C.2, find the value of $a$.

**C.7.** The Legendre polynomials and their zeros are important (Section 6.6). Using Exercise C.2, find at least two zeros of the Legendre polynomial of fifth degree

$$P_5(x) = \frac{63}{8}\left(x^5 - \frac{10}{9}x^3 + \frac{5}{21}x\right)$$

**C.8.** Using Exercise C.2, find a root of $x^3 - 2x - 5 = 0$. Newton illustrated the Newton method by solving this equation in 1669.

**C.9.** A streamline is a curve within a fluid whose tangent at any point is in the direction of the velocity at that point. The stream function $\psi$ is constant along a streamline. The stream function $\psi = U\,y(1 - (1/r^2))$ represents an ideal, two-dimensional flow around a circular cylinder of a unit radius in a uniform velocity $U$ in the $x$-direction far from the cylinder as shown in the figure. Here $r = \sqrt{x^2 + y^2}$ is the radial distance from the center of the cylinder. Write a computer program to compute the values of $y$ for $\psi = 1$ and $x = -5, -4, \ldots, 4, 5$. Use $U = 1$.

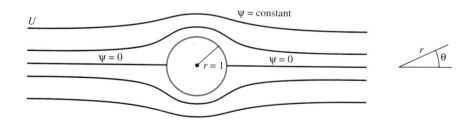

**C.10.** The solution of a partial differential equation by separation of variables often leads to finding the roots of $f(x) = 0$. The heat equation leads to the boundary problem $\phi'' + \lambda\phi = 0$; $\phi(0) = 0$, $\phi(1) = \phi'(1)$. The solution of this boundary value problem requires us to solve $\tan \lambda = \lambda$. Find at least two roots other than $x = 0$ of $\tan x - x = 0$.

**C.11.** Using Exercise C.1, find a root of $x^2 + 1 = 0$. Try to find a complex root using real arithmetic. Use various initial guesses including $1/\sqrt{3}$. For $r_0 = 1/\sqrt{3}$, we get chaos, not a root (Strang 1991).

## 2.9 FIXED-POINT ITERATION

This section will discuss fixed-point iteration of which the Newton–Raphson method is a special case. Let

$$g(r) = r - \frac{f(r)}{f'(r)} \tag{2.9.1}$$

The Newton–Raphson formula [Equation (2.8.4)] becomes

$$r_{n+1} = g(r_n) \quad \text{for } n = 0, 1, \ldots \tag{2.9.2}$$

If $g$ is continuous and the generated sequence $r_0, r_1, r_2, \ldots$ converges to $r$, then

$$r = \lim_{n\to\infty} r_{n+1} = \lim_{n\to\infty} g(r_n) = g(\lim_{n\to\infty} r_n) = g(r) \tag{2.9.3}$$

A root $r$ of $x = g(x)$ is called a **fixed point** of $g$. If $r$ is a fixed point of $g$, then from Equation (2.9.1), $f(r) = 0$. Thus a fixed point $r$ of $g$ is also a root of $f(x) = 0$. Contruct $g(x)$ for a given function $f(x)$ such that $r = g(r)$ whenever $f(r) = 0$. The idea is to rewrite $f(x) = 0$ in an equivalent form $x = g(x)$ using algebraic manipulations so that both equations have the same solution set.

**EXAMPLE 2.9.1**  The function $f(x) = x - x^2 = 0$ can be written as $x = g(x)$ in any one of the following ways:

1. $x = x^2$     thus     $g(x) = x^2$

2. $x = \sqrt{x}$     $g(x) = \sqrt{x}$

3. $x = x + 2(x - x^2)$     $g(x) = x + 2(x - x^2)$

4. $x = x - \dfrac{(x - x^2)}{(1 - 2x)}$     $g(x) = x - \dfrac{(x - x^2)}{(1 - 2x)}$     ■■■

The construction of $g(x)$ is not unique. Now if an efficient procedure can be found to find a fixed point for $g(x)$, $a \le x \le b$, then it is an efficient procedure to find a zero of $f(x)$, $a \le x \le b$. We start with $r_0$ and use the algorithm

$$r_{n+1} = g(r_n) \quad \text{for } n = 0, 1, \ldots \tag{2.9.2}$$

to find $r$.

Let $g(x) = x^2$. Then 0 and 1 are the fixed points of $x = g(x) = x^2$ and also zeros of $f(x) = x - x^2$. Geometrically, the fixed points 0 and 1 are the points at which the graphs of $y = x$ and $y = g(x) = x^2$ intersect as shown in the Figure 2.9.1. In order to find the intersection points, start with $r_0 = \frac{1}{2}$. Using Equation (2.9.2), $r_1 = g(r_0) = r_0^2 = 1/4$, $r_2 = g(r_1) = r_1^2 = 1/16$, $r_3 = g(r_2) = r_2^2 = 1/256, \ldots$, we get the sequence $\{1/2, 1/4, 1/16, 1/256, \ldots\}$, which converges to 0, a zero of $f$. If we start with $r_0 = 5/4$, then we get the sequence $\{5/4, 25/16, 625/256, \ldots\}$, which diverges (Figure 2.9.1).

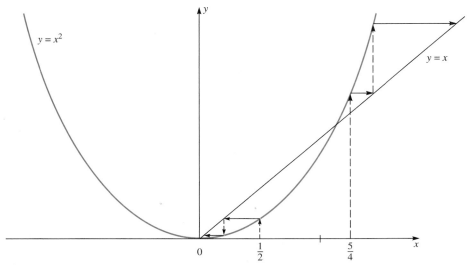

Fig. 2.9.1

Starting with $r_0 = 2$, the results of $r_{n+1} = g(r_n)$ for $n = 0, 1, \ldots$ are given in Table 2.9.1 for each case of Example 2.9.1.

Cases 1 and 3 diverge and Cases 2 and 4 converge to 1. Starting with $r_0 = 0.8$, the numerical results are given in Table 2.9.2.

Case 3 diverges, Case 1 converges to 0, and Cases 2 and 4 converge to 1, as in the previous case.

Again starting with $r_0 = 0.2$, the iterations are given in Table 2.9.3. Now Cases 1 and 4 converge to 0 and Cases 2 and 3 converge to 1.

It is natural to ask what makes the various iteration forms behave in the way they do, as shown in Tables 2.9.1, 2.9.2, and 2.9.3. A general theory is developed to explain this behavior.

**Table 2.9.1**    Iterations for $x - x^2 = 0$ starting with $r_0 = 2.0$

| | Case 1 | Case 2 | Case 3 | Case 4 |
|---|---|---|---|---|
| $n$ | $r_n$ | $r_n$ | $r_n$ | $r_n$ |
| 0 | 2.00000 | 2.00000 | 2.00000 | 2.00000 |
| 1 | 4.00000 | 1.41421 | −8.00000 | 1.33333 |
| 2 | 16.00000 | 1.18920 | −368.00000 | 1.06660 |
| 3 | 256.00000 | 1.09050 | −679328.00000 | 1.00390 |
| 4 | 65536.00000 | 1.04420 | | 1.00000 |
| 5 | | 1.02180 | | |

**Table 2.9.2** Iterations for $x - x^2 = 0$ starting with $r_0 = 0.8$

| n | Case 1 $r_n$ | Case 2 $r_n$ | Case 3 $r_n$ | Case 4 $r_n$ |
|---|---|---|---|---|
| 0 | 0.80000 | 0.80000 | 0.80000 | 0.80000 |
| 1 | 0.64000 | 0.89440 | 1.60000 | 1.06660 |
| 2 | 0.40960 | 0.94574 | $-3.20000$ | 1.00392 |
| 3 | 0.16777 | 0.97240 | $-70.40000$ | 1.00000 |
| 4 | 0.02810 | 0.98615 | $-25203.20000$ | 1.00000 |
| 5 | 0.00079 | 0.99305 | | 1.00000 |

**Table 2.9.3** Iterations for $x - x^2 = 0$ starting with $r_0 = 0.2$

| n | Case 1 $r_n$ | Case 2 $r_n$ | Case 3 $r_n$ | Case 4 $r_n$ |
|---|---|---|---|---|
| 0 | 0.20000 | 0.20000 | 0.20000 | 0.20000 |
| 1 | 0.04000 | 0.44720 | 1.00000 | $-0.06660$ |
| 2 | 0.00160 | 0.66870 | 1.00000 | $-0.00390$ |
| 3 | 0.00000 | 0.81770 | 1.00000 | 0.00000 |
| 4 | 0.00000 | 0.90430 | 1.00000 | 0.00000 |
| 5 | 0.00000 | 0.95090 | 1.00000 | 0.00000 |

**Algorithm**

**Subroutine** Fixed $(f, g, rn, \text{root}, \text{maxit}, \text{eps})$
Comment: $rn = r_0$; maxit is the maximum number of iterations to be computed; eps is tolerence.
1 **Print** $rn$, maxit, eps
2 $n = 0$
3 **Repeat** steps 4–8 **while** $n \leq$ maxit
    4 root $= g(rn)$
    5 **Print** root, $f(\text{root})$, and $n$
    6 If $|\text{root} - rn| <$ eps and $|rn - \text{root}| <$ eps $|\text{root}|$, **then print** 'finished' and exit.
    7 $rn = $ root
    8 $n = n + 1$
9 **Print** 'algorithm fails: no convergence' and exit. ◀

On a given interval $I = [a, b]$, $g(x)$ may have many fixed points or none at all. If $r$ is a fixed point in $I$, then $g(r) \in I$. Further $r_{n+1} = g(r_n)$ for $n = 0, 1, \ldots$, therefore, if $r_n \in I$, then $g(r_n) \in I$. Thus it is reasonable to assume that for each $x \in I$, $g(x) \in I$.

Further, if $g$ is discontinuous in $I$, then part of the graph of $g$ can be above the line $y = x$ and part below $y = x$. In such a case, $y = x$ may not intersect $y = g(x)$. Therefore $g$ must be continuous on $I$.

Let $g$ be continuous on $I = [a, b]$ and $g(x) \in I$ for all $x \in I$. Is there any fixed point in $I$? If $g(a) = a$ or $g(b) = b$, the endpoint is a fixed point. Assume that this is not the case. Then $g(a) > a$ and $g(b) < b$ (why?). Let $h(x) = g(x) - x$. Since $g$ is continuous on $I$, then $h$ is also continuous on $I$ and $h(a) > 0$ and $h(b) < 0$. Thus, by the Immediate Value Theorem (Theorem A.2), there exists at least one $r \in (a, b)$ such that $h(r) = 0$. So $g(r) = r$. Thus we have at least one fixed point in $I$.

Let $e_n = r_n - r$. Then $e_n = r_n - r = g(r_{n-1}) - g(r) = g'(\xi_{n-1})(r_{n-1} - r)$ where $\xi_{n-1}$ is between $r$ and $r_{n-1}$ because of the Mean Value Theorem (Theorem A.3) and $g(r) = r$. Hence we have

$$e_n = g'(\xi_{n-1}) \, e_{n-1} \tag{2.9.4}$$

Similarly, $e_{n-1} = g'(\xi_{n-2}) \, e_{n-2}$. Continuing, assuming $g'(\xi_i)$ exists for all $i$

$$e_n = g'(\xi_{n-1}) \, g'(\xi_{n-2}) \cdots g'(\xi_0) \, e_0$$

Thus

$$|e_n| = |g'(\xi_{n-1})| \, |g'(\xi_{n-2})| \cdots |g'(\xi_0)| \, |e_0|$$

If we assume that $|g'(x)| \leq k < 1$ for all $x$ in $(a, b)$, then $|e_n| \leq k^n |e_0|$ and $\lim_{n \to \infty} |e_n| = 0$ regardless of the initial error.

It should be pointed out that $|g'(x)| \leq k < 1$ for all $x$ in $(a, b)$ is an unnecessarily stringent requirement. It is not even necessary that $|g'(\xi_i)| < 1$ for every $\xi_i$. The only requirement is that the sequence $g'(\xi_0)$, $g'(\xi_0)g'(\xi_1)$, $g'(\xi_0)g'(\xi_1)g'(\xi_2)$, ... converges to zero. If $|g'(x)| \leq k < 1$ for all $x$ in $(a, b)$, then the sequence $\{r_0, r_1, r_2, \ldots\}$ generated by $r_{n+1} = g(r_n)$ for $n = 0, 1, \ldots$ converges to $r$.

If $|g'(x)| \leq k < 1$ for all $x \in (a, b)$, then we show that $r$ is unique. Let $r_A$ be another fixed point in $[a, b]$. Then

$$r - r_A = g(r) - g(r_A) = g'(\xi)(r - r_A)$$

where $\xi \in (r, r_A)$ by the Mean Value Theorem. Therefore

$$|r - r_A| = |g'(\xi)| \, |r - r_A| < |r - r_A|$$

which is a contradiction. Hence $r$ is unique. This leads to the next theorem.

**Theorem 2.9.1**    Let $g(x) \in [a, b]$ for all $x \in [a, b]$. Further let $g'$ exist on $(a, b)$ such that $|g'(x)| \leq k < 1$ for all $x$ in $(a, b)$ and $g$ is continuous on $[a, b]$. If $r_0 \in [a, b]$, then the sequence generated by

$$r_{n+1} = g(r_n) \quad \text{for } n = 0, 1, \ldots$$

converges to a unique fixed point $r$ in $[a, b]$.    ∎

This theorem gives sufficient conditions for the iterates to converge to the fixed point in $[a, b]$. This theorem remains valid if $[a, b]$ is replaced by $(-\infty, \infty)$.

**E X A M P L E   2.9.2**    Consider $x - x^2 = 0$. The roots of this equation are 0 and 1. Analyze the four cases that we considered in Example 2.9.1.

***Case 1***    When $g(x) = x^2$, $g$ is continuous for all $x$ and $g'(x) = 2x$. Thus $|g'(x)| < 1$ if $x \in (-\frac{1}{2}, \frac{1}{2})$. Further $g(x) \in [-\frac{1}{2}, \frac{1}{2}]$ for all $x$ in $[-\frac{1}{2}, \frac{1}{2}]$. So for any number $r_0$ in

$[-\frac{1}{2}, \frac{1}{2}]$, the iterative method converges. Geometrically it can be seen that if we pick $r_0$ in $(-1, 1)$, the iterates still converge as in Table 2.9.2 and Table 2.9.3. Keep in mind that the theorem does not give necessary conditions.

*Case 2*   Since $g(x) = \sqrt{x}$, $g'(x) = 1/(2\sqrt{x})$. It can be verified that $g$ is continuous and $|g'(x)| < 1$ if $x$ is in $(\frac{1}{4}, \infty)$. It can be verified that $g(x) \in [\frac{1}{4}, \infty)$ for all $x$ in $[\frac{1}{4}, \infty)$. Thus for $r_0$ in $[\frac{1}{4}, \infty)$, the iterative method converges. For $r_0 = 0.8$ and 2, the iterates converge to 1. It can be seen from Table 2.9.3 that the iterates converge to 1 with even $r_0 = 0.2$. In fact starting with any positive real number, the iterates converge to 1. There is no way to reach zero except by starting with $r_0 = 0$.

*Case 3*   In this case $g(x) = x + 2(x - x^2) = 3x - 2x^2$ is continuous for all $x$ and $g'(x) = 3 - 4x$. Thus $|g'(x)| < 1$ if $\frac{1}{2} < x < 1$. But $g(x) \notin [\frac{1}{2}, 1]$ for all $x$ in $[\frac{1}{2}, 1]$. Thus there is no guarantee that the iterates will converge for any $x$ in $[\frac{1}{2}, 1]$. However, starting with $x_0 = 0.2$, it converges to 1.

*Case 4*   $g(x) = x - [(x - x^2)/(1 - 2x)] = -x^2/(1 - 2x)$ is continuous if $x \neq \frac{1}{2}$ and $g'(x) = -2x(1 - x)/(1 - 2x)^2$. Then $|g'(x)| < 1$ if $x \in (-\infty, (3 - \sqrt{3})/6) \cup ((3 + \sqrt{3})/6, \infty)$. Further $g(x) \in (\frac{1}{2}, \infty)$ for all $x$ in $(\frac{1}{2}, \infty)$ and $g(x) \in (-\infty, \frac{1}{2})$ for all $x$ in $(-\infty, \frac{1}{2})$. Thus $g(x) \in [(3 + \sqrt{3})/6, \infty)$ for all $x$ in $[(3 + \sqrt{3})/6, \infty)$ and $|g'(x)| < 1$ for all $x$ in $((3 + \sqrt{3})/6, \infty)$. Thus for any $r_0$ in $[(3 + \sqrt{3})/6, \infty)$, the iterative method converges. For $r_0 = 0.8$ and 2, the iterates converge to 1. Also $g(x) \in (-\infty, (3 - \sqrt{3})/6]$ for all $x$ in $(-\infty, (3 - \sqrt{3})/6]$ and $|g'(x)| < 1$ for all $x$ in $(-\infty, (3 - \sqrt{3})/6)$. Thus for $r_0$ in $(-\infty, (3 - \sqrt{3})/6]$, the iterative method converges to 0. It can be seen from Table 2.9.3 that the iterates converge to 0 with $r_0 = 0.2$.   ■ ■ ■

Sometimes it is difficult to verify the first assumption of Theorem 2.9.1. In such circumstances, the following corollary gives sufficient conditions for convergence.

**Corollary 2.9.1**

If $g'(x)$ is continuous in a small open interval containing the fixed point $r$ and if $|g'(r)| < 1$, then there exists $\epsilon > 0$ such that

$$r_{n+1} = g(r_n) \quad \text{for } n = 0, 1, \ldots$$

converges whenever $|r_0 - r| \leq \epsilon$.   ■

## Rate of Convergence

Let $g$ be continuous on $[a, b]$ and $|g'(x)| \leq k < 1$ for all $x$ in $(a, b)$. Further assume $g(x) \in [a, b]$ for all $x \in [a, b]$. From Equation (2.9.4), we have $e_{n+1} = g'(\xi_n)e_n$ for some $\xi_n$ between $r$ and $r_n$. Hence

$$\lim_{n \to \infty} \left| \frac{e_{n+1}}{e_n} \right| = \lim_{n \to \infty} |g'(\xi_n)| = |g'(r)| \quad \text{if } e_n \neq 0 \quad \text{for all } n \quad \textbf{(2.9.5)}$$

If $|g'(r)| \neq 0$, and if the fixed-point iteration method converges, it converges linearly. If $|g'(r)| < 1$, then the iteration converges to $r$. When $|g'(r)| > 1$, the iteration converges to $r$ only by a lucky choice of the initial approximation $r_0$. When $|g'(r)| = 1$, the convergence is unpredictable without further information. In the special case when $g'(r) = 0$, the nature of convergence depends upon the behavior of the higher derivatives near $r$. For example

$$r_{n+1} = g(r_n) = g(r + (r_n - r)) = g(r) + g'(r)(r_n - r) + \frac{g''(r)}{2!}(r_n - r)^2$$

$$+ \cdots + \frac{g^{(k)}(r)}{k!}(r_n - r)^k + \frac{g^{(k+1)}(\xi)}{(k+1)!}(r_n - r)^{(k+1)} \tag{2.9.6}$$

where $\xi \in (r, r_n)$ by expanding $g(r_n)$ by Taylor's Theorem (Theorem A.8) about $r$.

Since $g(r) = r$, Equation (2.9.6) can be written as

$$e_{n+1} = g'(r)\,e_n + \frac{g''(r)}{2!}e_n^2 + \cdots + \frac{g^{(k)}(r)}{k!}e_n^k + \frac{g^{(k+1)}(\xi)}{(k+1)!}e_n^{k+1} \tag{2.9.7}$$

If $g'(r) = 0$ and $g''(x) \neq 0$ for all $x$ in $I$, then we get

$$\lim_{n\to\infty}\left|\frac{e_{n+1}}{e_n^2}\right| = \lim_{n\to\infty}\left|\frac{g''(\xi)}{2!}\right| = \left|\frac{g''(r)}{2}\right|$$

We get quadratic convergence. Similarly if $g'(r) = g''(r) = \cdots = g^{(k)}(r) = 0$ and $g^{(k+1)}(x) \neq 0$ for all $x$ in $I$, we have

$$\lim_{n\to\infty}\left|\frac{e_{n+1}}{e_n^{k+1}}\right| = \lim_{n\to\infty}\left|\frac{g^{(k+1)}(\xi)}{(k+1)!}\right| = \frac{|g^{(k+1)}(r)|}{(k+1)!} \tag{2.9.8}$$

The more the derivatives of $g$ vanish at $r$, the faster the rate of convergence.

Let $g(x) = x - (f(x)/f'(x))$. Then it can be verified (Exercise 14) that

$$g'(x) = \frac{f(x)f''(x)}{(f'(x))^2} \tag{2.9.9}$$

Since $f(r) = 0$, $g'(r) = 0$. Therefore the Newton–Raphson method converges at least quadratically in the case of a nonrepeated root.

## EXERCISES

1. Write each of the following equations as $x = g(x)$ in at least four different ways:
   (a) $x - \cos x = 0$     (b) $x^3 - x - 1 = 0$     (c) $e^{x/2} - 2x = 0$
2. By using $r_{n+1} = g(r_n)$ in Exercise 1(a), compute $r_1$, $r_2$, $r_3$, and $r_4$ starting with $r_0 = 1$ for each $g(x)$.
3. Verify that $x - \cos x = 0$ has a root between $x = 0$ and 1. Show that the iteration $r_{n+1} = \cos r_n$ converges to the solution of the equation $x - \cos x = 0$ for any $r_0 \in [0, 1]$. Further, show that the iteration $r_{n+1} = \cos r_n$ converges to the solution of the equation $x - \cos x = 0$ for any starting real number $r_0$.
4. Graph the curve $g(x) = \cos x$, the line $y = x$ and plot the fixed points of $g$. Start with $r_0 = \frac{1}{2}$ and compute $r_1$, $r_2$, $r_3$, and $r_4$ and show it as in Figure 2.9.1

5. Verify that $e^{x/2} - 2x = 0$ has a root between $x = 0$ and 1. Verify that the iteration $r_{n+1} = (e^{r_n/2})/2$ converges to this root if $r_0$ is chosen in the interval $[0, 1]$.

6. Graph the curve $g(x) = (e^{x/2})/2$, the line $y = x$, and plot the fixed points of $g$. Start with $r_0 = \frac{1}{2}$ and compute $r_1$, $r_2$, $r_3$, and $r_4$ and show it as in Figure 2.9.1.

7. Verify that $e^{x/2} - 2x = 0$ has a root between 4 and 5. Show that this root cannot be found using the iteration $r_{n+1} = (e^{r_n/2})/2$ by selecting $r_0$ in the interval $[4, 5]$.

8. Graph the curve $g(x) = (e^{x/2})/2$, the line $y = x$, and plot the fixed points of $g$. Start with $r_0 = \frac{9}{2}$ and compute $r_1$, $r_2$, $r_3$, and $r_4$ and show it as in Figure 2.9.1.

9. Determine an interval that contains a fixed point for
   (a) $g(x) = (x^2 + 2)/3$     (b) $g(x) = e^x$     (c) $g(x) = -\ln x$

10. Find an interval (if any) for which Theorem 2.9.1 guarantees convergence for the following iterations:
    (a) $r_{n+1} = (r_n^2 + 2)/3$     (b) $r_{n+1} = -\ln r_n$     (c) $r_{n+1} = \frac{1}{2}(r_n + (2/r_n))$

11. In order to find a root of $f(x) = 0$, rewrite it as $x = x + h(x) f(x) = g(x)$. Let $\alpha$ be a root of $f(x) = 0$ and $f'(\alpha) \neq 0$. The sequence $r_{n+1} = g(r_n)$ converges quadratically if $g'(\alpha) = 0$. Find $h(x)$ such that $g'(\alpha) = 0$. Verify that the sequence $r_{n+1} = g(r_n)$ is the Newton–Raphson formula.

12. Verify that $x(1 - a) + ax^2 = 0$ has roots 0 and $(a - 1)/a$ where $a$ is a constant. Using $g(x) = ax - ax^2$, find the values of $a$ for which $x_{n+1} = g(x_n)$ converges to 0 and $(a - 1)/a$.

13. Prove Corollary 2.9.1

14. Verify Equation (2.9.9)

15. Prove Theorem 2.8.1 using Theorem 2.9.1. (*Hint*: Use $g(x) = x - (f(x)/f'(x))$.)

16. Suppose $f$ is four times differentiable and has a simple zero $\alpha$. Define

$$r_{n+1} = \frac{1}{2}\left[ r_n - \frac{f(r_n)}{f'(r_n)} + r_n - \frac{\mu(r_n)}{\mu'(r_n)} \right]$$

where $\mu(r_n) = f(r_n)/f'(r_n)$. If the sequence $\{r_n\}$ converges to $\alpha$, then prove that the convergence is cubic.

17. The condition in Theorem 2.9.1 can be relaxed by a weaker Lipschitz condition. A function $g$ is said to satisfy the Lipschitz condition with Lipschitz constant $L$ on $[a, b]$ if for all $x_1, x_2 \in [a, b]$,

$$|g(x_1) - g(x_2)| \leq L |x_1 - x_2|$$

    (a) Show that $|g'(x)| \leq L$ for all $x \in [a, b]$ implies that $g$ satisfies the Lipschitz condition with Lipschitz constant $L$ on $[a, b]$.
    (b) Show that $g(x) = |x|/2$ satisfies the Lipschitz condition on $[-1, 1]$ yet $g'(x)$ does not exist for all $x \in [-1, 1]$.

18. Let $g(x) \in [a, b]$ for all $x \in [a, b]$. Also $g$ satisfies the Lipschitz condition given in Exercise 17 with a Lipschitz constant $L < 1$ on $[a, b]$. If $r_0 \in [a, b]$, prove that the sequence generated by

$$r_{n+1} = g(r_n), \quad \text{for } n = 0, 1, \ldots$$

converges to a unique fixed point $r$ in $[a, b]$.

**COMPUTER EXERCISES**

**C.1.** Write a subroutine that accepts as parameters a function $g$, the initial guess $r_0$, the maximum number of iterations maxit, tolerance eps, and carries out the computations $r_{n+1} = g(r_n)$ for $n = 0, 1, \ldots$.

**C.2.** Using Exercise C.1, write a computer program to find a root of $f(x) = 0$ on a given interval $[a, b]$ accurate to within $10^{-5}$. Print $n$, $r_n$, and $f(r_n)$.

**C.3.** Using Exercise C.2, find a root of the equation in Exercise 1(a), 1(b), or 1(c). Use various $r_0$. Analyze all cases using Theorem 2.9.1.

**C.4.** Kepler developed the mathematical description of the motion of a planet around the sun. It is given by $y = x - e \sin x$, where $x$ is its eccentric anomaly, $y$ is its mean anomaly and $e$ is the eccentricity of the orbit. Using Exercise C.2, find the root of $x - 0.05 \sin x = 2.18$.

**C.5.** Using Exercise C.2, find a root of the equation $x^3 - x^2 - x - 1 = 0$.

**C.6.** Using Exercise C.2, find at least two roots other than $x = 0$ of $\tan x - x = 0$.

**C.7.** An object weighing 4 lb is dropped from rest from a balloon which is 1500 ft above the ground. The forces acting on it are gravity $g$, which is approximately 32 ft/sec$^2$, and air resistance, which is numerically equal to $v/2$ (in other words, $k = \frac{1}{2}$). Determine the time it takes the object to reach the ground and the impact velocity.

**C.8.** Using Exercise C.2, find the coordinates of the points of intersection of the curves $y = x^3$ and $y = x^2 - 3x + 2$. (*Hint*: Sketch the curves.)

**C.9.** Using Exercise C.2, and $g(x) = ax - ax^2$, find the roots of $x(1 - a) + ax^2 = 0$ where $a$ is a constant for various values of $a$. For $a \geq 3$, convergence is not possible; however, cycling is possible. Use $a = 3.4, 3.5, 3.6,$ and 4.0 and observe cycling and chaos (Strang 1991).

## SUGGESTED READINGS

K. E. Atkinson, *An Introduction to Numerical Analysis*, 2nd ed., John Wiley & Sons, New York, 1989.

R. H. F. Pao, *Fluid Dynamics*, Charles E. Merrill Books, Inc., Columbus, Ohio, 1966.

A. Ralston, *A First Course in Numerical Analysis*, McGraw-Hill, New York, 1965.

G. Strang, ''A Chaotic Search for i,'' *The College Mathematics Journal*, **22** (1991), pp. 3–12.

# Chapter

# 3 ROOTS OF POLYNOMIAL EQUATIONS

$$\text{L}\,\text{et } A = \begin{bmatrix} 1 & 2 \\ 2 & 1 \end{bmatrix} \text{ and } I = \begin{bmatrix} 1 & 0 \\ 0 & 1 \end{bmatrix}. \text{ Find } \lambda \text{ such that}$$

$$\det(A - \lambda I) = \begin{vmatrix} 1 - \lambda & 2 \\ 2 & 1 - \lambda \end{vmatrix} = (1 - \lambda)^2 - 4 = \lambda^2 - 2\lambda - 3 = 0$$

Then $\lambda$ is called an eigenvalue of $A$ and $\det(A - \lambda I) = \lambda^2 - 2\lambda - 3 = 0$ is called the characteristic equation of $A$. To find the eigenvalues of $A$, we need to find the roots of

$$\lambda^2 - 2\lambda - 3 = 0$$

In this chapter, methods will be developed to find the roots of polynomial equations of this type.

## 3.1 INTRODUCTION

In the previous chapter we have not taken into account any particular property of $f$. Also, we have concentrated on finding a root in a given interval rather than finding all the roots of $f(x) = 0$. In this chapter we develop methods for finding all roots of a polynominal equation $P(x) = 0$ with real coefficients.

**Theorem 3.1.1**

**(Fundamental Theorem of Algebra)** If $P$ is a polynomial with real or complex coefficients of degree $N \geq 1$, then there exists at least one real or complex number $r$ such that $P(r) = 0$. ∎

This important theorem was proved by Gauss in his Ph.D. thesis in 1799.

Let $P(x)$ be a polynomial of degree $N \geq 1$. If $N = 1$, then solving $P(x) = 0$ results in one root. Let $N > 1$. By the Fundamental Theorem of Algebra, there is a real or complex number $r_1$ such that $P(r_1) = 0$. Divide $P(x)$ by $x - r_1$ and write $P(x) = (x - r_1)q_1(x)$ where $q_1(x)$ is a polynomial of degree $N - 1$. If $N - 1 \geq 1$, then again by the Fundamental Theorem of Algebra, $q_1(x)$ has a root $r_2$ such that $q_1(x) = (x - r_2)q_2(x)$. Thus $P(x) = (x - r_1)(x - r_2)q_2(x)$. If $N = 2$, then $q_2(x) = $ a constant. If $N > 2$, then continuing this process, we find $r_1, r_2, \ldots, r_N$ such that $P(r_k) = 0$ for $k = 1, 2, \ldots, N$ and

$$P(x) = a_N(x - r_1)(x - r_2) \cdots (x - r_N) \tag{3.1.1}$$

This means that every polynomial of degree $N$ can be factored into $N$ linear factors. These factors need not be distinct or real. Hence

**Corollary 3.1.1**    There are exactly $N$ zeros of a polynomial of degree $N$.    ∎

Finding the zeros of an $N$th degree polynomial is one of the oldest and most studied problems in mathematics. Let

$$P(x) = a_N x^N + a_{N-1} x^{N-1} + \cdots + a_1 x + a_0 \tag{3.1.2}$$

denote a polynomial of degree $N$ with real coefficients $a_0, a_1, a_2, \ldots, a_N$ where $a_N \neq 0$. Certainly for $N = 1$, $r = -a_0/a_1$. For $N = 2$, the zeros of $P(x) = a_2 x^2 + a_1 x + a_0$ are given by the familiar quadratic formula

$$r = \frac{-a_1 \pm \sqrt{a_1^2 + 4a_0 a_2}}{2a_2}$$

For $N = 3$, Geronimo Cardano (1501–1576) published in *Arts Magna* the formulas for the roots, and his student Ludovico Ferrari (1522–1565) gave the formulas for $N = 4$. For these formulas we refer the reader to Uspensky (1948). These formulas are not easy to evaluate and therefore are rarely used to compute the roots. Many attempts were made to find formulas for the roots of a fifth degree polynomial equation. Finally Abel (1802–1829) and Galois (1811–1832) proved that it is not possible to have the formulas for the zeros when $N \geq 5$. Numerical methods must be developed to approximate the zeros of a given polynomial.

Certain basic properties and basic algorithms are needed for the solutions of polynomial equations.

Let $P(x) = x^2 + 1 = (x - I)(x + I)$ where $I = \sqrt{-1}$. Note that the zeros can be complex even when the coefficients are real. $I$ and $-I$ are the zeros of the given $P(x)$. Also $I$ and $-I$ are complex conjugates of each other.

Let $z = x + Iy$ be any complex number. Then the **complex conjugate** of $z$, denoted by $\bar{z}$, is defined by $\bar{z} = x - Iy$. It can be verified (Exercise 1) that $\overline{z_1 + z_2} = \bar{z}_1 + \bar{z}_2$ and $\overline{z_1 z_2} = \bar{z}_1 \bar{z}_2$. Inductively

$$\overline{z_1 + z_2 + \cdots + z_n} = \bar{z}_1 + \bar{z}_2 + \cdots + \bar{z}_n \tag{3.1.3}$$

and

$$\overline{z_1 z_2 \cdots z_n} = \bar{z}_1 \bar{z}_2 \ldots \bar{z}_n \tag{3.1.4}$$

If $a$ is any real number, then $\bar{a} = a$. Let $r$ be a complex zero of $P$ with real coefficients given by Equation (3.1.2). Then

$$a_N r^N + a_{N-1} r^{N-1} + \cdots + a_1 r + a_0 = 0 \tag{3.1.5}$$

By taking the complex conjugate on both sides of Equation (3.1.5), using Equations (3.1.3) and (3.1.4), we conclude that

$$a_N \bar{r}^N + a_{N-1}\bar{r}^{N-1} + \cdots + a_1\bar{r} + a_0 = 0$$

Therefore $P(\bar{r}) = 0$. Thus

**Theorem 3.1.2**     The complex zeros of a polynomial with real coefficents occur in conjugate pairs.   ■

As a consequence of Corollary 3.1.1 and Theorem 3.1.2,

**Corollary 3.1.2**     A polynomial of odd degree with real coefficients has at least one real zero.   ■

Descartes' rule of signs is often helpful in determining the number of positive and negative zeros of a polynomial with real coefficients. Arrange the terms of a given polynomial in order of descending powers (ignoring the missing terms). If two successive terms have opposite signs, then a variation in sign has occurred.

**Theorem 3.1.3**     **(Descartes' Rule of Signs)**   The number of positive real roots of a polynomial with real coefficients is never greater than the number of variations in sign of it. If the number of positive real roots is less, then it is always less by an even number.   ■

**E X A M P L E   3.1.1**     Consider $P(x) = x^6 + x^4 - 2x^3 - x - 1$.
    This has one variation in sign, so from Descartes' rule of signs, $P(x)$ has exactly one positive zero (since "less by an even number" is less than 0). Note that $P(-x)$ has as its zeros the negatives of those of $P(x)$. Thus

$$P(-x) = x^6 + x^4 + 2x^3 + x - 1$$

has exactly one positive zero by Descartes' rules of signs. Hence $P(x)$ has one negative zero. Therefore $P(x)$ has one positive, one negative, and four complex zeros.

■ ■ ■

At times, it is very helpful to know the relationship between the zeros and the coefficients of a given polynomial. Consider $P(x) = a_2x^2 + a_1x + a_0$ with zeros $z_1$ and $z_2$. Then $P(x) = a_2(x - z_1)(x - z_2)$, so we have

$$a_2x^2 + a_1x + a_0 = a_2[x^2 - (z_1 + z_2)x + z_1z_2]$$

Equating the coefficients yields $z_1z_2 = a_0/a_2$ and $z_1 + z_2 = -(a_1/a_2)$. In order to find $z_1$ and $z_2$, solve the system of these two nonlinear equations.
    A given matrix has a corresponding characteristic polynomial (Appendix B). This relation also can be reversed since every polynomial is a characteristic polynomial of its corresponding companion matrix. Consider polynomials of the form

$$P(x) = x^4 + a_3x^3 + a_2x^2 + a_1x + a_0 \tag{3.1.6}$$

The **companion matrix** of $P(x)$ given by Equation (3.1.6) is defined to be the $4 \times 4$ matrix

$$A = \begin{bmatrix} 0 & 1 & 0 & 0 \\ 0 & 0 & 1 & 0 \\ 0 & 0 & 0 & 1 \\ -a_0 & -a_1 & -a_2 & -a_3 \end{bmatrix} \qquad (3.1.7)$$

It can be verified (Exercise 6) that the characteristic polynomial of Equation (3.1.7) is given by Equation (3.1.6). Thus the methods to find eigenvalues of a given matrix can also be used to find zeros of a given polynomial and vice versa.

## 3.2    THE HORNER METHOD

For a given polynomial, the real zeros can be found using the Newton–Raphson method. In order to use the Newton–Raphson method, it is necessary to find $P(\alpha)$ and $P'(\alpha)$ at a given number $\alpha$. As we will see, there is a better way to find $P(\alpha)$ and $P'(\alpha)$ than the direct substitution of $\alpha$ in $P(x)$ and $P'(x)$.

For $N = 4$, Equation (3.1.2) becomes

$$P(x) = a_4 x^4 + a_3 x^3 + a_2 x^2 + a_1 x + a_0 \qquad (3.2.1)$$

In order to evaluate $P(\alpha)$ by direct substitution in Equation (3.2.1), we need at least seven multiplications and four additions. Equation (3.2.1) can be written as nested multiplication as

$$P(x) = \{[(a_4 x + a_3)x + a_2]x + a_1\}x + a_0 \qquad (3.2.2)$$

$P(\alpha)$ in Equation (3.2.2) can be evaluated by only four multiplications and four additions. Thus with Equation (3.2.2) there is a considerable savings in multiplication. Equation (3.2.2) can be extended for any $N$ as

$$P(x) = \{[(a_N x + a_{N-1})x + a_{N-2}]x + \cdots + a_1\}x + a_0 \qquad (3.2.3)$$

Thus

$$P(\alpha) = \{[(a_N \alpha + a_{N-1})\alpha + a_{N-2}]\alpha + \cdots + a_1\}\alpha + a_0$$

This can be computed easily by computing $b_{N-1} = a_N \alpha + a_{N-1}$. Then $b_{N-2} = b_{N-1}\alpha + a_{N-2}$. We can continue and find $P(\alpha) = b_1 \alpha + a_0$. This can be written as an algorithm by

$$\begin{aligned} b_N &= a_N \\ b_i &= b_{i+1}\alpha + a_i \quad \text{for } i = N - 1, N - 2, \ldots, 0 \end{aligned} \qquad (3.2.4)$$

Then $b_0 = b_1 \alpha + a_0 = P(\alpha) = r_0$ from Equation (3.2.3). This algorithm is known as **Horner's method** and computes $P(\alpha)$ exactly analogous to Equation (3.2.3). This requires $N$ multiplications and $N$ additions. Moreover, this iterative scheme can be used to evaluate the derivatives $P^{(m)}(\alpha)$ for $0 \leq m \leq N$.

Let

$$q_1(x) = b_N x^{N-1} + b_{N-1} x^{N-2} + \cdots + b_2 x + b_1 \tag{3.2.5}$$

where $b_i$ is given by Equation (3.2.4). Then

$$
\begin{aligned}
(x - \alpha)q_1(x) + b_0 &= (x - \alpha)[b_N x^{N-1} + b_{N-1} x^{N-2} + \cdots + b_2 x + b_1] + b_0 \\
&= b_N x^N + (b_{N-1} - \alpha b_N)x^{N-1} + \cdots + (b_1 - \alpha b_2)x + b_0 - \alpha b_1 \\
&= a_N x^N + a_{N-1} x^{N-1} + \cdots + a_1 x + a_0 \quad \text{from Equation (3.2.4)} \\
&= P(x)
\end{aligned}
$$

Thus

$$P(x) = (x - \alpha)q_1(x) + b_0 = (x - \alpha)q_1(x) + r_0 \tag{3.2.6}$$

Differentiating both sides of Equation (3.2.6) with respect to $x$ yields

$$P'(x) = q_1(x) + (x - \alpha)q_1'(x) \tag{3.2.7}$$

Thus from Equation (3.2.7),

$$P'(\alpha) = q_1(\alpha)$$

In order to find $P'(\alpha)$, find $q_1(\alpha)$. Equation (3.2.5) can be written as

$$q_1(x) = \{[(b_N x + b_{N-1})x + b_{N-2}]x + \cdots + b_2\}x + b_1 \tag{3.2.8}$$

Also, $q_1(\alpha)$ can be computed using Horner's method. Thus

$$
\begin{aligned}
c_N &= b_N \\
c_i &= c_{i+1}\alpha + b_i \quad \text{for } i = N - 1, N - 2, \ldots, 1
\end{aligned} \tag{3.2.9}
$$

From Equation (3.2.8), $c_1 = q_1(\alpha) = P'(\alpha) = r_1$. Therefore, $q_1(x) = (x - \alpha)q_2(x) + r_1$. Hence

$$
\begin{aligned}
P(x) &= (x - \alpha)q_1(x) + r_0 \\
&= (x - \alpha)[(x - \alpha)q_2(x) + r_1] + r_0 \\
&= (x - \alpha)^2 q_2(x) + r_1(x - \alpha) + r_0
\end{aligned}
$$

By continuing this procedure and using synthetic division, we get Taylor's expansion of $P(x)$ about $x = \alpha$

$$P(x) = r_N(x - \alpha)^N + r_{N-1}(x - \alpha)^{N-1} + \cdots + r_2(x - \alpha)^2 + r_1(x - \alpha) + r_0$$

Thus $P^{(m)}(\alpha) = m! r_m$ and $r_m = P^{(m)}(\alpha)/m!$.

**Algorithm**

**Subroutine** Horner (**a**, **b**, **c**, $\alpha$, $N$)
Comment: $P(x) = a_N x^N + a_{N-1} x^{N-1} + \cdots + a_1 x + a_0$ is evaluated at $\alpha$ by $P(\alpha) = b_0$ and its derivative at $\alpha$ by $P'(\alpha) = c_1$ using Equations (3.2.4) and (3.2.9). We have $\mathbf{a} = [a_0, a_1, \ldots, a_N]$, $\mathbf{b} = [b_0, b_1, \ldots, b_N]$, and $\mathbf{c} = [c_1, c_2, \ldots, c_N]$.
1 **Print a**, $\alpha$, $N$
2 $b_N = a_N$
3 $c_N = b_N$
4 **Repeat** steps 5–6 **with** $i = N - 1, N - 2, \ldots, 1$
     5 $b_i = b_{i+1}\alpha + a_i$
     6 $c_i = c_{i+1}\alpha + b_i$
7 $b_0 = b_1\alpha + a_0$
8 **Print** $c_1$, $b_0$, and exit.

EXAMPLE 3.2.1

Evaluate $P(2)$ and $P'(2)$ using Horner's method if $P(x) = x^4 - 10x^3 + 35x^2 - 50x + 24$.

We have $a_4 = 1$, $a_3 = -10$, $a_2 = 35$, $a_1 = -50$, and $a_0 = 24$.
From Equation (3.2.4), $b_4 = a_4 = 1$; $b_i = b_{i+1}\alpha + a_i$ for $i = 3, 2, 1$, and 0.
For $i = 3$, $b_3 = b_4\alpha + a_3 = 1(2) - 10 = -8$.
For $i = 2$, $b_2 = b_3\alpha + a_2 = -8(2) + 35 = 19$.
For $i = 1$, $b_1 = b_2\alpha + a_1 = 19(2) - 50 = -12$ and for $i = 0$, $b_0 = b_1\alpha + a_0 = -12(2) + 24 = 0$. Thus $P(2) = b_0 = 0$. Hence $(x - 2)$ is a factor of $P(x)$.
So

$$q(x) = b_4x^3 + b_3x^2 + b_2x + b_1$$
$$= x^3 - 8x^2 + 19x - 12$$

From Equation (3.2.9), $c_4 = b_4 = 1$; $c_i = c_{i+1}\alpha + b_i$ for $i = 3, 2$, and 1.
For $i = 3$, $c_3 = c_4\alpha + b_3 = 1(2) - 8 = -6$.
For $i = 2$, $c_2 = c_3\alpha + b_2 = -6(2) + 19 = 7$ and for $i = 1$, $c_1 = c_2\alpha + b_1 = 7(2) - 12 = 2$. Thus $P'(2) = q(2) = c_1 = 2$. ∎

When we use hand calculations, it is easier to write the coefficients $a_i$ on a line and compute $a_i + \alpha b_{i+1}$ and $c_i = b_i + \alpha c_{i+1}$ as shown below in Table 3.2.1. This suggests the name **synthetic division** for this technique.

**Table 3.2.1**

| $a_N$ | $a_{N-1}$ | $a_{N-2}$ | ... | $a_i$ | ... | $a_2$ | $a_1$ | $a_0$ |
|---|---|---|---|---|---|---|---|---|
| | $\alpha b_N$ | $\alpha b_{N-1}$ | ... | $\alpha b_{i+1}$ | ... | $\alpha b_3$ | $\alpha b_2$ | $\alpha b_1$ |
| $b_N$ | $b_{N-1}$ | $b_{N-2}$ | ... | $b_i$ | ... | $b_2$ | $b_1$ | $b_0 = P(\alpha)$ |
| | $\alpha c_N$ | $\alpha c_{N-1}$ | ... | $\alpha c_{i+1}$ | ... | $\alpha c_3$ | $\alpha c_2$ | |
| $c_N$ | $c_{N-1}$ | $c_{N-2}$ | ... | $c_i$ | ... | $c_2$ | $c_1 = P'(\alpha)$ | |

EXAMPLE 3.2.2

Evaluate $P(2)$ and $P'(2)$ using synthetic division if $P(x) = x^4 - 10x^3 + 35x^2 - 50x + 24$.

Writing just the coefficients,

$$
\begin{array}{r|rrrrr}
 & 1 & -10 & 35 & -50 & 24 \\
\alpha = 2 & & 2 & -16 & 38 & -24 \\
\hline
 & 1 & -8 & 19 & -12 & 0 \quad = P(2) \\
 & & 2 & -12 & 14 & \\
\hline
 & 1 & -6 & 7 & 2 & = P'(2)
\end{array}
$$

Hence $P(2) = 0$ and $P'(2) = 2$. ∎

Therefore, when the Newton–Raphson method is used to find a zero of a given polynomial $P(x)$, use Horner's method or synthetic division to evaluate $P(\alpha)$ and $P'(\alpha)$.

**EXAMPLE 3.2.3**    Compute the first two iterates of the Newton–Raphson method starting with $r_0 = 0.5$ for the polynomial

$$P(x) = x^4 - 10x^3 + 35x^2 - 50x + 24$$

Use synthetic division to compute $P(r_n)$ and $P'(r_n)$.

Since $r_1 = r_0 - (P(r_0)/P'(r_0))$, we find $P(r_0)$ and $P'(r_0)$ using synthetic division. Using the scheme shown in Table 3.2.1, we have

|  |  | 1 | $-10$ | 35 | $-50$ | 24 |  |
|---|---|---|---|---|---|---|---|
| $\alpha = 0.5$ |  |  | 0.5 | $-4.75$ | 15.125 | $-17.4375$ |  |
|  |  | 1 | $-9.5$ | 30.25 | $-34.875$ | 6.5625 | $= P(0.5)$ |
|  |  |  | 0.5 | $-4.50$ | 12.875 |  |  |
|  |  | 1 | $-9.0$ | 25.75 | $-22.00$ |  | $= P'(0.5)$ |

Hence $r_1 = r_0 - (P(r_0)/P'(r_0)) = 0.5 - (6.5625/(-22.00)) = 0.5 + 0.2983 = 0.7983$. Repeating the procedure to find $r_2$,

|  |  | 1 | $-10$ | 35 | $-50$ | 24 |  |
|---|---|---|---|---|---|---|---|
| $\alpha = 0.7983$ |  |  | 0.7983 | $-7.34572$ | 22.07641 | $-22.2914$ |  |
|  |  | 1 | $-9.2017$ | 27.65428 | $-27.92359$ | 1.70860 | $= P(0.7983)$ |
|  |  |  | 0.7983 | $-6.70843$ | 16.72107 |  |  |
|  |  | 1 | $-8.40340$ | 20.94585 | $-11.20252$ |  | $= P'(0.7983)$ |

Hence $r_2 = r_1 - (P(r_1)/P'(r_1)) = 0.7983 - (1.70860/(-11.20252)) = 0.95082$. An actual zero is 1.0.    ■ ■ ■

If the $N$th iterate $r_N$ of the Newton–Raphson method is the zero of $P$, then

$$P(x) = (x - r_N)Q_1(x)$$

If the other zeros are to be found, then they must be the zeros of $Q_1(x)$. We refer to $Q_1(x)$ as the deflated polynomial. We can find a second zero of $P$ by applying the Newton–Raphson method to $Q_1(x)$, and this procedure can be repeated. This method sometimes leads to inaccurate approximations which can be improved by applying the Newton–Raphson method to the original polynomial $P$ by using the approximated zero as the initial value. If the $N$th degree polynomial $P$ has $N$ real zeros, then by using this procedure iteratively, $(N - 2)$ zeros can be found along with an approximate quadratic factor $Q_{N-2}(x)$. If desired, the quadratic formula can be used to solve $Q_{N-2}(x) = 0$ for the last two zeros of $P$.

## 3.3   THE BAIRSTOW METHOD

In order to find a complex root of a polynomial, the methods described so far are not of much help (except in Exercise 10, p. 45). If complex arithmetic capabilities are available, then the Newton–Raphson method can be used with a complex-valued initial estimate to find a complex root of a polynomial. However, for the polynomials with real-valued coefficients, the complex roots occur in conjugate pairs. The product of the linear factors of a pair of these complex conjugate roots is a quadratic polynomial with real coefficients. If we can extract the quadratic factors, then we can avoid complex arithmetic. To find the complex roots of a polynomial, first find the quadratic factors and then find the zeros of the quadratic factors by using the quadratic formula.

Let $x^2 - rx - s$ be a trial quadratic factor near the desired quadratic factor of a polynomial

$$P(x) = a_N x^N + a_{N-1} x^{N-1} + \cdots + a_1 x + a_0 \tag{3.3.1}$$

Then

$$P(x) = (x^2 - rx - s)(b_N x^{N-2} + b_{N-1} x^{N-3} + \cdots + b_3 x + b_2) + b_1(x - r) + b_0 \tag{3.3.2}$$

where $b_1$ and $b_0$ are both zero if $x^2 - rx - s$ is an exact divisor. Expanding Equation (3.3.2) in powers of $x$,

$$P(x) = b_N x^N + (b_{N-1} - rb_N)x^{N-1} + (b_{N-2} - rb_{N-1} - sb_N)x^{N-2}$$
$$+ \cdots + (b_i - rb_{i+1} - sb_{i+2})x^i$$
$$+ \cdots + (b_1 - rb_2 - sb_3)x + (b_0 - rb_1 - sb_2) \tag{3.3.3}$$

Comparing the coefficients of like powers of $x$ in Equations (3.3.1) and (3.3.3) and then solving for $b_i$,

$$b_N = a_N \qquad b_{N-1} = a_{N-1} + rb_N \qquad b_i = a_i + rb_{i+1} + sb_{i+2} \tag{3.3.4}$$

for $i = N - 2, N - 3, \ldots, 1, 0$.

For hand calculations, it is easier to write the coefficients $a_i$ on a line and then compute $b_i = a_i + rb_{i+1} + sb_{i+2}$ as shown below in Table 3.3.1.

**Table 3.3.1**

|  | $a_N$ | $a_{N-1}$ | $a_{N-2}$ | $a_{N-3}$ | $\ldots$ | $a_i$ | $\ldots$ | $a_2$ | $a_1$ | $a_0$ |
|---|---|---|---|---|---|---|---|---|---|---|
| $r$ |  | $rb_N$ | $rb_{N-1}$ | $rb_{N-2}$ | $\ldots$ | $rb_{i+1}$ | $\ldots$ | $rb_3$ | $rb_2$ | $rb_1$ |
| $s$ |  |  | $sb_N$ | $sb_{N-1}$ | $\ldots$ | $sb_{i+2}$ | $\ldots$ | $sb_4$ | $sb_3$ | $sb_2$ |
| $b_N$ | $b_{N-1}$ | $b_{N-2}$ | $b_{N-3}$ | $\ldots$ | $b_i$ | $\ldots$ | $b_2$ | $b_1$ | $b_0$ |

We would like $b_1$ and $b_0$ both to be zero; however, usually this will not happen. If $r$ and $s$ are changed properly, we can make $b_1$ and $b_0$ be zero. From Equation (3.3.4), $b_1$ and $b_0$ are functions of $r$ and $s$. Therefore,

$$b_1 = b_1(r, s) \quad \text{and} \quad b_0 = b_0(r, s) \tag{3.3.5}$$

Let $r_0$ and $s_0$ be close to $r$ and $s$ which makes $b_1(r, s) = 0$ and $b_0(r, s) = 0$. Find $r$ and $s$ by starting with $r_0$ and $s_0$ such that $b_1(r, s) = 0$ and $b_0(r, s) = 0$. Then by Taylor's Theorem in two dimensions (Theorem A.9),

$$b_1(r, s) = b_1(r_0 + (r - r_0), s_0 + (s - s_0)) = b_1(r_0, s_0)$$

$$+ \left[ (r - r_0) \frac{\partial b_1}{\partial r}(r_0, s_0) + (s - s_0) \frac{\partial b_1}{\partial s}(r_0, s_0) \right]$$

$$+ \frac{1}{2!} \left[ (r - r_0)^2 \frac{\partial^2 b_1}{\partial r^2}(\xi_1, \eta_1) \right.$$

$$+ 2(r - r_0)(s - s_0) \frac{\partial^2 b_1}{\partial r \partial s}(\xi_1, \eta_1) + (s - s_0)^2 \frac{\partial^2 b_1}{\partial s^2}(\xi_1, \eta_1) \left. \right]$$

where

$$\xi_1 = r_0 + \tau_1(r - r_0), \qquad \eta_1 = s_0 + \tau_1(s - s_0), \quad \text{and} \quad 0 \le \tau_1 \le 1 \quad \textbf{(3.3.6)}$$

and

$$b_0(r, s) = b_0(r_0 + (r - r_0), s_0 + (s - s_0)) = b_0(r_0, s_0)$$

$$+ \left[ (r - r_0) \frac{\partial b_0}{\partial r}(r_0, s_0) + (s - s_0) \frac{\partial b_0}{\partial s}(r_0, s_0) \right]$$

$$+ \frac{1}{2!} \left[ (r - r_0)^2 \frac{\partial^2 b_0}{\partial r^2}(\xi_2, \eta_2) \right.$$

$$+ 2(r - r_0)(s - s_0) \frac{\partial^2 b_0}{\partial r \partial s}(\xi_2, \eta_2) + (s - s_0)^2 \frac{\partial^2 b_0}{\partial s^2}(\xi_2, \eta_2) \left. \right]$$

where

$$\xi_2 = r_0 + \tau_2(r - r_0), \quad \eta_2 = s_0 + \tau_2(s - s_0), \quad \text{and} \quad 0 \le \tau_2 \le 1$$

Assuming that $r_0$ is very close to $r$ and $s_0$ is very close to $s$, the higher-order terms in Equation (3.3.6) can be ignored, obtaining

$$b_1(r, s) \approx b_1(r_0, s_0) + \left[ (r - r_0) \frac{\partial b_1}{\partial r}(r_0, s_0) + (s - s_0) \frac{\partial b_1}{\partial s}(r_0, s_0) \right]$$

and

$$b_0(r, s) \approx b_0(r_0, s_0) + \left[ (r - r_0) \frac{\partial b_0}{\partial r}(r_0, s_0) + (s - s_0) \frac{\partial b_0}{\partial s}(r_0, s_0) \right]$$

Since $b_1(r, s) = 0$ and $b_0(r, s) = 0$,

$$(r - r_0) \frac{\partial b_1}{\partial r}(r_0, s_0) + (s - s_0) \frac{\partial b_1}{\partial s}(r_0, s_0) \approx -b_1(r_0, s_0)$$

and

$$(r - r_0) \frac{\partial b_0}{\partial r} (r_0, s_0) + (s - s_0) \frac{\partial b_0}{\partial s} (r_0, s_0) \approx -b_0(r_0, s_0) \tag{3.3.7}$$

Now we want $\partial b_1/\partial r$, $\partial b_1/\partial s$, $\partial b_0/\partial r$, and $\partial b_0/\partial s$ in order to find $r$ and $s$ in Equation (3.3.7).

From Equation (3.3.4), $b_N = a_N$, so $\partial b_N/\partial r = \partial a_N/\partial r = 0$ and $\partial b_N/\partial s = \partial a_N/\partial s = 0$ since $a_N$ is a constant. Also $b_{N-1} = a_{N-1} + rb_N$, therefore $\partial b_{N-1}/\partial r = r(\partial b_N/\partial r) + b_N = b_N$ and $\partial b_{N-1}/\partial s = r(\partial b_N/\partial s) = 0$.

From Equation (3.3.4), we have

$$b_i = a_i + rb_{i+1} + sb_{i+2} \quad \text{for } i = N - 2, N - 1, \ldots, 0 \tag{3.3.8}$$

Differentiating Equation (3.3.8) with respect to $r$,

$$\frac{\partial b_i}{\partial r} = b_{i+1} + r \frac{\partial b_{i+1}}{\partial r} + s \frac{\partial b_{i+2}}{\partial r} \tag{3.3.9}$$

For convenience, define $c_{i+1} = \partial b_i/\partial r$. Then Equation (3.3.9) becomes

$$c_{i+1} = b_{i+1} + rc_{i+2} + sc_{i+3} \tag{3.3.10}$$

Differentiating Equation (3.3.8) with respect to $s$, we get

$$\frac{\partial b_i}{\partial s} = r \frac{\partial b_{i+1}}{\partial s} + b_{i+2} + s \frac{\partial b_{i+2}}{\partial s} \tag{3.3.11}$$

Again for convenience, define $d_{i+1} = \partial b_{i-1}/\partial s$. Then Equation (3.3.11) reduces to

$$d_{i+2} = rd_{i+3} + b_{i+2} + sd_{i+4}$$

This can be written as

$$d_{i+1} = b_{i+1} + rd_{i+2} + sd_{i+3} \tag{3.3.12}$$

which is exactly like Equation (3.3.10). Therefore

$$c_{i+1} = d_{i+1} = \frac{\partial b_i}{\partial r} = \frac{\partial b_{i-1}}{\partial s} \tag{3.3.13}$$

Hence

$$c_N = \frac{\partial b_{N-1}}{\partial r} = b_N$$

$$c_{N-1} = \frac{\partial b_{N-2}}{\partial r} = \frac{\partial}{\partial r} [a_{N-2} + rb_{N-1} + sb_N]$$

$$= b_{N-1} + r \frac{\partial b_{N-1}}{\partial r} + s \frac{\partial b_N}{\partial r} = b_{N-1} + rc_N$$

and

$$c_i = b_i + rc_{i+1} + sc_{i+2} \quad \text{for } i = N - 2, N - 1, \ldots, 1 \tag{3.3.14}$$

Using Equation (3.3.13) in Equation (3.3.7) yields

$$(r - r_0)c_2 + (s - s_0)c_3 \approx -b_1$$
$$(r - r_0)c_1 + (s - s_0)c_2 \approx -b_0 \tag{3.3.15}$$

where $c_1$, $c_2$, and $c_3$ are given by Equation (3.3.14) and $b_0$ and $b_1$ are given by Equation (3.3.8).

Solving Equation (3.3.15) for $(r - r_0)$ and $(s - s_0)$,

$$r - r_0 \approx \frac{b_1 c_2 - b_0 c_3}{c_1 c_3 - c_2^2} \quad \text{and} \quad s - s_0 \approx \frac{b_0 c_2 - b_1 c_1}{c_1 c_3 - c_2^2}$$

Thus the values of $r$ and $s$ are given by

$$r_{j+1} = r_j + \frac{b_1 c_2 - b_0 c_3}{c_1 c_3 - c_2^2} \quad \text{and} \quad s_{j+1} = s_j + \frac{b_0 c_2 - b_1 c_1}{c_1 c_3 - c_2^2} \qquad \textbf{(3.3.16)}$$

for $j = 0, 1, 2, \ldots$, where

$$b_N = a_N$$

$$b_{N-1} = a_{N-1} + r_j b_N$$

$$b_i = a_i + r_j b_{i+1} + s_j b_{i+2} \quad \text{for } i = N - 2, N - 3, \ldots, 1, 0 \qquad \textbf{(3.3.17)}$$

and

$$c_N = b_N$$

$$c_{N-1} = b_{N-1} + r_j c_N$$

$$c_i = b_i + r_j c_{i+1} + s_j c_{i+2} \quad \text{for } i = N - 2, N - 3, \ldots, 1 \qquad \textbf{(3.3.18)}$$

For a given $r_0$ and $s_0$, calculate $b_i$ from Equation (3.3.17) and $c_i$ from Equation (3.3.18) using $r_0$ and $s_0$. With $b_0$, $b_1$, and $c_1$, $c_2$, $c_3$, calculate $r_1$ and $s_1$ from Equation (3.3.16). Substituting $r_1$ and $s_1$, recalculate $b_i$ from Equation (3.3.17) and $c_i$ from Equation (3.3.18). Using these new values of $b_i$ and $c_i$ in Equation (3.3.16), calculate $r_2$ and $s_2$. Repeat this process until $b_1$ and $b_0$ become zero. This method is called the **Bairstow method**.

If the zeros of $x^2 - rx - s$ are distinct and $r_0$ and $s_0$ are sufficiently close to $r$ and $s$, then it can be shown that the Bairstow method will converge quadratically. The main drawback of this method is that for most polynomials even crude approximations to the quadratic factors are hard to obtain by mere inspection.

**Algorithm**

**Subroutine** Bair(**a**, **b**, **c**, $N$, $r$, $s$, maxit, eps)
Comment: This program determines a quadratic factor $x^2 - rx - s$ of the given polynomial $a_N x^N + a_{N-1} x^{N-1} + \cdots + a_1 x + a_0$ where $N \geq 3$ by providing initial values of $r$ and $s$. The term maxit is the maximum number of iterations; eps is tolerance. We have $\mathbf{a} = [a_0, a_1, \ldots, a_N]$, $\mathbf{b} = [b_0, b_1, \ldots, b_N]$, and $\mathbf{c} = [c_1, c_2, \ldots, c_N]$. Initial guesses of $r$ and $s$ are to be provided.
1 **Print** $N$, **a**, $r$, $s$, maxit, eps
2 $b_N = a_N$
3 $c_N = a_N$
4 $m = 0$
5 **Repeat** steps 6–18 **while** $m \leq$ maxit
   6 $b_{N-1} = a_{N-1} + r b_N$
   7 $c_{N-1} = b_{N-1} + r c_N$
   8 **Repeat** steps 9–10 **with** $i = N - 2, N - 3, \ldots, 1$
       9 $b_i = a_i + r b_{i+1} + s b_{i+2}$

$$10 \; c_i = b_i + rc_{i+1} + sc_{i+2}$$
11 $b_0 = a_0 + rb_1 + sb_2$
12 deno $= c_1 c_3 - c_2^2$
13 $r_1 = r + ((b_1 c_2 - b_0 c_3)/\text{deno})$
14 $s_1 = s + ((b_0 c_2 - b_1 c_1)/\text{deno})$
15 **If** $|s - s_1| <$ eps $|s_1|$ and $|r - r_1| <$ eps $|r_1|$, **then print** 'finished', $r_1$, and $s_1$, and exit.
16 $r = r_1$
17 $s = s_1$
18 $m = m + 1$
19 **Print** 'algorithm fails; start with different values of $r$ and $s$' and exit.  ◀

**EXAMPLE 3.3.1**

Find the quadratic factors of $x^4 - 10x^3 + 35x^2 - 50x + 24$.

We could start with any quadratic factor, but for hand calculations it is initially easy to start with $x^2 - x - 1$. Thus $r = 1$ and $s = 1$. Using the scheme shown in Table 3.3.1, we have

|  | 1 | $-10$ | 35 | $-50$ | 24 |
|---|---|---|---|---|---|
| $r_0 = 1$ |  | 1 | $-9$ | 27 | $-32$ |
| $s_0 = 1$ |  |  | 1 | $-9$ | 27 |
|  | 1 | $-9$ | 27 | $b_1 = -32$ | $b_0 = 19$ |

Thus we have $b_4 = 1$, $b_3 = -9$, $b_2 = 27$, $b_1 = -32$, and $b_0 = 19$. Since Equation (3.3.18) resembles Equation (3.3.17), the scheme used to find $b_i$ can also be used to find $c_i$. Therefore

|  | 1 | $-9$ | 27 | $-32$ | 19 |
|---|---|---|---|---|---|
| $r_0 = 1$ |  | 1 | $-8$ | 20 | $-20$ |
| $s_0 = 1$ |  |  | 1 | $-8$ | 20 |
|  | 1 | $-8$ | $c_2 = 20$ | $c_1 = -20$ | $c_0 = 19$ |

Thus for $j = 0$, Equation (3.3.16) becomes

$$r_1 = r_0 + \frac{b_1 c_2 - b_0 c_3}{c_1 c_3 - c_2^2} = 1 + \frac{(-32)(20) - 19(-8)}{(-20)(-8) - (20)^2} = 1 + \frac{-488}{-240} = 3.03333$$

$$s_1 = s_0 + \frac{b_0 c_2 - b_1 c_1}{c_1 c_3 - c_2^2} = 1 + \frac{(19)(20) - (-20)(-32)}{(-20)(-8) - (20)^2} = 1 + \frac{-260}{-240} = 2.08333$$

The computations with the values of $r_1$ and $s_1$ are given in the following table:

|  | 1 | $-10$ | 35 | $-50$ | 24 |
|---|---|---|---|---|---|
| $r_1 = 3.03333$ |  | 3.03333 | $-21.13221$ | 48.38507 | $-48.92415$ |
| $s_2 = 2.08333$ |  |  | 2.08333 | $-14.51387$ | 33.23145 |
|  | 1 | $-6.96667$ | 15.95112 | $-16.12886$ | 8.30729 |
|  |  | 3.03333 | $-11.93112$ | 18.51342 | $-17.62330$ |
|  |  |  | 2.08333 | $-8.19445$ | 12.71525 |
|  | 1 | $-3.93334$ | 6.10333 | $-5.80989$ | 3.39924 |

Thus we have $b_1 = -16.12886$, $b_0 = 8.30729$, $c_1 = -5.80981$, $c_2 = 6.10333$, and $c_3 = -3.93334$.

Thus for $j = 1$, Equation (3.3.16) becomes

$$r_2 = r_1 + \frac{b_1 c_2 - b_0 c_3}{c_1 c_3 - c_2^2}$$

$$= 3.03333 + \frac{(-16.12886)(6.10333) - (8.30729)(-3.93334)}{(-5.80989)(-3.93334) - (6.10333)^2}$$

$$= 3.03333 + \frac{-65.76436}{-14.39836} = 7.60082$$

$$s_2 = s_1 + \frac{b_0 c_2 - b_1 c_1}{c_1 c_3 - c_2^2}$$

$$= 2.08333 + \frac{(8.30729)(6.10333) - (-5.80989)(-16.12886)}{-14.39836}$$

$$= 2.08333 + \frac{-43.00477}{-14.39836} = 5.07011$$

Similarly, we continue and the computed values are given in Table 3.3.2.

**Table 3.3.2**

| $i$ | $r_i$ | $s_i$ |
|---|---|---|
| 0 | 1.00000 | 1.00000 |
| 1 | 3.03333 | 2.08333 |
| 2 | 7.60082 | 5.07011 |
| 3 | 5.71706 | 9.17668 |
| 4 | 5.34545 | 2.55475 |
| 5 | 5.16782 | -1.05846 |
| 6 | 5.07894 | -2.87953 |
| 7 | 5.03200 | -3.69951 |
| 8 | 5.00758 | -3.96511 |
| 9 | 5.00037 | -3.99945 |
| 10 | 5.00000 | -3.99999 |
| 11 | 5.00000 | -4.00000 |

One quadratic factor is $x^2 - 5x + 4 = (x - 4)(x - 1)$ and it can be verified that another quadratic factor is $x^2 - 5x + 6 = (x - 2)(x - 3)$. Thus $x^4 - 10x^3 + 35x^2 - 50x + 24 = (x^2 - 5x + 4)(x^2 - 5x + 6)$.  ■ ■ ■

---

**EXERCISES**

1. Let $z_1 = x_1 + I y_1$ and $z_2 = x_2 + I y_2$ be two complex numbers where $I = \sqrt{-1}$. Verify that (a) $\overline{z_1 + z_2} = \bar{z}_1 + \bar{z}_2$ and (b) $\overline{z_1 z_2} = \bar{z}_1 \bar{z}_2$.

2. Prove that $P(x) = x^4 + 2x^2 - 2x - 1$ has one positive, one negative, and two complex zeros without finding the zeros.

3. Prove that $P(x) = 2x^3 + x^2 + 1$ has one negative and two complex zeros without finding the zeros.

4. If $P(x) = a_3x^3 + a_2x^2 + a_1x + a_0$ with zeros $z_1$, $z_2$, and $z_3$, verify that $z_1z_2z_3 = -(a_0/a_3)$, $z_1z_2 + z_2z_3 + z_1z_3 = a_1/a_3$, and $z_1 + z_2 + z_3 = -(a_2/a_3)$. We solve the system of these three nonlinear equations numerically to find $z_1$, $z_2$, and $z_3$.

5. Find the companion matrix of $P(x) = x^3 - 3x^2 + 2$. Find the corresponding characteristic polynomial of the companion matrix.

6. Verify that the characteristic polynomial of Equation (3.1.7) is given by Equation (3.1.6).

7. Evaluate $P(1)$ and $P'(1)$ using synthetic division or Horner's method if $P(x) = x^5 + x^4 + x^3 + x^2 + x + 1$.

8. Evaluate $P'(2)$ and $P''(2)$ using synthetic division or Horner's method if $P(x) = x^4 + 3x^3 + 2x^2 + x + 1$.

In Exercises 9–11, find all the real zeros to within $10^{-2}$ of each of the polynomials using the Newton–Raphson method and synthetic division or Horner's method.

9. $P(x) = x^3 - 3x^2 + 2$

10. $P(x) = 16x^3 - 12x^2 - 4x + 3$

11. $P(x) = (35x^4 - 30x^2 + 3)/8$

For the polynomials in Exercises 12–14, compute the first three iterates of the Bairstow method with $r_0 = 1$ and $s_0 = 1$.

12. $P(x) = 4x^4 - 8x^3 + 5x^2 - x - 3$

13. $P(x) = 2x^3 - x^2 - 16x + 15$

14. $P(x) = x^4 - x^3 + 2x + 8$

## COMPUTER EXERCISES

**C.1.** Using Equations (3.2.4) and (3.2.9), write a subroutine to evaluate $P(\alpha)$ and $P'(\alpha)$.

**C.2.** Using the Newton–Raphson method, deflation, and Exercise C.1, write a computer program to find all real zeros of a given $N$th degree polynomial

$$P(x) = a_N x^N + a_{N-1}x^{N-1} + \cdots + a_1x + a_0$$

**C.3.** The Legendre polynomials are very useful and they are defined by

$$P_0(x) = 1, \qquad P_1(x) = x$$

and

$$P_{n+2}(x) = \frac{2n + 3}{n + 2}xP_{n+1}(x) - \frac{n + 1}{n + 2}P_n(x) \quad \text{for } n = 0, 1, \ldots$$

Determine $P_6$ and using Exercise C.2, find all real zeros of $P_6$ accurate to within $10^{-5}$.

**C.4.** Using the Bairstow method, write a subroutine to find a quadratic factor of a given polynomial

$$P(x) = a_N x^N + a_{N-1}x^{N-1} + \cdots + a_1x + a_0$$

Initial guesses for $r_0$ and $s_0$ are given in the main program. Assume that the convergence is reached when $|b_0|$ and $|b_1|$ are less than eps. Use Equations (3.3.16), (3.3.17), and (3.3.18). Put a limit on the number of iterations.

**C.5.** Using Equation (3.3.4), write a subroutine to divide a given polynomial

$$P(x) = a_N x^N + a_{N-1} x^{N-1} + \cdots + a_1 x + a_0$$

by its quadratic factor $x^2 - rx - s$.

**C.6.** Write a subroutine to solve a quadratic equation $x^2 - rx - s = 0$.

**C.7.** Using Exercises C.4, C.5, and C.6, write a computer program to find all zeros of a given $N$th degree polynomial

$$P(x) = a_N x^N + a_{N-1} x^{N-1} + \cdots + a_1 x + a_0$$

Read initial guesses $r_0$ and $s_0$, eps $= 10^{-5}$, $N$, and the coefficients $a_0$, $a_1$, $a_2$, ..., $a_N$.

**C.8.** Using Exercise C.7, find the roots of $x^6 + 5x^3 + 7x^2 + 1 = 0$. Use various $r_0$ and $s_0$.

**C.9.** Using Exercise C.7, find the eigenvalues of the matrix

$$A = \begin{bmatrix} 0 & 1 & 0 & 0 \\ 0 & 0 & 1 & 0 \\ 0 & 0 & 0 & 1 \\ -5 & 20 & -21 & 8 \end{bmatrix}$$

## 3.4  THE QUOTIENT-DIFFERENCE (QD) ALGORITHM*

The Newton–Raphson and Bairstow methods are very effective if close approximations to the zeros are already known. If such approximations are not known, then we run into a serious problem. If the zeros are real, then the bisection method can be used to find crude approximations most of the time. Unfortunately, a similar simple analog for complex zeros is not available. The QD algorithm provides the first approximations to all zeros. This algorithm is due to the Swiss mathematician Rutishauser who was a pioneer of numerical analysis for the computer age. For the proofs of convergence theorems, complex function theory is required; therefore, only the method will be presented. For the general discussion of the QD algorithm we refer the reader to Henrici (1964) and (for the classical papers on the QD algorithm by Rutishauser) Rutishauser (1956).

Let

$$P(x) = a_N x^N + a_{N-1} x^{N-1} + \cdots + a_1 x + a_0 \qquad (3.4.1)$$

We form an array of $q^{(0)}$ and $e^{(0)}$ in the following way:

| $e_0^{(0)}$ | $q_1^{(0)}$ | $e_1^{(0)}$ | $q_2^{(0)}$ | $e_2^{(0)}$ | $\cdots$ | $q_{N-1}^{(0)}$ | $e_{N-1}^{(0)}$ | $q_N^{(0)}$ | $e_N^{(0)}$ |
|---|---|---|---|---|---|---|---|---|---|
| | $-\dfrac{a_{N-1}}{a_N}$ | | 0 | | | 0 | | 0 | |
| 0 | | $\dfrac{a_{N-2}}{a_{N-1}}$ | | $\dfrac{a_{N-3}}{a_{N-2}}$ | | $\dfrac{a_0}{a_1}$ | | 0 | |

The first row consists only of the values of $q_i^{(0)}$ which are defined by

$$q_1^{(0)} = -\frac{a_{N-1}}{a_N} \quad \text{and} \quad q_2^{(0)} = q_3^{(0)} = \cdots = q_{N-1}^{(0)} = q_N^{(0)} = 0 \qquad \text{(3.4.2)}$$

The second row consists only of values of $e_i^{(0)}$ which are defined by

$$e_0^{(0)} = e_N^{(0)} = 0 \quad \text{and} \quad e_i^{(0)} = \frac{a_{N-i-1}}{a_{N-i}} \quad \text{for } i = 1, 2, \ldots, N-1 \qquad \text{(3.4.3)}$$

A new row of $q$ is formed by computing

$$q_i^{(k+1)} = (e_i^{(k)} - e_{i-1}^{(k)}) + q_i^{(k)} \quad \text{for } i = 1, 2, \ldots, N \quad \text{and } k = 0, 1, \ldots \quad \text{(3.4.4)}$$

Similarly a new row of $e$ is computed using

$$e_i^{(k+1)} = \left(\frac{q_{i+1}^{(k+1)}}{q_i^{(k+1)}}\right) e_i^{(k)} \quad \text{for } i = 1, 2, \ldots, N-1 \quad \text{and } k = 0, 1, \ldots \quad \text{(3.4.5)}$$

Suppose $a_i \neq 0$ for $i = 0, 1, \ldots, N$ and all roots are simple and real. Then it can be proved (Henrici 1964) that

$$\lim_{k \to \infty} q_i^{(k)} = r_i \quad \text{where } |r_1| > |r_2| \cdots |r_N| \quad \text{for } i = 1, 2, \ldots, N$$

and

$$\lim_{k \to \infty} e_i^{(k)} = 0 \quad \text{for } i = 1, 2, \ldots, N-1$$

Since the QD method converges slowly, the approximations are improved using the Newton–Raphson method.

**EXAMPLE 3.4.1**  Find the approximations of the zeros of the polynomial $P(x) = x^4 - 10x^3 + 35x^2 - 50x + 24$ using the QD method.

From Equations (3.4.2) and (3.4.3),

$$q_1^{(0)} = -\frac{a_{N-1}}{a_N} = -\frac{a_3}{a_4} = -\frac{(-10)}{(1)} = 10, \qquad q_2^{(0)} = q_3^{(0)} = q_4^{(0)} = 0$$

$$e_0^{(0)} = 0, \qquad e_1^{(0)} = \frac{a_{N-2}}{a_{N-1}} = \frac{a_2}{a_3} = \frac{35}{-10} = -3.5$$

$$e_2^{(0)} = \frac{a_{N-3}}{a_{N-2}} = \frac{a_1}{a_2} = \frac{-50}{35} = -1.42857$$

$$e_3^{(0)} = \frac{a_{N-4}}{a_{N-3}} = \frac{a_0}{a_1} = \frac{24}{-50} = -0.48, \qquad e_4^{(0)} = 0$$

For $k = 0$, Equations (3.4.4) and (3.4.5) give

$$q_1^{(1)} = (e_1^{(0)} - e_0^{(0)}) + q_1^{(0)} = (-3.5 - 0) + 10 = 6.5$$

$$q_2^{(1)} = (e_2^{(0)} - e_1^{(0)}) + q_2^{(0)} = (-1.42857 + 3.5) + 0.0 = 2.07143$$

$$q_3^{(1)} = (e_3^{(0)} - e_2^{(0)}) + q_3^{(0)} = (-0.48 + 1.42857) + 0.0 = 0.94857$$

$$q_4^{(1)} = (e_4^{(0)} - e_3^{(0)}) + q_4^{(0)} = (0.0 + 0.48) + 0.0 = 0.48$$

$$e_1^{(1)} = \left(\frac{q_2^{(1)}}{q_1^{(1)}}\right) e_1^{(0)} = \left(\frac{2.07143}{6.5}\right)(-3.5) = -1.11539$$

$$e_2^{(1)} = \left(\frac{q_3^{(1)}}{q_2^{(1)}}\right) e_2^{(0)} = \left(\frac{0.94851}{2.07143}\right)(-1.42857) = -0.65419$$

$$e_3^{(1)} = \left(\frac{q_4^{(1)}}{q_3^{(1)}}\right) e_3^{(0)} = \left(\frac{0.48}{0.94857}\right)(-0.48) = -0.24289$$

Continuing, we compute the values of $q^{(k)}$ and $e^{(k)}$ and these computed values are given in Table 3.4.1.

**Table 3.4.1**

| $k$ | $e_0^{(k)}$ | $q_1^{(k)}$ | $e_1^{(k)}$ | $q_2^{(k)}$ | $e_2^{(k)}$ | $q_3^{(k)}$ | $e_3^{(k)}$ | $q_4^{(k)}$ | $e_4^{(k)}$ |
|---|---|---|---|---|---|---|---|---|---|
| 0 | | 10.00000 | | 0.00000 | | 0.00000 | | 0.00000 | |
| | 0.00000 | | −3.50000 | | −1.42857 | | −0.48000 | | 0.00000 |
| 1 | | 6.50000 | | 2.07143 | | 0.94857 | | 0.48000 | |
| | 0.00000 | | −1.11539 | | −0.65419 | | −0.24289 | | 0.00000 |
| 2 | | 5.38462 | | 2.53263 | | 1.35987 | | 0.72289 | |
| | 0.00000 | | −0.52462 | | −0.35126 | | −0.12912 | | 0.00000 |
| 3 | | 4.86000 | | 2.70598 | | 1.58201 | | 0.85201 | |
| | 0.00000 | | −0.29210 | | −0.20536 | | −0.06954 | | 0.00000 |
| 4 | | 4.56790 | | 2.79272 | | 1.71783 | | 0.92155 | |
| | 0.00000 | | −0.17858 | | −0.12632 | | −0.03730 | | 0.00000 |
| 5 | | 4.38932 | | 2.84499 | | 1.80684 | | 0.95885 | |
| | 0.00000 | | −0.11575 | | −0.08022 | | −0.01980 | | 0.00000 |
| 6 | | 4.27357 | | 2.88052 | | 1.86727 | | 0.97865 | |
| | 0.00000 | | −0.07802 | | −0.05200 | | −0.01038 | | 0.00000 |
| 7 | | 4.19555 | | 2.90653 | | 1.90889 | | 0.98903 | |
| | 0.00000 | | −0.05405 | | −0.03415 | | −0.00538 | | 0.00000 |
| 8 | | 4.14150 | | 2.92643 | | 1.93767 | | 0.99440 | |
| | 0.00000 | | −0.03819 | | −0.02261 | | −0.00276 | | 0.00000 |
| 9 | | 4.10331 | | 2.94201 | | 1.95753 | | 0.99716 | |
| | 0.00000 | | −0.02733 | | −0.01505 | | −0.00141 | | 0.00000 |

All $e$ columns approach zero; therefore all the zeros of the polynomial are real. The $q$ columns converge to the zeros of $P(x)$, whose exact values are $r_1 = 4$, $r_2 = 3$, $r_3 = 2$, and $r_4 = 1$.    ■ ■ ■

If $P(x)$ has a pair of complex conjugate zeros, then one of the values of $e_i$ will not approach zero but will fluctuate in values. The sum of two $q$-values on either side of $e_i$ will approach $r$; that is, $q_i^{(k)} + q_{i+1}^{(k)} \approx r$. The product of the two $q$-values will approach $-s$; that is, $q_i^{(k-1)} q_{i+1}^{(k)} \approx -s$ of the quadratic factor $x^2 - rx - s$.

**EXAMPLE 3.4.2** Find the approximations of the zeros of the polynomial $P(x) = x^3 + 3x^2 + 3x + 2$. The computed values of $q^{(k)}$ and $e^{(k)}$ are given in Table 3.4.2.

**Table 3.4.2**

| $k$ | $e_0^{(k)}$ | $q_1^{(k)}$ | $e_1^{(k)}$ | $q_2^{(k)}$ | $e_2^{(k)}$ | $q_3^{(k)}$ | $e_3^{(k)}$ |
|---|---|---|---|---|---|---|---|
| 0 | | $-3.00000$ | | $0.00000$ | | $0.00000$ | |
| | $0.00000$ | | $1.00000$ | | $0.66667$ | | $0.00000$ |
| 1 | | $-2.00000$ | | $-0.33333$ | | $-0.66667$ | |
| | $0.00000$ | | $0.16667$ | | $1.33333$ | | $0.00000$ |
| 2 | | $-1.83333$ | | $0.83333$ | | $-2.00000$ | |
| | $0.00000$ | | $-0.07576$ | | $-3.20000$ | | $0.00000$ |
| 3 | | $-1.90909$ | | $-2.29091$ | | $1.20000$ | |
| | $0.00000$ | | $-0.09091$ | | $1.67619$ | | $0.00000$ |
| 4 | | $-2.00000$ | | $-0.52381$ | | $-0.47619$ | |
| | $0.00000$ | | $-0.02381$ | | $1.52381$ | | $0.00000$ |
| 5 | | $-2.02381$ | | $1.02381$ | | $-2.00000$ | |
| | $0.00000$ | | $0.01204$ | | $-2.97674$ | | $0.00000$ |
| 6 | | $-2.01176$ | | $-1.96498$ | | $0.97674$ | |
| | $0.00000$ | | $0.01176$ | | $1.47967$ | | $0.00000$ |
| 7 | | $-2.00000$ | | $-0.49708$ | | $-0.50292$ | |
| | $0.00000$ | | $0.00292$ | | $1.49708$ | | $0.00000$ |
| 8 | | $-1.99708$ | | $0.99708$ | | $-2.00000$ | |
| | $0.00000$ | | $-0.00146$ | | $-3.00293$ | | $0.00000$ |
| 9 | | $-1.99854$ | | $-2.00440$ | | $1.00293$ | |
| | $0.00000$ | | $-0.00146$ | | $1.50257$ | | $0.00000$ |

From the table, $e_1^{(k)}$ approaches zero but $e_2^{(k)}$ fluctuates; $q_1^{(k)}$ approaches $-2$ while $q_2^{(k)}$ and $q_3^{(k)}$ fluctuate. This indicates that there is a pair of complex conjugate zeros. Let

$$r^{(k)} = q_i^{(k)} + q_{i+1}^{(k)} = q_2^{(k)} + q_3^{(k)} \quad \text{and} \quad s^{(k)} = -q_i^{(k-1)} q_{i+1}^{(k)} = -q_2^{(k-1)} q_3^{(k)}$$

Then

$$r^{(1)} = q_2^{(1)} + q_3^{(1)} = -0.33333 - 0.66667 = -1.00000$$

$$s^{(1)} = -q_2^{(0)} q_3^{(1)} = -(0.0)(-0.66667) = 0.0$$

$$r^{(2)} = q_2^{(2)} + q_3^{(2)} = 0.83333 - 2.00000 = -1.16667$$

$$s^{(2)} = -q_2^{(1)} q_3^{(2)} = -(-0.33333)(-2.00000) = -0.66666$$

In the same way we compute $r^{(k)}$ and $s^{(k)}$. The computed values are given in Table 3.4.3.

**Table 3.4.3**

| $k$ | $r^{(k)}$ | $s^{(k)}$ |
|---|---|---|
| 1 | $-1.00000$ | $0.00000$ |
| 2 | $-1.16667$ | $-0.66666$ |
| 3 | $-1.09091$ | $-1.00000$ |
| 4 | $-1.00000$ | $-1.09091$ |
| 5 | $-0.97619$ | $-1.04762$ |
| 6 | $-0.98824$ | $-1.00000$ |
| 7 | $-1.00000$ | $-0.98823$ |
| 8 | $-1.00292$ | $-0.99416$ |
| 9 | $-1.00147$ | $-1.00000$ |

From the table, $r^{(k)}$ approaches $-1.0$ and $s^{(k)}$ approaches $-1.0$ as $k \to \infty$. Thus $x^2 + x + 1$ is a quadratic factor of $x^3 + 3x^2 + 3x + 2$. We have $-2$, $(-1 + I\sqrt{3})/2$, and $(-1 - I\sqrt{3})/2$ as zeros of the polynomial.  ■ ■ ■

It is possible that the scheme may break down because the denominator is zero or nearly zero. If $a_i = 0$, then $e_{N-i}$ is not defined and so our QD algorithm breaks down. A possible remedy is to introduce a new variable $y = x - c$ and consider the polynomial $P(y + c)$.

EXAMPLE 3.4.3

Let $P(x) = x^4 - x^3 + x + 1$. Then $a_2 = 0$ and therefore $e_2$ is not defined. Consider $x = y + 1$. Then the new polynomial

$$P^*(y) = P(y + 1) = (y + 1)^4 - (y + 1)^3 + (y + 1) + 1$$
$$= y^4 + 3y^3 + 3y^2 + 2y + 2$$

has all coefficients different than zero. Now use the QD algorithm to find the zeros of $P^*$. By using $x = y + 1$, the zeros of $P$ are found.  ■ ■ ■

The QD algorithm provides simultaneous approximations of all the zeros of a polynomial using only the information about the coefficients of a polynomial. It converges slowly; therefore, a large number of arithmetic operations may be needed. This may increase round-off errors. Consequently, we do not recommend using the QD algorithm to find the zeros of a polynomial $P$ to any degree of accuracy. Use the approximations of the zeros computed by the QD algorithm as the starting values for the Newton–Raphson and Bairstow methods. The changeover from the QD to the Newton–Raphson or Bairstow method is, to some extent, arbitrary. The combination of several methods has the advantage that the final values of the zeros are obtained from the original polynomial. It is challenging but rewarding to put all three algorithms into one working program.

## EXERCISES

For the polynomials in Exercises 1–4, compute the first three iterates of $q^{(k)}$ and $e^{(k)}$ of the QD method. Compare the real roots of $P(x) = 0$ with $q^{(3)}$.

1. $P(x) = x^2 + 3x + 2$
2. $P(x) = x^3 - x^2 + x - 1$
3. $P(x) = x^2 - 2x + 1$
4. $P(x) = (x - 1)^2(x^2 + 1)$
5. Transform the quadratic polynomial $P(x) = x^2 + 1$ into a corresponding quadratic polynomial in which all the coefficients are nonzero. Then compute the first three iterates of $q^{(k)}$ and $e^{(k)}$ of the QD method.

## COMPUTER EXERCISES

**C.1.** Write a subroutine to compute $q^{(k)}$ and $e^{(k)}$ of the QD algorithm by assuming that all zeros are real and simple. Stop the iterations if the difference between successive elements of the $q$-column is less than a prescribed parameter eps1 or if the number of iterations exceeds a prescribed parameter maxqd.

**C.2.** Write a subroutine to find a zero using the Newton–Raphson method or use the Newton–Raphson subroutine (Exercise C.1, p. 46). Use the values of $q^{(k)}$ obtained from Exercise C.1 as initial values for the zeros. If it does not converge to a specified tolerance eps2 within maxnr iterations, return to Exercise C.1 by redefining eps1 and maxqd of Exercise C.1.

**C.3.** Write a computer program to find all the zeros of a given $N$th degree polynomial whose zeros are simple and real. Use Exercise C.1 to find the initial values of the zeros for Exercise C.2.

In Exercises C.4–C.7, use Exercise C.3 to find all the roots of the polynomial equations.

**C.4.** $x^4 + x^3 - 13x^2 - x + 12 = 0$
**C.5.** $6x^3 - 7x^2 + 10x + 5 = 0$
**C.6.** $x^4 + 6x^3 + 5x^2 - 4x - 2 = 0$
**C.7.** $x^5 - 3x^4 + 6x^3 - 2x + 2 = 0$
**C.8.** Write a subroutine to compute $q^{(k)}$ and $e^{(k)}$ of the QD algorithm. If $e^{(k)}$ fluctuates, then compute $r^{(k)} = q_i^{(k)} + q_{i+1}^{(k)}$ and $s^{(k)} = -q_i^{(k-1)} q_{i+1}^{(k)}$. Stop the iterations if $r^{(k)}$ and $s^{(k)}$ are less than a prescribed parameter eps1 or if the number of iterations exceeds a prescribed parameter maxqd.

**C.9.** Write a subroutine to find a quadratic factor using the Bairstow method or use the Bairstow subroutine (Exercise C.4, p. 69). Use the values of $r^{(k)}$ and $s^{(k)}$ obtained from Exercise C.8 as initial guesses for the quadratic factors.

**C.10.** Write a subroutine to solve a quadratic equation $x^2 - rx - s = 0$ or use Exercise C.6, p. 70.

**C.11.** Use Exercises C.8, C.9, C.10, and C.2 to write a computer program to find all the zeros of a given $N$th degree polynomial.

In Exercises C.12–C.16, use Exercise C.11 to find all the roots of the polynomial equations.

**C.12.** $x^6 + 1 = 0$

**C.13.** $x^4 - 3x^3 + 2x^2 + 2x - 4 = 0$

**C.14.** $4x^5 - 16x^4 + 17x^3 + 6x^2 - 21x + 10 = 0$

**C.15.** $x^7 + x^6 + x^5 + x^4 + x^3 + x^2 + x + 1 = 0$

**C.16.** $x^4 - 8x^3 + 21x^2 - 20x + 5 = 0$

**C.17.** The dimensions of a rectangular box are 1, 2, and 3. Use Exercise C.11 to determine the dimensions of another rectangular box that has exactly twice the volume if each dimension is increased by the same amount.

**C.18.** In vibration analysis of a linear system it is necessary to determine the natural frequencies from a differential equation describing a system. While performing the dynamical analysis of the passenger compartment of an airplane, the following differential equation was obtained:

$$\frac{d^4x}{dt^4} + 0.5\frac{d^3x}{dt^3} + 0.12\frac{d^2x}{dt^2} + 0.22\frac{dx}{dt} + 0.22\,x = 0$$

Calculate the natural frequencies by finding the roots of its characteristic polynomial equation $r^4 + 0.5r^3 + 0.12r^2 + 0.22r + 0.22 = 0$.

## 3.5  ILL-CONDITIONED POLYNOMIALS*

Consider the roots of the polynomial equation

$$x^4 - 4x^3 + 6x^2 - 4x + 1 = (x - 1)^4 = 0 \qquad \textbf{(3.5.1)}$$

which has 1 as a root of multiplicity 4.

Consider another machine-represented polynomial

$$x^4 - 4x^3 + 6x^2 - 4x + 0.9999999 = 0 \qquad \textbf{(3.5.2)}$$

which is $(x - 1)^4 = 10^{-8}$.

The roots of Equation (3.5.2) are 1.01, 1.01, 0.99 and 0.99. Thus a change of $10^{-8}$ in the constant term has produced a change of $10^{-2}$ in the roots. A small change in a coefficient of a polynomial may produe a large change in the zeros. Such polynomials are called ill-conditioned polynomials. It is not only the polynomials with multiple zeros which are sometimes ill-conditioned; polynomials of higher degrees may also be ill-conditioned.

Let

$$P(x) = (x - 1)(x - 2)\cdots(x - 19)(x - 20)$$
$$= x^{20} - 210x^{19} + \cdots$$

This classic example is due to Wilkinson (1959). This polynomial has distinct roots 1, 2, . . . , 20. When the coefficient of $x^{19}$ was changed from $-210$ to $-210 + 0.1192$

$\times\ 10^{-8}$, Wilkinson computed the zeros of the new polynomial $P(x)\ +\ 0.1192\ \times\ 10^{-8}x^{19}$. These zeros, correct to nine decimal places, are

| | |
|---|---|
| 1.000 000 000 | $10.095\ 266\ 145\ \pm\ 0.643\ 500\ 904\ I$ |
| 2.000 000 000 | $11.793\ 633\ 881\ \pm\ 1.652\ 329\ 728\ I$ |
| 3.000 000 000 | $13.992\ 358\ 137\ \pm\ 2.518\ 830\ 070\ I$ |
| 4.000 000 000 | $16.730\ 737\ 466\ \pm\ 2.812\ 624\ 894\ I$ |
| 4.999 999 928 | $19.502\ 439\ 400\ \pm\ 1.940\ 330\ 347\ I$ |
| 6.000 006 944 | |
| 6.999 697 234 | |
| 8.007 267 603 | |
| 8.917 250 249 | |
| 20.846 908 101 | |

where $I\ =\ \sqrt{-1}$.

The small change in the coefficient of $x^{19}$ has caused ten of the zeros to become complex and six changed in size.

To minimize the effect of round-off for an ill-conditioned polynomial, double-precision arithmetic should be used. In order to achieve better accuracy for the solution of a high degree polynomial equation, double-precision arithmetic is highly desirable.

## SUGGESTED READINGS

P. Henrici, *Elements of Numerical Analysis,* John Wiley & Sons, New York, 1964.

H. Rutishauser, Der Quotienten-Differenzen-Algorithmus, *Mitteilungen aus dem Institut für angew,* Math. No. 7, Birkhaüser, Basal, and Stuttgart, 1956.

J. V. Uspensky, *Theory of Equations,* McGraw-Hill Book Company, Inc., New York, 1948.

J. H. Wilkinson, ''The evaluation of the zeros of ill-conditioned polynomials,'' *Numer. Math.* **1**(1959), pp. 150–180.

# Chapter

# 4 INTERPOLATION

T he following table gives the dynamic viscosity $\mu$ of water at various temperatures:

| Temperature (°C) | 50 | 55 | 60 | 65 | 70 |
|---|---|---|---|---|---|
| Dynamic Viscosity $\mu$ (Pa.s) | 0.00554 | 0.00508 | 0.00470 | 0.00437 | 0.00405 |

What is the dynamic viscosity of water at 63°C? Without repeating an expensive experiment, this information can be obtained accurately by interpolation, which is the subject of this chapter.

## 4.1 INTRODUCTION

Suppose $\{(x_0, y_0), (x_1, y_1), \ldots, (x_N, y_N)\}$ is a set of data points where $x_0 < x_1 < \cdots < x_N$. We do not have an analytic expression of a function $f$ whose graph contains these points, yet we want to calculate its value at an arbitrary point. We can estimate $f(x)$ at an arbitrary $x$ by drawing a curve through the given set of data points. Suppose we are given five points and we are looking for an approximating function $f$ whose graph passes through these five points. There are many ways to draw a curve that passes through the five points.

In Figure 4.1.1a, $f$ is not single-valued (it is not a function); in Figure 4.1.1b $f$ is not continuous. In Figure 4.1.1c, $f$ is not smooth; in other words, $f'$ is not continuous. In Figure 4.1.1d, $f_1$ and $f_2$ pass through the five points and both are smooth, yet they have different behaviors. Which is better? Obviously, something more needs to be said about the required $f$. We may want the $m$th derivative of $f$ to be continuous on $[a, b]$. One way of constructing $f$ that ensures the continuity of all the derivatives is to think of it as a polynomial $P(x)$ which passes through a given set of $(N + 1)$ points. The Weierstrass Approximation Theorem (published by Weierstrass in 1886) guarantees that any continuous function can be approximated by a polynomial.

**Theorem 4.1.1**   Let $f$ be continuous on $[a, b]$ and $\epsilon > 0$ be given. There exists a polynomial $P_N(x)$ of degree $N$ such that

$$|f(x) - P_N(x)| < \epsilon \quad \text{for all } x \text{ in } [a, b] \qquad \blacksquare$$

It is reasonable to consider a class of polynomials in order to approximate a continuous function $f$. If $f$ is smooth, then usually $|f(x) - P_N(x)|$ decreases fast with increasing $N$. In many cases $|f(x) - P_N(x)|$ decreases slowly so it is impractical to approximate $f$ with only one polynomial over the entire interval $[a, b]$. Some methods for polynomial approximations give approximations such that it is not certain that $|f(x) - P_N(x)| \to 0$ as $N \to \infty$ even when $f$ is quite smooth.

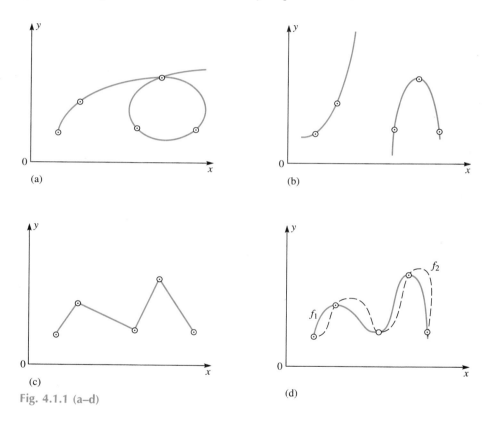

Fig. 4.1.1 (a–d)

Let $x_0, x_1, \ldots, x_N$ be distinct real numbers such that $x_0 < x_1 < \cdots < x_N$ and let $f(x_0), f(x_1), \cdots, f(x_N)$ be associated function values. If $x$ is any number in $[x_0, x_N]$ for which it is required to evaluate $f(x)$, then the approximation problem is called **interpolation**. If $x$ is outside of $[x_0, x_N]$, then the approximation problem is called **extrapolation**. Interpolation can be accomplished by fitting a polynomial through a set of tabulated data points whose argument values are close to the argument at which we desire to evaluate a function, and then evaluating the polynomial at the required argument. An approximating polynomial is also called an **interpolating polynomial**.

In practice, the given function is replaced by a simpler function for evaluation. For example, consider the problem of computing the values of the cosine function. The values of cos $x$ can be calculated exactly at few values of $x$. To calculate cos $x$, one can use its Maclaurin's series

$$\cos x = 1 - \frac{x^2}{2!} + \frac{x^4}{4!} - \frac{x^6}{6!} + \cdots$$

If $f$ is assumed to have many derivatives, then its approximating Taylor polynomial and the error committed in the approximation are given by Taylor's Theorem (Theorem A.8). The strength of the Weierstrass Theorem is that it requires a function $f$ only to be continuous in order to be approximated by a polynomial.

Before the use of computers, tables were used extensively to avoid repeating the calculations done by others. Interpolation formulas based on finite differences were used extensively for interpolation and a subject called finite difference calculus was used in numerical analysis and applied mathematics. The values of the most useful functions are now efficiently generated by computers and calculators at a desired point, reducing the need for interpolation formulas; however, we use interpolating polynomials to determine formulas for numerical differentiation and numerical integration.

## 4.2   UNDETERMINED COEFFICIENT METHOD

Let $x_0$, $x_1$, $x_2$, and $x_3$ be four distinct points. We are looking for a polynomial $P_3(x)$ such that $P_3(x_i) = f(x_i)$ for $i = 0, 1, 2,$ and 3.

Let

$$P_3(x) = a_3 x^3 + a_2 x^2 + a_1 x + a_0 \tag{4.2.1}$$

where $a_3$, $a_2$, $a_1$, and $a_0$ are to be determined such that $P_3(x_i) = f(x_i)$ for $i = 0, 1, 2,$ and 3. Thus $a_3$, $a_2$, $a_1$, and $a_0$ satisfy the following four linear equations:

$$a_3 x_0^3 + a_2 x_0^2 + a_1 x_0 + a_0 = f(x_0)$$

$$a_3 x_1^3 + a_2 x_1^2 + a_1 x_1 + a_0 = f(x_1)$$

$$a_3 x_2^3 + a_2 x_2^2 + a_1 x_2 + a_0 = f(x_2)$$

$$a_3 x_3^3 + a_2 x_3^2 + a_1 x_3 + a_0 = f(x_3)$$

This can also be written as

$$\begin{bmatrix} x_0^3 & x_0^2 & x_0 & 1 \\ x_1^3 & x_1^2 & x_1 & 1 \\ x_2^3 & x_2^2 & x_2 & 1 \\ x_3^3 & x_3^2 & x_3 & 1 \end{bmatrix} \begin{bmatrix} a_3 \\ a_2 \\ a_1 \\ a_0 \end{bmatrix} = \begin{bmatrix} f(x_0) \\ f(x_1) \\ f(x_2) \\ f(x_3) \end{bmatrix} \tag{4.2.2}$$

In other words, $V\mathbf{a} = \mathbf{y}$ where $V$ is a coefficient matrix, $\mathbf{a} = [a_3, a_2, a_1, a_0]'$, and $\mathbf{y} = [f(x_0), f(x_1), f(x_2), f(x_3)]'$. Matrix $V$ is called the **Vandermonde matrix**. Since $x_0$, $x_1$, $x_2$, and $x_3$ are distinct, $V$ is nonsingular. Hence Equation (4.2.2) has a unique solution.

The polynomial given by Equation (4.2.1) found by solving Equation (4.2.2) is a unique interpolating polynomial. This analysis can be extended to a polynomial of degree $N$ and the following theorem can be proven.

**Theorem 4.2.1**    There exists a unique polynomial $P_N(x)$ of degree $N$ or less such that $P(x_i) = f(x_i)$ for $i = 0, 1, 2, \ldots, N$.
■

However, we will determine the polynomial given by Equation (4.2.1) without solving the system in Equation (4.2.2) since for large systems this method is not efficient.

## 4.3  THE LAGRANGE INTERPOLATING POLYNOMIAL

We can also write a polynomial of degree three in the form

$$P_3(x) = f(x_0)L_0(x) + f(x_1)L_1(x) + f(x_2)L_2(x) + f(x_3)L_3(x) \qquad (4.3.1)$$

where $L_0(x)$, $L_1(x)$, $L_2(x)$, and $L_3(x)$ are polynomials of degree three called the **Lagrange polynomials**, which are to be determined such that $P_3(x_i) = f(x_i)$ for $i = 0, 1, 2$, and 3. Thus

$$P_3(x_0) = f(x_0)L_0(x_0) + f(x_1)L_1(x_0) + f(x_2)L_2(x_0) + f(x_3)L_3(x_0)$$

$$= f(x_0)$$

if $L_0(x_0) = 1, L_1(x_0) = 0, L_2(x_0) = 0$, and $L_3(x_0) = 0$. Similarly, $P_3(x_1) = f(x_1), P_3(x_2) = f(x_2)$, and $P_3(x_3) = f(x_3)$ if

$$\begin{array}{cccc} L_0(x_1) = 0 & L_1(x_1) = 1 & L_2(x_1) = 0 & L_3(x_1) = 0 \\ L_0(x_2) = 0 & L_1(x_2) = 0 & L_2(x_2) = 1 & L_3(x_2) = 0 \\ L_0(x_3) = 0 & L_1(x_3) = 0 & L_2(x_3) = 0 & L_3(x_3) = 1 \end{array}$$

Thus $L_i(x)$ is zero at every tabular argument except $x_i$ and 1 at $x_i$; in other words,

$$L_i(x_j) = \delta_{ij} = \begin{cases} 1 & \text{if } i = j \\ 0 & \text{if } i \neq j \end{cases}$$

for $i, j = 0, 1, 2$, and 3.

Since $L_0(x)$ is a polynomial of degree three such that $L_0(x_1) = 0, L_0(x_2) = 0$, and $L_0(x_3) = 0$, then $(x - x_1)$, $(x - x_2)$, and $(x - x_3)$ are the factors of $L_0(x)$. Therefore

$$L_0(x) = K(x - x_1)(x - x_2)(x - x_3)$$

where $K$ is a constant to be determined by the remaining condition $L_0(x_0) = 1$. Hence

$$L_0(x_0) = 1 = K(x_0 - x_1)(x_0 - x_2)(x_0 - x_3)$$

Therefore

$$K = \frac{1}{(x_0 - x_1)(x_0 - x_2)(x_0 - x_3)}$$

Thus

$$L_0(x) = \frac{(x - x_1)(x - x_2)(x - x_3)}{(x_0 - x_1)(x_0 - x_2)(x_0 - x_3)}$$

Observe the pattern. In the denominator and numerator, there is similarity. We get the denominator by replacing $x$ by $x_0$ in the numerator. We do not have $(x - x_0)$ as a factor for $L_0(x)$. Similarly $L_1(x)$, $L_2(x)$, and $L_3(x)$ can be obtained. Hence

$$\begin{aligned} P_3(x) = f(x_0) &\frac{(x - x_1)(x - x_2)(x - x_3)}{(x_0 - x_1)(x_0 - x_2)(x_0 - x_3)} \\ + f(x_1) &\frac{(x - x_0)(x - x_2)(x - x_3)}{(x_1 - x_0)(x_1 - x_2)(x_1 - x_3)} \\ + f(x_2) &\frac{(x - x_0)(x - x_1)(x - x_3)}{(x_2 - x_0)(x_2 - x_1)(x_2 - x_3)} \\ + f(x_3) &\frac{(x - x_0)(x - x_1)(x - x_2)}{(x_3 - x_0)(x_3 - x_1)(x_3 - x_2)} \end{aligned} \qquad (4.3.2)$$

It is convenient to use the product notation for $(x - x_1)(x - x_2)(x - x_3)$: $\prod_{i=1}^{3} (x - x_i) = (x - x_1)(x - x_2)(x - x_3)$. Here $\prod$ denotes the product notation; $i$ begins with 1 and ends at 3. The initial value and any other restrictions (if any) are denoted at the bottom of $\prod$ and the ending value is denoted at the top of $\prod$.

Therefore, $L_0(x)$ can be expressed as

$$\prod_{\substack{i=0 \\ i \neq 0}}^{3} \frac{(x - x_i)}{(x_0 - x_i)}$$

Thus Equation (4.3.2) can be written as

$$\begin{aligned} P_3(x) = f(x_0) \prod_{\substack{i=0 \\ i \neq 0}}^{3} \frac{(x - x_i)}{(x_0 - x_i)} &+ f(x_1) \prod_{\substack{i=0 \\ i \neq 1}}^{3} \frac{(x - x_i)}{(x_1 - x_i)} \\ + f(x_2) \prod_{\substack{i=0 \\ i \neq 2}}^{3} \frac{(x - x_i)}{(x_2 - x_i)} &+ f(x_3) \prod_{\substack{i=0 \\ i \neq 3}}^{3} \frac{(x - x_i)}{(x_3 - x_i)} \\ = \sum_{j=0}^{3} f(x_j) \prod_{\substack{i=0 \\ i \neq j}}^{3} \frac{(x - x_i)}{(x_j - x_i)} &= \sum_{j=0}^{3} f(x_j) L_j(x) \end{aligned} \qquad (4.3.3)$$

where

$$L_j(x) = \prod_{\substack{i=0 \\ i \neq j}}^{3} \frac{(x - x_i)}{(x_j - x_i)} \qquad (4.3.4)$$

Similarly a polynomial $P_N(x)$ passing through $(N + 1)$ points $(x_0, f(x_0))$, $(x_1, f(x_1))$, ..., $(x_N, f(x_N))$ is given by

$$P_N(x) = \sum_{j=0}^{N} f(x_j) L_j(x) \qquad (4.3.5)$$

where

$$L_j(x) = \prod_{\substack{i=0 \\ i \neq j}}^{N} \frac{(x - x_i)}{(x_j - x_i)} \tag{4.3.6}$$

The interpolating polynomial given by Equation (4.3.5) with $L_j(x)$ defined by Equation (4.3.6) is called a **Lagrange interpolating polynomial**.

Next, the question is whether the polynomials given by Equations (4.2.1) and (4.3.1) are really different. By Theorem 4.2.1, the polynomial of degree three through $(x_i, f(x_i))$, $i = 0, 1, 2,$ and 3 is unique. This uniqueness property is of practical use. Other formulas will be derived for the interpolating polynomial, but because of this uniqueness property the polynomial is the same. There is no particular reason to write Equation (4.3.1) in the form of Equation (4.2.1).

When $f(x)$ is approximated by a polynomial $P_3(x)$, how much error is left? There are two kinds of errors: round-off error and truncation error. The truncation error is due to using a finite degree polynomial. Let us analyze the truncation error. Since $P(x_i) = f(x_i)$ for $i = 0, 1, 2,$ and 3, the error is zero at $x_0, x_1, x_2,$ and $x_3$. Therefore

$$E(x) = f(x) - P_3(x) = (x - x_0)(x - x_1)(x - x_2)(x - x_3)g(x) \tag{4.3.7}$$

where $g(x)$ is to be determined. How is $g(x)$ determined? If we differentiate Equation (4.3.7) four times, then $P_3(x)$ is eliminated but a complicated equation results, involving the derivatives of $g(x)$ as well as $f(x)$ itself. This is of no help since $f(x)$ is not known and $g$ is not constant, because the truncation error depends on the argument. So to find $g(x)$ for a fixed value of $x$, define

$$\phi(t) = f(t) - P_3(t) - (t - x_0)(t - x_1)(t - x_2)(t - x_3)g(x) \tag{4.3.8}$$

Then $\phi(x_0) = \phi(x_1) = \phi(x_2) = \phi(x_3) = \phi(x) = 0$. By Generalized Rolle's Theorem (Corollary A.3.2), there is a number $\xi$ in the interval generated by $x_0, x_1, x_2, x_3,$ and $x$ such that $\phi^{(4)}(\xi) = 0$. Differentiating Equation (4.3.8) four times with respect to $t$ while treating $g(x)$ as a constant yields

$$\phi^{(4)}(t) = f^{(4)}(t) - 4!g(x)$$

Therefore

$$\phi^{(4)}(\xi) = f^{(4)}(\xi) - 4!g(x) = 0$$

Hence

$$g(x) = \frac{f^{(4)}(\xi)}{4!} \tag{4.3.9}$$

In Equation (4.3.9) the value of $\xi$ changes as $x$ changes and therefore $\xi$ can be regarded as a function of $x$. Substituting Equation (4.3.9) in Equation (4.3.7) yields

$$f(x) = P_3(x) + \frac{f^{(4)}(\xi(x))}{4!} \prod_{i=0}^{3} (x - x_i)$$

Hence, the following theorem can be proven.

**Theorem 4.3.1**     **(Lagrange Polynomial Approximation Theorem)**     Given a set $\{(x_i, f(x_i)) | i = 0, 1,$
$\ldots, N\}$ where $a = x_0 < x_1 < \ldots < x_N = b$ and the $(N + 1)$th derivative of $f$ is
continuous on $[a, b]$, then for any $x$ in $[a, b]$ there exists $\xi$ in $(a, b)$ such that

$$f(x) = \sum_{j=0}^{N} f(x_j)L_j(x) + \frac{f^{(N+1)}(\xi)}{(N + 1)!} \prod_{j=0}^{N} (x - x_j) \tag{4.3.10}$$

where

$$L_j(x) = \prod_{\substack{i=0 \\ i \neq j}}^{N} \frac{(x - x_i)}{(x_j - x_i)} \tag{4.3.11}$$

■

The truncation error is

$$E(x) = f(x) - P_N(x) = \frac{f^{(N+1)}(\xi(x))}{(N + 1)!} \prod_{i=0}^{N} (x - x_i) \tag{4.3.12}$$

Since we seldom know $f^{(N+1)}(\xi(x))$, the truncation error formula is of limited use.
However, it does give the order of the error if $|f^{(N+1)}(\xi)|$ is bounded. If $f$ is known
analytically, then an upper bound of $|f^{(N+1)}(x)|$ can be found over an entire interval
$[a, b]$; therefore, we can find an upper bound of the truncation error using Equation
(4.3.12).

**EXAMPLE 4.3.1**     The following table for $f(x) = \cos x$ is given. Approximate $f(0.23)$ using various
degrees of Lagrange interpolating polynomials.

| $x_i$ | 0.1 | 0.2 | 0.3 | 0.4 |
|---|---|---|---|---|
| $\cos x_i$ | 0.99500 | 0.98007 | 0.95534 | 0.92106 |

**1.**  For $N = 1$ (linear interpolation), let $x_0 = 0.2$ and $x_1 = 0.3$. Then

$$L_0(x) = \frac{(x - x_1)}{(x_0 - x_1)} = \frac{x - 0.3}{0.2 - 0.3} = -10(x - 0.3)$$

and

$$L_1(x) = \frac{(x - x_0)}{(x_1 - x_0)} = \frac{x - 0.2}{0.3 - 0.2} = 10(x - 0.2)$$

Hence

$$P_1(x) = f(x_0)L_0(x) + f(x_1)L_1(x)$$

$$= (-9.8007)(x - 0.3) + (9.5534)(x - 0.2)$$

$$= -0.2473x + 1.02953$$

Therefore $P_1(0.23) = 0.97265$.
Observe that $L_0(x) + L_1(x) = -10(x - 0.3) + 10(x - 0.2) = 3.0 - 2.0 = 1$.

We have $|\text{truncation error}| = |E(x)| = |f''(\xi)(x - 0.2)(x - 0.3)/(2!)|$ for some $\xi$ in $(0.2, 0.3)$. Hence $|E(0.23)| = |-\cos \xi(0.23 - 0.2)(0.23 - 0.3)/2|$. Since $|\cos \xi| \leq 1$, $|E(0.23)| \leq 0.00255$. We have $|\text{actual error}| = |\cos(0.23) - P(0.23)|$ $= |0.97367 - 0.97265| = 0.00102$. Thus the absolute value of the actual error is less than the estimated upper bound value of the truncation error.

2. For $N = 2$, let $x_0 = 0.1$, $x_1 = 0.2$, and $x_2 = 0.3$. Then $f(x_0) = 0.99500$, $f(x_1) = 0.98007$, and $f(x_2) = 0.95534$, and

$$L_0(x) = \frac{(x - x_1)(x - x_2)}{(x_0 - x_1)(x_0 - x_2)} = \frac{1}{0.02}(x^2 - 0.5x + 0.06)$$

$$L_1(x) = \frac{(x - x_0)(x - x_2)}{(x_1 - x_0)(x_1 - x_2)} = \frac{-1}{0.01}(x^2 - 0.4x + 0.03)$$

$$L_2(x) = \frac{(x - x_0)(x - x_1)}{(x_2 - x_0)(x_2 - x_1)} = \frac{1}{0.02}(x^2 - 0.3x + 0.02)$$

Hence

$$P_2(x) = \sum_{j=0}^{2} f(x_j)L_j(x)$$

$$= \left(\frac{0.99500}{0.02}\right)(x^2 - 0.5x + 0.06) - \frac{(0.98007)}{0.01}(x^2 - 0.4x + 0.03)$$

$$+ \left(\frac{0.95534}{0.02}\right)(x^2 - 0.3x + 0.02)$$

$$= -0.49x^2 - 0.0023x + 1.00013$$

Therefore $P_2(0.23) = 0.97368$.
Also observe that

$$L_0(x) + L_1(x) + L_2(x) = \frac{(x^2 - 0.5x + 0.06)}{0.01} + \frac{(x^2 - 0.4x + 0.03)}{0.01}$$

$$+ \frac{(x^2 - 0.3x + 0.02)}{0.02}$$

$$= -25x + 40x - 15x + 3 - 3 + 1 = 1$$

We have $|\text{truncation error}| = |E(x)| = |f'''(\xi)(x - 0.1)(x - 0.2)(x - 0.3)/(3!)|$ for some $\xi$ in $(0.1, 0.3)$. It can be verified that $|E(0.23)| \leq 0.00005$ and $|\text{actual error}| = |\cos(0.23) - P_2(0.23)| = 0.00001$.    ■ ■ ■

**Algorithm**

**Subroutine** Lag $(N, \mathbf{x}, \mathbf{y}, \text{xinter}, \text{yinter})$
Comment: $\mathbf{x}$ and $\mathbf{y}$ are vectors with components $x_i$ and $y_i = f(x_i)$, for $i = 0, 1,$ $\ldots, N$. This program computes yinter using the Lagrange interpolating polynomial [Equations (4.3.5) and (4.3.6)] when $(x_0, y_0), (x_1, y_1), \ldots, (x_N, y_N)$ and xinter are given.
  1 **Print** $N$, $\mathbf{x}$, $\mathbf{y}$, xinter
  2 sum $= 0.0$
  3 factor $= 1.$
  4 **Repeat** step 5 **with** $i = 0, 1, \ldots, N$
      5 factor $=$ factor (xinter $- x_i$)

6 **Repeat** steps 7–10 **with** $j = 0, 1, \ldots, N$
   7 deno $= 1$.
    8 **Repeat** step 9 **with** $i = 0, 1, \ldots, N$
      9 **If** $i \neq j$, **then** deno $=$ deno $*\ (x_j - x_i)$
   10 sum $=$ sum $+\ ((y_j * \text{factor})/((\text{xinter} - x_j) * \text{deno}))$
11 **Print** xinter, sum
Comment: sum $=$ yinter.

Let $f(x) = 1$ in Equation (4.3.10). Then the following theorem can be verified (Exercise 3).

**Theorem 4.3.2**

$$\sum_{j=0}^{N} L_j(x) = 1$$

This theorem is very helpful for checking calculations.

The degree of an interpolating polynomial is usually three or four. Using a higher degree interpolating polynomial does not guarantee a better approximation. A famous example due to Runge is $f(x) = 1/(1 + x^2)$ where $-5 \leq x \leq 5$. For $N = 10$, it can be verified (Exercise C.5) that $P_{10}$ wiggles up and down near the ends of the interval. When a table contains more than the required number of points for a desirable degree of the interpolating polynomial, we must use the right ''local'' points in the table. For example, consider the third degree interpolating polynomial. The largest values of

$$\left| (x - x_0)(x - x_1)(x - x_2)(x - x_3) \right|$$

occur in the outer intervals $[x_0, x_1]$ and $[x_2, x_3]$. It makes sense to choose for a given $x$ two interpolation points greater than $x$ and two interpolation points less than $x$.

## EXERCISES

1. Approximate $f(0.23)$ in Example 4.3.1 using $x_0 = 0.2$, $x_1 = 0.3$, and $x_2 = 0.4$.
2. Find $P_2(0.4)$ in Example 4.3.1 and the actual error. Also find an upper bound value of the truncation error at $x = 0.4$.
3. Prove Theorem 4.3.2.
4. Verify that the determinant of the Vandermonde matrix $V$ in Equation (4.2.2) is given by

$$(x_0 - x_1)(x_0 - x_2)(x_0 - x_3)(x_1 - x_2)(x_1 - x_3)(x_2 - x_3)$$

5. Derive the polynomial given by Equation (4.2.1) for the following table:

| $x$ | 0 | 1 | 2 | 4 |
|---|---|---|---|---|
| $f(x)$ | 0 | 0 | 1 | 0 |

Evaluate the polynomial at $x = 3$ as an estimate of $f(3)$.

**6.** Derive the Lagrange interpolating polynomial given by Equation (4.3.1) for the following table:

| $x$ | 0 | 1 | 2 | 4 |
|---|---|---|---|---|
| $f(x)$ | 0 | 0 | 1 | 0 |

Evaluate the polynomial at $x = 3$ as an estimate of $f(3)$. Compare the polynomial and its value at $x = 3$ with Exercise 5.

**7.** Use the method of undetermined coefficients to find the quadratic polynomial $Q_2(x)$ such that $Q_2(1) = 1$, $Q_2'(1) = 0$, and $Q_2(2) = 5$.

**8.** Let $f(x) = x^2 + 1$.

(a) Find the linear Lagrange interpolating polynomial using $x_0 = 2$ and $x_1 = 3$ and use it to approximate $f(2.5)$.

(b) What would be the maximum truncation error in (a)? Compare it with the actual error.

(c) Find the quadratic Lagrange interpolating polynomial using $x_0 = 1$, $x_1 = 2$, and $x_2 = 3$ and use it to approximate $f(2.5)$.

(d) What would be the maximum truncation error in (c)? Compare it with the actual error.

**9.** Let $f(x)$ be a polynomial of degree $\leq N$. Let $P_N(x)$ be a Lagrange interpolating polynomial of degree $\leq N$ based on $N + 1$ nodes $x_0, x_1, \ldots, x_N$ such that $P(x_i) = f(x_i)$ for $i = 0, 1, \ldots, N$. Show that the truncation error $E(x)$ is identically zero and thus $f(x) = P_N(x)$ for all $x$.

**10.** The third degree Lagrange polynomial

$$L_0(x) = \frac{(x - x_1)(x - x_2)(x - x_3)}{(x_0 - x_1)(x_0 - x_2)(x_0 - x_3)}$$

is 1 when $x = x_0$ and zero when $x = x_1, x_2,$ and $x_3$. Show that

$$L_0(x) = 1 + \frac{(x - x_0)}{(x_0 - x_1)} + \frac{(x - x_0)(x - x_1)}{(x_0 - x_1)(x_0 - x_2)} + \frac{(x - x_0)(x - x_1)(x - x_2)}{(x_0 - x_1)(x_0 - x_2)(x_0 - x_3)}$$

**11.** Verify that $\sum_{j=0}^{N} x_j L_j(x) = x$ where $N \geq 1$. (*Hint:* Use $f(x) = x$ in Equation (4.3.10).)

**12.** Verify that the two polynomials

$$P_1(x) = 2 - x \quad \text{and} \quad P_2(x) = x^2 - 4x + 4$$

interpolate

| $x$ | 1 | 2 |
|---|---|---|
| $f(x)$ | 1 | 0 |

Explain why this does not violate Theorem 4.2.1.

---

## COMPUTER EXERCISES

**C.1.** Write a subroutine to calculate the value of a Lagrange interpolating polynomial at a particular number xinter using Equations (4.3.5) and (4.3.6) for a given $N$ and $(x_i, y_i)$ for $i = 0, 1, \ldots, N$.

**C.2.** Write a computer program to calculate the value of a Lagrange interpolating polynomial at a particular number xinter using Exercise C.1. The inputs are $N$, $x_i$, $y_i = f(x_i)$ for $i = 0, 1, \ldots, N$, and xinter.

**C.3.** Use Exercise C.2 to calculate $e^{0.11}$, $e^{0.22}$, and $e^{0.33}$ from the following table:

| $x$ | 0 | 0.1 | 0.2 | 0.3 | 0.4 |
|-----|------|------|------|------|------|
| $e^x$ | 1.00000 | 1.10517 | 1.22140 | 1.34986 | 1.49182 |

Use $N = 2$ and 4. Compare the absolute value of actual errors to the upper bound value of the truncation errors.

**C.4.** Use Exercise C.2 to calculate the dynamic viscosity $\mu$ (Pa.s) of water at 52°C, 63°C, and 66°C from the following table:

| Temperature (°C) | 50 | 55 | 60 | 65 | 70 |
|------------------|------|------|------|------|------|
| Dynamic Viscosity $\mu$ (Pa.s) | 0.00554 | 0.00508 | 0.00470 | 0.00437 | 0.00405 |

Use $N = 2$ and 4.

**C.5.** Use Exercise C.1 to approximate $f(x) = 1/(1 + x^2)$ on $[-5, 5]$. Use $N = 4, 10$, and 20. Generate the values of $f(x_i)$ for $i = 0, 1, \ldots, N$ using $f(x) = 1/(1 + x^2)$. Compare the computed values with the exact values near the endpoints.

**C.6.** Use Exercise C.1 to approximate $f(x) = \sqrt{x}$ on $[0, 1]$. Use $N = 4, 10$, and 20. Generate the values of $f(x_i)$ for $i = 0, 1, \ldots, N$ using $f(x) = \sqrt{x}$. Compare the computed values with the exact values near the endpoints.

## 4.4 THE NEWTON DIVIDED DIFFERENCE FORMULAS

In practice, we do not know in advance how many points will provide either the best approximation or an approximation of desired accuracy because $f$ may not be known in analytical form. If $f$ is known in analytical form, it may be too complicated to evaluate higher derivatives to know an upper bound value of the truncation error. If we add an additional point in the Lagrange method, a great deal of new work is required. Each old Lagrange polynomial must be multiplied by a factor in both the numerator and denominator, and a new Lagrange polynomial must be added that corresponds to the additional point. The Newton divided difference formula avoids the recomputation; an addition of a term makes it a higher degree interpolating polynomial. We are looking for an $N$th degree interpolating polynomial when an $(N-1)$th degree interpolating polynomial and an additional point are known. Let $Q(x)$ be such an additional term of degree $N$ such that

$$P_N(x) = P_{N-1}(x) + Q(x)$$

Since we want $P_N(x_i) = P_{N-1}(x_i) = f(x_i)$ for $i = 0, 1, \ldots, N-1$, $Q(x_i) = 0$ for $i = 0, 1, \ldots, N-1$. Thus $Q(x)$ must contain $(x - x_0)(x - x_1) \cdots (x - x_{N-1})$. Therefore

$$Q(x) = a_N(x - x_0)(x - x_1) \cdots (x - x_{N-1})$$

Thus the interpolating polynomial we are looking for can be written in the form

$$P_N(x) = a_0 + a_1(x - x_0) + a_2(x - x_0)(x - x_1) + \cdots$$
$$+ a_N(x - x_0)(x - x_1) \cdots (x - x_{N-1}) \tag{4.4.1}$$

where $a_0, a_1, \ldots, a_N$ are constants to be determined.

Since $P_N(x_0) = f(x_0) = a_0$, $a_0 = f(x_0)$. Similarly $P_N(x_1) = f(x_1) = a_0 + a_1(x_1 - x_0)$ $= f(x_0) + a_1(x_1 - x_0)$. Thus

$$a_1 = \frac{f(x_1) - f(x_0)}{x_1 - x_0}$$

It makes sense to define some compact notations. Since the numerator and denominator are differences, define the **first divided difference** of $f$ at the points $x_0$ and $x_1$ as

$$f[x_1, x_0] = \frac{f(x_1) - f(x_0)}{x_1 - x_0}$$
$$= \frac{f(x_0) - f(x_1)}{x_0 - x_1} = \frac{f(x_0)}{x_0 - x_1} + \frac{f(x_1)}{x_1 - x_0} = f[x_0, x_1] \tag{4.4.2}$$

Thus $f$ is symmetric with respect to its arguments $x_0$ and $x_1$ and $a_1 = f[x_1, x_0]$.

Similarly,

$$P_N(x_2) = f(x_2) = a_0 + a_1(x_2 - x_0) + a_2(x_2 - x_0)(x_2 - x_1) \tag{4.4.3}$$

Substituting for $a_0$ and $a_1$ in Equation (4.4.3) and then solving for $a_2$ yields

$$a_2 = \frac{f(x_2) - f(x_0) - (x_2 - x_0) f[x_1, x_0]}{(x_2 - x_0)(x_2 - x_1)} \tag{4.4.4}$$

Further

$$f(x_2) - f(x_0) = f(x_2) - f(x_1) + f(x_1) - f(x_0)$$
$$= f(x_2) - f(x_1) + \frac{f(x_1) - f(x_0)}{x_1 - x_0}(x_1 - x_0)$$
$$= f(x_2) - f(x_1) + f[x_1, x_0](x_1 - x_0) \tag{4.4.5}$$

Substituting Equation (4.4.5) in Equation (4.4.4) yields

$$a_2 = \frac{f(x_2) - f(x_1) - (x_2 - x_1) f[x_1, x_0]}{(x_2 - x_0)(x_2 - x_1)}$$
$$= \frac{\dfrac{f(x_2) - f(x_1)}{x_2 - x_1}}{x_2 - x_0} - \frac{f[x_1, x_0]}{x_2 - x_0}$$
$$= \frac{f[x_2, x_1] - f[x_1, x_0]}{x_2 - x_0} \tag{4.4.6}$$

Define

$$f[x_2, x_1, x_0] = \frac{f[x_2, x_1] - f[x_1, x_0]}{x_2 - x_0} \tag{4.4.7}$$

Thus $a_2 = f[x_2, x_1, x_0]$. In addition, it can be verified [Exercise 6(b)] that

$$f[x_2, x_1, x_0] = \frac{f(x_2)}{(x_2 - x_0)(x_2 - x_1)} + \frac{f(x_1)}{(x_1 - x_0)(x_1 - x_2)} + \frac{f(x_0)}{(x_0 - x_1)(x_0 - x_2)}$$

In a similar way, the ***i*th order divided difference** is defined by

$$f[x_i, x_{i-1}, \ldots, x_1, x_0] = \frac{f[x_i, x_{i-1}, \ldots, x_1] - f[x_{i-1}, x_{i-2}, \ldots, x_0]}{(x_i - x_0)} \qquad (4.4.8)$$

It can be shown that

$$a_i = f[x_i, x_{i-1}, \ldots, x_1, x_0] \qquad (4.4.9)$$

So Equation (4.4.1) can be written as

$$\begin{aligned} P_N(x) &= f(x_0) + f[x_1, x_0](x - x_0) + f[x_2, x_1, x_0](x - x_0)(x - x_1) + \cdots \\ &\quad + f[x_N, x_{N-1}, \ldots, x_1, x_0](x - x_0)(x - x_1) \cdots (x - x_{N-1}) \\ &= f(x_0) + \sum_{i=1}^{N} f[x_i, x_{i-1}, \ldots, x_1, x_0](x - x_0)(x - x_1) \cdots (x - x_{i-1}) \end{aligned}$$

$$(4.4.10)$$

This is called the **Newton divided difference formula** for the interpolating polynomial. The computation of differences for Equation (4.4.10) can be generated in a simple manner as shown in Table 4.4.1.

**Table 4.4.1**

| $x$ | $f(x)$ | First Divided Differences | Second Divided Differences | Third Divided Differences |
|---|---|---|---|---|
| $x_0$ | $f(x_0)$ | | | |
| | | $f[x_1, x_0] = \dfrac{f(x_1) - f(x_0)}{x_1 - x_0}$ | | |
| $x_1$ | $f(x_1)$ | | $f[x_2, x_1, x_0]$ | |
| | | $f[x_2, x_1]$ | | $f[x_3, x_2, x_1, x_0]$ |
| $x_2$ | $f(x_2)$ | | $f[x_3, x_2, x_1]$ | |
| | | $f[x_3, x_2]$ | | |
| $x_3$ | $f(x_3)$ | | | |

Since $f[x_1, x_0] = f'(\xi)$ for some $\xi$ in $(x_0, x_1)$ by the Mean Value Theorem (Theorem A.3), the same type of relation exists between $f[x_N, x_{N-1}, \ldots, x_1, x_0]$ and the $N$th derivative. Assume that the $N$th derivative of $f$ is continuous on $[a, b]$ and $x_0, x_1, \ldots, x_N$ are distinct numbers in $[a, b]$. Define

$$g(x) = f(x) - P_N(x) \qquad (4.4.11)$$

Since $f(x_i) = P_N(x_i)$ for $i = 0, 1, \ldots, N$, then $g(x_i) = 0$ for $i = 0, 1, \ldots, N$. By the Generalized Rolle's Theorem (Corollary A.3.2), there exists $\xi$ in $(a, b)$ such that $g^{(N)}(\xi) = 0$. Thus $f^{(N)}(\xi) - P_N^{(N)}(\xi) = 0$. But $P_N^{(N)}(\xi) = N!a_N = N! f[x_N, x_{N-1}, \ldots, x_1, x_0]$. Therefore

$$f[x_N, x_{N-1}, \ldots, x_1, x_0] = \frac{P_N^{(N)}(\xi)}{N!} = \frac{f^{(N)}(\xi)}{N!} \qquad (4.4.12)$$

When we approximate $f(x)$ by $P_1(x)$, the truncation error $E(x)$ is given by

$$
\begin{aligned}
E(x) &= f(x) - P_1(x) \\
&= f(x) - f(x_0) - (x - x_0)f[x_1, x_0] \\
&= (x - x_0)\left(\frac{f(x) - f(x_0)}{x - x_0} - f[x_1, x_0]\right) \\
&= (x - x_0)(f[x, x_0] - f[x_0, x_1]) \\
&= (x - x_0)(x - x_1)\left(\frac{f[x, x_0] - f[x_0, x_1]}{(x - x_1)}\right) \\
&= (x - x_0)(x - x_1)f[x, x_0, x_1] \\
&= f[x, x_1, x_0](x - x_0)(x - x_1)
\end{aligned}
$$

If $f(x)$ is approximated by $P_N(x)$, then it can be verified that the truncation error $E(x)$ is given by

$$
E(x) = f(x) - P_N(x)
$$

$$
= f[x, x_N, x_{N-1}, \ldots, x_1, x_0](x - x_0)(x - x_1) \cdots (x - x_N)
$$

The following theorem can be proven.

**Theorem 4.4.1**    Given a set $\{(x_i, f(x_i)) \mid i = 0, 1, \ldots, N\}$ where $x_i$ are distinct points in $[a, b]$, then

$$
f(x) = f(x_0) + \sum_{i=1}^{N} f[x_i, x_{i-1}, \ldots, x_1, x_0](x - x_0)(x - x_1) \cdots (x - x_{i-1})
$$

$$
+ f[x, x_N, x_{N-1}, \ldots, x_1, x_0]\prod_{i=0}^{N}(x - x_i) \qquad \textbf{(4.4.13)}
$$

■

Since the interpolating polynomial is unique by Theorem 4.2.1, comparison of Equation (4.3.10) with Equation (4.4.13) yields

$$
f[x, x_N, x_{N-1}, \ldots, x_1, x_0] = \frac{f^{(N+1)}(\xi)}{(N+1)!}
$$

where $\xi$ is in $(a, b)$.

**Algorithm**    **Subroutine** Neinter $(N, \mathbf{x}, \mathbf{y}, \text{xinter}, \text{yinter})$
Comment: $\mathbf{x}$ and $\mathbf{y}$ are vectors with components $x_i$ and $y_i = f(x_i)$ for $i = 0, 1, \ldots,$
$N$. This program computes yinter using Equation (4.4.10) for given $(x_0, y_0)$, $(x_1, y_1)$,
$\ldots$, $(x_N, y_N)$ and xinter.
  1 **Print** $N, \mathbf{x}, \mathbf{y}$, xinter
Comment: This part computes the divided differences.
  2 **Repeat** step 3 **for** $i = 0, 1, \ldots, N$
      3 $d_i = y_i$
  4 **Repeat** steps 5–6 **for** $i = 1, 2, \ldots, N$
      5 **Repeat** step 6 **for** $j = N, N - 1, \ldots, i$
          6 $d_j = \dfrac{d_j - d_{j-1}}{x_j - x_{j-i}}$

Comment: We get $d_0 = f(x_0)$, $d_1 = f[x_1, x_0]$, ..., $d_N = f[x_N, x_{N-1}, ..., x_0]$. Compute Equation (4.4.10) for xinter using Horner's method.

    7   $t = d_N$
      8   **Repeat** step 9 **for** $i = N - 1, N - 2, ..., 0$
         9   $t = d_i + (\text{xinter} - x_i) * t$
   10   **Print** xinter, $t$                   ◀

**E X A M P L E   4.4.1**

Table 4.4.2 for $f(x) = \cos x$ is given. Approximate $f(0.13)$ using the Newton divided difference formula [Equation (4.4.10)]. Compare the absolute value of the actual error with an upper bound value of the truncation error.

**Table 4.4.2**

| $i$ | $x_i$ | $f(x_i) = \cos x_i$ | $f[x_{i+1}, x_i]$ | $f[x_{i+2}, x_{i+1}, x_i]$ | $f[x_{i+3}, x_{i+2}, x_{i+1}, x_i]$ |
|---|---|---|---|---|---|
| | | | $-0.14930$ | | |
| 1 | 0.2 | 0.98007 | | $-0.49000$ | |
| | | | $-0.24730$ | | 0.04167 |
| 2 | 0.3 | 0.95534 | | $-0.47750$ | |
| | | | $-0.34280$ | | |
| 3 | 0.4 | 0.92106 | | | |

**1.**   For $N = 1$ (linear), let $x_0 = 0.1$ and $x_1 = 0.2$. Then

$$P_1(x) = f(x_0) + f[x_1, x_0](x - x_0)$$

$$= 0.99500 + (-0.14930)(x - 0.1)$$

Therefore $P_1(0.13) = 0.99052$.
We have

$$|\text{actual error}| = |\cos(0.13) - P_1(0.13)| = |0.99156 - 0.99052| = 0.00104$$

and

$$|\text{truncation error}| = |E(x)| = |f''(\xi)(x - 0.1)(x - 0.2)/(2!)|$$

Since $|f''(\xi)| = |\cos\xi| \le 1$, $|E(0.13)| \le 0.00105$. Thus the absolute value of the actual error is less than the estimated upper bound value of the truncation error.

**2.**   For $N = 2$, let $x_0 = 0.1$, $x_1 = 0.2$, and $x_3 = 0.3$. Then

$$P_2(x) = f(x_0) + f[x_1, x_0](x - x_0) + f[x_2, x_1, x_0](x - x_0)(x - x_1)$$

$$= P_1(x) + (-0.49000)(x - 0.1)(x - 0.2)$$

Hence $P_2(0.13) = P_1(0.13) + (-0.49000)(0.03)(-0.07) = 0.99052 + 0.00103 = 0.99155$.

We have $|\text{truncation error}| = |E(x)| = |f^{(3)}(\xi)(x - 0.1)(x - 0.2)(x - 0.3)/(3!)|$. Since $|f^{(3)}(\xi)| = |\sin\xi| \le 1$, $|E(0.13) \le 0.00006$. Further, $|\text{actual error}| = |\cos(0.13) - P_2(0.13)| = |0.99156 - 0.99155| = 0.00001$ which is less than the estimated upper bound value of the truncation error.     ■ ■ ■

Suppose $\{(x_0, f(x_0)), (x_1, f(x_1)), \ldots, (x_N, f(x_N))\}$ is a set of data points. Often the value of $f(c)$ is known and we want to find the corresponding value of $c$. In order to find $c$, interchange the role of independent and dependent variables; in other words, the values of the function $f(x_i)$ become the abscissas and the corresponding $x_i$ become the corresponding ordinates. Since the functional values $f(x_i)$ are not necessarily uniformly spaced, the Newton divided difference formula is very convenient for this kind of a problem. This process is called an **inverse interpolation**.

**EXAMPLE 4.4.2**

The values of the Bessel function of order zero $J_0(x)$ is given in the following table:

| $x_i$ | 5.4 | 5.5 | 5.6 | 5.7 |
|---|---|---|---|---|
| $J_0(x_i)$ | $-0.04121$ | $-0.00684$ | $0.02697$ | $0.05992$ |

Find the value of $c$ for which $J_0(c) = 0$.

In Table 4.4.3 we construct the divided differences by interchanging the role of $x_i$ and $J_0(x_i)$.

**Table 4.4.3**

| $i$ | $J_0(x_i)$ | $x_i$ | $f[J_0(x_{i+1}), J_0(x_i)]$ | $f[J_0(x_{i+2}), J_0(x_{i+1}), J_0(x_i)]$ | $f[J_0(x_{i+3}), J_0(x_{x+2}), J_0(x_{i+1}), J_0(x_i)]$ |
|---|---|---|---|---|---|
| 0 | $-0.04121$ | 5.4 | | | |
| | | | 2.90951 | | |
| 1 | $-0.00684$ | 5.5 | | 0.70681 | |
| | | | 2.95770 | | 4.44547 |
| 2 | 0.02697 | 5.6 | | 1.15638 | |
| | | | 3.03490 | | |
| 3 | 0.05992 | 5.7 | | | |

The first element of $f[J_0(x_{i+1}), J_0(x_i)]$ in Table 4.4.3 is

$$f[J_0(x_1), J_0(x_0)] = \frac{x_1 - x_0}{(J_0(x_1) - J_0(x_0))} = \frac{5.5 - 5.4}{-0.00684 + 0.04121} = 2.90951$$

The remaining divided differences are similarly computed and are shown in Table 4.4.3. Then

$$\begin{aligned}
P(J_0(x)) &= f(J_0(x_0)) + f[J_0(x_1), J_0(x_0)](J_0(x) - J_0(x_0)) \\
&\quad + f[J_0(x_2), J_0(x_1), J_0(x_0)](J_0(x) - J_0(x_0))(J_0(x) - J_0(x_1)) \\
&\quad + f[J_0(x_3), J_0(x_2), J_0(x_1), J_0(x_0)](J_0(x) \\
&\quad - (J_0(x_0))(J_0(x) - (J_0(x_1))(J_0(x) - J_0(x_2)) \\
&= 5.4 + (2.90951)(J_0(x) + 0.04121) + (0.70681)(J_0(x) \\
&\quad + 0.04121)(J_0(x) \\
&\quad + 0.00684) + (4.4457)(J_0(x) \\
&\quad + 0.04121)(J_0(x) + 0.00684)(J_0(x) - 0.02697)
\end{aligned}$$

Consequently,

$$c \approx P(J_0(c)) = P(0) = 5.4 + (2.90951)(0.04121) + (0.70681)(0.04121)(0.00684)$$
$$+ (4.44547)(0.04121)(0.00684)(-0.02697)$$
$$= 5.52007$$ ■ ■ ■

---

## EXERCISES

**1.** The following table for $f(x) = \sin x$ is given:

| $x$ | 1 | 1.02 | 1.05 | 1.06 |
|---|---|---|---|---|
| $f(x)$ | 0.84147 | 0.85211 | 0.86742 | 0.87236 |

(a) Calculate the divided difference table.
(b) Find $P_2(x)$ and $P_3(x)$ using Equation (4.4.10).
(c) Find $P_2(1.01)$ and $P_3(1.01)$.
(d) Compare the absolute value of the actual errors in (c) to the upper bound value of the truncation errors.

**2.** The following table for $f(x) = x^2 + x + 1$ is given:

| $x$ | 0 | 2 | 3 | 5 |
|---|---|---|---|---|
| $f(x)$ | 1 | 7 | 13 | 31 |

(a) Calculate the divided difference table.
(b) Find $P_2(x)$ and $P_3(x)$ using Equation (4.4.10).
(c) Find $P_2(4)$ and $P_3(4)$.
(d) Compare the absolute value of the actual errors in (c) to the upper bound value of the truncation errors.

**3.** If $f(x)$ is a linear function, then prove that $f[x_1, x_0]$ is independent of $x_0$ and $x_1$.

**4.** If $f(x) = 1/(a - x)$ where $a$ is a constant, then show that
(a) $f[x_1, x_0] = 1/((a - x_1)(a - x_0))$
(b) $f[x_2, x_1, x_0] = 1/((a - x_2)(a - x_1)(a - x_0))$

**5.** If $f(x) = p(x)q(x)$, then show that $f[x_1, x_0] = p(x_1)q[x_1, x_0] + q(x_0)p[x_1, x_0]$.

**6.** Verify that
(a) $f[x_1, x_0] = (f(x_1)/(x_1 - x_0)) + (f(x_0)/(x_0 - x_1))$.
(b) $f[x_2, x_1, x_0] = (f(x_2)/((x_2 - x_0)(x_2 - x_1))) + (f(x_1)/((x_1 - x_0)(x_1 - x_2))) + (f(x_0)/((x_0 - x_1)(x_0 - x_2)))$.

**7.** Prove that $f[x_2, x_1, x_0] = f[x_0, x_1, x_2] = f[x_1, x_2, x_0]$.

**8.** Show that $a_3$ in Equation (4.4.1) is given by $a_3 = f[x_3, x_2, x_1, x_0]$.

**9.** Prove that $f[x_2, x_1, x_0] = f^{(2)}(\xi)/2!$ for some $\xi$ in $(x_0, x_2)$.

**10.** If a function $g$ interpolates $(x_0, f(x_0)), (x_1, f(x_1)), \ldots, (x_{N-1}, f(x_{N-1}))$ and a function $h$ interpolates $(x_1, f(x_1)), (x_2, f(x_2)), \ldots, (x_N, f(x_N))$, then show that $g(x) + ((x_0 - x)/(x_N - x_0))[g(x) - h(x)]$ interpolates $(x_0, f(x_0)), (x_1, f(x_1)), \ldots, (x_N, f(x_N))$.

**11.** The following table for $f(x) = \cosh x$ is given:

| $x$ | 1.0 | 1.1 | 1.2 | 1.3 | 1.4 |
|-----|-----|-----|-----|-----|-----|
| $f(x) = \cosh x$ | 1.54308 | 1.66852 | 1.81066 | 1.97091 | 2.15090 |

Find the value of $c$ such that $f(c) = \cosh c = 1.75$.

## COMPUTER EXERCISES

**C.1.** Write a subroutine to generate the difference table (similar to Table 4.4.2). Print the difference table (be happy with your achievements). Compute the value of the Newton divided difference formula [Equation (4.4.10)] at a given number xinter.

**C.2.** Write a computer program to calculate $P_N$ given by Equation (4.4.10) using Exercise C.1 at a given number xinter. The inputs are $N$, xinter, $x_i$ and $f(x_i)$, $i = 0, 1, \ldots, N$.

**C.3.** Use Exercise C.2 to calculate $\ln(1.2)$, $\ln(1.5)$, and $\ln(2.0)$ from the following table:

| $x$ | 1.0 | 1.1 | 1.4 | 1.6 | 1.9 |
|-----|-----|-----|-----|-----|-----|
| $\ln x$ | 0.00000 | 0.09531 | 0.33647 | 0.47000 | 0.64185 |

**C.4.** Use Exercise C.2 to calculate the density (kg/m$^3$) of water at 55°C, 70°C, and 78°C from the following table:

| Temperature (°C) | 50 | 60 | 65 | 75 | 80 |
|------------------|-----|-----|-----|-----|-----|
| Density $\rho$ (kg/m$^3$) | 988.0 | 985.7 | 980.5 | 974.8 | 971.6 |

**C.5.** Use Exercise C.2 to find the value $c$ such that $f(c) = \cosh c = 1.75$ in Exercise 11. Use $N = 3$ and 4.

**C.6.** The values of the Bessel function of order one $J_1(x)$ are given in the following table:

| $x_i$ | 6.8 | 6.9 | 7.0 | 7.1 | 7.2 |
|-------|-----|-----|-----|-----|-----|
| $J_1(x_i)$ | $-0.06522$ | $-0.03490$ | $-0.00468$ | 0.02515 | 0.05433 |

Use Exercise C.2 to find the value of $c$ such that $J_1(c) = 0$.

**C.7.** Use Exercise C.2 to calculate the population of Australia in 1960, 1970, and 1980 from the following table:

| Year | 1954 | 1961 | 1971 | 1976 | 1981 |
|------|------|------|------|------|------|
| Population (in millions) | 8.99 | 10.51 | 12.94 | 13.92 | 14.93 |

## 4.5  EQUAL INTERVAL METHODS

In the case of an equally spaced table, Equation (4.4.10) can be simplified. Let $f(x)$ be tabulated at $a = x_0, x_1, \ldots, x_N = b$ where $x_0, x_1, \ldots, x_N$ are arranged consecutively with equal spacing. Let $h = (b - a)/N = x_{i+1} - x_i$ for $i = 0, 1, \ldots, N - 1$.

Equation (4.4.10) can further be simplified by introducing the **forward difference operator** $\Delta$:

$$\Delta f(x) = f(x + h) - f(x) \tag{4.5.1}$$

It can be verified that [Exercise 3(a)]

$$\Delta(\alpha f(x) + \beta g(x)) = \alpha \Delta f(x) + \beta \Delta g(x)$$

where $\alpha$ and $\beta$ are constants; in other words, $\Delta$ is a linear operator.

**Higher order forward differences** are defined recursively by

$$\Delta^{i+1} f(x) = \Delta(\Delta^i f(x)) \text{ for } i = 0, 1, \ldots$$

For example,

$$
\begin{aligned}
\Delta^2 f(x) &= \Delta(\Delta f(x)) \\
&= \Delta(f(x + h) - f(x)) \\
&= \Delta f(x + h) - \Delta f(x) \\
&= f(x + 2h) - 2f(x + h) + f(x)
\end{aligned}
$$

Similarly

$$
\begin{aligned}
\Delta^3 f(x) &= \Delta(\Delta^2 f(x)) \\
&= \Delta(f(x + 2h) - 2f(x + h) + f(x)) \\
&= \Delta f(x + 2h) - 2\Delta f(x + h) + \Delta f(x) \\
&= f(x + 3h) - 3f(x + 2h) + 3f(x + h) - f(x) \\
&= \sum_{i=0}^{3} \binom{3}{i} (-1)^i f(x + (3 - i)h)
\end{aligned}
$$

Generalizing

$$\Delta^N f(x) = \sum_{i=0}^{N} \binom{N}{i} (-1)^i f(x + (N - i)h)$$

We can rewrite the divided differences in terms of the forward differences as follows:

$$f[x_1, x_0] = \frac{f(x_1) - f(x_0)}{x_1 - x_0} = \frac{f(x_0 + h) - f(x_0)}{h} = \frac{\Delta f(x_0)}{h}$$

$$
\begin{aligned}
f[x_2, x_1, x_0] &= \frac{f[x_2, x_1] - f[x_1, x_0]}{x_2 - x_0} = \frac{\dfrac{\Delta f(x_1)}{h} - \dfrac{\Delta f(x_0)}{h}}{2h} \\
&= \frac{\Delta(f(x_0 + h) - f(x_0))}{2h^2} = \frac{\Delta^2 f(x_0)}{2! \, h^2}
\end{aligned}
\tag{4.5.2}
$$

It can be proven by mathematical induction that

$$f[x_N, x_{N-1}, \ldots, x_1, x_0] = \frac{\Delta^N f(x_0)}{N! \, h^N} \tag{4.5.3}$$

Writing the divided differences in terms of the forward differences and $h$ in Equation (4.4.10) yields the **Newton forward difference formula**

$$P_N(x) = f(x_0) + \frac{\Delta f(x_0)}{h}(x - x_0) + \frac{\Delta^2 f(x_0)}{2! \, h^2}(x - x_0)(x - x_1) + \cdots$$
$$+ \frac{\Delta^N f(x_0)}{N! \, h^N}(x - x_0)(x - x_1) \cdots (x - x_{N-1}) \tag{4.5.4}$$

It is convenient to introduce a nondimensional variable $s$ as

$$s = \frac{x - x_0}{h}$$

and so

$$x = x_0 + sh \tag{4.5.5}$$

where $s$ is any real number. Then

$$x - x_i = x - x_0 + x_0 - x_i$$
$$= sh - (x_i - x_0)$$
$$= sh - ih$$
$$= (s - i)h \tag{4.5.6}$$

Substituting Equation (4.5.6) in Equation (4.5.4) yields

$$P_N(s) = f(x_0) + \Delta f(x_0)s + \frac{\Delta^2 f(x_0)}{2!}s(s - 1) + \cdots$$
$$+ \frac{\Delta^N f(x_0)}{N!}s(s - 1) \cdots (s - N + 1) \tag{4.5.7}$$

Using the binomial coefficient notation ($s$ is a positive real number)

$$\binom{s}{i} = \frac{s(s - 1) \cdots (s - i + 1)}{i!} \quad \text{and} \quad \binom{s}{0} = 1$$

in Equation (4.5.7), we get

$$P_N(s) = \sum_{i=0}^{N} \Delta^i f(x_0) \binom{s}{i} \tag{4.5.8}$$

Sometimes, this is also called the **Newton forward difference formula** for the interpolating polynomial. For $N = 1$, $P_1(s) = f(x_0) + \Delta f(x_0)s$, a linear interpolation formula. The computation of the forward differences in Equation (4.5.8) can be generated in a similar manner as divided differences as shown in Table 4.5.1.

**Table 4.5.1**

| $x$ | $f(x)$ | $\Delta f(x)$ | $\Delta^2 f(x)$ | $\Delta^3 f(x)$ |
|-----|--------|---------------|-----------------|-----------------|
| $x_0$ | $f(x_0)$ | | | |
| | | $\Delta f(x_0)$ | | |
| $x_1$ | $f(x_1)$ | | $\Delta^2 f(x_0)$ | |
| | | $\Delta f(x_1)$ | | $\Delta^3 f(x_0)$ |
| $x_2$ | $f(x_2)$ | | $\Delta^2 f(x_1)$ | |
| | | $\Delta f(x_2)$ | | |
| $x_3$ | $f(x_3)$ | | | |

**EXAMPLE 4.5.1**

The values of $f(x) = \cos x$ are given in Table 4.5.2. Approximate $f(0.13)$ using the Newton forward difference formula [Equation (4.5.8)] for various $N$.

**Table 4.5.2**

| $i$ | $x_i$ | $f(x_i)$ | $\Delta f(x_i)$ | $\Delta^2 f(x_i)$ | $\Delta^3 f(x_i)$ |
|-----|-------|----------|-----------------|-------------------|-------------------|
| 0 | 0.1 | 0.99500 | | | |
| | | | $-0.01493$ | | |
| 1 | 0.2 | 0.98007 | | $-0.00980$ | |
| | | | $-0.02473$ | | 0.00025 |
| 2 | 0.3 | 0.95534 | | $-0.00955$ | |
| | | | $-0.03428$ | | |
| 3 | 0.4 | 0.92106 | | | |

1. For $N = 1$, $P_1(s) = f(x_0) + \Delta f(x_0)s = 0.99500 + (-0.01493)s$.
   Since $s = (x - x_0)/h$, then for $x = 0.13$, $s = 0.3$. Therefore

$$P_1(0.3) = 0.99500 + (-0.01493)(0.3)$$

$$= 0.99052$$

2. For $N = 2$,

$$P_2(s) = f(x_0) + \Delta f(x_0)s + \frac{\Delta^2 f(x_0)}{2!} s(s - 1) = P_1(s) + \frac{\Delta^2 f(x_0)}{2!} s(s - 1)$$

Therefore

$$P_2(0.3) = 0.99052 + \frac{(-0.00980)}{2} (0.3)(-0.7)$$

$$= 0.99155$$

3. For $N = 3$,

$$P_3(s) = P_2(s) + \frac{\Delta^3 f(x_0)}{3!} s(s - 1)(s - 2)$$

Therefore

$$P_3(0.3) = 0.99155 + \frac{(0.00025)(0.3)(-0.7)(-1.7)}{6}$$
$$= 0.99156$$

The exact value to five figures of $\cos(0.13) = 0.99156$.    ■ ■ ■

For the derivatives and interpolations of the values near the end of the table, the backward difference is very convenient. We define

$$\nabla f(x) = f(x) - f(x - h) = \Delta f(x - h) \qquad \textbf{(4.5.9)}$$

This is called the **backward difference** of $f$ at $x$ and $\nabla$ is called the backward difference operator. It can be verified that $\nabla$ is a linear operator [Exercise 3(b)]. We extend the notion to higher differences by defining

$$\begin{aligned} \nabla^{i+1} f(x) &= \nabla^i(\nabla f(x)) \\ &= \nabla^i(f(x) - f(x - h)) \\ &= \nabla^i f(x) - \nabla^i f(x - h) \end{aligned}$$

For the backward difference interpolating formula, start from the bottom of the table. Replace $x_0$ by $x_M$, $x_1$ by $x_{M-1}$, ..., $x_i$ by $x_{M-i}$ in Equation (4.4.10). Then

$$\begin{aligned} P_N(x) &= f(x_M) + f[x_{M-1}, x_M](x - x_M) + f[x_{M-2}, x_{M-1}, x_M](x - x_M)(x - x_{M-1}) \\ &\quad + \cdots + f[x_{M-N}, x_{M-N+1}, \ldots, x_M](x - x_M)(x - x_{M-1}) \cdots (x - x_{M-N+1}) \quad \textbf{(4.5.10)} \end{aligned}$$

We write the divided differences in terms of the backward differences as follows:

$$f[x_{M-1}, x_M] = \frac{f(x_{M-1}) - f(x_M)}{x_{M-1} - x_M} = \frac{f(x_M) - f(x_{M-1})}{x_M - x_{M-1}} = \frac{\nabla f(x_M)}{h}$$

$$\begin{aligned} f[x_{M-2}, x_{M-1}, x_M] &= \frac{f[x_{M-2}, x_{M-1}] - f[x_{M-1}, x_M]}{(x_{M-2} - x_M)} \\[2mm] &= \frac{f[x_{M-1}, x_M] - f[x_{M-2}, x_{M-1}]}{(x_M - x_{M-2})} \\[2mm] &= \frac{\dfrac{\nabla f(x_M)}{h} - \dfrac{\nabla f(x_{M-1})}{h}}{2h} \\[2mm] &= \frac{\nabla(f(x_M) - f(x_{M-1}))}{2h^2} \\[2mm] &= \frac{\nabla^2 f(x_M)}{2! h^2} \end{aligned}$$

By mathematical induction it can be proven that

$$f[x_{M-i}, x_{M-i+1}, \ldots, x_{M-1}, x_M] = \frac{\nabla^i f(x_M)}{i! h^i} \quad \text{for } i = 0, 1, \ldots, M \qquad \textbf{(4.5.11)}$$

Substituting Equation (4.5.11) in Equation (4.5.10), we get the Newton backward difference formula

$$P_N(x) = f(x_M) + \frac{\nabla f(x_M)}{h}(x - x_M) + \frac{\nabla^2 f(x_M)}{2! \, h^2}(x - x_M)(x - x_{M-1})$$

$$+ \cdots + \frac{\nabla^N f(x_M)}{N! \, h^N}(x - x_M)(x - x_{M-1}) \cdots (x - x_{M-N+1}) \qquad \textbf{(4.5.12)}$$

Just as we introduced a nondimensional variable for the Newton forward difference formula, we introduce

$$t = \frac{(x - x_M)}{h}$$

Then

$$x = x_M + th \qquad \textbf{(4.5.13)}$$

and

$$x - x_{M-i} = x - x_M + x_M - x_{M-i}$$

$$= th + ih$$

$$= (t + i)h$$

Substituting Equation (4.5.13) in Equation (4.5.12), yields

$$P_N(t) = f(x_M) + \nabla f(x_M)t + \frac{\nabla^2 f(x_M)}{2!} t(t + 1) + \cdots$$

$$+ \frac{\nabla^N f(x_M)}{N!} t(t + 1) \cdots (t + N - 1) \qquad \textbf{(4.5.14)}$$

Sometimes this is also called the **Newton backward difference formula** for the interpolating polynomial. The computation of the backward differences in Equation (4.5.14) can be generated as shown in Table 4.5.3.

**Table 4.5.3**

| $x$ | $f(x)$ | $\nabla f(x)$ | $\nabla^2 f(x)$ | $\nabla^3 f(x)$ |
|---|---|---|---|---|
| | | | | |
| | | $\nabla f(x_{M-2})$ | | |
| $x_{M-2}$ | $f(x_{M-2})$ | | $\nabla^2 f(x_{M-1})$ | |
| | | $\nabla f(x_{M-1})$ | | $\nabla^3 f(x_M)$ |
| $x_{M-1}$ | $f(x_{M-1})$ | | $\nabla^2 f(x_M)$ | |
| | | $\nabla f(x_M)$ | | |
| $x_M$ | $f(x_M)$ | | | |

**EXAMPLE 4.5.2**

The values of $f(x) = \cos x$ are given in Table 4.5.2. Approximate $f(0.33)$ and $f(0.13)$ using the Newton backward difference formula [Equation (4.5.14)].

Since $\nabla f(x_3) = f(x_3) - f(x_2) = \Delta f(x_2)$, $\nabla f(0.4) = f(0.4) - f(0.3) = 0.92106 - 0.95534 = -0.03428 = \Delta f(0.3)$. Thus the backward differences are identical to the forward differences. There is a difference in notation to represent them.

1. For $N = 1$, $P_1(t) = f(x_3) + \nabla f(x_3)t = 0.92106 + (-0.03428)t$.
Since $t = (x - x_3)/h$, then for $x = 0.33$, $t = -0.7$. Therefore

$$P_1(-0.7) = 0.92106 + (-0.03428)(-0.7) = 0.94506$$

For $x = 0.13$, $t = -2.7$; therefore

$$P_1(-2.7) = 0.92106 + (-0.03428)(-2.7) = 1.01362$$

Further, $\cos(0.13) = 0.99156$, and yet we got 1.01362. Why is there so much error? Look at the truncation error:

$$|\text{truncation error}| = |E(x)| = |f''(\xi)(x - 0.4)(x - 0.3)/(2!)|$$
$$\leq \frac{(0.27)(0.17)}{2} = 0.02295$$

Far away base points lead to higher truncation error. Because of this, pick the base points closer to the point at which we want the value of the function.

2. For $N = 2$,

$$P_2(t) = P_1(t) + \frac{\nabla^2 f(x_3)}{2!} t(t + 1)$$
$$= P_1(t) + \frac{(-0.00955)}{2} t(t + 1)$$

For $x = 0.33$,

$$P_2(-0.7) = P_1(-0.7) + (-0.00478)(-0.7)(0.3)$$
$$= 0.94506 + 0.00100 = 0.94606$$

For $x = 0.13$,

$$P_2(-2.7) = P_1(-2.7) + (-0.00478)(-2.7)(-1.7)$$
$$= 1.01362 - 0.02192 = 0.99170$$

3. For $N = 3$,

$$P_3(t) = P_2(t) + \frac{\nabla^3 f(x_3)}{3!} t(t + 1)(t + 2)$$
$$= P_2(t) + \frac{(0.00025)}{6} t(t + 1)(t + 2)$$

For $x = 0.33$,

$$P_3(-0.7) = P_2(-0.7) + (0.00004)(-0.7)(0.3)(1.3)$$
$$= 0.94606 - 0.00001 = 0.94605$$

For $x = 0.13$,

$$P_3(-2.7) = P_2(-2.7) + (0.00025)(-2.7)(-1.7)(-0.7)$$
$$= 0.99170 - 0.00013 = 0.99157$$

Because of round-off errors, the approximate value of $f(0.13)$ differs slightly from the value we obtained by using the Newton forward difference formula. In fact, it can be proven that the forward and backward interpolating polynomials ending at the same difference entry are identical (Exercise 10). ∎ ∎ ∎

**EXERCISES**

1. The following table for $f(x) = \sin x$ is given:

| $x$ | 1.00 | 1.02 | 1.04 | 1.06 |
|---|---|---|---|---|
| $f(x)$ | 0.84147 | 0.85211 | 0.86240 | 0.87236 |

(a) Compute the forward difference table and note that the backward differences are identical; only the indices of the base points differ.
(b) Approximate $f(1.01)$ and $f(1.05)$ using Equation (4.5.7) with $N = 2$ and 3.
(c) Approximate $f(1.01)$ and $f(1.05)$ using Equation (4.5.14) with $N = 2$ and 3.
(d) Compare the approximated values for $f(1.01)$ and $f(1.05)$ to (b) and (c). Explain the results.

2. The following table for $f(x) = x^2 + x + 1$ is given:

| $x$ | 0 | 1 | 2 | 3 |
|---|---|---|---|---|
| $f(x)$ | 1 | 3 | 7 | 13 |

(a) Compute the forward difference table and note that the backward differences are identical; only the indices of the base points differ.
(b) Approximate $f(0.5)$ and $f(2.5)$ using Equation (4.5.7) with $N = 2$ and 3.
(c) Approximate $f(0.5)$ and $f(2.5)$ using Equation (4.5.14) with $N = 2$ and 3.
(d) Compare the approximated values for $f(0.5)$ and $f(2.5)$ to (b) and (c). Explain the results.

3. Prove that
(a) $\Delta(\alpha f(x) + \beta g(x)) = \alpha \Delta f(x) + \beta \Delta g(x)$
(b) $\nabla(\alpha f(x) + \beta g(x)) = \alpha \nabla f(x) + \beta \nabla g(x)$
where $\alpha$ and $\beta$ are constants. Hence $\Delta$ and $\nabla$ are linear operators.

4. Show that $\Delta^4 f(x) = f(x + 4h) - 4f(x + 3h) + 6f(x + 2h) - 4f(x + h) + f(x)$.

5. Show that
(a) $\nabla^2 f(x) = f(x) - 2f(x - h) + f(x - 2h)$
(b) $\nabla^3 f(x) = f(x) - 3f(x - h) + 3f(x - 2h) - f(x - 3h)$

6. Show that (a) $\nabla^2 f(x_0) = \Delta^2 f(x_0 - 2h)$ and (b) $\nabla^3 f(x_0) = \Delta^3 f(x_0 - 3h)$.

7. Prove that $\Delta(p(x)q(x)) = p(x)\Delta q(x) + q(x + h)\Delta p(x)$.

8. If $f(x) = cx + d$ where $c$ and $d$ are constants, then find (a) $\Delta f(x)$ and (b) $\Delta^2 f(x)$.

9. (a) $\begin{array}{c|c} x_0 & f(x_0) \\ & \quad\quad \Delta f(x_0) \\ x_1 & f(x_1) \end{array}$    (b) $\begin{array}{c|c} x_0 & f(x_0) \\ & \quad\quad \nabla f(x_1) \\ x_1 & f(x_1) \end{array}$

From table (a), the Newton forward interpolating polynomial of degree one is given by

$$P_1(s) = f(x_0) + \Delta f(x_0)s \quad \text{where } s = \frac{x - x_0}{h} \tag{9a}$$

From table (b), the Newton backward interpolating polynomial of degree one is given by

$$P_1(t) = f(x_1) + \nabla f(x_1)t \quad \text{where } t = \frac{x - x_1}{h} \tag{9b}$$

Verify that the polynomials given by Equations 9a and 9b are identical by writing $s$ and $t$ in terms of $x$.

10. (a)

| $x_0$ | $f(x_0)$ | | |
|---|---|---|---|
| | | $\Delta f(x_0)$ | |
| $x_1$ | $f(x_1)$ | | $\Delta^2 f(x_0)$ |
| | | $\Delta f(x_1)$ | |
| $x_2$ | $f(x_2)$ | | |

(b)

| $x_0$ | $f(x_0)$ | | |
|---|---|---|---|
| | | $\nabla f(x_1)$ | |
| $x_1$ | $f(x_1)$ | | $\nabla^2 f(x_2)$ |
| | | $\nabla f(x_2)$ | |
| $x_2$ | $f(x_2)$ | | |

From table (a), the Newton forward interpolating polynomial of second degree is given by

$$P_2(s) = f(x_0) + \Delta f(x_0)s + \frac{\Delta^2 f(x_0)}{2!} s(s - 1) \quad \text{where } s = \frac{x - x_0}{h} \tag{10a}$$

and from table (b), the Newton backward interpolating polynomial of second degree is given by

$$P_2(t) = f(x_2) + \nabla f(x_2)t + \frac{\nabla^2 f(x_2)}{2!} t(t + 1) \quad \text{where } t = \frac{x - x_2}{h} \tag{10b}$$

Verify that the polynomials given by Equations 10a and 10b are identical by writing $s$ and $t$ in terms of $x$.

11. The **central difference operator**, denoted by $\delta$, is defined by $\delta f(x) = f(x + (h/2)) - f(x - (h/2))$.
Show that

$$\frac{\delta^2 f(x_0 + h)}{2! \, h^2} = f[x_2, x_1, x_0]$$

Just as we derived Equation (4.5.8) in terms of the forward differences and Equation (4.5.14) in terms of the backward differences, we can derive a similar formula in terms of the central differences (Hilderbrand 1974).

## COMPUTER EXERCISES

C.1. Write a subroutine to generate the forward difference table (similar to Table 4.5.1). Print the forward difference table. Compute the value of the Newton forward difference formula [Equation (4.5.7)] at a given number $a$. Also, compute the value of the Newton backward difference formula [Equation (4.5.14)] at a given number $b$.

C.2. Write a computer program to calculate the value of $P_N$ given by Equation (4.5.7) using Exercise C.1 at $a$. Also, calculate the value of $P_N$ given by Equation (4.5.14) using Exercise C.1 at $b$. Convert a given number xinter in terms of $a$

and *b*. Your inputs are $N$, a particular number xinter, $x_i$, and $f(x_i)$, $i = 0$, 1, . . . , $N$.

**C.3.**    Use Exercise C.2 to compute log 1.05, log 1.23, and log 1.44 from the following table of the values of log $x$:

| $x$ | 1.0 | 1.1 | 1.2 | 1.3 | 1.4 |
|---|---|---|---|---|---|
| **log $x$** | 0.00000 | 0.04139 | 0.07918 | 0.11394 | 0.14613 |

Compare the computed values by both methods and then compare them with the exact values.

**C.4.**    Use Exercise C.2 to calculate the population of the United States in 1954, 1967, and 1978 from the following table:

| **Year** | 1940 | 1950 | 1960 | 1970 | 1980 | 1990 |
|---|---|---|---|---|---|---|
| **Population (in thousands)** | 132164 | 151325 | 179323 | 203302 | 226542 | 248709 |

**C.5.**    Use Exercise C.2 to calculate the surface tension ($N/m$) of water at 55°C, 70°C, and 78°C from the following table:

| **Temperature (°C)** | 40 | 50 | 60 | 70 | 80 |
|---|---|---|---|---|---|
| **Surface Tension ($N/m$)** | 7.01 | 6.82 | 6.68 | 6.50 | 6.30 |

## 4.6    HERMITE INTERPOLATION*

So far we have considered interpolating a function $f$ whose values $f(x_0), f(x_1), \ldots,$ $f(x_N)$ are known. Sometimes $f'(x_0), f'(x_1), \ldots, f'(x_N)$ are also known along with the values $f(x_0), f(x_1), \ldots, f(x_N)$. For simplicity, find a polynomial $P(x)$ such that $P(x_0) = f(x_0)$, $P(x_1) = f(x_1)$, $P(x_2) = f(x_2)$, $P'(x_0) = f'(x_0)$, $P'(x_1) = f'(x_1)$ and $P'(x_2) = f'(x_2)$. Therefore, we need a polynomial of degree five. Just as we expressed the interpolating polynomial in terms of the Lagrange polynomials, express

$$P_5(x) = f(x_0)H_0(x) + f(x_1)H_1(x) + f(x_2)H_2(x)$$
$$+ f'(x_0)\hat{H}_0(x) + f'(x_1)\hat{H}_1(x) + f'(x_2)\hat{H}_2(x) \qquad \textbf{(4.6.1)}$$

where $H_0(x)$, $H_1(x)$, $H_2(x)$, $\hat{H}_0(x)$, $\hat{H}_1(x)$, and $\hat{H}_2(x)$ are polynomials of degree five to be determined so that $P_5(x_i) = f(x_i)$ and $P'_5(x_i) = f'(x_i)$ for $i = 0$, 1, and 2. The polynomials $H_0(x)$, $H_1(x)$, $H_2(x)$, $\hat{H}_0(x)$, $\hat{H}_1(x)$, and $\hat{H}_2(x)$ are called the **Hermite polynomials**.

Thus

$$P_5(x_0) = f(x_0)H_0(x_0) + f(x_1)H_1(x_0) + f(x_2)H_2(x_0)$$
$$+ f'(x_0)\hat{H}_0(x_0) + f'(x_1)\hat{H}_1(x_0) + f'(x_2)\hat{H}_2(x_0)$$
$$= f(x_0)$$

if

$$H_0(x_0) = 1 \qquad H_1(x_0) = H_2(x_0) = \hat{H}_0(x_0) = \hat{H}_1(x_0) = \hat{H}_2(x_0) = 0 \qquad \textbf{(4.6.2)}$$

Differentiating Equation (4.6.1) with respect to $x$ yields

$$P_5'(x) = f(x_0)H_0'(x) + f(x_1)H_1'(x) + f(x_2)H_2'(x)$$
$$+ f'(x_0)\hat{H}_0'(x) + f'(x_1)\hat{H}_1'(x) + f'(x_2)\hat{H}_2'(x)$$

Therefore

$$P_5'(x_0) = f(x_0)H_0'(x_0) + f(x_1)H_1'(x_0) + f(x_2)H_2'(x_0) + f'(x_0)\hat{H}_0'(x_0)$$
$$+ f'(x_1)\hat{H}_1'(x_0) + f'(x_2)\hat{H}_2'(x_0)$$
$$= f'(x_0)$$

if

$$H_0'(x_0) = H_1'(x_0) = H_2'(x_0) = \hat{H}_1'(x_0) = \hat{H}_2'(x_0) = 0 \quad \text{and} \quad \hat{H}_0'(x_0) = 1 \quad \textbf{(4.6.3)}$$

Similarly $P_5(x_1) = f(x_1)$, $P_5'(x_1) = f'(x_1)$, $P_5(x_2) = f(x_2)$, and $P_5'(x_2) = f'(x_2)$ lead to other conditions similar to Equations (4.6.2) and (4.6.3).

From all these conditions,

$$H_0(x_1) = H_0'(x_1) = H_0(x_2) = H_0'(x_2) = 0 \qquad H_0(x_0) = 1 \quad \text{and} \quad H_0'(x_0) = 0$$
$$\textbf{(4.6.4)}$$

We have to determine $H_0(x)$ such that Equation (4.6.4) is satisfied. Since $H_0(x_1) = H_0'(x_1) = 0$, $H_0(x) = (x - x_1)^2 q(x)$ where $q(x)$ is to be determined. Also, $H_0(x_2) = H_0'(x_2) = 0$; therefore, $H_0(x) = (x - x_1)^2(x - x_2)^2 q_1(x)$ where $q_1(x)$ is to be determined. Since $L_0(x) = [(x - x_1)(x - x_2)]/[(x_0 - x_1)(x_0 - x_2)]$ and $H_0(x)$ is a polynomial of degree five,

$$H_0(x) = (ax + b)L_0^2(x) \qquad \textbf{(4.6.5)}$$

where $a$ and $b$ are to be determined so that $H_0(x_0) = 1$ and $H_0'(x_0) = 0$. From Equation (4.6.5), $H_0(x_0) = (ax_0 + b)L_0^2(x_0) = 1$. Since $L_0(x_0) = 1$,

$$ax_0 + b = 1 \qquad \textbf{(4.6.6)}$$

Differentiating Equation (4.6.5) with respect to $x$ yields

$$H_0'(x) = aL_0^2(x) + 2(ax + b)L_0(x)L_0'(x)$$

Hence

$$H_0'(x_0) = aL_0^2(x_0) + 2(ax_0 + b)L_0(x_0)L_0'(x_0)$$

Using Equation (4.6.6), $L_0(x_0) = 1$, and $H_0'(x_0) = 0$ in this equation, we get

$$a + 2L_0'(x_0) = 0 \qquad \textbf{(4.6.7)}$$

Solving Equations (4.6.6) and (4.6.7) for $a$ and $b$ and then substituting the values of $a$ and $b$ in Equation (4.6.5) yields

$$H_0(x) = [1 - 2(x - x_0)L_0'(x_0)]L_0^2(x) \qquad \textbf{(4.6.8)}$$

Similarly it can be verified that (Exercise 3)

$$H_1(x) = [1 - 2(x - x_1)L'_1(x_1)]L_1^2(x)$$

and

$$H_2(x) = [1 - 2(x - x_2)L'_2(x_2)]L_2^2(x) \qquad \textbf{(4.6.9)}$$

We have

$$\hat{H}_0(x_1) = \hat{H}'_0(x_1) = \hat{H}_0(x_2) = \hat{H}'_0(x_2) = 0 \qquad \hat{H}_0(x_0) = 0 \quad \text{and} \quad \hat{H}'_0(x_0) = 1 \quad \textbf{(4.6.10)}$$

for the fifth-degree polynomial $\hat{H}_0(x)$. Since $\hat{H}_0(x_1) = \hat{H}'_0(x_1) = 0$ and $\hat{H}_0(x_2) = \hat{H}'_0(x_2) = 0$,

$$\hat{H}_0(x) = (cx + d)L_0^2(x) \qquad \textbf{(4.6.11)}$$

where $c$ and $d$ are to be determined so that $\hat{H}_0(x_0) = 0$ and $\hat{H}'_0(x_0) = 1$. It is left to the reader (Exercise 4) to verify that

$$\hat{H}_0(x) = (x - x_0)L_0^2(x) \qquad \textbf{(4.6.12)}$$

Similarly

$$\hat{H}_1(x) = (x - x_1)L_1^2(x) \quad \text{and} \quad \hat{H}_2(x) = (x - x_2)L_2^2(x) \qquad \textbf{(4.6.13)}$$

Substituting Equations (4.6.8), (4.6.9), (4.6.12), and (4.6.13) in Equation (4.6.1) yields

$$P_5(x) = \sum_{j=0}^{2} \{ f(x_j)[1 - 2(x - x_j)L'_j(x_j)] + f'(x_j)(x - x_j) \} L_j^2(x) \qquad \textbf{(4.6.14)}$$

Since this is a fifth-degree polynomial, the error term is given by $(f^{(6)}(\xi)/6!)(x - x_0)^2 (x - x_1)^2(x - x_2)^2$ where $x_0 < \xi < x_2$. Similarly, consider a set $\{x_0, x_1, \ldots, x_N\}$ of $N + 1$ distinct points such that $a = x_0 < x_1 < \ldots < x_N = b$ and two sets $\{f(x_0), f(x_1), \ldots, f(x_N)\}$ and $\{f'(x_0), f'(x_1), \ldots, f'(x_N)\}$ of $N + 1$ values. Then it can be verified that there exists a unique polynomial of at most degree $2N + 1$ given by

$$P_{2N+1}(x) = \sum_{j=0}^{N} \{ f(x_j)[1 - 2(x - x_j)L'_j(x_j)] + f'(x_j)(x - x_j) \} L_j^2(x) \qquad \textbf{(4.6.15)}$$

such that $P_{2N+1}(x_j) = f(x_j)$ and $P'_{2N+1}(x_j) = f'(x_j)$ for $j = 0, 1, \ldots, N$.
We can generalize and prove the following theorem.

**Theorem 4.6.1**    **(Hermite Polynomial Approximation Theorem)**    Given a set $\{(x_i, f(x_i)), (x_i, f'(x_i)) | i = 0, 1, \ldots, N\}$ where $a = x_0 < x_1 < x_2 < \ldots < x_N = b$ and the $(2N + 2)$th derivative of $f$ is continuous on $[a, b]$, then for any $x$ in $[a, b]$ there exists $\xi$ in $(a, b)$ such that

$$f(x) = \sum_{j=0}^{N} \{ f(x_j)[1 - 2(x - x_j)L'_j(x_j)] + f'(x_j)(x - x_j) \} L_j^2(x)$$

$$+ \frac{f^{(2N+2)}(\xi)}{(2N + 2)!} (x - x_0)^2(x - x_1)^2 \cdots (x - x_N)^2 \qquad \textbf{(4.6.16)}$$

where

$$L_j(x) = \prod_{\substack{i=0 \\ i \neq j}}^{N} \frac{(x - x_i)}{(x_j - x_i)}$$

It can be verified that (Exercise 5)

$$L_j'(x_j) = \sum_{\substack{i=0 \\ i \neq j}}^{N} \frac{1}{(x_j - x_i)} \tag{4.6.17}$$

**EXAMPLE 4.6.1**   The following table for $f(x_i)$ and $f'(x_i)$ at points $x_i$ is given for $f(x) = e^x$. Approximate $f(0.23)$ using the Hermite interpolating polynomial of degree three. Compare the absolute values of the actual error to an upper bound value of the truncation error.

| $x$ | $f(x)$ | $f'(x)$ |
|-----|--------|---------|
| 0.2 | 1.22140 | 1.22140 |
| 0.3 | 1.34986 | 1.34986 |

From Example 4.3.1, $L_0(x) = -10(x - 0.3)$ and $L_1(x) = 10(x - 0.2)$. Therefore $L_0'(x_0) = -10$ and $L_1'(x_1) = 10$. Also, from Equation (4.6.17),

$$L_0'(x_0) = \frac{1}{x_0 - x_1} = \frac{1}{0.2 - 0.3} = -10 \quad \text{and} \quad L_1'(x_1) = \frac{1}{x_1 - x_0} = \frac{1}{0.3 - 0.2} = 10$$

Therefore

$$H_0(x) = [1 - 2(x - x_0)L_0'(x_0)]L_0^2(x) = [1 - 2(x - 0.2)(-10)](100)(x - 0.3)^2$$
$$= 100(20x - 3)(x - 0.3)^2$$
$$H_1(x) = [1 - 2(x - x_1)L_1'(x_1)]L_1^2(x) = [1 - 2(x - 0.3)(10)](100)(x - 0.2)^2$$
$$= 100(-20x + 7)(x - 0.2)^2$$
$$\hat{H}_0(x) = (x - x_0)L_0^2(x) = 100(x - 0.2)(x - 0.3)^2$$
$$\hat{H}_1(x) = (x - x_1)L_1^2(x) = 100(x - 0.3)(x - 0.2)^2$$

We have

$$P_3(x) = f(x_0)H_0(x) + f(x_1)H_1(x) + f'(x_0)\hat{H}_0(x) + f'(x_1)\hat{H}_1(x)$$
$$= 122.14(20x - 3)(x - 0.3)^2 + 134.986(-20x + 7)(x - 0.2)^2$$
$$+ 122.14(x - 0.2)(x - 0.3)^2 + 134.986(x - 0.3)(x - 0.2)^2$$

Hence $P_3(0.23) = 0.95758 + 0.29157 + 0.01795 - 0.00850 = 1.25860$.
Further

$$|\text{truncation error}| = |E(x)| = |f^{(4)}(\xi)(x - x_0)^2(x - x_1)^2/(4!)|$$

Since

$$|f^{(4)}(\xi)| \leq e^{0.3} \qquad |E(0.23)| \leq 0.00001$$

Also, we have $|\text{actual error}| = |e^{0.23} - P_3(0.23)| = 0.00000 \leq 0.00001$.  ■ ■ ■

**Algorithm**

**Subroutine** Heinter ($N$, **x**, **y**, **dy**, xinter)

Comment: **x**, **y** and **dy** are vectors with components $x_i$, $y_i = f(x_i)$, and $dy_i = f'(x_i)$ for $i = 0, 1, \ldots, N$. This program computes yinter $= P_{2N+1}$(xinter) using Equations (4.6.15) and (4.6.17).

  1 **Print** $N$, **x**, **y**, **dy**, xinter
  2 sum $= 0.0$
  3 **Repeat** steps 4–9 **for** $j = 0, 1, \ldots, N$
     4 sum1 $= 0.0$
     5 term $= 1.0$
     6 **Repeat** steps 7–8 **for** $i = 0, 1, \ldots N$
       7 **If** $i \neq j$, **then** sum1 $=$ sum1 $+ (1/(x_j - x_i))$
       8 **If** $i \neq j$, **then** term $=$ term $\ast$ (xinter $- x_i)/(x_j - x_i)$
     9 sum $=$ sum $+ \{y_j[1.0 - 2.0(\text{xinter} - x_j)\text{sum1}] + dy_j(\text{xinter} - x_j)\}(\text{term})^2$
  10 yinter $=$ sum
  11 **Print** xinter, yinter

---

**EXERCISES**

**1.** The following table for $f(x_i)$ and $f'(x_i)$ at points $x_i$ is given for $f(x) = xe^x$. Approximate $f(0.23)$ using the Hermite interpolating polynomial of degree three. Compare the absolute value of the actual error to an upper bound value of the truncation error.

| $x$ | $f(x)$ | $f'(x)$ |
|-----|--------|---------|
| 0.2 | 0.24428 | 1.46568 |
| 0.3 | 0.40496 | 1.75482 |

**2.** The following table for $f(x_i)$ and $f'(x_i)$ at points $x_i$ is given for $f(x) = e^{x^2}$. Approximate $f(0.23)$ using the Hermite interpolating polynomial of degree three. Compare the absolute value of the actual error to an upper bound value of the truncation error.

| $x$ | $f(x)$ | $f'(x)$ |
|-----|--------|---------|
| 0.2 | 1.04081 | 0.41632 |
| 0.3 | 1.09417 | 0.65650 |

**3.** Verify Equation (4.6.9).

**4.** Derive Equation (4.6.12) from Equation (4.6.11).

**5.** Prove Equation (4.6.17).

**6.** Prove that the polynomial given by Equation (4.6.14) is unique.

**7.** Find the polynomial $P(x)$ such that $P(0) = 0$, $P'(0) = 0$, $P(1) = 1$, and $P'(1) = 2$.

**8.** Determine the polynomial such that $P(a) = f(a)$, $P'(a) = f'(a)$, $P(b) = f(b)$, and $P'(b) = f'(b)$.

**9.** Find the formula similar to Equation (4.6.14) when $f(x_0), f'(x_0), f''(x_0), f(x_1), f'(x_1)$, and $f''(x_1)$ at two distinct points $x_0$ and $x_1$ are known.

## COMPUTER EXERCISES

**C.1.** Write a subroutine to calculate the value of $P_{2N+1}$ given by Equation (4.6.15) at a given number xinter.

**C.2.** Write a computer program to calculate $P_{2N+1}$ at a given number xinter using Exercise C.1. Your inputs are $N$, xinter, $f(x_i)$, and $f'(x_i)$ for $i = 0, 1, \ldots, N$.

**C.3.** Use Exercise C.2 to calculate $\ln(1.15)$, $\ln(1.28)$, and $\ln(1.32)$ from the following table:

| $x$ | 1.0 | 1.1 | 1.2 | 1.3 | 1.4 |
|-----|-----|-----|-----|-----|-----|
| ln $x$ | 0.0 | 0.09531 | 0.18232 | 0.26236 | 0.33647 |
| $1/x$ | 1.0 | 0.90909 | 0.83333 | 0.76923 | 0.71429 |

Use $N = 3$ and 4. Compare the absolute value of the actual errors to the upper bound value of the truncation errors.

**C.4.** Use Exercise C.2 to calculate $\sin(1.01)$, $\sin(1.03)$, and $\sin(1.05)$ from the following table:

| $x$ | 1.0 | 1.02 | 1.04 | 1.06 |
|-----|-----|------|------|------|
| sin $x$ | 0.84147 | 0.85211 | 0.86240 | 0.87236 |
| cos $x$ | 0.54030 | 0.52337 | 0.50622 | 0.48887 |

Use $N = 2$ and 3. Compare the absolute value of the actual errors to the upper bound value of the truncation errors.

**C.5.** Use Exercise C.2 to calculate $\tan(1.01)$, $\tan(1.03)$, and $\tan(1.05)$ from the following table:

| $x$ | 1.0 | 1.02 | 1.04 | 1.06 |
|-----|-----|------|------|------|
| tan $x$ | 1.55741 | 1.62813 | 1.70361 | 1.78442 |
| sec$^2$ $x$ | 3.42552 | 3.65081 | 3.90230 | 4.18417 |

Compare the absolute value of the actual errors to the upper bound value of the truncation errors.

**C.6.** An object weighing 4 lb is dropped from a balloon which is 1500 ft above the ground. The data is given in the following table:

| Time | 0 | 2 | 4 | 6 | 8 |
|---|---|---|---|---|---|
| **Distance** | 1500 | 1430 | 1195 | 748 | 29 |
| **Speed** | 0 | 73 | 166 | 286 | 440 |

Use Exercise C.2 to predict the position of the object and its speed at $t = 5$.

## 4.7 MINIMIZING THE MAXIMUM ERROR AND CHEBYSHEV POLYNOMIALS*

So far we assumed that a particular function has already been tabulated at given argument values. We discussed the methods for exact matching of a polynomial and the related errors. Assume that we can select $(N + 1)$ distinct tabular points as we like. How should we choose those points? Since we are interpolating a function $f$ by a polynomial of degree $N$, we could select those $(N + 1)$ distinct tabular points so that $\max_{a \leq x \leq b} |f(x) - P_N(x)|$ is minimum. We know that the truncation error $|E(x)| = |f(x) - P_N(x)| = |f^{(N+1)}(\xi)\Pi_{i=0}^{N}(x - x_i)/(N + 1)!|$. We want to minimize $\max_{a \leq x \leq b} |E(x)|$. Since $N + 1$ is fixed, we minimize $\max_{a \leq x \leq b} |f^{(N+1)}(\xi)||\Pi_{i=0}^{N}(x - x_i)|$. We do not know $f^{(N+1)}(\xi)$, therefore we cannot minimize $\max_{a \leq x \leq b} |f^{(N+1)}(\xi(x))|$. We do the next best thing by selecting $\hat{x}_0, \hat{x}_1, \ldots, \hat{x}_N$ so that $\max_{a \leq x \leq b} |\Pi_{i=0}^{N}(x - \hat{x}_i)|$ is minimized.

This seemingly difficult problem can be solved by the Chebyshev polynomials which are defined on $[-1, 1]$. Transform the interval $[a, b]$ in which the tabular points are given to the interval $[-1, 1]$ by producing for every $z$ in $[a, b]$ a corresponding point $x$ in $[-1, 1]$ by the transformation

$$x = \frac{2z - (a + b)}{b - a} \tag{4.7.1}$$

Our problem is to find $x_0, x_1, \ldots, x_N$ in $[-1, 1]$ so that $\max_{-1 \leq x \leq 1} |(x - x_0)(x - x_1) \cdots (x - x_N)|$ is minimized.

The Chebyshev polynomials are needed because their zeros provide the required points. The **Chebyshev polynomial** of degree $n$ is given by

$$T_n(x) = \cos(n \cos^{-1} x) \quad \text{for} \quad n = 0, 1, \ldots \tag{4.7.2}$$

It seems difficult to believe that Equation (4.7.2) is actually an algebraic polynomial. By direct calculation,

$$T_0(x) = \cos(0 \cos^{-1} x) = \cos(0) = 1$$

$$T_1(x) = \cos(\cos^{-1} x) = x \text{ (Keep in mind that } -1 \leq x \leq 1.)$$

Substitution of $\theta = \cos^{-1} x$, $0 \leq \theta \leq \pi$ in Equation (4.7.2) gives

$$T_n(\cos \theta) = \cos n\theta \tag{4.7.3}$$

Since $\cos(n + 1)\theta + \cos(n - 1)\theta = 2 \cos \theta \cos n\theta$, $T_{n+1} + T_{n-1} = 2T_1 T_n$. Therefore

$$T_{n+1}(x) = 2x T_n(x) - T_{n-1}(x) \tag{4.7.4}$$

where $x = \cos \theta$, $T_0(x) = 1$, $T_1(x) = x$ and $n = 1, 2, \ldots$.
For $n = 1$, Equation (4.7.4) gives

$$T_2(x) = 2x T_1(x) - T_0(x) = 2x^2 - 1$$

For $n = 2$, Equation (4.7.4) becomes

$$T_3(x) = 2x T_2(x) - T_1(x) = 2x(2x^2 - 1) - x = 4x^3 - 3x$$

Thus the Chebyshev polynomials of any desired degree can be generated from Equation (4.7.4). The first eight values of $T_i$ are given in Table 4.7.2. It can be proven by mathematical induction that $T_n(x)$ is a polynomial of degree $n$ with leading coefficient $2^{n-1}$. The graphs of $T_0(x)$, $T_1(x)$, $T_2(x)$, and $T_3(x)$ are shown in Figure 4.7.1.

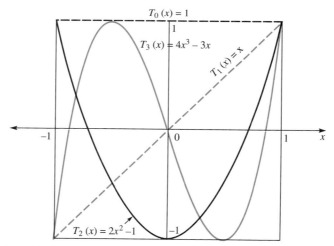

**Fig. 4.7.1**

Let us find the roots of $T_n(x) = 0$. Since $T_n(x) = T_n(\cos \theta) = \cos n\theta$, $\cos n\theta = 0$. Therefore

$$\theta_i = \frac{(2i + 1)\pi}{2n} \quad \text{for} \quad i = 0, 1, \ldots, n - 1$$

are $n$ zeros of $\cos n\theta$ in $[0, \pi]$. The graphs of $\cos \theta$, $\cos 2\theta$, $\cos 3\theta$, and $\cos 4\theta$ are shown in Figure 4.7.2.
The zeros of $T_n(x)$ are given by

$$x_i = \cos \frac{(2i + 1)\pi}{2n} \quad \text{for} \quad i = 0, 1, \ldots, n - 1 \tag{4.7.5}$$

Now let $\psi(x) = (x - x_0)(x - x_1)$. We want to select $x_0$ and $x_1$ such that $\max_{-1 \le x \le 1}|\psi(x)|$ is minimum.
Consider $\psi(x) = T_2(x)/2 = x^2 - \frac{1}{2}$ which has extreme values at $x_0 = -1$, $x_1 = 0$, and $x_2 = 1$ as shown in Figure 4.7.3.

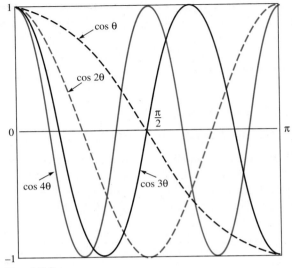

**Fig. 4.7.2**

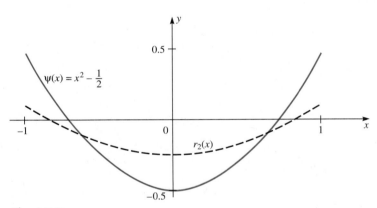

**Fig. 4.7.3**

Assume that there exists another monic polynomial $r_2(x)$ such that $\max_{-1 \leq x \leq 1} |r_2(x)| < \max_{-1 \leq x \leq 1} |\psi(x)|$. The graph of a proposed $r_2(x)$ is shown as a dashed curve in Figure 4.7.3.

Let $Q(x) = (T_2(x)/2) - r_2(x) = x^2 - (1/2) - r_2(x)$. Then $Q(x)$ is a first degree polynomial since $r_2(x)$ is a monic polynomial. Further $Q(x_0) > 0$, $Q(x_1) < 0$ and $Q(x_2) > 0$ because of the assumption about $r_2(x)$ on $[-1, 1]$. Thus $Q(x)$ changes sign twice on $[-1, 1]$; it has two roots. This is impossible, since $Q(x)$ is a first degree polynomial. Thus it is not possible to select $r_2(x)$ such that $\max_{-1 \leq x \leq 1} |r_2(x)| < \max_{-1 \leq x \leq 1} |\psi(x)|$. Hence $\psi(x)$ is the required polynomial. We can extend the discussion for $\psi(x) = T_{N+1}/2^N$ and prove the following theorem.

**Theorem 4.7.1**

Let $\psi(x) = \Pi_{i=0}^{N}(x - x_i)$. Then $\max_{-1 \le x \le 1}|\psi(x)|$ is minimized if $\psi(x) = T_{N+1}(x)/2^N$. Thus the zeros of $T_{N+1}(x)$ minimize $\max_{-1 \le x \le 1}|\psi(x)|$. ∎

To minimize the maximum error due to the term $\Pi_{i=0}^{N}(x - x_i)$, choose the zeros of $T_{N+1}(x)$ as the interpolating nodes.

**EXAMPLE 4.7.1**

Find the approximating polynomials of the second degree using (1) an equally spaced interval and (2) the zeros of $T_3(x)$ as the nodes for $f(x) = e^{-x}$ on $[-1, 1]$. Then compute the values of both polynomials and compare them with the exact values at various points in the interval $[-1, 1]$.

**1.** Let $x_0 = -1$, $x_1 = 0$, and $x_2 = 1$. Then the Lagrange interpolating polynomial

$$P_2(x) = f(x_0) \frac{(x - x_1)(x - x_2)}{(x_0 - x_1)(x_0 - x_2)} + f(x_1) \frac{(x - x_0)(x - x_2)}{(x_1 - x_0)(x_1 - x_2)}$$

$$+ f(x_2) \frac{(x - x_0)(x - x_1)}{(x_2 - x_0)(x_2 - x_1)}$$

$$= e^{-1} \frac{x(x - 1)}{(-1)(-2)} + e^0 \frac{(x + 1)(x - 1)}{(1)(-1)} + e^1 \frac{(x + 1)x}{2 \cdot 1}$$

$$= \frac{1}{e} \frac{(x^2 - x)}{2} - \frac{(x^2 - 1)}{1} + e \frac{x^2 + x}{2}$$

$$= \frac{x^2}{2}\left(e - 2 + \frac{1}{e}\right) + \frac{x}{2}\left(e - \frac{1}{e}\right) + 1$$

$$= \frac{x^2(e - 1)^2}{2e} + \frac{x(e^2 - 1)}{2e} + 1 \approx 0.54308 \, x^2 + 1.17520x + 1$$

**2.** Since $T_3(x) = 4x^3 - 3x = 0$, $x_0 = -\sqrt{3}/2$, $x_1 = 0$, and $x_2 = \sqrt{3}/2$ are the interpolating nodes. Then

$$\bar{P}_2(x) = e^{-\sqrt{3}/2} \frac{x(x - \sqrt{3}/2)}{(-\sqrt{3}/2)(-\sqrt{3})} + e^0 \frac{(x + \sqrt{3}/2)(x - \sqrt{3}/2)}{(\sqrt{3}/2)(-\sqrt{3}/2)}$$

$$+ e^{\sqrt{3}/2} \frac{(x + \sqrt{3}/2)x}{\sqrt{3}(\sqrt{3}/2)}$$

$$= \frac{2(e^{\sqrt{3}/2} - 1)^2}{3e^{\sqrt{3}/2}}x^2 + \frac{(e^{\sqrt{3}} - 1)}{\sqrt{3}e^{\sqrt{3}/2}}x + 1$$

$$\approx 0.53204x^2 + 1.12977x + 1$$

The values of $f(x)$, $\bar{P}_2(x)$, and $P_2(x)$ are listed for various values of $x$ along with $|f(x) - \bar{P}_2(x)|$ and $|f(x) - P_2(x)|$ in Table 4.7.1.

| $x$ | $f(x) = e^x$ | $\tilde{P}_2(x)$ | $|e^x - \tilde{P}_2(x)|$ | $P_2(x)$ | $|e^x - P_2(x)|$ |
|---|---|---|---|---|---|
| $-1.00$ | 0.36788 | 0.40227 | 0.03439 | 0.36788 | 0.00000 |
| $-0.75$ | 0.47237 | 0.45194 | 0.02043 | 0.42408 | 0.04829 |
| $-0.50$ | 0.60653 | 0.56812 | 0.03841 | 0.54817 | 0.05836 |
| $-0.25$ | 0.77880 | 0.75081 | 0.02799 | 0.74014 | 0.03866 |
| 0.00 | 1.00000 | 1.00000 | 0.00000 | 1.00000 | 0.00000 |
| 0.25 | 1.28403 | 1.31570 | 0.03167 | 1.32774 | 0.04371 |
| 0.50 | 1.64872 | 1.69790 | 0.04918 | 1.72337 | 0.07465 |
| 0.75 | 2.11700 | 2.14660 | 0.02960 | 2.18688 | 0.06988 |
| 1.00 | 2.71828 | 2.66181 | 0.05647 | 2.71828 | 0.00000 |

It can be seen from the Table 4.7.1 that the maximum error involved in using $\tilde{P}_2(x)$ is considerably less than that of using $P_2(x)$.    ■ ■ ■

Another useful property we need is the orthogonality of Chebyshev polynomials (Appendix E).
Let $w(x) = 1/\sqrt{1 - x^2}$. Then consider (Appendix E)

$$I = \langle T_i(x), T_j(x) \rangle = \int_{-1}^{1} w(x) T_i(x) T_j(x) dx \tag{4.7.6}$$

Substitute $x = \cos\theta$, $T_i(x) = \cos i\theta$, $T_j(x) = \cos j\theta$, and $dx = -\sin\theta\, d\theta$ in Equation (4.7.6), then

$$I = -\int_{\pi}^{0} \frac{1}{\sin\theta} \cos i\theta \cos j\theta \sin\theta\, d\theta = \int_{0}^{\pi} \cos i\theta \cos j\theta\, d\theta \tag{4.7.7}$$

It can be verified (Exercise 6) that

$$\langle T_i(x), T_j(x) \rangle = I = \begin{cases} 0 & \text{if } i \neq j \\ \pi/2 & \text{if } i = j \neq 0 \\ \pi & \text{if } i = j = 0 \end{cases} \tag{4.7.8}$$

Thus the set of **Chebyshev polynomials** forms an **orthogonal** set on $[-1, 1]$ with respect to the **weight function** $w(x) = 1/\sqrt{1 - x^2}$.
The orthogonal set of Chebyshev polynomials is also a set of linearly independent polynomials on $[-1, 1]$ with the weight function $w(x) = 1/\sqrt{1 - x^2}$ by Theorem E.1. Therefore we can represent any polynomial $P_N(x)$ of degree $N$ as a linear combination of the Chebyshev polynomials up to degree $N$. Thus

$$P_N(x) = \sum_{i=0}^{N} c_i T_i(x) \tag{4.7.9}$$

where (Appendix E)

$$c_i = \frac{\langle P_N(x), T_i(x) \rangle}{\langle T_i(x), T_i(x) \rangle}$$

$$= \frac{\displaystyle\int_{-1}^{1} P_N(x) \frac{T_i(x)}{\sqrt{1 - x^2}} \, dx}{\displaystyle\int_{-1}^{1} \frac{T_i^2(x)}{\sqrt{1 - x^2}} \, dx}$$   **(4.7.10)**

**EXAMPLE 4.7.2**    Let $P_N(x) = x^3$. Then $P_3(x) = x^3 = \sum_{i=0}^{3} c_i T_i(x)$.
From Equation (4.7.10)

$$c_0 = \frac{\displaystyle\int_{-1}^{1} P_3(x) \frac{T_0(x)}{\sqrt{1 - x^2}} \, dx}{\displaystyle\int_{-1}^{1} \frac{T_0^2(x)}{\sqrt{1 - x^2}} \, dx}$$

Using $x = \cos \theta$, it can be verified that $c_0 = 0$. Similarly, we can show that $c_1 = \frac{3}{4}$, $c_2 = 0$, and $c_3 = \frac{1}{4}$. Hence $P_3(x) = x^3 = \frac{3}{4} x + \frac{1}{4} (4x^3 - 3x)$.    ■ ■ ■

The first eight values of $T_i$ and the explicit representation of $x^i$ for $i = 0$ to 8 in terms of $T_i$ are given in Table 4.7.2.

**Table 4.7.2**

| | |
|---|---|
| $T_0(x) = 1$ | $1 = T_0$ |
| $T_1(x) = x$ | $x = T_1$ |
| $T_2(x) = 2x^2 - 1$ | $x^2 = \frac{1}{2} (T_0 + T_2)$ |
| $T_3(x) = 4x^3 - 3x$ | $x^3 = \frac{1}{4} (T_3 + 3T_1)$ |
| $T_4(x) = 8x^4 - 8x^2 + 1$ | $x^4 = \frac{1}{8} (T_4 + 4T_2 + 3T_0)$ |
| $T_5(x) = 16x^5 - 20x^3 + 5x$ | $x^5 = \frac{1}{16} (T_5 + 5T_3 + 10T_1)$ |
| $T_6(x) = 32x^6 - 48x^4 + 18x^2 - 1$ | $x^6 = \frac{1}{32} (T_6 + 6T_4 + 15T_2 + 10T_0)$ |
| $T_7(x) = 64x^7 - 112x^5 + 56x^3 - 7x$ | $x^7 = \frac{1}{64} (T_7 + 7T_5 + 21T_3 + 35T_1)$ |
| $T_8(x) = 128x^8 - 256x^6 + 160x^4 - 32x^2 + 1$ | $x^8 = \frac{1}{128} (T_8 + 8T_6 + 28T_4 + 56T_2 + 35T_0)$ |

## Economization

Suppose we have a power series expansion of a function given by

$$f(x) = a_0 + a_1 x + a_2 x^2 + \cdots + a_N x^N + E_N(x)$$   **(4.7.11)**

in the interval $[-1, 1]$. Also, it is known that $|E_N(x)| < \epsilon_1$ where $\epsilon_1$ is smaller than the prescribed tolerance $\epsilon$. Since $|a_N| + |\epsilon_1| > \epsilon$, we cannot discard the last term $a_N x^N$.

Thus the approximation is given by

$$f(x) \approx a_0 + a_1 x + a_2 x^2 + \cdots + a_N x^N \qquad \textbf{(4.7.12)}$$

When we substitute for $x, x^2, \ldots, x^N$ in terms of Chebyshev polynomials in Equation (4.7.12), we get

$$f(x) \approx b_0 T_0(x) + b_1 T_1(x) + \cdots + b_N T_N(x) \qquad \textbf{(4.7.13)}$$

Since $x^N = (T_N(x) + N T_{N-2}(x) + \cdots)/2^{N-1}$, $x^{N-1} = (T_{N-1}(x) + (N-1) T_{N-3}(x) + \cdots)/2^{N-2}$, $b_i$ of Equation (4.7.13) becomes small very rapidly compared to $a_i$ of Equation (4.7.11) as $i$ approaches $N$ for a broad class of functions. For large $N$ it may happen that

$$(|b_N| + |b_{N-1}| \cdots + |b_{N-m}|) + \epsilon_1 < \epsilon$$

Since $|T_i(x)| \le 1$ on $[-1, 1]$, the last $m$ terms of the right-hand side of Equation (4,.7.13) are negligible and we get

$$f(x) \approx b_0 T_0(x) + \cdots + b_{N-m} T_{N-m}(x) \qquad \textbf{(4.7.14)}$$

After substituting the values of $T_i(x)$ from Table 4.7.2,

$$f(x) \approx A_0 + A_1 x + \cdots + A_{N-m} x^{N-m} \qquad \textbf{(4.7.15)}$$

We get a polynomial approximation to $f(x)$ on $[-1, 1]$ involving fewer terms than would be required by the original approximation which provides the prescribed accuracy.

---

**EXAMPLE 4.7.3**     Compare the Maclaurin and Chebyshev expansions of $f(x) = \sinh x$ on $[-1, 1]$.
The Taylor Theorem A.8 gives

$$f(x) = \sinh x = x + \frac{x^3}{3!} + \frac{x^5}{5!} + R_5(x)$$

where $|R_5(x)| = |f^{(7)}(\xi(x)) x^7/7!| \le (e^1 + e^{-1})/(2 \cdot 7!) = 0.00031.$

Suppose we want $|P(x) - f(x)| \le 0.001$. We have $|R_3(x)| = |f^{(5)}(\xi) x^5/5!| \le 0.01286$. Therefore, we need $x^5/5!$. Thus

$$P_5(x) = x + \frac{x^3}{3!} + \frac{x^5}{5!}$$

$$= T_1(x) + \frac{1}{3!} \cdot \frac{1}{4}(T_3(x) + 3T_1(x)) + \frac{1}{5!} \cdot \frac{1}{16}(T_5(x) + 5T_3(x) + 10T_1(x))$$

$$= \left(1 + \frac{3}{3!\,4} + \frac{10}{5!\,16}\right) T_1(x) + \left(\frac{3}{3!\,4} + \frac{10}{5!\,16}\right) T_3(x) + \frac{1}{5!\,16} T_5(x)$$

$$= \frac{2170}{1920} T_1(x) + \frac{85}{1920} T_3(x) + \frac{1}{1920} T_5(x)$$

The last term will introduce additional error not exceeding $T_5(x)/1920 \le 0.00052$ for all $x$ in $[-1, 1]$. Thus, with a total error smaller than 0.00083,

$$\sinh x \approx \frac{2170}{1920} T_1(x) + \frac{85}{1920} T_3(x)$$

From Table 4.7.2,

$$\sinh x \approx \frac{2170}{1920} x + \frac{85}{1920}(4x^3 - 3x)$$

Therefore

$$\sinh x \approx Q_3(x) = \frac{1915}{1920} x + \frac{340}{1920} x^3$$

The computed values of $P_5(x)$, $Q_3(x)$, and $P_3(x)$ at various points are given in Table 4.7.3.

**Table 4.7.3**

| $x$ | $\sinh x$ | $P_5(x)$ | $|\sinh x - P_5(x)|$ | $Q_3(x)$ | $|\sinh x - Q_3(x)|$ | $P_3(x)$ | $|\sinh x - P_3(x)|$ |
|---|---|---|---|---|---|---|---|
| 1.00 | 1.17520 | 1.17500 | 0.00020 | 1.17448 | 0.00072 | 1.16667 | 0.00853 |
| 0.75 | 0.82232 | 0.82229 | 0.00003 | 0.82275 | 0.00043 | 0.82031 | 0.00201 |
| 0.25 | 0.25261 | 0.25261 | 0.00000 | 0.25212 | 0.00049 | 0.25260 | 0.00001 |

The tabulated entries for $Q_3(x)$ are well within the error bound we set from the beginning. Thus for a given maximum error bound, we can generally obtain a lower-order polynomial approximation by truncating terms of the Chebyshev expansion than by truncating the original polynomial expansion. This is called the **economization of power series**. ■ ■ ■

**EXERCISES**

1. Find the approximating linear polynomials using (a) $x_0 = -1$ and $x_1 = 1$, and (b) the zeros of $T_2(x)$ as the nodes for $f(x) = e^x$ on the closed interval $[-1, 1]$. Then compute the values of both polynomials at various points and compare them with the exact values at those points in the interval $[-1, 1]$.
2. The second degree Lagrange polynomial is $P_2(x) = f(x_0)L_0(x) + f(x_1)L_1(x) + f(x_2)L_2(x)$. Using the zeros of $T_3(x)$ as nodes, find $L_0(x)$, $L_1(x)$, and $L_2(x)$.
3. Find $T_4(x)$, $T_5(x)$, $T_6(x)$, $T_7(x)$, and $T_8(x)$ using Equation (4.7.4) and the values of $T_2(x)$ and $T_3(x)$ from Table 4.7.2.
4. Find the zeros of $T_4(x)$ and $T_5(x)$.
5. Express $x^2$ in terms of $T_0$, $T_1$, and $T_2$ using Equation (4.7.10).
6. Verify Equation (4.7.8).
7. Verify that $T_0(x)$, $T_1(x)$, and $T_2(x)$ are orthogonal polynomials with respect to the weight function $w(x) = 1/\sqrt{1 - x^2}$.
8. Verify that $T_{N+1}$ assumes its extreme values $N + 2$ times on $[-1, 1]$.
9. Sketch the graph of $T_3(x)$ on $[-1, 1]$. Find the extrema of $T_3(x)$.
10. Construct the second degree Lagrange interpolating polynomial using the zeros of $T_3(x)$ as the nodes for $f(x) = e^x$ on the interval $[2, 4]$.

**11.** Prove by mathematical induction that $T_N(x)$ is a polynomial of degree $N$ with leading coefficient $2^{N-1}$.

**12.** Show that $T_{n+m}(x) + T_{n-m}(x) = 2T_n(x)T_m(x)$ where $m$ and $n$ are integers and $n \geq m$.

**13.** Show that $(1 - x^2)T_n''(x) - xT_n'(x) + n^2T_n(x) = 0$.

**14.** For $f(x) = \sin x$, construct a polynomial of the lowest possible degree such that the error is less than 0.0001 on $[-1, 1]$.

**15.** For $f(x) = e^x$, construct a polynomial of the lowest possible degree such that the error is less than 0.0001 on $[-1, 1]$.

## COMPUTER EXERCISES

**C.1.** Compute the values of two approximating polynomials of second degree using (a) equally spaced nodes, and (b) the zeros of $T_3(x)$ as nodes for $f(x) = \sin x$ on $[-1, 1]$ at various points. Compare them with the exact values at those points in the interval $[-1, 1]$. Use Exercise C.1, p. 87.

**C.2.** Compute the values of two approximating polynomials of fourth degree using (a) equally spaced nodes, and (b) the zeros of $T_5(x)$ as nodes for $f(x) = xe^x$ on $[-1, 1]$ at various points. Compare them with the exact values at those points in the interval $[-1, 1]$. Use Exercise C.1, p. 87.

**C.3.** Find the roots of $T_{11}(x)$ using Exercise C.3, p. 75.

**C.4.** Compute the values of two approximating polynomials of tenth degree using (a) equally spaced nodes, and (b) the zeros of $T_{11}(x)$ (use Exercise C.3 or Equation (4.7.5)) as nodes for $f(x) = 1/(1 + 25x^2)$ on $[-1, 1]$ at various points and compare them with the exact values at those points in the interval $[-1, 1]$. Use Exercise C.1, p. 87.

## SUGGESTED READING

F. B. Hilderbrand, 1974. *Introduction to Numerical Analysis,* McGraw-Hill, New York.

# 5 NUMERICAL DIFFERENTIATION

A n object weighing 4 lb is dropped from a balloon that is 1500 ft above the ground. The following table gives the data from observation:

| Time (sec) | 0 | 1 | 2 | 3 | 4 | 5 |
|---|---|---|---|---|---|---|
| Distance (ft) | 1500 | 1483 | 1430 | 1336 | 1195 | 1002 |

Find the speed of the object at 0 sec, 3 sec, and 5 sec as accurately as possible. In this chapter we discuss methods to find derivatives at a given number $a$.

## 5.1 INTRODUCTION

Suppose the values of a function $f$ are given at $x_0, x_1, \ldots, x_N$. The object now is to find the derivative $f'$ at some arbitrary number $x$ in the interval generated by $x_0, x_1, \ldots, x_N$. The obvious solution is to approximate $f$ by another function that is simple to differentiate and then find the derivative of the replaced function. The natural choice for this approximating function is an interpolating polynomial which passes through $(x_0, f(x_0)), (x_1, f(x_1)), \ldots, (x_N, f(x_N))$. This interpolating polynomial could be a Lagrange interpolating polynomial, or a Newton forward or backward formula for the interpolating polynomial.

## 5.2 FIRST DERIVATIVES

Consider the representation of $f$ whose values at $x_0$ and $x_1$ are known, where $x_0 < x_1$. The Lagrange polynomial of degree one passing through $(x_0, f(x_0))$ and $(x_1, f(x_1))$ is given by

$$P_1(x) = f(x_0) \frac{x - x_1}{x_0 - x_1} + f(x_1) \frac{x - x_0}{x_1 - x_0} \qquad \textbf{(5.2.1)}$$

and therefore

$$f(x) = f(x_0)\frac{x - x_1}{x_0 - x_1} + f(x_1)\frac{x - x_0}{x_1 - x_0} + \frac{f''(\xi(x))}{2!}(x - x_0)(x - x_1) \quad \textbf{(5.2.2)}$$

where $x_0 < \xi(x) < x_1$ by the Lagrange Polynomial Approximation Theorem 4.3.1. Differentiating Equation (5.2.2) with respect to $x$ yields

$$f'(x) = \frac{f(x_0)}{x_0 - x_1} + \frac{f(x_1)}{x_1 - x_0}$$

$$+ \frac{f''(\xi)}{2!}[(x - x_1) + (x - x_0)]$$

$$+ \frac{(x - x_0)(x - x_1)}{2!}\frac{d}{dx}(f''(\xi(x))) \quad \textbf{(5.2.3)}$$

Since $\xi$ is a function of $x$, we need to find $d(f''(\xi(x)))/dx$.

In order to do this, divide Equation (5.2.2) by $P_2(x) = (x - x_0)(x - x_1)$ and get

$$\frac{f(x)}{P_2(x)} = \frac{f(x_0)}{(x - x_0)(x_0 - x_1)} + \frac{f(x_1)}{(x - x_1)(x_1 - x_0)} + \frac{f''(\xi(x))}{2!} \quad \textbf{(5.2.4)}$$

Differentiation of Equation (5.2.4) with respect to $x$ gives

$$\frac{d}{dx}\left[\frac{f(x)}{P_2(x)}\right] = -\frac{f(x_0)}{(x - x_0)^2(x_0 - x_1)}$$

$$- \frac{f(x_1)}{(x - x_1)^2(x_1 - x_0)} + \frac{1}{2!}\frac{d}{dx}[f''(\xi(x))] \quad \textbf{(5.2.5)}$$

Assume that $f(x_2)$ where $x_2$ is close to $x$ is also known. Then $f$ is given by

$$f(x) = f(x_0)\frac{(x - x_1)(x - x_2)}{(x_0 - x_1)(x_0 - x_2)} + f(x_1)\frac{(x - x_0)(x - x_2)}{(x_1 - x_0)(x_1 - x_2)}$$

$$+ f(x_2)\frac{(x - x_0)(x - x_1)}{(x_2 - x_0)(x_2 - x_1)}$$

$$+ \frac{f^{(3)}(\tau(x))}{3!}(x - x_0)(x - x_1)(x - x_2) \quad \textbf{(5.2.6)}$$

where $\tau(x)$ is in the open interval generated by $x_0$, $x$, $x_2$, and $x_1$. Equation (5.2.6) can be rewritten after dividing it by $(x - x_0)(x - x_1)(x - x_2)$ as

$$\frac{f(x)}{P_2(x)(x - x_2)} - \frac{f(x_2)}{P_2(x_2)(x - x_2)} = \frac{f(x_0)}{(x - x_0)(x_0 - x_1)(x_0 - x_2)}$$

$$+ \frac{f(x_1)}{(x - x_1)(x_1 - x_0)(x_1 - x_2)} + \frac{f^{(3)}(\tau(x))}{3!} \quad \textbf{(5.2.7)}$$

In order to find the limits of Equation (5.2.7) as $x_2 \rightarrow x$, write it as

$$\lim_{x_2 \rightarrow x}\frac{\dfrac{f(x_2)}{P_2(x_2)} - \dfrac{f(x)}{P_2(x)}}{(x_2 - x)} = \lim_{x_2 \rightarrow x}\left[-\frac{f(x_0)}{(x - x_0)(x_0 - x_1)(x_2 - x_0)}\right.$$

$$\left. - \frac{f(x_1)}{(x - x_1)(x_1 - x_0)(x_2 - x_1)} + \frac{f^{(3)}(\tau(x))}{3!}\right] \quad \textbf{(5.2.8)}$$

Using the definition of the derivative and taking limits, Equation (5.2.8) becomes

$$\frac{d}{dx}\left[\frac{f(x)}{P_2(x)}\right] = -\frac{f(x_0)}{(x - x_0)^2 (x_0 - x_1)}$$

$$-\frac{f(x_1)}{(x - x_1)^2 (x_1 - x_0)} + \lim_{x_2 \to x} \frac{f^{(3)}(\tau(x))}{3!} \qquad (5.2.9)$$

Comparing Equations (5.2.5) and (5.2.9) yields

$$\frac{1}{2!}\frac{d}{dx}\left(f''(\xi(x))\right) = \lim_{x_2 \to x} \frac{1}{3!} f^{(3)}(\tau(x)) = \frac{1}{3!} f^{(3)}(\eta(x)) \qquad (5.2.10)$$

where $x_0 < \eta(x) < x_1$. Similar arguments can be used to prove (Exercise 9)

$$\frac{1}{n!}\frac{d}{dx}\left(f^{(n)}(\xi(x))\right) = \frac{1}{(n + 1)!} f^{(n+1)}(\eta(x)) \qquad (5.2.11)$$

where $\xi(x)$ and $\eta(x)$ are defined as above.

By using Equation (5.2.10) in Equation (5.2.3), Equation (5.2.3) can be rewritten as

$$f'(x) = \frac{f(x_1) - f(x_0)}{x_1 - x_0} + \frac{f''(\xi)}{2!}(2x - x_0 - x_1)$$

$$+ \frac{f^{(3)}(\eta)}{3!}(x - x_0)(x - x_1) \qquad (5.2.12)$$

where $x_0 < \eta < x_1$.

When a table is given, usually the goal is to find the derivative at nodal points. Let us find the derivative at $x_0$. Substituting $x = x_0$ and $x_1 = x_0 + h$ in Equation (5.2.12),

$$f'(x_0) = \frac{f(x_0 + h) - f(x_0)}{h} - \frac{h}{2}f''(\xi_1) \qquad (5.2.13)$$

where $x_0 < \xi_1 < x_1$. Since $\xi_1$ is unknown, we cannot determine the exact value of $f''(\xi_1)$, but we can find an upper bound $M_2$ such that $|f''(x)| \le M_2$ for all $x$ in $[x_0, x_0 + h]$. Thus for small $h$, use the formula

$$f'(x_0) \approx \frac{f(x_0 + h) - f(x_0)}{h} \qquad (5.2.14)$$

with an error bounded by $hM_2/2$. This formula is called the **forward difference formula**.

Substitution of $x = x_1$ and $x_0 = x_1 - h$ in Equation (5.2.12) leads to

$$f'(x_1) = \frac{f(x_1) - f(x_0)}{x_1 - x_0} + \frac{h}{2}f''(\xi_2)$$

$$= \frac{f(x_1) - f(x_1 - h)}{h} + \frac{h}{2}f''(\xi_2) \qquad (5.2.15)$$

where $x_0 < \xi_2 < x_1$. For small $h$, use the formula

$$f'(x_1) \approx \frac{f(x_1) - f(x_1 - h)}{h} \qquad (5.2.16)$$

This formula is called the **backward difference formula**.

We know that $f'(x) = \lim_{h \to 0} [(f(x + h) - f(x))/h]$. If $h > 0$, then Equation (5.2.14) follows from the definition of $f'(x_0)$ and if $h < 0$, then Equation (5.2.16) follows from the definition of $f'(x_1)$. Thus Equations (5.2.14) and (5.2.16) fit with the intuitive definition of $f'(x)$.

**EXAMPLE 5.2.1**

Let $f(x) = \tan x$. Then find $f'(1)$ using Equations (5.2.14) and (5.2.16) with various $h$ and compare it with the exact value. Use five decimal places or six significant digits in all calculations.

We have from Equation (5.2.14),

$$f'(1) \approx \frac{f(1 + h) - f(1)}{h} = \frac{\tan(1 + h) - \tan 1}{h}$$

Carrying out calculations with five decimal places or six significant digits for $h = 0.1$ yields

$$\frac{\tan(1 + 0.1) - \tan 1}{h} = \frac{1.96476 - 1.55741}{0.1} = 4.07350$$

For various $h$, the computed results are given in Table 5.2.1. Since $f'(x) = \sec^2 x$, $f'(1) = \sec^2 1 \approx 3.42553$. The $|\text{actual error}| = |\sec^2 1 - [\tan(1 + h) - \tan 1/h]|$ is also given in Table 5.2.1.

**Table 5.2.1**

| $h$ | $\dfrac{\tan(1 + h) - \tan 1}{h}$ | \|Actual Error\| | $\dfrac{\tan 1 - \tan(1 - h)}{h}$ | \|Actual Error\| |
|---|---|---|---|---|
| 0.100000 | 4.07350 | 0.64797 | 2.97250 | 0.45303 |
| 0.010000 | 3.48000 | 0.05447 | 3.37300 | 0.05253 |
| 0.001000 | 3.43000 | 0.00447 | 3.42000 | 0.00553 |
| 0.000100 | 3.40000 | 0.02553 | 3.40000 | 0.02553 |
| 0.000010 | 3.00000 | 0.42553 | 4.00000 | 0.57447 |
| 0.000001 | 0.00000 | 3.42553 | 0.00000 | 3.42553 |

From Equation (5.2.16),

$$f'(1) \approx \frac{f(1) - f(1 - h)}{h} = \frac{\tan 1 - \tan(1 - h)}{h}$$

For $h = 0.1$, we get

$$f'(1) = \frac{\tan 1 - \tan(1 - h)}{h} = \frac{1.55741 - 1.26016}{0.1} = 2.97250$$

For $h = 0.000001$, $\tan(1 + h)$, $\tan(1 - h)$, and $\tan 1$ are equal since we use five decimal places; subtraction cancels them. This can also be seen in Table 5.2.1. If $f(x)$ can be evaluated only up to five decimal places, it does not pay to have $h$ smaller than five decimal places. Also, observe that as $h$ decreases, $|\text{actual error}|$ does not decrease.

■ ■ ■

Subtraction of almost equal numbers are involved in Equations (5.2.14) and (5.2.16) and therefore large round-off errors are introduced. Let us first examine the round-off errors when using the forward difference formula [Equation (5.2.14)]. Let us assume that the round-off errors $e(x_0 + h)$ and $e(x_0)$ are introduced in computing $f(x_0 + h)$ and $f(x_0)$ when we use the computed values $\bar{f}(x_0 + h)$ and $\bar{f}(x_0)$. The relationships between them are given by

$$f(x_0 + h) = \bar{f}(x_0 + h) + e(x_0 + h) \quad \text{and} \quad f(x_0) = \bar{f}(x_0) + e(x_0) \qquad \textbf{(5.2.17)}$$

The error in using approximations $\bar{f}(x_0 + h)$ and $\bar{f}(x_0)$ in Equation (5.2.14) is

$$f'(x_0) - \frac{\bar{f}(x_0 + h) - \bar{f}(x_0)}{h} = f'(x_0) - \frac{f(x_0 + h) - f(x_0)}{h} + \frac{e(x_0 + h) - e(x_0)}{h}$$

$$= -\frac{h}{2}f''(\xi) + \frac{e(x_0 + h) - e(x_0)}{h} \qquad \textbf{(5.2.18)}$$

The total error in Equation (5.2.18) is made up of truncation and round-off errors. Suppose that the round-off errors $e(x_0 + h)$ and $e(x_0)$ are bounded by $\epsilon > 0$; that is, $|e(x_0 + h)| \le \epsilon$ and $|e(x_0)| \le \epsilon$. Then from Equation (5.2.18),

$$\left| f'(x_0) - \frac{\bar{f}(x_0 + h) - \bar{f}(x_0)}{h} \right| \le \frac{hM_2}{2} + \frac{2\epsilon}{h}$$

The maximum truncation error $hM_2/2$ and maximum rounding error $2\epsilon/h$ as a function of $h$ are shown in Figure 5.2.1.

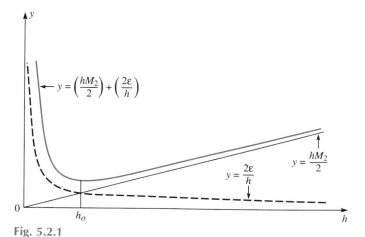

Fig. 5.2.1

As $h$ becomes small, the truncation error $hM_2/2$ becomes small while the round-off error $2\epsilon/h$ becomes large. When $h$ becomes large, the truncation error $hM_2/2$ becomes large and the round-off error $2\epsilon/h$ becomes small. We want the value of $h$ for which the sum of these errors is the smallest. This value $h_o$ is called the **optimal step size** and it can be obtained by minimizing the quantity

$$g(h) = \frac{hM_2}{2} + \frac{2\epsilon}{h}$$

Setting $g'(h) = 0$ results in $h = 2\sqrt{\epsilon/M_2}$. It can be verified that $g''(h) > 0$ for this value of $h$; therefore we obtain the optimal value

$$h_o = 2\sqrt{\frac{\epsilon}{M_2}} \qquad (5.2.19)$$

It can be verified that Equation (5.2.19) also gives the optimal step size $h_o$ for the backward difference formula [Equation (5.2.16)]. Thus, there is no sense in choosing $h$ much smaller than $h_o$.

**EXAMPLE 5.2.2**

Find the optimal step size $h_o$ for Example 5.2.1.

Since $f''(x) = 2\sec^2 x \tan x$, it can be verified that $M_2 = \max_{x\in[1,1.1]}|f''(x)| \approx 19.09851$. $M_2$ depends on the selection of the interval. Further, the values of $f$ are given to five decimal places, therefore it is reasonable to assume that $\epsilon = 0.000005$. Substituting the values of $M_2$ and $\epsilon$ in Equation (5.2.19) yields

$$h_o = 2\sqrt{\frac{0.000005}{19.09851}} \approx 0.00102$$

The step size $h = 0.001$ is close to the optimal value $h_o$ for the forward and backward difference formulas and it gives the best approximation of $f'(1)$ for the various values of $h$.   ∎ ∎ ∎

The truncation error could be reduced further by using higher-degree polynomials. Consider the representation of $f$ whose values are known at equally-spaced nodal points $x_0, x_1 = x_0 + h$, and $x_2 = x_1 + h = x_0 + 2h$. One can extend the previous work using the quadratic Lagrange interpolating polynomial as an approximation function of $f$. There is another simple procedure by which higher-order formulas can be derived at a given nodal point. Expanding $f(x_1) = f(x_0 + h)$ and $f(x_2) = f(x_0 + 2h)$ about $x_0$ by Taylor's Theorem (Theorem A.8) results in

$$f(x_0 + h) = f(x_0) + hf'(x_0) + \frac{h^2}{2!}f''(x_0) + \frac{h^3}{3!}f^{(3)}(\xi_1) \qquad (5.2.20)$$

and

$$f(x_0 + 2h) = f(x_0) + 2hf'(x_0) + \frac{4h^2}{2!}f''(x_0) + \frac{8h^3}{3!}f^{(3)}(\xi_2) \qquad (5.2.21)$$

where $x_0 < \xi_1 < x_0 + h$ and $x_0 < \xi_2 < x_0 + 2h$. We are expanding at $x_1 = x_0 + h$ and $x_2 = x_0 + 2h$ since the values of $f$ are known at those nodal points. We want the formula for $f'(x_0)$ and therefore we include the higher-order term $f''(x_0)$ in Equations (5.2.20) and (5.2.21) with the hope that the term in $f''(x_0)$ will be eliminated between Equation (5.2.20) and (5.2.21). If we can manage to eliminate more higher-order terms, then we get higher-order formulas. Multiplying Equation (5.2.20) by 4 and Equation

(5.2.21) by $-1$ and then adding,

$$4f(x_0 + h) - f(x_0 + 2h) = 3f(x_0) + 2hf'(x_0)$$
$$+ \frac{h^3}{3!}(4f^{(3)}(\xi_1) - 8f^{(3)}(\xi_2)) \tag{5.2.22}$$

Assume that $f^{(3)}$ is continuous on $[x_0, x_0 + 2h]$. Then using Corollary A.2.2

$$4f^{(3)}(\xi_1) + 4f^{(3)}(\xi) = 8f^{(3)}(\xi_2) \tag{5.2.23}$$

where $x_0 < \xi < x_0 + 2h$. Using Equation (5.2.23) in Equation (5.2.22) yields

$$4f(x_0 + h) - f(x_0 + 2h) = 3f(x_0) + 2hf'(x_0) - \frac{4h^3}{3!}f^{(3)}(\xi) \tag{5.2.24}$$

Solving Equation (5.2.24) for $f'(x_0)$,

$$f'(x_0) = \frac{-f(x_0 + 2h) + 4f(x_0 + h) - 3f(x_0)}{2h} + \frac{h^2}{3}f^{(3)}(\xi) \tag{5.2.25}$$

Since $x_0 + h$ and $x_0 + 2h$ are forward points from $x_0$, Equation (5.2.25) is called the **three-point forward difference formula**. We could also get the four- or five-point forward difference formula.

Assume that the values of $f$ at $x_0$, $x_0 - h$, and $x_0 - 2h$ are known. Then we can represent $f$ by a quadratic Lagrange interpolating polynomial and get the **three-point backward difference formula**

$$f'(x_0) = \frac{f(x_0 - 2h) - 4f(x_0 - h) + 3f(x_0)}{2h} + \frac{h^2}{3}f^{(3)}(\xi) \tag{5.2.26}$$

using the Lagrange Polynomial Approximation Theorem 4.3.1. Assume that $f$ is known at $x_0 - h$, $x_0$, and $x_0 + h$. Then expanding $f(x_0 - h)$ and $f(x_0 + h)$ about $x_0$ by Taylor's Theorem (Theorem A.8), we get

$$f(x_0 - h) = f(x_0) - hf'(x_0) + \frac{h^2}{2!}f''(x_0) - \frac{h^3}{3!}f^{(3)}(\xi_1) \tag{5.2.27}$$

and

$$f(x_0 + h) = f(x_0) + hf'(x_0) + \frac{h^2}{2!}f''(x_0) + \frac{h^3}{3!}f^{(3)}(\xi_2) \tag{5.2.28}$$

where $x_0 - h < \xi_1 < x_0$ and $x_0 < \xi_2 < x_0 + h$. Subtracting Equation (5.2.27) from Equation (5.2.28) results in

$$f(x_0 + h) - f(x_0 - h) = 2hf'(x_0) + \frac{h^3}{3!}(f''(\xi_1) + f''(\xi_2)) \tag{5.2.29}$$

Using Corollary A.2.2 in Equation (5.2.29) and then solving Equation (5.2.29) for $f'(x_0)$, yields

$$f'(x_0) = \frac{f(x_0 + h) - f(x_0 - h)}{2h} - \frac{h^2}{6}f^{(3)}(\xi) \tag{5.2.30}$$

where $x_0 - h < \xi < x_0 + h$. This is called the **central difference formula**. Thus for small $h$, use the formula

$$f'(x_0) \approx \frac{f(x_0 + h) - f(x_0 - h)}{2h} \tag{5.2.31}$$

with an error bound $h^2 M_3/6$, where $\left| f^{(3)}(x) \right| \leq M_3$ for all $x$ in $[x_0 - h, x_0 + h]$. If neither $f''(x)$ nor $f^{(3)}(x)$ is large, then for $h < 1$ the truncation error for the formula of Equation (5.2.31) will be considerably smaller than the one given by Equation (5.2.14) or Equation (5.2.16).

Since we subtract almost equal numbers in Equation (5.2.31), round-off error becomes dominant. Like the forward difference formula given by Equation (5.2.14), error analysis can be carried out for Equation (5.2.31). It can be verified that the optimal step size $h_o$ for the central difference formula is given by (Exercise 13)

$$h_o = \sqrt[3]{\frac{3\epsilon}{M_3}} \tag{5.2.32}$$

EXAMPLE 5.2.3    Let $f(x) = \tan x$. Find $f'(1)$ using Equation (5.2.31) and compare it with the forward difference values in Table 5.2.1. Use five decimal places or six significant digits in all calculations.

From Equation (5.2.31),

$$f'(1) \approx \frac{f(1 + h) - f(1 - h)}{2h} = \frac{\tan(1 + h) - \tan(1 - h)}{2h}$$

For $h = 0.1$, using five decimal places in our calculations, we get

$$f'(1) \approx \frac{\tan(1.0 + 0.1) - \tan(1.0 - 0.1)}{2(0.1)} = \frac{1.96476 - 1.26016}{0.2} = 3.52300$$

The results are computed using five decimal places in all calculations and are given in Table 5.2.2.

**Table 5.2.2**

| $h$ | $\dfrac{\tan(1 + h) - \tan(1 - h)}{2h}$ | \|Actual Error\| | $\dfrac{\tan(1 + h) - \tan 1}{h}$ | \|Actual Error\| |
|---|---|---|---|---|
| 0.100000 | 3.52300 | 0.09747 | 4.07350 | 0.64797 |
| 0.010000 | 3.42650 | 0.00097 | 3.48000 | 0.05447 |
| 0.001000 | 3.42500 | 0.00053 | 3.43000 | 0.00447 |
| 0.000100 | 3.40000 | 0.02553 | 3.40000 | 0.02553 |
| 0.000010 | 3.50000 | 0.07447 | 3.00000 | 0.42553 |
| 0.000001 | 0.00000 | 3.42553 | 0.00000 | 3.42553 |

■ ■ ■

## EXERCISES

1. Let $f(x) = \sin x$.
   (a) Approximate $f'(1)$ using Equations (5.2.14) and (5.2.31) with $h = 0.1, 0.01,$ $0.001, 0.0001,$ and $0.00001$. Use five decimal places in all calculations.
   (b) Compute the absolute value of the actual error in each case.
2. Find the optimal step size $h_o$ for Exercise 1(a) and check it with your calculations.
3. Let $f(x) = e^x$.
   (a) Approximate $f'(1)$ using Equations (5.2.16) and (5.2.31) with $h = 0.1, 0.01,$ $0.001, 0.0001,$ and $0.00001$. Use five decimal places in all calculations.
   (b) Compute the absolute value of the actual error in each case.
4. Find the optimal step size $h_o$ for Exercise 3(a) and check it with your calculations.
5. Let $f(x) = x^3$.
   (a) Approximate $f'(1)$ using Equations (5.2.14) and (5.2.31) with $h = 0.1, 0.01,$ $0.001, 0.0001,$ and $0.00001$. Use five decimal places in all calculations.
   (b) Compute the absolute value of the actual error in each case.
6. Find the optimal step size $h_o$ for Exercise 5(a) and check it with your calculations.
7. The following table gives the values of the Bessel function of order zero:

| $x_i$ | 5.3 | 5.4 | 5.5 | 5.6 | 5.7 | 5.8 |
|---|---|---|---|---|---|---|
| $J_0(x_i)$ | $-0.07580$ | $-0.04121$ | $-0.00684$ | 0.02697 | 0.05972 | 0.09170 |

   Approximate $J_0'(x)$ for $x = 5.3, 5.5,$ and $5.8$ as accurately as possible.
8. An object weighing 4 lb is dropped from a balloon that is 1500 ft above the ground. The following table gives the data from observation:

| Time (t) | 0 | 1 | 2 | 3 | 4 | 5 |
|---|---|---|---|---|---|---|
| Distance (ft) | 1500 | 1483 | 1430 | 1336 | 1195 | 1002 |

   Find the speed of the object at time $t = 0, 3,$ and $5$ as accurately as possible.
9. Prove Equation (5.2.11).
10. Derive Equation (5.2.13) using Taylor's Theorem (Theorem A.8).
11. Derive Equation (5.2.26) using Taylor's Theorem (Theorem A.8).
12. Derive Equations (5.2.26) and (5.2.30) using the quadratic Lagrange interpolating polynomial.
13. Verify that the optimal step size for the central difference formula

$$f'(x_0) = \frac{f(x_0 + h) - f(x_0 - h)}{2h} - \frac{h^2}{6}f^{(3)}(\xi)$$

is given by

$$h_o = \sqrt[3]{\frac{3\epsilon}{M_3}}$$

where $|f^{(3)}(\xi)| \le M_3$ for all $x$ in $[x_0 - h, x_0 + h]$ and $f(x)$ can be evaluated within $\pm\, \epsilon$.

14. Derive the five-point formula

$$f'(x_0) = \frac{1}{12h} [f(x_0 - 2h) - 8f(x_0 - h)$$

$$+ 8f(x_0 + h) - f(x_0 + 2h)] + \frac{h^4}{30} f^{(5)}(\xi)$$

where $\xi$ lies between $x_0 - 2h$ and $x_0 + 2h$. (*Hint*: Expand $f(x_0 - 2h), f(x_0 - h)$, $f(x_0 + h)$, and $f(x_0 + 2h)$ about $x_0$ using Taylor's Theorem.)

15. The partial derivatives of $f(x, y)$ with respect to $x$ and $y$ are the functions $f_x(x, y)$ and $f_y(x, y)$, defined by the limits

$$f_x(x, y) = \lim_{h \to 0} \frac{f(x + h, y) - f(x, y)}{h} \quad \text{and} \quad f_y(x, y) = \lim_{h \to 0} \frac{f(x, y + h) - f(x, y)}{h}$$

Equation (5.2.31) can be adapted to partial derivatives as

$$f_x(x, y) \approx \frac{f(x + h, y) - f(x - h, y)}{2h} \quad \text{and}$$

$$f_y(x, y) \approx \frac{f(x, y + h) - f(x, y - h)}{2h} \quad \text{(5.2.33)}$$

(a) Let $f(x, y) = x^2 e^y$. Approximate $f_x(1, 2)$ and $f_y(1, 2)$ using Equation (5.2.33) with $h = 0.1, 0.01, 0.001$, and $0.00001$. Use five decimal places in all calculations. Compare the actual error in each case.

(b) Let $f(x, y) = \cos (x/y)$. Approximate $f_x(1, 2)$ and $f_y(1, 2)$ using Equation (5.2.33) with $h = 0.1, 0.01, 0.001$, and $0.00001$. Use five decimal places in all calculations. Compare the actual error in each case.

16. Divide Equation (5.2.6) by $(x - x_0)(x - x_1)$ and then compare it with Equation (5.2.4). Verify that

$$\frac{f''(\xi(x))}{2!} - \frac{f(x_0)}{(x_0 - x_1)(x_0 - x_2)} - \frac{f(x_1)}{(x_1 - x_0)(x_1 - x_2)}$$

$$- \frac{f(x_2)}{(x_2 - x_0)(x_2 - x_1)} = (x - x_2) \frac{f^{(3)}(\eta(x))}{3!}$$

17. Prove that $\dfrac{f^{(n)}(\xi(x))}{n!} + c = (x - x_n) \dfrac{f^{(n+1)}(\eta(x))}{(n + 1)!}$ where $c$ is a fixed constant.

    (*Hint*: This is similar to Exercise 16.)

18. Find the optimal step size $h_o$ for the three-point forward difference formula [Equation (5.2.25)].

**COMPUTER EXERCISES**

C.1. Write a subroutine to find the first derivative at a number $a$ using (a) the forward difference formula [Equation (5.2.14)], (b) the backward difference formula

[Equation (5.2.16)], and (c) the central difference formula [Equation (5.2.31)] for various $h$.

**C.2.**  Write a computer program to compute the first derivative at a number $a$ using (a) the forward difference formula [Equation 5.2.14)], (b) the backward difference formula [Equation (5.2.16)], and (c) the central difference formula [Equation (5.2.31)] using Exercise C.1.

**C.3.**  Let $f(x) = \sin x$. Find $f'(1)$ using Exercise C.2. Use $h$ as small as possible. Compute the actual error in each case. Find $\epsilon$ of your computer. Then use your $\epsilon$ to find (a) the optimal step size $h$ for the forward difference formula and the backward difference formula and (b) the optimal step size for the central difference formula. Compare your calculations for different $h$.

**C.4.**  Let $f(x) = e^x$. Find $f'(1)$ using Exercise C.2. Use $h$ as small as possible. Compute the actual error in each case. Find $\epsilon$ of your computer. Then use your $\epsilon$ to find (a) the optimal step size $h$ for the forward difference formula and the backward difference formula and (b) the optimal step size $h$ for the central difference formula. Compare your calculations for different $h$.

## 5.3    HIGHER DERIVATIVES

Formulas for higher derivatives can be obtained using the techniques employed to derive the formulas for the first derivative. For the second derivative, we need at least a second-degree polynomial since the second derivative of a first-degree polynomial is zero. Consider the representation of $f$ whose values are known at equally spaced nodal points $x_0$, $x_1 = x_0 + h$, and $x_2 = x_1 + h$. Then the Newton forward difference formula reduces to

$$f(x) = f(x_0) + \frac{\Delta f(x_0)}{1!\,h}(x - x_0) + \frac{\Delta^2 f(x_0)}{2!\,h^2}(x - x_0)(x - x_1)$$

$$+ \frac{f^{(3)}(\xi)}{3!}(x - x_0)(x - x_1)(x - x_2) \qquad (5.3.1)$$

where $x_0 < \xi < x_2$. Differentiating Equation (5.3.1) with respect to $x$ and using Equation (5.2.11) yields

$$f'(x) = \frac{\Delta f(x_0)}{h} + \frac{\Delta^2 f(x_0)}{2!\,h^2}[(x - x_0) + (x - x_1)]$$

$$+ \frac{f^{(3)}(\xi)}{3!}[(x - x_1)(x - x_2) + (x - x_0)(x - x_2)$$

$$+ (x - x_0)(x - x_1)]$$

$$+ \frac{f^{(4)}(\eta(x))}{4!}(x - x_0)(x - x_1)(x - x_2) \qquad (5.3.2)$$

where $x_0 < \eta < x_2$. We want the formula for the second derivative; therefore, differentiate Equation (5.3.2) one more time with respect to $x$ and use Equation (5.2.11) to get

$$f''(x) = \frac{\Delta^2 f(x_0)}{h^2} + \frac{f^{(3)}(\xi)}{3}[(x - x_0) + (x - x_1) + (x - x_2)]$$

$$+ \frac{(f^{(4)}(\eta) + f^{(4)}(\xi_1))}{4!}[(x - x_1)(x - x_2) + (x - x_0)(x - x_2)$$

$$+ (x - x_0)(x - x_1)]$$

$$+ \frac{f^{(5)}(\eta_1(x))}{5!}(x - x_0)(x - x_1)(x - x_2) \qquad \textbf{(5.3.3)}$$

where $x_0 < \xi_1, \eta_1 < x_2$.

Since we want the derivative at the nodal point $x_0$, substituting $x = x_0$ in Equation (5.3.3) yields

$$f''(x_0) = \frac{\Delta^2 f(x_0)}{h^2} - hf^{(3)}(\xi) + \frac{2h^2}{4!}(f^{(4)}(\eta) + f^{(4)}(\xi_1)) \qquad \textbf{(5.3.4)}$$

Similarly substituting $x = x_1$ in Equation (5.3.3) results in

$$f''(x_1) = \frac{\Delta^2 f(x_0)}{h^2} - \frac{h^2}{4!}(f^{(4)}(\eta) + f^{(4)}(\xi_1)) \qquad \textbf{(5.3.5)}$$

If $f^{(4)}$ is continuous on $[x_0, x_0 + 2h]$, then by Corollary A.2.2, Equation (5.3.5) reduces to

$$f''(x_1) = \frac{\Delta^2 f(x_0)}{h^2} - \frac{h^2}{12}f^{(4)}(\xi) \qquad \textbf{(5.3.6)}$$

where $\xi$ is in $(x_0, x_0 + 2h)$. Substituting for $\Delta^2 f(x_0)$ from Section 4.5 in Equation (5.3.6) yields

$$f''(x_1) = \frac{f(x_0 + 2h) - 2f(x_0 + h) + f(x_0)}{h^2} - \frac{h^2}{12}f^{(4)}(\xi) \qquad \textbf{(5.3.7)}$$

In addition, we want to find the formula for $f''(x_0)$ and Equation (5.3.4) is not convenient because of the complicated truncation error terms. Hence replace $x_0 + h$ for $x_0 + 2h$, $x_0$ for $x_0 + h$, and $x_0 - h$ for $x_0$ in Equation (5.3.7). This leads to

$$f''(x_0) = \frac{f(x_0 + h) - 2f(x_0) + f(x_0 - h)}{h^2} - \frac{h^2}{12}f^{(4)}(\xi) \qquad \textbf{(5.3.8)}$$

where $\xi$ is in $(x_0 - h, x_0 + h)$. Higher derivative formulas can likewise be developed. We derive Equation (5.3.8) using Taylor's Theorem (Theorem A.8). Expanding $f(x_0 + h)$ and $f(x_0 - h)$ about $x_0$ by Taylor's Theorem,

$$f(x_0 + h) = f(x_0) + hf'(x_0) + \frac{h^2}{2!}f''(x_0) + \frac{h^3}{3!}f^{(3)}(x_0) + \frac{h^4}{4!}f^{(4)}(\xi_1) \qquad \textbf{(5.3.9)}$$

and

$$f(x_0 - h) = f(x_0) - hf'(x_0) + \frac{h^2}{2!}f''(x_0) - \frac{h^3}{3!}f^{(3)}(x_0) + \frac{h^4}{4!}f^{(4)}(\xi_2) \qquad \textbf{(5.3.10)}$$

where $x_0 < \xi_1 < x_0 + h$ and $x_0 - h < \xi_2 < x_0$. Adding Equations (5.3.9) and (5.3.10) results in

$$f(x_0 + h) + f(x_0 - h) = 2f(x_0) + h^2 f''(x_0) + \frac{h^4}{4!} (f^{(4)}(\xi_1) + f^{(4)}(\xi_2)) \qquad \textbf{(5.3.11)}$$

If $f^{(4)}$ is continuous on $[x_0 - h, x_0 + h]$, then by using Corollary A.2.2, Equation (5.3.11) can be written as

$$f''(x_0) = \frac{f(x_0 + h) - 2f(x_0) + f(x_0 - h)}{h^2} - \frac{h^2}{12} f^{(4)}(\xi)$$

where $\xi$ is in $(x_0 - h, x_0 + h)$. For small $h$, use the formula

$$f''(x_0) \approx \frac{f(x_0 + h) - 2f(x_0) + f(x_0 - h)}{h^2} \qquad \textbf{(5.3.12)}$$

with an error bound $h^2 M_4/12$ where $|f^{(4)}(x)| \leq M_4$ for all $x$ in $[x_0 - h, x_0 + h]$. For convenience, the formulas for the derivatives are given in Table 5.3.1. For more formulas, we refer the reader to Collatz (1966).

**Table 5.3.1**

$$f'(x_0) = \frac{f(x_0 + h) - f(x_0)}{h} - \frac{h}{2} f''(\xi)$$

$$f'(x_0) = \frac{f(x_0) - f(x_0 - h)}{h} + \frac{h}{2} f''(\xi)$$

$$f'(x_0) = \frac{f(x_0 + h) - f(x_0 - h)}{2h} - \frac{h^2}{6} f^{(3)}(\xi)$$

$$f'(x_0) = \frac{-3f(x_0) + 4f(x_0 + h) - f(x_0 + 2h)}{2h} + \frac{h^2}{3} f^{(3)}(\xi)$$

$$f'(x_0) = \frac{f(x_0 - 2h) - 8f(x_0 - h) + 8f(x_0 + h) - f(x_0 + 2h)}{12h} + \frac{h^4}{30} f^{(5)}(\xi)$$

$$f''(x_0) = \frac{f(x_0 - h) - 2f(x_0) + f(x_0 + h)}{h^2} - \frac{h^2}{12} f^{(4)}(\xi)$$

$$f''(x_0) = \frac{-f(x_0 - 2h) + 16f(x_0 - h) - 30f(x_0) + 16f(x_0 + h) - f(x_0 + 2h)}{12h^2} + \frac{h^4}{90} f^{(6)}(\xi)$$

$$f''(x_0) = \frac{2f(x_0) - 5f(x_0 + h) + 4f(x_0 + 2h) - f(x_0 + 3h)}{h^2} + \frac{11h^2}{12} f^{(4)}(\xi)$$

$$f^{(3)}(x_0) = \frac{-f(x_0 - 2h) + 2f(x_0 - h) - 2f(x_0 + h) + f(x_0 + 2h)}{2h^3} - \frac{h^2}{4} f^{(5)}(\xi)$$

$$f^{(4)}(x_0) = \frac{f(x_0 - 2h) - 4f(x_0 - h) + 6f(x_0) - 4f(x_0 + h) + f(x_0 + 2h)}{h^4} - \frac{h^2}{6} f^{(6)}(\xi)$$

**EXAMPLE 5.3.1**

For $f(x) = \tan x$, find $f''(1)$ using Equation (5.3.12) and compare it with the exact value. Use five decimal places or six significant digits in all calculations.

From Equation (5.3.12), for $h = 0.1$ we get

$$f''(1) \approx \frac{f(1 + h) - 2f(1) + f(1 - h)}{h^2} = \frac{\tan(1.1) - 2\tan(1) + \tan(0.9)}{(0.1)^2}$$

$$= \frac{1.96476 - 2(1.55741) + 1.26016}{0.01} = 11.01000$$

The calculations are carried out using five decimal places in all calculations and are shown in Table 5.3.2. Since $f'(x) = \sec^2 x$, $f''(x) = 2\sec^2 x \tan x$, then the exact value of $f''(1) = 2\sec^2 1 \tan 1 \approx 10.66986$. Also

$$|\text{actual error}| = \left| f''(1) - \frac{f(1 + h) - 2f(1) + f(1 - h)}{h^2} \right|$$

is calculated for various $h$ and is also given in Table 5.3.2.

**Table 5.3.2**

| $h$ | $\dfrac{\tan(1 + h) - 2\tan 1 + \tan(1 - h)}{h^2}$ | \|Actual Error\| |
|---|---|---|
| 0.1000 | 11.01000 | 0.34014 |
| 0.0250 | 10.69052 | 0.02066 |
| 0.0100 | 10.70000 | 0.03014 |
| 0.0010 | 10.00000 | 0.66986 |

Again, round-off errors are important. Assume that the round-off errors $e(x_0 + h)$, $e(x_0)$, and $e(x_0 - h)$ are introduced in computing $f(x_0 + h)$, $f(x_0)$, and $f(x_0 - h)$ when we use the computed values $\bar{f}(x_0 + h)$, $\bar{f}(x_0)$, and $\bar{f}(x_0 - h)$. These relationships are given by

$$f(x_0 + h) = \bar{f}(x_0 + h) + e(x_0 + h) \qquad f(x_0) = \bar{f}(x_0) + e(x_0)$$

$$f(x_0 - h) = \bar{f}(x_0 - h) + e(x_0 - h) \tag{5.3.13}$$

The error is computed using Equations (5.3.13) and (5.3.8):

$$\left| f''(x_0) - \frac{\bar{f}(x_0 + h) - 2\bar{f}(x_0) + \bar{f}(x_0 - h)}{h^2} \right|$$

$$= \left| f''(x_0) - \frac{f(x_0 + h) - 2f(x_0) + f(x_0 - h)}{h^2} \right.$$

$$\left. + \frac{e(x_0 + h) - 2e(x_0) + e(x_0 - h)}{h^2} \right|$$

$$= \left| -\frac{h^2}{12} f^{(4)}(\xi) + \frac{e(x_0 + h) - 2e(x_0) + e(x_0 - h)}{h^2} \right| \tag{5.3.14}$$

If we assume that the round-off errors $e(x_0 + h)$, $e(x_0)$, and $e(x_0 - h)$ are bounded by $\epsilon$ and $|f^{(4)}(x)| \leq M_4$ for all $x \in [x_0 - h, x_0 + h]$, then we get from Equation (5.3.14)

$$\left| f''(x_0) - \frac{\tilde{f}(x_0 + h) - 2\tilde{f}(x_0) + \tilde{f}(x_0 - h)}{h^2} \right| \leq \frac{M_4 h^2}{12} + \frac{4\epsilon}{h^2}$$

The optimum step size $h_o = 2\sqrt[4]{3\epsilon/M_4}$ will minimize the quantity $(M_4 h^2/12) + (4\epsilon/h^2)$.

**EXAMPLE 5.3.2**

Find the optimal step size $h_0$ for Example 5.3.1.

Since $f^{(4)}(x) = 2 \sec^2 x\, (2 \tan x + \sec^2 x + 4 \sec^2 \tan x)$, it can be verified that $M_4 = \max_{x \in [1,1.1]} |f^{(4)}(x)| \approx 456.74$. The value of $M_4$ depends on the interval. Since we are carrying five decimal places, it is reasonable to assume that $\epsilon = 0.000005$. Therefore, $h_o$ is given by

$$h_o = 2\sqrt[4]{\frac{3\epsilon}{M_4}} = 2\sqrt[4]{\frac{(3)(0.000005)}{456.74}} \approx 0.02692$$

The step size $h = 0.025$ is close to $h_o$ among the various values of $h$ and it gives the best approximation to $f''(1)$, which is not very accurate. ■ ■ ■

Small values of $h$ needed to reduce truncation error cause the round-off error to grow. Thus, we have encountered the process where the reduction of $h$ does not mean better accuracy. If possible, avoid numerical differentiation. We derived numerical differentiation formulas since we need them to approximate the solutions of ordinary and partial differential equations.

## 5.4  RICHARDSON'S EXTRAPOLATION

In general, it is not possible to reduce the total error in numerical differentiation by simply decreasing the size of $h$. However, it is possible to improve the accuracy by Richardson's extrapolation procedure. This is similar to Aitken's $\Delta^2$ process but does not resort to a more complicated formula. We compute several estimates of the required quantity and then combine these estimates to provide a better estimate.

Consider Equation (5.2.13), which can be written as

$$\frac{f(x_0 + h) - f(x_0)}{h} = f'(x_0) + \frac{h}{2}f''(\xi) \tag{5.4.1}$$

where $x_0 < \xi < x_0 + h$. We are approximating the unknown quantity $f'(x_0)$ by $D(h) = [(f(x_0 + h) - f(x_0))/h]$. Thus Equation (5.4.1) can be written as

$$D(h) = a + bh \tag{5.4.2}$$

where $a = f'(x_0)$ and $b = f''(\xi(h))/2$.

The basic idea is to compute two different approximations $D(h_1)$ and $D(h_2)$ and use them to eliminate $b$ to obtain a better approximation of $f'(x_0)$. We assume that $b$ is independent of $h$ for small values of $h$. Let

$$D(h_1) = a + bh_1 \quad \text{and} \quad D(h_2) = a + bh_2 \tag{5.4.3}$$

Elimination of $b$ between these two equations gives

$$a = \frac{h_2 D(h_1) - h_1 D(h_2)}{h_2 - h_1} \tag{5.4.4}$$

If we let $h_1 = h$ and $h_2 = rh$, then Equation (5.4.4) reduces to

$$a = \frac{rD(h) - D(rh)}{r - 1} \tag{5.4.5}$$

If $r = \frac{1}{2}$, then Equation (5.4.5) further simplifies to

$$a = 2D(h/2) - D(h) \tag{5.4.6}$$

Define $D_1(h) = 2D(h/2) - D(h)$. Then $D_1(h)$ may be expected to be a better approximation of $a$ than either $D(h)$ or $D(h/2)$. Since $b$ is not generally independent of $h$, $D_1(h)$ does not give the exact value of $f'(x_0)$. If $h$ is too small, then the computed values of $D(h)$ and $D(h/2)$ will be inaccurate and so will $D_1(h)$.

**EXAMPLE 5.4.1**

Approximate $f'(1.0)$ from the given table of $f(x) = \tan x$ using Richardson's extrapolation technique. Use five decimal places or six significant digits in all calculations.

$$D(0.4) = \frac{f(1.4) - f(1)}{0.4} = \frac{\tan(1.4) - \tan 1}{0.4} = \frac{5.79788 - 1.55741}{0.4}$$

$$= \frac{4.24047}{0.4} = 10.60118$$

Similarly $D(0.2)$, $D(0.1)$, $D(0.05)$, $D(0.025)$, $D(0.0125)$, and $D(0.00625)$ are computed and shown in Table 5.4.1.

Since $D_1(h) = 2D(h/2) - D(h)$, $D_1(0.4) = 2D(0.2) - D(0.4) = 2(5.07370) - 10.60118 = -0.45378$. Similarly the values $D_1(0.2)$, $D_1(0.1)$, $D_1(0.05)$, $D_1(0.025)$, and $D_1(0.0125)$ shown in Table 5.4.1 are also computed. Also $|\text{actual error}| = |\sec^2 1 -$

**Table 5.4.1**

| $h$ | $x$ | $f(x) = \tan x$ | $D(h)$ | $D_1(h)$ | $\vert$Actual Error$\vert$ |
|---|---|---|---|---|---|
| | 1.00000 | 1.55741 | | | |
| 0.40000 | 1.40000 | 5.79788 | 10.60118 | $-0.45378$ | 3.87931 |
| 0.20000 | 1.20000 | 2.57215 | 5.07370 | 3.07330 | 0.35223 |
| 0.10000 | 1.10000 | 1.96476 | 4.07350 | 3.36290 | 0.06263 |
| 0.05000 | 1.05000 | 1.74332 | 3.71820 | 3.41140 | 0.01413 |
| 0.02500 | 1.02500 | 1.64653 | 3.56480 | 3.42240 | 0.00313 |
| 0.01250 | 1.01250 | 1.60108 | 3.49360 | 3.42414 | 0.00139 |
| 0.00625 | 1.00625 | 1.57903 | 3.45887 | | |

$D_1(h)|$ where $\sec^2 1 = 3.42553$ to five places is given in Table 5.4.1. Compare the values in Table 5.4.1 with those in Table 5.2.1.   ■ ■ ■

This technique can also be used to derive higher-order formulas for the derivative. Consider

$$f(x_0 + h) = f(x_0) + hf'(x_0) + \frac{h^2}{2!}f''(x_0) + \frac{h^3}{3!}f^{(3)}(\xi_1(h)) \qquad (5.4.7)$$

where $x_0 < \xi_1 < x_0 + h$. Solving Equation (5.4.7) for $f'(x_0)$ gives

$$f'(x_0) = \frac{f(x_0 + h) - f(x_0)}{h} - \frac{h}{2}f''(x_0) - \frac{h^2}{3!}f^{(3)}(\xi_1) \qquad (5.4.8)$$

Since $h/2$ is not appropriate for a nodal point, replace $h$ by $2h$ in Equation (5.4.8). This gives

$$f'(x_0) = \frac{f(x_0 + 2h) - f(x_0)}{2h} - hf''(x_0) - \frac{4h^2}{3!}f^{(3)}(\xi_2) \qquad (5.4.9)$$

where $x_0 < \xi_2 < x_0 + 2h$. Eliminating $f''(x_0)$ between Equations (5.4.8) and (5.4.9), we get

$$f'(x_0) = \frac{-f(x_0 + 2h) + 4f(x_0 + h) - 3f(x_0)}{2h}$$
$$+ \frac{h^3}{3}(2f^{(3)}(\xi_2) - f^{(3)}(\xi_1)) \qquad (5.4.10)$$

If $f^{(3)}$ is continuous on $[x_0, x_0 + 2h]$, then it can be shown that $f^{(3)}(\xi_1) + f^{(3)}(\xi) = 2f^{(3)}(\xi_2)$ where $x_0 < \xi < x_0 + 2h$ using Corollary A.2.2. This reduces Equation (5.4.10) to

$$f'(x_0) = \frac{-f(x_0 + 2h) + 4f(x_0 + h) - 3f(x_0)}{2h} + \frac{h^3}{3}f^{(3)}(\xi) \qquad (5.4.11)$$

This is the three-point forward difference formula. Other formulas can be derived in a similar manner. The concept of the Richardson method will also be used in approximating integrals in Section 6.5.

## EXERCISES

1. Let $f(x) = \sin x$.
   (a) Use Equation (5.3.12) with $h = 0.1, 0.01$, and $0.005$ to approximate $f''(1)$ by carrying five decimal places in all calculations.
   (b) Compute the actual error in each case.
2. Find the optimal step size $h_o$ for Exercise 1(a) and check it with your calculations.
3. Let $f(x) = e^x$.
   (a) Use Equation (5.3.12) with $h = 0.1, 0.01$, and $0.005$ to approximate $f''(0)$ by carrying five decimal places in all calculations.
   (b) Compute the actual error in each case.

4. Find the optimal step size $h_o$ for Exercise 3(a) and check it with your calculations.
5. Let $f(x) = x^3$.
   (a) Use Equation (5.3.12) with $h = 0.1, 0.01,$ and $0.005$ to approximate $f''(0)$ by carrying five decimal places in all calculations.
   (b) Compute the actual error in each case.
6. Find the optimal step size $h_o$ for Exercise 5(a) and check it with your calculations.
7. The following table gives the values of the Bessel function of order zero:

| $x_i$ | 5.3 | 5.4 | 5.5 | 5.6 | 5.7 | 5.8 |
|---|---|---|---|---|---|---|
| $J_0(x_i)$ | −0.07580 | −0.04121 | −0.00684 | 0.02697 | 0.05972 | 0.09170 |

Approximate $J_0''(x)$ for $x = 5.3, 5.5,$ and $5.8$ as accurately as possible.

8. Approximate $f'(1)$ from the given table of $f(x) = \sin x$ using the Richardson extrapolation technique. The inherent round-off error has the bound $|\epsilon| \le 5 \times 10^{-6} = 0.000005$. Use the rounded values in your calculations. Compare the approximated values with the exact value of $f'(1)$.

| $x$ | $f(x) = \sin x$ |
|---|---|
| 1.4000 | 0.98545 |
| 1.2000 | 0.93204 |
| 1.1000 | 0.89121 |
| 1.0500 | 0.86742 |
| 1.0250 | 0.85471 |
| 1.0125 | 0.84816 |
| 1.0000 | 0.84147 |

9. Approximate $f'(3)$ from the given table of $f(x) = \ln x$ using the Richardson extrapolation technique. The inherent round-off error has the bound $|\epsilon| \le 5 \times 10^{-6} = 0.000005$. Use the rounded values in your calculations. Compare the approximated values with the exact value of $f'(3)$.

| $x$ | $f(x) = \ln x$ |
|---|---|
| 3.4000 | 1.22378 |
| 3.2000 | 1.16315 |
| 3.1000 | 1.13140 |
| 3.0500 | 1.11514 |
| 3.0250 | 1.10691 |
| 3.0125 | 1.10277 |
| 3.0000 | 1.09861 |

**10.** Use Taylor's Theorem to derive

$$f''(x_0) = \frac{-f(x_0 - 2h) + 16f(x_0 - h) - 30f(x_0) + 16f(x_0 + h) - f(x_0 + 2h)}{12h^2}$$

$$+ \frac{h^4}{90} f^{(6)}(\xi)$$

**11.** Find the optimal step size $h_o$ for the formula in Exercise 10.

**12.** Use the Richardson extrapolation technique to derive the five-point formula

$$f'(x_0) = \frac{1}{12h} [f(x_0 - 2h) - 8f(x_0 - h) + 8f(x_0 + h) - f(x_0 + 2h)]$$

$$+ \frac{h^4}{30} f^{(5)}(\xi)$$

**13.** Find the truncation error if we use the approximation

$$f^{(3)}(x_0) \approx \frac{f(x_0 + 3h) - 3f(x_0 + 2h) + 3f(x_0 + h) - f(x_0)}{h^3}$$

**14.** Verify that the sum of the coefficients is zero in all approximation formulas given in Table 5.3.1.

## COMPUTER EXERCISES

**C.1.** Write a subroutine to approximate the second derivative at a number $a$ using Equation (5.3.12) for various $h$.

**C.2.** Write a computer program to compute the second derivative at a number $a$ using Exercise C.1.

**C.3.** Let $f(x) = \sin x$. Find $f''(1)$ using Exercise C.2. Use $h$ as small as possible. Compute the actual error in each case. Find $\epsilon$ for your computer. Then use your $\epsilon$ to find the optimal step size for Equation (5.3.12). Compare your calculations for different $h$.

**C.4.** Let $f(x) = e^x$. Find $f''(1)$ using Exercise C.2. Use $h$ as small as possible. Compute the actual error in each case. Find $\epsilon$ of your computer. Then use your $\epsilon$ to find the optimal step size for Equation (5.3.12). Compare your calculations for different $h$.

## SUGGESTED READING

L. Collatz, *The Numerical Treatment of Differential Equations*, 3rd ed., Springer–Verlag, New York, 1966.

# Chapter

# 6 NUMERICAL INTEGRATION

The normalized cumulative distribution function

$$\Phi(x) = \int_{-\infty}^{x} e^{-(t^2/2)} \, dt$$

is very important in statistics. Since $\int e^{-(t^2/2)} \, dt$ is not expressible in terms of elementary functions, $\Phi(x)$ is obtained from tables. In this chapter, we discuss methods that compute $\Phi(x)$.

## 6.1  INTRODUCTION

Sometimes it is not possible to evaluate a definite integral exactly by well-known techniques; therefore numerical methods are used. Also, sometimes $f(x)$ is not known analytically; in other words, it is given by a table. Numerical integration, sometimes called quadrature, does not have the instability problem that numerical differentiation has. We seek numerical procedures to evaluate the definite integral given by

$$I(f, a, b) = \int_{a}^{b} f(x) \, dx \tag{6.1:1}$$

where $f$ is continuous or Riemann integrable on $[a, b]$. We look for $\Sigma_{i=0}^{N} c_i f(x_i)$ to approximate Equation (6.1.1). The $(N + 1)$ distinct points $a = x_0 < x_1 < \cdots < x_N = b$ are called the quadrature points or nodes, and the quantities $c_i$ are called the weights or coefficients. The basic problem is the selection of the nodes and coefficients so that

$$\left| \int_{a}^{b} f(x) \, dx - \sum_{i=o}^{n} c_i f(x_i) \right|$$

will be minimum for a large class of functions.

The classical method is to replace the function $f$ by a polynomial that can be easily integrated. The use of the interpolating polynomial as the replacement function leads to a variety of integration formulas, called the **Newton–Cotes formulas**. The interpolating polynomial can be obtained using a Newton divided difference formula for the interpolating polynomial or Lagrange interpolating polynomial. We use a Lagrange interpolating polynomial for our derivation.

## 6.2    THE TRAPEZOIDAL RULE

One of the simplest formulas to use is the first-degree Lagrange interpolating polynomial. Let $x_0 = a$ and $x_1 = b$. Then by the Lagrange Polynomial Approximation Theorem 4.3.1,

$$f(x) = f(a)\frac{x - b}{a - b} + f(b)\frac{x - a}{b - a} + \frac{f''(\xi)}{2!}(x - a)(x - b) \qquad \textbf{(6.2.1)}$$

where $a < \xi(x) < b$. Integration of Equation (6.2.1) from $a$ to $b$ gives

$$\int_a^b f(x)\,dx = -\frac{f(a)}{(b - a)}\int_a^b (x - b)\,dx + \frac{f(b)}{(b - a)}\int_a^b (x - a)\,dx$$

$$+ \frac{1}{2!}\int_a^b f''(\xi)(x - a)(x - b)\,dx$$

$$= -\frac{f(a)}{(b - a)}\frac{(x - b)^2}{2}\bigg|_a^b + \frac{f(b)}{(b - a)}\frac{(x - a)^2}{2}\bigg|_a^b$$

$$+ \frac{1}{2}\int_a^b f''(\xi)(x - a)(x - b)\,dx$$

$$= \frac{(b - a)}{2}[f(a) + f(b)] + \frac{1}{2}\int_a^b f''(\xi)(x - a)(x - b)\,dx \qquad \textbf{(6.2.2)}$$

The last term in Equation (6.2.2) is the error term. In the error term, $(x - a)(x - b) \le 0$ for all $x$ in $[a, b]$. Therefore, using the Second Mean Value Theorem for the Definite Integrals (Theorem A.5), the error term can be written as

$$\int_a^b f''(\xi)(x - a)(x - b)\,dx = f''(c)\int_a^b (x - a)(x - b)\,dx \qquad \textbf{(6.2.3)}$$

where $a < c < b$. Let $u = x - a$ and $h = b - a$. Then

$$\int_a^b (x - a)(x - b)\,dx = \int_0^h u(u - h)\,du$$

$$= \left[\frac{u^3}{3} - h\frac{u^2}{2}\right]\bigg|_0^h = \frac{h^3}{3} - \frac{h^3}{2} = -\frac{h^3}{6} \qquad \textbf{(6.2.4)}$$

Substituting the value of the integral from Equation (6.2.4) in Equation (6.2.3) yields

$$\int_a^b f''(\xi)(x - a)(x - b)\, dx = -\frac{h^3}{6} f''(c) \tag{6.2.5}$$

Substitution of Equation (6.2.5) in Equation (6.2.2) gives the **trapezoidal rule with error term**

$$\int_a^b f(x)\, dx = \frac{h}{2}[f(a) + f(b)] - \frac{h^3}{12} f''(c) \tag{6.2.6}$$

where $a < c < b$.

The error term in the trapezoidal rule is zero when $f''(c) = 0$. If $f(x)$ is linear, then the trapezoidal rule gives the exact value. If the error term is small, then the approximation of the definite integral is given by

$$\int_a^b f(x)\, dx \approx \frac{(b - a)}{2}[f(a) + f(b)] \tag{6.2.7}$$

Geometrically, when $f$ is a function with positive values, $(b - a)[f(a) + f(b)]/2$ represents the area of the trapezoid as shown in Figure 6.2.1. Therefore Equation (6.2.7) is called the **trapezoidal rule**.

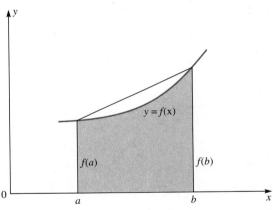

**Fig. 6.2.1**

If $h = b - a$ is not sufficiently small, then Equation (6.2.7) is not of much use because the truncation error can be large. In such cases, divide the integral into a sum of integrals over small intervals and then apply Equation (6.2.7) to each of these smaller intervals. Let us divide a closed interval $[a, b]$ into $N$ equally-spaced subintervals of the form $[x_0, x_1], [x_1, x_2], \cdots, [x_{N-1}, x_N]$ such that $a = x_0 < x_1 < \cdots < x_{N-1} < x_N = b$ as shown in Figure 6.2.2.

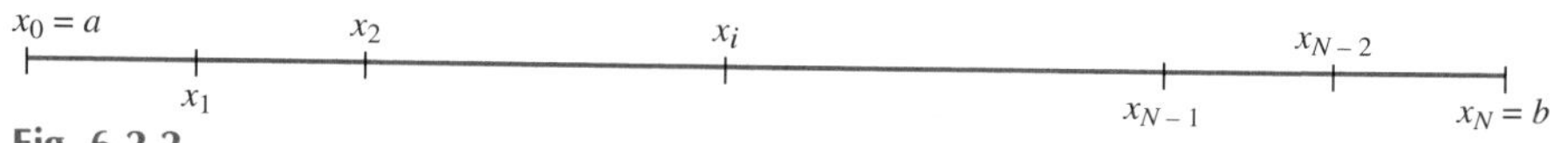

**Fig. 6.2.2**

The length of each subinterval $[x_{i-1}, x_i]$ is denoted by $h$; therefore $h = (b - a)/N$ and $x_i = a + ih$ for $i = 0, 1, ..., N$. Then

$$\int_a^b f(x)\, dx = \int_{x_0}^{x_1} f(x)\, dx + \int_{x_1}^{x_2} f(x)\, dx + \cdots + \int_{x_{N-1}}^{x_N} f(x)\, dx$$

$$= \frac{h}{2}[f(x_0) + f(x_1) + f(x_1) + f(x_2) + \cdots + f(x_{N-1}) + f(x_N)]$$

$$- \frac{h^3}{12}[f''(c_1) + f''(c_2) + \cdots + f''(c_N)]$$

$$= \frac{h}{2}\left[f(a) + 2\sum_{i=1}^{N-1} f(x_i) + f(b)\right] - \frac{h^3}{12}\sum_{i=1}^{N} f''(c_i) \qquad \textbf{(6.2.8)}$$

where $x_{i-1} < c_i < x_i$. Let $f''(x)$ be continuous on $[a, b]$. Then by Corollary A.2.2, there exists $\mu$ in $[a, b]$ such that $\sum_{i=1}^{N} f''(c_i) = Nf''(\mu)$. Thus the last term of Equation (6.2.8) simplifies to

$$\frac{h^3}{12}\sum_{i=1}^{N} f''(c_i) = \frac{h^3}{12}Nf''(\mu) = \frac{h^2(b - a)}{12}f''(\mu)$$

Equation (6.2.8) reduces to

$$\int_a^b f(x)\, dx = \frac{h}{2}\left[f(a) + 2\sum_{i=1}^{N-1} f(x_i) + f(b)\right] - \frac{h^2(b - a)}{12}f''(\mu) \qquad \textbf{(6.2.9)}$$

where $a < \mu < b$. Equation (6.2.9) is called the **composite trapezoidal rule with error term**. If the error term is reasonably small, then

$$\int_a^b f(x)\, dx \approx \frac{h}{2}\left[f(a) + 2\sum_{i=1}^{N-1} f(x_i) + f(b)\right] \qquad \textbf{(6.2.10)}$$

Sometimes the composite trapezoidal rule given by Equation (6.2.10) is also referred to as the *trapezoidal rule*.

**EXAMPLE 6.2.1**    Use the composite trapezoidal rule to approximate the definite integral $\int_0^1 dx/(1 + x^2)$.
Divide the interval $[0, 1]$ into two equal intervals. Then $N = 2$, so $h = (1 - 0)/2 = \frac{1}{2}$, $x_0 = 0$, $x_1 = \frac{1}{2}$, and $x_2 = 1$. For $N = 2$, Equation (6.2.10) becomes

$$\int_0^1 f(x)\, dx \approx \frac{h}{2}[f(x_0) + 2f(x_1) + f(x_2)]$$

and therefore

$$\int_0^1 \frac{dx}{1 + x^2} \approx \frac{1/2}{2}\left[\frac{1}{1 + 0} + 2\frac{1}{1 + 1/4} + \frac{1}{1 + 1}\right] = 0.775$$

For $N = 4$, we get $h = (1 - 0)/4 = \frac{1}{4}$, $x_0 = 0$, $x_1 = \frac{1}{4}$, $x_2 = \frac{1}{2}$, $x_3 = \frac{3}{4}$, and $x_4 = 1$. For $N = 4$, Equation (6.2.10) becomes

$$\int_0^1 f(x)\,dx \approx \frac{h}{2}[f(x_0) + 2f(x_1) + 2f(x_2) + 2f(x_3) + f(x_4)]$$

and therefore

$$\int_0^1 \frac{dx}{1 + x^2} \approx \frac{1/4}{2}\left[\frac{1}{1 + 0} + \frac{2}{1 + (1/16)} + \frac{2}{1 + (1/4)} + \frac{2}{1 + (9/16)} + \frac{1}{1 + 1}\right]$$
$$\approx 0.78279$$

The calculations using Equation (6.2.10) for various values of $N$ are given in Table 6.2.1 along with the absolute value of the actual errors.

**Table 6.2.1**

| $N$ | Using Equation (6.2.10) | \|Actual Error\| | $E_N$ |
|-----|-------------------------|------------------|-------|
| 2   | 0.77500                 | 0.01039          | 0.04166 |
| 4   | 0.78279                 | 0.00260          | 0.01041 |
| 8   | 0.78474                 | 0.00065          | 0.00260 |
| 16  | 0.78523                 | 0.00016          | 0.00065 |
| 32  | 0.78535                 | 0.00004          | 0.00016 |
| 64  | 0.78538                 | 0.00001          | 0.00004 |
| 128 | 0.78539                 | 0.00000          | 0.00001 |

The exact value of the integral

$$\int_0^1 \frac{dx}{1 + x^2} = \tan^{-1} x \Big|_0^1 = \tan^{-1} 1 = \frac{\pi}{4} \approx 0.78539$$

The error in using the composite trapezoidal rule is $-(b - a)^3 f''(\mu)/12N^2$, where $0 < \mu < 1$. Since we do not know $\mu$, the most we can say is the absolute error is less than or equal to

$$E_N = \max_{0 \le x \le 1} \frac{|f''(x)|}{12N^2}$$

Since $f(x) = 1/(1 + x^2)$, it can be verified that

$$f'(x) = -\frac{2x}{(1 + x^2)^2} \qquad f''(x) = -\frac{2(1 - 3x^2)}{(1 + x^2)^3} \qquad f^{(3)}(x) = \frac{24x\,(1 - x^2)}{(1 + x^2)^4}$$

The maximum value of $|f''(x)|$ occurs at the critical points of $f''(x)$, which are given by $f^{(3)}(x) = 0$. Thus 0, 1, and $-1$ are the critical points of $f''(x)$. Hence max $|f''(x)| =$ max $\{2, \frac{1}{2}\} = 2$. The absolute value of the actual error is less than or equal to

$$E_N = \frac{2}{12N^2} = \frac{1}{6N^2} \tag{6.2.11}$$

The values of $E_N$ for various values of $N$ are also given in Table 6.2.1.    ■ ■ ■

**Algorithm**

**Subroutine** trap $(a, b, f, N)$
Comment: This program computes the definite integral $\int_a^b f(x)\, dx$ using the composite trapezoidal rule given by Equation (6.2.10) with $N \geq 2$.
1 **Print** $a, b, N$
2 $h = (b - a)/N$
3 sum $= 0.0$
4 **Repeat** steps 5–6 **with** $i = 1, 2, \ldots, N - 1$
    5 $x = a + ih$
    6 sum $=$ sum $+ f(x)$
7 deint $= h\,(f(a) + 2$ sum $+ f(b))/2$
8 **Print** deint, $a, b, N, h$    ◀

## The Method of Undetermined Coefficients

We can determine the coefficients $c_i$ such that

$$\int_a^b f(x)\, dx \cong \sum_{i=0}^{N} c_i f(x_i) \tag{6.2.12}$$

using the method of undetermined coefficients.
    For $N = 1$,

$$\int_a^b f(x)\, dx \cong c_0 f(a) + c_1 f(b) \tag{6.2.13}$$

Equation (6.2.13) is exact for any polynomial of degree one or less. Therefore by successively using $f(x) = 1$ and $x$ in Equation (6.2.13), we get

$$\int_a^b 1\, dx = x \Big|_a^b = b - a = c_0 f(a) + c_1 f(b) = c_0 + c_1$$

and

$$\int_a^b x\, dx = \frac{x^2}{2} \Big|_a^b = \frac{1}{2}(b^2 - a^2) = c_0 f(a) + c_1 f(b) = c_0\, a + c_1\, b \tag{6.2.14}$$

Solving these equations for $c_0$ and $c_1$, it can be verified that $c_0 = c_1 = (b - a)/2$.

Thus we get

$$\int_a^b f(x)\, dx \approx \frac{(b-a)}{2}[f(a) + f(b)] \tag{6.2.15}$$

The truncation error is given by

$$E_2(f) = \int_a^b f(x)\, dx - \frac{b-a}{2}[f(a) + f(b)]$$

If it is known that the integration formula has an error term of the form $E_m(f) = Af^{(m)}(\xi)$, then it can be determined. Consider $f(x) = x^2$. Then

$$E_2(f) = \int_a^b x^2\, dx - \frac{b-a}{2}[a^2 + b^2] = \frac{(b^3 - a^3)}{3} - \frac{(a^2b - a^3 + b^3 - ab^2)}{2}$$

$$= -\frac{(b-a)^3}{6}$$

Since this is nonzero,

$$-\frac{(b-a)^3}{6} = Af^{(2)}(\xi) = 2A$$

Therefore

$$A = -\frac{(b-a)^3}{12}$$

and

$$E_2(f) = -\frac{(b-a)^3}{12}f''(\xi) = \int_a^b f(x)\, dx - \frac{(b-a)}{2}[f(a) + f(b)]$$

Thus, we get the familiar trapezoidal rule

$$\int_a^b f(x)\, dx = \frac{(b-a)}{2}[f(a) + f(b)] - \frac{(b-a)^3}{12}f''(\xi) \tag{6.2.16}$$

## EXERCISES

1. Use the trapezoidal rule with $N = 2$ to approximate each of the following integrals. Compare each approximate value with the exact value:
   (a) $\int_0^1 x\, dx$     (b) $\int_0^1 e^x\, dx$     (c) $\int_0^{\pi/2} \sin x\, dx$

2. Determine an upper bound value of the truncation error for the approximations in each case of Exercise 1 and compare it with the absolute value of the actual error.

3. Derive the simple rectangle rule

$$\int_a^b f(x)\, dx = (b - a)f(a) + \frac{(b-a)^2}{2}f'(\xi)$$

where $a < \xi < b$.

4. Derive Equation (6.2.6) by approximating $f$ with the Newton forward difference formula for the interpolating polynomial.

5. Use the trapezoidal rule with $N = 2$ to approximate each of the following integrals. Compare each approximate value with the exact value.

   (a) $\int_0^1 x^2\, dx$　　(b) $\int_0^1 x^3\, dx$

6. The truncation error in the trapezoidal rule can be expressed differently.

   (a) Show that $\frac{1}{2} \int_a^b f''(\xi_2(x))(x - a)(x - b)\, dx = \frac{1}{2} \int_0^h t\, (t - h)\, f''(\xi_2(t))\, dt$ using $x - a = t$ and $b - a = h$. Also verify that $\int_0^h t\, (t - h)\, dt = -h^3/6$.

   (b) Show that

$$\int_0^h t\, (t - h)\, f''(\xi_2(t))\, dt = \int_0^h \left[ -\frac{h^2}{6} + \left( \frac{h^2}{6} + t\, (t - h) \right) \right] f''(\xi_2(t))\, dt$$

$$= -\frac{h^2}{3} [f'(\xi(h)) - f'(\xi(0))]$$

$$- \frac{1}{18} \int_0^h f^{(3)}(\xi_3(t))\, [2t^3 - 3t^2 h + h^2 t]\, dt$$

   (c) Show that

$$\int_0^h f^{(3)}(\xi_3(t))(2t^3 - 3t^2 h + h^2 t)\, dt = -\frac{1}{8} \int_0^h f^{(4)}(\xi_4(t))\, t^2\, (t - h)^2\, dt$$

   Also verify that $\int_0^h t^2(t - h)^2\, dt = h^5/30$.

   (d) Show that

$$\int_0^h t^2\, (t - h)^2\, f^{(4)}(\xi_4(t))\, dt = \int_0^h \left[ \frac{h^4}{30} + t^2\, (t - h)^2 - \frac{h^4}{30} \right] f^{(4)}(\xi(t))\, dt$$

$$= \frac{2h^4}{15} \left[ f^{(3)}(\xi_4(h)) - f^{(3)}(\xi_4(0)) \right]$$

$$+ \int_0^h f^{(4)}(\xi_4(t)) \left[ t^2\, (t - h)^2 - \frac{h^4}{30} \right] dt$$

   (e) Using (a), (b), (c), and (d), verify that the truncation error in the trapezoidal rule is given by

$$\frac{1}{2} \int_a^b f''(\xi_2(x))(x - a)(x - b)\, dx = -\frac{h^2}{6} [f'(\xi_2(h)) - f'(\xi_2(0))]$$

$$+ \frac{h^4}{2160} \left[ f^{(3)}(\xi_4(h)) - f^{(3)}(\xi_4(0)) \right]$$

$$+ \frac{1}{288} \int_0^h f^{(4)}(\xi_4(t)) \left[ t^2\, (t - h)^2 - \frac{h^4}{30} \right] dt$$

This can continue indefinitely. Therefore

$$\frac{b-a}{2}[f(a) + f(b)] = \int_a^b f(x)\,dx + A_2 h^2 + A_4 h^4 + A_6 h^6 + \cdots$$

where $A_2, A_4, A_6, \ldots$ are constants.

## COMPUTER EXERCISES

**C.1.** Write a subroutine that accepts $a$, $b$, $f$, and $N$ and carries out the composite trapezoidal rule procedure using Equation (6.2.10).

**C.2.** Write a computer program to evaluate $\int_a^b f(x)\,dx$ accurate within $10^{-5}$ using Exercise C.1.

**C.3.** Evaluate the definite integral in Exercise 1(b) or 1(c) using Exercise C.2 with $N = 4, 8, 16,$ and $32$ and compare these values with the exact value.

## 6.3 SIMPSON'S RULE

To improve the trapezoidal rule, use the quadratic Lagrange interpolating polynomial to approximate $f$ on $[a, b]$. Let $h = (b - a)/2$. Then $x_0 = a$, $x_1 = a + h$, and $x_2 = a + 2h = b$ as shown in Figure 6.3.1.

Then by the Lagrange Polynomial Approximation Theorem 4.3.1,

$$f(x) = f(x_0)\frac{(x - x_1)(x - x_2)}{2h^2} - f(x_1)\frac{(x - x_0)(x - x_2)}{h^2}$$

$$+ f(x_2)\frac{(x - x_0)(x - x_1)}{2h^2} + \frac{f^{(3)}(\xi(x))}{3!}(x - x_0)(x - x_1)(x - x_2) \quad \textbf{(6.3.1)}$$

where $x_0 < \xi < x_2$.

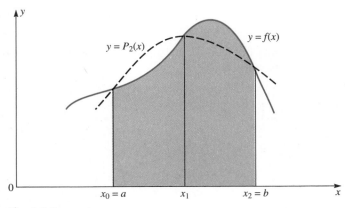

**Fig. 6.3.1**

Integrating both sides of Equation (6.3.1) from $x_0$ to $x_2$,

$$\int_a^b f(x)\,dx = \int_{x_0}^{x_2} f(x)\,dx = \frac{f(x_0)}{2h^2}\int_{x_0}^{x_2}(x-x_1)(x-x_2)\,dx$$

$$-\frac{f(x_1)}{h^2}\int_{x_0}^{x_2}(x-x_0)(x-x_2)\,dx$$

$$+\frac{f(x_2)}{2h^2}\int_{x_0}^{x_2}(x-x_0)(x-x_1)\,dx$$

$$+\frac{1}{3!}\int_{x_0}^{x_2} f^{(3)}(\xi(x))(x-x_0)(x-x_1)(x-x_2)\,dx \qquad \textbf{(6.3.2)}$$

In order to evaluate the right-hand side of Equation (6.3.2), integrate each term separately. Substitute $x = x_0 + ht$. Then

$$\int_{x_0}^{x_2}(x-x_1)(x-x_2)\,dx = h^3\int_0^2(t-1)(t-2)\,dt = h^3\int_0^2(t^2-3t+2)\,dt$$

$$= h^3\left[\frac{t^3}{3}-\frac{3t^2}{2}+2t\right]\Bigg|_0^2 = h^3\left(\frac{8}{3}-6+4\right)$$

$$= \frac{2h^3}{3} \qquad \textbf{(6.3.3)}$$

$$\int_{x_0}^{x_2}(x-x_0)(x-x_2)\,dx = h^3\int_0^2 t(t-2)\,dt = h^3\int_0^2(t^2-2t)\,dt$$

$$= h^3\left[\frac{t^3}{3}-t^2\right]\Bigg|_0^2 = h^3\left(\frac{8}{3}-4\right) = -\frac{4h^3}{3} \qquad \textbf{(6.3.4)}$$

and

$$\int_{x_0}^{x_2}(x-x_0)(x-x_1)\,dx = h^3\int_0^2 t(t-1)\,dt = h^3\int_0^2(t^2-t)\,dt$$

$$= h^3\left[\frac{t^3}{3}-\frac{t^2}{2}\right]\Bigg|_0^2 = h^3\left(\frac{8}{3}-2\right) = \frac{2h^3}{3} \qquad \textbf{(6.3.5)}$$

Let us integrate

$$\int_{x_0}^{x_2} f^{(3)}(\xi(x))(x-x_0)(x-x_1)(x-x_2)\,dx \qquad \textbf{(6.3.6)}$$

Since $\xi(x)$ is unknown and $(x-x_0)(x-x_1)(x-x_2)$ changes its sign at $x_1 = (x_0 + x_2)/2 = (a+b)/2$ in the open interval $(x_0, x_2)$, we cannot use the Second Mean Value

Theorem for the Definite Integrals (Theorem A.5) as we did for the trapezoidal rule. In order to integrate Equation (6.3.6), define

$$q(x) = \int_{x_0}^{x} (t - x_0)(t - x_1)(t - x_2)\, dt \tag{6.3.7}$$

It can be verified (Exercise 7) that

$$q(x) = \frac{(x - x_0)^2(x - x_2)^2}{4} \tag{6.3.8}$$

Further $q(x_0) = 0$, $q(x_2) = 0$, and $q(x) \geq 0$ for all $x$. Thus

$$\int_{x_0}^{x_2} f^{(3)}(\xi(x))(x - x_0)(x - x_1)(x - x_2)\, dx = \int_{x_0}^{x_2} f^{(3)}(\xi(x)) \frac{dq}{dx}\, dx$$

$$= f^{(3)}(\xi(x))\, q(x) \Big|_{x_0}^{x_2} - \int_{x_0}^{x_2} q(x) \frac{d}{dx} [f^{(3)}(\xi(x))]\, dx \qquad \text{(integration by parts)}$$

$$= -\int_{x_0}^{x_2} q(x) \frac{f^{(4)}(\eta(x))}{4}\, dx \qquad \left( q(x_0) = 0,\, q(x_2) = 0 \text{ and Equation (5.2.11)} \right)$$

$$= -\frac{f^{(4)}(\mu)}{4} \int_{x_0}^{x_2} q(x)\, dx \qquad (x_0 < \mu < x_2) \qquad (q(x) \geq 0)$$

$$= -\frac{f^{(4)}(\mu)}{4} \left\{ xq(x) \Big|_{x_0}^{x_2} - \int_{x_0}^{x_2} x \frac{dq}{dx}\, dx \right\} \qquad \text{(integration by parts)}$$

$$= \frac{f^{(4)}(\mu)}{4} \int_{x_0}^{x_2} x(x - x_0)(x - x_1)(x - x_2)\, dx \tag{6.3.9}$$

$$= -\frac{h^5}{15} f^{(4)}(\mu) \tag{6.3.10}$$

Substitution of Equations (6.3.3), (6.3.4), (6.3.5), and (6.3.10) in Equation (6.3.2) leads to

$$\int_{a}^{b} f(x)\, dx = \int_{x_0}^{x_2} f(x)dx = \frac{h}{3}[f(x_0) + 4f(x_1) + f(x_2)] - \frac{h^5}{90} f^{(4)}(\mu(x))$$

$$= \frac{h}{3}[f(a) + 4f(a + h) + f(b)]$$

$$- \frac{h^5}{90} f^{(4)}(\mu(x)) \tag{6.3.11}$$

where $a < \mu(x) < b$. This is the well-known **Simpson rule with error term**.

In general $h = (b - a)/2$ is not small; therefore, the truncation error can be large. As with the trapezoidal rule, a composite formula is obtained by dividing $[a, b]$ into an even number (why?) of subintervals $[x_i, x_{i+1}]$ where $x_i = a + ih$, and then applying

Equation (6.3.11) to each pair of consecutive subintervals. Let $N$ be an even number of intervals such that $h = (b - a)/N$ and $a = x_0 < x_1 < \cdots < x_N = b$. Then

$$\int_a^b f(x)\,dx = \int_{x_0}^{x_2} f(x)\,dx + \int_{x_2}^{x_4} f(x)\,dx + \cdots + \int_{x_{N-2}}^{x_N} f(x)\,dx$$

$$= \frac{h}{3}\left\{ f(x_0) + 4f(x_1) + f(x_2) + f(x_2) + 4f(x_3) + f(x_4) + \cdots \right.$$

$$\left. + f(x_{N-2}) + 4f(x_{N-1}) + f(x_N) \right\}$$

$$- \frac{h^5}{90} \sum_{i=1}^{N/2} f^{(4)}(\mu_i) \qquad (x_{2i-2} < \mu_i < x_{2i})$$

$$= \frac{h}{3}\left[ f(a) + 4 \sum_{i=1}^{N/2} f(x_{2i-1}) + 2 \sum_{i=1}^{(N/2)-1} f(x_{2i}) + f(b) \right]$$

$$- \frac{h^5}{90} \sum_{i=1}^{N/2} f^{(4)}(\mu_i) \tag{6.3.12}$$

As with the composite trapezoidal rule, by Corollary A.2.2 there exists $\eta$ in $(a, b)$ such that

$$\sum_{i=1}^{N/2} f^{(4)}(\mu_i) = \frac{N}{2} f^{(4)}(\eta) \tag{6.3.13}$$

Using Equation (6.3.13) in Equation (6.3.12) yields

$$\int_a^b f(x)\,dx = \frac{h}{3}\left[ f(a) + 4 \sum_{i=1}^{N/2} f(x_{2i-1}) + 2 \sum_{i=1}^{(N/2)-1} f(x_{2i}) + f(b) \right]$$

$$- \frac{h^4(b - a)}{180} f^{(4)}(\eta) \tag{6.3.14}$$

This is the **Simpson composite rule with error term**. If the truncation error is reasonably small, then

$$\int_a^b f(x)\,dx \approx \frac{h}{3}\left[ f(a) + 4 \sum_{i=1}^{N/2} f(x_{2i-1}) + 2 \sum_{i=1}^{(N/2)-1} f(x_{2i}) + f(b) \right] \tag{6.3.15}$$

Sometimes the composite Simpson rule is referred to as the **Simpson rule**. From Equation (6.3.14) we see that the truncation error is zero if $f$ is a polynomial of degree three or less, even though a quadratic interpolating polynomial is used in Equation (6.3.1). This results in better accuracy and therefore Equation (6.3.15) is widely used. It is probably the most frequently used formula of all the numerical integration formulas.

**EXAMPLE 6.3.1**

Use the Simpson rule to approximate the definite integral $\int_0^1 dx/(1 + x^2)$.

Divide $[0, 1]$ into two equal intervals. Then $N = 2$, $h = \frac{1}{2}$, $x_0 = 0$, $x_1 = \frac{1}{2}$, and $x_2 = 1$. For $N = 2$, Equation (6.3.15) becomes

$$\int_0^1 f(x)\,dx \approx \frac{h}{3}[f(x_0) + 4f(x_1) + f(x_2)]$$

Therefore

$$\int_0^1 \frac{dx}{1 + x^2} \approx \frac{1/2}{3}\left[\frac{1}{1 + 0} + \frac{4}{1 + (1/4)} + \frac{1}{1 + 1}\right] \approx 0.78333$$

For $N = 4$, we get $h = \frac{1}{4}$, $x_0 = 0$, $x_1 = \frac{1}{4}$, $x_2 = \frac{1}{2}$, $x_3 = \frac{3}{4}$, and $x_4 = 1$. In the case of $N = 4$, Equation (6.3.15) becomes

$$\int_0^1 f(x)\,dx \approx \frac{h}{3}[f(x_0) + 4f(x_1) + 2f(x_2) + 4f(x_3) + f(x_4)]$$

Therefore

$$\int_0^1 \frac{dx}{1 + x^2} \approx \frac{1/4}{3}\left[\frac{1}{1 + 0} + \frac{4}{1 + (1/16)} + \frac{2}{1 + (1/4)} + \frac{4}{1 + (9/16)} + \frac{1}{1 + 1}\right]$$
$$\approx 0.78539$$

Similar calculations are carried out for various $N$ using Equation (6.3.15) and are given in Table 6.3.1 along with the absolute value of the actual errors.

**Table 6.3.1**

| $N$ | Using Equation (6.3.15) | \|Actual Error\| | $E_N$ | Trapezoidal Rule | \|Actual Error\| |
|---|---|---|---|---|---|
| 2 | 0.78333 | 0.00206 | 0.00833 | 0.77500 | 0.01039 |
| 4 | 0.78539 | 0.00000 | 0.00052 | 0.78279 | 0.00260 |
| 8 | 0.78539 | 0.00000 | 0.00003 | 0.78474 | 0.00085 |

The error in the Simpson composite rule is given by $-(b - a)^5 f^{(4)}(\eta)/180\, N^4$, where $0 < \eta < 1$. Again, as in the trapezoidal rule, $\eta$ is not known; however,

$$E_N = \max_{0 \le x \le 1} \frac{|f^{(4)}(x)|}{180\, N^4}$$

where $E_N$ is the maximum error. It can be verified that

$$f^{(4)}(x) = \frac{24(1 - 10x^2 + 5x^4)}{(1 + x^2)^5} \qquad f^{(5)}(x) = \frac{-240\, x\, (3 - 10x^2 + 3x^4)}{(1 + x^2)^6}$$

and

$$\max_{0 \le x \le 1} f^{(4)}(x) = \max\left\{|f^{(4)}(0)|, \left|f^{(4)}\left(\pm\frac{1}{\sqrt{3}}\right)\right|, \left|f^{(4)}\left(\pm\sqrt{3}\right)\right|, |f^{(4)}(1)|\right\}$$
$$= \max\{24, 10.125, 0.375, 3\} = 24$$

Thus

$$E_N = \frac{24}{180\, N^4}$$

The values of $N$ and $E_N$ are also given in Table 6.3.1. For comparison, the values computed by the composite trapezoidal rule from Table 6.2.1 are also given in Table 6.3.1.    ■ ■ ■

**Algorithm**

**Subroutine** Simpson $(a, b, f, N)$
Comment: This program computes the definite integral $\int_a^b f(x)\,dx$ using the composite Simpson rule given by Equation (6.3.15). $N$ must be a positive even integer.
1  **Print** $a$, $b$, $N$
2  $h = (b - a)/N$
3  $M = N/2$
4  sumeven $= 0.$
5  sumodd $= 0.$
6  **Repeat** steps 7–8 **with** $i = 1, 2, ..., M - 1$
   7  $x = a + 2ih$
    8  sumeven $=$ sumeven $+ f(x)$
9  **Repeat** steps 10–11 **with** $i = 1, 2, ..., M$
   10  $x = a + (2i - 1)\,h$
    11  sumodd $=$ sumodd $+ f(x)$
12  deint $= h\,(f(a) + 4\text{ sumodd} + 2\text{ sumeven} + f(b))/3$
13  **Print** deint, $a$, $b$, $N$, $h$    ◀

**E X E R C I S E S**

1.  Use the Simpson rule with $N = 2$ to approximate each of the following integrals. Compare each approximate value with the exact value:
   (a)  $\int_0^1 x\,dx$    (b)  $\int_0^1 e^x\,dx$    (c)  $\int_0^{\pi/2} \sin x\,dx$
2.  Determine an upper bound value of the truncation error for the approximations in each case of Exercise 1 and compare it with the actual error.
3.  Use the Newton forward difference formula for the interpolating polynomial to approximate $f$ to derive Equation (6.3.11).
4.  Derive the Simpson rule using the method of undetermined coefficients.
5.  Using the method of undetermined coefficients, determine $w_0$, $w_1$, $w_2$, and $w_3$ such that

$$\int_a^b f(x)\,dx \cong w_0 f(a) + w_1 f(b) + w_2 f'(a) + w_3 f'(b)$$

6.  Use the Simpson rule with $N = 2$ to approximate each of the following integrals. Compare each approximate value with the exact value.
   (a)  $\int_0^1 x^2\,dx$    (b)  $\int_0^1 x^3\,dx$    (c)  $\int_0^1 x^4\,dx$.
7.  Derive Equation (6.3.8) from Equation (6.3.7).

■■■■■■■■ COMPUTER EXERCISES ■■■■■■■■

**C.1.** Write a subroutine that accepts $a$, $b$, $f$, and $N$ and carries out the composite trapezoidal rule procedure using Equation (6.2.10).

**C.2.** Write a subroutine that accepts $a$, $b$, $f$, and $N$ and carries out the composite Simpson rule procedure using Equation (6.3.15).

**C.3.** Write a computer program to evaluate $\int_a^b f(x)\, dx$ accurate within $10^{-5}$ using Exercise C.1 or Exercise C.1, p. 146 and Exercise C.2.

**C.4.** Evaluate the definite integral in Exercise 1(b) or 1(c) using Exercise C.3 with $N = 4$, 8, 16 and 32 and compare these values with the exact value.

**C.5.** The **Gauss error function**, erf($x$), is defined by

$$\operatorname{erf}(x) = \frac{2}{\sqrt{\pi}} \int_0^x e^{-t^2}\, dt$$

Compute erf($x$) for $0 \leq x \leq 1$ using Exercise C.3. Print erf($x$) for $x = 0., 0.1,$ ...., 1.

**C.6.** The period of oscillation of a simple pendulum is given by

$$P(\theta_0) = 4\sqrt{\frac{l}{g}} \int_0^{\pi/2} \frac{dx}{\sqrt{1 - k^2 \sin^2 x}} \qquad \textbf{(6.3.16)}$$

where $l$ is the length of pendulum, $g$ is the acceleration due to gravity, and $k = \sin(\theta_0/2)$. The angle $\theta_0$ is the initial position of the pendulum when it is released from rest at time $t = 0$. Determine the period of oscillation for $\theta_0 = \frac{\pi i}{180}$ for $i = 1, 3, ..., 45$. Use Exercise C.3 and $l/g = 1$. The integral given by Equation (6.3.16) is called an **elliptic integral**.

**C.7.** The **Bessel function of order zero** is defined by

$$J_0(x) = \frac{1}{\pi} \int_0^{\pi} \cos(x \sin \theta)\, d\theta$$

Compute $J_0(x)$ for $x = 2, 2.1, ...., 3$ using Exercise C.3.

## 6.4   NEWTON–COTES FORMULAS

We require a general approach of generating coefficients that arise from the integration of the Lagrangian polynomials, since for $N = 3, 4, 5, ....$ the algebra involved in calculating the coefficients becomes more and more involved. If the endpoints $a$ and $b$ of the interval $[a, b]$ are included as nodal points and the remaining nodal points are strictly within the interval, then the intergrating formulas are called the **closed**

**Newton–Cotes integrating formulas**. The trapezoidal rule given by Equation (6.2.6) and the Simpson rule given by Equation (6.3.11) are examples of closed Newton–Cotes integrating formulas. Sometimes we want to integrate a function whose value at one endpoint or both endpoints is not known. In this case we use equally-spaced nodal points within the interval $[a, b]$ as shown in Figure 6.4.1 to develop the integrating formulas. These integrating formulas are called the **open Newton–Cotes integrating formulas**.

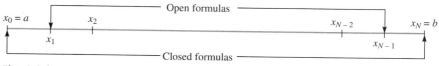

Fig. 6.4.1

## Closed Newton–Cotes Integrating Formulas

Let $N + 1$ distinct, equally spaced points be denoted by $x_i = a + ih$ where $h = (b - a)/N$ for $i = 0, 1,\ldots, N - 1$ and $x_N = b$ as shown in Figure 6.4.1. We form an interpolating polynomial $P_N(x)$ of at most degree $N$ such that $P_N(x_i) = f(x_i)$ for $i = 0, 1,\ldots, N$. The Newton–Cotes integration formulas are derived by replacing the integrand $f(x)$ by a suitable interpolating polynomial $P_N(x)$.

Using the Lagrange interpolating polynomial given by Equation (4.3.5),

$$P_N(x) = \sum_{j=0}^{N} f(x_j)\, L_j(x)$$

where

$$L_j(x) = \prod_{\substack{i=0 \\ i \neq j}}^{N} \frac{(x - x_i)}{(x_j - x_i)}$$

Thus

$$\int_a^b P_N(x)\, dx = \sum_{j=0}^{N} c_j f(x_j) \tag{6.4.1}$$

where

$$c_j = \int_a^b \prod_{\substack{i=0 \\ i \neq j}}^{N} \frac{(x - x_i)}{(x_j - x_i)}\, dx$$

$N = 1$ gives the trapezoidal rule and $N = 2$ gives the Simpson rule. Thus the trapezoidal rule and the Simpson rule are the first two cases of the closed Newton–Cotes integration formulas.

For $N = 3$, we have

$$\int_a^b P_3(x)\,dx = \int_{x_0}^{x_3} P_3(x)\,dx = f(x_0) \int_{x_0}^{x_3} \frac{(x - x_1)(x - x_2)(x - x_3)}{(x_0 - x_1)(x_0 - x_2)(x_0 - x_3)}\,dx$$

$$+ f(x_1) \int_{x_0}^{x_3} \frac{(x - x_0)(x - x_2)(x - x_3)}{(x_1 - x_0)(x_1 - x_2)(x_1 - x_3)}\,dx$$

$$+ f(x_2) \int_{x_0}^{x_3} \frac{(x - x_0)(x - x_1)(x - x_3)}{(x_2 - x_0)(x_2 - x_1)(x_2 - x_3)}\,dx$$

$$+ f(x_3) \int_{x_0}^{x_3} \frac{(x - x_0)(x - x_1)(x - x_2)}{(x_3 - x_0)(x_3 - x_1)(x_3 - x_2)}\,dx$$

For integration, substitute $x = x_0 + ht$. Then $dx = h\,dt$ and $x - x_i = h(t - t_i)$ and so

$$\int_{x_0}^{x_3} \frac{(x - x_1)(x - x_2)(x - x_3)}{(x_0 - x_1)(x_0 - x_2)(x_0 - x_3)}\,dx = -\frac{h}{6} \int_0^3 (t - 1)(t - 2)(t - 3)\,dt = \frac{3h}{8}$$

$$\int_{x_0}^{x_3} \frac{(x - x_0)(x - x_2)(x - x_3)}{(x_1 - x_0)(x_1 - x_2)(x_1 - x_3)}\,dx = \frac{h}{2} \int_0^3 t(t - 2)(t - 3)\,dt = \frac{9h}{8}$$

$$\int_{x_0}^{x_3} \frac{(x - x_0)(x - x_1)(x - x_3)}{(x_2 - x_0)(x_2 - x_1)(x_2 - x_3)}\,dx = -\frac{h}{2} \int_0^3 t(t - 1)(t - 3)\,dt = \frac{9h}{8}$$

and

$$\int_{x_0}^{x_3} \frac{(x - x_0)(x - x_1)(x - x_3)}{(x_3 - x_0)(x_3 - x_1)(x_3 - x_2)}\,dx = \frac{h}{2} \int_0^3 t(t - 1)(t - 2)\,dt = \frac{3h}{8}$$

Thus

$$\int_a^b P_3(x)\,dx = \int_{x_0}^{x_3} P_3(x)\,dx = \frac{3h}{8}\,[f(x_0) + 3f(x_1) + 3f(x_2) + f(x_3)]$$

$$= \frac{3h}{8}\,[f(a) + 3f(a + h) + 3f(a + 2h)$$

$$+ f(b)] \tag{6.4.2}$$

The truncation error is given by

$$E_3(f) = \int_{x_0}^{x_3} [f(x) - P_3(x)]\,dx$$

$$= \int_{x_0}^{x_3} \frac{f^{(4)}(\xi(x))}{4!}\,(x - x_0)(x - x_1)(x - x_2)(x - x_3)\,dx \tag{6.4.3}$$

where $x_0 = a < \xi(x) < x_3 = b$. Since $(x - x_0)(x - x_1)(x - x_2)(x - x_3)$ changes sign in $[a, b]$, we cannot use the Second Mean Value Theorem for Definite Integrals (Theorem A.5) directly. We cannot use the technique that we used for the Simpson rule because we cannot construct $q(x) \geq 0$ on $[x_0 = a, x_3 = b]$. Instead, divide the interval $[x_0, x_3]$ into two intervals, $[x_0, x_2]$ and $[x_2, x_3]$. Therefore

$$E_3(f) = \int_{x_0}^{x_2} \frac{f^{(4)}(\xi(x))}{4!} (x - x_3)(x - x_0)(x - x_1)(x - x_2) \, dx$$

$$+ \int_{x_2}^{x_3} \frac{f^{(4)}(\xi(x))}{4!} \prod_{i=0}^{3} (x - x_i) \, dx \qquad \textbf{(6.4.4)}$$

Since we know (Exercise 17, p. 128)

$$\frac{f^{(4)}(\xi(x))}{4!} (x - x_3) = \frac{f^{(3)}(\eta(x))}{3!} + c$$

where $a < \eta(x), \xi(x) < b$ and $\prod_{i=0}^{3} (x - x_i)$ does not vanish on $(x_2, x_3)$, Equation (6.4.4) can be rewritten using the Second Mean Value Theorem of Definite Integrals (Theorem A.5) as

$$E_3(f) = \int_{x_0}^{x_2} \left[ \frac{f^{(3)}(\eta(x))}{3!} + c \right] (x - x_0)(x - x_1)(x - x_2) \, dx$$

$$+ \frac{f^{(4)}(\xi_1)}{4!} \int_{x_2}^{x_3} \prod_{i=0}^{3} (x - x_i) \, dx \qquad \textbf{(6.4.5)}$$

where $\xi_1 \in [a, b]$. Using the technique developed for the error term of the Simpson rule, it can be verified (Exercise 4) that

$$\int_{x_0}^{x_2} \left[ \frac{f^{(3)}(\eta(x))}{3!} + c \right] (x - x_0)(x - x_1)(x - x_2) \, dx$$

$$= \frac{f^{(4)}(\eta_1)}{4!} \int_{x_0}^{x_2} \prod_{i=0}^{3} (x - x_i) \, dx \qquad \textbf{(6.4.6)}$$

where $\eta_1 \in [x_0, x_2]$. Substituting Equation (6.4.6) in Equation (6.4.5) yields

$$E_3(f) = \frac{f^{(4)}(\eta_1)}{4!} \int_{x_0}^{x_2} \prod_{i=0}^{3} (x - x_i) \, dx + \frac{f^{(4)}(\xi_1)}{4!} \int_{x_2}^{x_3} \prod_{i=0}^{3} (x - x_i) \, dx \qquad \textbf{(6.4.7)}$$

Since

$$\int_{x_0}^{x_2} \prod_{i=0}^{3} (x - x_i) \, dx \quad \text{and} \quad \int_{x_2}^{x_3} \prod_{i=0}^{3} (x - x_i) \, dx$$

are constants of the same sign, by Corollary A.2.2 there exists $\xi$ in $[a, b]$ such that

$$E_3(f) = \frac{f^{(4)}(\xi)}{4!} \left[ \int_{x_0}^{x_2} \prod_{i=0}^{3} (x - x_i) \, dx + \int_{x_2}^{x_3} \prod_{i=0}^{3} (x - x_i) \, dx \right]$$

$$= \frac{f^{(4)}(\xi)}{4!} \int_{x_0}^{x_3} \prod_{i=0}^{3} (x - x_i) \, dx = -\frac{3h^5}{80} f^{(4)}(\xi) \tag{6.4.8}$$

From Equations (6.4.2) and (6.4.8),

$$\int_a^b f(x) \, dx = \int_{x_0}^{x_3} f(x) \, dx = \frac{3h}{8} \left[ f(x_0) + 3f(x_1) + 3f(x_2) + f(x_3) \right] - \frac{3h^5}{80} f^{(4)}(\xi)$$

$$= \frac{3h}{8} \left[ f(a) + 3f(a + h) + 3f(a + 2h) \right.$$

$$\left. + f(b) \right] - \frac{3h^5}{80} f^{(4)}(\xi) \tag{6.4.9}$$

By using the techniques we developed for $N = 2$ and $N = 3$, the following theorem can be proven.

**Theorem 6.4.1**    If $f^{(N+2)}$ is continuous on $[a, b]$, then there exists $\xi$ in $[a, b]$ such that

$$\int_a^b f(x) \, dx = \sum_{j=0}^{N} f(x_j) \int_a^b \prod_{\substack{i=0 \\ i \neq j}}^{N} \frac{(x - x_i)}{(x_j - x_i)} \, dx + E_N(f) \tag{6.4.10}$$

where

$$E_N(f) = \begin{cases} \dfrac{f^{(N+2)}(\xi)}{(N + 2)!} \displaystyle\int_a^b x \prod_{i=0}^{N} (x - x_i) \, dx & \text{if } N \text{ is even} \\[4mm] \dfrac{f^{(N+1)}(\xi)}{(N + 1)!} \displaystyle\int_a^b \prod_{i=0}^{N} (x - x_i) \, dx & \text{if } N \text{ is odd} \end{cases} \tag{6.4.11}$$

and $x_0 = a$, $x_i = x_0 + ih$ for $i = 0, 1, ..., N$, and $x_N = b$.    ∎

For the proof, see Isaacson and Keller (1966).

For reference, the most common closed Newton–Cotes formulas with their error terms are given in Table 6.4.1 where $h = (b - a)/N$.

## Open Newton–Cotes Integral Formulas

In Figure 6.4.1, $(N + 1)$ distinct, equally spaced points are shown. We form the interpolation polynomial $P_{N-2}(x)$ of at most degree $N - 2$ such that $P_{N-2}(x_i) = f(x_i)$ for $i = 1, 2, ..., N - 1$. Then $\int_a^b P_{N-2}(x) \, dx$ is taken as an approximation to $\int_a^b f(x) \, dx$.

**Table 6.4.1**

1. $N = 1$

$$\int_a^b f(x)\, dx = \frac{h}{2}\,[f(a) + f(b)] - \frac{h^3}{12} f^{(2)}(\xi) \qquad \text{(trapezoidal rule)}$$

2. $N = 2$

$$\int_a^b f(x)\, dx = \frac{h}{3}\,[f(a) + 4f(a + h) + f(b)] - \frac{h^5}{90} f^{(4)}(\xi) \qquad \text{(Simpson's rule)}$$

3. $N = 3$

$$\int_a^b f(x)\, dx = \frac{3h}{8}\,[f(a) + 3f(a + h) + 3f(a + 2h) + f(b)] - \frac{3h^5}{80} f^{(4)}(\xi)$$

4. $N = 4$

$$\int_a^b f(x)\, dx = \frac{2h}{45}\,[7f(a) + 32f(a + h) + 12f(a + 2h)$$
$$+\, 32f(a + 3h) + 7f(b)] - \frac{8h^7}{975} f^{(6)}(\xi)$$

5. $N = 5$

$$\int_a^b f(x)\, dx = \frac{5h}{288}\,[19f(a) + 75f(a + h) + 50f(a + 2h) + 50f(a + 3h) + 75f(a + 4h)$$
$$+\, 19f(b)] - \frac{275h^7}{12096} f^{(6)}(\xi)$$

6. $N = 6$

$$\int_a^b f(x)\, dx = \frac{h}{140}\,[41f(a) + 216f(a + h) + 27f(a + 2h) + 272f(a + 3h) + 27f(a + 4h)$$
$$+\, 216f(a + 5h) + 41f(b)] - \frac{9h^9}{1400} f^{(8)}(\xi)$$

7. $N = 7$

$$\int_a^b f(x)\, dx = \frac{7h}{17280}\,[751f(a) + 3577f(a + h) + 1323f(a + 2h) + 2989f(a + 3h)$$
$$+\, 2989f(a + 4h) + 1323f(a + 5h) + 3577f(a + 6h) + 751f(b)] - \frac{8183h^9}{518400} f^{(8)}(\xi)$$

1. Let $N = 2$. Then $x_0 = a$, $x_1 = (a + b)/2$, $x_2 = b$, and $h = (b - a)/2$. We do not use $x_0 = a$ and $x_2 = b$ but instead use the point $x_1 = (a + b)/2$. Then

$$f(x) = f(x_1) + (x - x_1)f'(\eta(x)) \tag{6.4.12}$$

where $\eta(x)$ is between $x$ and $x_1$. Integrating both sides of Equation (6.4.12) yields

$$\int_a^b f(x)\,dx = f(x_1)\int_a^b dx + \int_a^b (x - x_1)f'(\eta(x))\,dx \tag{6.4.13}$$

Since $(x - x_1)$ changes its signs at $x = x_1$, use the technique developed to integrate the truncation error and verify that (Exercise 7)

$$\int_a^b (x - x_1)f'(\eta(x))\,dx = \frac{(b - a)^3}{24}f''(\xi) \tag{6.4.14}$$

where $a \le \xi \le b$. Using Equation (6.4.14) and $b - a = 2h$ in Equation (6.4.13),

$$\int_a^b f(x)\,dx = 2hf\left(\frac{a + b}{2}\right) + \frac{(b - a)^3}{24}f''(\xi) \tag{6.4.15}$$

2. Consider $N = 3$. Then $x_0 = a$, $x_1 = a + h$, $x_2 = a + 2h$, $x_3 = b$, and $h = (b - a)/3$. We use $x_1$ and $x_2$. Using the Lagrange interpolating polynomial,

$$f(x) = f(x_1)\frac{(x - x_2)}{(x_1 - x_2)} + f(x_2)\frac{x - x_1}{x_2 - x_1} + f''(\eta(x))\frac{(x - x_1)(x - x_2)}{2!} \tag{6.4.16}$$

By integrating both sides of Equation (6.4.16), it can be verified (Exercise 8) that

$$\int_a^b f(x)\,dx = \frac{3h}{2}\left[f\left(\frac{2a + b}{3}\right) + f\left(\frac{a + 2b}{3}\right)\right] + \frac{(b - a)^3}{36}f''(\xi) \tag{6.4.17}$$

where $a \le \xi \le b$. Higher-order formulas can be derived similarly and are given by the following theorem.

**Theorem 6.4.2**    If $f^{(N+2)}$ is continuous on $[a, b]$, then there exists $\xi$ in $[a, b]$ such that

$$\int_a^b f(x)\,dx = \sum_{j=1}^{N-1} f(x_j) \int_a^b \prod_{\substack{i=1 \\ i \ne j}}^{N-1} \frac{(x - x_i)}{(x_j - x_i)}\,dx + E_N(f) \tag{6.4.18}$$

where

$$E_N(f) = \begin{cases} \dfrac{f^{(N+2)}(\xi)}{(N + 2)!} \displaystyle\int_a^b x \prod_{i=1}^{N-1} (x - x_i)\,dx & \text{if } N \text{ is even} \\[4ex] \dfrac{f^{(N+1)}(\xi)}{(N + 1)!} \displaystyle\int_a^b \prod_{i=1}^{N-1} (x - x_i)\,dx & \text{if } N \text{ is odd} \end{cases} \tag{6.4.19}$$

and $x_0 = a$, $x_i = x_0 + ih$ for $i = 0, 1, \dots, N$, and $x_N = b$.    ∎

For the proof, see Isaacson and Keller (1966).

The formula given by Equation (6.4.18) is used in deriving formulas for the solution of ordinary differential equations.

---

**EXERCISES**

1. Approximate each of the following integrals using the closed Newton–Cotes formulas for $N = 3, 4$, and 5:
   (a) $\int_0^1 e^x \, dx$    (b) $\int_0^1 \sin x \, dx$

2. Approximate each of the following integrals using the closed Newton–Cotes formulas for $N = 3, 4$, and 5:
   (a) $\int_2^4 x^2 e^x \, dx$    (b) $\int_0^1 e^{-x^2} \, dx$

3. Find an upper bound for the truncation error in each case of Exercise 1 and compare it with the actual error.

4. Verify Equation (6.4.6).

5. Verify Equation (6.4.8).

6. Prove Theorem 6.4.1.

7. Prove that $\int_a^b (x - x_1) f'(\eta(x)) \, dx = [(b - a)^3/24] f''(\xi)$, where $\xi \in (a, b)$, $\eta \in (a, b)$, and $x_1 = (a + b)/2$.

8. Derive Equation (6.4.17).

9. Approximate each of the following integrals using the open Newton–Cotes formulas for $N = 2$ and 3:
   (a) $\int_0^1 e^x \, dx$    (b) $\int_0^1 \sin x \, dx$

10. Approximate each of the following integrals using the open Newton–Cotes formulas for $N = 2$ and 3:
    (a) $\int_2^4 x^2 e^x \, dx$    (b) $\int_0^1 e^{-x^2} \, dx$

11. Find an upper bound for the truncation error in each case of Exercise 9 and compare it with the actual error.

12. Derive the open Newton–Cotes formula for $N = 4$ using Theorem 6.4.2.

---

## 6.5    ROMBERG INTEGRATION

In order to reduce the truncation error, we can decrease the value of $h$ or use a higher-order formula. In this section we will discuss the method of Romberg integration that approximates the integral automatically with progressively higher-order formulas.

The trapezoidal rule can be written as (Exercise 6, p. 145).

$$T_0(h) = T_0(h/2^0) = \frac{h}{2} [f(a) + f(b)]$$

$$= \int_a^b f(x) \, dx + A_2 h^2 + A_4 h^4 + A_6 h^6 + \cdots \qquad \textbf{(6.5.1)}$$

where $h = b - a$ and $A_2, A_4, \ldots$ are constants. One can use the Richardson extrapolation procedure to obtain a better approximation for $I = \int_a^b f(x)\, dx$ by eliminating the lowest-order error term in Equation (6.5.1). Evaluating the trapezoidal rule with the interval length $h/2$ yields

$$T_0(h/2) = \frac{h}{4}\left[f(a) + 2f\left(a + \frac{h}{2}\right) + f(b)\right]$$

$$= I + A_2\left(\frac{h}{2}\right)^2 + A_4\left(\frac{h}{2}\right)^4 + A_6\left(\frac{h}{2}\right)^6 + \cdots \qquad (6.5.2)$$

In order to eliminate $A_2$ between Equations (6.5.1) and (6.5.2), multiply Equation (6.5.1) by $(-\frac{1}{4})$ and add it to Equation (6.5.2). This gives

$$T_0(h/2) - \frac{1}{4}T_0(h/2^0) = I\left(1 - \frac{1}{4}\right) + A_4\left(\frac{1}{2^4} - \frac{1}{2^2}\right)h^4 + A_6\left(\frac{1}{2^6} - \frac{1}{2^2}\right)h^6 + \cdots$$

Multiplying this equation by 4/3 yields

$$\frac{4^1\, T_0(h/2) - T_0(h/2^0)}{4^1 - 1} = I + A_4^{(1)}\, h^4 + A_6^{(1)}\, h^6 + \cdots \qquad (6.5.3)$$

where

$$A_4^{(1)} = \frac{A_4}{3}\left(\frac{1}{4} - 1\right) \qquad A_6^{(1)} = \frac{A_6}{3}\left(\frac{1}{4^2} - 1\right), \ldots$$

By taking a linear combination of two trapezoidal results, we are able to increase the order from two to four. Denoting this linear combination by $T_1(h/2)$, we have

$$T_1(h/2) = \frac{4^1\, T_0(h/2) - T_0(h/2^0)}{4^1 - 1} = I + A_4^{(1)}\, h^4 + A_6^{(1)}\, h^6 + \cdots \qquad (6.5.4)$$

In order to eliminate $A_4^{(1)}$, evaluate Equation (6.5.4) with the interval length $h/2$. This leads to

$$T_1(h/2^2) = \frac{4^1\, T_0(h/2^2) - T_0(h/2)}{4^1 - 1} = I + A_4^{(1)}\left(\frac{h}{2}\right)^4 + A_6^{(1)}\left(\frac{h}{2}\right)^6 + \cdots \qquad (6.5.5)$$

Eliminate $A_4^{(1)}$ between Equations (6.5.4) and (6.5.5) by multiplying Equation (6.5.4) with $(-1/4^2)$ and adding it to Equation (6.5.5). This leads to

$$T_1(h/2^2) - \frac{1}{4^2}T_1(h/2) = I\left(1 - \frac{1}{4^2}\right) + A_6^{(1)}\left(\frac{1}{2^6} - \frac{1}{4^2}\right)h^6 + \cdots$$

This equation simplifies to

$$\frac{4^2\, T_1(h/2^2) - T_1(h/2)}{4^2 - 1} = I + A_6^{(1)}\frac{\left(\frac{1}{2^2} - 1\right)}{(4^2 - 1)}h^6 + \cdots$$

Denoting this linear combination by $T_2(h/2^2)$, $A_6^{(1)} \dfrac{(\frac{1}{2^2} - 1)}{(4^2 - 1)} = A_6^{(2)}$, and so on, results in

$$T_2(h/2^2) = \frac{4^2 \, T_1(h/2^2) - T_1(h/2)}{4^2 - 1} = I + A_6^{(2)} \, h^6 + A_8^{(2)} \, h^8 + \cdots \qquad \textbf{(6.5.6)}$$

At the $i$th step we get

$$T_i(h/2^i) = \frac{4^i \, T_{i-1}(h/2^i) - T_{i-1}(h/2^{i-1})}{4^i - 1} = I + A_{2i+2}^{(i)} \, h^{2i+2} + \cdots \qquad \textbf{(6.5.7)}$$

for $i = 1, 2, ..., N$. This application of the Richardson extrapolation to the trapezoidal rule is called the **Romberg integration method**. In order to show this in a more systematic way, we construct Table 6.5.1.

**Table 6.5.1**

| | | | |
|---|---|---|---|
| $T_0(h/2^0) = T_{0,0}$ | | | |
| $T_0(h/2^1) = T_{1,0}$ | $T_1(h/2^1) = T_{1,1}$ | | |
| $T_0(h/2^2) = T_{2,0}$ | $T_1(h/2^2) = T_{2,1}$ | $T_2(h/2^2) = T_{2,2}$ | |
| $T_0(h/2^3) = T_{3,0}$ | $T_1(h/2^3) = T_{3,1}$ | $T_2(h/2^3) = T_{3,2}$ | $T_3(h/2^3) = T_{3,3}$ |

We start with the trapezoidal rule with $h = b - a$ as

$$T_0(h) = T_0(h/2^0) = T_{0,0} = \frac{(b - a)}{2} [f(a) + f(b)]$$

Then we divide the interval $h = b - a$ into two equal intervals and use the trapezoidal rule to get

$$T_{1,0} = T_0(h/2^1) = \left(\frac{b - a}{2}\right) \cdot \frac{1}{2} \left[ f(a) + 2f\left(a + \frac{b - a}{2}\right) + f(b) \right]$$

$$= \frac{1}{2} \left[ \frac{(b - a)}{2} \left(f(a) + f(b)\right) + (b - a)f\left(a + \frac{b - a}{2}\right) \right]$$

$$= \frac{1}{2} \left[ T_0(h) + hf\left(a + \frac{h}{2}\right) \right]$$

Divide the two intervals into four intervals and get

$$T_{2,0} = T_0(h/2^2) = T_0(h/4)$$

$$= \frac{h}{4} \cdot \frac{1}{2} \left[ f(a) + 2f\left(a + \frac{h}{4}\right) + 2f\left(a + 2\left(\frac{h}{4}\right)\right) + 2f\left(a + 3\left(\frac{h}{4}\right)\right) + f(b) \right]$$

$$= \frac{1}{2} \left[ \frac{h}{4} \left( \left[ f(a) + 2f\left(a + \frac{h}{2}\right) + f(b) \right] + 2f\left(a + \frac{h}{4}\right) + 2f\left(a + \frac{3h}{4}\right) \right) \right]$$

$$= \frac{1}{2} \left[ T_0(h/2) + \frac{h}{2}f\left(a + \frac{h}{4}\right) + \frac{h}{2}f\left(a + \frac{3h}{4}\right) \right]$$

In general, the $i$th element of the first column is given by

$$T_{i,0} = T_0(h/2^i) = \frac{1}{2}\left[ T_0(h/2^{i-1}) + \frac{h}{2^{i-1}} \sum_{j=1}^{2^{i-1}} f\left( a + \left( j - \frac{1}{2}\right) \frac{h}{2^{i-1}}\right)\right] \qquad (6.5.8)$$

for $i = 1, 2, ..., N$. Once the first two elements of the first column of Table 6.5.1 are computed, then compute $T_1(h/2)$ from Equation (6.5.4) by

$$T_{1,1} = T_1(h/2^1) = \frac{4^1 T_0(h/2^1) - T_0(h/2^0)}{4^1 - 1} = \frac{4^1 T_{1,0} - T_{0,0}}{4^1 - 1} \qquad (6.5.9)$$

Next compute $T_{2,0} = T_0(h/2^2)$, $T_{2,1} = T_1(h/2^2)$, and $T_{2,2} = T_2(h/2^2)$. Actually, we compute row by row rather than column by column. $T_1(h/2^2)$ is computed from Equation (6.5.9) by replacing $h$ by $h/2$ and is given by

$$T_{2,1} = T_1(h/2^2) = \frac{4^1 T_0(h/2^2) - T_0(h/2^1)}{4^1 - 1} = \frac{4^1 T_{2,0} - T_{1,0}}{4^1 - 1} \qquad (6.5.10)$$

In general, $T_1(h/2^i)$, the $i$th element of the second column, is given by

$$T_{i,1} = T_1(h/2^i) = \frac{4^1 T_0(h/2^i) - T_0(h/2^{i-1})}{4^1 - 1} = \frac{4^1 T_{i,0} - T_{i-1,0}}{4^1 - 1} \qquad (6.5.11)$$

for $i = 1, 2, ..., N$.

The approximations given in the second column of Table 6.5.1 are actually the approximations given by the Simpson rule starting with interval size $h/2$ (Exercise 9). From Equation (6.5.6), it can be seen that the third column can be computed by

$$T_{i,2} = T_2(h/2^i) = \frac{4^2 T_1(h/2^i) - T_1(h/2^{i-1})}{4^2 - 1} = \frac{4^2 T_{i,1} - T_{i-1,1}}{4^2 - 1} \qquad (6.5.12)$$

for $i = 2, 3, ...N$. In general, any $k + 1$ column is given by

$$T_{i,k} = T_k(h/2^i) = \frac{4^k T_{k-1}(h/2^i) - T_{k-1}(h/2^{i-1})}{4^k - 1} = \frac{4^k T_{i,k-1} - T_{i-1,k-1}}{4^k - 1} \qquad (6.5.13)$$

for $i = k, k + 1, ..., N$. Also, it can be verified that the approximation in the third column corresponds to the approximation given by the five-point closed Newton–Cotes formula ($N = 4$) in Table 6.5.1. The sequences $T_k(h/2^i)$ for $k = 3, 4, ...$ do not correspond to closed Newton–Cotes formulas as do those for $k = 0, 1,$ and 2.

**EXAMPLE 6.5.1**    Use the Romberg integration method to approximate the definite integral $\int_0^1 dx/(1 + x^2)$.
Here $h = b - a = 1 - 0 = 1$. Thus

$$T_{0,0} = T_0(h/2^0) = \frac{h}{2}[f(a) + f(b)] = \frac{1}{2}\left[ \frac{1}{1 + 0^2} + \frac{1}{1 + 1^2}\right] = \frac{1}{2}\left[ 1 + \frac{1}{2}\right] = \frac{3}{4}$$

$$= 0.75$$

Compute the elements of the second row ($i = 1$)

$$T_{1,0} = T_0(h/2^1) = \frac{1}{2}\left[T_0(h) + hf\left(a + \frac{h}{2}\right)\right] = \frac{1}{2}\left[0.75 + 1\frac{1}{1 + (\frac{1}{2})^2}\right] = 0.775$$

$$T_{1,1} = T_1(h/2^1) = \frac{4\,T_{1,0} - T_{0,0}}{4^1 - 1}$$

$$= \frac{4\,T_0(h/2^1) - T_0(h)}{4^1 - 1} = \frac{4 \times 0.775 - 0.75}{3} = 0.78333$$

Now compute the elements of the third row ($i = 2$)

$$T_{2,0} = T_0(h/2^2)$$

$$= \frac{1}{2}\left[T_0(h/2) + \frac{h}{2}\sum_{j=1}^{2^{2-1}} f\left(a + \left(j - \frac{1}{2}\right)\frac{h}{2^{2-1}}\right)\right]$$

$$= \frac{1}{2}\left[T_0(h/2) + \frac{h}{2}\sum_{j=1}^{2} f\left(a + \left(j - \frac{1}{2}\right)\frac{h}{2}\right)\right]$$

$$= \frac{1}{2}\left[0.775 + \frac{1}{2}\left\{f\left(a + \frac{h}{4}\right) + f\left(a + \frac{3h}{4}\right)\right\}\right]$$

$$= \frac{1}{2}\left[0.775 + \frac{1}{2}\left\{f\left(\frac{1}{4}\right) + f\left(\frac{3}{4}\right)\right\}\right]$$

$$= \frac{1}{2}\left[0.775 + 0.5\,(0.94118 + 0.64000)\right] = 0.78279$$

$$T_{2,1} = T_1(h/2^2) = \frac{4^1\,T_{2,0} - T_{1,0}}{4^1 - 1} = \frac{4(0.78279) - 0.775}{3} = 0.78539$$

$$T_{2,2} = T_2(h/2^2) = \frac{4^2\,T_{2,1} - T_{1,1}}{4^2 - 1} = \frac{16(0.78539) - 0.78333}{15} = 0.78552$$

The other elements shown in Table 6.5.2 are similarly computed.

**Table 6.5.2**

| | | | | |
|---|---|---|---|---|
| 0.75000 | | | | |
| 0.77500 | 0.78333 | | | |
| 0.78279 | 0.78539 | 0.78552 | | |
| 0.78474 | 0.78539 | 0.78539 | 0.78539 | |
| 0.78523 | 0.78539 | 0.78539 | 0.78539 | 0.78539 |

For Romberg integration, compute $T_{0,0}$ using the trapezoidal rule with $h = b - a$. Then the elements $T_{1,0}$ and $T_{1,1}$ are computed using Equations (6.5.8) and (6.5.13), respectively. Next the elements $T_{2,0}$, $T_{2,1}$, and $T_{2,2}$ are computed. $T_{2,0}$ is computed using the trapezoidal rule given by Equation (6.5.8). $T_{2,1}$ and $T_{2,2}$ are computed using Equation (6.5.13). Then compute the elements $T_{3,0}$, $T_{3,1}$, $T_{3,2}$, and $T_{3,3}$ of the fourth row. Compute the array row by row until the last two values in the same row agree to the desired accuracy. We continued computations to show the convergence of the diagonal terms.

Since the trapezoidal rule converges to $I$ as $h \to 0$ and the error in the Romberg integration is smaller than the trapezoidal rule for any $h$, $T_k(h/2^i)$ converges to $I$ for each $k$ as $h \to 0$. Also, the diagonal sequence $T_{0,0}, T_{1,1}, \dots, T_{k,k}, \dots$ converges much more rapidly than the trapezoidal sequence $T_{0,0}, T_{1,0}, \dots, T_{k,0}, \dots$. Many results concerning convergence of the Romberg integration can be found in Bauer et al. (1963).

**Algorithm**

**Subroutine** Romb($N$, $a$, $b$, $f$, eps)
Comment: This program approximates a given definite integral $\int_a^b F(x)\,dx$ using the Romberg method. We denote $T_0(h) = T_0(h/2^0) = T_{0,0}$, $T_0(h/2^1) = T_{1,0}, \dots, T_0(h/2^i) = T_{i,0}, \dots, T_0(h/2^N) = T_{N,0}$; $T_1(h/2^1) = T_{1,1}$, $T_1(h/2^2) = T_{2,1}, \dots, T_1(h/2^i) = T_{i,1}; \dots, T_N(h/2^N) = T_{N,N}$. $N$ is the maximum number of rows.
  1 **Print** $N$, $a$, $b$, eps
  2 $h = b - a$
  3 $T_{0,0} = h[f(a) + f(b)]/2$
  4 **Repeat** steps 5–13 **with** $i = 1, 2, \dots, N$
     5 sum $= 0.$
     6 $M = 2^{i-1}$
     7 $h1 = h/M$
     8 **Repeat** step 9 **with** $j = 1, 2, \dots, M$
       9 sum $=$ sum $+ f(a + (j - 0.5)\, h1)$
     10 $T_{i,0} = [T_{i-1,0} + (h1 * \text{sum})]/2$
Comment: This part computes the elements in the $(i + 1)$ row. We use Equation (6.5.13) to compute $T_{i,k}$.
    11 **Repeat** step 12 **with** $k = 1, 2, \dots, i$
      12 $T_{i,k} = (4^k\, T_{i,k-1} - T_{i-1,k-1})/(4^k - 1)$
    13 **If** $|\, T_{i,i} - T_{i,i-1}\,| <$ eps, **then** print deint $= T_{i,i}$ and exit.
14 **Print** 'The Romberg method is not converging for a given $N$'. ◀

**EXERCISES**

**1.** The definite integral $\int_0^1 e^x\,dx$ is approximated by $T_{2,0} = 1.72722$. Compute $T_{3,0}$.
**2.** The definite integral $\int_0^1 e^{-x^2}\,dx$ is approximated by $T_{3,0} = 0.74658$. Compute $T_{4,0}$.
**3.** The values of $T_{0,0}$, $T_{1,0}$, $T_{2,0}$, and $T_{3,0}$ are given for the definite integral $\int_1^2 dx/x$ $= \ln 2 \approx 0.69315$ in the following table:

| $k$ | $T_{k,0}$ | $T_{k,1}$ | $T_{k,2}$ | $T_{k,3}$ |
|---|---|---|---|---|
| 0 | 0.75000 | | | |
| 1 | 0.70833 | | | |
| 2 | 0.69702 | | | |
| 3 | 0.69412 | | | |

(a) Compute $T_{1,1}$     (b) Compute $T_{2,1}$ and $T_{2,2}$     (c) Compute $T_{3,1}$, $T_{3,2}$, and $T_{3,3}$.

4. The values of $T_{0,0}$, $T_{1,0}$, $T_{2,0}$, and $T_{3,0}$ are given for the definite integral $\int_0^\pi \sin x \, dx = 2$ in the following table:

| $k$ | $T_{k,0}$ | $T_{k,1}$ | $T_{k,2}$ | $T_{k,3}$ |
|---|---|---|---|---|
| 0 | 0.00000 | | | |
| 1 | 1.57080 | | | |
| 2 | 1.89612 | | | |
| 3 | 1.97423 | | | |

   (a) Compute $T_{1,1}$    (b) Compute $T_{2,1}$ and $T_{2,2}$    (c) Compute $T_{3,1}$, $T_{3,2}$, and $T_{3,3}$.
5. To approximate the definite integral $\int_0^1 e^{-x^2} \, dx$, Romberg integration is used. If $T_{1,1} = 0.74685$ and $T_{2,2} = 0.74682$, find $T_{2,1}$.
6. To approximate the definite integral $\int_0^2 e^x \, dx$, Romberg integration is used. If $T_{1,1} = 6.42073$ and $T_{2,2} = 6.38924$, find $T_{2,1}$.
7. Use Romberg integration through the third row of Table 6.5.1 to approximate each of the following definite integrals and compare each approximate value with the exact value:
   (a) $\int_0^1 x \, dx$    (b) $\int_0^1 e^x \, dx$
8. Use Romberg integration through the third row of Table 6.5.1 to approximate each of the following definite integrals:
   (a) $\int_0^1 \sin x \, dx$    (b) $\int_0^1 e^{-x^2} \, dx$
9. Show that $T_1(h/2)$ in Table 6.5.1 is the same as the Simpson rule with step size $h/2$.
10. Show that $T_2(h/2^2)$ in Table 6.5.1 is the same as the closed Newton–Cotes formula for $N = 4$ with step size $h/4$.
11. Show that $T_3(h/2^3)$ in Table 6.5.1 does not correspond to any of the closed Newton–Cotes formulas.

## COMPUTER EXERCISES

C.1. Write a subroutine that uses the trapezoidal rule to evaluate $\int_a^b f(x) \, dx$ over an arbitrary number of subintervals $(b - a)/2^i$ in length, where $i$ is a parameter of the subroutine. Use Equation (6.5.8).
C.2. Write a computer program to evaluate $\int_a^b f(x) \, dx$ accurate within $10^{-5}$ using the Romberg method with Exercise C.1.
C.3. Use Exercise C.2 to compute the definite integral

$$J_2(1) = \frac{1}{3\pi} \int_{-1}^{1} (1 - x^2)^{3/2} \cos x \, dx$$

C.4. Sometimes integrals are used to define special mathematical functions. The error function erf$(x)$ is defined by

$$\text{erf}(x) = \frac{2}{\sqrt{\pi}} \int_0^x e^{-t^2} \, dt$$

Use Exercise C.2 to find erf$(0.1)$, erf$(0.5)$, and erf$(1.0)$.

**C.5.** The Bessel function of order zero is defined by

$$J_0(x) = \frac{1}{\pi} \int_0^\pi \cos(x \sin \theta) \, d\theta$$

Use Exercise C.2 to find $J_0(0.1)$, $J_0(0.5)$, and $J_0(1.0)$.

**C.6.** The **Frensel integral** is defined by

$$C(x) = \int_0^x \cos\left(\frac{\pi}{2} t^2\right) dt$$

Use Exercise C.2 to find $C(\frac{1}{2})$, $C(1)$, and $C(2)$.

---

## 6.6 GAUSSIAN INTEGRATION

All integration formulas developed are of the form

$$\int_a^b f(x) \, dx \approx a_0 f(x_0) + a_1 f(x_1) + \ldots + a_N f(x_N) \tag{6.6.1}$$

The nodes $x_0, x_1, \ldots, x_N$ have been specified to be equally spaced so there is no choice in the selection of the base points. To integrate a function that is in an equally-spaced tabulated form, these methods are clearly preferable. However, if a function $f$ is known analytically, there is no need to require equally-spaced nodes for the integrating formulas. If $x_0, x_1, \ldots, x_N$ are not fixed in advance and if there are no other restrictions on them, then there are $2N + 2$ unknowns or parameters in Equation (6.6.1) that should satisfy $2N + 2$ equations. Thus it seems reasonable to expect that we can obtain a formula that is exact whenever $f$ is a polynomial of degree $k \leq 2N + 1$. Gauss showed that by selecting $x_0, x_1, \ldots, x_N$ properly it is possible to construct formulas far more accurate than the corresponding Newton–Cotes formulas. The formulas based on this principle are called **Gaussian integration formulas**.

Let us determine the parameters in the case of two points. It is convenient to determine the parameters if the integral involved is of the form $\int_{-1}^1 f(x) \, dx$. Let us determine four parameters $a_0$, $a_1$, $x_0$, and $x_1$ such that

$$\int_{-1}^1 f(x) \, dx \approx a_0 f(x_0) + a_1 f(x_1) \tag{6.6.2}$$

This formula gives the exact value whenever $f$ is a polynomial of degree three or less. In order to get four equations, let $f(x) = 1$, $x$, $x^2$, and $x^3$ in Equation (6.6.2). First consider $f(x) = 1$. Then $\int_{-1}^1 f(x) \, dx = \int_{-1}^1 1 \, dx = x|_{-1}^1 = 2$ and $a_0 f(x_0) + a_1 f(x_1) = a_0 \cdot 1 + a_1 \cdot 1$. Thus we have $a_0 + a_1 = 2$. Let $f(x) = x$. Then $\int_{-1}^1 f(x) \, dx = \int_{-1}^1 x \, dx = x^2/2|_{-1}^1 = 0$, and $a_0 f(x_0) + a_1 f(x_1) = a_0 x_0 + a_1 x_1$. Hence $a_0 x_0 + a_1 x_1 = 0$. Similarly considering $f(x) = x^2$ and $x^3$, we get $a_0 x_0^2 + a_1 x_1^2 = \frac{2}{3}$ and $a_0 x_0^3 + a_1 x_1^3 = 0$. We have

$$a_0 + a_1 = 2$$
$$a_0 x_0 + a_1 x_1 = 0$$
$$a_0 x_0^2 + a_1 x_1^2 = 2/3$$
$$a_0 x_0^3 + a_1 x_1^3 = 0 \tag{6.6.3}$$

Solving these four nonlinear equations,

$$a_0 = a_1 = 1 \qquad x_0 = -x_1 = -\sqrt{3}/3$$

Therefore the integration formula is given by

$$\int_{-1}^{1} f(x)\, dx \approx f(-\sqrt{3}/3) + f(\sqrt{3}/3) \tag{6.6.4}$$

This is called the two-point Gaussian integration formula. It is remarkable that by adding two values, we get the exact value of an integral of any polynomial of degree three or less.

In order to use Equation (6.6.4), write $\int_a^b f(x)\, dx$ in the form $\int_{-1}^{1} F(t)\, dt$. Let $x = \alpha t + \beta$ where $\alpha$ and $\beta$ are to be determined so that when $t = 1$, $x = b$ and when $t = -1$, $x = a$. Thus we have $\alpha + \beta = b$ and $-\alpha + \beta = a$. Solving these two equations

$$\alpha = \frac{(b - a)}{2} \quad \text{and} \quad \beta = \frac{(b + a)}{2}$$

Thus if

$$x = \frac{(b - a)t + (b + a)}{2} \tag{6.6.5}$$

then

$$\int_a^b f(x)\, dx = \frac{(b - a)}{2} \int_{-1}^{1} f\left(\frac{(b - a)t + (a + b)}{2}\right) dt$$

**EXAMPLE 6.6.1**

Integrate $\int_0^1 e^{2x}\, dx$ using the two-point Gaussian integration formula.

$$\int_0^1 e^{2x}\, dx = \frac{1}{2} \int_{-1}^{1} e^{t+1}\, dt \approx \frac{1}{2}[e^{(-\sqrt{3}/3)+1} + e^{(\sqrt{3}/3)+1}] \approx 3.18405$$

The exact value to five places is

$$\int_0^1 e^{2x}\, dx = \left.\frac{e^{2x}}{2}\right|_0^1 = \frac{1}{2}(e^2 - 1) = 3.19453$$

The error is $0.1048 \times 10^{-1}$.    ■ ■ ■

Formulas containing more terms can be derived using the same technique that was used to derive Equation (6.6.4). The solutions to the corresponding nonlinear systems are difficult to obtain, so we present an alternative derivation of these formulas. Since $x_i$ are unknowns, use the Lagrange interpolating polynomial, which allows arbitrarily-spaced base points. Using the Lagrange Polynomial Approximation Theorem 4.3.1,

$$f(x) = \sum_{j=0}^{N} f(x_j)\, L_j(x) + \frac{f^{(N+1)}(\xi(x))}{(N + 1)!} \prod_{j=0}^{N} (x - x_j) \tag{6.6.6}$$

where

$$L_j(x) = \prod_{\substack{i=0 \\ i \neq j}}^{N} \frac{(x - x_i)}{(x_j - x_i)} \quad \text{and} \quad -1 < \xi < 1$$

For Equation (6.6.1), $a_0, a_1, \ldots, a_N, x_0, x_1, \ldots, x_N$ are $2N + 2$ unknowns, and therefore Equation (6.6.1) gives the exact value whenever $f(x)$ is a polynomial of degree $2N + 1$ or less. Now if we assume that $f(x)$ is a polynomial of degree $2N + 1$, then the term $f^{(N+1)}(\xi(x))/(N + 1)!$ is a polynomial of degree $N$ at most. Let

$$\frac{f^{(N+1)}(\xi(x))}{(N + 1)!} = q_N(x) \qquad \textbf{(6.6.7)}$$

where $q_N(x)$ is a polynomial of degree $N$.

Substituting Equation (6.6.7) in Equation (6.6.6) and integrating between $-1$ and $1$ result in

$$\int_{-1}^{1} f(x) \, dx = \sum_{j=0}^{N} f(x_j) \int_{-1}^{1} L_j(x) \, dx + \int_{-1}^{1} q_N(x) \prod_{j=0}^{N} (x - x_j) \, dx \qquad \textbf{(6.6.8)}$$

We want to select $x_j$ in such a way that the error term in Equation (6.6.8) vanishes since $f(x)$ is a polynomial of degree $\leq 2N + 1$. We want

$$\int_{-1}^{1} q_N(x) \prod_{j=0}^{N} (x - x_j) \, dx = 0 \qquad \textbf{(6.6.9)}$$

$\prod_{j=0}^{N} (x - x_j)$ is a polynomial of degree $N + 1$ and $q_N(x)$ is a polynomial of degree $N$ or less; therefore, Equation (6.6.9) is satisfied if we choose polynomial $\prod_{j=0}^{N} (x - x_j)$ of degree $N + 1$ orthogonal to all polynomials of degree $N$ or less on the interval $[-1, 1]$.

The **Legendre polynomials** defined by

$$P_0(x) = 1$$

$$P_1(x) = x$$

$$P_i(x) = \frac{1}{i} [(2i - 1) x P_{i-1}(x) - (i - 1) P_{i-2}(x)] \quad \text{for } i = 2, 3, \ldots \qquad \textbf{(6.6.10)}$$

are orthogonal polynomials over $[-1, 1]$ with respect to the weight function $w(x) = 1$. Orthogonal polynomials are also linearly independent (Appendix E) and, therefore, $q_N(x)$ in Equation (6.6.9) can be written as a linear combination of Legendre polynomials $P_i(x)$, $i = 0, 1, \ldots, N$. If we pick $x_j$, $j = 0, 1, \ldots, N$ of $\prod_{j=0}^{N} (x - x_j)$ as the zeros of the $(N + 1)$th degree Legendre polynomial $P_{N+1}(x)$, then Equation (6.6.9) will be satisfied. Also, it is known that the zeros of the Legendre polynomial of any degree $\geq 1$ are all real and distinct (Appendix E). By selecting the zeros of Legendre polynomial $P_{N+1}$ as the nodes for Equation (6.6.1), Equation (6.6.8) reduces to

$$\int_{-1}^{1} f(x) \, dx = \sum_{j=0}^{N} f(x_j) \int_{-1}^{1} L_j(x) \, dx \qquad \textbf{(6.6.11)}$$

whenever $f(x)$ is a polynomial of degree $2N + 1$ or less. Therefore

$$\int_{-1}^{1} f(x)\,dx = \sum_{j=0}^{N} f(x_j) \int_{-1}^{1} L_j(x)\,dx + \frac{f^{(2N+2)}(\eta)}{(2N+2)!} \int_{-1}^{1} \sum_{i=0}^{N} (x - x_i)^2\,dx \quad \textbf{(6.6.12)}$$

Thus

$$\int_{-1}^{1} f(x)\,dx \approx \sum_{j=0}^{N} a_j f(x_j) \qquad \textbf{(6.6.13)}$$

where $a_j = \int_{-1}^{1} L_j(x)\,dx$ and $x_j$, $j = 0, 1, \ldots, N$ are the zeros of a Legendre polynomial $P_{N+1}(x)$.

EXAMPLE 6.6.2

Find $a_0$, $a_1$, and $a_2$ for $N = 2$.

The zeros of $P_3(x) = 5(x^3 - (3/5)x)/2$ are $-\sqrt{3/5}$, 0, and $\sqrt{3/5}$. Since $x_0 = -\sqrt{3/5}$, $x_1 = 0$, and $x_2 = \sqrt{3/5}$,

$$L_0(x) = \frac{(x - 0)(x - \sqrt{3/5})}{(-\sqrt{3/5} - 0)(-\sqrt{3/5} - \sqrt{3/5})} = \frac{5}{6} x\,(x - \sqrt{3/5})$$

$$L_1(x) = \frac{5}{3}(x + \sqrt{3/5})(x - \sqrt{3/5})$$

$$L_2(x) = \frac{5}{6} x(x + \sqrt{3/5})$$

Hence

$$a_0 = \int_{-1}^{1} L_0(x)\,dx = \frac{5}{6} \int_{-1}^{1} [x^2 - (\sqrt{3/5})x]\,dx = \frac{5}{9}$$

Similarly, it can be verified that $a_1 = \int_{-1}^{1} L_1(x)\,dx = 8/9$ and $a_2 = \int_{-1}^{1} L_2\,dx = 5/9$.

■ ■ ■

A short table of Legendre polynomials, zeros of Legendre polynomials, and the values of the coefficients $a_j$ are given in Table 6.6.1.

EXAMPLE 6.6.3

Use Gaussian quadrature with $N = 3$ and 4 to evaluate $\int_0^1 e^{2x}\,dx$.

Using Equation (6.6.5), we get $\int_0^1 e^{2x}\,dx = \frac{1}{2} \int_{-1}^{1} e^{1+t}\,dt$.

From Table 6.6.1 for $N = 3$,

$$\frac{1}{2} \int_{-1}^{1} e^{1+t}\,dt \approx \frac{1}{2}\left[\frac{5}{9} e^{1-\sqrt{3/5}} + \frac{8}{9} e + \frac{5}{9} e^{1+\sqrt{3/5}}\right] \approx 3.19444$$

**Table 6.6.1**

| Legendre Polynomials | Zeros | Weights $a_j$ |
|---|---|---|
| $P_1(x) = x$ | 0 | 2 |
| $P_2(x) = \frac{3}{2}\left(x^2 - \frac{1}{3}\right)$ | $-\frac{1}{\sqrt{3}} = -0.57735$ | 1.0 |
| | $\frac{1}{\sqrt{3}} = \phantom{-}0.57735$ | 1.0 |
| $P_3(x) = \frac{5}{2}\left(x^3 - \frac{3}{5}x\right)$ | $-\sqrt{\frac{3}{5}} = -0.77460$ | $\frac{5}{9} = 0.55556$ |
| | $0.00000$ | $\frac{8}{9} = 0.88889$ |
| | $\sqrt{\frac{3}{5}} = 0.77460$ | $\frac{5}{9} = 0.55556$ |
| $P_4(x) = \frac{35}{8}\left(x^4 - \frac{6}{7}x^2 + \frac{3}{35}\right)$ | $-0.86114$ | $0.34785$ |
| | $-0.33998$ | $0.65214$ |
| | $0.33998$ | $0.65214$ |
| | $0.86114$ | $0.34785$ |
| $P_5(x) = \frac{63}{8}\left(x^5 - \frac{10}{9}x^3 + \frac{5}{21}x\right)$ | $-0.90618$ | $0.23692$ |
| | $-0.53847$ | $0.47863$ |
| | $0.00000$ | $0.56889$ |
| | $0.53847$ | $0.47863$ |
| | $0.90618$ | $0.23692$ |

From Table 6.6.1 for $N = 4$,

$$\frac{1}{2}\int_{-1}^{1} e^{1+t}\, dt \approx \frac{1}{2}[(0.34785)\, e^{1-0.86114} + (0.65214)\, e^{1-0.33998}$$
$$+ (0.65214)\, e^{1+0.33998} + (0.34785)\, e^{1+0.86114}] \approx 3.19450$$

while the exact value of the integral is $(e^2 - 1)/2 = 3.19453$ to five places.  ∎∎∎

---

### EXERCISES

1.  Use the two-point and three-point Gaussian integration formulas to approximate:

    (a) $\int_{-1}^{1} e^x\, dx$    (b) $\int_0^\pi \sin x\, dx$    (c) $\int_2^3 dx/x$

    Compare each approximated value with the exact value.

2.  Find the truncation error formulas for the two-point and three-point Gaussian integration formulas and compare them with the truncation error term for the Simpson rule.

3.  One can derive the composite rule for the Gaussian integration formulas by dividing the interval $[a, b]$ into $M$ equal subintervals. Then use the Gaussian integration formula on each subinterval. Approximate $\int_2^3 dx/x$ by dividing $[2, 3]$

into two subintervals, and on each subinterval use the two-point Gaussian integration formula and Equation (6.6.5).

4. Using $f(x) = 1$ and $x$, derive the one-point Gaussian integration formula

$$\int_{-1}^{1} f(x)\,dx \approx a_0 f(x_0)$$

5. Solve the system of equations given by Equation (6.6.3).
6. Derive the three-point Gaussian integration formula

$$\int_{-1}^{1} f(x)\,dx \approx a_0 f(-\alpha) + a_1 f(0) + a_2 f(\alpha)$$

by solving the system of equations.
7. Find $a_0$, $x_0$, $x_1$, and $x_2$ such that

$$\int_{-1}^{1} f(x)\,dx \approx a_0\,[f(x_0) + f(x_1) + f(x_2)]$$

Use this to evaluate Exercise 1(a).
8. Using Equation (6.6.10), determine $P_2(x)$, $P_3(x)$, $P_4(x)$, and $P_5(x)$. Find the zeros of $P_5(x)$.
9. Verify the entries in Table 6.6.1 for $N = 2, 3$, and 4.
10. Verify that the Legendre polynomials $P_1(x)$, $P_2(x)$, $P_3(x)$, and $P_4(x)$ are orthogonal over the interval $[-1, 1]$ with respect to the weight function $w(x) = 1$.
11. Expand the second-degree polynomial $P(x) = x^2 + 8x + 1$ in terms of the Legendre polynomials. Verify that $\int_{-1}^{1} (x^2 + 8x + 1) \prod_{i=0}^{2} (x - x_i)\,dx = 0$ where $x_0$, $x_1$, and $x_2$ are the zeros of $P_3(x)$.
12. The Legendre polynomials are also given by

$$P_i(x) = \frac{1}{2^i\,i!}\frac{d^i}{dx^i}(x^2 - 1)^i$$

for $i = 1, 2, \ldots$, with $P_0(x) = 1$. Find $P_1(x)$, $P_2(x)$, and $P_3(x)$.
13. Verify that $P_1(x)$, $P_2(x)$, and $P_3(x)$ satisfy the Legendre differential equation

$$(1 - x^2)\frac{d^2 y}{dx^2} - 2x\frac{dy}{dx} + i(i + 1)y = 0$$

with $i = 1, 2$, and 3 respectively.

## COMPUTER EXERCISES

C.1. Write a subroutine to find all real and simple zeros of the $N$th degree Legendre polynomial using the Newton–Raphson method.
C.2. Write a computer program to approximate $\int_2^3 x^x\,dx$ using the Gaussian integration formula with $N = 4$. Use Equation (6.6.5) and Table 6.6.1 or Exercise C.1 or C.3, page 75.

**C.3.** Write a computer program to approximate an integral $\int_a^b f(x)\,dx$ using the Gaussian composite rule. Use Equation (6.6.5) and Table 6.6.1 or Exercises C.1 or C.3, page 75, to approximate $\int_2^3 dx/x$ by dividing [2, 3] into four equal subintervals.

## 6.7  IMPROPER INTEGRALS

We may not be able to use any of the standard numerical methods we have discussed to approximate an improper integral. Sometimes an improper integral can be approximated by first modifying it and then applying a standard numerical method. In this section we consider a variety of techniques through illustration.

**E X A M P L E   6.7.1**

Approximate $\int_0^a f(x)/\sqrt{x}\,dx$ where $f$ is differentiable on $[0, a]$ as many times as necessary.

Since $f(x)/\sqrt{x}$ is undefined at $x = 0$, we cannot use any numerical method directly. However, by substituting $t = \sqrt{x}$

$$\int_0^a \frac{f(x)}{\sqrt{x}}\,dx = 2\int_0^{\sqrt{a}} f(t^2)\,dt$$

The right-hand side can be approximated by any standard numerical technique since the integrand is smooth.

This integral can also be modified by using integration by parts. Let $u = f(x)$ and $dv = dx/\sqrt{x}$. Then

$$\int_0^a \frac{f(x)}{\sqrt{x}}\,dx = 2\sqrt{x}f(x)\,\Big|_0^a - 2\int_0^a \sqrt{x}f'(x)\,dx$$

$$= 2\sqrt{a}f(a) - 2\int_0^a \sqrt{x}f'(x)\,dx$$

Since $\sqrt{x}\,f'(x)$ is smooth on $[0, a]$, $\int_0^a \sqrt{x}\,f'(x)\,dx$ can be approximated by any standard numerical technique.  ■ ■ ■

**E X A M P L E   6.7.2**

Approximate $\int_1^\infty e^{-x}/x^2\,dx$.

The upper limit can be converted to 0 by using $x = 1/y$. Thus

$$\int_1^\infty \frac{e^{-x}}{x^2}\,dx = -\int_1^0 e^{-(1/y)}\,dy = \int_0^1 e^{-(1/y)}\,dy$$

$\lim_{y \to 0^+} e^{-1/y} = 0$ and therefore $f(0^+) = 0$. Use any standard numerical technique to approximate $\int_0^1 e^{-(1/y)}\,dy$ with $f(0) = 0$.  ■ ■ ■

**EXAMPLE 6.7.3**

Approximate $\int_{-\infty}^{\infty} e^{-x^2}\, dx$.

Sometimes in order to integrate from $-\infty$ to $\infty$, we integrate over the interval $[-R_1, R_2]$ assuming that $f(x)$ is not contributing significantly for $x \leq -R_1$ and $x \geq R_2$. In this case for $x = \pm 5$, the integrand $e^{-25}$ is small. Therefore the integral can be approximated by evaluating $\int_{-5}^{5} e^{-x^2}\, dx$ using the Simpson rule. We can estimate the error involved in replacing $(-\infty, \infty)$ by $(-5, 5)$. ■ ■ ■

**EXERCISES**

Approximate each of the following integrals using any numerical technique:

**1.** $\displaystyle \int_0^{\pi} \frac{\sin x}{\sqrt{x}}\, dx$

**2.** $\displaystyle \int_0^{\pi} \frac{\sin x}{x^{3/2}}\, dx$

**3.** $\displaystyle \int_0^{\infty} \frac{dx}{e^x + e^{-x}}$

**4.** $\displaystyle \int_0^{\pi/2} x^2 \sin \frac{1}{x}\, dx$

**5.** $\displaystyle \int_1^{\infty} \frac{\sin x}{x^2}\, dx$

**COMPUTER EXERCISES**

**C.1.** The **gamma function** is defined by

$$\Gamma(t) = \int_0^{\infty} x^{t-1} e^{-x}\, dx \qquad t > 0$$

If $t$ is a positive integer, then $\Gamma(t) = (t - 1)!$. Using integration by parts, verify that $\Gamma(t + 1) = t\Gamma(t)$. For this reason, we need a table for $1 \leq t \leq 2$. Compute $\Gamma(t)$ for $t = 1.0, 1.1, \ldots, 2$.

**C.2.** The **sine integral** is defined by

$$\text{Si}(x) = \int_0^x \frac{\sin t}{t}\, dt$$

Compute $\text{Si}(x)$ for $x = 0.0, 0.1, \ldots, 1.0$. (*Note*: At $t = 0$, $(\sin t)/t$ is undefined.)

## SUGGESTED READINGS

E. Isaacson and H. B. Keller, *Analysis of Numerical Methods*, John Wiley & Sons, Inc., New York, 1966.

F. L. Bauer, H. Rutishauser, and E. Stiefel, *New Aspects in Numerical Quadrature,* Proc. Symp. Appl. Math., Am. Math. Soc. **15** (1963), 199–218.

# 7 SYSTEMS OF LINEAR EQUATIONS

L et

$$4x_1 + 2x_2 + 3x_3 = 7$$
$$2x_1 - 4x_2 - x_3 = 1$$
$$-x_1 + x_2 + 4x_3 = -5$$

be three planes in 3-dimensional space. We want to know whether the planes intersect. If so, we want to find the point of intersection. In this chapter and in Chapters 10 and 11, we develop methods to find the solutions of systems of linear equations of this type.

## 7.1 INTRODUCTION

We consider the problem of solving a system of $N$ linear equations in $N$ unknowns in this chapter. Systems of linear equations arise in virtually all areas of the physical, biological, and social sciences. Also, the numerical solutions of boundary-value problems in ordinary differential equations, partial differential equations, integral equations, optimization problems, and various other problems require the solving of systems of linear equations.

We develop direct methods to solve

$$A\mathbf{x} = \mathbf{b}$$

where $A$ is a given $N \times N$ matrix, $\mathbf{b}$ is a given $N$-column vector, and the $N$-column vector $\mathbf{x}$ is to be determined. The direct methods discussed are Gaussian elimination and factorization methods.

The term **direct method** refers to a procedure that, if all round-off errors are ignored, would lead to an exact solution in a finite number of steps. All direct methods involve some kind of elimination procedure.

## 7.2  GAUSSIAN ELIMINATION AND PIVOTING

Consider the simple example given at the beginning of the chapter. The system of equations

$$\begin{aligned} 4x_1 + 2x_2 + 3x_3 &= 7 \\ 2x_1 - 4x_2 - x_3 &= 1 \\ -x_1 + x_2 + 4x_3 &= -5 \end{aligned}$$  (7.2.1)

can be written in matrix form as

$$\begin{bmatrix} 4 & 2 & 3 \\ 2 & -4 & -1 \\ -1 & 1 & 4 \end{bmatrix} \begin{bmatrix} x_1 \\ x_2 \\ x_3 \end{bmatrix} = \begin{bmatrix} 7 \\ 1 \\ -5 \end{bmatrix}$$  (7.2.2)

The method starts by eliminating the first variable $x_1$ from the second and third equations. This requires that we subtract $\frac{1}{2}$ times the first equation from the second, and subtract $-\frac{1}{4}$ times the first equation from the third. The coefficient 4 of $x_1$ in the first equation is known as the **pivot** in this first elimination step. The result is an equivalent system of equations

$$\begin{bmatrix} 4 & 2 & 3 \\ 0 & -5 & -5/2 \\ 0 & 3/2 & 19/4 \end{bmatrix} \begin{bmatrix} x_1 \\ x_2 \\ x_3 \end{bmatrix} = \begin{bmatrix} 7 \\ -5/2 \\ -13/4 \end{bmatrix}$$  (7.2.3)

At the second stage, ignore the first equation. The second and third equations contain only two unknowns, $x_2$ and $x_3$. Eliminate $x_2$ from the third equation by subtracting $-3/10$ times the second equation from the third. The pivot for this step is $-5$. This leads to

$$\begin{bmatrix} 4 & 2 & 3 \\ 0 & -5 & -5/2 \\ 0 & 0 & 4 \end{bmatrix} \begin{bmatrix} x_1 \\ x_2 \\ x_3 \end{bmatrix} = \begin{bmatrix} 7 \\ -5/2 \\ -4 \end{bmatrix}$$  (7.2.4)

This method is based on the fact that replacing any equation by a linear combination of itself and another equation of the system does not change the solution of the original system (this is proven in an introductory linear algebra course).

The second part of the method is to solve Equation (7.2.4) by **back substitution**. The third row in the matrix Equation (7.2.4) states $4x_3 = -4$; hence, $x_3 = -1$. When $x_3 = -1$ is substituted into the second row equation $-5x_2 - (5/2)x_3 = -5/2$, the result is $x_2 = 1$. Substitution of these values of $x_2$ and $x_3$ into the first row equation $4x_1 + 2x_2 + 3x_3 = 7$ gives $x_1 = 2$. This example illustrates the method of **Gaussian elimination**.

The idea of elimination can be extended to $N$ equations in $N$ unknowns, no matter how large $N$ may be. In the first step, use multiples of the first equation to eliminate the first variable from the other equations. This leaves a system of $(N - 1)$ equations in $(N - 1)$ unknowns. The process is repeated until we are left with one equation and one unknown. By using back substitution, the upper triangular system of linear equations is solved. Let us formalize the procedure. Consider a system of $N$ linear equations in $N$ unknowns

$$
\begin{aligned}
a_{11}x_1 + a_{12}x_2 + \cdots + a_{1N}x_N &= b_1 \\
a_{21}x_1 + a_{22}x_2 + \cdots + a_{2N}x_N &= b_2 \\
\vdots \qquad \vdots \qquad \cdots \qquad \vdots \qquad \vdots & \\
a_{N1}x_1 + a_{N2}x_2 + \cdots + a_{NN}x_N &= b_N
\end{aligned}
$$

(7.2.5)

This can be written as

$$
A\mathbf{x} = \mathbf{b}
$$

(7.2.6)

where

$$
A = \begin{bmatrix} a_{11} & a_{12} & \cdots & a_{1N} \\ a_{21} & a_{22} & \cdots & a_{2N} \\ \vdots & \vdots & \ddots & \vdots \\ a_{N1} & a_{N2} & \cdots & a_{NN} \end{bmatrix}
\qquad
\mathbf{x} = \begin{bmatrix} x_1 \\ x_2 \\ \vdots \\ x_N \end{bmatrix}
\qquad
\mathbf{b} = \begin{bmatrix} b_1 \\ b_2 \\ \vdots \\ b_N \end{bmatrix}
$$

Let us augment the matrix $A$ by $\mathbf{b}$ to get a new matrix

$$
[A \mid \mathbf{b}] = \left[ \begin{array}{cccc|c} a_{11} & a_{12} & \cdots & a_{1N} & b_1 \\ a_{21} & a_{22} & \cdots & a_{2N} & b_2 \\ \vdots & \vdots & \ddots & \vdots & \vdots \\ a_{N1} & a_{N2} & \cdots & a_{NN} & b_N \end{array} \right]
$$

(7.2.7)

The first step is to eliminate all elements in the first column of $[A \mid \mathbf{b}]$ except $a_{11}$. Suppose the pivot $a_{11} \neq 0$. For convenience, define the row multipliers by

$$
m_{i1} = \frac{a_{i1}}{a_{11}} \quad \text{for} \quad i = 2, 3, \ldots, N
$$

Then by subtracting $m_{i1}$ times the first row from the $i$th row for $i = 2, 3, \ldots, N$, we get

$$
\begin{bmatrix} a_{11} & a_{12} & \cdots & a_{1N} & b_1 \\ 0 & a_{22} - m_{21}a_{12} & \cdots & a_{2N} - m_{21}a_{1N} & b_2 - m_{21}b_1 \\ \vdots & \vdots & \ddots & \vdots & \vdots \\ 0 & a_{N2} - m_{N1}a_{12} & \cdots & a_{NN} - m_{N1}a_{1N} & b_N - m_{N1}b_1 \end{bmatrix}
$$

Rewrite the new matrix as

$$
[A \mid \mathbf{b}]^{(1)} = \left[ \begin{array}{ccccc} a_{11} & a_{12} & \cdots & a_{1N} & b_1 \\ 0 & a_{22}^{(1)} & \cdots & a_{2N}^{(1)} & b_2^{(1)} \\ \vdots & \vdots & \ddots & \vdots & \vdots \\ 0 & a_{N2}^{(1)} & \cdots & a_{NN}^{(1)} & b_N^{(1)} \end{array} \right]
$$

(7.2.8)

where the superscript (1) indicates that one elimination has been carried out and

$$
a_{ij}^{(1)} = a_{ij} - m_{i1}a_{1j} \quad \text{for } i, j = 2, 3, \ldots, N
$$

and

$$
b_i^{(1)} = b_i - m_{i1}b_1 \quad \text{for } i = 2, 3, \ldots, N
$$

Assume the pivot $a_{22}^{(1)} \neq 0$. In the second step, eliminate all elements in the second column of $[A \mid \mathbf{b}]^{(1)}$ below $a_{22}^{(1)}$ by defining $m_{i2} = a_{i2}^{(1)}/a_{22}^{(1)}$ for $i = 3, 4, \ldots, N$ and by subtracting $m_{i2}$ times the second row from the $i$th row for $i = 3, 4, \ldots, N$. Then we get

$$[A \mid \mathbf{b}]^{(2)} = \begin{bmatrix} a_{11} & a_{12} & a_{13} & \ldots & a_{1N} & b_1 \\ 0 & a_{22}^{(1)} & a_{23}^{(1)} & \ldots & a_{2N}^{(1)} & b_2^{(1)} \\ 0 & 0 & a_{33}^{(2)} & \ldots & a_{3N}^{(2)} & b_3^{(2)} \\ \vdots & \vdots & \vdots & \ddots & \vdots & \vdots \\ 0 & 0 & a_{N3}^{(2)} & \ldots & a_{NN}^{(2)} & b_N^{(2)} \end{bmatrix}$$

where

$$a_{ij}^{(2)} = a_{ij}^{(1)} - m_{i2} a_{2j}^{(1)} \quad \text{for } i, j = 3, 4, \ldots, N$$

and

$$b_i^{(2)} = b_i^{(1)} - m_{i2} b_2^{(1)} \quad \text{for } i = 3, 4, \ldots, N$$

Repeating this procedure $(N - 1)$ times yields

$$[A \mid \mathbf{b}]^{(N-1)} = \begin{bmatrix} a_{11} & a_{12} & a_{13} & \ldots & a_{1N} & b_1 \\ 0 & a_{22}^{(1)} & a_{23}^{(1)} & \ldots & a_{2N}^{(1)} & b_2^{(1)} \\ 0 & 0 & a_{33}^{(2)} & \ldots & a_{3N}^{(2)} & b_3^{(2)} \\ \vdots & \vdots & \vdots & \ddots & \vdots & \vdots \\ 0 & 0 & 0 & \ldots & a_{NN}^{(N-1)} & b_N^{(N-1)} \end{bmatrix} \qquad \textbf{(7.2.9)}$$

provided that at the $i$th stage the pivots $a_{ii}^{(i-1)} \neq 0$ for $i = 1, 2, \ldots, N - 1$, where

$$a_{ij}^{(k)} = a_{ij}^{(k-1)} - m_{ik} a_{kj}^{(k-1)}$$

$$b_i^{(k)} = b_i^{(k-1)} - m_{ik} b_k^{(k-1)}$$

$$m_{ik} = \frac{a_{ik}^{(k-1)}}{a_{kk}^{(k-1)}}$$

for $k = 1, 2, \ldots, N - 1$, and $i, j = k + 1, k + 2, \ldots, N$, and

$$a_{ij}^{(0)} = a_{ij} \quad \text{for } i, j = 1, 2, \ldots, N \quad \text{and} \quad b_i^{(0)} = b_i \quad \text{for } i = 1, 2, \ldots, N$$
$$\textbf{(7.2.10)}$$

From Equation (7.2.9),

$$x_N = \frac{b_N^{(N-1)}}{a_{NN}^{(N-1)}} \quad \text{if } a_{NN}^{(N-1)} \neq 0$$

and

$$x_i = \frac{1}{a_{ii}^{(i-1)}} \left[ b_i^{(i-1)} - \sum_{j=i+1}^{N} a_{ij}^{(i-1)} x_j \right] \quad \text{for } i = N - 1, N - 2, \ldots, 1 \qquad \textbf{(7.2.11)}$$

**EXAMPLE 7.2.1**    Solve

$$4x_1 + 2x_2 + 3x_3 = 7$$

$$2x_1 - 4x_2 - x_3 = 1$$

$$-x_1 + x_2 + 4x_3 = -5$$

using Gaussian elimination with back substitution.

The augmented matrix is given by

$$[A \mid \mathbf{b}] = \begin{bmatrix} 4 & 2 & 3 & \mid & 7 \\ 2 & -4 & -1 & \mid & 1 \\ -1 & 1 & 4 & \mid & -5 \end{bmatrix}$$

We want to eliminate all elements in the first column of $[A \mid \mathbf{b}]$ except the diagonal element 4. Define $m_{21} = a_{21}/a_{11} = 2/4 = \frac{1}{2}$, then multiply the first row by $\frac{1}{2}$ and subtract it from the second row. In order to eliminate the third element in the first column, define $m_{31} = a_{31}/a_{11} = -\frac{1}{4}$, then multiply the first row by $-\frac{1}{4}$ and subtract it from the third row. This gives

$$[A \mid \mathbf{b}]^{(1)} = \begin{bmatrix} 4 & 2 & 3 & \mid & 7 \\ 0 & -5 & -5/2 & \mid & -5/2 \\ 0 & 3/2 & 19/4 & \mid & -13/4 \end{bmatrix}$$

Now we want to eliminate $3/2$ in the second column of $[A \mid \mathbf{b}]^{(1)}$ below $-5$. Define $m_{32} = a_{32}^{(1)}/a_{22}^{(1)} = (3/2)/(-5) = -3/10$. Multiply the second row by $-3/10$ and subtract it from the third row. This leads to

$$[A \mid \mathbf{b}]^{(2)} = \begin{bmatrix} 4 & 2 & 3 & \mid & 7 \\ 0 & -5 & -5/2 & \mid & -5/2 \\ 0 & 0 & 4 & \mid & -4 \end{bmatrix}$$

The equation $4x_3 = -4$ gives $x_3 = -1$. From the second row, we have $-5x_2 - (5/2)x_3 = -5/2$. Substituting $x_3 = -1$, we get $x_2 = 1$. From the first row, we have $4x_1 + 2x_2 + 3x_3 = 7$. Using $x_2 = 1$ and $x_3 = -1$ yields $x_1 = 2$.    ■ ■ ■

The method described above assumes that at each stage the pivotal element $a_{ii}^{(i-1)} \neq 0$. It is simple to construct examples that have pivotal zero elements. Let $a_{ii}^{(i-1)} = 0$. For such examples when this method is used, the procedure breaks down since we cannot define $m_{ik}$. To ensure that zero pivots are not used to compute the multipliers at stage $i$, a search is made in column $i$ for a nonzero element from rows $i + 1, \ldots, N$. If some entry is nonzero, say in row $k$, then we interchange rows $k$ and $i$. Again, this interchange of rows does not change the solution of the system. If all elements from the pivot down in column $i$ including the pivot are zero, then the matrix is **singular** (Exercise 8). The elimination method will pinpoint the singularity of the matrix.

In exact arithmetic, the Gaussian elimination method gives the solution of a linear system provided the matrix of it is nonsingular. Row exchanges may be necessary to avoid zero pivots. In practice, the round-off errors affect the computed solution. For a system of moderate size, say $N = 100$, the elimination procedure will require the order of $10^6$ operations (we will see this later). Although the round-off error may be small in each individual operation, the accumulated round-off errors could be large. We shall see that this potential accumulation of round-off errors is not as serious as we might expect if we take proper precautions. The other possibility is one of catastrophic round-off errors. Consider an example of a $2 \times 2$ system to see how it occurs.

Consider the system

$$0.0001x_1 + x_2 = 1$$

$$x_1 - x_2 = 0 \tag{7.2.12}$$

which has the exact solution $x_1 = x_2 = 10000/10001$.

Let us assume that we have a decimal computer with a word length of three digits; that is, numbers are represented in the form $0.*** \times 10^p$. Also, assume that it chops the resulting number. Let us carry out the Gaussian elimination procedure on this hypothetical computer. First note $a_{11} \neq 0$ so no interchange of rows is required. Then

$$m_{21} = \frac{a_{21}}{a_{11}} = \frac{0.1 \times 10^1}{0.1 \times 10^{-3}} = 0.1 \times 10^5$$

which is exact. Then

$$a_{22}^{(1)} = a_{22} - m_{21}a_{12} = -0.1 \times 10^1 - (0.1 \times 10^5)(0.1 \times 10^1)$$

$$= -0.1 \times 10^1 - 0.1 \times 10^5 = -0.00001 \times 10^5 - 0.1 \times 10^5 \approx -0.1 \times 10^5$$

The exact value of $a_{22}^{(1)}$ is $-0.10001 \times 10^5$, but the computer has a word length of only three digits, therefore it is represented by $-0.100 \times 10^5$. This is the first round-off error. Then

$$b_2^{(1)} = b_2 - m_{21}b_1 = 0 - (0.1 \times 10^5)(0.1 \times 10^1) = -0.1 \times 10^5$$

Thus

$$x_2 = \frac{b_2^{(1)}}{a_{22}^{(1)}} = \frac{-0.1 \times 10^5}{-0.1 \times 10^5} = 0.1 \times 10^1 = 1$$

and

$$x_1 = \frac{b_1 - a_{12}x_2}{a_{11}} = \frac{0.1 \times 10^1 - (0.1 \times 10^1)(0.1 \times 10^1)}{0.1 \times 10^{-3}} = 0$$

No round-off errors occurred in the calculations of $b_2^{(1)}$, $x_2$, and $x_1$. The computed value of $x_2$ agrees with the exact solution; however, $x_1$ is no longer a reasonable approximation of the exact solution. The only error made is in the calculation of $a_{22}^{(1)}$ and that is in the fifth decimal place. Every other operation was exact. How can this error cause the computed $x_1$ to deviate so drastically from its exact value?

Note that the quantity $-0.00001 \times 10^5$ dropped from $a_{22}^{(1)}$ is simply $a_{22}$. Since this is the only place where $a_{22}$ enters the calculation, the computed solution would have been the same if $a_{22}$ were zero. In other words, we have computed on our hypothetical computer the solution of the equations

$$0.0001\, x_1 + x_2 = 1$$

$$x_1 \qquad = 0 \tag{7.2.13}$$

Indeed, we expect the two systems of Equations (7.2.12) and (7.2.13) to have rather different solutions. Why did this occur? The large value of $m_{21}$ made it impossible to include $a_{22}$ in the sum on our hypothetical computer. The large value of $m_{21}$ was due to the smallness of $a_{11}$ relative to $a_{21}$. The remedy is to keep the absolute value of the multipliers less than or equal to one. This can be achieved by an interchange of equations. Consider

$$x_1 - x_2 = 0$$

$$0.0001x_1 + x_2 = 1 \qquad (7.2.14)$$

On our hypothetical three digits computer,

$$m_{21} = \frac{a_{21}}{a_{11}} = \frac{0.1 \times 10^{-3}}{0.1 \times 10^1} = 0.1 \times 10^{-3}$$

Thus

$$a_{22}^{(1)} = a_{22} - m_{21}a_{12} = 0.1 \times 10^1 - (0.1 \times 10^{-3})(-0.1 \times 10^1) = 0.1 \times 10^1$$

$$b_2^{(1)} = b_2 - m_{21}b_1 = (0.1 \times 10^1) - (0.1 \times 10^{-3})(0) \approx 0.1 \times 10^1$$

Therefore

$$x_2 = \frac{b_2^{(1)}}{a_{22}^{(1)}} = \frac{0.1 \times 10^1}{0.1 \times 10^1} = 0.1 \times 10^1 = 1$$

and

$$x_1 = \frac{b_1 - a_{12}x_2}{a_{11}} = \frac{0 - (-0.1 \times 10^1)(0.1 \times 10^1)}{0.1 \times 10^1} = 0.1 \times 10^1 = 1$$

Now the computed values of $x_1$ and $x_2$ agree well with the exact values. Thus, from the practical point of view, we choose the element with the largest magnitude from the $i$th column below or on the main diagonal and exchange that row with the $i$th row. This keeps the absolute value of the multipliers less than or equal to one. This procedure is known as **partial pivoting**.

**Algorithm**

**Subroutine** Solver ($N$, $A$, **b**, **x**, eps)
Comment: This subroutine solves the system of equations given by $A\mathbf{x} = \mathbf{b}$ using the Gaussian elimination method with partial pivoting. $A$ is a square matrix of order $N$. If the absolute value of any pivot element is less than eps, then the subroutine assumes that $A$ is singular. We denote $m_{ik}$ by $zm_{ik}$.
  1 **Print** $N$, $A$, **b**, eps
  2 **Repeat** steps 3–11 **with** $k = 1, 2, \ldots, N - 1$
      3 Jcol $= k$
      4 Call **subroutine** Pivot ($N$, $A$, **b**, Jcol)
      Comment: The subroutine Pivot locates Irow where the absolute value of the element is the largest in the column Jcol and then interchanges all elements in row Irow with those in row Jcol.
      5 If $|a_{kk}| <$ eps **then print** 'Algorithm fails because $|a_{kk}| <$ eps' and exit

6 **Repeat** steps 7–11 **with** $i = k + 1, k + 2, \ldots, N$
    7 $zm_{ik} = a_{ik}/a_{kk}$
    8 $a_{ik} = 0$
    9 $b_i = b_i - zm_{ik}b_k$
    10 **Repeat** step 11 **with** $j = k + 1, k + 2, \ldots, N$
        11 $a_{ij} = a_{ij} - zm_{ik}a_{kj}$
Comment: This part is for back substitution.
12 $x_N = b_N/a_{NN}$
13 **Repeat** steps 14–17 **with** $i = N - 1, N - 2, \ldots, 1$
    14 sum $= 0.0$
    15 **Repeat** step 16 **with** $j = i + 1, i + 2, \ldots, N$
        16 sum $=$ sum $+ a_{ij}x_j$
    17 $x_i = (b_i - \text{sum})/a_{ii}$
18 Print **x**

**Subroutine** Pivot ($N$, A, **b**, Jcol)
Comment: This subroutine locates in column Jcol the element whose absolute value is maximum at Irow. Then it interchanges the elements of the row Jcol with those of the row Irow.
1 **Print** $N$, A, **b**, Jcol
2 Apivot $= |a_{(\text{Jcol, Jcol})}|$
3 Ipivot $=$ Jcol
4 $J1 =$ Jcol $+ 1$
5 **Repeat** steps 6–7 Irow $= J1, J1 + 1, \ldots, N$
    6 Amax $= |a_{(\text{Irow, Jcol})}|$
    7 **If** (Amax. Gt. Apivot) **then** set Ipivot $=$ Irow and Apivot $=$ Amax
8 **If** (Ipivot. eq. Jcol) **then print** 'no need to interchange rows' and exit.
9 **Repeat** steps 10–12 **with** $i =$ Jcol, Jcol $+ 1, \ldots, N$
    10 save $= a_{(\text{Jcol}, i)}$
    11 $a_{(\text{Jcol}, i)} = a_{(\text{Ipivot}, i)}$
    12 $a_{(\text{Ipivot}, i)} =$ save
13 save $= b_{\text{Jcol}}$
14 $b_{\text{Jcol}} = b_{\text{Ipivot}}$
15 $b_{\text{Ipivot}} =$ save

There is no need to move the elements of a matrix in computer memory in order to use partial pivoting. We could define an index vector $\mathbf{l} = [l_1, l_2, \ldots, l_N]$. Initially $\mathbf{l} = [l_1, l_2, \ldots, l_N] = [1, 2, \ldots, N]$ and in the elimination process we refer to the $i$th row via the $i$th element of $\mathbf{l}$. We treat $l(i)$ as the $i$th row and use $a_{l(i),j}$ instead of $a_{i,j}$. To switch two rows, $j$ and $i$, we simply exchange elements $j$ and $i$ of $\mathbf{l}$. The computer program for Gaussian elimination with partial pivoting could be written using $\mathbf{l}$ without physically switching rows (Exercise C.5 or Atkinson, Harley, and Hudson 1989).

Total pivoting, an extension of partial pivoting, is sometimes used. Determine the element with the largest magnitude in the submatrix $A_k = [a_{ij}^{(k)}]$ where $i, j = k, k + 1, \ldots, N$ and then use this element as a pivot. With complete pivoting, not only a row but a column interchange is needed to move the element into the pivot. Because of the time involved in searching for the element with the largest magnitude and the complications involved in column and row exchanges, complete pivoting is rarely used. Normally partial pivoting is quite adequate. Therefore, use partial pivoting except in those rare cases where partial pivoting is inadequate.

## Scaling

If the elements of the coefficient matrix $A$ vary greatly in size, then during computation it is likely that a loss of significant digits will occur. To avoid this problem, the coefficient matrix $A$ should be scaled so that the maximum element in each row and each column is of the same order of magnitude. This is called **equilibration** of the matrix. This can be done by multiplying the rows and columns by suitable constants before the elimination starts. This can be done in many ways (Wilkinson 1965).

**E X A M P L E   7.2.2**     Solve

$$10x_1 + 10^5 x_2 = 10^5$$

$$x_1 - x_2 = 0 \qquad\qquad (7.2.15)$$

using partial pivoting on a hypothetical decimal computer with a word length of three digits (1) without scaling and (2) with scaling. Assume that the hypothetical decimal computer chops the resulting number. The exact solution is given by $x_1 = x_2 = 10000/10001$.

1. Since 10, the first element in the first row, is the largest element in the first column, there is no need to interchange rows for the partial pivoting strategy. Then on our hypothetical computer, we get

$$m_{21} = \frac{a_{21}}{a_{11}} = \frac{1}{10} = 0.1$$

Thus

$$a_{22}^{(1)} = a_{22} - m_{21}a_{12} = -0.1 \times 10^1 - (0.1 \times 10^0)(0.1 \times 10^6)$$

$$= 0.00001 \times 10^5 - 0.1 \times 10^5 \approx -0.1 \times 10^5$$

$$b_2^{(1)} = b_2 - m_{21}b_1 = 0 - (0.1) \times 10^5 = -0.1 \times 10^5$$

Hence

$$x_2 = \frac{b_2^{(1)}}{a_{22}^{(1)}} = \frac{-0.1 \times 10^5}{-0.1 \times 10^5} = 0.1 \times 10^1 = 1$$

and

$$x_1 = \frac{b_1 - a_{12}x_2}{a_{11}} = \frac{0.1 \times 10^6 - (0.1 \times 10^6)(0.1 \times 10^1)}{0.1 \times 10^2} = 0$$

The computed value of $x_2$ agrees with the exact solution; however, $x_1$ is no longer a reasonable approximation to the exact solution. We encounter the same problem with Equation (7.2.12). In fact, Equation (7.2.15) is obtained by multiplying the first equation of Equation (7.2.12) by $10^5$.

2. Scale the first row of Equation (7.2.15) so that the maximum element is approximately one. We equilibrate Equation (7.2.15) by multiplying the first equation of Equation (7.2.15) by $10^{-5}$ and get the system

$$0.0001x_1 + x_2 = 1$$

$$x_1 - x_2 = 0$$

that is Equation (7.2.12). Using the partial pivoting strategy, we get $x_1 = 1$ and $x_2 = 1$.    ∎∎∎

Unfortunately, equilibration of a matrix involves certain difficulties. Consider the system

$$x_1 + x_2 + x_3 = 2$$

$$10^4x_1 + x_2 - x_3 = 0$$

$$10^4x_1 - x_2 + x_3 = 0 \qquad \textbf{(7.2.16)}$$

If we multiply the first column by $10^{-4}$ (our new $x_1 = 10^{-4}x_1$), then

$$0.0001x_1 + x_2 + x_3 = 2$$

$$x_1 + x_2 - x_3 = 0$$

$$x_1 - x_2 + x_3 = 0 \qquad \textbf{(7.2.17)}$$

On the other hand, multiplying the second and third rows by $10^{-4}$ gives

$$x_1 + x_2 + x_3 = 2$$

$$x_1 + 0.0001x_2 - 0.0001x_3 = 0$$

$$x_1 - 0.0001x_2 + 0.0001x_3 = 0 \qquad \textbf{(7.2.18)}$$

We have equilibrated the systems of Equations (7.2.17) and (7.2.18). The system given by Equation (7.2.17) will behave properly with partial pivoting strategy on our hypothetical decimal computer with a word length of three digits while on the same computer Equation (7.2.18) produces an erroneous solution.

It is difficult to find a foolproof scaling algorithm, therefore it is recommended that the scaling of unknowns and equations be done on a problem-by-problem basis (Golub and Van Loan 1983, Forsyth, Malcolm, and Moler 1977, and Watkins 1991).

## EXERCISES

1. Solve the following linear systems using Gaussian elimination with partial pivoting. Check your answers using Gaussian elimination without partial pivoting.

(a)    $x_1 - x_2 + x_3 = 1$          (b)    $2x_1 + x_2 - x_3 = 2$
        $2x_1 + 3x_2 - x_3 = 4$                  $x_1 - x_2 + 3x_3 = 3$
       $-3x_1 + x_2 + x_3 = -1$                $-5x_1 + 2x_2 + 7x_3 = 0$

2. Use Gaussian elimination with and without partial pivoting on a hypothetical decimal computer with a word length of three digits. Assume that the hypothetical decimal computer chops the resulting number. Solve

$$0.0001x_1 + x_2 = 1$$
$$x_1 + x_2 = 2$$

3. Solve the linear systems in Exercise 1 using Cramer's rule.
4. Solve the linear system

$$x_1 + \quad x_2 - \quad x_3 = 0.5$$
$$10000x_1 + 5000x_2 + 5000x_3 = 15{,}000$$
$$0.002x_1 + 0.001x_2 + 0.003x_3 = 0.005$$

(*Hint:* First scale it properly.)

5. Using Gaussian elimination with partial pivoting and without partial pivoting on a hypothetical computer with a word length of three digits, solve

$$x_1 + \quad x_2 = 1$$
$$x_1 + 1.0001x_2 = 0$$

6. Prove that $A$ is singular where

$$A = \begin{bmatrix} a_{11} & a_{12} & a_{13} & a_{14} \\ 0 & 0 & a_{23} & a_{24} \\ 0 & 0 & a_{33} & a_{34} \\ 0 & 0 & a_{43} & a_{44} \end{bmatrix}$$

7. Solve the following linear system using partial pivoting on a hypothetical decimal computer with a word length of three digits (a) without scaling and (b) with scaling. Assume that the hypothetical decimal computer chops the resulting number.

$$2x_1 + 20000x_2 = 20000$$
$$x_1 - \quad x_2 = 0$$

8. If all elements from the pivot down in column $i$ including the pivot are zero, then prove that the matrix is singular.

9. Compute the solution of the linear system in Exercise 1(a) using Gaussian elimination with total pivoting.

10. At what step will Gaussian elimination without partial pivoting discover that $A$ is singular?

$$A = \begin{bmatrix} 1 & 2 & 3 & 4 \\ 2 & 3 & 4 & 5 \\ 3 & 4 & 5 & 6 \\ 3 & 4 & 5 & 6 \end{bmatrix}$$

11. At what step will Gaussian elimination without partial pivoting discover that $A$ is singular?

$$A = \begin{bmatrix} 1 & 2 & 3 & 3 \\ 2 & 3 & 4 & 4 \\ 3 & 4 & 5 & 5 \\ 4 & 5 & 6 & 6 \end{bmatrix}$$

**12.** Find an example that shows that there may be more than one way to equilibrate a matrix.

## COMPUTER EXERCISES

**C.1.** Write a subroutine that locates the largest element in absolute value from the $i$th column below or on the main diagonal and exchange that row with the $i$th row.

**C.2.** Write a subroutine that reduces $[A \mid \mathbf{b}]$ to Equation (7.2.9) using Exercise C.1.

**C.3.** Write a subroutine that solves Equation (7.2.9) by back substitution. Use Equation (7.2.11).

**C.4.** Write a program that reads the values of $A$ and $\mathbf{b}$ in order to solve a linear system of equations $A\mathbf{x} = \mathbf{b}$. Print the values of $A$ and $\mathbf{b}$. Use subroutines from Exercises C.1, C.2, and C.3 to solve $A\mathbf{x} = \mathbf{b}$ by Gaussian elimination with partial pivoting. Print the values of $\mathbf{x}$. Multiply $A$ and $\mathbf{x}$ and print those values along with $\mathbf{b}$.

**C.5.** Write a program that reads the values of $A$ and $\mathbf{b}$ in order to solve a linear system of equations $A\mathbf{x} = \mathbf{b}$. Print the values of $A$ and $\mathbf{b}$. Use Gaussian elimination with partial pivoting without physically switching rows by using index vector $\mathbf{l}$. Print the values of $\mathbf{x}$. Multiply $A$ and $\mathbf{x}$ and print those values along with $\mathbf{b}$.

**C.6.** Using Exercises C.4 or C.5, solve the following linear systems, if possible.

(a)
$$x_1 + \tfrac{1}{2}x_2 + \tfrac{1}{3}x_3 + \tfrac{1}{4}x_4 = \tfrac{1}{8}$$
$$\tfrac{1}{2}x_1 + \tfrac{1}{3}x_2 + \tfrac{1}{4}x_3 + \tfrac{1}{5}x_4 = \tfrac{1}{9}$$
$$\tfrac{1}{3}x_1 + \tfrac{1}{4}x_2 + \tfrac{1}{5}x_3 + \tfrac{1}{6}x_4 = \tfrac{1}{10}$$
$$\tfrac{1}{4}x_1 + \tfrac{1}{5}x_2 + \tfrac{1}{6}x_3 + \tfrac{1}{7}x_4 = \tfrac{1}{11}$$

(b)
$$x_1 + 0.50000x_2 + 0.33333x_3 + 0.25000x_4 = 0.12500$$
$$0.50000x_1 + 0.33333x_2 + 0.25000x_3 + 0.20000x_4 = 0.11111$$
$$0.33333x_1 + 0.25000x_2 + 0.20000x_3 + 0.16667x_4 = 0.10000$$
$$0.25000x_1 + 0.20000x_2 + 0.16667x_3 + 0.14286x_4 = 0.09091$$

**C.7.** An electric network is shown in the following diagram. The voltage applied at node $i$ is denoted by $v_i$ volts and the current flowing between node $i$ and node $j$ is denoted by $I_{ij}$ amps. To obtain voltage at each node, we use

(a) Ohm's law which states that the current $I_{ij}$ flowing between nodes $i$ and $j$ is given by

$$I_{ij} = \frac{v_i - v_j}{R_{ij}}$$

where $R_{ij}$, in ohms, is the resistance between nodes $i$ and $j$

(b) Kirchhoff's current law which states that the sum of the current arriving at each node is zero.

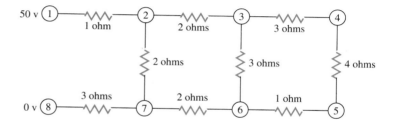

Applying these two laws at node two, we get

$$I_{12} + I_{32} + I_{72} = \frac{v_1 - v_2}{1} + \frac{v_3 - v_2}{2} + \frac{v_7 - v_2}{2} = 0$$

Since $v_1 = 50$ volts, we get

$$-4v_2 + v_3 + v_7 = -100$$

Similarly writing an equation for each node, we get the following system of six simultaneous equations in six unknowns:

$$
\begin{aligned}
-4v_2 + \phantom{7}v_3 \phantom{+ 2v_4 + 2v_6} + \phantom{3}v_7 &= -100 \\
3v_2 - 7v_3 + 2v_4 \phantom{+} + \phantom{1}2v_6 \phantom{+ 3v_7} &= 0 \\
4v_3 - 7v_4 + 3v_5 \phantom{+ 11v_6 + 3v_7} &= 0 \\
v_4 - 5v_5 + \phantom{1}4v_6 \phantom{+ 3v_7} &= 0 \\
2v_3 \phantom{- 7v_4} + 6v_5 - 11v_6 + 3v_7 &= 0 \\
3v_2 \phantom{- 7v_3 + 2v_4 + 6v_5} + \phantom{1}3v_6 - 8v_7 &= 0
\end{aligned}
$$

Find the voltages at all nodes and the current $I_{ij}$ in each resistor using Exercises C.4 or C.5.

**C.8.**  An electric network is shown in the following diagram.

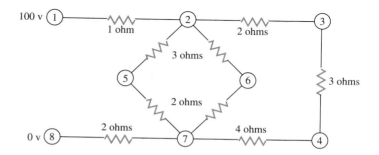

Find the voltages at all nodes and the current $I_{ij}$ in each resistor using Exercises C.4 or C.5.

**C.9.**  Wooden strips are said to be pin-jointed if they are fastened by a single nail. This allows us to ignore any torques transmitted by the members. As a result, the members have forces along their lengths. Determine the forces in the 7-member plane truss as shown in the accompanying diagram using Exercises C.4 or C.5.

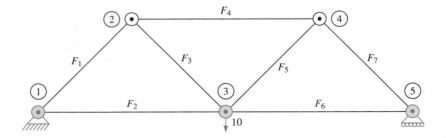

Since the number of joints $j$ is related to the number of members $m$ by $2j - 3 = m$, the truss is statically determinant. In other words, the forces are determined by the conditions of static equilibrium at each of the joints.

## 7.3   GAUSS–JORDAN ELIMINATION

A popular variant of Gaussian elimination is Gauss–Jordan elimination. In this method, we proceed as before to get Equation (7.2.8). Now assume that $a_{22}^{(1)} \neq 0$ in Equation (7.2.8). We eliminate all elements in the second column of $[A \mid \mathbf{b}]^{(1)}$ in Equation (7.2.8) except the diagonal element $a_{22}^{(1)}$ by defining $m_{i2} = a_{i2}^{(1)}/a_{22}^{(1)}$ for $i = 1, 3, 4, \ldots, N$ where $a_{12}^{(1)} = a_{12}$ and subtracting $m_{i2}$ times the second row from the $i$th row for $i = 1, 3, 4, \ldots, N$. This reduces Equation (7.2.8) to

$$
\begin{bmatrix}
a_{11} & 0 & a_{13}^{(2)} & \cdots & a_{1N}^{(2)} & b_1^{(2)} \\
0 & a_{22}^{(1)} & a_{23}^{(1)} & \cdots & a_{2N}^{(1)} & b_2^{(1)} \\
0 & 0 & a_{33}^{(2)} & \cdots & a_{3N}^{(2)} & b_3^{(2)} \\
\vdots & \vdots & \vdots & \ddots & \vdots & \vdots \\
0 & 0 & a_{N3}^{(2)} & \cdots & a_{NN}^{(2)} & b_N^{(2)}
\end{bmatrix}
\tag{7.3.1}
$$

The elements in the first row are given by

$$
a_{1j}^{(2)} = a_{1j} - \left( \frac{a_{12}}{a_{22}^{(1)}} \right) a_{2j}^{(1)} \quad \text{for } j = 3, 4, \ldots, N
$$

and

$$
b_1^{(2)} = b_1 - \left( \frac{a_{12}}{a_{22}^{(1)}} \right) b_2^{(1)}
$$

The idea is to reduce all elements in a column to zero except the diagonal element. Repeating this procedure, we get

$$\begin{bmatrix} a_{11} & 0 & 0 & \dots & 0 & b_1^{(N-1)} \\ 0 & a_{22}^{(1)} & 0 & \dots & 0 & b_2^{(N-1)} \\ 0 & 0 & a_{33}^{(2)} & \dots & 0 & b_3^{(N-1)} \\ \vdots & \vdots & \vdots & \ddots & \vdots & \vdots \\ 0 & 0 & 0 & \dots & a_{NN}^{(N-1)} & b_N^{(N-1)} \end{bmatrix} \tag{7.3.2}$$

where

$$a_{ij}^{(k)} = a_{ij}^{(k-1)} - m_{ik} a_{kj}^{(k-1)}$$

$$b_i^{(k)} = b_i^{(k-1)} - m_{ik} b_k^{(k-1)}$$

$$m_{ik} = \frac{a_{ik}^{(k-1)}}{a_{kk}^{(k-1)}}$$

for $k = 1, 2, \dots, N - 1$, $i, j = 1, 2, \dots, k - 1, k + 1, \dots, N$, and $a_{ij}^{(0)} = a_{ij}$ for $i, j = 1, 2, \dots, N$ and $b_i^{(0)} = b_i$ for $i = 1, 2, \dots, N$.

The solution of Equation (7.3.2) is given by

$$x_i = \frac{b_i^{(N-1)}}{a_{ii}^{(i-1)}} \quad \text{for } i = 1, 2, \dots, N$$

**EXAMPLE 7.3.1**   Using the Gauss–Jordan elimination method, solve

$$4x_1 + 2x_2 + 3x_3 = 7$$
$$2x_1 - 4x_2 - x_3 = 1$$
$$-x_1 + x_2 + 4x_3 = -5$$

By the Gaussian elimination method in Example 7.2.1, we had

$$[A \mid \mathbf{b}]^{(1)} = \begin{bmatrix} 4 & 2 & 3 & \mid & 7 \\ 0 & -5 & -5/2 & \mid & -5/2 \\ 0 & 3/2 & 19/4 & \mid & -13/4 \end{bmatrix}$$

Now eliminate all elements in the second column except the diagonal element $-5$. We want to eliminate 2 from the second column. Therefore, define $m_{12} = a_{12}^{(1)}/a_{22}^{(1)} = 2/-5$ and multiply the second row by $-2/5$ and subtract it from the first row. In order to eliminate 3/2 from the third column, define $m_{32} = a_{32}^{(1)}/a_{22}^{(1)} = (3/2)/(-5) = -(3/10)$. Multiply the second row by $-3/10$ and subtract it from the third row. This gives

$$[A \mid \mathbf{b}]^{(2)} = \begin{bmatrix} 4 & 0 & 2 & \mid & 6 \\ 0 & -5 & -5/2 & \mid & -5/2 \\ 0 & 0 & 4 & \mid & -4 \end{bmatrix}$$

We want to eliminate 2 and $-5/2$ from the third column. Define $m_{31} = a_{13}^{(2)}/a_{33}^{(2)} = \frac{2}{4} = \frac{1}{2}$ and $m_{32} = a_{23}^{(2)}/a_{33}^{(2)} = -(5/2)/4 = -5/8$. Multiply the third row by $\frac{1}{2}$ and subtract it from the first row. In order to eliminate $-5/2$ from the third column, multiply the third row by $-5/8$ and subtract it from the second row. This leads to

$$[A \mid \mathbf{b}]^{(3)} = \begin{bmatrix} 4 & 0 & 0 & \mid & 8 \\ 0 & -5 & 0 & \mid & -5 \\ 0 & 0 & 4 & \mid & -4 \end{bmatrix}$$

Thus

$$x_1 = \frac{b_{11}^{(3)}}{a_{11}^{(3)}} = \frac{8}{4} = 2$$

$$x_2 = \frac{b_{22}^{(3)}}{a_{22}^{(3)}} = \frac{-5}{-5} = 1$$

$$x_3 = \frac{b_{33}^{(3)}}{a_{33}^{(3)}} = \frac{-4}{4} = -1$$

■ ■ ■

It may seem that the Gauss–Jordan elimination is preferable to Gaussian elimination at first sight but we will show in the next section that the Gaussian elimination method is more efficient.

## 7.4   COMPUTATIONAL CONSIDERATIONS

Let us analyze the efficiencies of the Gaussian elimination method and the Gauss–Jordan method by finding the number of arithmetic operations needed to compute the solution vector $\mathbf{x}$. Division takes more time than multiplication and multiplication takes more time than addition or subtraction on many computers. The amount of time required to perform addition or subtraction is the same. Therefore we will keep count of division, multiplication, and addition/subtraction separately.

Let us look at the first step:

$$[A \mid \mathbf{b}] = \begin{bmatrix} a_{11} & a_{12} & \dots & a_{1N} & \mid & b_1 \\ a_{21} & a_{22} & \dots & a_{2N} & \mid & b_2 \\ \vdots & \vdots & \ddots & \vdots & \mid & \vdots \\ a_{N1} & a_{N2} & \dots & a_{NN} & \mid & b_N \end{bmatrix} \qquad \textbf{(7.4.1)}$$

Multiply the first row by $m_{21} = (a_{21})/a_{11}$ and subtract it from the second row. We need one division to get $m_{21}$ and $N$ multiplications to get $m_{21}a_{12}, m_{21}a_{13}, \dots, m_{21}a_{1N}$, and $m_{21}b_1$. Then we subtract this product from the second row. This takes $N$ additions/subtractions. Since we know that the first element in the second row is going to be zero, there is no need to form $m_{21}a_{11}$ and subtract it from $a_{21}$. It takes one division, $N$ multiplications and $N$ additions/subtractions to get the second row of Equation (7.4.2) from Equation (7.4.1). To get the 3rd, 4th, $\dots$, $N$th row of Equation (7.4.2) from Equation (7.4.1), we need one division, $N$ multiplications, and $N$ additions/subtractions for each row. Thus we need $N - 1$ divisions, $(N - 1)N$ multiplications, and $(N - 1)N$ additions/subtractions in order to reduce Equation (7.4.1) to

$$[A \mid \mathbf{b}]^{(1)} = \begin{bmatrix} a_{11} & a_{12} & a_{13} & \dots & a_{1N} & \big| & b_1 \\ 0 & a_{22}^{(1)} & a_{23}^{(1)} & \dots & a_{2N}^{(1)} & \big| & b_2^{(1)} \\ 0 & a_{32}^{(1)} & a_{33}^{(1)} & \dots & a_{3N}^{(1)} & \big| & b_3^{(1)} \\ \vdots & \vdots & \vdots & \ddots & \vdots & \big| & \vdots \\ 0 & a_{N2}^{(1)} & a_{N3}^{(1)} & \dots & a_{NN}^{(1)} & \big| & b_N^{(1)} \end{bmatrix} \qquad \textbf{(7.4.2)}$$

Similar analysis shows that we need $N - 2$ divisions, $(N - 2)(N - 1)$ multiplications, and $(N - 2)(N - 1)$ additions/subtractions to reduce Equation (7.4.2) to

$$[A \mid \mathbf{b}]^{(2)} = \begin{bmatrix} a_{11} & a_{12} & a_{13} & \dots & a_{1N} & \big| & b_1 \\ 0 & a_{22}^{(1)} & a_{23}^{(1)} & \dots & a_{2N}^{(1)} & \big| & b_2^{(1)} \\ 0 & 0 & a_{33}^{(2)} & \dots & a_{3N}^{(2)} & \big| & b_3^{(2)} \\ \vdots & \vdots & \vdots & \ddots & \vdots & \big| & \vdots \\ 0 & 0 & a_{N3}^{(2)} & \dots & a_{NN}^{(2)} & \big| & b_N^{(2)} \end{bmatrix} \qquad \textbf{(7.4.3)}$$

Continuing this procedure to reduce Equation (7.4.1) to

$$[A \mid \mathbf{b}]^{(N-1)} = \begin{bmatrix} a_{11} & a_{12} & a_{13} & \dots & a_{1N} & \big| & b_1 \\ 0 & a_{22}^{(1)} & a_{23}^{(1)} & \dots & a_{2N}^{(1)} & \big| & b_2^{(1)} \\ 0 & 0 & a_{33}^{(2)} & \dots & a_{3N}^{(2)} & \big| & b_3^{(2)} \\ \vdots & \vdots & \vdots & \ddots & \vdots & \big| & \vdots \\ 0 & 0 & 0 & \dots & a_{NN}^{(N-1)} & \big| & b_N^{(N-1)} \end{bmatrix} \qquad \textbf{(7.4.4)}$$

we have performed $(N - 1) + (N - 2) + \cdots + 1$ divisions, $(N - 1) \cdot N + (N - 2) \cdot (N - 1) + \cdots + 1 \cdot 2$ multiplications, and $(N - 1) \cdot N + (N - 2) \cdot (N - 1) + \cdots + 1 \cdot 2$ additions/subtractions. We need an equal number of multiplications and additions/subtractions to reduce Equation (7.4.1) to Equation (7.4.4). To evaluate the total number of multiplications, let us rewrite them as

$$1 \cdot 2 + 2 \cdot 3 + \cdots + (N - 1) \cdot N = \sum_{i=1}^{N} (i - 1) \cdot i$$

$$= \sum_{i=1}^{N} i^2 - \sum_{i=1}^{N} i$$

$$= \frac{N(N + 1)(2N + 1)}{6} - \frac{N(N + 1)}{2}$$

$$= \frac{N(N + 1)}{6} [2N + 1 - 3] = \frac{N(N - 1)(N + 1)}{3}$$

Thus, in order to reduce Equation (7.4.1) to Equation (7.4.4), we need $N(N - 1)(N + 1)/3$ multiplications and $N(N - 1)(N + 1)/3$ additions/subtractions. The total number of divisions is given by $1 + 2 + \cdots + (N - 1) = \sum_{i=1}^{N-1} i = N(N - 1)/2$.

We have yet to count the number of operations for back substitutions. From Equation (7.4.4), $a_{NN}^{(N-1)} x_N = b_N^{(N-1)}$. Therefore $x_N = b_N^{(N-1)}/a_{NN}^{(N-1)}$, and thus we need just one division. Solving $a_{N-1,N-1}^{(N-2)} x_{N-1} + a_{N-1,N}^{(N-2)} x_N = b_{N-1}^{(N-2)}$ for $x_{N-1}$, we get

$$x_{N-1} = \frac{b_{N-1}^{(N-2)} - a_{N-1,N}^{(N-2)} x_N}{a_{N-1,N-1}^{(N-2)}}$$

We need one multiplication to get $a_{N-1,N}^{(N-2)} x_N$, one addition/subtraction to get the numerator, and then one division. Continuing we get

$$x_1 = \frac{b_1 - a_{12}x_2 - \cdots - a_{1N}x_N}{a_{11}}$$

by solving $a_{11}x_1 + a_{12}x_2 + \cdots + a_{1N}x_N = b_1$ for $x_1$. Thus we need $(N-1)$ additions/subtractions, $(N-1)$ multiplications, and one division. For back substitution, we have performed $N$ divisions, $1 + 2 + \cdots + (N-1)$ multiplications, and $1 + 2 + \cdots + (N-1)$ additions/subtractions. Adding, we need $N(N-1)/2$ additions/subtractions, $N(N-1)/2$ multiplications, and $N$ divisions for back substitution. Thus for the complete process we need

$$\frac{N(N-1)}{2} + N = \frac{N(N+1)}{2} \quad \text{divisions}$$

$$\frac{N(N-1)(N+1)}{3} + \frac{N(N-1)}{2} = \frac{N(N-1)(2N+5)}{6} \quad \text{multiplications}$$

$$\frac{N(N-1)(2N+5)}{6} \quad \text{additions/subtractions}$$

It can be shown that we require $N^2$ divisions, $N(N^2-1)/2$ multiplications, and $N(N^2-1)/2$ additions/subtractions to compute $\mathbf{x}$ when we use the Gauss–Jordan elimination method (Exercise 5).

In Table 7.4.1, the total number of required divisions, multiplications, and additions/subtractions are given for the Gaussian elimination method and Gauss–Jordan method for comparison for different $N$.

**Table 7.4.1**

| | **Gaussian Elimination** | | | **Gauss–Jordan** | | |
|---|---|---|---|---|---|---|
| $N$ | Divisions | Multiplications | Additions/ Subtractions | Divisions | Multiplications | Additions/ Subtractions |
| 3 | 6 | 11 | 11 | 9 | 12 | 12 |
| 5 | 15 | 50 | 50 | 25 | 60 | 60 |
| 10 | 55 | 375 | 375 | 100 | 495 | 495 |
| 100 | 5050 | 338,250 | 338,250 | 10,000 | 499,950 | 499,950 |

The Gauss–Jordan method requires more operations than the Gaussian elimination method and therefore, the Gaussian elimination method is recommended. Ordinarily, we are not concerned about the cost when $N$ is small.

For large $N$, keeping the terms of higher order, the number of operations for Gaussian elimination is proportional to $N^3$ (Exercise 6). We say that Gaussian elimination is of order $N^3$ $(O(N^3))$ (Exercise 6). There exist algorithms (Pan 1984, Strassen 1969, and Coppersmith and Winograd 1982) that are of order $N^\alpha$ where $\alpha < 3$. The proportionality constant in front of $N^\alpha$ is large and coding is complicated so the new algorithms are largely of theoretical interest.

For large $N$, it is traditional to count the number of arithmetic operations in terms of **flop** (short term for floating-point operation). A flop consists of work associated with the statement

$$\text{sum} = \text{sum} + a_i * b_i$$

Thus a flop consists of one floating-point multiplication and one floating-point addition or subtraction. This terminology reminds us that various overheads like fetching, subscripting, and so forth are associated. In most linear algebra the division is not used extensively and therefore does not contribute significantly (Table 7.4.1). Thus, we say that the cost of Gaussian elimination is $N^3/3$ flops while the cost of Gauss–Jordan elimination is $N^3/2$ flops by ignoring lesser costs.

# EXERCISES

1. Solve the following linear systems using the Gauss–Jordan method with partial pivoting. Check your answers using the Gauss–Jordan method without partial pivoting.

   (a)
   $$\begin{aligned} x_1 - x_2 + x_3 &= 1 \\ 2x_1 + 3x_2 - x_3 &= 4 \\ -3x_1 + x_2 + x_3 &= -1 \end{aligned}$$

   (b)
   $$\begin{aligned} 2x_1 + x_2 - x_3 &= 2 \\ x_1 - x_2 + 3x_3 &= 3 \\ -5x_1 + 2x_2 + 7x_3 &= 0 \end{aligned}$$

2. Count the number of divisions, multiplications, and additions/subtractions used to solve Exercise 1(a).

3. Let $A$ and $B$ be $N \times N$ matrices and $\mathbf{x} \in R^N$.
   (a) Find the number of multiplications and additions necessary to compute $A\mathbf{x}$.
   (b) Find the number of multiplications and additions necessary to compute $AB$.

4. Derive or prove by mathematical induction

   (a) $\displaystyle\sum_{i=1}^{N} i = \frac{N(N + 1)}{2}$      (b) $\displaystyle\sum_{i=1}^{N} i^2 = \frac{N(N + 1)(2N + 1)}{6}$

5. Verify that Gauss–Jordan elimination requires $N^2$ divisions, $N(N^2 - 1)/2$ multiplications, and $N(N^2 - 1)/2$ additions/subtractions to solve a system of $N$ linear equations in $N$ unknowns.

6. Verify that the number of operations for Gaussian elimination is proportional to $2N^3/3$ while the number of operations for Gauss–Jordan elimination is proportional to $N^3$ for large $N$.

7. Make a table for the required number of divisions, multiplications, and additions/subtractions for Gaussian elimination and Gauss–Jordan elimination for $N = 20$, 40, and 1000.

**8.** An IBM 370 takes 0.26 $\mu$s ($\mu$s = microsecond = $10^{-6}$ second) for division, 0.07 $\mu$s for multiplication, and 0.04 $\mu$s for addition/subtraction. Find the required time to solve $A\mathbf{x} = \mathbf{b}$ of order 600 ($N = 600$) on an IBM 370.

## COMPUTER EXERCISES

**C.1.** Verify that Equation (7.2.10) can be written as

```
DO 5 K = 1,N − 1
DO 5 I = K + 1,N
    AM = A(I,K)/A(K,K)
    B(I) = B(I) − AM*B(K)
DO 5 J = K + 1,N
5 A(I,J) = A(I,J) − AM*A(K,J)
```

Also, count the number of operations in this code.

**C.2.** Write a subroutine that locates the largest element in absolute value from the $i$th column below or on the main diagonal and exchange that row with the $i$th row.

**C.3.** Using Exercise C.2, write a subroutine that reduces $[A \mid \mathbf{b}]$ to Equation (7.3.2). Count the number of divisions, multiplications, and additions/subtractions it uses by having proper counters at appropriate places.

**C.4.** Write a program that reads the values of $A$ and $\mathbf{b}$ in order to solve a linear system of equations $A\mathbf{x} = \mathbf{b}$. Print the values of $A$ and $\mathbf{b}$. Use subroutines C.2 and C.3 to solve $A\mathbf{x} = \mathbf{b}$ by Gauss–Jordan elimination with partial pivoting. Print the values of $\mathbf{x}$. Multiply $A$ and $\mathbf{x}$ and print those values along with $\mathbf{b}$.

**C.5.** Using Exercise C.4, solve the following linear systems, if possible.

(a)
$$x_1 + \tfrac{1}{2}x_2 + \tfrac{1}{3}x_3 + \tfrac{1}{4}x_4 = \tfrac{1}{8}$$

$$\tfrac{1}{2}x_1 + \tfrac{1}{3}x_2 + \tfrac{1}{4}x_3 + \tfrac{1}{5}x_4 = \tfrac{1}{9}$$

$$\tfrac{1}{3}x_1 + \tfrac{1}{4}x_2 + \tfrac{1}{5}x_3 + \tfrac{1}{6}x_4 = \tfrac{1}{10}$$

$$\tfrac{1}{4}x_1 + \tfrac{1}{5}x_2 + \tfrac{1}{6}x_3 + \tfrac{1}{7}x_4 = \tfrac{1}{11}$$

(b)
$$x_1 + 0.50000x_2 + 0.33333x_3 + 0.25000x_4 = 0.12500$$

$$0.50000x_1 + 0.33333x_2 + 0.25000x_3 + 0.20000x_4 = 0.11111$$

$$0.33333x_1 + 0.25000x_2 + 0.20000x_3 + 0.16667x_4 = 0.10000$$

$$0.25000x_1 + 0.20000x_2 + 0.16667x_3 + 0.14286x_4 = 0.09091$$

## 7.5 DIRECT FACTORIZATION METHODS

In the Gaussian elimination method, the system was reduced to a triangular form and then solved by back substitution. It is much easier to solve triangular systems. Let us exploit this idea and assume that a given $N \times N$ matrix $A$ can be written as a product of two matrices $L$ and $U$ so that

$$A = LU \qquad (7.5.1)$$

where $L$ is an $N \times N$ lower-triangular matrix and $U$ is an $N \times N$ upper-triangular matrix. The factorization in Equation (7.5.1) is called an $LU$ **decomposition** of $A$. Then $A\mathbf{x} = \mathbf{b}$ is equivalent to $LU\mathbf{x} = \mathbf{b}$. Further $L(U\mathbf{x}) = \mathbf{b}$ decomposes into two triangular systems

$$U\mathbf{x} = \mathbf{y} \quad \text{and} \quad L\mathbf{y} = \mathbf{b} \qquad (7.5.2)$$

Both systems are triangular and therefore easy to solve. What we need is a procedure to generate factorization. Before we develop the procedure, let us look very closely at the solved Example 7.2.1.

$$[A \mid \mathbf{b}] = \begin{bmatrix} 4 & 2 & 3 & \mid & 7 \\ 2 & -4 & -1 & \mid & 1 \\ -1 & 1 & 4 & \mid & -5 \end{bmatrix}$$

is reduced by the Gauss elimination method to

$$[A \mid \mathbf{b}]^{(2)} = \begin{bmatrix} 4 & 2 & 3 & \mid & 7 \\ 0 & -5 & -5/2 & \mid & -5/2 \\ 0 & 0 & 4 & \mid & -4 \end{bmatrix}$$

Thus for

$$A = \begin{bmatrix} 4 & 2 & 3 \\ 2 & -4 & -1 \\ -1 & 1 & 4 \end{bmatrix}$$

we have

$$U = \begin{bmatrix} 4 & 2 & 3 \\ 0 & -5 & -5/2 \\ 0 & 0 & 4 \end{bmatrix}$$

We construct $L$ from multipliers by defining

$$L = \begin{bmatrix} 1 & 0 & 0 \\ m_{21} & 1 & 0 \\ m_{31} & m_{32} & 1 \end{bmatrix} = \begin{bmatrix} 1 & 0 & 0 \\ 1/2 & 1 & 0 \\ -1/4 & -3/10 & 1 \end{bmatrix}$$

Then

$$LU = \begin{bmatrix} 1 & 0 & 0 \\ 1/2 & 1 & 0 \\ -1/4 & -3/10 & 1 \end{bmatrix} \begin{bmatrix} 4 & 2 & 3 \\ 0 & -5 & -5/2 \\ 0 & 0 & 4 \end{bmatrix} = \begin{bmatrix} 4 & 2 & 3 \\ 2 & -4 & -1 \\ -1 & 1 & 4 \end{bmatrix} = A$$

Thus we observe that we obtain $L$ and $U$ by Gaussian elimination. In fact, we prove the following theorem.

**Theorem 7.5.1**    If Gaussian elimination can be carried out on the system $A\mathbf{x} = \mathbf{b}$ without interchanging rows or columns, then $A = LU$ where

$$L = \begin{bmatrix} 1 & 0 & 0 & \ldots & 0 \\ m_{21} & 1 & 0 & \ldots & 0 \\ m_{31} & m_{32} & 1 & \ldots & 0 \\ \vdots & \vdots & \vdots & \ddots & \vdots \\ m_{N1} & m_{N2} & m_{N3} & \ldots & 1 \end{bmatrix} \quad U = \begin{bmatrix} a_{11} & a_{12} & a_{13} & \ldots & a_{1N} \\ 0 & a_{22}^{(1)} & a_{23}^{(1)} & \ldots & a_{2N}^{(1)} \\ 0 & 0 & a_{33}^{(2)} & \ldots & a_{3N}^{(2)} \\ \vdots & \vdots & \vdots & \ddots & \vdots \\ 0 & 0 & 0 & \ldots & a_{NN}^{(N-1)} \end{bmatrix} \quad \textbf{(7.5.3)}$$

*Proof*    Let us find $(LU)_{ij}$. From Equation (7.5.3),

$$(LU)_{ij} = [m_{i1}\ m_{i2}\ \ldots\ m_{i,i-1}\ 1\ 0\ \ldots\ 0] \begin{bmatrix} a_{1j} \\ a_{2j}^{(1)} \\ a_{3j}^{(2)} \\ \vdots \\ a_{j-1,j}^{(j-2)} \\ a_{jj}^{(j-1)} \\ 0 \\ \vdots \\ 0 \end{bmatrix}$$

We have two cases,

*Case 1*    For $i \leq j$,

$$(LU)_{ij} = m_{i1}a_{1j} + m_{i2}a_{2j}^{(1)} + \cdots + m_{i,i-1}a_{i-1,j}^{(i-2)} + 1a_{i,j}^{(i-1)}$$

$$= \sum_{k=1}^{i-1} m_{ik}a_{kj}^{(k-1)} + a_{i,j}^{(i-1)}$$

$$= \sum_{k=1}^{i-1} (a_{ij}^{(k-1)} - a_{ij}^{(k)}) + a_{i,j}^{(i-1)} \quad \text{from Equation (7.2.10)}$$

$$= (a_{ij} - a_{ij}^{(1)}) + (a_{ij}^{(1)} - a_{ij}^{(2)}) + \cdots + (a_{ij}^{(i-2)} - a_{ij}^{(i-1)}) + a_{ij}^{(i-1)}$$

$$= a_{ij}$$

*Case 2*    For $i > j$,

$$(LU)_{ij} = m_{i1}a_{1j} + m_{i2}a_{2j}^{(1)} + \cdots + m_{ij}a_{jj}^{(j-1)}$$

$$= \sum_{k=1}^{j-1} m_{ik}a_{kj}^{(k-1)} + m_{ij}a_{jj}^{(j-1)}$$

$$= \sum_{k=1}^{j-1} (a_{ij}^{(k-1)} - a_{ij}^{(k)}) + \frac{a_{ij}^{(j-1)}}{a_{jj}^{(j-1)}} a_{jj}^{(j-1)} \quad \text{from Equation (7.2.10)}$$

$$= (a_{ij} - a_{ij}^{(1)}) + (a_{ij}^{(1)} - a_{ij}^{(2)}) + \cdots + (a_{ij}^{(j-2)} - a_{ij}^{(j-1)}) + a_{ij}^{(j-1)}$$

$$= a_{ij}$$

Thus $LU = A$.    ∎

Since $A = LU$, $\det(A) = \det(L)\det(U)$. Further $L$ and $U$ are triangular matrices; therefore, their determinants are the product of their diagonal elements. Thus $\det(L) = 1$; so, $\det(A) = \det(U)$.

We are looking for an $LU$ decomposition of $A$ without resorting to Gaussian elimination. It is convenient to determine $L$ and $U$ directly from $A$. Before we develop methods, we would like to know under what conditions $A$ can have $LU$ factors.

Consider

$$A = \begin{bmatrix} 0 & 1 \\ 1 & 1 \end{bmatrix} = \begin{bmatrix} 1 & 0 \\ l_{12} & 1 \end{bmatrix} \begin{bmatrix} u_{11} & u_{12} \\ 0 & u_{22} \end{bmatrix} = \begin{bmatrix} u_{11} & u_{12} \\ l_{12}u_{11} & l_{12}u_{12} + u_{22} \end{bmatrix}$$

Hence $u_{11} = 0$ and $l_{12}u_{11} = 1$. There is no $l_{12}$ such that $l_{12} \times 0 = 1$. Thus $A = \begin{bmatrix} 0 & 1 \\ 1 & 1 \end{bmatrix}$ does not have an $LU$ decomposition even though $A$ is nonsingular. On the other hand, the singular matrix $\begin{bmatrix} 1 & 1 \\ 1 & 1 \end{bmatrix}$ has $LU$ factors $\begin{bmatrix} 1 & 0 \\ 1 & 1 \end{bmatrix}$ and $\begin{bmatrix} 1 & 1 \\ 0 & 0 \end{bmatrix}$. The question of existence for $LU$ decomposition can be answered in terms of the leading principal submatrices.

The **leading principal submatrices** of

$$A = \begin{bmatrix} 4 & 2 & 3 \\ 2 & -4 & -1 \\ -1 & 1 & 4 \end{bmatrix}$$

are $A_1 = [4]$, $A_2 = \begin{bmatrix} 4 & 2 \\ 2 & -4 \end{bmatrix}$ and $A_3 = A$. The leading principal submatrices of $A = \begin{bmatrix} 0 & 1 \\ 1 & 1 \end{bmatrix}$ are $A_1 = [0]$ and $A_2 = A$.

If we carry out Gaussian elimination on an augmented matrix $[A \mid \mathbf{b}]$ given by Equation (7.2.7) and reduce it to the form $[A \mid \mathbf{b}]^{(N-1)}$ given by Equation (7.2.9), then we have

$$\det(A_k) = a_{11}a_{22}^{(1)} \ldots a_{kk}^{(k-1)} \quad \text{for } k = 1, 2, \ldots, N$$

that is, the product of the first $k$ pivots. Therefore, Gaussian elimination can be carried out without row interchanges if and only if $\det(A_k) \neq 0$ for $k = 1, 2, \ldots, N - 1$. Using this, the following theorem can be proven (Exercise 7).

**Theorem 7.5.2**    Let $A$ be an $N \times N$ matrix. If $\det(A_k) \neq 0$ for $k = 1, 2, \ldots, N - 1$, then there exists a unique lower-triangular matrix $L$ with $l_{ii} = 1$ and a unique upper-triangular matrix $U$ such that $LU = A$.    ∎

So far we used $l_{ii} = 1$ in $L$. If we require $L$ and $U$ to be lower- and upper-triangular matrices, then we can select $L$ and $U$ in many ways. Let us consider a $4 \times 4$ matrix

$$A = \begin{bmatrix} a_{11} & a_{12} & a_{13} & a_{14} \\ a_{21} & a_{22} & a_{23} & a_{24} \\ a_{31} & a_{32} & a_{33} & a_{34} \\ a_{41} & a_{42} & a_{43} & a_{44} \end{bmatrix} \tag{7.5.4}$$

Then the factors of $A$ are given by (assuming it is factorable)

$$L = \begin{bmatrix} l_{11} & 0 & 0 & 0 \\ l_{21} & l_{22} & 0 & 0 \\ l_{31} & l_{32} & l_{33} & 0 \\ l_{41} & l_{42} & l_{43} & l_{44} \end{bmatrix} \quad \text{and} \quad U = \begin{bmatrix} u_{11} & u_{12} & u_{13} & u_{14} \\ 0 & u_{22} & u_{23} & u_{24} \\ 0 & 0 & u_{33} & u_{34} \\ 0 & 0 & 0 & u_{44} \end{bmatrix} \tag{7.5.5}$$

We have 16 known elements for $A$ in Equation (7.5.4) and 20 unknowns for $L$ and $U$ in Equation (7.5.5). For a unique solution of a system having 20 unknowns, we need 20 equations while we have 16 equations since $LU = A$. So we specify four additional conditions on the unknowns in the following well-known ways:

1. **Crout's method:** Let $u_{11} = u_{22} = u_{33} = u_{44} = 1$.
2. **Doolittle's method:** Let $l_{11} = l_{22} = l_{33} = l_{44} = 1$.

Let us consider Crout's method. In this method, we want

$$\begin{bmatrix} a_{11} & a_{12} & a_{13} & a_{14} \\ a_{21} & a_{22} & a_{23} & a_{24} \\ a_{31} & a_{32} & a_{33} & a_{34} \\ a_{41} & a_{42} & a_{43} & a_{44} \end{bmatrix} = \begin{bmatrix} l_{11} & 0 & 0 & 0 \\ l_{21} & l_{22} & 0 & 0 \\ l_{31} & l_{32} & l_{33} & 0 \\ l_{41} & l_{42} & l_{43} & l_{44} \end{bmatrix} \begin{bmatrix} 1 & u_{12} & u_{13} & u_{14} \\ 0 & 1 & u_{23} & u_{24} \\ 0 & 0 & 1 & u_{34} \\ 0 & 0 & 0 & 1 \end{bmatrix}$$

$$= \begin{bmatrix} l_{11} & l_{11}u_{12} & l_{11}u_{13} & l_{11}u_{14} \\ l_{21} & l_{21}u_{12} + l_{22} & l_{21}u_{13} + l_{22}u_{23} & l_{21}u_{14} + l_{22}u_{24} \\ l_{31} & l_{31}u_{12} + l_{32} & l_{31}u_{13} + l_{32}u_{23} + l_{33} & l_{31}u_{14} + l_{32}u_{24} + l_{33}u_{34} \\ l_{41} & l_{41}u_{12} + l_{42} & l_{41}u_{13} + l_{42}u_{23} + l_{43} & l_{41}u_{14} + l_{42}u_{24} + l_{43}u_{34} + l_{44} \end{bmatrix} \tag{7.5.6}$$

Comparing each element of the first column and first row of Equation (7.5.6) we get

$$l_{i1} = a_{i1} \quad \text{for } i = 1, 2, 3, \text{ and } 4$$

$$u_{1j} = \frac{a_{1j}}{l_{11}} \quad \text{for } j = 2, 3, \text{ and } 4$$

Comparing the last three elements of the second column,

$$l_{i2} = -l_{i1}u_{12} + a_{i2} \quad \text{for } i = 2, 3, \text{ and } 4$$

Comparing the last two elements of the second row,

$$l_{21}u_{13} + l_{22}u_{23} = a_{23} \quad \text{and} \quad l_{21}u_{14} + l_{22}u_{24} = a_{24}$$

Therefore

$$u_{2i} = \frac{a_{2i} - l_{21}u_{1i}}{l_{22}} \quad \text{for } i = 3 \text{ and } 4$$

The comparison of the last two elements of the third column yields

$$l_{31}u_{13} + l_{32}u_{23} + l_{33} = a_{33} \quad \text{and} \quad l_{41}u_{13} + l_{42}u_{23} + l_{43} = a_{43}$$

Therefore

$$l_{i3} = a_{i3} - l_{i1}u_{13} - l_{i2}u_{23} \quad \text{for } i = 3 \text{ and } 4$$

The comparison of the last element of the third row yields

$$u_{34} = \frac{a_{34} - l_{31}u_{14} - l_{32}u_{24}}{l_{33}}$$

Similarly, the last element of the fourth column yields

$$l_{44} = a_{44} - l_{41}u_{14} - l_{42}u_{24} - l_{43}u_{34}$$

For $k = 1, 2, \ldots, N$, the elements of the decomposition matrices $L$ and $U$ of an $N \times N$ matrix $A$ are given by

$$u_{kk} = 1$$

$$l_{ik} = a_{ik} - \sum_{m=1}^{k-1} l_{im}u_{mk} \quad \text{for } i = k, k+1, \ldots, N$$

$$u_{kj} = \frac{1}{l_{kk}}\left[ a_{kj} - \sum_{m=1}^{k-1} l_{km}u_{mj} \right] \quad \text{for } j = k+1, k+2, \ldots, N \qquad (7.5.7)$$

We assume that whenever the lower limit of summation is greater than the upper limit of summation, the sum is zero. Hence

$$l_{11} = a_{11} - \sum_{m=1}^{0} l_{1m}u_{m1} = a_{11} - 0 = a_{11}$$

Similarly, for $k = 1, 2, \ldots, N$, the elements of the decomposition matrices $L$ and $U$ of an $N \times N$ matrix $A$ by the Doolittle method are given by (Exercise 8)

$$l_{kk} = 1$$

$$u_{kj} = a_{kj} - \sum_{m=1}^{k-1} l_{km}u_{mj} \quad \text{for } i = k, k+1, \ldots, N$$

$$l_{ik} = \frac{1}{u_{kk}}\left[ a_{ik} - \sum_{m=1}^{k-1} l_{im}u_{mk} \right] \quad \text{for } j = k+1, k+2, \ldots, N \qquad (7.5.8)$$

The solution of the system of equations $L\mathbf{y} = \mathbf{b}$ is given by

$$y_i = \frac{1}{l_{ii}}\left[ b_i - \sum_{j=1}^{i-1} l_{ij}y_j \right] \quad \text{for } i = 1, 2, \ldots, N \qquad (7.5.9)$$

and the solution of the system of equations $U\mathbf{x} = \mathbf{y}$ is given by

$$x_i = \frac{1}{u_{ii}}\left[ y_i - \sum_{j=i+1}^{N} u_{ij}x_j \right] \quad \text{for } i = N, N-1, \ldots, 1 \qquad (7.5.10)$$

**EXAMPLE 7.5.1**     Using the Crout factorization method, solve

$$4x_1 + 2x_2 + 3x_3 = 7$$
$$2x_1 - 4x_2 - x_3 = 1$$
$$-x_1 + x_2 + 4x_3 = -5$$

First let us find $L$ and $U$ such that

$$A = \begin{bmatrix} 4 & 2 & 3 \\ 2 & -4 & -1 \\ -1 & 1 & 4 \end{bmatrix} = \begin{bmatrix} l_{11} & 0 & 0 \\ l_{21} & l_{22} & 0 \\ l_{31} & l_{32} & l_{33} \end{bmatrix} \begin{bmatrix} 1 & u_{12} & u_{13} \\ 0 & 1 & u_{23} \\ 0 & 0 & 1 \end{bmatrix}$$

Using Equation (7.5.7) for $k = 1$, we get $l_{i1} = a_{i1} - \sum_{m=1}^{0} l_{im}u_{m1} = a_{i1}$ for $i = 1, 2,$ and 3. Therefore $l_{11} = a_{11} = 4$, $l_{21} = a_{21} = 2$ and $l_{31} = a_{31} = -1$.

For $k = 1$

$$u_{1j} = \left[ a_{1j} - \sum_{m=1}^{0} l_{1m}u_{mj} \right] / (l_{11}) = (1/l_{11})a_{1j} \quad \text{for } j = 2 \text{ and } 3$$

Hence $u_{12} = a_{12}/l_{11} = \frac{2}{4} = \frac{1}{2}$, and $u_{13} = a_{13}/l_{11} = \frac{3}{4}$.

From Equation (7.5.7) for $k = 2$, we get $l_{i2} = a_{i2} - l_{i1}u_{12}$ for $i = 2$ and 3. Hence $l_{22} = a_{22} - l_{21}u_{12} = -4 - (2)(\frac{1}{2}) = -5$ and $l_{32} = a_{32} - l_{31}u_{12} = 1 - (-1)(\frac{1}{2}) = \frac{3}{2}$.

Also, $u_{2j} = [a_{2j} - l_{21}u_{1j}]/l_{22}$ for $j = 3$. Therefore $u_{23} = [a_{23} - l_{21}u_{13}]/l_{22} = [(-1) - (2)(\frac{3}{4})]/(-5) = (1 + \frac{3}{2})/5 = \frac{1}{2}$.

For $k = 3$, we get from Equation (7.5.7), $l_{i3} = a_{i3} - (l_{i1}u_{13} + l_{i2}u_{23})$ for $i = 3$. Hence $l_{33} = a_{33} - (l_{31}u_{13} + l_{32}u_{23}) = 4 - [(-1)(\frac{3}{4}) + (\frac{3}{2})(\frac{1}{2})] = 4$.

Therefore

$$A = \begin{bmatrix} 4 & 2 & 3 \\ 2 & -4 & -1 \\ -1 & 1 & 4 \end{bmatrix} = \begin{bmatrix} 4 & 0 & 0 \\ 2 & -5 & 0 \\ -1 & 3/2 & 4 \end{bmatrix} \begin{bmatrix} 1 & 1/2 & 3/4 \\ 0 & 1 & 1/2 \\ 0 & 0 & 1 \end{bmatrix}$$

Now we want to solve $A\mathbf{x} = \mathbf{b} = L(U\mathbf{x})$. Let $U\mathbf{x} = \mathbf{y}$. Then we have $L\mathbf{y} = \mathbf{b}$. Thus

$$\begin{bmatrix} 4 & 0 & 0 \\ 2 & -5 & 0 \\ -1 & 3/2 & 4 \end{bmatrix} \begin{bmatrix} y_1 \\ y_2 \\ y_3 \end{bmatrix} = \begin{bmatrix} 7 \\ 1 \\ -5 \end{bmatrix}$$

Solving, we get $y_1 = \frac{7}{4}$, $y_2 = (1 - 2y_1)/-5 = (1 - \frac{7}{2})/-5 = \frac{1}{2}$, and $y_3 = (-5 + y_1 - (\frac{3}{2})y_2)/4 = (-5 + (\frac{7}{4}) - (\frac{3}{4}))/4 = -1$. Now we have to solve $U\mathbf{x} = \mathbf{y}$, which is

$$\begin{bmatrix} 1 & 1/2 & 3/4 \\ 0 & 1 & 1/2 \\ 0 & 0 & 1 \end{bmatrix} \begin{bmatrix} x_1 \\ x_2 \\ x_3 \end{bmatrix} = \begin{bmatrix} 7/4 \\ 1/2 \\ -1 \end{bmatrix}$$

Solving these equations, we get $x_1 = 2$, $x_2 = 1$, and $x_3 = -1$. Also we can find $x_1$, $x_2$, and $x_3$ using Equations (7.5.9) and (7.5.10).  ■ ■ ■

**Algorithm**

**Subroutine** Factor ($N$, $A$, **b**, **x**)
Comment: This subroutine solves the system of equations $A\mathbf{x} = \mathbf{b}$ using the factorization method. Matrix $A$ is written as a product of two matrices $L$ and $U$ such that $A = LU$ where $L$ is a lower-triangular matrix and $U$ is an upper-triangular matrix with $u_{kk} = 1$ for $k = 1, 2, \ldots, N$ using Equation (7.5.7). We denote the elements of $L$ by $zl_{ij}$.

1 **Print** $N$, $A$, **b**
2 **Repeat** step 3 with $k = 1, 2, \ldots, N$
   3  $u_{kk} = 1$
4 **Repeat** steps 5–15 with $k = 1, 2, \ldots, N$
      5 **Repeat** steps 6–9 with $i = k, k + 1, \ldots, N$
         6  sum $= 0.0$
         7 **Repeat** step 8 with $m = 1, 2, \ldots, k - 1$
            8  sum $=$ sum $+ zl_{im} u_{mk}$
         9  $zl_{ik} = a_{ik} -$ sum
      10 **If** $|zl_{kk}| <$ eps, **then print** 'A cannot be decomposed by Crout's method'
      11 **Repeat** steps 12–15 with $j = k + 1, k + 2, \ldots, N$
         12  sum $= 0.0$
         13 **Repeat** step 14 with $m = 1, 2, \ldots, k - 1$
            14  sum $=$ sum $+ zl_{km} u_{mj}$
         15  $u_{kj} = (a_{kj} -$ sum$)/zl_{kk}$
Comment: This solves $A\mathbf{x} = (LU)\mathbf{x} = L(U\mathbf{x}) = \mathbf{b}$ by solving $L\mathbf{y} = \mathbf{b}$ and $U\mathbf{x} = \mathbf{y}$ using Equations (7.5.9) and (7.5.10).
16  $y_1 = b_1/zl_{11}$
17 **Repeat** steps 18–21 with $i = 2, 3, \ldots, N$
      18  sum $= 0.0$
      19 **Repeat** step 20 with $j = 1, 2, \ldots, i - 1$
         20  sum $=$ sum $+ zl_{ij} y_j$
      21  $y_i = (b_i -$ sum$)/zl_{ii}$
22  $x_N = y_N$
23 **Repeat** steps 24–27 with $i = N - 1, N - 2, \ldots, 1$
      24  sum $= 0.0$
      25 **Repeat** step 26 with $j = i + 1, i + 2, \ldots, N$
         26  sum $=$ sum $+ u_{ij} x_j$
      27  $x_i = y_i -$ sum
28 **Print** **x**, $A$, $L$, $U$

This method is very useful when we want to solve many linear systems with the same coefficient matrix $A$ but with different right-hand **b** such as $A\mathbf{x}_1 = \mathbf{b}_1$, $A\mathbf{x}_2 = \mathbf{b}_2, \ldots, A\mathbf{x}_n = \mathbf{b}_n$.

Equation (7.5.7) shows that $LU$ decomposition fails if any of $l_{kk}$ is zero. Besides, if $l_{kk}$ is small, then round-off errors are magnified by $1/l_{kk}$. In order to maximize $l_{kk}$, row interchanges can be introduced in the $LU$ decomposition algorithm similar to partial pivoting in Gaussian elimination.

To rearrange or permute rows of a given matrix, we define an $N \times N$ **permutation matrix** $P = [\mathbf{e}_{j_1}, \mathbf{e}_{j_2}, \ldots, \mathbf{e}_{j_N}]$ where $j_1, j_2, \ldots, j_N$ is a permutation of the integers 1, 2, $\ldots, N$ and $\mathbf{e}_i = [0, 0, \ldots, 1, \ldots, 0]'$ where 1 is in the $i$th position and 0 in the remaining positions.

**EXAMPLE 7.5.2**    The following are permutation matrices:

$$\begin{bmatrix} 0 & 1 \\ 1 & 0 \end{bmatrix} = [\mathbf{e}_2, \mathbf{e}_1], \quad \begin{bmatrix} 0 & 1 & 0 \\ 0 & 0 & 1 \\ 1 & 0 & 0 \end{bmatrix} = [\mathbf{e}_3, \mathbf{e}_1, \mathbf{e}_2], \quad \begin{bmatrix} 1 & 0 & 0 \\ 0 & 1 & 0 \\ 0 & 0 & 1 \end{bmatrix} = [\mathbf{e}_1, \mathbf{e}_2, \mathbf{e}_3] \quad ∎∎∎$$

**EXAMPLE 7.5.3**    Solve the following system using $LU$ decomposition with $|l_{11}|$, $|l_{22}|$, and $|l_{33}|$ as large as possible using row interchanges.

$$\begin{aligned} 2x_1 - 4x_2 - x_3 &= 1 \\ -x_1 + x_2 + 4x_3 &= -5 \\ 4x_1 + 2x_2 + 3x_3 &= 7 \end{aligned}$$

First find the triangular factors of the matrix

$$A = \begin{bmatrix} 2 & -4 & -1 \\ -1 & 1 & 4 \\ 4 & 2 & 3 \end{bmatrix}$$

Since $l_{11} = a_{11}$, interchange row 1 with row 3, which has the largest element in the first column. This will maximize $|l_{11}|$. Thus we have

$$\begin{bmatrix} 4 & 2 & 3 \\ -1 & 1 & 4 \\ 2 & -4 & -1 \end{bmatrix} = \begin{bmatrix} l_{11} & 0 & 0 \\ l_{21} & l_{22} & 0 \\ l_{31} & l_{32} & l_{33} \end{bmatrix} \begin{bmatrix} 1 & u_{12} & u_{13} \\ 0 & 1 & u_{23} \\ 0 & 0 & 1 \end{bmatrix}$$

$$= \begin{bmatrix} l_{11} & l_{11}u_{12} & l_{11}u_{13} \\ l_{21} & l_{21}u_{12} + l_{22} & l_{21}u_{13} + l_{22}u_{23} \\ l_{31} & l_{31}u_{12} + l_{32} & l_{31}u_{13} + l_{32}u_{23} + l_{33} \end{bmatrix}$$

Comparing elements of the first column and first row, we get

$$l_{11} = 4 \qquad l_{21} = -1 \qquad l_{31} = 2 \qquad u_{12} = \frac{2}{l_{11}} = \frac{1}{2} \qquad u_{13} = \frac{3}{l_{11}} = \frac{3}{4}$$

and so

$$\begin{bmatrix} 4 & 2 & 3 \\ -1 & 1 & 4 \\ 2 & -4 & -1 \end{bmatrix} = \begin{bmatrix} 4 & 0 & 0 \\ -1 & l_{22} & 0 \\ 2 & l_{32} & l_{33} \end{bmatrix} \begin{bmatrix} 1 & 1/2 & 3/4 \\ 0 & 1 & u_{23} \\ 0 & 0 & 1 \end{bmatrix}$$

$$= \begin{bmatrix} 4 & 2 & 3 \\ -1 & (-1/2) + l_{22} & (-3/4) + l_{22}u_{23} \\ 2 & 1 + l_{32} & (3/2) + l_{32}u_{23} + l_{33} \end{bmatrix}$$

Comparing the elements at $(2, 2)$, we get $(-\frac{1}{2}) + l_{22} = 1$. Hence $l_{22} = \frac{3}{2}$. We want to maximize $|l_{22}|$, therefore interchanging rows 2 and 3 may result in a different $l_{22}$.

Interchange rows 2 and 3 to get

$$\begin{bmatrix} 4 & 2 & 3 \\ 2 & -4 & -1 \\ -1 & 1 & 4 \end{bmatrix} = \begin{bmatrix} 4 & 0 & 0 \\ 2 & \hat{l}_{22} & 0 \\ -1 & \hat{l}_{32} & \hat{l}_{33} \end{bmatrix} \begin{bmatrix} 1 & 1/2 & 3/4 \\ 0 & 1 & \hat{u}_{23} \\ 0 & 0 & 1 \end{bmatrix}$$

$$= \begin{bmatrix} 4 & 2 & 3 \\ 2 & 1 + \hat{l}_{22} & (3/2) + \hat{l}_{22}\hat{u}_{23} \\ -1 & (-1/2) + \hat{l}_{32} & (-3/4) + \hat{l}_{32}\hat{u}_{23} + \hat{l}_{33} \end{bmatrix}$$

Comparisons of the elements at (2, 2) give $1 + \hat{l}_{22} = -4$. Hence $\hat{l}_{22} = -5$. This gives $|\hat{l}_{22}| = 5 > \frac{3}{2}$; therefore we accept these factors. Further comparing elements, we get $\hat{u}_{23} = \frac{1}{2}$, $\hat{l}_{32} = -1$, and $\hat{l}_{33} = 4$. Hence

$$A' = \begin{bmatrix} 4 & 2 & 3 \\ 2 & -4 & -1 \\ -1 & 1 & 4 \end{bmatrix} = \begin{bmatrix} 4 & 0 & 0 \\ 2 & -5 & 0 \\ -1 & 3/2 & 4 \end{bmatrix} \begin{bmatrix} 1 & 1/2 & 3/4 \\ 0 & 1 & 1/2 \\ 0 & 0 & 1 \end{bmatrix}$$

which is the matrix $A$ with rearranged rows. We have

$$PA = \begin{bmatrix} 0 & 0 & 1 \\ 1 & 0 & 0 \\ 0 & 1 & 0 \end{bmatrix} \begin{bmatrix} 2 & -4 & -1 \\ -1 & 1 & 4 \\ 4 & 2 & 3 \end{bmatrix} = \begin{bmatrix} 4 & 2 & 3 \\ 2 & -4 & -1 \\ -1 & 1 & 4 \end{bmatrix} = A' = LU$$

Thus

$$A = P^{-1}LU = \begin{bmatrix} 0 & 1 & 0 \\ 0 & 0 & 1 \\ 1 & 0 & 0 \end{bmatrix} \begin{bmatrix} 4 & 0 & 0 \\ 2 & -5 & 0 \\ -1 & 3/2 & 4 \end{bmatrix} \begin{bmatrix} 1 & 1/2 & 3/4 \\ 0 & 1 & 1/2 \\ 0 & 0 & 1 \end{bmatrix}$$

$$= \begin{bmatrix} 2 & -5 & 0 \\ -1 & 3/2 & 4 \\ 4 & 0 & 0 \end{bmatrix} \begin{bmatrix} 1 & 1/2 & 3/4 \\ 0 & 1 & 1/2 \\ 0 & 0 & 1 \end{bmatrix}$$

The interchange of rows produces $L$ and $U$ with rows of $L$ out of order. In order to solve $A\mathbf{x} = \mathbf{b}$, we solve

$$(P^{-1}LU)\mathbf{x} = \mathbf{b}$$

Let $U\mathbf{x} = \mathbf{y}$. Then $P^{-1}L\mathbf{y} = \mathbf{b}$ and so $L\mathbf{y} = P\mathbf{b}$.

Actually it is easy to find $P\mathbf{b}$ by keeping a record of any row interchanges made when $A$ is factored and then interchanging the elements of $\mathbf{b}$ or exchanging them at the same time. Thus

$$\begin{bmatrix} 4 & 0 & 0 \\ 2 & -5 & 0 \\ -1 & 3/2 & 4 \end{bmatrix} \begin{bmatrix} y_1 \\ y_2 \\ y_3 \end{bmatrix} = \begin{bmatrix} 0 & 0 & 1 \\ 1 & 0 & 0 \\ 0 & 1 & 0 \end{bmatrix} \begin{bmatrix} 1 \\ -5 \\ 7 \end{bmatrix} = \begin{bmatrix} 7 \\ 1 \\ -5 \end{bmatrix}$$

Solving these equations, we get $y_1 = \frac{7}{4}$, $y_2 = -\frac{1}{2}$, and $y_3 = -1$. Then $U\mathbf{x} = \mathbf{y}$ becomes

$$\begin{bmatrix} 1 & 1/2 & 3/4 \\ 0 & 1 & 1/2 \\ 0 & 0 & 1 \end{bmatrix} \begin{bmatrix} x_1 \\ x_2 \\ x_3 \end{bmatrix} = \begin{bmatrix} 7/4 \\ 1/2 \\ -1 \end{bmatrix}$$

Solving, we get $x_1 = 2$, $x_2 = 1$, and $x_3 = -1$.    ■ ■ ■

It seems that the introduction of row interchanges into the *LU* decomposition algorithm increases the total amount of work, but it can be shown that the total number of arithmetic operations required to solve an $N \times N$ system of equations using the *LU* decomposition method and Gaussian elimination is the same with or without row interchanges (Fox 1964).

## EXERCISES

1. Find an *LU* decomposition of each of the following matrices:

   (a) $$\begin{bmatrix} 2 & 1 & 1 \\ -4 & -5 & -1 \\ 6 & -3 & 7 \end{bmatrix}$$    (b) $$\begin{bmatrix} 2 & 4 & 2 & 4 \\ -1 & -4 & 1 & 2 \\ 2 & 5 & 2 & 4 \\ 1 & 4 & 0 & 2 \end{bmatrix}$$

2. Show that each of the following matrices do not have an *LU* decomposition:

   (a) $$\begin{bmatrix} 0 & 2 \\ 1 & 3 \end{bmatrix}$$    (b) $$\begin{bmatrix} 1 & 1 & 2 \\ 1 & 1 & 3 \\ 2 & 3 & 4 \end{bmatrix}$$

3. Find an *LU* decomposition of each of the following matrices using Gaussian elimination:

   (a) $$\begin{bmatrix} 2 & 8 & 6 \\ 4 & 2 & -2 \\ 3 & -1 & 1 \end{bmatrix}$$    (b) $$\begin{bmatrix} 1 & 1 & 1 \\ 2 & 3 & 1 \\ 1 & -1 & -2 \end{bmatrix}$$

4. Solve the following linear systems using an *LU* decomposition:

   (a)  $x_1 + 4x_2 + 3x_3 = 10$        (b) $2x_1 + 4x_2 - 6x_3 = 22$
   
     $2x_1 + x_2 - x_3 = -1$           $3x_1 + 6x_2 - 8x_3 = 32$
     
     $6x_1 - 2x_2 + 2x_3 = 22$          $2x_1 + x_2 - x_3 = 8$

5. Find the determinant of the following matrices using an *LU* decomposition:

   (a) $$\begin{bmatrix} 1 & 0 \\ 2 & 3 \end{bmatrix}$$    (b) $$\begin{bmatrix} 1 & 4 & 0 \\ 0 & 1 & 0 \\ 0 & 4 & 1 \end{bmatrix}$$

6. Solve the linear system in Exercise 4 by an *LU* decomposition with $|l_{11}|$, $|l_{22}|$, and $|l_{33}|$ as large as possible using row interchanges.
7. Prove Theorem 7.5.2.
8. Verify Equation (7.5.8).
9. Verify Equations (7.5.9) and (7.5.10).
10. If $P$ is a permutation matrix, then prove that $P$ is nonsingular and $P^{-1} = P^t$.

**COMPUTER EXERCISES**

**C.1.** Write a subroutine that factors an $N \times N$ matrix $A$ into two factors $L$ and $U$ using Equation (7.5.7).

**C.2.** Using Equations (7.5.9) and (7.5.10), write a subroutine that solves the systems $L\mathbf{y} = \mathbf{b}$ and $U\mathbf{x} = \mathbf{y}$ where matrices $L$ and $U$ are given by Exercise C.1.

**C.3.** Write a computer program that reads the values of $A$ and $\mathbf{b}$ to solve the linear system of equations $A\mathbf{x} = \mathbf{b}$ by Crout's method. Print the values of $A$ and $\mathbf{b}$. Use subroutines C.1 and C.2. Print the values of $\mathbf{x}$, $L$, and $U$. Multiply $L$ and $U$ to show that it is actually $A$ by printing the values $LU$ along with $A$. Multiply $U$ and $\mathbf{x}$ and then $L$ and $U\mathbf{x}$; print the values of $L(U\mathbf{x})$ along with $\mathbf{b}$.

**C.4.** Using Exercise C.3, solve the following linear system of equations:

$$x_1 + 3x_2 + 5x_3 - x_4 = 1.8125$$
$$x_1 + 4x_2 + 11x_3 \qquad = 2.875$$
$$3x_1 + 8x_2 + 11x_3 + 6x_4 = 5.25$$
$$2x_1 + 6x_2 + 9x_3 - 4x_4 = 3.375$$

**C.5.** Using Exercise C.3, solve the system in Exercise C.7, page 186.

**C.6.** Using Exercise C.3, find the cubic polynomial $y = Ax^3 + Bx^2 + Cx + D$ that passes through the four points $(-2, -33)$, $(-1, -7)$, $(0, 1)$ and $(1, 9)$.

**C.7.** Using Exercise C.3, solve the following linear system of equations (if possible):

$$0.5x_1 + 0.6x_2 + 0.7x_3 + 0.8x_4 = 2.6$$
$$0.6x_1 + 0.7x_2 + 0.8x_3 + 0.9x_4 = 3.0$$
$$0.7x_1 + 0.8x_2 + 0.9x_3 + x_4 = 3.4$$
$$0.8x_1 + 0.9x_2 + x_3 + 1.1x_4 = 3.8$$

**C.8.** Using Exercise C.3, solve the following linear system of equations (if possible):

$$x_1 + \tfrac{1}{2}x_2 + \tfrac{1}{3}x_3 + \tfrac{1}{4}x_4 = \tfrac{1}{8}$$
$$\tfrac{1}{2}x_1 + \tfrac{1}{3}x_2 + \tfrac{1}{4}x_3 + \tfrac{1}{5}x_4 = \tfrac{1}{9}$$
$$\tfrac{1}{3}x_1 + \tfrac{1}{4}x_2 + \tfrac{1}{5}x_3 + \tfrac{1}{6}x_4 = \tfrac{1}{10}$$
$$\tfrac{1}{4}x_1 + \tfrac{1}{5}x_2 + \tfrac{1}{6}x_3 + \tfrac{1}{7}x_4 = \tfrac{1}{11}$$

**C.9.** Using Exercise C.3, solve the following system of equations (if possible):

$$x_1 + 0.50000x_2 + 0.33333x_3 + 0.25000x_4 = 0.12500$$
$$0.50000x_1 + 0.33333x_2 + 0.25000x_3 + 0.20000x_4 = 0.11111$$
$$0.33333x_1 + 0.25000x_2 + 0.20000x_3 + 0.16667x_4 = 0.10000$$
$$0.25000x_1 + 0.20000x_2 + 0.16667x_3 + 0.14286x_4 = 0.09091$$

## 7.6   POSITIVE DEFINITE SYSTEMS*

For solving a system of linear equations, it is advisable to use partial pivoting strategy in the interest of numerical stability or to avoid failure. An important class of matrices for which interchanges of rows are not necessary is positive definite matrices. In the case of positive matrices, Gaussian elimination can be used without interchanging rows.

Consider the symmetric matrix

$$A = \begin{bmatrix} 1 & 1/2 & 1/3 \\ 1/2 & 1/3 & 1/4 \\ 1/3 & 1/4 & 1/5 \end{bmatrix}$$

It can be verified that the $LU$ factorization of $A$ is

$$\begin{bmatrix} 1 & 1/2 & 1/3 \\ 1/2 & 1/3 & 1/4 \\ 1/3 & 1/4 & 1/5 \end{bmatrix} = \begin{bmatrix} 1 & 0 & 0 \\ 1/2 & 1/12 & 0 \\ 1/3 & 1/12 & 1/180 \end{bmatrix} \begin{bmatrix} 1 & 1/2 & 1/3 \\ 0 & 1 & 1 \\ 0 & 0 & 1 \end{bmatrix} = L_2 U_2$$

$$= \begin{bmatrix} 1 & 0 & 0 \\ 1/2 & 1 & 0 \\ 1/3 & 1 & 1 \end{bmatrix} \begin{bmatrix} 1 & 0 & 0 \\ 0 & 1/12 & 0 \\ 0 & 0 & 1/180 \end{bmatrix} \begin{bmatrix} 1 & 1/2 & 1/3 \\ 0 & 1 & 1 \\ 0 & 0 & 1 \end{bmatrix}$$

$$= L_3 D U_2 = L_3 D L_3^t$$

Thus the factorization of a symmetric matrix has the form $LDL^t$ where $D$ is a diagonal matrix with every diagonal element positive.

A symmetric matrix is called **positive definite** if it has an $LDL^t$ decomposition where every diagonal element of $D$ is positive.

Positive definite matrices arise in the numerical solution of partial differential equations, least squares problems, elastic structure problems and other applications. It can also be proven that a symmetric matrix $A$ is positive definite if and only if $\mathbf{x}^t A \mathbf{x} > 0$ for every $\mathbf{x} \neq \mathbf{0}$ (Exercise 6). For more details about the properties of positive definite matrices, refer to Strang (1988).

Let $A$ be a positive definite $N \times N$ matrix. Then $A$ is symmetric and $A = LDL^t$ where every diagonal element $d_i$ for $i = 1, 2, \ldots, N$ of diagonal matrix $D$ is positive. Further

$$A = LDL^t = L\sqrt{D}\sqrt{D}L^t = (L\sqrt{D})(L\sqrt{D})^t$$

where

$$\sqrt{D} = \begin{bmatrix} \sqrt{d_1} & 0 & \cdots & 0 \\ 0 & \sqrt{d_2} & \cdots & 0 \\ \vdots & \vdots & \ddots & \vdots \\ 0 & 0 & \cdots & \sqrt{d_N} \end{bmatrix}$$

Let $G = L\sqrt{D}$. Then $G$ is a lower triangular matrix. Since $l_{ii} = 1$ for $i = 1, 2, \ldots,$ $N$ and every diagonal element of $\sqrt{D}$ is positive, then every diagonal element of $G$ is positive. Thus $A = GG'$. This decomposition is known as **Cholesky's decomposition**.

**Theorem 7.6.1**

**(Cholesky Decomposition Theorem)**    If $A$ is a positive definite matrix, then $A$ can be factored in the form

$$A = GG' \tag{7.6.1}$$

where $G$ is a lower triangular matrix with positive diagonal elements. ∎

This factorization is unique (Exercise 10).

By equating elements of both sides of Equation (7.6.1), we get the elements of $G$ in terms of the elements of $A$. Consider

$$
\begin{bmatrix} a_{11} & a_{21} & a_{31} \\ a_{21} & a_{22} & a_{32} \\ a_{31} & a_{32} & a_{33} \end{bmatrix} = \begin{bmatrix} g_{11} & 0 & 0 \\ g_{21} & g_{22} & 0 \\ g_{31} & g_{32} & g_{33} \end{bmatrix} \begin{bmatrix} g_{11} & g_{21} & g_{31} \\ 0 & g_{22} & g_{32} \\ 0 & 0 & g_{33} \end{bmatrix}
$$

$$
= \begin{bmatrix} g_{11}^2 & g_{11}g_{21} & g_{11}g_{31} \\ g_{21}g_{11} & g_{21}^2 + g_{22}^2 & g_{21}g_{31} + g_{22}g_{32} \\ g_{31}g_{11} & g_{31}g_{21} + g_{32}g_{22} & g_{31}^2 + g_{32}^2 + g_{33}^2 \end{bmatrix}
$$

Comparing each element of the first column yields

$$g_{11} = \sqrt{a_{11}}$$

$$g_{21} = \frac{a_{21}}{g_{11}} = \frac{a_{21}}{\sqrt{a_{11}}}$$

$$g_{31} = \frac{a_{31}}{g_{11}} = \frac{a_{31}}{\sqrt{a_{11}}}$$

Comparing the last two elements of the second column yields

$$g_{22} = \sqrt{a_{22} - g_{21}^2}$$

$$g_{32} = \frac{(a_{32} - g_{31}g_{21})}{g_{22}}$$

The comparison of the last element of the third column gives

$$g_{33} = \sqrt{a_{33} - (g_{31}^2 + g_{32}^2)}$$

For a positive definite $N \times N$ matrix, the elements of factor $G$ of $GG'$ are given by for $k = 1, 2, \ldots, N$

$$g_{kk} = \sqrt{a_{kk} - \sum_{m=1}^{k-1} (g_{km})^2}$$

$$g_{ik} = \frac{1}{g_{kk}} \left( a_{ik} - \sum_{m=1}^{k-1} g_{im} g_{km} \right) \quad \text{for } i = k + 1, k + 2, \ldots, N \tag{7.6.2}$$

**EXAMPLE 7.6.1**

Using the Cholesky factorization method, solve

$$4x_1 - 2x_2 + 4x_3 = 6$$

$$-2x_1 + 10x_2 - 2x_3 = 6$$

$$4x_1 - 2x_2 + 8x_3 = 10$$

We have

$$A = \begin{bmatrix} 4 & -2 & 4 \\ -2 & 10 & -2 \\ 4 & -2 & 8 \end{bmatrix}$$

$A$ is symmetric and positive definite if we can find $G$ with positive diagonal elements such that $A = GG'$. Then we solve $A\mathbf{x} = (GG')\mathbf{x} = \mathbf{b}$.

From Equation (7.6.2) for $k = 1$

$$g_{11} = \sqrt{a_{11} - \sum_{m=1}^{0} (g_{1m})^2} = \sqrt{a_{11}} = \sqrt{4} = 2$$

$$g_{i1} = \frac{(a_{i1} - \sum_{m=1}^{0} g_{im} g_{1m})}{g_{11}} = \frac{a_{i1}}{g_{11}} \quad \text{for } i = 2 \text{ and } 3$$

For $i = 2$, $g_{21} = a_{21}/g_{11} = -2/2 = -1$ and for $i = 3$, $g_{31} = a_{31}/g_{11} = 4/2 = 2$.
For $k = 2$, Equation (7.6.2) becomes

$$g_{22} = \sqrt{a_{22} - \sum_{m=1}^{1} (g_{2m})^2} = \sqrt{a_{22} - g_{21}^2} = \sqrt{10 - 1} = 3$$

$$g_{i2} = \frac{(a_{i2} - \sum_{m=1}^{1} g_{im} g_{2m})}{g_{22}} = \frac{(a_{i2} - g_{i1} g_{21})}{g_{22}} \quad \text{for } i = 3$$

For $i = 3$, $g_{32} = (a_{32} - g_{31} g_{21})/g_{22} = (-2 - (2)(-1))/3 = 0$.
From Equation (7.6.2) for $k = 3$

$$g_{33} = \sqrt{a_{33} - \sum_{m=1}^{2} (g_{3m})^2} = \sqrt{a_{33} - (g_{31})^2 - (g_{32})^2}$$

$$= \sqrt{8 - (2)^2 - (0)^2} = \sqrt{4} = 2$$

Thus

$$G = \begin{bmatrix} 2 & 0 & 0 \\ -1 & 2 & 0 \\ 2 & 0 & 2 \end{bmatrix}$$

We want to solve $A\mathbf{x} = (GG')\mathbf{x} = G(G'\mathbf{x}) = \mathbf{b}$.
Let $\mathbf{y} = G'\mathbf{x}$. Then $G\mathbf{y} = \mathbf{b}$ and solving

$$G\mathbf{y} = \begin{bmatrix} 2 & 0 & 0 \\ -1 & 2 & 0 \\ 2 & 0 & 2 \end{bmatrix} \begin{bmatrix} y_1 \\ y_2 \\ y_3 \end{bmatrix} = \begin{bmatrix} 6 \\ 6 \\ 10 \end{bmatrix}$$

we get $y_1 = 3$, $y_2 = 3$, and $y_3 = 2$. Further, solving $G'\mathbf{x} = \mathbf{y}$, we get $x_1 = 1$, $x_2 = 1$, and $x_3 = 1$. ■ ■ ■

**Algorithm**

**Subroutine** Choles ($N$, $A$, **b**, **x**)

Comment: This subroutine solves the system of equations $A\mathbf{x} = \mathbf{b}$ where $A$ is an $N \times N$ symmetric positive definite matrix. If $A$ is not positive definite, the program prints "$A$ is not positive definite" and exits. First $G$ with positive diagonal elements is computed such that $A = GG'$. Then $A\mathbf{x} = (GG')\mathbf{x} = G(G'\mathbf{x}) = \mathbf{b}$ is solved by first solving $G\mathbf{y} = \mathbf{b}$ and then $G'\mathbf{x} = \mathbf{y}$ using Equations (7.5.9) and (7.5.10).

  1  **Print** $N$, $A$, **b**
  2  **Repeat** steps 3–12 **with** $k = 1, 2, \ldots, N$
      3  sum = 0
      4  **Repeat** step 5 **with** $m = 1, 2, \ldots, k - 1$
         5  sum = sum + $g_{km}g_{km}$
      6  **If** $a_{kk} - $ sum $\leq 0$, **then** print '$A$ is not positive definite' and exit
      7  $g_{kk} = \sqrt{a_{kk} - \text{sum}}$
      8  **Repeat** steps 9–12 **with** $i = k + 1, k + 2, \ldots, N$
         9  sum1 = 0.0
         10  **Repeat** step 11 **with** $m = 1, 2, \ldots, k - 1$
            11  sum1 = sum1 + $g_{im}g_{mk}$
         12  $g_{ik} = (a_{ik} - \text{sum1})/g_{kk}$
13  **Print** $N$, $G$

Comment: This part solves $G\mathbf{y} = \mathbf{b}$ and then $G'\mathbf{x} = \mathbf{y}$ using Equations (7.5.9) and (7.5.10).

14  $y_1 = b_1/g_{11}$
15  **Repeat** steps 16–19 **with** $i = 2, 3, \ldots, N$
      16  sum = 0.0
      17  **Repeat** step 18 **with** $j = 1, 2, \ldots, i - 1$
         18  sum = sum + $g_{ij}y_j$
      19  $y_i = (b_i - \text{sum})/g_{ii}$
20  $x_N = y_N/g_{NN}$
21  **Repeat** steps 22–25 **with** $i = N - 1, N - 2, \ldots, 1$
      22  sum = 0.0
      23  **Repeat** step 24 **with** $j = i + 1, i + 2, \ldots, N$
         24  sum = sum + $g_{ij}x_j$
      25  $x_i = (y_i - \text{sum})/g_{ii}$
26  **Print** **x**, $A$, $G$

      The Cholesky algorithm (Exercise 11) requires $N^3/6$ flops. From Equation (7.6.2), it can be verified that

$$|g_{kj}| \leq \sqrt{a_{kk}} \quad \text{for } j = 1, 2, \ldots, k, k = 1, 2, \ldots, N$$

Thus the elements of $G$ remain bounded.

      If available storage space is limited, then the Cholesky algorithm has a clear edge over many algorithms. We need to store $a_{ij}$ for which $i \geq j$; in other words, the main diagonal and entries below it. Each element $a_{ij}$ is used to compute $g_{ij}$. Thus $g_{ij}$ can be stored over $a_{ij}$. However, this compact scheme makes programming difficult.

      If a real symmetric matrix $A$ is suspected to be positive definite, then it is relatively straightforward to obtain factors $G$ and $G'$ using the Cholesky factorization algorithm given by Equation (7.6.2). If a square root of a negative number occurs at any stage, then $A$ is not a positive definite matrix.

The Gaussian elimination method and factorization methods are very useful for moderately sized systems of linear equations. Systems of the order of 10,000 are not uncommon and are of a special type. In Chapter 11, we will discuss some methods used to solve large systems.

## EXERCISES

1. Let $A = \begin{bmatrix} 1 & 0 \\ 0 & 4 \end{bmatrix}$. Find four lower-triangular matrices $L$ such that $A = LL'$. Prove that $A$ is positive definite by finding the Cholesky factors.

Determine whether each of the matrices in Exercises 2–4 is positive definite.

2. $\begin{bmatrix} 1 & 2 \\ 2 & 3 \end{bmatrix}$    3. $\begin{bmatrix} 2 & 1 & 1 \\ 1 & 1 & 3 \\ 1 & 3 & 1 \end{bmatrix}$    4. $\begin{bmatrix} 2 & -1 & -1 \\ -1 & 2 & -1 \\ -1 & -1 & 2 \end{bmatrix}$

5. If $A = LL'$ where $L$ is a real nonsingular lower-triangular matrix, prove that $A$ is positive definite.

6. If $A$ is positive definite, then prove that $\mathbf{x}'A\mathbf{x} > 0$ for every $\mathbf{x} \neq \mathbf{0}$ (*Hint:* Substitute $A = LDL'$ and $\mathbf{y} = L'\mathbf{x}$). If $A$ is symmetric and $\mathbf{x}'A\mathbf{x} > 0$ for every $\mathbf{x} \neq \mathbf{0}$, then prove that $A$ is positive definite.

7. If $\mathbf{x}'A\mathbf{x} > 0$ for every $\mathbf{x} \neq \mathbf{0}$, then show that each diagonal element of $A$ is positive.

8. If $A$ is positive definite and $P$ is nonsingular, prove that $B = P'AP$ is also positive definite.

9. Find an upper-triangular matrix $R$ such that $A = RR'$ where

$$A = \begin{bmatrix} a_{11} & a_{21} & a_{31} \\ a_{21} & a_{22} & a_{32} \\ a_{31} & a_{32} & a_{33} \end{bmatrix}$$

is positive definite.

10. Prove that $G$ is unique in Theorem (7.6.1) (Cholesky Decomposition Theorem).

11. Verify that the Cholesky algorithm requires $N^3/6$ flops.

12. Prove that (a) $|a_{ij}| \leq (a_{ij} + a_{ji})/2$    (b) $|a_{ij}| \leq \sqrt{a_{ii}a_{jj}}$ for $i \neq j$, if $a_{ij}$ are the elements of a positive definite matrix $A$.

## COMPUTER EXERCISES

C.1. Write a subroutine that factors an $N \times N$ positive definite matrix $A$ into two factors $G$ and $G'$ using Equation (7.6.2).

**C.2.** Write a subroutine that solves the systems $G\mathbf{y} = \mathbf{b}$ and $G'\mathbf{x} = \mathbf{y}$ where matrices $G$ and $G'$ are given by Exercise C.1 using Equations (7.5.9) and (7.5.10).

**C.3.** Using Exercises C.1 and C.2, solve the following linear system of equations (if possible):

$$0.5x_1 + 0.6x_2 + 0.7x_3 + 0.8x_4 = 2.6$$
$$0.6x_1 + 0.7x_2 + 0.8x_3 + 0.9x_4 = 3.0$$
$$0.7x_1 + 0.8x_2 + 0.9x_3 + \quad x_4 = 3.4$$
$$0.8x_1 + 0.9x_2 + \quad x_3 + 1.1x_4 = 3.8$$

**C.4.** Using Exercises C.1 and C.2, solve the following linear system of equations (if possible):

$$x_1 + \tfrac{1}{2}x_2 + \tfrac{1}{3}x_3 + \tfrac{1}{4}x_4 = \tfrac{1}{8}$$
$$\tfrac{1}{2}x_1 + \tfrac{1}{3}x_2 + \tfrac{1}{4}x_3 + \tfrac{1}{5}x_4 = \tfrac{1}{9}$$
$$\tfrac{1}{3}x_1 + \tfrac{1}{4}x_2 + \tfrac{1}{5}x_3 + \tfrac{1}{6}x_4 = \tfrac{1}{10}$$
$$\tfrac{1}{4}x_1 + \tfrac{1}{5}x_2 + \tfrac{1}{6}x_3 + \tfrac{1}{7}x_4 = \tfrac{1}{11}$$

**C.5.** Using Exercises C.1 and C.2, solve the following linear system of equations (if possible):

$$x_1 + 0.50000x_2 + 0.33333x_3 + 0.25000x_4 = 0.12500$$
$$0.50000x_1 + 0.33333x_2 + 0.25000x_3 + 0.20000x_4 = 0.11111$$
$$0.33333x_1 + 0.25000x_2 + 0.20000x_3 + 0.16667x_4 = 0.10000$$
$$0.25000x_1 + 0.20000x_2 + 0.16667x_3 + 0.14286x_4 = 0.09091$$

## 7.7   TRIDIAGONAL SYSTEMS*

Finite difference or finite element approximations of a boundary-value problem lead to a system of equations whose coefficient matrix has a band structure. $A$ is called a **band matrix** if all nonzero elements are located on the main and some adjacent diagonals. Outside the band, all elements are zero; in other words

$$a_{ij} = 0 \quad \text{if } i - j > m \quad \text{and} \quad j - i < n$$

where $m$ and $n$ are nonnegative integers. When $m = n = 0$, we get a diagonal matrix and when $m = n = 1$, we get a **tridiagonal matrix**.

Consider a system of the form

$$
\begin{aligned}
b_1 x_1 + c_1 x_2 && = s_1 \\
a_2 x_1 + b_2 x_2 + c_2 x_3 && = s_2 \\
a_3 x_2 + b_3 x_3 + c_3 x_4 && = s_3 \\
&\vdots& \vdots \\
a_N x_{N-1} + b_N x_N &&= s_N
\end{aligned}
\tag{7.7.1}
$$

In matrix form, Equation (7.7.1) becomes

$$
A\mathbf{x} = \mathbf{s} \tag{7.7.2}
$$

where

$$
A = \begin{bmatrix}
b_1 & c_1 & 0 & \cdots & 0 & 0 & 0 \\
a_2 & b_2 & c_2 & \cdots & 0 & 0 & 0 \\
0 & a_3 & b_3 & \cdots & 0 & 0 & 0 \\
\vdots & \vdots & \vdots & & \vdots & \vdots & \vdots \\
0 & 0 & 0 & \cdots & a_{N-1} & b_{N-1} & c_{N-1} \\
0 & 0 & 0 & \cdots & 0 & a_N & b_N
\end{bmatrix}
\quad
\mathbf{x} = \begin{bmatrix}
x_1 \\ x_2 \\ x_3 \\ \vdots \\ x_{N-1} \\ x_N
\end{bmatrix}
\quad
\mathbf{s} = \begin{bmatrix}
s_1 \\ s_2 \\ s_3 \\ \vdots \\ s_{N-1} \\ s_N
\end{bmatrix}
$$

The matrix $A$ has nonzero elements only on the diagonal and in the positions adjacent to the diagonal. Hence $A$ is called a tridiagonal matrix and Equation (7.7.1) a **tridiagonal system**. Consider a $10 \times 10$ matrix. There are ten elements on the diagonal, nine elements on the superdiagonal, and nine elements on the subdiagonal. There is a total of $10 + 9 + 9 = 28$ nonzero elements. Since $10 \times 10$ has 100 elements and 28 nonzero elements, there are $100 - 28 = 72$ zero elements. Thus the matrix is 72% zero. These zeros bring tremendous simplification.

A tridiagonal system can be solved very efficiently by the factorization method. Let us assume that $A$ is nonsingular and can be factored as the product

$$
A = LU \tag{7.7.3}
$$

where

$$
L = \begin{bmatrix}
\beta_1 & 0 & 0 & \cdots & 0 & 0 & 0 \\
\alpha_2 & \beta_2 & 0 & \cdots & 0 & 0 & 0 \\
0 & \alpha_3 & \beta_3 & \cdots & 0 & 0 & 0 \\
\vdots & \vdots & \vdots & & \vdots & \vdots & \vdots \\
0 & 0 & 0 & \cdots & \alpha_{N-1} & \beta_{N-1} & 0 \\
0 & 0 & 0 & \cdots & 0 & \alpha_N & \beta_N
\end{bmatrix}
$$

$$
U = \begin{bmatrix}
1 & \gamma_1 & 0 & \cdots & 0 & 0 & 0 \\
0 & 1 & \gamma_2 & \cdots & 0 & 0 & 0 \\
0 & 0 & 1 & \cdots & 0 & 0 & 0 \\
\vdots & \vdots & \vdots & & \vdots & \vdots & \vdots \\
0 & 0 & 0 & \cdots & 0 & 1 & \gamma_{N-1} \\
0 & 0 & 0 & \cdots & 0 & 0 & 1
\end{bmatrix}
$$

Then comparing the elements of

$$LU = \begin{bmatrix} \beta_1 & 0 & 0 & \cdots & 0 & 0 \\ \alpha_2 & \beta_2 & 0 & \cdots & 0 & 0 \\ 0 & \alpha_3 & \beta_3 & \cdots & 0 & 0 \\ \vdots & \vdots & \vdots & \ddots & \vdots & \vdots \\ 0 & 0 & 0 & \vdots & \beta_{N-1} & 0 \\ 0 & 0 & 0 & \cdots & \alpha_N & \beta_N \end{bmatrix} \begin{bmatrix} 1 & \gamma_1 & 0 & \cdots & 0 & 0 \\ 0 & 1 & \gamma_2 & \cdots & 0 & 0 \\ 0 & 0 & 1 & \cdots & 0 & 0 \\ \vdots & \vdots & \vdots & \ddots & \vdots & \vdots \\ 0 & 0 & 0 & \cdots & 1 & \gamma_{N-1} \\ 0 & 0 & 0 & \cdots & 0 & 1 \end{bmatrix}$$

$$= \begin{bmatrix} \beta_1 & \beta_1\gamma_1 & 0 & \cdots & 0 & 0 \\ \alpha_2 & \alpha_2\gamma_1 + \beta_2 & \beta_2\gamma_2 & \cdots & 0 & 0 \\ 0 & \alpha_3 & \alpha_3\gamma_2 + \beta_3 & \cdots & 0 & 0 \\ \vdots & \vdots & \vdots & \ddots & \vdots & \vdots \\ 0 & 0 & 0 & \cdots & \alpha_{N-1}\gamma_{N-2} + \beta_{N-1} & \beta_{N-1}\gamma_{N-1} \\ 0 & 0 & 0 & \cdots & \alpha_N & \alpha_N\gamma_{N-1} + \beta_N \end{bmatrix}$$

$$= \begin{bmatrix} b_1 & c_1 & 0 & \cdots & 0 & 0 \\ a_2 & b_2 & c_2 & \cdots & 0 & 0 \\ 0 & a_3 & b_3 & \cdots & 0 & 0 \\ \vdots & \vdots & \vdots & \ddots & \vdots & \vdots \\ 0 & 0 & 0 & \cdots & b_{N-1} & c_{N-1} \\ 0 & 0 & 0 & \cdots & a_N & b_N \end{bmatrix} = A$$

we get

$$\beta_1 = b_1 \qquad \gamma_1 = \frac{c_1}{\beta_1}; \qquad \alpha_i = a_i \qquad \beta_i = b_i - a_i\gamma_{i-1} \qquad \gamma_i = \frac{c_i}{\beta_i}$$

for $i = 2, 3, \ldots, N - 1$ and

$$\alpha_N = a_N \qquad \beta_N = b_N - a_N\gamma_{N-1} \tag{7.7.4}$$

Equation (7.7.2) becomes

$$LU\mathbf{x} = \mathbf{s} \tag{7.7.5}$$

Let $U\mathbf{x} = \mathbf{z}$. Then Equation (7.7.5) becomes $L\mathbf{z} = \mathbf{s}$; in other words

$$\begin{bmatrix} \beta_1 & 0 & 0 & \cdots & 0 & 0 \\ a_2 & \beta_2 & 0 & \cdots & 0 & 0 \\ 0 & a_3 & \beta_3 & \cdots & 0 & 0 \\ \vdots & \vdots & \vdots & \ddots & \vdots & \vdots \\ 0 & 0 & 0 & \cdots & \beta_{N-1} & 0 \\ 0 & 0 & 0 & \cdots & a_N & \beta_N \end{bmatrix} \begin{bmatrix} z_1 \\ z_2 \\ z_3 \\ \vdots \\ z_{N-1} \\ z_N \end{bmatrix} = \begin{bmatrix} s_1 \\ s_2 \\ s_3 \\ \vdots \\ s_{N-1} \\ s_N \end{bmatrix}$$

Solving for $z_i$, we get

$$z_1 = \frac{s_1}{\beta_1}$$

$$z_i = \frac{(s_i - \alpha_i z_{i-1})}{\beta_i} \quad \text{for } i = 2, 3, \ldots, N \tag{7.7.6}$$

Once $\mathbf{z}$ is known, then we solve $U\mathbf{x} = \mathbf{z}$; in other words

$$\begin{bmatrix} 1 & \gamma_1 & 0 & \cdots & 0 & 0 \\ 0 & 1 & \gamma_2 & \cdots & 0 & 0 \\ 0 & 0 & 1 & \cdots & 0 & 0 \\ \vdots & \vdots & \vdots & \ddots & \vdots & \vdots \\ 0 & 0 & 0 & \cdots & 1 & \gamma_{N-1} \\ 0 & 0 & 0 & \cdots & 0 & 1 \end{bmatrix} \begin{bmatrix} x_1 \\ x_2 \\ x_3 \\ \vdots \\ x_{N-1} \\ x_N \end{bmatrix} = \begin{bmatrix} z_1 \\ z_2 \\ z_3 \\ \vdots \\ z_{N-1} \\ z_N \end{bmatrix}$$

and get

$$x_N = z_N$$

$$x_i = z_i - \gamma_1 x_{i+1} \quad \text{for } i = N - 1, N - 2, \ldots, 1 \tag{7.7.7}$$

In summary, the following steps are involved in solving Equation (7.7.1):

1.  Calculate $\alpha_i$, $\beta_i$, and $\gamma_i$ from Equation (7.7.4).
2.  Calculate $z_i$ from Equation (7.7.6).
3.  Calculate $x_i$ from Equation (7.7.7).

**EXAMPLE 7.7.1**

Using Equations (7.7.4), (7.7.6), and (7.7.7), solve

$$-4x_1 + x_2 \qquad\qquad\quad = -1$$
$$x_1 - 4x_2 + x_3 \qquad\quad = 0$$
$$x_2 - 4x_3 + x_4 = 0$$
$$x_3 - 4x_4 = 0$$

We decompose $A$ into $L$ and $U$ factors using Equation (7.7.4). The elements of $L$ and $U$ are given by

$$\beta_1 = b_1 = -4 \qquad \gamma_1 = \frac{c_1}{\beta_1} = -\frac{1}{4}$$

$$\alpha_2 = a_2 = 1 \qquad \beta_2 = b_2 - a_2\gamma_1 = -4 - (1)\left(-\frac{1}{4}\right) = -\frac{15}{4}$$

$$\gamma_2 = \frac{c_2}{\beta_2} = \frac{1}{(-15/4)} = -\frac{4}{15}$$

$$\alpha_3 = a_3 = 1 \qquad \beta_3 = b_3 - a_3\gamma_2 = -4 - 1\left(-\frac{4}{15}\right) = -\frac{56}{15}$$

$$\gamma_3 = \frac{c_3}{\beta_3} = \frac{1}{(-56/15)} = -\frac{15}{56}$$

$$\alpha_4 = a_4 = 1 \qquad \beta_4 = b_4 - a_4\gamma_3 = -4 - (1)\left(-\frac{15}{56}\right) = -\frac{209}{56}$$

Thus we have

$$
\begin{bmatrix}
-4 & 1 & 0 & 0 \\
1 & -4 & 1 & 0 \\
0 & 1 & -4 & 1 \\
0 & 0 & 1 & -4
\end{bmatrix}
$$

$$
= \begin{bmatrix}
-4 & 0 & 0 & 0 \\
1 & -15/4 & 0 & 0 \\
0 & 1 & -56/15 & 0 \\
0 & 0 & 1 & -209/56
\end{bmatrix}
\begin{bmatrix}
1 & -1/4 & 0 & 0 \\
0 & 1 & -4/15 & 0 \\
0 & 0 & 1 & -15/56 \\
0 & 0 & 0 & 1
\end{bmatrix}
$$

Then solving $L\mathbf{z} = \mathbf{s}$ using Equation (7.7.6),

$$
z_1 = \frac{s_1}{\beta_1} = \frac{-1}{-4} = \frac{1}{4} \qquad
z_2 = \frac{s_2 - \alpha_2 z_1}{\beta_2} = \frac{0 - 1(1/4)}{(-15/4)} = \frac{1}{15}
$$

$$
z_3 = \frac{s_3 - \alpha_3 z_2}{\beta_3} = \frac{0 - 1(1/15)}{(-56/15)} = \frac{1}{56} \qquad
z_4 = \frac{s_4 - \alpha_4 z_3}{\beta_4} = \frac{0 - 1(1/56)}{(-209/56)} = \frac{1}{209}
$$

Further, we find $\mathbf{x}$ by solving $U\mathbf{x} = \mathbf{z}$ using Equation (7.7.7). Thus

$$
x_4 = z_4 = \frac{1}{209} \qquad
x_3 = z_3 - \gamma_3 x_4 = \frac{1}{56} + \frac{15}{56} \cdot \frac{1}{209} = \frac{4}{209}
$$

$$
x_2 = z_2 - \gamma_2 x_3 = \frac{1}{15} + \frac{4}{15} \cdot \frac{4}{209} = \frac{15}{209} \qquad
x_1 = z_1 - \gamma_1 x_2 = \frac{1}{4} + \frac{1}{4} \cdot \frac{15}{209} = \frac{56}{209}
$$

■ ■ ■

**Algorithm**

**Subroutine** Tridi $(\mathbf{a}, \mathbf{b}, \mathbf{c}, \mathbf{s}, N, \mathbf{x})$
Comment: This subroutine solves a tridiagonal system using Equations (7.7.4), (7.7.6), and (7.7.7). We have $\mathbf{a} = [a_2, a_3, \ldots, a_N]$, $\mathbf{b} = [b_1, b_2, \ldots, b_N]$, $\mathbf{c} = [c_1, c_2, \ldots, c_{N-1}]$ and $\mathbf{s} = [s_1, s_2, \ldots, s_N]$.
  1 **Print** $N$, $\mathbf{a}$, $\mathbf{b}$, $\mathbf{c}$, $\mathbf{s}$
  2 $\beta_1 = b_1$
  3 $\gamma_1 = c_1/\beta_1$
  4 **Repeat** steps 5–7 **with** $i = 2, 3, \ldots, N - 1$
      5 $\alpha_i = a_i$
      6 $\beta_i = b_i - a_i \gamma_{i-1}$
      7 $\gamma_i = c_i/\beta_i$
  8 $\alpha_N = a_N$
  9 $\beta_N = b_N - a_N \gamma_{N-1}$
10 $z_1 = s_1/\beta_1$
11 **Repeat** step 12 **with** $i = 2, 3, \ldots, N$
    12 $z_i = (s_i - \alpha_i z_{i-1})/\beta_i$
13 $x_N = z_N$
14 **Repeat** step 15 **with** $i = N - 1, N - 2, \ldots, 1$
    15 $x_i = z_i - \gamma_i x_{i+1}$
16 **Print** $\mathbf{x}$ and 'finished' and exit.

We can show (Exercise 3) that the number of operations for an *LU* decomposition method needed to solve a tridiagonal system $A\mathbf{x} = \mathbf{b}$ is proportional to $N$, not to a higher power of $N$, where $A$ is an $N \times N$ matrix. This is almost instantaneous. In many applications $N > 1000$ and there is a significant savings in using Equations (7.7.4), (7.7.6), and (7.7.7). Also, the storage necessary to solve a tridiagonal system is minimal. There is no need to use a two-dimensional array for the array since it would require $N^2$ storage spaces for $3N - 2$ elements. The nonzero elements can be stored in a one-dimensional array.

## 7.8  MATRIX INVERSION

The inverse of a matrix $A$ is defined in terms of its cofactors and determinant (Appendix B). However, this is not a practical way to find an inverse. Gaussian elimination or the factorization methods can be used to compute the inverse of $A$. Since $AA^{-1} = I$, it follows that the $i$th column of $A^{-1}$ is the solution of the linear system of equations $A\mathbf{x} = \mathbf{e}_i$. Hence, we obtain $A^{-1}$ by solving the systems of linear equations

$$A\mathbf{x} = \mathbf{e}_i \quad \text{for } i = 1, 2, \dots, N \tag{7.8.1}$$

The solution vectors $\mathbf{x}_1, \mathbf{x}_2, \dots, \mathbf{x}_N$ will be the columns of $A^{-1}$.

It is straightforward to use *LU* factorization of $A$ since Equation (7.8.1) represents a situation in which the same matrix $A$ appears with the different right-hand $\mathbf{e}_i$. Hence, by solving the systems

$$A\mathbf{x} = (LU)\mathbf{x} = \mathbf{e}_i \quad \text{for } i = 1, 2, \dots, N \tag{7.8.2}$$

we get

$$A^{-1} = [\mathbf{x}_1, \mathbf{x}_2, \dots, \mathbf{x}_N] \tag{7.8.3}$$

**E X A M P L E   7.8.1**    Find the inverse of the matrix

$$A = \begin{bmatrix} 4 & 2 & 3 \\ 2 & -4 & -1 \\ -1 & 1 & 4 \end{bmatrix}$$

To find $A^{-1}$, we must solve

$$
\begin{array}{lll}
4x_1 + 2x_2 + 3x_3 = 1 & 4x_1 + 2x_2 + 3x_3 = 0 & 4x_1 + 2x_2 + 3x_3 = 0 \\
2x_1 - 4x_2 - x_3 = 0 & 2x_1 - 4x_2 - x_3 = 1 & 2x_1 - 4x_2 - x_3 = 0 \\
-x_1 + x_2 + 4x_3 = 0 & -x_1 + x_2 + 4x_3 = 0 & -x_1 + x_2 + 4x_3 = 1
\end{array}
\tag{7.8.4}
$$

Equation (7.8.4) can be easily solved by the *LU* decomposition method. Since

$$A = \begin{bmatrix} 4 & 2 & 3 \\ 2 & -4 & -1 \\ -1 & 1 & 4 \end{bmatrix} = \begin{bmatrix} 1 & 0 & 0 \\ 1/2 & 1 & 0 \\ -1/4 & -3/10 & 1 \end{bmatrix} \begin{bmatrix} 4 & 2 & 3 \\ 0 & -5 & -5/2 \\ 0 & 0 & 4 \end{bmatrix} = LU$$

we need to solve $A\mathbf{x} = (LU)\mathbf{x} = \mathbf{e}_i$ for $i = 1, 2,$ and 3. Solving

$$(LU)\mathbf{x} = \begin{bmatrix} 1 & 0 & 0 \\ 1/2 & 1 & 0 \\ -1/4 & -3/10 & 1 \end{bmatrix} \begin{bmatrix} 4 & 2 & 3 \\ 0 & -5 & -5/2 \\ 0 & 0 & 4 \end{bmatrix} \begin{bmatrix} x_1 \\ x_2 \\ x_3 \end{bmatrix} = \mathbf{e}_1 = \begin{bmatrix} 1 \\ 0 \\ 0 \end{bmatrix}$$

we get

$$\mathbf{x}_1 = \begin{bmatrix} 3/16 \\ 7/80 \\ 1/40 \end{bmatrix}$$

Similarly solving $(LU)\mathbf{x} = \mathbf{e}_2$ and $(LU)\mathbf{x} = \mathbf{e}_3$, we get

$$\mathbf{x}_2 = \begin{bmatrix} 1/16 \\ -19/80 \\ 3/40 \end{bmatrix} \quad \text{and} \quad \mathbf{x}_3 = \begin{bmatrix} -1/8 \\ -1/8 \\ 1/4 \end{bmatrix}$$

Thus

$$A^{-1} = [\mathbf{x}_1, \mathbf{x}_2, \mathbf{x}_3] = \begin{bmatrix} 3/16 & 1/16 & -1/8 \\ 7/80 & -19/80 & -1/8 \\ 1/40 & 3/40 & 1/4 \end{bmatrix}$$

Alternatively, the Gaussian elimination method can be conveniently used on the larger augmented matrix

$$\begin{bmatrix} 4 & 2 & 3 & | & 1 & 0 & 0 \\ 2 & -4 & -1 & | & 0 & 1 & 0 \\ -1 & 1 & 4 & | & 0 & 0 & 1 \end{bmatrix}$$

Eliminating elements in column one below the diagonal element 4 gives

$$\begin{bmatrix} 4 & 2 & 3 & | & 1 & 0 & 0 \\ 0 & -5 & -5/2 & | & -1/2 & 1 & 0 \\ 0 & 3/2 & 19/4 & | & 1/4 & 0 & 1 \end{bmatrix}$$

Eliminating the elements below the diagonal element $-5$ in column 2 gives

$$\begin{bmatrix} 4 & 2 & 3 & | & 1 & 0 & 0 \\ 0 & -5 & -5/2 & | & -1/2 & 1 & 0 \\ 0 & 0 & 4 & | & 1/10 & 3/10 & 1 \end{bmatrix} \tag{7.8.5}$$

The solutions of Equation (7.8.5) are found by using the backward substitution on each of the three augmented matrices

$$\begin{bmatrix} 4 & 2 & 3 & | & 1 \\ 0 & -5 & -5/2 & | & -1/2 \\ 0 & 0 & 4 & | & 1/10 \end{bmatrix} \quad \begin{bmatrix} 4 & 2 & 3 & | & 0 \\ 0 & -5 & -5/2 & | & 1 \\ 0 & 0 & 4 & | & 1/10 \end{bmatrix} \quad \begin{bmatrix} 4 & 2 & 3 & | & 0 \\ 0 & -5 & -5/2 & | & 0 \\ 0 & 0 & 4 & | & 1 \end{bmatrix}$$

$$\tag{7.8.6}$$

The solutions of Equation (7.8.5) are given by

$$
\begin{array}{lll}
x_1 = 3/16 & x_1 = 1/16 & x_1 = -1/8 \\
x_2 = 7/80 & x_2 = -19/80 & x_2 = -1/8 \\
x_3 = 1/40 & x_3 = 3/40 & x_3 = 1/4
\end{array}
$$

and so

$$
A^{-1} = \begin{bmatrix}
3/16 & 1/16 & -1/8 \\
7/80 & -19/80 & -1/8 \\
1/40 & 3/40 & 1/4
\end{bmatrix}
$$

■ ■ ■

## EXERCISES

1. Solve the following linear systems using Equations (7.7.4), (7.7.6), and (7.7.7):

   (a) $\begin{aligned} -4x_1 + x_2 \quad\quad &= -1 \\ x_1 - 4x_2 + x_3 &= 0 \\ x_2 - 4x_3 &= 0 \end{aligned}$

   (b) $\begin{aligned} -2x_1 + x_2 \quad\quad &= -2 \\ x_1 - 2x_2 + x_3 &= 0 \\ x_2 - 2x_3 &= 0 \end{aligned}$

2. Count the number of divisions, multiplications, and additions/subtractions you used to solve Exercise 1(a).

3. Find the number of divisions, multiplications, and additions/subtractions we need to solve a tridiagonal system $A\mathbf{x} = \mathbf{b}$ where $A$ is an $N \times N$ matrix when we use Equations (7.7.4), (7.7.6), and (7.7.7).

4. Make a table for the required number of divisions, multiplications, and additions/subtractions we need to solve a tridiagonal system $A\mathbf{x} = \mathbf{b}$ of order 3, 5, 10, and 100 when we use Equations (7.7.4), (7.7.6), and (7.7.7). Compare these values with the values given in Table 7.4.1.

5. Let

$$
A = \begin{bmatrix}
2 & -1 & 0 & 0 & 0 & 0 \\
-1 & 2 & -1 & 0 & 0 & 0 \\
0 & -1 & 2 & -1 & 0 & 0 \\
0 & 0 & -1 & 2 & -1 & 0 \\
0 & 0 & 0 & -1 & 2 & -1 \\
0 & 0 & 0 & 0 & -1 & 2
\end{bmatrix}
$$

   be the triangular matrix. Find the matrices $L$ and $U$ using Equation (7.7.4). Generalize the matrices $L$ and $U$ for such an $N \times N$ matrix $A$.

6. The decomposition in Equation (7.7.3) is not unique. We can use

$$
L = \begin{bmatrix}
1 & 0 & 0 & \cdots & 0 & 0 \\
\alpha_2 & 1 & 0 & \cdots & 0 & 0 \\
0 & \alpha_3 & 1 & \cdots & 0 & 0 \\
\vdots & \vdots & \vdots & & \vdots & \vdots \\
0 & 0 & 0 & \cdots & 1 & 0 \\
0 & 0 & 0 & \cdots & \alpha_N & 1
\end{bmatrix}
$$

   Derive recursive relations similar to Equations (7.7.4), (7.7.5), and (7.7.6) to solve a tridiagonal system $A\mathbf{x} = \mathbf{b}$ with the above $L$ and $U$.

7. Find the inverse of the following matrices and check your answer by computing $AA^{-1}$:

(a) $\begin{bmatrix} 1 & 0 \\ 3 & 3 \end{bmatrix}$
(b) $\begin{bmatrix} 1 & 2 & \frac{3}{2} \\ 0 & 1 & -1 \\ 3 & 5 & 7 \end{bmatrix}$

8. For solving $A\mathbf{x} = \mathbf{b}$, we can find $A^{-1}$ and then find $\mathbf{x} = A^{-1}\mathbf{b}$. Why it is better to solve the linear system of equations $A\mathbf{x} = \mathbf{b}$ than to find $A^{-1}$ and $\mathbf{x}$ by $\mathbf{x} = A^{-1}\mathbf{b}$?

## COMPUTER EXERCISES

**C.1.** Using Equation (7.7.4), write a subroutine that factors an $N \times N$ tridiagonal matrix $A$ given by Equation (7.7.2) into two factors $L$ and $U$.

**C.2.** Using Equations (7.7.6) and (7.7.7), write a subroutine that solves the systems $L\mathbf{z} = \mathbf{s}$ and $U\mathbf{x} = \mathbf{z}$ where matrices $L$ and $U$ are given by Exercise C.1.

**C.3.** Using Exercises C.1 and C.2, write a computer program that reads the values of a tridiagonal matrix $A$ and $\mathbf{s}$ to solve the linear system of equations $A\mathbf{x} = \mathbf{s}$. Print the values of $A$ and $\mathbf{s}$. Print the values of $\mathbf{x}$, $L$, and $U$. Multiply $L$ and $U$ to show that that it is actually $A$ by printing the values $LU$ along with $A$. Multiply $U$ and $\mathbf{x}$ and then $L$ and $U\mathbf{x}$; print the values of $L(U\mathbf{x})$ and $\mathbf{s}$.

**C.4.** Using Exercise C.3, solve the following linear system of equations:

$$\begin{aligned}
-2x_1 + x_2 &= -2 \\
x_1 - 2x_2 + x_3 &= 0 \\
x_2 - 2x_3 + x_4 &= 0 \\
x_3 - 2x_4 + x_5 &= 0 \\
x_4 - 2x_5 &= -3
\end{aligned}$$

**C.5.** Using Exercise C.3 for $N = 10$ and 20 and $h = \pi/(2N)$, solve the following linear system of equations

$$\left(1 + \frac{h^2}{12}\right) Y_{i-1} + \left(\frac{10h^2}{12} - 2\right) Y_i + \left(1 + \frac{h^2}{12}\right) Y_{i+1} = 0$$

for $i = 1, 2, \ldots, N - 1$ where $Y_0 = 1$ and $Y_N = 0$.

**C.6.** The system of equations for an electric ladder is given by

$$\begin{aligned}
10x_1 - 5x_2 &= 20 \\
-5x_{i-1} + 10x_i - 5x_{i+1} &= 0 \quad \text{for } i = 2, 3, \ldots, N - 1 \\
-5x_{N-1} + 10x_N &= 0
\end{aligned}$$

Using Exercise C.3 with $N = 10$ and 20, solve the system.

**C.7.** Using Equation (7.5.7), write a subroutine that factors an $N \times N$ matrix $A$ given by Equation (7.2.6) into two factors $L$ and $U$.

**C.8.** Using Equations (7.5.9) and (7.5.10), write a subroutine that solves the systems $L\mathbf{y} = \mathbf{b}$ and $U\mathbf{x} = \mathbf{y}$ where matrices $L$ and $U$ are given by Exercise C.7.

**C.9.** Write a computer program that reads the values of a matrix $A$ to find $A^{-1}$ using Exercises C.7 and C.8 or Exercises C.1 and C.2, page 205. Multiply $A$ and $A^{-1}$ and verify that $AA^{-1} = I$. Print the values of $A$, $A^{-1}$, and $AA^{-1}$.

**C.10.** Using Exercise C.9 or the Gaussian elimination method, find the inverse of the matrix

$$\begin{bmatrix} 1 & 1/2 & 1/3 & 1/4 \\ 1/2 & 1/3 & 1/4 & 1/5 \\ 1/3 & 1/4 & 1/5 & 1/6 \\ 1/4 & 1/5 & 1/6 & 1/7 \end{bmatrix}$$

**C.11.** Using Exercise C.9 or the Gaussian elimination method, find the inverse of the matrix

$$\begin{bmatrix} 10 & 7 & 8 & 7 \\ 7 & 5 & 6 & 5 \\ 8 & 6 & 10 & 9 \\ 7 & 5 & 9 & 10 \end{bmatrix}$$

## SUGGESTED READINGS

L. V. Atkinson, P. J. Harley, and J. D. Hudson, *Numerical Methods with Fortran 77, A Practical Introduction,* Addison-Wesley Publishing Company, Workingham, England, 1989.

D. Coppersmith and S. Winograd, *On the asymptotic complexity of matrix multiplications,* SIAM J. Comput **11** (1982), 472–492.

G. E. Forsyth, M. A. Malcolm, and C. B. Moler, *Computer Methods for Mathematical Computations,* Prentice-Hall, Inc., Englewood Cliffs, N.J., 1977.

L. Fox, *An Introduction to Numerical Algebra,* Oxford University Press, Oxford, 1964.

G. H. Golub and C. F. Van Loan, *Matrix Computations,* 2nd ed., The Johns Hopkins University Press, Baltimore, Maryland, 1989.

V. Pan, *How can we speed up matrix multiplication?* SIAM Rev **26** (1984), 393–415.

G. Strang, *Linear Algebra and its Applications,* 3rd ed., Harcourt Brace Jovanovich, Publishers, San Diego, 1988.

V. Strassen, *Gaussian elimination is not optimal,* Numer. Math. **13** (1969), 354–356.

D. S. Watkins, *Fundamental of Matrix Computations,* John Wiley & Sons, New York, 1991.

J. H. Wilkinson, *The Algebraic Eigenvalue Problem,* Oxford University Press, Oxford, 1965.

# Chapter

# 8 INITIAL-VALUE PROBLEMS: SINGLE-STEP METHODS

I f a planet of mass $m$ revolves around the sun of mass $M$, then the equations of motion in polar coordinates $(r, \theta)$ are given by (McCuskey 1962)

$$\frac{d^2r}{dt^2} - r\left(\frac{d\theta}{dt}\right)^2 = -\frac{\mu}{r^2} \tag{1}$$

and

$$\frac{d}{dt}\left(r^2\frac{d\theta}{dt}\right) = 0 \tag{2}$$

where $\mu = G(M + m)$ and $G$ is the constant of gravitation.

Integrating Equation (2), we get

$$r^2\frac{d\theta}{dt} = h \tag{3}$$

where $h$ is a constant. Substituting Equation (3) in Equation (1) gives

$$\frac{d^2r}{dt^2} - \frac{h^2}{r^3} = -\frac{\mu}{r^2} \tag{4}$$

Substituting $u = 1/r$ in Equation (4) yields

$$\frac{d^2u}{d\theta^2} + u = \frac{\mu}{h^2} \tag{5}$$

Using the general theory of relativity, it can be shown (Tolman 1958) that the path of a planet is given by

$$\frac{d^2u}{d\theta^2} + u = \frac{\mu}{h^2} + \frac{3\mu}{c^2}u^2 \tag{6}$$

where $c$ is the speed of light. The last term in Equation (6) is a very small correction.

The solution of Equation (5) is given by

$$u = \frac{1}{r} = \frac{\mu}{h^2}[1 + e\cos(\theta - \omega)] \tag{7}$$

where $e$ is the eccentricity of the orbit and $\omega$ is the longitude of perihelion. Equation (7) represents an ellipse in polar coordinates. The vertex of an ellipse that is nearest to the focus where the sun is located is called a perihelion; the other vertex is called an aphelion. The approximate solution of Equation (6) suggests an advance in the longitude of its perihelion. For most planets this amount is too small to be observed. In the case of the planet Mercury, there was a discrepancy between the predicted value given by Equation (7) and actual observations. The advancement predicted by Equation (6) was actually observed in the longitude of perihelion of Mercury. This gave a boost to the new theory of relativity at that time. In Chapters 8, 9, and 14, we develop methods to approximate the solutions of ordinary differential equations.

## 8.1    INTRODUCTION

Differential equations are divided into two classes, **ordinary** and **partial**, according to the number of independent variables present in the differential equations; one for ordinary and more than one for partial. The **order** of the differential equation is the order of its highest derivative. The general solution of the $N$th-order ordinary differential equation contains $N$ independent arbitrary constants. To determine these $N$ arbitrary constants, we need $N$ conditions. If these $N$ conditions are prescribed at one point, then these conditions are called **initial conditions**. A differential equation with initial conditions is called an **initial-value problem**. If these $N$ conditions are prescribed at different points, then these conditions are called **boundary conditions**. A differential equation with boundary conditions is called a **boundary-value problem**.

These two types of problems have different properties. An initial-value problem can be identified with a time-dependent problem. The solution of a time-dependent problem depends only on what has happened. A boundary-value problem can be identified with a steady-state problem. The solution of a steady-state problem at different spatial points may be interdependent. In this chapter and the next chapter, we develop methods to solve ordinary differential equations with given initial conditions.

We shall start with a first-order ordinary differential equation

$$\frac{dy}{dx} = f(x, y)$$

with the initial condition

$$y(x_0) = y_0 \tag{8.1.1}$$

where $y(x)$ is the unknown and $f$ is a given function of the independent variable $x$ and unknown $y$.

Let us consider a set of $M$ first-order simultaneous equations in the form

$$\frac{dy_1}{dx} = f_1(x, y_1, y_2, \ldots, y_M)$$

$$\frac{dy_2}{dx} = f_2(x, y_1, y_2, \ldots, y_M)$$

$$\vdots$$

$$\frac{dy_M}{dx} = f_M(x, y_1, y_2, \ldots, y_M)$$

with the initial conditions

$$y_1(x_0) = \alpha_1, y_2(x_0) = \alpha_2, \ldots, y_M(x_0) = \alpha_M \qquad \textbf{(8.1.2)}$$

where $y_1, y_2, \ldots, y_M$ are unknowns and $f_1, f_2, \ldots, f_M$ are given functions of the independent variable $x$ and unknowns $y_1, y_2, \ldots, y_M$.

Let

$$\mathbf{y}(x) = \begin{bmatrix} y_1(x) \\ y_2(x) \\ \vdots \\ y_M(x) \end{bmatrix} \qquad \mathbf{f}(x, \mathbf{y}) = \begin{bmatrix} f_1(x, y_1, y_2, \ldots, y_M) \\ f_2(x, y_1, y_2, \ldots, y_M) \\ \vdots \\ f_M(x, y_1, y_2, \ldots, y_M) \end{bmatrix} \qquad \mathbf{y}_0 = \begin{bmatrix} \alpha_1 \\ \alpha_2 \\ \vdots \\ \alpha_M \end{bmatrix}$$

Then the system of Equation (8.1.2) can be written as

$$\frac{d\mathbf{y}}{dx} = \mathbf{f}(x, \mathbf{y})$$

with the initial condition

$$\mathbf{y}(x_0) = \mathbf{y}_0 \qquad \textbf{(8.1.3)}$$

It will later be shown that any numerical method that is applicable to Equation (8.1.1) can be generalized to solve Equation (8.1.3).

An $M$th-order initial-value problem has the form

$$\frac{d^M y}{dx^M} = f\left(x, y, \frac{dy}{dx}, \frac{d^2 y}{dx^2}, \ldots, \frac{d^{M-1} y}{dx^{M-1}}\right) \qquad \textbf{(8.1.4a)}$$

with the initial conditions

$$y(x_0) = \alpha_1, \frac{dy}{dx}(x_0) = \alpha_2, \frac{d^2 y}{dx^2}(x_0) = \alpha_3, \ldots, \frac{d^{M-1} y}{dx^{M-1}} = \alpha_M \qquad \textbf{(8.1.4b)}$$

where $y(x)$ is unknown and $f$ is a given function of the independent variable $x$ and unknown $y$ and its derivatives. We can convert Equation (8.1.4a) to a set of $M$ simultaneous equations by defining

$$y_1(x) = y(x)$$
$$y_2(x) = \frac{dy_1}{dx} = \frac{dy}{dx}$$
$$y_3(x) = \frac{dy_2}{dx} = \frac{d^2 y}{dx^2}$$
$$\vdots$$
$$y_M(x) = \frac{dy_{M-1}}{dx} = \frac{d^{M-1} y}{dx^{M-1}}$$

Then Equation (8.1.4a) can be written as

$$\frac{dy_1}{dx} = y_2$$
$$\frac{dy_2}{dx} = y_3$$
$$\vdots$$
$$\frac{dy_{M-1}}{dx} = y_M$$

$$\frac{dy_M}{dx} = f(x, y_1, y_2, \ldots, y_M) \tag{8.1.5a}$$

and the initial conditions given by Equation (8.1.4b) become

$$y(x_0) = \alpha_1, \ y_2(x_0) = \alpha_2, \ldots, y_M(x_0) = \alpha_M \tag{8.1.5b}$$

---

**EXAMPLE 8.1.1**     To illustrate these ideas, write

$$\frac{d^3y}{dx^3} + y\frac{d^2y}{dx^2} + xy^2\frac{dy}{dx} + y = 8x$$

with initial conditions $y(0) = 1$, $dy/dx\,(0) = 2$, and $d^2y/dx^2\,(0) = 5$ as a system of first-order ordinary differential equations.

This is equivalent to

$$\frac{dy_1}{dx} = y_2$$

$$\frac{dy_2}{dx} = y_3$$

$$\frac{dy_3}{dx} = -y_1\,y_3 - xy_1^2\,y_2 - y_1 + 8x$$

with $y_1(0) = 1$, $y_2(0) = 2$, and $y_3(0) = 5$.     ■ ■ ■

Thus Equation (8.1.1) is basic for the numerical solutions of ordinary differential equations with given initial conditions. So far we have discussed initial-value problems. Many times boundary conditions are specified rather than initial conditions. Consider for example,

$$\frac{d^2y}{dx^2} + y = 0 \tag{8.1.6a}$$

together with the boundary conditions

$$y(0) = 0 \quad \text{and} \quad y\left(\frac{\pi}{2}\right) = 1 \tag{8.1.6b}$$

The general solution of Equation (8.1.6a) is given by

$$y(x) = A\sin x + B\cos x \tag{8.1.7}$$

where $A$ and $B$ are constants to be determined from Equation (8.1.6b). Since $y(0) = 0$, $B = 0$. Thus the solution of Equation (8.1.6a) and the first condition of (8.1.6b) is

$$y(x) = A\sin x \tag{8.1.8}$$

This gives rise to a family of different solutions as shown in Figure 8.1.1. Since $y(\pi/2) = 1$, this serves to select the particular member of the family. However, if $y(\pi) = 0$ had been specified, then there are infinitely many solutions; if $y(\pi) \neq 0$, then there are no solutions.

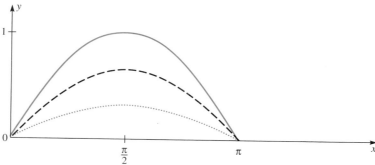

Fig. 8.1.1

Before attempting to solve an initial-value problem, we would like to know whether a solution exists and, if so, whether it is unique. Fortunately, it is possible to state conditions under which an initial-value problem will have a unique solution.

We need $f(x, y)$ to be continuous in a region $R = \{(x, y)| |x - x_0| \le a, |y - y_0| \le b\}$ and to satisfy a Lipschitz condition in $R$ which is defined below.

A function $f(x, y)$ is said to satisfy a **Lipschitz condition** in the variable $y$ in $R$ if there exists a constant $L > 0$ such that

$$|f(x, y_1) - f(x, y_2)| \le L |y_1 - y_2| \tag{8.1.9}$$

for every $(x, y_1)$ and $(x, y_2)$ belonging to $R$. The constant $L$ is called a **Lipschitz constant**.

Let $f$ be such that $\partial f/\partial y$ exists and is bounded for all $(x, y)$ in $R$. Applying the Mean Value Theorem (Theorem A.3) for derivatives, for any $(x, y_1)$ and $(x, y_2)$ in $R$ there exists $\xi$ between $y_1$ and $y_2$ such that

$$\frac{f(x, y_1) - f(x, y_2)}{y_1 - y_2} = \frac{\partial f}{\partial y}(x, \xi)$$

Thus

$$|f(x, y_1) - f(x, y_2)| = \left|\frac{\partial f}{\partial y}(x, \xi)\right| |y_1 - y_2|$$

$$\le L |y_1 - y_2|$$

where $|\partial f/\partial y| \le L$ for all $(x, y)$ in $R$. Thus if $f$ is defined in $R$ and if there exists $L$ such that $|\partial f/\partial y \, (x, y)| \le L$ for all $(x, y)$ in $R$, then $f$ satisfies the Lipschitz condition (8.1.9) in $R$ in $y$ with the Lipschitz constant $L$.

The Lipschitz condition (8.1.9) is weaker than the existence of $\partial f/\partial y$; in other words, there are functions whose partial derivatives with respect to $y$ do not exist and yet they satisfy the Lipschitz condition. Consider the function $f$ defined by $f(x, y) = x|y|$ in $R$ where $R = \{(x, y) \mid |x| \le a, |y| \le b\}$. Although $\partial f/\partial y$ does not exist for $(x, 0)$ in $R$,

$$|f(x, y_1) - f(x, y_2)| \le |x| \, |y_1 - y_2| \le a \, |y_1 - y_2|$$

for all $(x, y_1)$ and $(x, y_2)$ in $R$. Thus $f(x, y) = x |y|$ satisfies the Lipschitz condition in the variable $y$ in $R$.

We now state the Existence and Uniqueness Theorems.

**Theorem 8.1.1**    **(Existence Theorem)**    Let $f(x, y)$ be continuous on a closed rectangle

$$R = \{(x, y) \mid |x - x_0| \le a, |y - y_0| \le b\}$$

and $|f(x, y)| \le M$ for all $(x, y)$ in $R$. Then the initial-value problem given by Equation (8.1.1) has at least one solution on the interval $|x - x_0| \le h$ where $h$ is the smaller of the two numbers $a$ and $b/M$.    ∎

**Theorem 8.1.2**    **(Uniqueness Theorem)**    Let $f(x, y)$ be continuous on a closed rectangle

$$R = \{(x, y) \mid |x - x_0| \le a, |y - y_0| \le b\}$$

bounded; in other words, $|f(x, y)| \le M$ for all $(x, y)$ in $R$, and $f(x, y)$ satisfies a Lipschitz condition in $R$. Then the initial-value problem given by Equation (8.1.1) has a unique solution on the interval $|x - x_0| \le h$, where $h$ is the smaller of the two numbers $a$ and $b/M$.    ∎

Complete details can be found in most of the books on ordinary differential equations (for example, see Ross 1974).

**EXAMPLE 8.1.2**    Discuss the existence and uniqueness of the initial-value problem $dy/dx = y^{1/3}$, $y(0) = 0$.

Since $f(x, y) = y^{1/3}$ is continuous in $R = \{(x, y) \mid |x| \le a, |y| \le b\}$ and $|f(x, y)| \le b^{1/3}$, then by the Existence Theorem there exists at least one solution on $|x| < h$ where $h = \min (a, b/b^{1/3})$.

Let us examine the question of uniqueness. Since $\partial f/\partial y = \frac{1}{3} y^{-2/3}$, $\partial f/\partial y$ does not exist on the line $y = 0$. Let us check whether $f(x, y)$ satisfies the Lipschitz condition in $R$ or not. We have

$$
\begin{aligned}
\frac{|f(x, y_1) - f(x, y_2)|}{|y_1 - y_2|} &= \frac{|y_1^{1/3} - y_2^{1/3}|}{|y_1 - y_2|} \\
&= \frac{|y_1^{1/3} - y_2^{1/3}|}{|(y_1^{1/3} - y_2^{1/3})(y_1^{2/3} + y_1^{1/3} y_2^{1/3} + y_2^{2/3})|} \\
&= \frac{1}{|y_1^{2/3} + y_1^{1/3} y_2^{1/3} + y_2^{2/3}|}
\end{aligned}
$$

By selecting $y_1$ and $y_2$ arbitrarily small, this ratio becomes arbitrarily large and so we cannot find $L$. Thus $f(x, y)$ does not satisfy a Lipschitz condition on any domain that contains the line $y = 0$ and therefore not on $R$. We cannot apply the Uniqueness Theorem. However, do not conclude that uniqueness is impossible just because the

Lipschitz condition is not satisfied; rather, no conclusion can be drawn about uniqueness in this problem.

In fact, it can be shown that there are two solutions that are given by

$$y_1(x) = 0 \quad \text{for all } x \qquad y_2(x) = \begin{cases} (\tfrac{2}{3}x)^{3/2} & \text{for } x \geq 0 \\ 0 & \text{for } x < 0 \end{cases}$$  ∎∎∎

## EXERCISES

1.  Write each of the following initial-value problems as a system of first-order ordinary differential equations:

    (a) $\dfrac{d^2y}{dx^2} - \left(\dfrac{dy}{dx}\right)^2 + y = x^2 \qquad y(0) = 1 \qquad \dfrac{dy}{dx}(0) = 2$

    (b) $\dfrac{d^3y}{dx^3} - 3\dfrac{d^2y}{dx^2} - \dfrac{dy}{dx} + y^2 = e^x \qquad y(0) = 1 \qquad \dfrac{dy}{dx}(0) = 1 \qquad \dfrac{d^2y}{dx^2}(0) = 0$

    (c) $\dfrac{d^2y}{dx^2} + p(x)\dfrac{dy}{dx} + q(x)\,y = g(x) \qquad y(0) = \alpha_1 \qquad \dfrac{dy}{dx}(0) = \alpha_2$

2.  Show that each of the following functions satisfies a Lipschitz condition in $y$ in the rectangle $R = \{(x, y) \mid |x| \leq a, |y| \leq b\}$:

    (a) $f(x, y) = 2x^2 + y^2$      (b) $f(x, y) = y \sin x + x \sin y$

    (c) $f(x, y) = (x + y)e^{x^2+y^2}$

3.  For each of the following initial-value problems find the largest interval $|x - x_0| \leq h$ on which the Uniqueness Theorem guarantees the existence of a unique solution.

    (a) $\dfrac{dy}{dx} = y^2 \qquad y(-1) = -1$      (b) $\dfrac{dy}{dx} = x^2 + y^2 \qquad y(0) = 0$

4.  For each of the following initial-value problems show that there exists a unique solution of the problem if $y_0 \neq 0$. Discuss the existence and uniqueness of the solution if $y_0 = 0$.

    (a) $\dfrac{dy}{dx} = y^{2/3} \qquad y(x_0) = y_0$      (b) $\dfrac{dy}{dx} = y^{4/3} \qquad y(x_0) = y_0$

5.  Show that if $a_{11}, a_{12} \neq 0, a_{21} \neq 0$, and $a_{22}$ are constants and if the functions $f_1$ and $f_2$ are differentiable, then the initial-value problem

    $$\dfrac{dy_1}{dx} = a_{11}\,y_1 + a_{12}\,y_2 + f_1(x)$$

    $$\dfrac{dy_2}{dx} = a_{21}\,y_1 + a_{22}\,y_2 + f_2(x)$$

    with $y_1(0) = \alpha_1$, $y_2(0) = \alpha_2$ can be transformed into a second-order initial-value problem.

6.  Show that $\int_a^b f(x)\,dx$ can be computed by solving the initial-value problem $dy/dx = f(x)$ for $a \leq x \leq b$ with $y(a) = 0$.

7.  Derive Equation (5).

## 8.2 FINITE-DIFFERENCE METHODS

We want to compute approximate values of the solution of an initial-value problem

$$\frac{dy}{dx} = f(x, y) \qquad y(x_0) = y_0 \tag{8.2.1}$$

at selected discrete points

$$x_0 = a \qquad x_i = x_{i-1} + h_i = x_0 + ih_i \quad \text{for } i$$
$$= 1, 2, \ldots, N - 1 \quad \text{and} \quad x_N = b \tag{8.2.2}$$

where $h_i$ is the local interval or step size as shown in Figure 8.2.1.

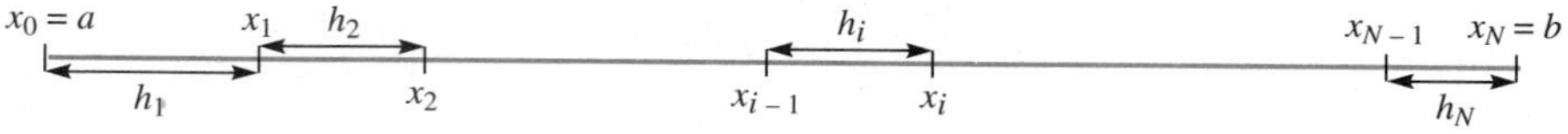

**Fig. 8.2.1**

Often, we take $h_i$ to be a constant $h$ independent of $i$. Then Equation (8.2.2) reduces to

$$x_0 = a \qquad x_i = x_{i-1} + h = x_0 + i h \quad \text{for } i$$
$$= 1, 2, \ldots, N - 1 \quad \text{and} \quad x_N = b \tag{8.2.3}$$

The points $x_i$ are referred to as mesh points. At $x = x_i > x_0$, the unknown $y$ satisfies Equation (8.2.1)

$$\frac{dy}{dx}(x_i) = f(x_i, y(x_i)) \tag{8.2.4}$$

The simplest way to approximate Equation (8.2.4) is to replace the derivative $dy/dx$ at $x_i$ by

$$\frac{dy}{dx}(x_i) = \frac{y(x_i + h) - y(x_i)}{h} - \frac{h}{2} y''(\xi_i) \tag{8.2.5}$$

where $x_i < \xi_i < x_i + h$ from Table 5.3.1. Substituting Equation (8.2.5) in Equation (8.2.4) yields

$$y(x_i + h) - y(x_i) - hf(x_i, y(x_i)) = \frac{h^2}{2} y''(\xi_i) \tag{8.2.6}$$

for $i = 0, 1, \ldots, N - 1$. If we choose $h$ small enough, we can ignore the right-hand side of Equation (8.2.6). Denoting the approximate value of $y$ at $x_i$ by $Y_i$ [i.e., $Y_i \cong y(x_i)$] and the approximate value of $y$ at $x_i + h$ by $Y_{i+1}$ [i.e., $Y_{i+1} \cong y(x_i + h)$], Equation (8.2.6) reduces to

$$Y_{i+1} - Y_i - hf(x_i, Y_i) = 0$$

or

$$Y_{i+1} = Y_i + hf(x_i, Y_i) \tag{8.2.7}$$

for $i = 0, 1, \ldots, N - 1$. Equation (8.2.7), known as **Euler's formula** or **Euler's forward formula**, provides a means of obtaining $Y_{i+1}$ from a knowledge of $Y_i$. Starting with $Y_0 \cong y(x_0)$, we first compute $Y_1$ as an approximation of $y(x_1)$, then $Y_2, Y_3$, and so on.

Similarly, substituting

$$\frac{dy}{dx}(x_i) = \frac{y(x_i) - y(x_i - h)}{h} + \frac{h}{2}y''(\eta_i)$$

where $x_i - h < \eta_i < x_i$ from Table 5.3.1 in Equation (8.2.4), we get

$$y(x_i) - y(x_i - h) - hf(x_i, y(x_i)) = -\frac{h^2}{2}y''(\eta_i) \qquad \textbf{(8.2.8)}$$

for $i = 1, 2, \ldots, N$. Choosing $h$ small, we can ignore the right-hand side of Equation (8.2.8). Denoting the approximate value of $y$ at $x_i$ by $Y_i$ and the approximate value of $y$ at $x_i - h$ by $Y_{i-1}$, Equation (8.2.8) reduces to

$$Y_i - Y_{i-1} - hf(x_i, Y_i) = 0 \quad \text{for } i = 1, 2, \ldots, N$$

or

$$Y_{i+1} = Y_i + hf(x_{i+1}, Y_{i+1}) \quad \text{for } i = 0, 1, \ldots, N - 1 \qquad \textbf{(8.2.9)}$$

Since $Y_{i+1}$ appears on both sides of Equation (8.2.9), this formula is called the **implicit** or **backward Euler formula**. We have to solve Equation (8.2.9) to get $Y_{i+1}$.

Further substituting

$$\frac{dy}{dx}(x_i) = \frac{y(x_i + h) - y(x_i - h)}{2h} - \frac{h^2}{6}y^{(3)}(\zeta_i)$$

where $x_i - h < \zeta_i < x_i + h$ from Table 5.3.1 in Equation (8.2.4) and choosing $h$ small, we get

$$Y_{i+1} - Y_{i-1} - 2hf(x_i, Y_i) = 0$$

or

$$Y_{i+1} = Y_{i-1} + 2hf(x_i, Y_i) \qquad \textbf{(8.2.10)}$$

for $i = 1, 2, \ldots, N - 1$. This is known as the **midpoint formula**. The value of $Y_1$ must be computed by another method.

**Algorithm**

**Subroutine** Euler $(a, b, N, y0, f, Y)$
Comment: This program approximates the solution of an initial-value problem $y' = f(x, y)$, $y(x_0) = y0 = Y_0$ on a given interval $[a, b]$ where $x_0 = a$ and $x_N = b$ using Equation (8.2.7). The output consists of the values of $Y$ at various $x$.
  1 **Print** $a, b, N, y0$
  2 $h = (b - a)/N$
  3 $x = a$
  4 $Y = y0$
  5 **Print** $h, x, Y$
  6 **Repeat** steps 7–9 **with** $i = 1, 2, \ldots, N$
      7 $Y = Y + hf(x, Y)$
      8 $x = x + h$
      9 **Print** $i, x, Y$
10 **Print** 'finished' and exit.

**EXAMPLE 8.2.1**    Solve the initial-value problem

$$\frac{dy}{dx} = -5y \qquad y(0) = 1$$

for $0 \leq x \leq 1$ using Equations (8.2.7), (8.2.9), and (8.2.10) and compare it with the exact solution $y = e^{-5x}$. Use $h = 0.1$.

Since $f(x, y) = -5y$, Equation (8.2.7) reduces to

$$Y_{i+1} = Y_i - 5hY_i = (1 - 5h) Y_i \quad \text{for } i = 0, 1, \ldots, 9 \qquad \textbf{(8.2.11)}$$

Then for $i = 0$, Equation (8.2.11) reduces to

$$Y_1 = (1 - 5h) Y_0 = (1 - 5h) = 0.5$$

For $i = 1$, Equation (8.2.11) reduces to

$$Y_2 = (1 - 5h) Y_1 = (1 - 5h)^2 = (0.5)^2 = 0.25$$

Similarly, we continue and the computed results are presented in Table 8.2.1.

**Table 8.2.1**

| $i$ | $x_i$ | $Y_i$ Using Eq. (8.2.11) | \|Actual Error\| for Eq. (8.2.11) | $Y_i$ Using Eq. (8.2.12) | \|Actual Error\| for Eq. (8.2.12) | $Y_i$ Using Eq. (8.2.13) | \|Actual Error\| for Eq. (8.2.13) |
|---|---|---|---|---|---|---|---|
| 0 | 0.0 | 1.00000 | 0.00000 | 1.00000 | 0.00000 | 1.00000 | 0.00000 |
| 1 | 0.1 | 0.50000 | 0.10653 | 0.66667 | 0.06014 | 0.60653 | 0.00000 |
| 2 | 0.2 | 0.25000 | 0.11788 | 0.44444 | 0.07657 | 0.39347 | 0.02559 |
| 3 | 0.3 | 0.12500 | 0.09813 | 0.29630 | 0.07317 | 0.21306 | 0.01007 |
| 4 | 0.4 | 0.06250 | 0.07284 | 0.19753 | 0.06220 | 0.18041 | 0.04507 |
| 5 | 0.5 | 0.03125 | 0.05083 | 0.13169 | 0.04960 | 0.03265 | 0.04943 |
| 6 | 0.6 | 0.01563 | 0.03416 | 0.08779 | 0.03800 | 0.14775 | 0.09797 |
| 7 | 0.7 | 0.00781 | 0.02238 | 0.05853 | 0.02833 | 0.11510 | 0.14530 |
| 8 | 0.8 | 0.00391 | 0.01441 | 0.03902 | 0.02070 | 0.26286 | 0.24454 |
| 9 | 0.9 | 0.00195 | 0.00916 | 0.02601 | 0.01490 | 0.37796 | 0.38907 |
| 10 | 1.0 | 0.00098 | 0.00576 | 0.01734 | 0.01060 | 0.64081 | 0.63408 |

Substituting $f(x, y) = -5y$, Equation (8.2.9) reduces to $Y_{i+1} = Y_i - 5hY_{i+1}$. Solving this implicit equation for $Y_{i+1}$ yields

$$Y_{i+1} = \frac{1}{(1 + 5h)} Y_i \quad \text{for } i = 0, 1, \ldots, 9 \qquad \textbf{(8.2.12)}$$

For $i = 0$, Equation (8.2.12) reduces to $Y_1 = 1/(1 + 5h) = 0.66667$. For $i = 1$, Equation (8.2.12) gives $Y_2 = (1/(1 + 5h)) Y_1 = 1/(1 + 5h)^2 = 4/9 = 0.44444$. In Table 8.2.1, we list the computed values $Y_i$ obtained using Equation (8.2.12).

Further, using $f(x, y) = -5y$, Equation (8.2.10) reduces to

$$Y_{i+1} = Y_{i-1} - 10hY_i \quad \text{for } i = 1, 2, \ldots, 9 \qquad \textbf{(8.2.13)}$$

For $i = 0$, we get $Y_{-1}$ which is not defined; therefore $Y_1$ cannot be computed using Equation (8.2.13). We use $Y_1 = e^{-5h}$. For $i = 1$, Equation (8.2.13) reduces to $Y_2 = Y_0 - 10hY_1 = 1 - 10\,he^{-0.5} = 1 - e^{-0.5} = 0.39347$. For $i = 2$, $Y_3 = Y_1 - 10\,hY_2 = e^{-0.5} - 0.39347 = 0.21306$. In the last two columns of Table 8.2.1, the computed values $Y_i$ obtained using Equation (8.2.13), and the absolute value of the actual errors are given. $|y(x_i) - Y_i|$ is the absolute value of the actual error at the $i$th step.

The error in the values of $Y$ computed using the midpoint formula [Equation (8.2.10)] in the last column of Table 8.2.1 is considerably smaller than the error by the other two methods for the first few steps and then it grows rapidly. To understand this phenomenon, let us look closely at Equation (8.2.13). The solution of this difference equation is given by (Appendix D) $Y_i = r^i$. Substitution of $Y_i = r^i$ in Equation (8.2.13) leads to the characteristic equation

$$r^2 + 10hr - 1 = 0$$

whose roots are given by $r_1 = -5h + \sqrt{1 + 25h^2}$ and $r_2 = -5h - \sqrt{1 + 25h^2} = -1/r_1$. Thus the general solution of Equation (8.2.13) is given by

$$Y_i = c_1 r_1^i + c_2 r_2^i \quad \text{for } i = 0, 1, \dots$$

where the constants $c_1$ and $c_2$ are to be determined by the initial conditions $Y_0 = 1$ and $Y_1 = e^{-5h}$. This leads to the following system

$$c_1 + c_2 = Y_0 = 1$$

$$c_1 r_1 + c_2 r_2 = Y_1 = e^{-5h}$$

whose solution is given by

$$c_1 = \frac{e^{-5h} - r_2}{r_1 - r_2} \quad \text{and} \quad c_2 = \frac{r_1 - e^{-5h}}{r_1 - r_2} \tag{8.2.14}$$

If $h$ is small, then

$$r_1 = -5h + \sqrt{1 + 25h^2} \approx 1 - 5h \quad \text{and} \quad r_2 \approx -(1 + 5h)$$

Since $\lim_{h \to 0} (1 + h)^{1/h} = e$ and $x_i = ih$,

$$\lim_{i \to \infty} r_1^i = \lim_{h \to 0} (1 - 5h)^{x_i/h} = \lim_{h \to 0} (1 - 5h)^{(-5x_i)/(-5h)} = e^{-5x_i}$$

Similarly

$$\lim_{h \to 0} (1 + 5h)^{x_i/h} = e^{5x_i}$$

Hence the solution of Equation (8.2.13) approaches

$$Y_i = c_1 e^{-5x_i} + c_2 e^{5x_i}(-1)^i \tag{8.2.15}$$

as $h \to 0$.

The first term in Equation (8.2.15) approximates the analytical solution $e^{-5x_i}$. The second term is a "parasitic solution" that arises because we replaced the first-order ordinary differential equation by the second-order difference equation. Using Taylor's series, it can be verified that $c_1 \approx 1$ and $c_2 \approx (125/12)\,h^3$. Thus a small error is introduced and is multiplied by the exponential factor $(-1)^i\,e^{5x_i}$. Since the first term is exponentially decreasing and the second term (the extraneous solution) is exponentially increasing, the error term introduced by the extraneous solution overshadows the first term. This leads to an incorrect solution.    ■ ■ ■

A method is considered to be **unstable** if errors introduced into the calculations grow as the computation proceeds. The midpoint formula is unstable. We will discuss stability further in Section 9.4.

### EXERCISES

In Exercises 1–4, (a) use the Euler formula [Equation (8.2.7)] to approximate the solution of the given initial-value problem on the given interval, (b) solve the initial-value problem exactly, and (c) compare (a) and (b) at three different points in the given interval including the right endpoint of the interval. Use $h = 0.1$.

1. $\dfrac{dy}{dx} = x^2 \qquad y(0) = 0 \quad$ on $\quad [0, 0.4]$

2. $\dfrac{dy}{dx} = x + y \qquad y(0) = 0 \quad$ on $\quad [0, 0.4]$

3. $(x^2 + 1)\dfrac{dy}{dx} + xy = 0 \qquad y(0) = 1 \quad$ on $\quad [0, 0.4]$

4. $\dfrac{dy}{dx} = 1 - x + 4y \qquad y(0) = 1 \quad$ on $\quad [0, 0.4]$

5. Use the implicit Euler formula [Equation (8.2.9)] to approximate the solution of the initial-value problem

$$\frac{dy}{dx} = x + y \qquad y(0) = 0 \quad \text{on} \quad [0, 0.4]$$

and compare it with the exact solution at three different points including $x = 0.4$. Use $h = 0.1$.

6. Use the implicit Euler method [Equation (8.2.9)] to approximate the solution of the initial-value problem

$$\frac{dy}{dx} = 1 - x + 4y \qquad y(0) = 1 \quad \text{on} \quad [0, 0.4]$$

and compare it with the exact solution at three different points including $x = 0.4$. Use $h = 0.1$.

7. Use the midpoint formula [Equation (8.2.10)] to approximate the solution of the initial-value problem

$$\frac{dy}{dx} = x^2 \qquad y(0) = 0 \quad \text{on} \quad [0, 0.4]$$

and compare it with the exact solution at three different points including $x = 0.4$. Use $h = 0.1$ and $Y_1 = y(x_1)$ where $y(x)$ is the exact solution of the given initial-value problem.

8.  Use the midpoint formula [Equation (8.2.10)] to approximate the solution of the initial-value problem

$$(x^2 + 1)\frac{dy}{dx} + xy = 0 \qquad y(0) = 1 \quad \text{on} \quad [0, 0.4]$$

and compare it with the exact solution at three different points including $x = 0.4$. Use $h = 0.1$ and $Y_1 = y(x_1)$ where $y(x)$ is the exact solution of the given initial-value problem.

9.  Find the general solution of the difference equation

$$Y_{i+1} = \frac{1}{(1 + 5h)} Y_i \quad \text{for } i = 0, 1, \ldots$$

Analyze the solution.

10  Find the general solution of the difference equation

$$Y_{i+1} = (1 - 5h) Y_i \quad \text{for } i = 0, 1, \ldots$$

Analyze the solution.

## COMPUTER EXERCISES

C.1.  Write a subroutine to compute the Euler forward formula $Y_{i+1} = Y_i + hf(x_i, Y_i)$ where $a$, $b$, $N$, $y0$, and $f$ are parameters. Compute $h = (b - a)/N$.

C.2.  Write a subroutine to compute the midpoint formula $Y_{i+1} = Y_{i-1} + 2hf(x_i, Y_i)$ where $a$, $b$, $N$, $y0$, $Y_1$, and $f$ are parameters. Compute $h = (b - a)/N$.

C.3.  Using Exercises C.1 and C.2, write a computer program to approximate the solution of an initial-value problem

$$\frac{dy}{dx} = f(x, y) \qquad y(x_0) = Y_0 \quad \text{for } x_0 = a \le x \le x_N = b$$

C.4.  Approximate the solution of the initial-value problem $dy/dx = 1 - x - 4y$, $y(0) = 1$ for $0 \le x \le 1$ with step sizes $h = 0.1, 0.05, 0.01, 0.005$, and $0.001$ using Exercise C.3. Print the exact solution, approximate solution, error, and relative error at 11 nodes including $x_0 = 0$ and $x_N = 1$.

C.5.  Approximate the solution of the initial-value problem $dy/dx = 1 - x - 4y$, $y(0) = 1$ for $0 \le x \le 1$ with step sizes $h = 0.1, 0.05, 0.01, 0.005$, and $0.001$ using the backward Euler formula. Print the exact solution, approximate solution, error, and relative error at 11 nodes including $x_0 = 0$ and $x_N = 1$.

C.6.  Approximate the solution of the initial-value problem $(x^2 + 1)(dy/dx) + xy = 0$, $y(0) = 1$ for $0 \le x \le 1$ with step sizes $h = 0.1, 0.05, 0.01, 0.005$, and $0.001$ using Exercise C.3. Print the exact solution, approximate solution, error, and relative error at 11 nodes including $x_0 = 0$ and $x_N = 1$.

C.7.  Approximate the solution of the initial-value problem $(x^2 + 1)(dy/dx) + xy = 0$, $y(0) = 1$ for $0 \le x \le 1$ with step sizes $h = 0.1, 0.05, 0.01, 0.005$, and $0.001$ using the backward Euler formula. Print the exact solution, approximate solution, error, and relative error at 11 nodes including $x_0 = 0$ and $x_N = 1$.

## 8.3  TAYLOR SERIES METHOD

If the unique solution of an initial-value problem

$$\frac{dy}{dx} = f(x, y) \qquad y(x_0) = y_0 \tag{8.3.1}$$

exists, then

$$y(x) = y(x_0 + (x - x_0)) = y(x_0) + y'(x_0)(x - x_0)$$
$$+ \frac{y''(x_0)}{2!}(x - x_0)^2 + \cdots$$
$$+ \frac{y^{(n)}(x_0)}{n!}(x - x_0)^n + \frac{y^{(n+1)}(\xi)}{(n+1)!}(x - x_0)^{(n+1)} \tag{8.3.2}$$

where $|x - x_0| < \xi$, provided the solution $y(x)$ has derivatives of all orders.

Let us denote $h = x - x_0$ in Equation (8.3.2). Then Equation (8.3.2) becomes

$$y(x) = y(x_0) + y'(x_0)\,h + \frac{y''(x_0)}{2!}\,h^2 + \cdots + \frac{y^{(n)}(x_0)}{n!}\,h^n$$
$$+ \frac{y^{(n+1)}(\xi)}{(n+1)!}\,h^{n+1} \tag{8.3.3}$$

We know $y(x_0) = y_0$. We want to find $y'(x_0), y''(x_0), \ldots, y^{(n)}(x_0)$, and $y^{(n+1)}(\xi)$. In order to find them, let us first find $y'(x), y''(x), \ldots, y^{(n+1)}(x)$.

Since $dy/dx = f(x, y)$,

$$y'(x) = \frac{dy}{dx} = f(x, y) \tag{8.3.4}$$

$$y''(x) = \frac{dy'}{dx} = \frac{d}{dx}[f(x, y(x))] = f'(x, y(x))$$
$$= \frac{\partial f}{\partial x} + \frac{\partial f}{\partial y}\frac{dy}{dx} = f_x(x, y) + f_y(x, y)f(x, y) \tag{8.3.5}$$

$$y^{(3)}(x) = f''(x, y) = f_{xx}(x, y) + 2f(x, y)f_{xy}(x, y)$$
$$+ f^2(x, y)f_{yy}(x, y) + f_x(x, y)f_y(x, y) + f(x, y)f_y^2(x, y) \tag{8.3.6}$$

Continuing,

$$y^{(n)}(x) = f^{(n-1)}(x, y)$$

Thus Equation (8.3.3) can be written as

$$y(x) = y(x_0) + f(x_0, y_0)\,h + \frac{f'(x_0, y_0)}{2!}\,h^2 + \frac{f''(x_0, y_0)}{3!}\,h^3 + \cdots$$
$$+ \frac{f^{(n-1)}(x_0, y_0)}{n!}\,h^n + \frac{f^{(n)}(\xi, y(\xi))}{(n+1)!}\,h^{n+1} \tag{8.3.7}$$

Since $f(x, y)$ is complicated (otherwise, we would solve the equation analytically), evaluation of higher derivatives is time consuming. Therefore instead of using a high-degree Taylor series over a relatively large distance, we divide the interval $[x_0, b]$ into small subintervals and use a lower-degree Taylor series over each subinterval.

Let us partition the interval $[x_0 = a, b]$ into $N$ equally-spaced subintervals by introducing grid points $a = x_0 < x_1 < x_2 \cdots < x_N = b$ as shown in Figure 8.3.1.

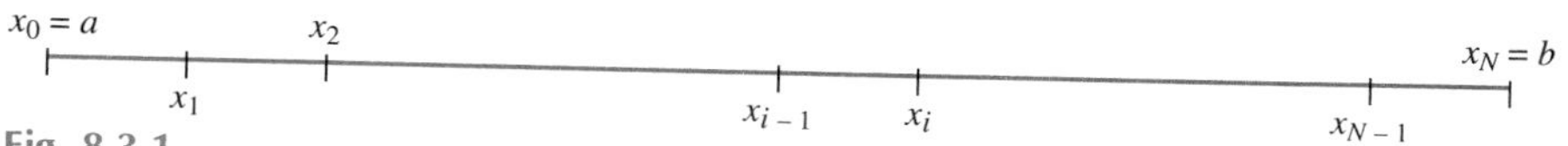

**Fig. 8.3.1**

If we denote the interval length by $h$, then $h = (b - a)/N$ and $x_i = x_0 + ih$ for $i = 0, 1, \ldots, N$. From Equation (8.3.7),

$$y(x_1) = y(x_0) + f(x_0, y_0)\, h + \frac{f'(x_0, y_0)}{2!} h^2 + \frac{f''(x_0, y_0)}{3!} h^3 + \cdots$$

$$+ \frac{f^{(n-1)}(x_0, y_0)}{n!} h^n + \frac{f^{(n)}(\xi, y(\xi))}{(n+1)!} h^{n+1} \qquad \text{(8.3.8)}$$

For convenience, denote

$$T_n(x, y, h) = f(x, y) + \frac{f'(x, y)}{2!} h + \cdots + \frac{f^{(n-1)}(x, y)}{n!} h^{n-1} \qquad \text{(8.3.9)}$$

Then Equation (8.3.8) becomes

$$y(x_1) = y(x_0) + h T_n(x_0, y_0, h) + \frac{f^{(n)}(\xi, y(\xi))}{(n+1)!} h^{n+1} \qquad \text{(8.3.10)}$$

Since we do not know $\xi$, we cannot compute the last term; however, since $h$ is small, we may ignore the last term in Equation (8.3.10). Using $\cong$ to denote "approximately equal to," we approximate $y(x_1)$ by

$$y(x_1) \cong y_0 + h T_n(x_0, y_0, h) \qquad \text{(8.3.11)}$$

Since this value is an approximate value of $y(x_1)$, denote it by $Y_1 \cong y(x_1)$. The approximate value of $y$ at $x_1$ is now known, so we can find the approximate value of $y$ at $x_2$ by computing

$$T_n(x_1, Y_1, h) = f(x_1, Y_1) + \frac{h}{2!} f'(x_1, Y_1) + \frac{h^2}{3!} f''(x_1, Y_1) + \cdots + \frac{h^{n-1}}{n!} f^{(n-1)}(x_1, Y_1)$$

and

$$Y_2 = Y_1 + h T_n(x_1, Y_1, h)$$

Similarly, we find the approximate values of $y$ at $x_3, x_4, \ldots, x_N$. In general, we have

$$Y_0 = y(x_0)$$

$$Y_{i+1} = Y_i + h\, T_n(x_i, Y_i, h) \quad \text{for } i = 0, 1, \ldots, N - 1 \qquad \text{(8.3.12)}$$

where

$$T_n(x_i, Y_i, h) = f(x_i, Y_i) + \frac{h}{2!} f'(x_i, Y_i) + \frac{h^2}{3!} f''(x_i, Y_i) + \cdots$$

$$+ \frac{h^{n-1}}{n!} f^{(n-1)}(x_i, Y_i) \qquad \text{(8.3.13)}$$

This method is called the **Taylor Series method of order $n$**. The Taylor Series method of order one is also called Euler's method.

**EXAMPLE 8.3.1**   Solve the initial-value problem

$$\frac{dy}{dx} = y - x \qquad y(0) = 0$$

for $0 \le x \le 1$ using the Taylor Series method of orders one, two, and three. Use $h = 0.05$.

For $n = 1$, Equation (8.3.13) gives $T_1(x_i, Y_i, h) = f(x_i, Y_i)$, and therefore Equation (8.3.12) becomes

$$
\begin{aligned}
Y_{i+1} &= Y_i + hT_1(x_i, Y_i, h) \\
&= Y_i + hf(x_i, Y_i) \\
&= Y_i + h(Y_i - x_i) \\
&= Y_i(1 + h) - hx_i
\end{aligned}
\qquad \textbf{(8.3.14)}
$$

For $i = 0$, we get from Equation (8.3.14)

$$Y_1 = Y_0(1 + h) - hx_0 = (0) \cdot (1 + h) - (h)(0) = 0.0$$

For $i = 1$, Equation (8.3.14) gives

$$Y_2 = Y_1(1 + h) - hx_1 = (0.0)(1.05) - (0.05)(0.05) = -0.00250$$

and similarly we compute the values at $x_3, x_4, \ldots$, and $x_{20} = 1.0$. The computed values are given in Table 8.3.1 along with the absolute value of the actual errors.

For $n = 2$, Equation (8.3.13) gives

$$T_2(x_i, Y_i, h) = f(x_i, Y_i) + \frac{h}{2}f'(x_i, Y_i)$$

Since $f(x, y) = y - x$,

$$f'(x, y) = \frac{d}{dx}(y - x) = \frac{dy}{dx} - 1 = y - x - 1$$

Hence

$$T_2(x_i, Y_i, h) = (Y_i - x_i) + \frac{h}{2}(Y_i - x_i - 1)$$

$$= (Y_i - x_i)\left(1 + \frac{h}{2}\right) - \frac{h}{2}$$

Therefore Equation (8.3.12) becomes

$$Y_{i+1} = Y_i + h\left[(Y_i - x_i)\left(1 + \frac{h}{2}\right) - \frac{h}{2}\right]$$

$$= Y_i\left(1 + h + \frac{h^2}{2}\right) - x_i\left(h + \frac{h^2}{2}\right) - \frac{h^2}{2} \qquad \textbf{(8.3.15)}$$

For $i = 0$, we get from Equation (8.3.15)

$$Y_1 = Y_0 \left( 1 + h + \frac{h^2}{2} \right) - x_0 \left( h + \frac{h^2}{2} \right) - \frac{h^2}{2}$$

$$= 0 - 0 - \frac{1}{2}(0.0025) = -0.00125$$

For $i = 1$, Equation (8.3.15) gives

$$Y_2 = Y_1 \left( 1 + h + \frac{h^2}{2} \right) - x_1 \left( h + \frac{h^2}{2} \right) - \frac{h^2}{2} = -0.00513$$

We can continue to compute $Y_3, Y_4, \ldots, Y_{20}$. Table 8.3.1 lists these values along with the absolute value of the actual errors.

For $n = 3$, we get

$$T_3(x_i, Y_i, h) = f(x_i, Y_i) + \frac{h}{2!} f'(x_i, Y_i) + \frac{h^2}{3!} f''(x_i, Y_i)$$

$$= T_2(x_i, Y_i, h) + \frac{h^2}{3!} f''(x_i, Y_i)$$

We have

$$f''(x, y) = \frac{d}{dx} f'(x, y) = \frac{d}{dx}(y - x - 1)$$

$$= \frac{dy}{dx} - 1 = y - x - 1$$

therefore

$$T_3(x_i, Y_i, h) = T_2(x_i, Y_i, h) + \frac{h^2}{6}(Y_i - x_i - 1)$$

$$= (Y_i - x_i) \left( 1 + \frac{h}{2} \right) - \frac{h}{2} + \frac{h^2}{6}(Y_i - x_i - 1)$$

$$= Y_i \left( 1 + \frac{h}{2} + \frac{h^2}{6} \right) - x_i \left( 1 + \frac{h}{2} + \frac{h^2}{6} \right) - \frac{h}{2} - \frac{h^2}{6}$$

Equation (8.3.12) becomes

$$Y_{i+1} = Y_i \left( 1 + h + \frac{h^2}{2} + \frac{h^3}{6} \right) - x_i \left( h + \frac{h^2}{2} + \frac{h^3}{6} \right) - \frac{h^2}{2} - \frac{h^3}{6} \qquad \textbf{(8.3.16)}$$

For $i = 0$, we get from Equation (8.3.16)

$$Y_1 = Y_0 \left( 1 + h + \frac{h^2}{2} + \frac{h^3}{6} \right) - x_0 \left( h + \frac{h^2}{2} + \frac{h^3}{6} \right) - \frac{h^2}{2} - \frac{h^3}{6}$$

$$= 0 - 0 - 0.00125 - 0.00002 = -0.00127$$

For $i = 1$, Equation (8.3.16) gives

$$Y_2 = Y_1 \left( 1 + h + \frac{h^2}{2} + \frac{h^3}{6} \right) - x_1 \left( h + \frac{h^2}{2} + \frac{h^3}{6} \right) - \frac{h^2}{2} - \frac{h^3}{6}$$

$$= -0.00127 \, (1 + 0.05 + 0.00125 + 0.00002)$$

$$- 0.05 \, (0.05 + 0.00125 + 0.00002) - 0.00127$$

$$= -0.00517$$

For $n = 3$, the computed values $Y_i$ and the absolute value of the actual errors are given in Table 8.3.1 along with the actual values of the exact solution $y(x) = 1 + x - e^x$.

**Table 8.3.1**

| $i$ | $x_i$ | $Y_i$ for $n = 1$ | \|Actual Error\| $n = 1$ | $Y_i$ for $n = 2$ | \|Actual Error\| $n = 2$ | $Y_i$ for $n = 3$ | \|Actual Error\| $n = 3$ | Exact Value |
|---|---|---|---|---|---|---|---|---|
| 0 | 0.00 | 0.00000 | 0.00000 | 0.00000 | 0.00000 | 0.00000 | 0.00000 | 0.00000 |
| 1 | 0.05 | −0.00000 | 0.00127 | −0.00125 | 0.00002 | −0.00127 | 0.00000 | −0.00127 |
| 2 | 0.10 | −0.00250 | 0.00267 | −0.00513 | 0.00004 | −0.00517 | 0.00000 | −0.00517 |
| 3 | 0.15 | −0.00763 | 0.00420 | −0.01176 | 0.00007 | −0.01183 | 0.00000 | −0.01183 |
| 4 | 0.20 | −0.01551 | 0.00589 | −0.02130 | 0.00010 | −0.02140 | 0.00000 | −0.02140 |
| 5 | 0.25 | −0.02628 | 0.00775 | −0.03390 | 0.00013 | −0.03402 | 0.00001 | −0.03403 |
| 6 | 0.30 | −0.04010 | 0.00976 | −0.04970 | 0.00016 | −0.04986 | 0.00000 | −0.04986 |
| 8 | 0.40 | −0.07746 | 0.01436 | −0.09159 | 0.00023 | −0.09182 | 0.00000 | −0.09182 |
| 10 | 0.50 | −0.12889 | 0.01983 | −0.14839 | 0.00033 | −0.14872 | 0.00000 | −0.14872 |
| 12 | 0.60 | −0.19586 | 0.02626 | −0.22168 | 0.00044 | −0.22211 | 0.00001 | −0.22212 |
| 14 | 0.70 | −0.27993 | 0.03382 | −0.31319 | 0.00056 | −0.31375 | 0.00000 | −0.31375 |
| 16 | 0.80 | −0.38287 | 0.04267 | −0.42483 | 0.00071 | −0.42553 | 0.00001 | −0.42554 |
| 18 | 0.90 | −0.50662 | 0.05298 | −0.55871 | 0.00089 | −0.55959 | 0.00001 | −0.55960 |
| 20 | 1.00 | −0.65330 | 0.06498 | −0.71719 | 0.00109 | −0.71827 | 0.00001 | −0.71828 |

The computed values for $n = 3$ are very close to the actual values as can be seen in Table 8.3.1. The Taylor Series method requires major effort to find the derivatives of $f(x, y)$ and it may be costly to evaluate the necessary derivatives of $f(x, y)$ for $n > 3$. Symbol-manipulating programs are now available and therefore differentiation of a complicated expression can be turned over to a computer. However, the Runge–Kutta methods replace the derivatives of $f(x, y)$ with suitable combinations of values of $f(x, y)$. We will discuss the Runge–Kutta methods in Section 8.5.

EXERCISES

1.  Let $dy/dx = f(x, y(x)) = y'$. Verify that

    $$y^{(3)}(x) = f_{xx}(x, y) + 2f(x, y)f_{xy}(x, y) + (f(x, y))^2 f_{yy}(x, y)$$
    $$+ f_x(x, y)f_y(x, y) + f(x, y)(f_y(x, y))^2$$

2.  Find the formulas for $Y_{i+1}$ of the Taylor Series method of orders one, two, and three to solve the following initial-value problems:
    (a)  $dy/dx = y + x$ $\quad$ $y(0) = 0$ $\qquad$ (b)  $dy/dx = x^2 + y^2$ $\quad$ $y(0) = 0$
3.  Use the Taylor Series method of orders one, two, and three with $h = 0.1$ to approximate the initial-value problem

    $$\frac{dy}{dx} = y + x \qquad y(0) = 0 \quad \text{on} \quad [0, 0.4]$$

4.  Find $f^{(3)}(x, y(x))$ if
    (a)  $dy/dx = f(x, y) = e^x + y$ $\qquad$ (b)  $dy/dx = f(x, y) = \sin x \cos y$

COMPUTER EXERCISES

C.1.  Approximate the solution of the initial-value problem $dy/dx = x^2 + y^2$, $y(0) = 0$ for $0 \le x \le 1$ with step sizes $h = 0.1, 0.05, 0.01, 0.005,$ and $0.001$ using the Taylor Series method of orders one, two, and three. Print the exact solution, approximate solution, error, and relative error at 11 nodes including $x_0 = 0$ and $x_N = 1$.
C.2.  Approximate the solution of the initial-value problem $dy/dx = 1 - x + 4y$, $y(0) = 1$ for $0 \le x \le 1$ with step sizes $h = 0.1, 0.05, 0.01, 0.005,$ and $0.001$ using the Taylor Series method of orders one, two, and three. Print the exact solution, approximate solution, error, and relative error at 11 nodes including $x_0 = 0$ and $x_N = 1$.

## 8.4  THE EULER METHOD*

The Taylor Series method of order one is given by

$$Y_{i+1} = Y_i + hf(x_i, Y_i) \quad \text{for } i = 0, 1, \ldots \qquad \textbf{(8.4.1)}$$

for solving the initial-value problem

$$\frac{dy}{dx} = f(x, y) \qquad y(x_0) = Y_0 \qquad \textbf{(8.4.2)}$$

Equation (8.4.1) is also called the **Euler formula** or the **Euler forward formula**. The Euler method is probably the simplest of all numerical methods for solving ordinary differential equations; however, it is computationally inefficient. Nevertheless the study of its properties is very instructive and serves as a basis for the analysis of more accurate methods. Before discussing its properties, let us look at its geometrical interpretation.

The solution given by Equation (8.4.1) across $[x_0, x_1]$ follows the tangent line at $x_0$. In general, the larger $x_1 - x_0$, the larger the deviation between the solution and tangent line. If $f(x, y)$ is well-behaved and if $Y_1$ is near $y(x_1)$, then we can hope that $f(x_1, y(x_1)) \cong f(x_1, Y_1)$. Next we follow the line from $(x_1, Y_1)$ in the direction of the slope $f(x_1, Y_1)$. We compute $Y_2 = Y_1 + hf(x_1, Y_1)$ as an approximation of $y(x_2)$. It is possible that the line joining $(x_1, Y_1)$ and $(x_2, Y_2)$ may not be parallel to the tangent line to the solution curve $y(x)$ at $(x_1, y(x_1))$ since $y(x_1)$ is not known exactly. Continuing this process we obtain a polygon line, often referred to as an Euler polygon, as shown in Figure 8.4.1.

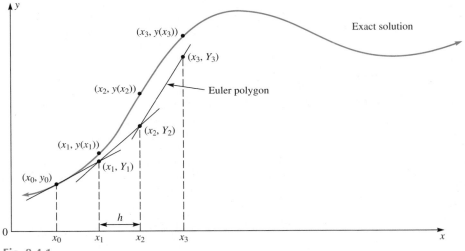

Fig. 8.4.1

In order to understand the local truncation error, suppose $y(x_i)$ is known. Then apply the Euler formula to compute $Y_{i+1}$:

$$Y_{i+1} = y(x_i) + hf(x_i, y(x_i)) \tag{8.4.3}$$

The correct value of $y(x_{i+1})$ is given by

$$y(x_{i+1}) = y(x_i) + hf(x_i, y(x_i)) + \frac{h^2}{2} y''(\xi_i) \tag{8.4.4}$$

where $x_i < \xi_i < x_{i+1}$.

For simplicity, assume that our computer computes with complete accuracy; that is, we retain an infinite number of places. Then $Y_{i+1}$ is computed accurately without any round-off error. The **local truncation error** in the $(i + 1)$th step is the difference between the exact value $y(x_{i+1})$ and $Y_{i+1}$ and is given by subtracting Equation (8.4.3) from Equation (8.4.4) as

$$y(x_{i+1}) - Y_{i+1} = \frac{h^2}{2} y''(\xi_i) \tag{8.4.5}$$

Since we do not know $\xi_i$, we cannot find the local truncation error unless $y(x_{i+1})$ and $Y_{i+1}$ are known. However, we can find an upper bound. Let $|y''(x)| \leq M_2$ for all $x$ in a given interval. Then our local truncation error, denoted by $L_T$, is given by

$$L_T = \frac{h^2}{2}|y''(\xi_i)| \leq \frac{M_2 h^2}{2} \qquad (8.4.6)$$

Thus the local truncation error of the Euler method is of second order in $h$. To keep the local truncation error small keep the interval length $h$ small. This requires a large number of steps to reach the final goal, causing more computational error and more propagation. Choice methods have the local truncation error of at least fifth order in $h$. The local truncation error given by Equation (8.4.6) is only in one step. When we add up all truncation errors, how far is our computed value $Y_N$ from the exact value $y(x_N)$ at $x_N = b$? The accumulation of all the local truncation errors is called the **global error** of the method. Thus the global error in $Y_{i+1}$ is the error when all previous values $Y_i, Y_{i-1}, \ldots, Y_1$ are also determined by the Euler method. Let us denote the error $y(x_{i+1}) - Y_{i+1}$ by $e_{i+1}$. Then subtracting Equation (8.4.1) from Equation (8.4.4), we get

$$\begin{aligned} e_{i+1} &= y(x_{i+1}) - Y_{i+1} \\ &= y(x_i) + hf(x_i, y(x_i)) + \frac{h^2}{2!}y''(\xi_i) - Y_i - hf(x_i, Y_i) \\ &= y(x_i) - Y_i + h[f(x_i, y(x_i)) - f(x_i, Y_i)] + \frac{h^2}{2!}y''(\xi_i) \qquad (8.4.7) \end{aligned}$$

Assume that $f(x, y)$ satisfies a Lipschitz condition. Then

$$|f(x_i, y(x_i)) - f(x_i, Y_i)| \leq L|y(x_i) - Y_i| = L\,|e_i| \qquad (8.4.8)$$

Using Equations (8.4.6) and (8.4.8) in Equation (8.4.7),

$$|e_{i+1}| \leq |e_i| + hL|e_i| + \frac{M_2 h^2}{2} = |e_i|(1 + hL) + \frac{M_2 h^2}{2}$$

Thus

$$|e_{i+1}| \leq (1 + hL)|e_i| + \frac{M_2 h^2}{2} \qquad (8.4.9)$$

Let $i = 0$ in Equation (8.4.9). Then $|e_1| \leq (1 + hL)|e_0| + (M_2 h^2/2)$. Since $|y(x_0) - Y_0| = |e_0|$ and our computer is completely accurate by our assumption, $|e_0| = 0$. We will consider round-off errors later. Thus $|e_1| \leq (M_2 h^2/2)$. For $i = 1$, Equation (8.4.9) becomes

$$|e_2| \leq (1 + hL)|e_1| + \frac{M_2 h^2}{2}$$

Since $|e_1| \leq (M_2 h^2/2)$,

$$|e_2| \leq \{1 + (1 + hL)\}\frac{M_2 h^2}{2}$$

Similarly continuing, we get

$$|e_{i+1}| \leq \{1 + (1 + hL) + (1 + hL)^2 + \cdots + (1 + hL)^i\}\frac{M_2 h^2}{2} \qquad (8.4.10)$$

Further, $1 + (1 + hL) + (1 + hL)^2 + \cdots + (1 + hL)^i = [(1 + hL)^{i+1} - 1]/(hL)$ because it is a geometric series. Hence Equation (8.4.10) becomes

$$|e_{i+1}| \leq \frac{M_2 h}{2L} [(1 + hL)^{i+1} - 1] \tag{8.4.11}$$

Since $e^x = 1 + x + (x^2/2) e^{\xi}$ where $0 < \xi < x$, $1 + x \leq e^x$. Hence $1 + hL \leq e^{hL}$ and therefore Equation (8.4.11) can be simplified to

$$|e_{i+1}| \leq \frac{M_2 h}{2L} [e^{hL(i+1)} - 1]$$

Further $x_{i+1} = x_0 + (i + 1) h$, therefore

$$|e_{i+1}| \leq \frac{M_2 h}{2L} [e^{(x_{i+1} - x_0)L} - 1] \tag{8.4.12}$$

Thus the accumulated truncation error or global truncation error is proportional to the step size $h$. This is expressed by saying that the Euler method is of the first order. The error bound given by Equation (8.4.12) provides only qualitative information since in reality we do not even know $y(x)$ and therefore $M_2$ is not known.

A method is said to be of **order $p$** if there is a constant $A$ such that

$$|y(x_{i+1}) - Y_{i+1}| = |e_{i+1}| \leq Ah^p$$

A rule of thumb is that for a method of order $p$, a local truncation error is of order $p + 1$.

---

**EXAMPLE 8.4.1**

Determine an upper bound of the global error of Euler's method in solving the initial-value problem

$$\frac{dy}{dx} = y - x \qquad y(0) = 0$$

from $x_0 = 0$ to 1. Also, compare the results of Euler's method for various $h$.

Table 8.4.1 compares the values for $h = 0.1$, 0.05, and 0.025 at various $x$. The absolute value of the actual errors for various $h$ and the values of the exact solution

**Table 8.4.1**    Comparison of the results of Euler's method with different $h$

| $i$ | $x_i$ | $h = 0.1$ | \|Actual Error\| | $h = 0.05$ | \|Actual Error\| | $h = 0.025$ | \|Actual Error\| | $y(x)$ |
|---|---|---|---|---|---|---|---|---|
| 0 | 0.00 | 0.00000 | 0.00000 | 0.00000 | 0.00000 | 0.00000 | 0.00000 | 0.00000 |
| 1 | 0.10 | 0.00000 | 0.00517 | −0.00250 | 0.00267 | −0.00381 | 0.00136 | −0.00517 |
| 2 | 0.20 | −0.01000 | 0.01140 | −0.01551 | 0.00589 | −0.01840 | 0.00300 | −0.02140 |
| 3 | 0.30 | −0.03100 | 0.01886 | −0.04010 | 0.00976 | −0.04489 | 0.00497 | −0.04986 |
| 4 | 0.40 | −0.06410 | 0.02772 | −0.07746 | 0.01436 | −0.08451 | 0.00731 | −0.09182 |
| 5 | 0.50 | −0.11051 | 0.03821 | −0.12889 | 0.01983 | −0.13862 | 0.01010 | −0.14872 |
| 6 | 0.60 | −0.17156 | 0.05056 | −0.19586 | 0.02626 | −0.20873 | 0.01339 | −0.22212 |
| 7 | 0.70 | −0.24872 | 0.06503 | −0.27993 | 0.03382 | −0.29650 | 0.01725 | −0.31375 |
| 8 | 0.80 | −0.34359 | 0.08195 | −0.38287 | 0.04267 | −0.40376 | 0.02178 | −0.42554 |
| 9 | 0.90 | −0.45795 | 0.10165 | −0.50662 | 0.05298 | −0.53254 | 0.02706 | −0.55960 |
| 10 | 1.00 | −0.59374 | 0.12454 | −0.65330 | 0.06498 | −0.68506 | 0.03322 | −0.71828 |

$y(x) = 1 + x - e^x$ at various $x$ are also given. It can be seen from Table 8.4.1 that as the step size $h$ decreases, the error decreases.

Since $y(x) = 1 + x - e^x$, $y'(x) = 1 - e^x$ and $y''(x) = -e^x$. Hence $|y''(x)| = |-e^x| \le e$ and therefore $M_2 = e$ on $[0, 1]$. Further $f(x, y) = y - x$, therefore $|f(x, y_1) - f(x, y_2)| = |y_1 - y_2|$ and so $L = \partial f/\partial y = 1$. Using $M_2 = e$ and $L = 1$ in Equation (8.4.12), the accumulated error bound is given by

$$|y(x_{i+1}) - Y_{i+1}| = |e_{i+1}| \le \frac{eh}{2}(e^{x_{i+1}} - 1) \tag{8.4.13}$$

In Table 8.4.2, the values of Equation (8.4.13) and the absolute value of the actual errors are given at various $x$.

**Table 8.4.2**

| $x$ | 0.2 | 0.4 | 0.6 | 0.8 | 1.0 |
|---|---|---|---|---|---|
| For $h = 0.1$ | | | | | |
| \|Actual Error\| | 0.01140 | 0.02772 | 0.05056 | 0.08195 | 0.12454 |
| Error bound | 0.03009 | 0.06685 | 0.11174 | 0.16657 | 0.23354 |
| For $h = 0.05$ | | | | | |
| \|Actual Error\| | 0.00590 | 0.01437 | 0.02626 | 0.04267 | 0.06498 |
| Error bound | 0.01505 | 0.03342 | 0.05587 | 0.08328 | 0.11677 |
| For $h = 0.025$ | | | | | |
| \|Actual Error\| | 0.00300 | 0.00732 | 0.01339 | 0.02178 | 0.03322 |
| Error bound | 0.00752 | 0.01671 | 0.02793 | 0.04164 | 0.05838 |

From Table 8.4.2, it can be seen that the error bound given by Equation (8.4.13) is considerably higher than the actual error. It must be emphasized that Equation (8.4.12) provides an upper bound rather than a realistic bound.　■ ■ ■

The importance of Equation (8.4.12) is to show that the error tends to zero as $h \to 0$; thus Equation (8.4.12) establishes the convergence of the method. However as $h$ becomes smaller, more calculations are necessary. Since we use finite digit arithmetic, more round-off errors are expected. Therefore, let us consider round-off errors involved in Euler's method.

In practice, due to the limitations of computers, it is generally difficult to compute $Y_{i+1}$ exactly from the formula. Just as we defined local truncation error, we define local round-off error. We have

$$Y_{i+1} = Y_i + hf(x_i, Y_i)$$

and $Y_i$ is not known exactly but $\hat{Y}_i$ is known. Also we cannot exactly compute $Y_{i+1}$ even if $Y_i$ is known exactly. Let us denote the computed value of $Y_{i+1}$ by $\hat{Y}_{i+1}$ which is given by

$$\hat{Y}_{i+1} = \hat{Y}_i + hf(x_i, \hat{Y}_i) + \rho_i$$

where $\rho_i$ is the local round-off error introduced by the inexact computation of $\hat{Y}_i + hf(x_i, \hat{Y}_i)$. In practice, $\hat{Y}_i$ will differ from $Y_i$ because of round-off errors. We denote

$$r_i = Y_i - \hat{Y}_i \tag{8.4.14}$$

where $Y_i$ is the exact solution of Equation (8.4.1) and $\hat{Y}_i$ is the computed value at $x = x_i$. Let $\rho_0$ be the initial round-off error. Then $\hat{Y}_0 - y_0 = \hat{Y}_0 - Y_0 = \rho_0$. The total error in the $(i + 1)$th step is

$$
\begin{aligned}
E_{i+1} &= y(x_{i+1}) - \hat{Y}_{i+1} \\
&= y(x_{i+1}) - Y_{i+1} + Y_{i+1} - \hat{Y}_{i+1} \\
&= e_{i+1} + r_{i+1}
\end{aligned}
$$

Further,

$$
\begin{aligned}
E_{i+1} &= y(x_{i+1}) - \hat{Y}_{i+1} \\
&= y(x_i) + hf(x_i, y(x_i)) + \frac{h^2}{2} y''(\xi_i) \\
&\quad - (\hat{Y}_i + hf(x_i, \hat{Y}_i) + \rho_i) \\
&= y(x_i) - \hat{Y}_i + h[f(x_i, y(x_i)) - f(x_i, \hat{Y}_i)] \\
&\quad - \rho_i + \frac{h^2}{2} y''(\xi_i)
\end{aligned}
\tag{8.4.15}
$$

Let $\rho = \max_{0 \le i \le n} |\rho_i|$ and $|y''(x_i)| \le M_2$ for all $x$ in the given interval. Then Equation (8.4.15) gives

$$
\begin{aligned}
|E_{i+1}| &\le |E_i| + hL |E_i| + \rho + \frac{M_2 h^2}{2} \\
&= (1 + hL) |E_i| + \rho + \frac{M_2 h^2}{2}
\end{aligned}
\tag{8.4.16}
$$

Just as we derived Equation (8.4.12) from Equation (8.4.9), follow the same procedure (Exercise 5) to get

$$
|E_{i+1}| \le e^{(1+i)hL} |E_0| + \frac{1}{L} \left( \frac{M_2 h}{2} + \frac{\rho}{h} \right) [e^{(1+i)hL} - 1]
\tag{8.4.17}
$$

## EXERCISES

Determine an upper bound using Equation (8.4.6) for the local truncation error of Euler's method for each of the following exercises:

1. $\dfrac{dy}{dx} = x^2 \qquad y(0) = 0 \quad \text{on} \quad [0, \tfrac{1}{2}]$

2. $(x^2 + 1) \dfrac{dy}{dx} + xy = 0 \qquad y(0) = 1 \quad \text{on} \quad [0, \tfrac{1}{2}]$

Determine an upper bound using Equation (8.4.12) for the global truncation error of Euler's method for each of the following exercises:

3. $\dfrac{dy}{dx} = x + y \qquad y(0) = 0 \quad \text{on} \quad [0, \tfrac{1}{2}]$

4. $\dfrac{dy}{dx} = 1 - x + 4y \qquad y(0) = 1 \quad \text{on} \quad [0, \tfrac{1}{2}]$

5. Derive Equation (8.4.17).

Determine an upper bound using Equation (8.4.17) for the global errors of Euler's method for each of the following exercises. (Use $\rho = (\frac{1}{2}) \times 10^{-5}$.)

**6.** $\dfrac{dy}{dx} = x^2 \qquad y(0) = 0 \quad$ on $\quad [0, \frac{1}{2}]$

**7.** $(x^2 + 1)\dfrac{dy}{dx} + xy = 0 \qquad y(0) = 1 \quad$ on $\quad [0, \frac{1}{2}]$

**8.** Derive Euler's formula by integrating $dy/dx = f(x, y)$ from $x_i$ to $x_{i+1}$ using the left-hand rectangular rule; in other words, $\int_{x_i}^{x_{i+1}} f(x, y)\, dx \cong f(x_i, y(x_i))(x_{i+1} - x_i)$.

**9.** Verify that $y_i = (1 + h)^i$, for $i = 0, 1, \ldots$ approximates the initial-value problem $dy/dx = y(x)$, $y(0) = 1$ on $[0, 1]$ by the Euler method. Also, verify that the local truncation error $L_T \le (h^2 e)/2$ and the global truncation error $\le he\,(e - 1)/2$.

**10.** Find the optimal step size $h$ for which the bound given by Equation (8.4.17) is minimum.

## COMPUTER EXERCISES

**C.1.** Write a subroutine to compute $Y_{i+1} = Y_i + hf(x_i, Y_i)$ where $a$, $b$, $y_0$, $f$, and $N$ are parameters.

**C.2.** Use your program of Exercise C.1 to approximate the solution of the initial-value problem $dy/dx = 1 - x + 4y$, $y(0) = 1$ for $0 \le x \le 1$ with step sizes $h = 0.1, 0.05, 0.01, 0.005$, and $0.001$. Print the exact solution, approximate solution, error and relative error at 11 nodes including $x_0 = 0$ and $x_N = 1$. Find the optimal step size $h$ using Exercise 10 and knowing $\rho$ of your computer. Analyze your output for errors by finding an upper bound for the global error. Use Example 8.4.1 for your model.

**C.3.** To get some idea of the possible dangers in the initial conditions, such as those due to round-off, suppose that in the initial condition a mistake is made and 1.0001 is used instead of 1. Using your program of Exercise C.1, solve $dy/dx = 1 - x - 4y$, $y(0) = 1.0001$ with step sizes $h = 0.1, 0.05, 0.01, 0.005$, and $0.001$. Print the exact solution, approximate solution, error and relative error at 11 nodes including $x_0 = 0$ and $x_N = 1$. Compare your numerical solution with the numerical solution in Exercise C.2.

**C.4.** Approximate the solution of the initial-value problem $dy/dx = \sqrt{y}$, $y(0) = 0$ using Exercise C.1. This initial-value problem has the nontrivial solution $y(x) = (x/2)^2$. Are you getting the approximate nontrivial solution? Explain this paradox.

## 8.5  THE RUNGE–KUTTA METHODS

In order to solve an initial-value problem

$$\frac{dy}{dx} = f(x, y) \qquad y(x_0) = y_0 \qquad\qquad \textbf{(8.5.1)}$$

we can use a Taylor Series method of higher order for better accuracy; however, in order to use higher orders, we need to evaluate higher derivatives of $f(x, y)$. The Runge–Kutta methods do not require the evaluation of the derivatives of $f(x, y)$ and at the same time keep the desirable property of higher-order local truncation error.

Let us start with the Taylor Series method of order two by considering Taylor's expansion

$$y(x_{i+1}) = y(x_i) + hy'(x_i) + \frac{h^2}{2!} y''(x_i) + \frac{h^3}{3!} y^{(3)}(\xi_i) \qquad \textbf{(8.5.2)}$$

where $x_i < \xi_i < x_{i+1}$. From Equation (8.5.1) we have $y'(x_i) = f(x_i, y(x_i))$, therefore Equation (8.5.2) becomes

$$y(x_{i+1}) = y(x_i) + hf(x_i, y(x_i)) + \frac{h^2}{2!} f'(x_i, y(x_i)) + \frac{h^3}{3!} f''(\xi_i, y(\xi_i)) \qquad \textbf{(8.5.3)}$$

The idea is to avoid $f'(x_i, y(x_i))$ which is equal to $f_x(x_i, y(x_i)) + f(x_i, y(x_i)) f_y(x_i, y(x_i))$. This can be done by approximating $f'(x_i, y_i)$ by the forward difference $[f(x_i + h, y(x_i + h)) - f(x_i, y(x_i))]/h$. Thus Equation (8.5.3) becomes

$$y(x_{i+1}) = y(x_i) + hf(x_i, y(x_i))$$

$$+ \frac{h^2}{2!} \left\{ \frac{(f(x_i + h, y(x_i + h)) - f(x_i, y(x_i)))}{h} - \frac{h}{2} f''(\zeta_i, y(\zeta_i)) \right\}$$

$$+ \frac{h^3}{3!} f''(\xi_i, y(\xi_i))$$

$$= y(x_i) + \frac{h}{2} [f(x_i, y(x_i)) + f(x_i + h, y(x_i + h))] - \frac{h^3}{12} f''(\eta_i, y(\eta_i))$$

This yields

$$Y_{i+1} = Y_i + \frac{h}{2} [f(x_i, Y_i) + f(x_{i+1}, Y_{i+1})] \qquad \textbf{(8.5.4)}$$

In this equation, the unknown $Y_{i+1}$ on the right-hand side is replaced using Euler's formula $Y_{i+1} = Y_i + hf(x_i, Y_i)$, yielding

$$Y_{i+1} = Y_i + \frac{h}{2} [f(x_i, Y_i) + f(x_i + h, Y_i + hf(x_i, Y_i))] \qquad \textbf{(8.5.5)}$$

This formula is called the **Runge–Kutta formula of second order**. Comparing it with the Taylor Series method of second order

$$Y_{i+1} = Y_i + hf(x_i, Y_i) + \frac{h^2}{2} f'(x_i, Y_i) \qquad \textbf{(8.5.6)}$$

the difference is clear. In Equation (8.5.5), we do not need $f'(x_i, Y_i)$ and yet it has the same truncation order as Equation (8.5.6).

We want to develop a general procedure to derive Equation (8.5.5) from Equation (8.5.6) and other similar higher-order methods. We are looking at the formula

$$Y_{i+1} = Y_i + w_1 k_1 + w_2 k_2 \qquad \textbf{(8.5.7)}$$

where

$$k_1 = hf(x_i, Y_i)$$
$$k_2 = hf(x_i + \alpha_2 h, Y_i + \beta_{21} k_1)$$

and the constants $w_1$, $w_2$, $\alpha_2$, and $\beta_{21}$ are to be determined so that Equations (8.5.6) and (8.5.7) are equal. This is also called a **two-stage Runge–Kutta formula**. Our object is to make the coefficients of $h$ and $h^2$ identical in the Taylor series expansion of both sides of Equation (8.5.7) about $(x_i, Y_i)$. The expansion of the left-hand side of Equation (8.5.7) is

$$Y_{i+1} = Y_i + hf(x_i, Y_i) + \frac{h^2}{2!} f'(x_i, Y_i) + \frac{h^3}{3!} f''(x_i, Y_i)$$
$$+ \text{ higher-order terms} \tag{8.5.8}$$

Substituting the values of $f'(x_i, Y_i)$ and $f''(x_i, Y_i)$ from Equations (8.3.5) and (8.3.6) in Equation (8.5.8), and using

$$f = f(x_i, Y_i) \qquad f_x = f_x(x_i, Y_i) \qquad f_y = f_y(x_i, Y_i)$$
$$f_{xx} = f_{xx}(x_i, Y_i) \qquad f_{xy} = f_{xy}(x_i, Y_i) \qquad f_{yy} = f_{yy}(x_i, Y_i) \tag{8.5.9}$$

we get

$$Y_{i+1} = Y_i + hf + \frac{h^2}{2!} [f_x + ff_y] + \frac{h^3}{3!} [f_{xx} + 2ff_{xy} + f^2 f_{yy}$$
$$+ f_y (f_x + ff_y)] + \text{ higher-order terms} \tag{8.5.10}$$

To get the expansion of the right-hand of Equation (8.5.7), we use the Taylor series expansion of a function of two variables (Theorem A.9) for $k_2$ through the second derivative terms, which gives

$$k_2 = hf(x_i + \alpha_2 h, Y_i + \beta_{21} k_1)$$

$$= h\left\{ f(x_i, Y_i) + \left( \alpha_2 h \frac{\partial}{\partial x} + \beta_{21} k_1 \frac{\partial}{\partial y} \right) f(x_i, Y_i) + \frac{1}{2!} \left( \alpha_2 h \frac{\partial}{\partial x} + \beta_{21} k_1 \frac{\partial}{\partial y} \right)^2 f(x_i, Y_i) \right\}$$

$$+ \text{ higher-order terms}$$

$$= h\left\{ f + \alpha_2 h f_x + \beta_{21} hff_y + \frac{h^2}{2} (\alpha_2^2 f_{xx} + 2 \alpha_2 \beta_{21} ff_{xy} + \beta_{21}^2 f^2 f_{yy}) \right\}$$

$$+ \text{ higher-order terms}$$

Substituting for $k_1$ and $k_2$ in Equation (8.5.7), we get

$$Y_{i+1} = Y_i + h (w_1 + w_2) f + w_2 h^2 [\alpha_2 f_x + \beta_{21} ff_y]$$
$$+ \frac{w_2 h^3}{2} [\alpha_2^2 f_{xx} + 2 \alpha_2 \beta_{21} ff_{xy} + \beta_{21}^2 f^2 f_{yy}]$$
$$+ \text{ higher-order terms} \tag{8.5.11}$$

Comparing the coefficients of $h$ and $h^2$ on the right-hand of Equations (8.5.10) and (8.5.11), we get

$$w_1 + w_2 = 1 \qquad \alpha_2 w_2 = \frac{1}{2} \qquad \beta_{21} w_2 = \frac{1}{2} \tag{8.5.12}$$

We have four unknowns and only three equations, therefore, we have one degree of freedom in the solution of Equation (8.5.12). Solving Equation (8.5.12) in terms of $\alpha_2$, we get

$$w_2 = \frac{1}{2\alpha_2} \qquad w_1 = 1 - w_2 = 1 - \frac{1}{2\alpha_2} = \frac{2\alpha_2 - 1}{2\alpha_2}$$

$$\beta_{21} = \frac{1}{2w_2} = \alpha_2 \tag{8.5.13}$$

We can use one degree of freedom to select the truncation error as small as possible that produced $h^3$ terms in the expansions of Equations (8.5.10) and (8.5.11).

The asymptotic form of the error term is found by taking the difference between the $h^3$ terms in Equations (8.5.10) and (8.5.11) and is given by

$$\frac{h^3}{6}[f_{xx} + 2ff_{xy} + f^2f_{yy} + f_y(f_x + ff_y)]$$

$$- \frac{w_2 h^3}{2}[\alpha_2^2 f_{xx} + 2\alpha_2 \beta_{21} ff_{xy} + \beta_{21}^2 f^2 f_{yy}]$$

$$= h^3\left\{\left(\frac{1}{6} - \frac{\alpha_2}{4}\right)[f_{xx} + 2ff_{xy} + f^2f_{yy}] + \frac{1}{6}f_y[f_x + ff_y]\right\} \tag{8.5.14}$$

If we select $\alpha_2$ such that Equation (8.5.14) is zero, then we get a third-order method. However, there is no value of $\alpha_2$ that will make Equation (8.5.14) vanish for all $x$ and $y$. We can select $\alpha_2$ such that Equation (8.5.14) is as small as possible or any Runge–Kutta formula reduces to a quadrature formula of the same order or higher for $f(x, y)$ independent of $y$.

In order to find a bound on Equation (8.5.14), let us define as suggested by Lotkin (1951)

$$|f(x, y)| < M \qquad \left|\frac{\partial^{i+j} f}{\partial x^i \partial y^j}\right| \leq \frac{L^{i+j}}{M^{j-1}} \quad \text{for } i + j \leq n$$

Therefore we have

$$|f(x, y)| < M \qquad |f_y| \leq L \qquad |f_x| < LM \qquad |f_{xy}| < L^2$$

$$|f_{xx}| < L^2 M \qquad |f_{yy}| < \frac{L^2}{M}$$

so that Equation (8.5.14) is bounded by

$$|\text{error}| = h^3\left|\left\{\left(\frac{1}{6} - \frac{\alpha_2}{4}\right)[f_{xx} + 2ff_{xy} + f^2f_{yy}] + \frac{1}{6}f_y[f_x + ff_y]\right\}\right|$$

$$\leq h^3 ML^2\left[4\left|\frac{1}{6} - \frac{\alpha_2}{4}\right| + \frac{1}{3}\right] \tag{8.5.15}$$

This is minimum when $\alpha_2 = \frac{2}{3}$ and in that case

$$|\text{error}| \leq \frac{ML^2 h^3}{3}$$

The well-known solutions of Equation (8.5.12) are:

1.  $\alpha_2 = \beta_{21} = 1$ and $w_1 = w_2 = \frac{1}{2}$. Then Equation (8.5.7) becomes

$$Y_{i+1} = Y_i + \frac{h}{2}[f(x_i, Y_i) + f(x_i + h, Y_i + hf(x_i, Y_i)] \qquad \textbf{(8.5.5)}$$

which is known as the **Improved Euler method** or **Heun method**. This can be written as

$$\hat{Y}_{i+1} = Y_i + hf(x_i, Y_i)$$

$$Y_{i+1} = Y_i + \frac{h}{2}\left[f(x_i, Y_i) + f(x_i + h, \hat{Y}_{i+1})\right] \qquad \textbf{(8.5.16)}$$

The first equation predicts $\hat{Y}_{i+1}$, a preliminary value, while the second equation gives the corrected value $Y_{i+1}$ by using the preliminary value $\hat{Y}_{i+1}$. This can be viewed as a **predictor and corrector method**. If $f(x, y)$ is independent of $y$, then Equation (8.5.16) reduces to the Trapezoidal rule. Therefore, Equation (8.5.16) is also known as the **Trapezoidal method**.

2.  $\alpha_2 = \beta_{21} = \frac{1}{2}$, $w_1 = 0$, and $w_2 = 1$. Then Equation (8.5.7) reduces to

$$Y_{i+1} = Y_i + hf\left(x_i + \frac{h}{2}, Y_i + \frac{h}{2}f(x_i, Y_i)\right)$$

This corresponds to the Newton–Cotes one-point open formula when $f(x, y)$ is independent of $y$. This can be written as

$$\hat{Y}_{i+1/2} = Y_i + \frac{h}{2}f(x_i, Y_i)$$

$$Y_{i+1} = Y_i + hf\left(x_i + \frac{h}{2}, \hat{Y}_{i+1/2}\right) \qquad \textbf{(8.5.17)}$$

which is called the **Modified Euler method** or **Improved Polygon method**.

3.  $\alpha_2 = \beta_{21} = \frac{2}{3}$, $w_1 = \frac{1}{4}$, and $w_2 = \frac{3}{4}$. In this case Equation (8.5.7) reduces to

$$Y_{i+1} = Y_i + \frac{h}{4}\left[f(x_i, Y_i) + 3f\left(x_i + \frac{2}{3}h, Y_i + \frac{2}{3}hf(x_i, Y_i)\right)\right]$$

This can be written as

$$\hat{Y}_{i+2/3} = Y_i + \frac{2}{3}hf(x_i, Y_i)$$

$$Y_{i+1} = Y_i + \frac{h}{4}\left[f(x_i, Y_i) + 3f\left(x_i + \frac{2}{3}h, \hat{Y}_{i+2/3}\right)\right] \qquad \textbf{(8.5.18)}$$

The error bound given by Equation (8.5.15) is minimum in this case.

**Algorithm**

**Subroutine** Runge2 $(a, b, N, y0, f, Y)$
Comment: This program approximates the solution of an initial-value problem $y' = f(x, y)$, $y(x_0) = y0 = Y_0$ on a given interval $[a, b]$ where $x_0 = a$ and $x_N = b$ using the Improved Euler method given by Equation (8.5.16).

```
1 Print a, b, N, y0
2 h = (b − a)/N
3 x = a
4 Y = y0
5 Print h, x, Y
6 Repeat steps 7–11 with i = 1, 2, . . ., N
      7 oldf = f(x, Y)
      8 Yhat = Y + h * oldf
      9 x = a + ih
     10 Y = Y + h(oldf + f(x, Yhat))/2
     11 Print i, x, Y
12 Print 'finished' and exit.
```

◀

**EXAMPLE 8.5.1**    Using Equations (8.5.16), (8.5.17), and (8.5.18), solve the initial-value problem

$$\frac{dy}{dx} = -xy^2 \qquad y(0) = 2 \quad \text{for } 0 \le x \le 1$$

and compare it with the exact solution $y = 2/(1 + x^2)$. Use $h = 0.1$.
    Since $f(x, y) = -xy^2$, Equation (8.5.16) reduces to

$$\hat{Y}_{i+1} = Y_i + hf(x_i, Y_i) = Y_i + h(-x_i Y_i^2) = Y_i(1 - hx_i Y_i)$$

$$Y_{i+1} = Y_i + \frac{h}{2}[f(x_i, Y_i) + f(x_i + h, \hat{Y}_{i+1})]$$

$$= Y_i + \frac{h}{2}[-x_i Y_i^2 - (x_i + h)\hat{Y}_{i+1}^2]$$

$$= Y_i - \frac{h}{2}[x_i Y_i^2 + (x_i + h)\hat{Y}_{i+1}^2] \qquad\qquad (8.5.19)$$

**Table 8.5.1**

| $x_i$ | $Y_i$ Using Eq. (8.5.16) | \|Actual Error\| for Eq. (8.5.16) | $Y_i$ Using Eq. (8.5.17) | \|Actual Error\| for Eq. (8.5.17) | $Y_i$ Using Eq. (8.5.18) | \|Actual Error\| for Eq. (8.5.18) | Exact $y(x)$ |
|---|---|---|---|---|---|---|---|
| 0.0 | 2.00000 | 0.00000 | 2.00000 | 0.00000 | 2.00000 | 0.00000 | 2.00000 |
| 0.1 | 1.98000 | 0.00020 | 1.98000 | 0.00020 | 1.98000 | 0.00020 | 1.98020 |
| 0.2 | 1.92273 | 0.00035 | 1.92235 | 0.00072 | 1.92248 | 0.00060 | 1.92308 |
| 0.3 | 1.83449 | 0.00037 | 1.83348 | 0.00138 | 1.83382 | 0.00104 | 1.83486 |
| 0.4 | 1.72391 | 0.00023 | 1.72221 | 0.00193 | 1.72278 | 0.00136 | 1.72414 |
| 0.5 | 1.60007 | 0.00007 | 1.59777 | 0.00223 | 1.59855 | 0.00145 | 1.60000 |
| 0.6 | 1.47105 | 0.00047 | 1.46836 | 0.00223 | 1.46927 | 0.00132 | 1.47059 |
| 0.7 | 1.34317 | 0.00089 | 1.34029 | 0.00199 | 1.34126 | 0.00102 | 1.34228 |
| 0.8 | 1.22080 | 0.00129 | 1.21790 | 0.00161 | 1.21888 | 0.00063 | 1.21951 |
| 0.9 | 1.10658 | 0.00161 | 1.10381 | 0.00116 | 1.10475 | 0.00022 | 1.10497 |
| 1.0 | 1.00184 | 0.00184 | 0.99928 | 0.00072 | 1.00015 | 0.00015 | 1.00000 |

For $i = 0$, Equation (8.5.19) gives

$$\hat{Y}_1 = Y_0 (1 - hx_0, Y_0) = 2$$

$$Y_1 = Y_0 - \frac{h}{2} [x_0 Y_0^2 + (x_0 + h) \hat{Y}_1^2]$$

$$= 2 - \frac{h}{2} [h \hat{Y}_1^2]$$

$$= 2 - \frac{h^2}{2} \hat{Y}_1^2$$

$$= 2 - 2h^2 = 1.98$$

Similarly we continue and the computed results are presented in Table 8.5.1. Equation (8.5.17) reduces to

$$\hat{Y}_{i+1/2} = Y_i + \frac{h}{2} f(x_i, Y_i) = Y_i + \frac{h}{2} (-x_i Y_i^2) = Y_i \left(1 - \frac{hx_i Y_i}{2}\right)$$

$$Y_{i+1} = Y_i + hf \left(x_i + \frac{h}{2}, \hat{Y}_{i+1/2}\right)$$

$$= Y_i - h \left(x_i + \frac{h}{2}\right) \hat{Y}_{i+1/2}^2 \qquad (8.5.20)$$

For $i = 0$, Equation (8.5.20) gives

$$\hat{Y}_{1/2} = Y_0 \left(1 - \frac{hx_0 Y_0}{2}\right) = 2$$

$$Y_1 = Y_0 - h \left(x_0 + \frac{h}{2}\right) \cdot \hat{Y}_{1/2}^2$$

$$= Y_0 - 2 h^2 = 1.98$$

Similarly we continue. The computed values are presented in Table 8.5.1 along with the computed values using Equation (8.5.18).

Observe the |Actual Error| for Equations (8.5.16), (8.5.17), and (8.5.18) and verify that |Actual Error| for Equation (8.5.18) is minimum. ∎ ∎ ∎

## Third-Order Methods

The second-order methods were derived from Equation (8.5.7). In order to derive the third-order methods, we add one term or stage $w_3 k_3$ to Equation (8.5.7) and get the **three-stage Runge–Kutta formula**

$$Y_{i+1} = Y_i + w_1 k_1 + w_2 k_2 + w_3 k_3 \qquad (8.5.21)$$

where

$$k_1 = hf(x_i, Y_i)$$
$$k_2 = hf(x_i + \alpha_2 h, Y_i + \beta_{21} k_1)$$
$$k_3 = hf(x_i + \alpha_3 h, Y_i + \beta_{31} k_1 + \beta_{32} k_2)$$

We have eight unknowns and we determine them by expanding Equation (8.5.21) in a Taylor series around $(x_i, Y_i)$ and compare it with the corresponding terms of the Taylor Series method of order three given by

$$Y_{i+1} = Y_i + hf(x_i, Y_i) + \frac{h^2}{2!}f'(x_i, Y_i) + \frac{h^3}{3!}f''(x_i, Y_i)$$

$$+ \text{ higher-order terms} \qquad \qquad \textbf{(8.5.22)}$$

Expanding $k_2$ and $k_3$ in Taylor series (Theorem A.9) around $(x_i, Y_i)$ and keeping terms up to $h^3$, we get

$$k_2 = h\{f + h\,[\alpha_2 f_x + \beta_{21} ff_y]$$

$$+ \frac{h^2}{2}[\alpha_2^2 f_{xx} + 2\,\alpha_2\,\beta_{21} ff_{xy} + \beta_{21}^2 f^2 f_{yy}]$$

$$+ \text{ higher-order terms}\}$$

$$k_3 = h\{f + h\,[\alpha_3 f_x + (\beta_{31} + \beta_{32})ff_y] + \frac{h^2}{2}[2\,\beta_{32}\,(\alpha_2 f_x + \beta_{21} ff_y)f_y$$

$$+ \alpha_3^2 f_{xx} + 2\alpha_3\,(\beta_{31} + \beta_{32})ff_{xy} + (\beta_{31} + \beta_{32})^2 f^2 f_{yy}]$$

$$+ \text{ higher-order terms}\}$$

It can be verified (Exercise 9) that substitution of $k_1$, $k_2$, and $k_3$ in Equation (8.5.21) leads to

$$Y_{i+1} = Y_i + h\,(w_1 + w_2 + w_3)f + h^2\,\{(w_2\,\alpha_2 + w_3\,\alpha_3)f_x$$

$$+ ff_y\,[w_2\,\beta_{21} + w_3\,(\beta_{31} + \beta_{32})]\}$$

$$+ h^3\left\{\frac{1}{2}\,(w_2\,\alpha_2^2 + w_3\,\alpha_3^2)f_{xx} + [w_2\,\alpha_2\,\beta_{21} + w_3\,\alpha_3\,(\beta_{31} + \beta_{32})]ff_{xy}\right.$$

$$+ \frac{1}{2}\,[w_2\,\beta_{21}^2 + w_3\,(\beta_{31} + \beta_{32})^2]f^2 f_{yy}$$

$$\left. + w_3\,\alpha_2\,\beta_{32}f_x f_y + w_3\,\beta_{21}\,\beta_{32}ff_y^2\right\}$$

$$+ \text{ higher-order terms} \qquad \qquad \textbf{(8.5.23)}$$

Substituting for $f'(x_i, Y_i)$ and $f''(x_i, Y_i)$ from Equations (8.3.5) and (8.3.6) in Equation (8.5.22),

$$Y_{i+1} = Y_i + hf + \frac{h^2}{2}\,[f_x + ff_y]$$

$$+ \frac{h^3}{3!}\,\{f_{xx} + 2ff_{xy} + f^2 f_{yy} + f_x f_y + ff_y^2\}$$

$$+ \text{ higher-order terms} \qquad \qquad \textbf{(8.5.24)}$$

Comparing the coefficients of $h$, $h^2$, and $h^3$ of Equations (8.5.23) and (8.5.24), we get

$$w_1 + w_2 + w_3 = 1 \qquad w_2\,\alpha_2\,\beta_{21} + w_3\,\alpha_3\,(\beta_{31} + \beta_{32}) = \frac{1}{3}$$

$$w_2\,\alpha_2 + w_3\,\alpha_3 = \frac{1}{2}$$

$$w_2\,\beta_{21}^2 + w_3\,(\beta_{31} + \beta_{32})^2 = \frac{1}{3}$$

$$w_2\,\beta_{21} + w_3\,(\beta_{31} + \beta_{32}) = \frac{1}{2}$$

$$w_3\,\alpha_2\,\beta_{32} = \frac{1}{6}$$

$$w_2\,\alpha_2^2 + w_3\,\alpha_3^2 = \frac{1}{3} \qquad \qquad w_3\,\beta_{21}\,\beta_{32} = \frac{1}{6} \qquad \textbf{(8.5.25)}$$

Since $w_3\,\alpha_2\,\beta_{32} = \frac{1}{6} = w_3\,\beta_{21}\,\beta_{32}$, $\alpha_2 = \beta_{21}$.

Further $w_2\,\alpha_2^2 + w_3\,\alpha_3^2 = \frac{1}{3} = w_2\,\alpha_2\,\beta_{21} + w_3\,\alpha_3\,(\beta_{31} + \beta_{32})$ and $\alpha_2 = \beta_{21}$; therefore, $\alpha_3 = \beta_{31} + \beta_{32}$. Thus Equation (8.5.25) reduces to

$$w_1 + w_2 + w_3 = 1$$

$$w_2\,\beta_{21} + w_3\,(\beta_{31} + \beta_{32}) = \frac{1}{2}$$

$$w_2\,\beta_{21}^2 + w_3\,(\beta_{31} + \beta_{32})^2 = \frac{1}{3}$$

$$w_3\,\beta_{21}\,\beta_{32} = \frac{1}{6} \tag{8.5.26}$$

The system of Equation (8.5.26) has six unknowns and four equations. Thus we obtain a two parameter family of such methods. The classical method is given by $w_1 = w_3 = \frac{1}{6}$, $w_2 = \frac{2}{3}$, $\alpha_2 = \frac{1}{2}$, $\alpha_3 = 1$, $\beta_{21} = \frac{1}{2}$, $\beta_{31} = -1$, and $\beta_{32} = 2$. Thus we have

$$Y_{i+1} = Y_i + \frac{1}{6}(k_i + 4k_2 + k_3) \tag{8.5.27}$$

where

$$k_1 = hf(x_i, Y_i)$$

$$k_2 = hf\left(x_i + \frac{h}{2}, Y_i + \frac{k_1}{2}\right)$$

$$k_3 = hf(x_i + h, Y_i - k_1 + 2k_2)$$

If $f(x, y)$ is a function of $x$ only, then Equation (8.5.27) is Simpson's rule.

## Fourth-Order Methods

The higher-order Runge–Kutta formulas can be derived in the same way; however, as the order increases the complexity increases very rapidly. The best known **Runge–Kutta method of four stage and fourth order** is given by

$$Y_{i+1} = Y_i + \frac{1}{6}(k_1 + 2k_2 + 2k_3 + k_4) \tag{8.5.28}$$

where

$$k_1 = hf(x_i, Y_i)$$

$$k_2 = hf\left(x_i + \frac{h}{2}, Y_i + \frac{k_1}{2}\right)$$

$$k_3 = hf\left(x_i + \frac{h}{2}, Y_i + \frac{k_2}{2}\right)$$

$$k_4 = hf(x_i + h, Y_i + k_3)$$

The local truncation error is of fifth order in $h$. We pay for this favorable truncation error by evaluating $f$ four times at each step. At first sight, these formulas seem to be complicated, but they are easy to program.

Let $s$ be the number of stages and $p$ be the order. Then Butcher (1987) has shown that (1) a Runge–Kutta method with $s$ stages has order not exceeding $s$ and (2) there is no $p$ stage $p$th order Runge–Kutta method for $p \geq 5$. Thus the Runge–Kutta method of order $n$ requires at least $(n + 1)$ evaluations of $f$ per step if $n \geq 5$. Consequently, we lose computational efficiency if the Runge–Kutta method of order five or greater is used. For this reason Equation (8.5.28) is possibly the most popular Runge–Kutta method.

**Algorithm**

**Subroutine** Runge4 $(a, b, N, y0, f, Y)$
Comment: This program approximates the solution of an initial-value problem $y' = f(x, y)$, $y(x_0) = y0 = Y_0$ on a given interval $[a, b]$ where $x_0 = a$ and $x_N = b$ using the classical fourth-order Runge–Kutta method given by Equation (8.5.28).
  1 **Print** $a, b, N, y0$
  2 $h = (b - a)/N$
  3 $x = a$
  4 $Y = y0$
  5 **Print** $h, x, Y$
  6 **Repeat** steps 7–13 **with** $i = 1, 2, \ldots, N$
      7 $ak1 = hf(x, Y)$
      8 $ak2 = hf(x + 0.5\ h, Y + 0.5\ ak1)$
      9 $ak3 = hf(x + 0.5\ h, y + 0.5\ ak2)$
  10 $x = x + h$
  11 $ak4 = hf(x, Y + ak3)$
  12 $Y = Y + (ak1 + 2ak2 + 2ak3 + ak4)/6$
  13 **Print** $i, x, Y$
14 **Print** 'finished' and exit.  ◀

**EXAMPLE 8.5.2**

Solve the initial-value problem

$$\frac{dy}{dx} = y - x \qquad y(0) = 0 \quad \text{for } 0 \leq x \leq 1$$

using the fourth-order Runge–Kutta method. Compare it with the first- and second-order Runge–Kutta methods and the exact solution. Use $h = 0.1$.
Since $f(x, y) = y - x$,

$$Y_{i+1} = Y_i + \frac{1}{6}(k_1 + 2k_2 + 2k_3 + k_4) \qquad \textbf{(8.5.29)}$$

where

$$k_1 = h(Y_i - x_i)$$

$$k_2 = h\left[Y_i + \frac{k_1}{2} - \left(x_i + \frac{h}{2}\right)\right]$$

$$k_3 = h\left[Y_i + \frac{k_2}{2} - \left(x_i + \frac{h}{2}\right)\right]$$

$$k_4 = h[Y_i + k_3 - (x_i + h)]$$

For $i = 0$, Equation (8.5.29) gives

$$Y_1 = Y_0 + \frac{1}{6}(k_1 + 2k_2 + 2k_3 + k_4)$$

$$= \frac{1}{6}(k_1 + 2k_2 + 2k_3 + k_4)$$

where

$$k_1 = h(Y_0 - x_0) = 0$$

$$k_2 = h\left[Y_0 + \frac{k_1}{2} - \left(x_0 + \frac{h}{2}\right)\right] = -\frac{h^2}{2}$$

$$k_3 = h\left[Y_0 + \frac{k_2}{2} - \left(x_0 + \frac{h}{2}\right)\right] = h\left[0 - \frac{h^2}{4} - \frac{h}{2}\right]$$

$$= -\frac{h^2}{2}\left(\frac{h}{2} + 1\right)$$

$$k_4 = h[Y_0 + k_3 - (x_0 + h)] = h\left[-\frac{h^2}{2}\left(\frac{h}{2} + 1\right) - h\right]$$

$$= -h^2\left[\frac{h}{2}\left(\frac{h}{2} + 1\right) + 1\right]$$

Therefore

$$Y_1 = \frac{1}{6}\left\{-h^2 - h^2\left(\frac{h}{2} + 1\right) - h^2\left(\frac{h^2}{4} + \frac{h}{2} + 1\right)\right\}$$

$$= -\frac{h^2}{6}\left(3 + h + \frac{h^2}{4}\right) = -0.00517$$

Similarly, we continue and the computed results are given in Table 8.5.2.

The Improved Euler method is given by

$$\hat{Y}_{i+1} = Y_i + hf(x_i, Y_i)$$

$$Y_{i+1} = Y_i + \frac{h}{2}\left[f(x_i, Y_i) + f(x_i + h, \hat{Y}_{i+1})\right]$$

For $i = 0$, we get

$$\hat{Y}_1 = Y_0 + hf(x_0, Y_0) = 0 + h(Y_0 - x_0) = 0$$

$$Y_1 = Y_0 + \frac{h}{2}\left[f(x_0, Y_0) + f(x_0 + h, \hat{Y}_1)\right]$$

$$= \frac{h}{2}\left[Y_0 - x_0 + (\hat{Y}_1 - (x_0 + h))\right] = \frac{h}{2}(-h) = -\frac{h^2}{2} = -0.00500$$

**Table 8.5.2**

| $x_i$ | $Y_i$ using Euler | \|Actual Error\| for Euler | $Y_i$ using Improved Euler | \|Actual Error\| for Improved Euler | $Y_i$ using Fourth-Order | \|Actual Error\| for Fourth-Order |
|------|------|------|------|------|------|------|
| 0.0 | 0.00000 | 0.00000 | 0.00000 | 0.00000 | 0.00000 | 0.00000 |
| 0.1 | 0.00000 | 0.00517 | −0.00500 | 0.00017 | −0.00517 | 0.00000 |
| 0.2 | −0.01000 | 0.01140 | −0.02103 | 0.00037 | −0.02140 | 0.00000 |
| 0.3 | −0.03100 | 0.01886 | −0.04923 | 0.00063 | −0.04986 | 0.00000 |
| 0.4 | −0.06410 | 0.02772 | −0.09090 | 0.00092 | −0.09182 | 0.00000 |
| 0.5 | −0.11051 | 0.03821 | −0.14745 | 0.00127 | −0.14872 | 0.00000 |
| 0.6 | −0.17156 | 0.05056 | −0.22043 | 0.00169 | −0.22212 | 0.00000 |
| 0.7 | −0.24872 | 0.06503 | −0.31157 | 0.00218 | −0.31375 | 0.00000 |
| 0.8 | −0.34359 | 0.08195 | −0.42279 | 0.00275 | −0.42554 | 0.00000 |
| 0.9 | −0.45795 | 0.10165 | −0.55618 | 0.00342 | −0.55960 | 0.00000 |
| 1.0 | −0.59374 | 0.12454 | −0.71408 | 0.00420 | −0.71828 | 0.00000 |

In Table 8.5.2, we list the computed values $Y_i$ obtained using the Improved Euler method. For comparison the results of Euler's method are also given from Table 8.4.1 along with the absolute value of the actual errors.

The fourth-order Runge–Kutta requires four evaluations of $f$ per step $h$, the second-order Runge–Kutta requires two evaluations of $f$ per step $h$, while Euler's method requires only one evaluation of $f$ per step $h$. Therefore, let us compare the results of the Euler method, second-order, and fourth-order Runge–Kutta methods by using $h/4$, $h/2$, and $h$ respectively. ∎ ∎ ∎

**EXAMPLE 8.5.3**    Using the Euler method, the Improved Euler method, and the fourth-order Runge–Kutta method with $h = 0.025, 0.05$, and $0.1$ respectively, solve the initial-value problem

$$\frac{dy}{dx} = y - x \qquad y(0) = 0 \quad \text{for } 0 \le x \le 1$$

**Table 8.5.3**

| $x$ | Euler $h = 0.025$ | \|Actual Error\| for Euler | Improved Euler $h = 0.05$ | \|Actual Error\| for Improved Euler | Fourth-Order $h = 0.1$ | \|Actual Error\| for Fourth-Order |
|------|------|------|------|------|------|------|
| 0.0 | 0.00000 | 0.00000 | 0.00000 | 0.00000 | 0.00000 | 0.00000 |
| 0.1 | −0.00381 | 0.00136 | −0.00513 | 0.00004 | −0.00517 | 0.00000 |
| 0.2 | −0.01840 | 0.00300 | −0.02130 | 0.00010 | −0.02140 | 0.00000 |
| 0.3 | −0.04489 | 0.00497 | −0.04970 | 0.00016 | −0.04986 | 0.00000 |
| 0.4 | −0.08451 | 0.00731 | −0.09159 | 0.00023 | −0.09182 | 0.00000 |
| 0.5 | −0.13862 | 0.01010 | −0.14839 | 0.00033 | −0.14872 | 0.00000 |
| 0.6 | −0.20873 | 0.01339 | −0.22168 | 0.00044 | −0.22212 | 0.00000 |
| 0.7 | −0.29650 | 0.01725 | −0.31319 | 0.00056 | −0.31375 | 0.00000 |
| 0.8 | −0.40376 | 0.02178 | −0.42483 | 0.00071 | −0.42554 | 0.00000 |
| 0.9 | −0.53254 | 0.02706 | −0.55871 | 0.00089 | −0.55960 | 0.00000 |
| 1.0 | −0.68506 | 0.03322 | −0.71719 | 0.00109 | −0.71828 | 0.00000 |

The results are given in Table 8.5.3.

From Table 8.5.3, the Runge–Kutta fourth-order method gives more accurate answers than the Runge–Kutta second-order method with one-half the mesh size. This is generally true but not always. Usually, you can go a long way with the fourth-order Runge–Kutta method.

■ ■ ■

## EXERCISES

In Exercises 1–4, (a) use the second-order Runge–Kutta formulas (8.5.16), (8.5.17), and (8.5.18) to approximate the solution of the given initial-value problem on the given interval and (b) compare (a) with the exact solution. Use $h = 0.1$.

1. $dy/dx = -y^2$, $y(0) = 1$ on [0, 0.3]; exact solution: $y(x) = 1/(1 + x)$.
2. $dy/dx = 2xy$, $y(0) = 1$ on [0, 0.3]; exact solution: $y(x) = e^{x^2}$.
3. $dy/dx = -y + x^2$, $y(0) = 1$ on [0, 0.3]; exact solution: $y(x) = 2 - 2x + x^2 - e^{-x}$.
4. $dy/dx = y - (2x/y)$, $y(0) = 1$ on [0, 0.3]; exact solution: $y(x) = \sqrt{2x + 1}$.
5. Use the third-order Runge–Kutta method to approximate the solution of the initial-value problem on the given interval in Exercise 1 and compare it with the exact solution.
6. Use the third-order Runge–Kutta method to approximate the solution of the initial-value problem on the given interval in Exercise 4 and compare it with the exact solution.
7. Use the fourth-order Runge–Kutta method to approximate the solution of the initial-value problem on the given interval in Exercise 2 and compare it with the exact solution.
8. Use the fourth-order Runge–Kutta method to approximate the solution of the initial-value problem on the given interval in Exercise 3 and compare it with the exact solution.
9. Verify Equation (8.5.23).
10. Verify that the two-parameter system (8.5.25) may be solved to give

$$w_1 = 1 + \frac{2 - 3(\alpha_2 + \alpha_3)}{6 \alpha_2 \alpha_3} \qquad \beta_{21} = \alpha_2$$

$$w_2 = \frac{3\alpha_3 - 2}{6 \alpha_2 (\alpha_3 - \alpha_2)} \qquad \beta_{31} = \frac{3\alpha_2 \alpha_3 (1 - \alpha_2) - \alpha_3^2}{\alpha_2 (2 - 3\alpha_2)}$$

$$w_3 = \frac{2 - 3\alpha_2}{6 \alpha_3 (\alpha_3 - \alpha_2)} \qquad \beta_{32} = \frac{\alpha_3 (\alpha_3 - \alpha_2)}{\alpha_2 (2 - 3\alpha_2)}$$

Show that when $\alpha_2 = 0$ there are no third-order Runge–Kutta methods.

11. The fourth-order Runge–Kutta formula (8.5.28) can be derived by starting with

$$Y_{i+1} = Y_i + w_1 k_1 + w_2 k_2 + w_3 k_3 + w_4 k_4$$

where

$$k_1 = hf(x_i, Y_i)$$
$$k_2 = hf(x_i + \alpha_2 h, Y_i + \beta_{21} k_1)$$
$$k_3 = hf(x_i + \alpha_3 h, Y_i + \beta_{31} k_1 + \beta_{32} k_2)$$
$$k_4 = hf(x_i + \alpha_4 h, Y_i + \beta_{41} k_1 + \beta_{42} k_2 + \beta_{43} k_3)$$

(a) Show that the parameters must satisfy the system of equations

$$w_1 + w_2 + w_3 + w_4 = 1 \qquad w_2 \alpha_2 \beta_{32} + w_4 (\alpha_2 \beta_{42} + \alpha_3 \beta_{43}) = \frac{1}{6}$$

$$w_2 \alpha_2 + w_3 \alpha_3 + w_4 \alpha_4 = \frac{1}{2} \qquad w_3 \alpha_2^2 \beta_{32} + w_4 (\alpha_2^2 \beta_{42} + \alpha_3^2 \beta_{43}) = \frac{1}{12}$$

$$w_3 \alpha_2^2 + w_3 \alpha_3^2 + w_4 \alpha_4^2 = \frac{1}{3} \qquad w_3 \alpha_2 \alpha_3 \beta_{32} + w_4 \alpha_4 (\alpha_2 \beta_{42} + \alpha_3^2 \beta_{43}) = \frac{1}{8}$$

$$w_2 \alpha_2^3 + w_3 \alpha_3^3 + w_4 \alpha_4^3 = \frac{1}{4} \qquad w_4 \alpha_2 \beta_{32} \beta_{43} = \frac{1}{24}$$

(b) Verify that $w_1 = w_4 = \frac{1}{6}$, $w_2 = w_3 = \frac{1}{3}$, $\alpha_2 = \alpha_3 = \frac{1}{2}$, $\alpha_4 = 1$, $\beta_{21} = \frac{1}{2}$, $\beta_{31} = 0$, $\beta_{32} = \frac{1}{2}$, $\beta_{41} = \beta_{42} = 0$, and $\beta_{43} = 1$ satisfy the system of equations in (a).

**12.**  Show that Equations (8.5.16), (8.5.17), and (8.5.18) give the same approximation to the initial-value problem

$$\frac{dy}{dx} = y + x \qquad y(0) = 1 \quad \text{for } 0 \le x \le 1$$

Why?

## COMPUTER EXERCISES

For Exercises C.1 and C.2, write a subroutine to approximate the solution of an initial-value problem

$$\frac{dy}{dx} = f(x, y) \qquad y(x_0) = Y_0 = y0$$

**C.1.**  Use the second-order Runge–Kutta method.

**C.2.**  Use the fourth-order Runge–Kutta method.

**C.3.**  Write a program to approximate the solution of an initial-value problem

$$\frac{dy}{dx} = f(x, y) \qquad y(x_0) = Y_0 = y0$$

on a given interval $[x_0 = a, b]$ using Exercises C.1 and C.2. Read $N$, $a$, $b$, and $y_0$. Use various $h$.

**C.4.**  Using Exercise C.3, approximate the solution of the initial-value problem $dy/dx = 1 - x + 4y$, $y(0) = 1$ for $0 \le x \le 1$. Compare your results with the exact solution and analyze your output.

**C.5.** Using Exercise C.3, approximate the solution of the initial-value problem $dy/dx = 100 (\sin x - y)$, $y(0) = 0$ for $0 \le x \le 1$. Use $h = 0.1, 0.05, 0.01, 0.005$, and $0.001$. Compare your solutions with the exact solution.

**C.6.** By solving an appropriate initial-value problem, prepare a table of values of the function $f(x)$ defined by

$$f(x) = \frac{1}{\sqrt{2\pi}} \int_0^x e^{(-t^2/2)} dt$$

for $0 \le x \le 2$. In this way we generate the table of areas for a standard normal distribution.

**C.7.** The logistic equation for a population growth is given by $dp/dt = ap - bp^2$, $p(0) = p_0$, where $p(t)$ is the population at time $t$, and $a$ and $b$ are constants. Use $a = b = 5$ and $p_0 = 20$ to study the population growth at various time $t$. Use Exercise C.3 with various $h$.

## 8.6    ADAPTIVE STEP SIZE AND RUNGE–KUTTA–FEHLBERG METHODS*

When we compute the numerical solution using the step size $h$, we would like to know whether $h$ is large or small. If $h$ is large, then we accumulate truncation errors. If $h$ is too small, then we accumulate round-off errors and waste computing resources. For these reasons, we develop methods to adjust step size $h$ automatically for a given initial-value problem. There is no reason why the step size $h$ should be kept fixed over the entire interval. We can move with a few big steps through a flat plateau, while we need many small steps through a mountainous range. For a given $x_i$, $Y_i$, we want to determine a step size $h$ as large as possible such that the magnitude of the local truncation error does not exceed a predetermined constant tolerance $\epsilon$ per unit step after one step with this step size. The tolerance $\epsilon$ should be greater than $k \cdot$ eps where eps is a machine epsilon and $k = \max \{y(x) \mid x \in [a, b]\}$.

First consider a simple procedure of interval halving. Assume that $Y_i$ within a given tolerance at $x_i$ is available with $h = x_i - x_{i-1}$. Further assume that we are using a method of order $p$. We want to compute the solution from $x_i$ to $x_i + h$. We compute the solution twice, first by using the current step size $h$ and again using two steps each of size $h/2$. Thus we have two estimates $Y_{i+1}^h$ and $Y_{i+2}^{h/2}$ of the value $y(x_i + h)$.

The local truncation errors for the method of order $p$ are given by

$$y(x_i + h) - Y_{i+1}^h = A(x_i) h^{p+1} + \text{higher-order terms} \qquad \textbf{(8.6.1)}$$

and

$$y(x_i + h) - Y_{i+2}^{h/2} = A(x_i) \left(\frac{h}{2}\right)^{p+1} + \text{higher-order terms} \qquad \textbf{(8.6.2)}$$

Subtracting Equation (8.6.2) from (8.6.1), we get

$$Y_{i+2}^{h/2} - Y_{i+1}^h = A(x_i) (2^{p+1} - 1) \left(\frac{h}{2}\right)^{p+1} + \text{higher-order terms} \qquad \textbf{(8.6.3)}$$

$A(x_i)$ of the principal part of the error can be estimated from Equation (8.6.3) as

$$A(x_i) \approx \frac{Y_{i+2}^{h/2} - Y_{i+1}^h}{2^{p+1} - 1} \left(\frac{2}{h}\right)^{p+1} \tag{8.6.4}$$

Thus the local truncation error

$$L_T = \left|y(x_i + h) - Y_{i+1}^h\right| \approx \frac{\left|Y_{i+2}^{h/2} - Y_{i+1}^h\right|}{(2^{p+1} - 1)} 2^{p+1} \tag{8.6.5}$$

provides an estimate of the local truncation error. The estimate $L_T$ can be used to decide whether the step size $h$ is just right or too small or too big.

A common policy is to keep the local truncation error $L_T$ per unit step below a given tolerance $\epsilon$; in other words, we want

$$\frac{L_T}{h} \le \epsilon \tag{8.6.6}$$

We need $L_T/h \ge \epsilon'$ where $\epsilon'$ is a lower bound. Since halving the step size reduces the truncation error by approximately $1/(2^{p+1})$, we set the lower bound $\epsilon' = \epsilon/(2^{p+1})$ for the method of order $p$. We have

**1.**

$$\epsilon' \le \frac{L_T}{h} \le \epsilon$$

In this case accept the value $Y_{i+2}^{h/2}$ and continue the integration from $x_i + h$ using the same step size $h$.

**2.**

$$\frac{L_T}{h} < \epsilon'$$

In this case accept $Y_{i+2}^{h/2}$. Since the local truncation error $L_T$ is small, continue the integration from $x_i + h$ using the step size $2h$.

**3.**

$$\frac{L_T}{h} > \epsilon$$

In this case the local truncation error $L_T$ is large and so reduce the step size to $h/2$ to integrate from $x_i$.

The method of interval halving to control errors requires substantial efforts and therefore is not recommended.

## Fehlberg's Method

Instead of comparing two approximations of the same method with different step sizes, Fehlberg (1969) suggested comparing two approximations with the same step size of two different methods of orders $p$ and $p + 1$. This gives rise to the **Runge–Kutta–Fehlberg method**.

Consider the second-order Runge–Kutta method

$$Y_{i+1} = Y_i + w_1 k_1 + w_2 k_2 \tag{8.6.7}$$

where

$$k_1 = hf(x_i, Y_i)$$
$$k_2 = hf(x_i + \alpha_2 h, Y_i + \beta_{21} k_1)$$

and the third-order Runge–Kutta method

$$\hat{Y}_{i+1} = Y_i + \hat{w}_1 k_1 + \hat{w}_2 k_2 + \hat{w}_3 k_3 \tag{8.6.8}$$

where

$$k_1 = hf(x_i, Y_i)$$
$$k_2 = hf(x_i + \alpha_2 h, Y_i + \beta_{21} k_1)$$
$$k_3 = hf(x_i + \alpha_3 h, Y_i + \beta_{31} k_1 + \beta_{32} k_2)$$

Let us assume that $Y_i$ within a given tolerance is available with $h = x_i - x_{i-1}$. The idea is to use the same $k_1$ and $k_2$ to evaluate Equations (8.6.7) and (8.6.8). In order to evaluate Equation (8.6.8), we need only one additional evaluation of $k_3$. We determine constants $\alpha_k$, $\beta_{kl}$, $w_k$, and $\hat{w}_k$ in Equations (8.6.7) and (8.6.8) as in Equations (8.5.7) and (8.5.21). One possible solution is given by the second-order Runge–Kutta method

$$Y_{i+1} = Y_i + (0k_1 + k_2) \tag{8.6.9}$$

where

$$k_1 = hf(x_i, Y_i)$$

and

$$k_2 = hf\left(x_i + \frac{h}{2}, Y_i + \frac{k_1}{2}\right)$$

and the third-order Runge–Kutta method

$$\hat{Y}_{i+1} = Y_i + \frac{1}{6}(k_1 + 4k_2 + k_3) \tag{8.6.10}$$

where

$$k_1 = hf(x_i, Y_i)$$
$$k_2 = hf\left(x_i + \frac{h}{2}, Y_i + \frac{k_1}{2}\right)$$
$$k_3 = hf(x_i + h, Y_i - k_1 + 2k_2)$$

The local truncation error for Equation (8.6.9) is given by

$$y(x_i + h) - Y_{i+1} = B(x_i) h^3 + \text{higher-order terms} \tag{8.6.11}$$

and for Equation (8.6.10) it is given by

$$y(x_i + h) - \hat{Y}_{i+1} = C(x_i) h^4 + \text{higher-order terms} \tag{8.6.12}$$

Subtracting Equation (8.6.12) from Equation (8.6.11), we get

$$\hat{Y}_{i+1} - Y_{i+1} = B(x_i) h^3 + \text{higher-order terms} \tag{8.6.13}$$

From Equation (8.6.13), we have

$$|B(x_i)| \approx \frac{|\hat{Y}_{i+1} - Y_{i+1}|}{h^3} \qquad \textbf{(8.6.14)}$$

Thus the local truncation error

$$L_T = |y(x_i + h) - Y_{i+1}| \approx |B(x_i)| \, h^3 \approx |\hat{Y}_{i+1} - Y_{i+1}|$$

The local truncation error of $Y_{i+1}$ is estimated using just one additional evaluation of $k_3$. We want to keep the local truncation error $L_T$ per unit step below a given tolerence $\epsilon$; in other words, we want $L_T/h \le \epsilon$. If $|\hat{Y}_{i+1} - Y_{i+1}| \le h\epsilon$, then the integration from $x_i$ to $x_i + h$ is successful. In this case, we want to find the right step size for the next step. But if $|\hat{Y}_{i+1} - Y_{i+1}| > h\epsilon$, then the integration from $x_i$ to $x_i + h$ is not successful. In this case, we must go back to $x_i$ and use a smaller $h$. In either case, we need to have a new step size. If we want the "new" step size $h_{\text{new}}$ to be successful, then we must have

$$L_T \approx |B(x_i) \, h_{\text{new}}^3| \le h_{\text{new}} \, \epsilon \qquad \textbf{(8.6.15)}$$

Substituting for $|B(x_i)|$ from Equation (8.6.14) in Equation (8.6.15), we get the approximation relation

$$\frac{|\hat{Y}_{i+1} - Y_{i+1}|}{h^3} \, |h_{\text{new}}^2| \le \epsilon$$

This can be used to estimate

$$h_{\text{new}} \approx h \sqrt{\frac{h\epsilon}{|\hat{Y}_{i+1} - Y_{i+1}|}} \qquad \textbf{(8.6.16)}$$

Since approximations are involved in Equation (8.6.16), we can use

$$h_{\text{new}} \approx \alpha h \sqrt{\frac{h\epsilon}{|\hat{Y}_{i+1} - Y_{i+1}|}} \qquad \textbf{(8.6.17)}$$

where $\alpha$ is a suitable adjustment factor. Usually we select $\alpha = 0.9$. The new step size $h_{\text{new}}$ is used to compute $Y_{i+2}$ at $x_{i+2}$ if $|\hat{Y}_{i+1} - Y_{i+1}| \le h\epsilon$. It is also used to compute $Y_{i+1}$ at $x_{i+1}$ if $|\hat{Y}_{i+1} - Y_{i+1}| > h\epsilon$ where $x_{i+1} = x_i + h_{\text{new}}$.

If we compute $Y_{i+1}$ and $\hat{Y}_{i+1}$ using the methods of order $p$ and $p + 1$ respectively, then Equation (8.6.17) is replaced by

$$h_{\text{new}} \approx \alpha h \sqrt[p]{\frac{h\epsilon}{|\hat{Y}_{i+1} - Y_{i+1}|}} \qquad \textbf{(8.6.18)}$$

The idea to use a pair of Runge–Kutta formulas in which some of the evaluations of $f$ can also be used in the other formulas to estimate the local truncation error was used by Fehlberg (1969) to derive other formulas. The **most widely used Runge–Kutta–Fehlberg formulas** are

$$Y_{i+1} = Y_i + \frac{25}{216} k_1 + \frac{1408}{2565} k_3 + \frac{2197}{4104} k_4 - \frac{1}{5} k_5 \qquad \textbf{(8.6.19)}$$

and

$$\hat{Y}_{i+1} = Y_i + \frac{16}{135} k_1 + \frac{6656}{12825} k_3 + \frac{28561}{56430} k_4 - \frac{9}{50} k_5 + \frac{2}{55} k_6 \qquad \textbf{(8.6.20)}$$

where

$$k_1 = hf(x_i, Y_i)$$

$$k_2 = hf\left(x_i + \frac{1}{4} h, Y_i + \frac{1}{4} k_1\right)$$

$$k_3 = hf\left(x_i + \frac{3}{8} h, Y_i + \frac{3}{32} k_1 + \frac{9}{32} k_2\right)$$

$$k_4 = hf\left(x_i + \frac{12}{13} h, Y_i + \frac{1932}{2197} k_1 - \frac{7200}{2197} k_2 + \frac{7296}{2197} k_3\right)$$

$$k_5 = hf\left(x_i + h, Y_i + \frac{439}{216} k_1 - 8k_2 + \frac{3680}{513} k_3 - \frac{845}{4104} k_4\right)$$

$$k_6 = hf\left(x_i + \frac{1}{2} h, Y_i - \frac{8}{27} k_1 + 2k_2 - \frac{3544}{2565} k_3 + \frac{1859}{4104} k_4 - \frac{11}{40} k_5\right) \textbf{(8.6.21)}$$

Here $Y_{i+1}$ is a fourth-order formula using five stages and $\hat{Y}_{i+1}$ is a fifth-order formula using six stages. The fact that $Y_{i+1}$ uses one more stage than necessary to achieve fourth-order accuracy is not important since we need to evaluate $f$ six times for $\hat{Y}_{i+1}$ anyway. It is of questionable value to use Equations (8.6.19) and (8.6.20) separately. Subtracting Equation (8.6.19) from Equation (8.6.20) yields

$$\hat{Y}_{i+1} - Y_{i+1} = \frac{1}{360} k_1 - \frac{128}{4275} k_3 - \frac{2197}{75240} k_4 + \frac{1}{50} k_5 + \frac{2}{55} k_6 \qquad \textbf{(8.6.22)}$$

**Algorithm**

**Subroutine** RKF45 ($a$, $b$, $y0$, $f$, $Y$, hmax, hmin, eps, maxit, $h$, $\alpha$)
Comment: This program computes the solution of an initial-value problem $y' = f(x, y)$, $y(x_0) = y0$ on a given interval $[a, b]$ where $x_0 = a$ using the Runge–Kutta–Fehlberg method of order four [Equation (8.6.19)] and five [Equation (8.6.20)]. Here hmax is the maximum step size; hmin is the minimum step size; eps is the tolerance; maxit is the maximum number of steps to be used in going from $a$ to $b$; $h$ is the initial step size; $\alpha$ is the suitable adjustment factor in Equation (8.6.18).

1 **Print** $a$, $b$, $y0$, hmax, hmin, eps, maxit, $h$, $\alpha$
2 $x = a$
3 $Y = y0$
4 **Repeat** steps 5–25 **with** $i = 1, 2, \ldots,$ maxit
   5 **If** ($b - x < h$), **then** $h = b - x$
   6 $ak1 = hf(x, Y)$
   7 $ak2 = hf(x + (h/4), Y + (ak1/4))$
   8 $ak3 = hf(x + (3/8)h, Y + ((3 \ (ak1) + 9 \ (ak2))/32))$
   9 $ak4 = hf(x + (12/13)h, Y + ((1932 \ (ak1) - 7200 \ (ak2) + 7296 \ (ak3))/2197))$

10 $ak5 = hf(x + h, Y + (439/216) (ak1) - 8 (ak2) + (3680/513) (ak3) - (845/4104) (ak4))$

11 $ak6 = hf(x + (1/2) h, Y - (8/27) (ak1) + 2 (ak2) - (3544/2565) (ak3) + (1859/4104) (ak4) - (11/40) (ak5))$

12 $TE = |(1/360)(ak1) - (128/4275)(ak3) - (2197/75240)(ak4) + (1/50)(ak5) + (2/55) (ak6)|$

13 $h_{new} = \alpha h (h \ eps/TE)^{0.25}$

14 **If** $TE \leq h$ eps, **then** do steps 15–21

   15 $Y = Y + (16/135) (ak1) + (6656/12825) (ak3) + (28561/56430) (ak4) - (9/50) (ak5) + (2/55) (ak6)$

  16 **Print** $h$, $x + h$, $Y$

  17 $h = h_{new}$

  18 **If** $h <$ hmin, **then** $h =$ hmin

  19 $x = x + h$

  20 **If** $x > b$, **then print** 'finished' and exit.

  21 **Print** $h$ and $x$ and return to 5

22 **If** $TE > h$ eps, **then** do steps 23–25

  23 $h = h_{new}$

  24 **If** $h >$ hmax, **then** h $=$ hmax

25 **Print** $h$, $x$, $Y$ and return to 5 ◀

**EXAMPLE 8.6.1** Using the RKF45 procedure, approximate the solution of the initial-value problem

$$\frac{dy}{dx} = -32xy^2 \qquad y\left(-\frac{1}{2}\right) = \frac{1}{5} \quad \text{for } -\frac{1}{2} \leq x \leq 0$$

**Table 8.6.1**

| $i$ | $x_i$ | $Y_i$ | Exact $y(x_i)$ | $|Y_i - y(x_i)|$ | $h_i = x_i - x_{i-1}$ |
|---|---|---|---|---|---|
| 0 | $-0.50000$ | 0.20000 | 0.20000 | — | 0.10000 |
| 1 | $-0.44489$ | 0.23999 | 0.23999 | 0.00000 | 0.05511 |
| 2 | $-0.39669$ | 0.28426 | 0.28426 | 0.00000 | 0.04819 |
| 3 | $-0.35333$ | 0.33361 | 0.33361 | 0.00000 | 0.04336 |
| 4 | $-0.30997$ | 0.39411 | 0.39411 | 0.00000 | 0.04336 |
| 5 | $-0.27307$ | 0.45598 | 0.45598 | 0.00000 | 0.03690 |
| 6 | $-0.23616$ | 0.52844 | 0.52844 | 0.00000 | 0.03690 |
| 7 | $-0.19926$ | 0.61152 | 0.61152 | 0.00000 | 0.03690 |
| 8 | $-0.16235$ | 0.70336 | 0.70336 | 0.00000 | 0.03690 |
| 9 | $-0.12545$ | 0.79885 | 0.79885 | 0.00000 | 0.03690 |
| 10 | $-0.08854$ | 0.88854 | 0.88854 | 0.00000 | 0.03690 |
| 11 | $-0.05728$ | 0.95012 | 0.95012 | 0.00000 | 0.03127 |
| 12 | $-0.02601$ | 0.98928 | 0.98928 | 0.00000 | 0.03127 |
| 13 | 0.00000 | 1.00000 | 1.00000 | 0.00000 | 0.02601 |

The exact solution $y(x) = 1/(1 + 16 x^2)$ rises rapidly from $\frac{1}{5}$ to 1 in the interval $[-(\frac{1}{2}), 0]$. Therefore our adaptive procedure is expected to choose a small step size near $x = 0$ and a larger step size elsewhere. An approximate solution is computed using the step-size control procedure described above. The results shown in Table 8.6.1 were obtained using eps $= 10^{-5}$, hmax $= 0.1$, hmin $= 0.001$, and initial $h = 0.1$.

∎ ∎ ∎

A step-size control for the combination of three- and four-stage Runge–Kutta methods is not possible since the corresponding systems of equations do not admit solutions so that both methods have the same $k_1$, $k_2$, and $k_3$.

### EXERCISES

1. Replace $w_1$, $w_2$, and $w_3$ by $\hat{w}_1$, $\hat{w}_2$, and $\hat{w}_3$ respectively in Equation (8.5.25). Using Equation (8.5.13), $\alpha_2 = \frac{1}{2}$, and $\beta_{21} = \frac{1}{2}$, verify Equations (8.6.9) and (8.6.10) by finding $w_1$, $w_2$, $w_3$, $\hat{w}_1$, $\hat{w}_2$, $\hat{w}_3$, $\alpha_3$, $\beta_{31}$, and $\beta_{32}$.

2. Replace $w_1$, $w_2$, and $w_3$ by $\hat{w}_1$, $\hat{w}_2$, and $\hat{w}_3$ respectively in Equation (8.5.25). Using Equation (8.5.12), $\alpha_2 = 1$, and $\beta_{21} = 1$, verify that the combination of Runge–Kutta methods of two stage

$$Y_{i+1} = Y_i + \frac{1}{2}(k_1 + k_2)$$

and three stage

$$\hat{Y}_{i+1} = Y_i + \frac{1}{6}(k_1 + 2k_2 + k_3)$$

where

$$k_1 = hf(x_i, Y_i)$$
$$k_2 = hf(x_i + h, Y_i + k_1)$$
$$k_3 = hf\left(x_i + \frac{h}{2}, Y_i + \frac{1}{4}(k_1 + k_2)\right)$$

gives a step-size control method.

3. Verify that the combination of three- and four-stage Runge–Kutta methods is not possible for a step-size control method.

In Exercises 4–5, use two steps of the step-size control method with the Runge–Kutta–Fehlberg formulas (8.6.9) and (8.6.10) to approximate the solution of the given initial-value problem on the interval. Use eps $= 0.01$.

4. $dy/dx = -xy^2$, $y(0) = 2$ on [0, 0.4]; exact solution: $y(x) = 2/(1 + x^2)$.
5. $dy/dx = y - x$, $y(0) = 0$ on [0, 0.4]; exact solution: $y(x) = 1 + x - e^x$.

■■■■■■■■■■■■ **COMPUTER EXERCISES** ■■■■■■■■■■■■

For Exercises C.1 and C.2, write a subroutine to approximate the solution of an initial-value problem

$$\frac{dy}{dx} = f(x, y) \qquad y(x_0) = Y_0 = y0$$

**C.1.** Use the step-size control method with the Runge–Kutta–Fehlberg formulas (8.6.9) and (8.6.10).

**C.2.** Use the step-size control method with Runge–Kutta–Fehlberg formulas (8.6.19) and (8.6.20).

**C.3.** Using Exercises C.1 and C.2, approximate the solution of the initial-value problem

$$\frac{dy}{dx} = \frac{10}{1 + 100 \, x^2} \qquad y(0) = 0 \quad \text{for } 0 \le x \le 1$$

Also use Exercise C.2, p. 258 to solve it. Compare your results with the exact solution.

**C.4.** Using Exercises C.1 and C.2, approximate the solution of the initial-value problem

$$\frac{dy}{dx} = 1 + y^2 \qquad y(0) = 0 \quad \text{for } 0 \le x \le 1.5$$

Compare your results with the exact solution.

**C.5.** Using Exercises C.1 and C.2, approximate the solution of the initial-value problem

$$\frac{dy}{dx} = -50 \, xy^2 \qquad y(-1) = \frac{1}{26} \quad \text{for } -1 \le x \le 0$$

Compare your results with the exact solution.

---

## 8.7    SYSTEMS OF DIFFERENTIAL EQUATIONS AND HIGHER-ORDER DIFFERENTIAL EQUATIONS

Let us consider a system of three first-order differential equations

$$\frac{dy_1}{dx} = f_1(x, y_1, y_2, y_3)$$

$$\frac{dy_2}{dx} = f_2(x, y_1, y_2, y_3)$$

$$\frac{dy_3}{dx} = f_3(x, y_1, y_2, y_3) \qquad \textbf{(8.7.1)}$$

with initial conditions

$$y_1(x_0) = \alpha_1 \qquad y_2(x_0) = \alpha_2 \qquad y_3(x_0) = \alpha_3$$

This can be expressed in a vector equation

$$\frac{d\mathbf{y}}{dx} = \mathbf{f}(x, \mathbf{y}) \qquad \mathbf{y}(x_0) = \boldsymbol{\alpha} \tag{8.7.2}$$

where

$$\mathbf{y} = \begin{bmatrix} y_1(x) \\ y_2(x) \\ y_3(x) \end{bmatrix} \qquad \mathbf{f}(x, \mathbf{y}) = \mathbf{f}(x, y_1, y_2, y_3) = \begin{bmatrix} f_1(x, y_1, y_2, y_3) \\ f_2(x, y_1, y_2, y_3) \\ f_3(x, y_1, y_2, y_3) \end{bmatrix} \qquad \boldsymbol{\alpha} = \begin{bmatrix} \alpha_1 \\ \alpha_2 \\ \alpha_3 \end{bmatrix}$$

In order to solve Equation (8.7.2), we could apply any of the methods for a single first-order ordinary differential equation with the vector quantities rather than scalar quantities. For example, the Euler method gives

$$\mathbf{Y}_{i+1} = \mathbf{Y}_i + h\mathbf{f}(x_i, \mathbf{Y}_i) \quad \text{for } i = 0, 1, \ldots, N - 1 \tag{8.7.3}$$

with

$$\mathbf{Y}(x_0) = \mathbf{Y}_0 = \boldsymbol{\alpha}$$

where

$$\mathbf{Y}_{i+1} = \begin{bmatrix} Y_{1,i+1} \\ Y_{2,i+1} \\ Y_{3,i+1} \end{bmatrix} \qquad \mathbf{f}(x_i, \mathbf{Y}_i) = \begin{bmatrix} f_1(x_i, Y_{1,i}, Y_{2,i}, Y_{3,i}) \\ f_2(x_i, Y_{1,i}, Y_{2,i}, Y_{3,i}) \\ f_3(x_i, Y_{1,i}, Y_{2,i}, Y_{3,i}) \end{bmatrix}$$

$$\mathbf{Y}_0 = \boldsymbol{\alpha} = \begin{bmatrix} Y_{1,0} \\ Y_{2,0} \\ Y_{3,0} \end{bmatrix} = \begin{bmatrix} y_1(x_0) \\ y_2(x_0) \\ y_3(x_0) \end{bmatrix}$$

Similarly the Runge–Kutta method of order four is given by

$$\mathbf{Y}'_{i+1} = \mathbf{Y}_i + \frac{1}{6}(\mathbf{k}_{1,i} + 2\mathbf{k}_{2,i} + 2\mathbf{k}_{3,i} + \mathbf{k}_{4,i}) \tag{8.7.4}$$

where

$$\mathbf{k}_{1,i} = h\mathbf{f}(x_i, \mathbf{Y}_i)$$

$$\mathbf{k}_{2,i} = h\mathbf{f}\left(x_i + \frac{h}{2}, \mathbf{Y}_i + \frac{1}{2}\mathbf{k}_{1,i}\right)$$

$$\mathbf{k}_{3,i} = h\mathbf{f}\left(x_i + \frac{h}{2}, \mathbf{Y}_i + \frac{1}{2}\mathbf{k}_{2,i}\right)$$

$$\mathbf{k}_{4,i} = h\mathbf{f}(x_i + h, \mathbf{Y}_i + \mathbf{k}_{3,i}) \quad \text{for } i = 0, 1, \ldots, N - 1$$

**Algorithm**

**Subroutine** System($a$, $b$, $N$, $M$, $\boldsymbol{\alpha}$, $\mathbf{Y}$, $\mathbf{f}$)
Comment: This program approximates the solution of a system of $M$ first-order ordinary differential equations $dy_j/dx = f_j(x, y_1, y_2, \ldots, y_M)$ with initial conditions $y_j = \alpha_j$ for $j = 1, 2, \ldots, M$ on a given interval $[a, b]$ where $x_0 = a$ and $x_N = b$ using the classical fourth-order Runge–Kutta method given by Equation (8.7.4).
  1 **Print** $a$, $b$, $N$, $M$, $\alpha_1$, $\alpha_2$, $\ldots$, $\alpha_M$
  2 $h = (b - a)/N$

    3  $x = a$
    4  **Repeat** step 5 **with** $j = 1, 2, \ldots, M$
        5  $Y_j = \alpha_j$
    6  **Print** $h, x, Y_1, Y_2, \ldots, Y_M$
    7  **Repeat** steps 8–19 **with** $i = 1, 2, \ldots, N$
        8  **Repeat** step 9 **with** $j = 1, 2, \ldots, M$
            9  $ak_{1,j} = hf_j(x, Y_1, Y_2, \ldots, Y_M)$
        10 **Repeat** step 11 **with** $j = 1, 2, \ldots, M$
            11 $ak_{2,j} = hf_j(x + \frac{h}{2}, Y_1 + \frac{1}{2} ak_{1,1}, Y_2 + \frac{1}{2} ak_{1,2}, \ldots, Y_M + \frac{1}{2} ak_{1,M})$
        12 **Repeat** step 13 **with** $j = 1, 2, \ldots, M$
            13 $ak_{3,j} = hf_j(x + \frac{h}{2}, Y_1 + \frac{1}{2} ak_{2,1}, Y_2 + \frac{1}{2} ak_{2,2}, \ldots, Y_M + \frac{1}{2} ak_{2,M})$
        14 **Repeat** step 15 **with** $j = 1, 2, \ldots, M$
            15 $ak_{4,j} = hf_j(x + h, Y_1 + ak_{3,1}, Y_2 + ak_{3,2}, \ldots, Y_M + ak_{3,M})$
        16 **Repeat** step 17 **with** $j = 1, 2, \ldots, M$
            17 $Y_j = Y_j + (ak_{1,j} + 2ak_{2,j} + 2ak_{3,j} + ak_{4,j})/6$
        18 $x = a + ih$
        19 **Print** $x, Y_1, Y_2, \ldots, Y_M$
    20 **Print** 'finished' and exit.    ◀

**EXAMPLE 8.7.1**    Solve

$$\frac{dy_1}{dx} = \frac{1}{2}(y_1 + 5y_2)$$

$$\frac{dy_2}{dx} = \frac{1}{2}(5y_1 + y_2)$$

with initial conditions $y_1(0) = 2$ and $y_2(0) = 0$ for $0 \leq x \leq 1$ using (1) the Euler method and (2) the fourth-order Runge–Kutta method.

1.  Divide $[0, 1]$ into ten equal intervals such that $x_0 = 0, x_1 = 0.1, \ldots, x_{10} = 1.0$ with $h = 0.1$. Also $f_1(x, y_1, y_2) = (y_1 + 5y_2)/2$ and $f_2(x, y_1, y_2) = (5y_1 + y_2)/2$. Then Equation (8.7.3) becomes

$$\begin{bmatrix} Y_{1,i+1} \\ Y_{2,i+1} \end{bmatrix} = \begin{bmatrix} Y_{1,i} + hf_1(x_i, Y_{1,i}, Y_{2,i}) \\ Y_{2,i} + hf_2(x_i, Y_{1,i}, Y_{2,i}) \end{bmatrix}$$

$$= \begin{bmatrix} Y_{1,i} + \dfrac{h}{2}(Y_{1,i} + 5Y_{2,i}) \\ Y_{2,i} + \dfrac{h}{2}(5Y_{1,i} + Y_{2,i}) \end{bmatrix} \tag{8.7.5}$$

where $Y_{1,i} \approx y_1(x_i)$ and $Y_{2,i} \approx y_2(x_i)$.
For $i = 0$, Equation (8.7.5) reduces to

$$\begin{bmatrix} Y_{1,1} \\ Y_{2,1} \end{bmatrix} = \begin{bmatrix} Y_{1,0} + \dfrac{h}{2}(Y_{1,0} + 5Y_{2,0}) \\ Y_{2,0} + \dfrac{h}{2}(5Y_{1,0} + Y_{2,0}) \end{bmatrix} = \begin{bmatrix} y_1(0) + \dfrac{h}{2}(y_1(0) + 5y_2(0)) \\ y_2(0) + \dfrac{h}{2}(5y_1(0) + y_2(0)) \end{bmatrix} = \begin{bmatrix} 2.1 \\ 0.5 \end{bmatrix}$$

For $i = 1$, Equation (8.7.5) reduces to

$$\begin{bmatrix} Y_{1,2} \\ Y_{2,2} \end{bmatrix} = \begin{bmatrix} Y_{1,1} + \dfrac{h}{2}(Y_{1,1} + 5Y_{2,1}) \\ Y_{2,1} + \dfrac{h}{2}(5Y_{1,1} + Y_{2,1}) \end{bmatrix} = \begin{bmatrix} 2.33 \\ 1.05 \end{bmatrix}$$

Similarly continuing, $Y_{1,i}$ and $Y_{2,i}$ are computed and are given in Table 8.7.1. For comparison, the values of the exact solution $y_1(x) = e^{3x} + e^{-2x}$ and $y_2(x) = e^{3x} - e^{-2x}$ are also given in Table 8.7.1 along with the absolute value of the actual errors. Errors are large due to large $h$ and low order of the Euler method.

2. For $i = 0$, Equation (8.7.4) becomes

$$\mathbf{Y}_1 = \mathbf{Y}_0 + \frac{1}{6}(\mathbf{k}_{1,0} + 2\mathbf{k}_{2,0} + 2\mathbf{k}_{3,0} + \mathbf{k}_{4,0})$$

where

$$\mathbf{k}_{1,0} = h\mathbf{f}(x_0, \mathbf{Y}_0) = h\begin{bmatrix} f_1(x_0, Y_{1,0}, Y_{2,0}) \\ f_2(x_0, Y_{1,0}, Y_{2,0}) \end{bmatrix} = \frac{h}{2}\begin{bmatrix} Y_{1,0} + 5Y_{2,0} \\ 5Y_{1,0} + Y_{2,0} \end{bmatrix} = \begin{bmatrix} 0.1 \\ 0.5 \end{bmatrix}$$

$$\mathbf{k}_{2,0} = h\mathbf{f}\left(x_0 + \frac{h}{2}, \mathbf{Y}_0 + \frac{1}{2}\mathbf{k}_{1,0}\right) = h\begin{bmatrix} f_1\left(x_0 + \dfrac{h}{2}, Y_{1,0} + 0.05, Y_{2,0} + 0.25\right) \\ f_2\left(x_0 + \dfrac{h}{2}, Y_{1,0} + 0.05, Y_{2,0} + 0.25\right) \end{bmatrix}$$

$$= \frac{h}{2}\begin{bmatrix} (Y_{1,0} + 0.5) + 5(Y_{2,0} + 0.25) \\ 5(Y_{1,0} + 0.5) + (Y_{2,0} + 0.25) \end{bmatrix} = \begin{bmatrix} 0.165 \\ 0.525 \end{bmatrix}$$

$$\mathbf{k}_{3,0} = h\mathbf{f}\left(x_0 + \frac{h}{2}, \mathbf{Y}_0 + \frac{1}{2}\mathbf{k}_{2,0}\right) = h\begin{bmatrix} f_1\left(x_0 + \dfrac{h}{2}, Y_{1,0} + 0.0825, Y_{2,0} + 0.2625\right) \\ f_2\left(x_0 + \dfrac{h}{2}, Y_{1,0} + 0.0825, Y_{2,0} + 0.2625\right) \end{bmatrix}$$

$$= \frac{h}{2}\begin{bmatrix} (Y_{1,0} + 0.0825) + 5(Y_{2,0} + 0.2625) \\ 5(Y_{1,0} + 0.0825) + (Y_{2,0} + 0.2625) \end{bmatrix} = \begin{bmatrix} 0.16975 \\ 0.53375 \end{bmatrix}$$

$$\mathbf{k}_{4,0} = h\mathbf{f}(x_0 + h, \mathbf{Y}_0 + \mathbf{k}_{3,0}) = h\begin{bmatrix} f_1(x_0 + h, Y_{1,0} + 0.16975, Y_{2,0} + 0.53375) \\ f_2(x_0 + h, Y_{1,0} + 0.16975, Y_{2,0} + 0.53375) \end{bmatrix}$$

$$= \frac{h}{2}\begin{bmatrix} (Y_{1,0} + 0.16975) + 5(Y_{2,0} + 0.53375) \\ 5(Y_{1,0} + 0.16975) + (Y_{2,0} + 0.53375) \end{bmatrix} = \begin{bmatrix} 0.24193 \\ 0.56913 \end{bmatrix}$$

**Table 8.7.1**

| $i$ | $x_i$ | $y_1(x) =$ $e^{3x} + e^{-2x}$ | $Y_{1,i}$ | \|Actual Error\| in $Y_1$ | $y_2(x) =$ $e^{3x} - e^{-2x}$ | $Y_{2,i}$ | \|Actual Error\| in $Y_2$ |
|---|---|---|---|---|---|---|---|
| 0 | 0.0 | 2.00000 | 2.00000 | 0.00000 | 0.00000 | 0.00000 | 0.00000 |
| 1 | 0.1 | 2.16859 | 2.10000 | 0.06859 | 0.53113 | 0.50000 | 0.03113 |
| 2 | 0.2 | 2.49244 | 2.33000 | 0.16244 | 1.15180 | 1.05000 | 0.10180 |
| 3 | 0.3 | 3.00842 | 2.70900 | 0.29942 | 1.91080 | 1.68500 | 0.22580 |
| 4 | 0.4 | 3.76945 | '3.26570 | 0.50375 | 2.87079 | 2.44650 | 0.42429 |
| 5 | 0.5 | 4.84957 | 4.04061 | 0.80896 | 4.11381 | 3.38525 | 0.72856 |
| 6 | 0.6 | 6.35084 | 5.08895 | 1.26189 | 5.74845 | 4.56466 | 1.18379 |
| 7 | 0.7 | 8.41277 | 6.48457 | 1.92820 | 7.91957 | 6.06513 | 1.85444 |
| 8 | 0.8 | 11.22507 | 8.32508 | 2.89999 | 10.82128 | 7.98954 | 2.83174 |
| 9 | 0.9 | 15.04503 | 10.73872 | 4.30631 | 14.71443 | 10.47028 | 4.24415 |
| 10 | 1.0 | 20.22087 | 13.89322 | 6.32765 | 19.95020 | 13.67848 | 6.27172 |

Hence

$$\mathbf{Y}_1 = \mathbf{Y}_0 + \frac{1}{6}(\mathbf{k}_{1,0} + 2\mathbf{k}_{2,0} + 2\mathbf{k}_{3,0} + \mathbf{k}_{4,0})$$

$$= \begin{bmatrix} 2 \\ 0 \end{bmatrix} + \frac{1}{6}\left( \begin{bmatrix} 0.1 \\ 0.5 \end{bmatrix} + \begin{bmatrix} 0.33 \\ 1.05 \end{bmatrix} + \begin{bmatrix} 0.3395 \\ 1.0675 \end{bmatrix} + \begin{bmatrix} 0.24193 \\ 0.56913 \end{bmatrix} \right)$$

$$= \begin{bmatrix} 2.16857 \\ 0.53110 \end{bmatrix} = \begin{bmatrix} Y_{1,1} \\ Y_{2,1} \end{bmatrix}$$

**Table 8.7.2**

| $i$ | $x_i$ | $y_1(x) =$ $e^{3x} + e^{-2x}$ | $Y_{1,i}$ | \|Actual Error\| for $Y_1$ | $y_2(x) =$ $e^{3x} - e^{-2x}$ | $Y_{2,i}$ | \|Actual Error\| for $Y_2$ |
|---|---|---|---|---|---|---|---|
| 0 | 0.0 | 2.00000 | 2.00000 | 0.00000 | 0.00000 | 0.00000 | 0.00000 |
| 1 | 0.1 | 2.16859 | 2.16857 | 0.00002 | 0.53113 | 0.53110 | 0.00003 |
| 2 | 0.2 | 2.49244 | 2.49239 | 0.00005 | 1.15180 | 1.15174 | 0.00006 |
| 3 | 0.3 | 3.00842 | 3.00830 | 0.00012 | 1.91080 | 1.91067 | 0.00013 |
| 4 | 0.4 | 3.76945 | 3.76924 | 0.00021 | 2.87079 | 2.87057 | 0.00022 |
| 5 | 0.5 | 4.84957 | 4.84922 | 0.00035 | 4.11381 | 4.11345 | 0.00036 |
| 6 | 0.6 | 6.35084 | 6.35027 | 0.00057 | 5.74845 | 5.74788 | 0.00043 |
| 7 | 0.7 | 8.41277 | 8.41187 | 0.00090 | 7.91957 | 7.91866 | 0.00091 |
| 8 | 0.8 | 11.22507 | 11.22369 | 0.00138 | 10.82128 | 10.81988 | 0.00140 |
| 9 | 0.9 | 15.04503 | 15.04292 | 0.00211 | 14.71443 | 14.71231 | 0.00212 |
| 10 | 1.0 | 20.22087 | 20.21770 | 0.00317 | 19.95020 | 19.94703 | 0.00317 |

The values of $Y_{1,i}$ and $Y_{2,i}$ given in Table 8.7.2 are computed in a similar manner. The values of the exact solutions and the absolute value of the actual errors are also given in Table 8.7.2.

Compared to Euler's method, the fourth-order Runge–Kutta method provides accurate results; however reducing $h$ further gives more accurate results. ■ ■ ■

An $M$th-order initial-value problem (8.1.4a) with (8.1.4b) can be written as a system of first-order ordinary differential equations given by Equation (8.1.5a) with Equation (8.1.5b). The system is solved by using Equation (8.7.3) or Equation (8.7.4).

**EXAMPLE 8.7.2**

Solve $(d^2u/d\theta^2) + u = 1 + 0.01\ u^2$ for $0 \le \theta \le \pi$ with initial conditions $u(0) = 0$ and $du/d\theta\ (0) = 0$. This is equivalent to

$$\frac{dy_1}{d\theta} = y_2$$

$$\frac{dy_2}{d\theta} = 1 - y_1 + 0.01\ y_1^2$$

along with $y_1(0) = y_{1,0} = 0$, $dy_1/d\theta\ (0) = y_2(0) = y_{2,0} = 0$ where $y_1 = u$.

Divide $[0, \pi]$ into 30 equal intervals such that $\theta_0 = 0$, $\theta_1 = \pi/30 = 0.10472, \ldots,$ $\theta_{30} = \pi = 3.14159$ with $h = 0.10472$. We have $f_1(\theta, y_1, y_2) = y_2$ and $f_2(\theta, y_1, y_2) =$

**Table 8.7.3**

| $i$ | $\theta_i$ | $Y_{1,i}$ | $Y_{2,i}$ |
|---|---|---|---|
| 0 | 0.00000 | 0.00000 | 0.00000 |
| 1 | 0.10472 | 0.00548 | 0.10453 |
| 2 | 0.20944 | 0.02185 | 0.20791 |
| 3 | 0.31416 | 0.04894 | 0.30902 |
| 4 | 0.41888 | 0.08645 | 0.40674 |
| 5 | 0.52360 | 0.13398 | 0.50002 |
| 10 | 1.04720 | 0.50010 | 0.86656 |
| 15 | 1.57080 | 1.00096 | 1.00334 |
| 20 | 2.09440 | 1.50437 | 0.87653 |
| 25 | 2.61799 | 1.87870 | 0.52162 |
| 30 | 3.14159 | 2.02689 | 0.03192 |

$1 - y_1 + 0.01\, y_1^2$. For $i = 0$, Equation (8.7.4) becomes

$$\mathbf{Y}_1 = \mathbf{Y}_0 + \frac{1}{6}(\mathbf{k}_{1,0} + 2\mathbf{k}_{2,0} + 2\mathbf{k}_{3,0} + \mathbf{k}_{4,0})$$

where

$$\mathbf{k}_{1,0} = h\mathbf{f}(\theta_0, \mathbf{Y}_0) = h\begin{bmatrix} f_1(\theta_0, Y_{1,0}, Y_{2,0}) \\ f_2(\theta_0, Y_{1,0}, Y_{2,0}) \end{bmatrix} = \begin{bmatrix} Y_{2,0} \\ 1 - Y_{1,0} + 0.01\, Y_{1,0}^2 \end{bmatrix} = \begin{bmatrix} 0 \\ h \end{bmatrix}$$

$$\mathbf{k}_{2,0} = h\mathbf{f}\left(\theta_0 + \frac{h}{2}, \mathbf{Y}_0 + \frac{1}{2}\mathbf{k}_{1,0}\right) = h\begin{bmatrix} f_1\left(\theta_0 + \dfrac{h}{2}, Y_{1,0} + 0, Y_{2,0} + \dfrac{h}{2}\right) \\ f_2\left(\theta_0 + \dfrac{h}{2}, Y_{1,0} + 0, Y_{2,0} + \dfrac{h}{2}\right) \end{bmatrix}$$

$$= h\begin{bmatrix} \dfrac{h}{2} \\ 1 \end{bmatrix} = \begin{bmatrix} \dfrac{h^2}{2} \\ h \end{bmatrix}$$

$$\mathbf{k}_{3,0} = h\mathbf{f}\left(\theta_0 + \frac{h}{2}, \mathbf{Y}_0 + \frac{1}{2}\mathbf{k}_{2,0}\right) = h\begin{bmatrix} f_1\left(\theta_0 + \dfrac{h}{2}, Y_{1,0} + \dfrac{h^2}{4}, Y_{2,0} + \dfrac{h}{2}\right) \\ f_2\left(\theta_0 + \dfrac{h}{2}, Y_{1,0} + \dfrac{h^2}{2}, Y_{2,0} + \dfrac{h}{2}\right) \end{bmatrix}$$

$$= h\begin{bmatrix} \dfrac{h}{2} \\ 1 - \dfrac{h^2}{4} + 0.01\left(\dfrac{h^2}{4}\right)^2 \end{bmatrix} = \begin{bmatrix} \dfrac{h^2}{2} \\ h - \dfrac{h^3}{4} + 0.01\,\dfrac{h^5}{16} \end{bmatrix}$$

$$\mathbf{k}_{4,0} = h\mathbf{f}(\theta_0 + h, \mathbf{Y}_0 + \mathbf{k}_{3,0}) = h\begin{bmatrix} f_1\left(\theta_0 + h, Y_{1,0} + \dfrac{h^2}{2}, Y_{2,0} + h - \dfrac{h^3}{4} + 0.01\,\dfrac{h^5}{16}\right) \\ f_2\left(\theta_0 + h, Y_{1,0} + \dfrac{h^2}{2}, Y_{2,0} + h - \dfrac{h^3}{4} + 0.01\,\dfrac{h^5}{16}\right) \end{bmatrix}$$

$$= h\begin{bmatrix} h - \dfrac{h^3}{4} + 0.01\,\dfrac{h^5}{16} \\ 1 - \dfrac{h^2}{2} + 0.01\,\dfrac{h^5}{4} \end{bmatrix} = \begin{bmatrix} h^2 - \dfrac{h^4}{4} + 0.01\,\dfrac{h^6}{16} \\ h - \dfrac{h^3}{2} + 0.01\,\dfrac{h^5}{4} \end{bmatrix}$$

Hence

$$\mathbf{Y}_1 = \mathbf{Y}_0 + \frac{1}{6}(\mathbf{k}_{1,0} + 2\mathbf{k}_{2,0} + 2\mathbf{k}_{3,0} + \mathbf{k}_{4,0})$$

$$= \begin{bmatrix} 0 \\ 0 \end{bmatrix}$$

$$+ \frac{1}{6}\left( \begin{bmatrix} 0 \\ h \end{bmatrix} + 2\begin{bmatrix} \dfrac{h^2}{2} \\ h \end{bmatrix} + 2\begin{bmatrix} \dfrac{h^2}{2} \\ h - \dfrac{h^3}{4} + 0.01\dfrac{h^5}{16} \end{bmatrix} + \begin{bmatrix} h^2 - \dfrac{h^4}{4} + 0.01\dfrac{h^6}{16} \\ h - \dfrac{h^3}{2} + 0.01\dfrac{h^5}{4} \end{bmatrix} \right)$$

$$= \frac{1}{6}\begin{bmatrix} h^2\left(3 - \dfrac{h^2}{4} + 0.01\dfrac{h^4}{16}\right) \\ h\left(6 - h^2 + 0.01\left(\dfrac{3h^4}{8}\right)\right) \end{bmatrix} = \begin{bmatrix} 0.00548 \\ 0.10453 \end{bmatrix}$$

The values of $Y_{1,i}$ and $Y_{2,i}$ given in Table 8.7.3 are computed in a similar manner.

■ ■ ■

---

EXERCISES

1.  Use the Euler method to approximate the solution of the following initial-value problem at $x = 1$ with $h = 0.25$.

$$\frac{dy_1}{dx} = 4y_2 \qquad y_1(0) = 2$$

$$\frac{dy_2}{dx} = y_1 \qquad y_2(0) = 0$$

2.  Use the second-order Runge–Kutta method to approximate the solution of the following initial-value problem at $x = 0.5$ with $h = 0.25$.

$$\frac{dy_1}{dx} = 4y_2 \qquad y_1(0) = 2$$

$$\frac{dy_2}{dx} = y_1 \qquad y_2(0) = 0$$

3.  Use the fourth-order Runge–Kutta method to approximate the solution of the following initial-value problem at $x = 0.5$ with $h = 0.25$.

$$\frac{dy_1}{dx} = 4y_2 \qquad y_1(0) = 2$$

$$\frac{dy_2}{dx} = y_1 \qquad y_2(0) = 0$$

**4.** Use the Euler method to approximate the solution of the following initial-value problem at $x = 1$ with $h = 0.25$.

$$\frac{d^2y}{dx^2} - 3\frac{dy}{dx} + 2y = x \qquad y(0) = 0 \qquad y'(0) = \frac{1}{2}$$

**5.** Use the second-order Improved Euler method to approximate the solution of the following initial-value problem at $x = 0.5$ with $h = 0.25$.

$$\frac{d^2y}{dx^2} - 3\frac{dy}{dx} + 2y = x \qquad y(0) = 0 \qquad y'(0) = \frac{1}{2}$$

**6.** Use the fourth-order Runge–Kutta method to approximate the solution of the following initial-value problem at $x = 0.5$ with $h = 0.25$.

$$\frac{d^2y}{dx^2} - 3\frac{dy}{dx} + 2y = x \qquad y(0) = 0 \qquad y'(0) = \frac{1}{2}$$

## COMPUTER EXERCISES

For Exercises C.1–C.3, write a subroutine to approximate the solution of a system of $M$ first-order differential equations

$$\frac{dy_i}{dx} = f_i(x, y_1, y_2, \dots, y_M)$$

with initial conditions

$$y_i(x_0) = \alpha_i \quad \text{for } i = 1, 2, \dots, M$$

**C.1.** Use the Euler method.
**C.2.** Use the second-order Runge–Kutta method.
**C.3.** Use the fourth-order Runge–Kutta method.
**C.4.** Write a program to approximate the solution of a system of $M$ first-order differential equations

$$\frac{dy_i}{dx} = f_i(x, y_1, y_2, \dots, y_M)$$

with initial conditions

$$y_i(x_0) = \alpha_i \quad \text{for } i = 1, 2, \dots, M$$

on a given interval $[a, b]$ using Exercises C.1 and C.3 or Exercises C.2 and C.3. Read $a$, $b$, $N$, and $\alpha_i$. Use various $h = (b - a)/N$.
**C.5.** Use Exercise C.4 to approximate the solution at $x = 1$ of the system of three

first-order differential equations given by

$$\frac{dy_1}{dx} = y_1 - y_2 - y_3$$

$$\frac{dy_2}{dx} = y_1 + 3y_2 + y_3$$

$$\frac{dy_3}{dx} = -3y_1 + y_2 - y_3$$

with initial conditions $y_1(0) = 1$, $y_2(0) = 2$, $y_3(0) = 0$.

**C.6.**   Use Exercise C.4 to approximate the solution of the initial-value problem at $x = 1$.

$$\frac{d^2y}{dx^2} - 3\frac{dy}{dx} + 2y = x \qquad y(0) = 0 \qquad y'(0) = \frac{1}{2}$$

**C.7.**   Predator–prey problem: Consider the population of two interacting species. Let $y_1 = y_1(t)$ and $y_2 = y_2(t)$ denote the number of preys and predators respectively at time $t$. Using several simplifying assumptions (Ortega and Poole 1981), we have the system of two nonlinear ordinary differential equations

$$\frac{dy_1}{dt} = ay_1 + by_1 y_2$$

$$\frac{dy_2}{dt} = cy_2 + dy_1 y_2$$

with $a > 0$, $b < 0$, $c < 0$, and $d > 0$. These equations are known as the Lotka–Voltera equations. Use Exercise C.4 to approximate the solution at $t = 1$ of the system given by

$$\frac{dy_1}{dt} = 3y_1 - y_1 y_2$$

$$\frac{dy_2}{dt} = -3y_2 + y_1 y_2$$

with initial conditions:

(a) $y_1(0) = 100$,    $y_2(0) = 200$
(b) $y_1(0) = 100$,    $y_2(0) = 500$
(c) $y_1(0) = 100$,    $y_2(0) = 700$

**C.8.**   The equation of motion of a simple pendulum consisting of mass $m$ suspended by a rod of length $l$ having negligible mass (Nagle and Saff 1989) is given by

$$\frac{d^2\theta}{dt^2} + \frac{g}{l}\sin\theta = 0$$

with initial condition $\theta(0) = \alpha$, $\theta'(0) = 0$.

For small angles, $\sin\theta \approx \theta$, the problem reduces to $(d^2\theta/dt^2) + (g/l)\theta = 0$, which can be solved by classical methods. The period of oscillation $P$ is given by $P = 2\pi\sqrt{l/g}$, which is independent of $\alpha$.

For large $\theta$, you might suspect that the period of a simple pendulum depends on the length $l$ of the rod and the initial displacement $\alpha$. Compare the solutions obtained from solving

$$\frac{d^2\theta}{dt^2} + \frac{g}{l}\sin\theta = 0 \qquad \theta(0) = \alpha \qquad \theta'(0) = 0$$

with those of the linearized problem

$$\frac{d^2\theta}{dt^2} + \frac{g}{l}\theta = 0 \qquad \theta(0) = \alpha \qquad \theta'(0) = 0$$

for $\alpha = 1°$, $5°$, $30°$, $60°$, and $90°$. Use $g = 9.8$ m/s$^2$ and $l = 1$ m. Use Newton's method to find the period of oscillation.

**C.9.** Consider the flow of an incompressible fluid over a semi-infinite plate. By limiting the viscous effects in the region close to the plate, the governing equation of motions and continuity are reduced to

$$2\frac{d^3y}{dt^3} + y\frac{d^2y}{dt^2} = 0$$

Use Exercise C.4 to approximate the solution that satisfies $y(0) = 0$, $y'(0) = 0$, and $y''(0) = 1$ on the closed interval $[0,1]$.

**C.10.** The Bessel equation of order zero is given by

$$x^2\frac{d^2y}{dx^2} + x\frac{dy}{dx} + x^2y = 0$$

Use Exercise C.4 to approximate the solution that satisfies $y(0) = 1$ and $y'(0) = 0$ on the interval $[0, 2]$. Also, using Newton's method, find the value of $x$ such that $y(x) = 0$.

**C.11.** The Lorenz equations (1963) given by

$$\frac{dx}{dt} = -Px + Py$$

$$\frac{dy}{dt} = -y + rx - xz$$

$$\frac{dz}{dt} = -bz + xy$$

where $P$, $r$, and $b$ are constants, played an important role in the development of ideas about chaos. Use Exercise C.4 to approximate its solution with
(a) $x(0) = 1$, $y(0) = 0$, $z(0) = 0$, $r = 131$, $b = 3/8$ and $P = 4$
(b) $x(0) = 1$, $y(0) = 0$, $z(0) = 0$, $r = 28$, $b = 8/3$ and $P = 10$

## SUGGESTED READINGS

J. C. Butcher, *The Numerical Analysis of Ordinary Differential Equations*, John Wiley, Chichester, 1987.

E. Fehlberg, *Klassiche Runge–Kutta formeln funfter und siebenter ordnung mit schritt-weitenkontrolle*, Computing **4** (1965), 93–106.

E. N. Lorenz, *Deterministic non-periodic flow*, J. Atmos. Sci., **20** (1963), 130–141.

M. Lotkin, *On the accuracy of Runge–Kutta's method*, Math. Comp. **5** (1951), 128–133.

S. W. McCuskey, *An Introduction to Advanced Dynamics*, Addison-Wesley Publishing Co., Inc., Reading, MA, 1962.

R. K. Nagle and E. B. Saff, *Fundamentals of Differential Equations*, 2nd ed., The Benjamin/Cummings Publishing Company, Inc., Redwood City, CA, 1989.

J. M. Ortega and W. G. Poole Jr., *An Introduction to Numerical Methods for Differential Equations*, Pitman Publishing Inc., Marshfield, MA, 1981.

S. L. Ross, *Differential Equations*, 2nd ed., Xerox College Publishing, Lexington, MA, 1974.

R. C. Tolman, *Relativity Thermodynamics and Cosmology*, Oxford at the Clarendon Press, 1958.

# 9 INITIAL-VALUE PROBLEMS: MULTISTEP METHODS

## 9.1 INTRODUCTION

The methods discussed in the previous chapter depended only on $Y_i$ to obtain the value of $Y_{i+1}$; therefore those methods are called **single-step methods**. As we move away from $x_0$, the error $|y(x_{i+1}) - Y_{i+1}|$ increases. Since the approximate values at $x_0$, $x_1, \ldots, x_i$ are available, it seems reasonable to use those values to approximate $Y_{i+1}$ accurately. A method that uses $k$ approximate values of $y$ to compute $Y_{i+1}$ is called a **$k$-step method**. So far we have derived methods based on the idea of expanding the exact solution in a Taylor series, while the multistep methods are derived using the idea of numerical integration.

A one-step method uses $y(x_0) = y_0 = Y_0$ to compute $Y_1$. Then $Y_1$ is used to compute $Y_2$ and so on. However, a two-step method requires $Y_0$ and $Y_1$ to compute $Y_2$. Clearly it is not possible to use a two-step method to compute $Y_1$ since not enough information is available. $Y_0, Y_1, \ldots, Y_{k-1}$ must be available to get started with a $k$-step method. These starting values must be computed by other methods.

## 9.2 MULTISTEP METHODS

Let us develop multistep methods to solve an initial-value problem

$$\frac{dy}{dx} = f(x, y) \qquad y(x_0) = Y_0 \tag{9.2.1}$$

Integrate Equation (9.2.1) from $x_i$ to $x_{i+1}$ to get

$$\int_{x_i}^{x_{i+1}} \frac{dy}{dx}\, dx = \int_{x_i}^{x_{i+1}} f(x, y)\, dx$$

Thus

$$y(x_{i+1}) = y(x_i) + \int_{x_i}^{x_{i+1}} f(x, y(x))\, dx \tag{9.2.2}$$

Since $y(x)$ is not known and $f(x, y(x))$ cannot be integrated exactly, we approximate $f(x, y(x))$ by an interpolating polynomial that uses the previously obtained data points $(x_i, f(x_i, y(x_i)))$, $(x_{i-1}, f(x_{i-1}, y(x_{i-1})))$, ..., $(x_{i-k}, f(x_{i-k}, y(x_{i-k})))$.

Let $k = 0$. Then for some $\eta_i$ in $(x_i, x_{i+1})$, Equation (9.2.2) becomes

$$y(x_{i+1}) = y(x_i) + \int_{x_i}^{x_{i+1}} [f(x_i, y(x_i)) + (x - x_i)f'(\eta_i, y(\eta_i(x)))]dx$$

$$= y(x_i) + hf(x_i, y(x_i)) + \frac{h^2}{2}f'(\xi_i, y(\xi_i)) \tag{9.2.3}$$

where $h = x_{i+1} - x_i$ and $x_i < \xi_i < x_{i+1}$. This gives the one-step Euler method

$$Y_{i+1} = Y_i + hf(x_i, Y_i) \tag{9.2.4}$$

Let $k = 1$. Although any interpolating polynomial through $(x_i, f(x_i, y(x_i)))$ and $(x_{i-1}, f(x_{i-1}, y(x_{i-1})))$ can be used, it is very convenient to use the Newton backward difference formula [Equation (4.5.12)]. Let $h = x_{i+1} - x_i = x_i - x_{i-1}$. Then Equation (9.2.2) becomes

$$y(x_{i+1}) = y(x_i) + \int_{x_i}^{x_{i+1}} \{ f(x_i, y(x_i)) + (x - x_i) \frac{\nabla f(x_i, y(x_i))}{h}$$

$$+ \frac{(x - x_i)(x - x_{i-1})}{2!} f''(\xi_i, y(\xi_i)) \}dx$$

$$= y(x_i) + hf(x_i, y(x_i))$$

$$+ \frac{h}{2} \nabla f(x_i, y(x_i)) + \frac{f''(\eta_i, y(\eta_i))}{2!} \int_{x_i}^{x_{i+1}} (x - x_i)(x - x_{i-1})dx$$

$$= y(x_i) + \frac{h}{2} \left\{ 3f(x_i, y(x_i)) - f(x_{i-1}, y(x_{i-1})) \right\} + \frac{5}{12} h^3 f''(\eta_i, y(\eta_i)) \tag{9.2.5}$$

where $x_i < \xi_i$, $\eta_i < x_{i+1}$. This two-step method that uses the information at the points $x_i$ and $x_{i-1}$ is called the **second-order Adams–Bashforth method** and is given by

$$Y_{i+1} = Y_i + \frac{h}{2} \left[ 3f(x_i, Y_i) - f(x_{i-1}, Y_{i-1}) \right] \tag{9.2.6}$$

Similarly for $k = 2$, using three points $(x_i, f(x_i, y(x_i)))$, $(x_{i-1}, f(x_{i-1}, y(x_{i-1})))$, and $(x_{i-2}, f(x_{i-2}, y(x_{i-2})))$, we get (Exercise 4) from Equation (9.2.2)

$$y(x_{i+1}) = y(x_i) + \frac{h}{12} \left\{ 23f(x_i, y(x_i)) - 16f(x_{i-1}, y(x_{i-1})) \right.$$

$$\left. + 5f(x_{i-2}, y(x_{i-2})) \right\} + \frac{3}{8} h^4 f^{(3)}(\eta_i, y(\eta_i)) \tag{9.2.7}$$

For $k = 3$,

$$y(x_{i+1}) = y(x_i) + \frac{h}{24} \left\{ 55f(x_i, y(x_i)) - 59f(x_{i-1}, y(x_{i-1})) + 37f(x_{i-2}, y(x_{i-2})) \right.$$

$$\left. - 9f(x_{i-3}, y(x_{i-3})) \right\} + \frac{251}{720} h^5 f^{(4)}(\xi_i, y(\xi_i)) \tag{9.2.8}$$

For $k = 4$

$$y(x_{i+1}) = y(x_i) + \frac{h}{720} \left\{ 1901 f(x_i, y(x_i)) - 2774 f(x_{i-1}, y(x_{i-1})) + 2616 f(x_{i-2}, y(x_{i-2})) \right.$$

$$\left. - 1274 f(x_{i-3}, y(x_{i-3})) + 251 f(x_{i-4}, y(x_{i-4})) \right\} + \frac{95}{288} h^6 f^{(5)} (\zeta_i, y(\zeta_i)) \quad \textbf{(9.2.9)}$$

In principle, the preceding procedure can be continued to obtain higher-order Adams–Bashforth formulas but the formulas become complex as $k$ increases.

Multistep methods need help getting started. Generally, a $k$-step method must have starting values $Y_0, Y_1, \ldots, Y_{k-1}$. These starting values must be computed by other methods. However, keep in mind that the starting values obtained must be as accurate as those produced by the final method. If a starting method is of lower order, then use a smaller step size to generate accurate starting values.

By using $(x_i, f(x_i, y(x_i)))$, $(x_{i-1}, f(x_{i-1}, y(x_{i-1})))$, $\ldots$, $(x_{i-k}, f(x_{i-k}, y(x_{i-k})))$, we derived the Adams–Bashforth methods. We can also use $(x_{i+1}, f(x_{i+1}, y(x_{i+1})))$, $(x_{i+2}, f(x_{i+2}, y(x_{i+2})))$, $\ldots$ to form an interpolating polynomial. An interpolating polynomial through $(x_{i+1}, f(x_{i+1}, y(x_{i+1})))$, $(x_i, f(x_i, y(x_i)))$, $\ldots$, $(x_{i-k}, f(x_{i-k}, y(x_{i-k})))$ that satisfies $P(x_j) = f(x_j, y(x_j))$ for $j = i + 1, i, \ldots, i - k$ generates a class of methods known as the **Adams–Moulton methods**.

Let $k = 0$. Then replacing $f(x, y(x))$ in Equation (9.2.2) by the interpolating polynomial through $(x_{i+1}, f(x_{i+1}, y(x_{i+1})))$ and $(x_i, f(x_i, y(x_i)))$, we get

$$y(x_{i+1}) = y(x_i) + \int_{x_i}^{x_{i+1}} \left\{ f(x_i, y(x_i)) \frac{x - x_{i+1}}{x_i - x_{i+1}} + f(x_{i+1}, y(x_{i+1})) \frac{x - x_i}{x_{i+1} - x_i} \right.$$

$$\left. + \frac{(x - x_i)(x - x_{i+1})}{2!} f'' (\xi_i(x), y(\xi_i(x))) \right\} dx$$

$$= y(x_i) + \frac{h}{2} \left[ f(x_i, y(x_i)) + f(x_{i+1}, y(x_{i+1})) \right] - \frac{h^3}{12} f'' (\eta_i, y(\eta_i)) \quad \textbf{(9.2.10)}$$

The **second-order Adams–Moulton formula** is given by

$$Y_{i+1} = Y_i + \frac{h}{2} [f(x_i, Y_i) + f(x_{i+1}, Y_{i+1})] \quad \textbf{(9.2.11)}$$

This is also known as the **Trapezoidal method**.

For $k = 1$, using the cubic interpolating polynomial through $(x_{i+1}, f(x_{i+1}, y(x_{i+1})))$, $(x_i, f(x_i, y(x_i)))$, and $(x_{i-1}, f(x_{i-1}, y(x_{i-1})))$ in Equation (9.2.2) gives

$$y(x_{i+1}) = y(x_i) + \frac{h}{12} \left[ 5 f(x_{i+1}, y(x_{i+1})) + 8 f(x_i, y(x_i)) \right.$$

$$\left. - f(x_{i-1}, y(x_{i-1})) \right] - \frac{h^4}{24} f^{(3)} (\zeta_i, y(\zeta_i)) \quad \textbf{(9.2.12)}$$

Similarly for $k = 2$

$$y(x_{i+1}) = y(x_i) + \frac{h}{24} \left[ 9 f(x_{i+1}, y(x_{i+1})) + 19 f(x_i, y(x_i)) - 5 f(x_{i-1}, y(x_{i-1})) \right.$$

$$\left. + f(x_{i-2}, y(x_{i-2})) \right] - \frac{19}{720} h^5 f^{(4)} (\xi_i, y(\xi_i)) \quad \textbf{(9.2.13)}$$

For $k = 3$

$$y(x_{i+1}) = y(x_i) + \frac{h}{720}\left[251f(x_{i+1}, y(x_{i+1})) + 646f(x_i, y(x_i)) - 264f(x_{i-1}, y(x_{i-1}))\right.$$
$$\left. + 106f(x_{i-2}, y(x_{i-2})) - 19f(x_{i-3}, y(x_{i-3}))\right] - \frac{3}{160}h^6 f^{(5)}(\alpha_i, y(\alpha_i)) \quad \textbf{(9.2.14)}$$

Note that in Equation (9.2.11), $f(x_{i+1}, Y_{i+1})$ is not known since $Y_{i+1}$ is not known. Thus $Y_{i+1}$ is defined implicitly in Equation (9.2.11). Similarly, $Y_{i+1}$ is defined implicitly in other Adams–Moulton formulas, and therefore they are called **implicit methods** while Adams–Bashforth methods define $Y_{i+1}$ **explicitly**.

**EXAMPLE 9.2.1**   Approximate the solution of the initial-value problem

$$\frac{dy}{dx} = y - x \qquad y(0) = 0$$

using the fourth-order (a) Adams–Bashforth method and (b) Adams–Moulton method. Use $h = 0.1$.

(a) Since $f(x, y) = y - x$, the fourth-order Adams–Bashforth formula [Equation (9.2.8)] reduces to

$$Y_{i+1} = Y_i + \frac{h}{24}\{55(Y_i - x_i) - 59(Y_{i-1} - x_{i-1}) + 37(Y_{i-2} - x_{i-2}) - 9(Y_{i-3} - x_{i-3})\}$$

$$\textbf{(9.2.15)}$$

for $i = 3, 4, \ldots$.

The fourth-order Runge–Kutta method was used to obtain the starting values $Y_1 = -0.00517$, $Y_2 = -0.02140$, and $Y_3 = -0.04986$. These values are given in Table 8.5.3.

For $i = 3$ and $h = 0.1$, Equation (9.2.15) reduces to

$$Y_4 = Y_3 + \frac{h}{24}\left\{55(Y_3 - x_3) - 59(Y_2 - x_2) + 37(Y_1 - x_1) - 9(Y_0 - x_0)\right\}$$
$$= -0.04986 + \frac{0.1}{24}\left\{55(-0.04986 - 0.3) - 59(-0.02140 - 0.2)\right.$$
$$\left. + 37(-0.00517 - 0.1) - 9(0 - 0)\right\} = -0.09182$$

Similarly we compute the other values given in Table 9.2.1. The absolute value of the actual errors using the exact solution $y(x) = 1 + x - e^x$ are also given in Table 9.2.1.

(b) Since $f(x, y) = y - x$, the fourth-order Adams–Moulton formula [Equation (9.2.13)] reduces to

$$Y_{i+1} = Y_i + \frac{h}{24}\{9(Y_{i+1} - x_{i+1}) + 19(Y_i - x_i) - 5(Y_{i-1} - x_{i-1}) + (Y_{i-2} - x_{i-2})\}$$

$$\textbf{(9.2.16)}$$

for $i = 2, 3, \ldots$.

**Table 9.2.1**

| $x_i$ | Fourth-Order Adams–Bashforth | \|Actual Error\| | Fourth-Order Adams–Moulton | \|Actual Error\| |
|---|---|---|---|---|
| 0.4 | $-0.09182$ | 0.00000 | $-0.09183$ | 0.00001 |
| 0.5 | $-0.14871$ | 0.00001 | $-0.14872$ | 0.00000 |
| 0.6 | $-0.22210$ | 0.00002 | $-0.22212$ | 0.00000 |
| 0.7 | $-0.31373$ | 0.00002 | $-0.31376$ | 0.00001 |
| 0.8 | $-0.42551$ | 0.00003 | $-0.42555$ | 0.00001 |
| 0.9 | $-0.55956$ | 0.00004 | $-0.55961$ | 0.00001 |
| 1.0 | $-0.71823$ | 0.00005 | $-0.71829$ | 0.00001 |

Solving Equation (9.2.16) for $Y_{i+1}$ yields

$$Y_{i+1} = \frac{1}{\left(1 - \dfrac{9h}{24}\right)} \left\{ Y_i + \frac{h}{24} \left[ -9x_{i+1} + 19(Y_i - x_i) \right. \right.$$

$$\left. \left. - 5(Y_{i-1} - x_{i-1}) + (Y_{i-2} - x_{i-2}) \right] \right\} \qquad \textbf{(9.2.17)}$$

for $i = 2, 3, \ldots$.

We are able to solve Equation (9.2.16) for $Y_{i+1}$, since $f(x, y) = y - x$ is a linear function. Normally it is difficult to solve implicit formulas for $Y_{i+1}$.

The starting values $Y_1 = -0.00517$, $Y_2 = -0.02140$, and $Y_3 = -0.04986$ obtained using the fourth-order Runge–Kutta method are used in order to compare results with (a), although we need only the values of $Y_0$, $Y_1$, and $Y_2$ to start. For $i = 3$ and $h = 0.1$, Equation (9.2.17) reduces to

$$Y_4 = \frac{1}{\left(1 - \dfrac{9h}{24}\right)} \left\{ Y_3 + \frac{h}{24} \left[ -9x_4 + 19(Y_3 - x_3) - 5(Y_2 - x_2) + (Y_1 - x_1) \right] \right\}$$

$$= 1.03896 \left\{ -0.04986 + \frac{0.1}{24} \left[ -9(0.4) + 19(-0.04986 - 0.3) \right. \right.$$

$$\left. \left. - 5(-0.0214 - 0.2) + (-0.00517 - 0.1) \right] \right\} = -0.09183$$

In Table 9.2.1, we list the computed values $Y_i$ obtained using the fourth-order Adams–Moulton method along with the absolute value of the actual errors. ∎ ∎ ∎

The truncation error of the Adams–Bashforth method is $(251/720)\, h^5 f^{(4)}(\eta_i, y(\eta_i))$, while the truncation error of the Adams–Moulton method is $-(19/720)\, h^5 f^{(4)}(\xi_i, y(\xi_i))$. Thus the absolute truncation error of the Adams–Moulton method is 0.076 times that of the Adams–Bashforth method. This can also be verified from Table 9.2.1. This is one of the main reasons for using the Adams–Moulton method, although it is an implicit method.

Instead of integrating Equation (9.2.1) from $x_i$ to $x_{i+1}$, we integrate it from $x_{i-k}$ to $x_{i+1}$, where $k$ is a positive integer, to derive a number of other multistep formulas. Thus

$$y(x_{i+1}) = y(x_{i-k}) + \int_{x_{i-k}}^{x_{i+1}} f(x, y)\,dx \qquad (9.2.18)$$

For $k = 0$, Equation (9.2.18) reduces to

$$y(x_{i+1}) = y(x_i) + \int_{x_i}^{x_{i+1}} f(x, y)\,dx \qquad (9.2.19)$$

By approximating $f(x, y)$ in Equation (9.2.19) by an interpolating polynomial passing through $(x_i, f(x_i, y(x_i)))$ and $(x_{i+1}, f(x_{i+1}, y(x_{i+1})))$, it can be verified that Equation (9.2.19) reduces to Equation (9.2.10).

Further for $k = 1$, Equation (9.2.18) reduces to

$$y(x_{i+1}) = y(x_{i-1}) + \int_{x_{i-1}}^{x_{i+1}} f(x, y)\,dx \qquad (9.2.20)$$

By approximating $f(x, y)$ in Equation (9.2.20) by an interpolating polynomial passing through $(x_{i-1}, f(x_{i-1}, y(x_{i-1})))$, $(x_i, f(x_i, y(x_i)))$, and $(x_{i+1}, f(x_{i+1}, y(x_{i+1})))$, it can be verified (Exercise 12) that Equation (9.2.20) reduces to

$$y(x_{i+1}) = y(x_{i-1}) + \frac{h}{3}\,[f(x_{i+1}, y(x_{i+1})) + 4f(x_i, y(x_i))$$
$$+ f(x_{i-1}, y(x_{i-1}))] - \frac{h^5}{90} f^{(4)}(\eta_i, y(\eta_i)) \qquad (9.2.21)$$

This is known as **Simpson's formula**.

For $k = 3$, Equation (9.2.18) reduces to

$$y(x_{i+1}) = y(x_{i-3}) + \int_{x_{i-3}}^{x_{i+1}} f(x, y)\,dx \qquad (9.2.22)$$

We approximate $f(x, y)$ in Equation (9.2.22) by an interpolating polynomial that passes through $(x_{i-2}, f(x_{i-2}, y(x_{i-2})))$, $(x_{i-1}, f(x_{i-1}, y(x_{i-1})))$, and $(x_i, f(x_i, y(x_i)))$. It can be verified (Exercise 13) that Equation (9.2.22) becomes

$$y(x_{i+1}) = y(x_{i-3}) + \frac{4h}{3}\,[2f(x_i, y(x_i)) - f(x_{i-1}, y(x_{i-1}))$$
$$+ 2f(x_{i-2}, y(x_{i-2}))] + \frac{28}{90}\,h^5 f^{(4)}(\xi_i, y(\xi_i)) \qquad (9.2.23)$$

This is known as **Milne's formula**.

All formulas derived so far in this section are special cases of

$$Y_{i+1} = \alpha_1 Y_i + \alpha_2 Y_{i-1} + \cdots + \alpha_k Y_{i-k+1} + h[\beta_0 f(x_{i+1}, Y_{i+1}) + \beta_1 f(x_i, Y_i)$$
$$+ \cdots + \beta_k f(x_{i-k+1}, Y_{i-k+1})] \quad \text{for } i = k, k+1, \ldots \qquad (9.2.24)$$

Since Equation (9.2.24) is a linear combination of $Y_i$ and $f(x_i, Y_i)$, Equation (9.2.24) is called a **linear $k$-step method**. If $\beta_0 \neq 0$, then Equation (9.2.24) is said to be **implicit**; if $\beta_0 = 0$, then Equation (9.2.24) is said to be **explicit**. Thus Adams–Moulton methods are implicit while Adams–Bashforth methods are explicit.

## EXERCISES

In Exercises 1–3, (a) use the second-order Adams–Bashforth method to approximate the solution of the given initial-value problem on the given interval and (b) compare (a) with the exact solution. Use the exact solution for the starting values. Use $h = 0.1$.

1. $dy/dx = y + x$, $y(0) = 0$ on $[0, 0.4]$; exact solution: $y(x) = -1 - x + e^x$.
2. $dy/dx = -xy$, $y(0) = 1$ on $[0, 0.4]$; exact solution: $y(x) = e^{-x^2/2}$.
3. $dy/dx = (x - y)/x$, $y(1) = 5/2$ on $[1, 1.4]$; exact solution: $y(x) = (x/2) + (2/x)$.
4. Derive the third-order Adams–Bashforth formula [Equation (9.2.7)].
5. Use the second-order Adams–Moulton method to approximate the solution of the initial-value problem $dy/dx = y + x$, $y(0) = 0$ on $[0, 0.4]$. Use $h = 0.1$. Use the exact solution $y(x) = -1 - x + e^x$ for starting values.
6. Use the second-order Adams–Moulton method to approximate the solution of the initial-value problem $dy/dx = y^2$, $y(1) = -1$ on $[1, 1.4]$. Use $h = 0.1$. Use the exact solution $y(x) = -1/x$ for starting values.
7. Verify Equation (9.2.9).
8. Verify Equation (9.2.13).
9. Derive Equation (9.2.5) using the Lagrange Polynomial Approximation Theorem 4.3.1.
10. An alternate method of deriving the Adams–Bashforth method is by the method of undetermined coefficients. Assume the form $Y_{i+1} = Y_i + af(x_i, Y_i) + bf(x_{i-1}, Y_{i-1})$ for the derivation of Equation (9.2.5). Expand $y'(x_{i-1}) = f(x_{i-1}, y(x_{i-1}))$ and $y_{i+1}$ in a Taylor series about $x_i$ and compare the coefficients of $h$ and $h^2$ to obtain $a$ and $b$.
11. Using the method of undetermined coefficients outlined in Exercise 10, derive the third-order Adams–Bashforth method

$$Y_{i+1} = Y_i + \frac{h}{24}\{23f(x_i, Y_i) - 16(x_{i-1}, Y_{i-1}) + 5f(x_{i-2}, Y_{i-2})\}$$

12. Verify Equation (9.2.21).
13. Verify Equation (9.2.23).

## COMPUTER EXERCISES

For Exercises C.1 and C.2, write a subroutine to approximate a solution of an initial-value problem

$$\frac{dy}{dx} = f(x, y) \qquad y(x_0) = y_0$$

C.1. Use the second-order Adams–Bashforth method.
C.2. Use the fourth-order Adams–Bashforth method.
C.3. Using Exercises C.1 and C.2, write a program to approximate a solution of an initial-value problem $dy/dx = f(x, y)$, $y(x_0) = y_0$ on a given interval $[x_0 = a,$

$b$]. Read $a$, $b$, $N$, and $y_0$.Use various $N$. Use Exercise C.2, p. 258 for starting values.

**C.4.**  Using Exercise C.3, approximate the solution of the initial-value problem $dy/dx = 1 - x + 4y$, $y(0) = 1$ for $0 \le x \le 1$. Compare your results with the exact solution and with the second-order and fourth-order Runge–Kutta methods.

**C.5.**  Using Exercise C.3, approximate the solution of the initial-value problem $dy/dx = -50y + 100$, $y(0) = 0$ for $0 \le x \le 1$. Compare your results with the exact solution and analyze your output.

**C.6.**  Using the second- and fourth-order Adams–Moulton methods, approximate the solution of the initial-value problem $dy/dx = 1 - x + 4y$, $y(0) = 1$ for $0 \le x \le 1$. Compare your results with the exact solution and with the second-order and fourth-order Runge–Kutta methods.

**C.7.**  Using the second- and fourth-order Adams–Moulton methods, approximate the solution of the initial-value problem $dy/dx = -50y + 100$, $y(0) = 0$ for $0 \le x \le 1$. Compare your results with the exact solution and analyze your output.

## 9.3  PREDICTOR-CORRECTOR METHODS

We derived multistep methods in Section 9.2. The absolute local truncation error of the fourth-order Adams–Moulton method

$$Y_{i+1} = Y_i + \frac{h}{24} [9f(x_{i+1}, Y_{i+1}) + 19f(x_i, Y_i) - 5f(x_{i-1}, Y_{i-1}) + f(x_{i-2}, Y_{i-2})]$$

**(9.3.1)**

for $i = 2, 3, \ldots$ is less than that of the fourth-order Adams–Bashforth method

$$Y_{i+1} = Y_i + \frac{h}{24} [55f(x_i, Y_i) - 59f(x_{i-1}, Y_{i-1}) + 37f(x_{i-2}, Y_{i-2}) - 9f(x_{i-3}, Y_{i-3})]$$

**(9.3.2)**

for $i = 3, 4, \ldots$ . Between Equations (9.3.1) and (9.3.2), Equation (9.3.1) is preferable, but it is an implicit formula. If $f(x, y)$ is nonlinear, then generally it is difficult to solve Equation (9.3.1) explicitly for $Y_{i+1}$. However, Equation (9.3.1) is a nonlinear equation with a root $Y_{i+1}$ and can be solved by the techniques of Chapter 2. For a fixed $i$, $Y_{i+1}$ is the solution of

$$y = g(y) \qquad\qquad \text{(9.3.3)}$$

where

$$g(y) = Y_i + \frac{h}{24} [9f(x_{i+1}, y) + 19f(x_i, Y_i) - 5f(x_{i-1}, Y_{i-1}) + f(x_{i-2}, Y_{i-2})]$$

To solve Equation (9.3.3), it is very convenient to use the fixed-point iteration method

$$y^{(k+1)} = g(y^{(k)}) \quad \text{for } k = 0, 1, \ldots \qquad\qquad \text{(9.3.4)}$$

since $Y_{i+1}$ is a fixed point of $g$. If $|g'(y)| < 1$ for all $y$ with $|y - Y_{i+1}| \le |y^{(0)} - Y_{i+1}|$, then Equation (9.3.4) converges. Since $g'(y) = (9h/24) \, \partial f/\partial y$, Equation (9.3.4) converges if $h < 1/(\frac{9}{24} \mid \frac{\partial f}{\partial y} \mid)$ and $y^{(0)}$ is sufficiently close to $Y_{i+1}$. Thus by selecting $y^{(0)}$ properly, Equation (9.3.4) converges without using many iterations. We calculate $y^{(0)}$ using Equation (9.3.2)

$$Y_{i+1}^{(0)} = Y_i + \frac{h}{24}[55f(x_i, Y_i) - 59f(x_{i-1}, Y_{i-1}) + 37f(x_{i-2}, Y_{i-2}) - 9f(x_{i-3}, Y_{i-3})]$$
$$\textbf{(9.3.5)}$$

This approximation is improved using Equation (9.3.1)

$$Y_{i+1}^{(1)} = Y_i + \frac{h}{24}[9f(x_{i+1}, Y_{i+1}^{(0)}) + 19f(x_i, Y_i) - 5f(x_{i-1}, Y_{i-1}) + f(x_{i-2}, Y_{i-2})]$$
$$\textbf{(9.3.6)}$$

This corrected value can be corrected again using Equation (9.3.1), but it will not improve $Y_{i+1}^{(1)}$ as an approximation of $y(x_{i+1})$. We use Equation (9.3.5) to predict a value of $Y_{i+1}$ and therefore Equation (9.3.5) is known as a predictor. We substitute the predicted value $Y_{i+1}^{(0)}$ into the right-hand side of the implicit formula (9.3.6) to get a corrected value $Y_{i+1}^{(1)}$ and therefore Equation (9.3.6) is known as a corrector. A combination of an explicit method to predict and an implicit method to correct is known as a predictor-corrector method. It has been shown (Henrici 1962) that if the predictor method has at least the order of the corrector method, then one iteration is sufficient to preserve the asymptotic accuracy of the corrector method.

A commonly used predictor-corrector method is the combination of the fourth-order Adams–Bashforth formula as a predictor and the fourth-order Adams–Moulton formula as a corrector. Thus

$$Y_{i+1}^{(p)} = Y_i + \frac{h}{24}\{55f(x_i, Y_i) - 59f(x_{i-1}, Y_{i-1}) + 37f(x_{i-2}, Y_{i-2}) - 9f(x_{i-3}, Y_{i-3})\}$$

$$Y_{i+1} = Y_i + \frac{h}{24}\{9f(x_{i+1}, Y_{i+1}^{(p)}) + 19f(x_i, Y_i) - 5f(x_{i-1}, Y_{i-1}) + f(x_{i-2}, Y_{i-2})\}$$
$$\textbf{(9.3.7)}$$

Equation (9.3.7) is widely used in combination with the fourth-order Runge–Kutta method as a starter. Like the fourth-order Runge–Kutta method, Equation (9.3.7) is one of the most reliable and widely used for the numerical solution of an initial-value problem.

**Algorithm**

**Subroutine** Precor $(a, b, y0, f, \mathbf{Y}, N)$
Comment: This program approximates the solution of an initial-value problem $dy/dx = f(x, y), y(x_0) = y0$ on a given interval $[x_0 = a, b]$ using the predictor-corrector method given by Equation (9.3.7). Starting values are computed by the subroutine Runge4 (fourth-order Runge–Kutta method) given in Section 8.5. $N$ is the number of subintervals of $[a, b]$ and $\mathbf{Y} = [Y_0, Y_1, \ldots, Y_N]$.
  1 **Print** $a, b, y0, N$
  2 $h = (b - a)/N$
  3 $x = a$
  4 $Y_0 = y0$
Comment: The next six steps are for calling the **subroutine** Runge4 from Section 8.5 which computes $Y_1, Y_2,$ and $Y_3$.
  5 **Repeat** steps 6–10 **with** $i = 1, 2, 3$
     6 $b2 = a + ih$

```
 7 b1 = b2 − h
 8 Call Subroutine Runge4(b1, b2, 1, y0, f, D)
 9 Y_i = D
10 y0 = D
11 Repeat steps 12–18 with i = 3, 4, . . . , N − 1
   12 x = a + ih
   13 temp = f(x, Y_i)
   14 temp1 = f(x − h, Y_{i−1})
   15 temp2 = f(x − 2h, Y_{i−2})
   16 P = Y_i + (h/24)[55 temp − 59 temp1 + 37 temp2 − 9f(x − 3h, Y_{i−3})]
   17 Y_{i+1} = Y_i + (h/24)[9f(x + h, P) + 19 temp − 5 temp1 + temp2]
   18 Print i + 1, x + h, Y_{i+1}
19 Print 'finished' and exit.
```
◀

In practice, we do not solve Equation (9.2.11) for $Y_{i+1}$ but rather combine it with the first-order Adams–Bashforth formula [Equation (9.2.4)] as a predictor. Thus

$$Y_{i+1}^{(p)} = Y_i + h f(x_i, Y_i)$$

$$Y_{i+1} = Y_i + \frac{h}{2}\{f(x_i, Y_i) + f(x_{i+1}, Y_{i+1}^{(p)})\} \tag{9.3.8}$$

**EXAMPLE 9.3.1**

Approximate the solution of the initial-value problem

$$\frac{dy}{dx} = y - x \qquad y(0) = 0$$

using the fourth-order predictor-corrector method. Use $h = 0.1$.
Since $f(x, y) = y - x$, Equation (9.3.7) reduces to

$$Y_{i+1}^{(p)} = Y_i + \frac{h}{24}\left\{55(Y_i - x_i) - 59(Y_{i-1} - x_{i-1}) + 37(Y_{i-2} - x_{i-2}) - 9(Y_{i-3} - x_{i-3})\right\}$$

$$Y_{i+1} = Y_i + \frac{h}{24}\left\{9(Y_{i+1}^{(p)} - x_{i+1}) + 19(Y_i - x_i) - 5(Y_{i-1} - x_{i-1}) + (Y_{i-2} - x_{i-2})\right\}$$
$$\tag{9.3.9}$$

The fourth-order Runge–Kutta method was used to obtain the starting values $Y_1 = -0.00517, Y_2 = -0.02140$, and $Y_3 = -0.04986$. These values are given in Table 8.5.2.
For $i = 3$, Equation (9.3.9) reduces to

$$Y_4^{(p)} = Y_3 + \frac{h}{24}\left\{55(Y_3 - x_3) - 59(Y_2 - x_2) + 37(Y_1 - x_1) - 9(Y_0 - x_0)\right\}$$

$$= -0.04986 + \frac{0.1}{24}\left\{55(-0.04968 - 0.3) - 59(-0.02140 - 0.2)\right.$$

$$\left. + 37(-0.00517 - 0.1) - 9(0 - 0)\right\} = -0.09182$$

$$Y_4 = Y_3 + \frac{h}{24}\left\{9(Y_4^{(p)} - x_4) + 19(Y_3 - x_3) - 5(Y_2 - x_2) + (Y_1 - x_1)\right\}$$

$$= -0.04986 + \frac{0.1}{24}\left\{9(-0.09182 - 0.4) + 19(-0.04968 - 0.3)\right.$$

$$\left. - 5(-0.0214 - 0.2) + (-0.00517 - 0.1)\right\} = -0.09182$$

**Table 9.3.1**

| $x_i$ | $Y_i$ Using Predictor-Corrector | \|Actual Error\| | $Y_i$ Using Runge–Kutta | \|Actual Error\| |
|---|---|---|---|---|
| 0.0 | 0.00000 | 0.00000 | 0.00000 | 0.00000 |
| 0.1 | | | −0.00517 | 0.00000 |
| 0.2 | | | −0.02140 | 0.00000 |
| 0.3 | | | −0.04986 | 0.00000 |
| 0.4 | −0.09182 | 0.00000 | −0.09182 | 0.00000 |
| 0.5 | −0.14872 | 0.00000 | −0.14872 | 0.00000 |
| 0.6 | −0.22212 | 0.00000 | −0.22212 | 0.00000 |
| 0.7 | −0.31375 | 0.00000 | −0.31375 | 0.00000 |
| 0.8 | −0.42554 | 0.00000 | −0.42554 | 0.00000 |
| 0.9 | −0.55960 | 0.00000 | −0.55960 | 0.00000 |
| 1.0 | −0.71828 | 0.00000 | −0.71828 | 0.00000 |

The difference between $Y_4^{(p)}$ and $Y_4$ is negligible in this case. Similarly we compute the other values. The computed values of $Y_i$ along with the absolute value of the actual errors are given in Table 9.3.1. For comparison, the values computed by using the fourth-order Runge–Kutta method are also given.

If we compare the absolute value of the actual errors in Tables 9.3.1 and 9.2.1, Table 9.3.1 shows that the predictor-corrector method provides better accuracy without a substantial amount of extra work.   ■ ■ ■

## Error Estimation

We adjusted step size $h$ to control the local truncation error in a single-step method by two approximations at $x_{i+1}$. When we use the predictor-corrector method, two approximations are available at $x_{i+1}$. We can use these approximations to adjust the step size $h$ to control the local truncation error. The local truncation error from Equations (9.2.8) and (9.2.13) in $Y_{i+1}^{(p)}$ and $Y_{i+1}$, respectively, is given by

$$y(x_{i+1}) - Y_{i+1}^{(p)} = \frac{251}{720} h^5 f^{(4)}(\xi_i, y(\xi_i)) \tag{9.3.10}$$

and

$$y(x_{i+1}) - Y_{i+1} = -\frac{19}{720} h^5 f^{(4)}(\zeta_i, y(\zeta_i)) \tag{9.3.11}$$

where $\xi_i \in (x_{i-3}, x_i)$ and $\zeta_i \in (x_{i-2}, x_{i+1})$. In general, $\xi_i$ and $\zeta_i$ are not equal, but if $h$ is small then we can assume that $f^{(4)}$ is not likely to vary much. We can assume $f^{(4)}(\xi_i, y(\xi_i)) \approx f^{(4)}(\zeta_i, y(\zeta_i)) \approx f^{(4)}(\xi, y(\xi))$. Subtracting Equation (9.3.11) from Equation (9.3.10) yields

$$Y_{i+1} - Y_{i+1}^{(p)} \approx \frac{3}{8} h^5 f^{(4)}(\xi, y(\xi))$$

and so

$$f^{(4)}(\xi, y(\xi)) \approx \frac{8(Y_{i+1} - Y_{i+1}^{(p)})}{3h^5} \tag{9.3.12}$$

If we want the "new" step size $h_{\text{new}}$ to be as large as possible so that the local truncation error per unit step does not exceed a predetermined constant tolerance $\epsilon$, then

$$\frac{\left|y(x_i + h_{\text{new}}) - Y_{i+h_{\text{new}}}\right|}{h_{\text{new}}} = \frac{19}{720}\left|f^{(4)}(\xi, y(\xi))\right|h_{\text{new}}^4 \approx \frac{19\left|Y_{i+1} - Y_{i+1}^{(p)}\right|}{270\, h^5} h_{\text{new}}^4 \leq \epsilon$$

Consequently

$$h_{\text{new}} \leq h \sqrt[4]{\frac{270\, \epsilon h}{19\left|Y_{i+1} - Y_{i+1}^{(p)}\right|}}$$

Since a number of approximations are used in this derivation, we choose

$$h_{\text{new}} = \alpha h \sqrt[4]{\frac{270\, \epsilon h}{19\left|Y_{i+1} - Y_{i+1}^{(p)}\right|}} \tag{9.3.13}$$

where $\alpha$ is a suitable adjustment factor. Usually $\alpha = 0.9$.

A change in the step size $h$ requires restarting the algorithm with four new points with the new spacing. This is expensive, therefore it is a common practice to ignore step size change as long as the local truncation error per unit step is between $\epsilon/5$ and $\epsilon$. Thus

$$\frac{\epsilon}{5} \leq \frac{\left|y(x_{i+1}) - Y_{i+1}\right|}{h} \approx \frac{19\left|Y_{i+1} - Y_{i+1}^{(p)}\right|}{270\, h} < \epsilon \tag{9.3.14}$$

## Systems of Differential Equations and Higher-Order Differential Equations

We could apply the predictor-corrector methods for a single first-order ordinary differential equation to approximate the solution of a system

$$\frac{d\mathbf{y}}{dx} = \mathbf{f}(x, \mathbf{y}) \qquad \mathbf{y}(\mathbf{x}_0) = \boldsymbol{\alpha} \tag{9.3.15}$$

where

$$\mathbf{y}(x) = \begin{bmatrix} y_1(x) \\ y_2(x) \\ \vdots \\ y_M(x) \end{bmatrix} \qquad \mathbf{f}(x, \mathbf{y}) = \begin{bmatrix} f_1(x, y_1, y_2, \ldots, y_M) \\ f_2(x, y_1, y_2, \ldots, y_M) \\ \vdots \\ f_M(x, y_1, y_2, \ldots, y_M) \end{bmatrix} \qquad \boldsymbol{\alpha} = \begin{bmatrix} \alpha_1 \\ \alpha_2 \\ \vdots \\ \alpha_M \end{bmatrix}$$

with vector quantities rather than scalar quantities; for example, Equation (9.3.8)

$$\mathbf{Y}_{i+1}^{(p)} = \mathbf{Y}_i + h\mathbf{f}(x_i, \mathbf{Y}_i)$$

$$\mathbf{Y}_{i+1} = \mathbf{Y}_i + \frac{h}{2}[\mathbf{f}(x_i, \mathbf{Y}_i) + \mathbf{f}(x_{i+1}, \mathbf{Y}_{i+1}^{(p)})] \tag{9.3.16}$$

for $i = 0, 1, \ldots$ with $\mathbf{Y}(x_0) = \mathbf{Y}_0 = \boldsymbol{\alpha}$ gives the approximate solution of Equation (9.3.15). The fourth-order predictor-corrector method is given by

$$\mathbf{Y}_{i+1}^{(p)} = \mathbf{Y}_i + \frac{h}{24}\{55\mathbf{f}(x_i, \mathbf{Y}_i) - 59\mathbf{f}(x_{i-1}, \mathbf{Y}_{i-1}) + 37\mathbf{f}(x_{i-2}, \mathbf{Y}_{i-2}) - 9\mathbf{f}(x_{i-3}, \mathbf{Y}_{i-3})\}$$

$$\mathbf{Y}_{i+1} = \mathbf{Y}_i + \frac{h}{24}\{9\mathbf{f}(x_{i+1}, \mathbf{Y}_{i+1}^{(p)}) + 19\mathbf{f}(x_i, \mathbf{Y}_i) - 5\mathbf{f}(x_{i-1}, \mathbf{Y}_{i-1}) + \mathbf{f}(x_{i-2}, \mathbf{Y}_{i-2})\}$$

**(9.3.17)**

for $i = 3, 4, \ldots$.

**Algorithm**

**Subroutine** Sypc $(a, b, N, M, \boldsymbol{\alpha}, \mathbf{Y}_N, \mathbf{f})$
Comment: This program approximates the solution of a system of $M$ first-order ordinary differential equations $dy_j/dx = f_j(x, y_1, y_2, \ldots, y_M)$ with initial conditions $y_j = \alpha_j$ for $j = 1, 2, \ldots, M$ on a given interval $[a, b]$ where $x_0 = a$ and $x_N = b$ using the fourth-order predictor-corrector method given by Equation (9.3.17). We have $\mathbf{Y} = [\mathbf{Y}_1, \mathbf{Y}_2, \ldots, \mathbf{Y}_M]$, $\mathbf{P} = [P_1, P_2, \ldots, P_M]$, $\boldsymbol{\alpha} = [\alpha_1, \alpha_2, \ldots, \alpha_M]$, and $\mathbf{f} = [f_1(x, \mathbf{Y}), f_2(x, \mathbf{Y}), \ldots, f_M(x, \mathbf{Y})]$. We denote the value of variable $\mathbf{Y}_j$ at $x_i$ by $Y_{j,i}$.
  1 **Print** $a, b, N, M, \boldsymbol{\alpha}$
  2 $h = (b - a)/N$
  3 $x = a$
  4 **Repeat** step 5 **with** $j = 1, 2, \ldots, M$
        5 $Y_{j,0} = \alpha_j$
Comment: The next seven steps are for calling **Subroutine** System from Section 8.7 which computes $Y_{j,i}$ for $i = 1, 2, 3$ and $j = 1, 2, \ldots, M$.
  6 **Repeat** steps 7–12 **with** $i = 1, 2,$ and 3
        7 $x1 = a + (i - 1)h$
        8 $x2 = a + ih$
        9 **Call** System $(x1, x2, 1, M, \boldsymbol{\alpha}, \mathbf{P}, \mathbf{f})$
       10 **Repeat** steps 11–12 **with** $j = 1, 2, \ldots, M$
             11 $Y_{j,i} = P_j$
             12 $\alpha_j = P_j$
 13 **Repeat** steps 14–21 **with** $i = 3, 4, \ldots, N - 1$
       14 $x = a + ih$
       15 **Repeat** steps 16–21 **with** $j = 1, 2, \ldots, M$
             16 temp $= f_j(x, Y_{1,i}, Y_{2,i}, \ldots, Y_{M,i})$
             17 temp1 $= f_j(x - h, Y_{1,i-1}, Y_{2,i-1}, \ldots, Y_{M,i-1})$
             18 temp2 $= f_j(x - 2h, Y_{1,i-2}, Y_{2,i-2}, \ldots, Y_{M,i-2})$
             19 $Pr = Y_{j,i} + (h/24)[55\,\text{temp} - 59\,\text{temp1} + 37\,\text{temp2} - 9f_j(x - 3h, Y_{j,i-3})]$
             20 $Y_{j,i+1} = Y_{j,i} + (h/24)[9f_j(x + h, Pr) + 19\,\text{temp} - 5\,\text{temp1} + \text{temp2}]$
             21 **Print** $i + 1, x + h, Y_{j,i+1}$    ◀
 22 **Print** 'finished' and exit.

**EXAMPLE 9.3.2**

Approximate the solution of

$$\frac{dy_1}{dx} = \frac{1}{2}(y_1 + 5y_2)$$

$$\frac{dy_2}{dx} = \frac{1}{2}(5y_1 + y_2)$$

with initial conditions $y_1(0) = 2$ and $y_2(0) = 0$ for $0 \le x \le 1$ using (a) Equation (9.3.16) and (b) Equation (9.3.17). Use $h = 0.1$.

Divide [0, 1] into ten equal intervals such that $x_0 = 0$, $x_1 = 0.1, \ldots, x_{10} = 1.0$. (a) Equation (9.3.16) becomes

$$
\begin{bmatrix} Y^{(p)}_{1,i+1} \\ Y^{(p)}_{2,i+1} \end{bmatrix} = \begin{bmatrix} Y_{1,i} + h\,f_1(x_i, Y_{1,i}, Y_{2,i}) \\ Y_{2,i} + h\,f_2(x_i, Y_{1,i}, Y_{2,i}) \end{bmatrix} = \begin{bmatrix} Y_{1,i} + \dfrac{h}{2}\,(Y_{1,i} + 5Y_{2,i}) \\ Y_{2,i} + \dfrac{h}{2}\,(5Y_{1,i} + Y_{2,i}) \end{bmatrix} \quad (9.3.18)
$$

and

$$
\begin{bmatrix} Y_{1,i+1} \\ Y_{2,i+1} \end{bmatrix} = \begin{bmatrix} Y_{1,i} + \dfrac{h}{2}\,[f_1(x_i, Y_{1,i}, Y_{2,i}) + f_1(x_{i+1}, Y^{(p)}_{1,i+1}, Y^{(p)}_{2,i+1})] \\ Y_{2,i} + \dfrac{h}{2}\,[f_2(x_i, Y_{1,i}, Y_{2,i}) + f_2(x_{i+1}, Y^{(p)}_{1,i+1}, Y^{(p)}_{2,i+1})] \end{bmatrix}
$$

$$
= \begin{bmatrix} Y_{1,i} + \dfrac{h}{4}\,[Y_{1,i} + 5Y_{2,i} + Y^{(p)}_{1,i+1} + 5Y^{(p)}_{2,i+1}] \\ Y_{2,i} + \dfrac{h}{4}\,[5Y_{1,i} + Y_{2,i} + 5Y^{(p)}_{1,i+1} + Y^{(p)}_{2,i+1}] \end{bmatrix} \quad (9.3.19)
$$

For $i = 0$, Equation (9.3.18) reduces to

$$
\begin{bmatrix} Y^{(p)}_{1,1} \\ Y^{(p)}_{2,1} \end{bmatrix} = \begin{bmatrix} \left(1 + \dfrac{h}{2}\right) Y_{1,0} + \dfrac{5h}{2}\,Y_{2,0} \\ \dfrac{5h}{2}\,Y_{1,0} + \left(1 + \dfrac{h}{2}\right) Y_{2,0} \end{bmatrix} = \begin{bmatrix} \left(1 + \dfrac{h}{2}\right) \cdot 2 + \dfrac{5h}{2} \cdot 0 \\ \dfrac{5h}{2} \cdot 2 + \left(1 + \dfrac{h}{2}\right) \cdot 0 \end{bmatrix} = \begin{bmatrix} 2.1 \\ 0.5 \end{bmatrix}
$$

and Equation (9.3.19) reduces to

$$
\begin{bmatrix} Y_{1,1} \\ Y_{2,1} \end{bmatrix} = \begin{bmatrix} Y_{1,0} + \dfrac{h}{4}\,[Y_{1,0} + 5Y_{2,0} + Y^{(p)}_{1,1} + 5Y^{(p)}_{2,1}] \\ Y_{2,0} + \dfrac{h}{4}\,[5Y_{1,0} + Y_{2,0} + 5Y^{(p)}_{1,1} + Y^{(p)}_{2,1}] \end{bmatrix} = \begin{bmatrix} 2.165 \\ 0.525 \end{bmatrix}
$$

Similarly continuing, $Y_{1,i}$ and $Y_{2,i}$ are computed. The computed values of $Y_{1,i}$ and $Y_{2,i}$ along with the absolute value of the actual errors are given in Table 9.3.2. Errors are large due to large $h$. Of course, we do not expect much from the lower-order methods. (b) Substituting $f_1(x, y_1, y_2) = (y_1 + 5y_2)/2$ and $f_2(x, y_1, y_2) = (5y_1 + y_2)/2$ in Equation (9.3.17) leads to

$$
Y^{(p)}_{1,i+1} = Y_{1,i} + \frac{h}{48}\,[55(Y_{1,i} + 5Y_{2,i}) - 59(Y_{1,i-1} + 5Y_{2,i-1})
$$
$$
+ 37(Y_{1,i-2} + 5Y_{2,i-2}) - 9(Y_{1,i-3} + 5Y_{2,i-3})]
$$

and

$$
Y^{(p)}_{2,i+1} = Y_{2,i} + \frac{h}{48}\,[55(5Y_{1,i} + Y_{2,i}) - 59(5Y_{1,i-1} + Y_{2,i-1})
$$
$$
+ 37(5Y_{1,i-2} + Y_{2,i-2}) - 9(5Y_{1,i-3} + Y_{2,i-3})] \quad (9.3.20)
$$

**Table 9.3.2**

| $i$ | $x_i$ | $Y_{1,i}$ | \|Actual Error\| in $Y_1$ | $Y_{2,i}$ | \|Actual Error\| in $Y_2$ |
|---|---|---|---|---|---|
| 0 | 0.0 | 2.00000 | 0.00000 | 0.00000 | 0.00000 |
| 1 | 0.1 | 2.16500 | 0.00359 | 0.52500 | 0.00613 |
| 2 | 0.2 | 2.48143 | 0.01101 | 1.13663 | 0.01517 |
| 3 | 0.3 | 2.98451 | 0.02391 | 1.88177 | 0.02902 |
| 4 | 0.4 | 3.72469 | 0.04475 | 2.82045 | 0.05034 |
| 5 | 0.5 | 4.77235 | 0.07722 | 4.03087 | 0.08294 |
| 6 | 0.6 | 6.22417 | 0.12667 | 5.61616 | 0.13230 |
| 7 | 0.7 | 8.21191 | 0.20086 | 7.71333 | 0.20624 |
| 8 | 0.8 | 10.91414 | 0.31093 | 10.50531 | 0.31597 |
| 9 | 0.9 | 14.57220 | 0.47283 | 14.23696 | 0.47747 |
| 10 | 1.0 | 19.51161 | 0.70927 | 19.23671 | 0.71349 |

For $i = 3$, Equation (9.3.20) reduces to

$$Y_{1,4}^{(p)} = Y_{1,3} + \frac{h}{48}[55(Y_{1,3} + 5Y_{2,3}) - 59(Y_{1,2} + 5Y_{2,2})$$
$$+ 37(Y_{1,1} + 5Y_{2,1}) - 9(Y_{1,0} + 5Y_{2,0})]$$

and

$$Y_{2,4}^{(p)} = Y_{2,3} + \frac{h}{48}[55(5Y_{1,3} + Y_{2,3}) - 59(5Y_{1,2} + Y_{2,2})$$
$$+ 37(5Y_{1,1} + Y_{2,1}) - 9(5Y_{1,0} + Y_{2,0})] \qquad \textbf{(9.3.21)}$$

Using $y_1(x) = e^{3x} + e^{-2x}$ and $y_2(x) = e^{3x} - e^{-2x}$ to compute $Y_{1,1}$, $Y_{1,2}$, $Y_{1,3}$, $Y_{2,1}$, $Y_{2,2}$, and $Y_{2,3}$ and then substituting these values in Equation (9.3.21), we get $\begin{bmatrix} Y_{1,4}^{(p)} \\ Y_{2,4}^{(p)} \end{bmatrix} =$ $\begin{bmatrix} 3.76799 \\ 2.86918 \end{bmatrix}$. Using these values in Equation (9.3.17) yields $\begin{bmatrix} Y_{1,4} \\ Y_{2,4} \end{bmatrix} = \begin{bmatrix} 3.76941 \\ 2.87077 \end{bmatrix}$.

The computed values of $Y_{1,i}$ and $Y_{2,i}$ along with the absolute value of the actual errors are given in Table 9.3.3. In comparison to (a), the fourth-order method provides accurate results; however, by reducing $h$ further, more accurate results could be obtained.

**Table 9.3.3**

| $i$ | $x_i$ | $Y_{1,i}$ | \|Actual Error\| in $Y_1$ | $Y_{2,i}$ | \|Actual Error\| in $Y_2$ |
|---|---|---|---|---|---|
| 4 | 0.4 | 3.76941 | 0.00004 | 2.87077 | 0.00002 |
| 5 | 0.5 | 4.84948 | 0.00009 | 4.11375 | 0.00006 |
| 6 | 0.6 | 6.35067 | 0.00018 | 5.74832 | 0.00013 |
| 7 | 0.7 | 8.41247 | 0.00030 | 7.91932 | 0.00025 |
| 8 | 0.8 | 11.22458 | 0.00049 | 10.82083 | 0.00045 |
| 9 | 0.9 | 15.04425 | 0.00078 | 14.71369 | 0.00074 |
| 10 | 1.0 | 20.21965 | 0.00122 | 19.94902 | 0.00118 |

■ ■ ■

**EXERCISES**

In Exercises 1–3, (a) use the second-order predictor-corrector formula

$$Y_{i+1}^{(p)} = Y_i + \frac{h}{2}\left\{3f(x_i, Y_i) - f(x_{i-1}, Y_{i-1})\right\}$$

$$Y_{i+1} = Y_i + \frac{h}{2}\left\{f(x_i, Y_i) + f(x_{i+1}, Y_{i+1}^{(p)})\right\}$$

to approximate the solution of the given initial-value problem on the given interval by using the exact solution for starting values and (b) compare (a) with the exact solution. Use $h = 0.1$.

1.  $dy/dx = y + x$, $y(0) = 0$ on $[0, 0.4]$; exact solution: $y(x) = -1 - x + e^x$.
2.  $dy/dx = -xy$, $y(0) = 1$ on $[0, 0.4]$; exact solution: $y(x) = e^{-x^2/2}$.
3.  $dy/dx = (x - y)/x$, $y(1) = 5/2$ on $[1, 1.4]$; exact solution: $y(x) = (x/2) + (2/x)$.
4.  Use the fourth-order predictor-corrector formula [Equation (9.3.7)] to approximate the solution of the initial-value problem $dy/dx = y + x$, $y(0) = 0$ on $[0, 0.4]$. Use $h = 0.1$. Use the exact solution $y(x) = -1 - x + e^x$ for starting values.
5.  Use the fourth-order predictor-corrector formula [Equation (9.3.7)] to approximate the solution of the initial-value problem $dy/dx = y^2$, $y(1) = -1$ on $[1, 0.4]$. Use $h = 0.1$. Use the exact solution $y(x) = -1/x$ for starting values.
6.  Use the predictor-corrector formula [Equation (9.3.16)] to approximate the solution of the initial-value problem

$$\frac{dy_1}{dx} = 4y_2 \qquad y_1(0) = 2$$

$$\frac{dy_2}{dx} = y_1 \qquad y_2(0) = 0$$

at $x = 0.5$ with $h = 0.1$.

7.  Use the fourth-order predictor-corrector formula [Equation (9.3.17)] to approximate the solution of the initial-value problem

$$\frac{dy_1}{dx} = 4y_2 \qquad y_1(0) = 2$$

$$\frac{dy_2}{dx} = y_1 \qquad y_2(0) = 0$$

at $x = 0.4$ with $h = 0.1$.

**COMPUTER EXERCISES**

For Exercises C.1 and C.2, write a subroutine to approximate a solution of an initial-value problem

$$\frac{dy}{dx} = f(x, y) \qquad y(x_0) = y0$$

**C.1.** Use the second-order predictor-corrector method.

**C.2.** Use the fourth-order predictor-corrector method.

**C.3.** Using Exercises C.1 and C.2, write a program to approximate a solution of an initial-value problem

$$\frac{dy}{dx} = f(x, y) \qquad y(x_0) = y0$$

on a given interval $[x_0 = a, b]$. Read $a$, $b$, $N$, and $y0$. Use various $N$. Use Exercise C.2, p. 258 for starting values.

**C.4.** Using Exercise C.3, approximate the solution of the initial-value problem $dy/dx = 1 - x + 4y$, $y(0) = 1$ for $0 \le x \le 1$. Compare your results with the exact solution and with the second-order and fourth-order Runge–Kutta methods.

**C.5.** Using Exercise C.3, approximate the solution of the initial-value problem $dy/dx = -50y + 100$, $y(0) = 0$ for $0 \le x \le 1$. Compare your results with the exact solution and analyze your output.

**C.6.** Using Equation (9.3.17), write a subroutine to approximate the solution of a system of $M$ first-order differential equations

$$\frac{dy_j}{dx} = f_j(x, y_1, y_2, \dots, y_M)$$

with initial conditions $y_j(x_0) = \alpha_j$ for $j = 1, 2, \dots, M$. Use Exercise C.3, p. 274 for starting values,

**C.7.** Use Exercise C.6 to approximate the solution at $x = 1$ of the system given by

$$\frac{dy_1}{dx} = y_1 - y_2 - y_3$$

$$\frac{dy_2}{dx} = y_1 + 3y_2 + y_3$$

$$\frac{dy_3}{dx} = 3y_1 + y_2 - y_3$$

with initial conditions $y_1(0) = 1$, $y_2(0) = 2$, and $y_3(0) = 0$.

**C.8.** Use Exercise C.6 to approximate the solution at $t = 1$ of the system given by

$$\frac{dy_1}{dt} = 3y_1 - y_1 y_2$$

$$\frac{dy_2}{dt} = -3y_2 + y_1 y_2$$

with initial conditions $y_1(0) = 100$ and $y_2(0) = 200$.

**C.9.** The motion of a moon in a planar orbit around a planet is given by

$$\frac{d^2x}{dt^2} = -\frac{Gmx}{r^3} \quad \text{and} \quad \frac{d^2y}{dt^2} = -\frac{Gmy}{r^3}$$

where $G$ is the gravitational constant, $m$ is the mass of this moon, and $r^2 = x^2 + y^2$. Compute the orbit of this moon by assuming $Gm = 1$, $x(0) = 1$, $y(0) = 0$, $x'(0) = 0$, and $y'(0) = 1$.

## 9.4   INSTABILITY AND STIFF DIFFERENTIAL EQUATIONS

We developed many numerical methods in this chapter to solve an initial-value problem

$$\frac{dy}{dx} = f(x, y) \qquad y(x_0) = y_0 \tag{9.4.1}$$

Let us consider a general multistep method of $k$ step of the form

$$Y_{i+1} = \alpha_1 Y_i + \alpha_2 Y_{i-1} + \cdots + \alpha_k Y_{i-k+1} + h[\beta_0 f(x_{i+1}, Y_{i+1}) + \beta_1 f(x_i, Y_i)$$
$$+ \cdots + \beta_k f(x_{i-k+1}, Y_{i-k+1}) \quad \text{for } i = k, k + 1, \ldots \tag{9.4.2}$$

Let initial conditions $Y_i$ for $i = 0, 1, \ldots, k$ used in solving Equation (9.4.2) be such that

$$\lim_{h \to 0} Y_i(h) = y_0 \quad \text{for } i = 0, 1, \ldots, k \tag{9.4.3}$$

where $y_0$ is the initial condition in Equation (9.4.1). Then the method defined by Equation (9.4.2) is convergent if $f$ satisfies the Lipschitz condition and

$$\lim_{\substack{i+1 \to \infty \\ (h \to 0)}} Y_{i+1} = y(x) \quad (i + 1)h = x - a \tag{9.4.4}$$

for all $x \in [a, b]$. Since $Y_i$ for $i = 0, 1, \ldots, k$ is not an exact solution of $dy/dx = f(x, y)$ in Equation (9.4.1); we need Equation (9.4.3). Loosely speaking, the concept of convergence means that any desired degree of accuracy can be achieved for Equation (9.4.1) by Equation (9.4.2) using a small enough $h$.

The local truncation error is defined to be the quantity

$$T_i(y; h) = y(x_{i+1}) - Y_{i+1}$$
$$= y(x_{i+1}) - [\alpha_1 Y_i + \alpha_2 Y_{i-1} + \cdots + \alpha_k Y_{i-k+1} + h(\beta_1 f(x_i, Y_i)$$
$$+ \beta_2 f(x_{i-1}, Y_{i-1}) + \cdots + \beta_k f(x_{i-k}, Y_{i-k+1})]$$

For convenience, let us define the normalized local truncation error $\tau_i(y; h) = T_i(y; h)/h$.

In order for Equation (9.4.2) to approximate the solution of Equation (9.4.1), it is necessary that as the step size approaches zero, the normalized local truncation error must approach zero at each step; in other words, $\lim_{h \to 0} |\tau(y; h)| = 0$ where $|\tau(y; h)| = \max_{a \le x_i \le b} |\tau_i(y; h)|$. This is often called the **consistency condition**.

EXAMPLE 9.4.1   Verify that the midpoint formula

$$Y_{i+1} = Y_{i-1} + 2hf(x_i, Y_i) \quad \text{for } i = 1, 2, \ldots$$

is consistent.

We have

$$
\begin{aligned}
T_i(y; h) &= y(x_i + h) - Y_{i+1} \\
&= y(x_i + h) - [y(x_i - h) + 2hf(x_i, y(x_i))] \\
&= y(x_i) + hy'(x_i) + \frac{h^2}{2!} y''(x_i) + \frac{h^3}{3!} y^{(3)}(\xi_1) \\
&\quad - [y(x_i) - hy'(x_i) + \frac{h^2}{2!} y''(x_i) - \frac{h^3}{3!} y^{(3)}(\xi_2) + 2hy'(x_i)] \\
&= \frac{h^3}{3!} y^{(3)}(\xi_1) + \frac{h^3}{3!} y^{(3)}(\xi_2) \\
&= \frac{h^3}{3} y^{(3)}(\eta_i) \qquad \text{where } \xi_1, \xi_2, \text{ and } \eta_i \in (x_i - h, x_i + h)
\end{aligned}
$$

Therefore $\tau_i(y; h) = (1/h) \, T_i(y; h) = (h^2/3) \, y^{(3)}(\eta_i)$. Hence

$$
|\tau(y; h)| = \max_{a \le x_i \le b} |\tau_i(y; h)| = \frac{h^2}{3} \max_{a \le x_i \le b} |y^{(3)}(\eta_i)| \le \frac{M_3 h^2}{3}
$$

where $|y^{(3)}(x)| \le M_3$ for all $x$ in $[a, b]$. Since $\lim_{h \to 0} M_3 h^2/3 = 0$, $\lim_{h \to 0} |\tau(y; h)| = 0$ using the Sandwich Theorem [Theorem A.0]. Hence the midpoint formula is consistent and of second order.  ■ ■ ■

When we used the midpoint formula in Section 8.2, we noticed that the computed values $Y_i$ did not converge to the values $y(x_i)$ of the exact solution. Thus consistency does not imply convergence. Stability is also necessary for convergence. We are concerned with instabilities of numerical methods in this section. We consider a method to be **absolutely stable** for a given step size $h$ and a given differential equation if a change due to perturbation $\delta$ at one of the mesh values $Y_i$ does not produce a change larger than $\delta$ in all subsequent values $Y_{i+j}$ for $j = 1, 2, \ldots$. This is dependent on a given differential equation, which is unfortunate.

The example we used to check instabilities in Section 8.2 can be modified and used to study instabilities of numerical methods. The idea is to apply a numerical method to a simple test equation

$$
\frac{dy}{dx} = \lambda y \qquad y(0) = y_0 = 1 \tag{9.4.5}
$$

where $\lambda$ is a parameter. The exact solution of Equation (9.4.5) is $y(x) = e^{\lambda x}$, so that

1.  if $\lambda > 0$, then $y(x) \to \infty$ as $x \to \infty$,
2.  if $\lambda < 0$, then $y(x) \to 0$ as $x \to \infty$, and
3.  if $\lambda$ is complex, then $y(x)$ oscillates.

Thus a wide range of general behavior can be represented by this simple test problem.

The **region of absolute stability** of a given method is the set of values of $h > 0$ and $\lambda$ (which may be complex) for which a perturbation $\delta$ introduced at $Y_i$ in the numerical solution of Equation (9.4.5) will produce a change not larger than $\delta$ in the subsequent values $Y_{i+j}$ for $j = 1, 2, \ldots$.

Let us investigate the stability of the second-order Adams–Bashforth formula

$$
Y_{i+1} = Y_i + \frac{h}{2} [3f(x_i, Y_i) - f(x_{i-1}, Y_{i-1})] \quad \text{for } i = 1, 2, \ldots \tag{9.4.6}
$$

We solve Equation (9.4.5) using Equation (9.4.6). By substituting $f(x, y) = \lambda y$ in Equation (9.4.6), we obtain the difference equations

$$Y_{i+1} = Y_i + \frac{\lambda h}{2}[3Y_i - Y_{i-1}] \quad \text{for } i = 1, 2, \ldots \tag{9.4.7}$$

To get the solution of Equation (9.4.7), substitute $Y_i = r^i$ (Appendix D) in Equation (9.4.7) and simplify to get the characteristic polynomial

$$r^2 - r - \frac{\lambda h}{2}(3r - 1) = 0$$

or

$$r^2 - r\left(1 + \frac{3\lambda h}{2}\right) + \frac{\lambda h}{2} = 0 \tag{9.4.8}$$

The roots of the characteristic polynomial (9.4.8) are

$$r_1 = \frac{1}{2}\left[1 + \frac{3}{2}\lambda h + \sqrt{1 + h\lambda + \frac{9}{4}h^2\lambda^2}\right]$$

$$r_2 = \frac{1}{2}\left[1 + \frac{3}{2}\lambda h - \sqrt{1 + h\lambda + \frac{9}{4}h^2\lambda^2}\right] \tag{9.4.9}$$

Then the general solution of Equation (9.4.7) is

$$Y_i = c_1 r_1^i + c_2 r_2^i \quad \text{for } i = 0, 1, \ldots \tag{9.4.10}$$

where $c_1$ and $c_2$ are to be determined such that $Y_0$ and $Y_1$ agree with $y(x_0)$ and $y(x_1)$ (which is not given). The original differential equation has only one parameter $y(x_0) = y_0$ while Equation (9.4.7) has two parameters $Y_0$ and $Y_1$. Thus from Equation (9.4.10), we have

$$\begin{aligned} Y_0 &= c_1 + c_2 = y_0 = 1 \\ Y_1 &= c_1 r_1 + c_2 r_2 = e^{\lambda h} \end{aligned} \tag{9.4.11}$$

Solving Equation (9.4.11) for $c_1$ and $c_2$ leads to

$$c_1 = \frac{Y_1 - r_2 Y_0}{r_1 - r_2} = \frac{e^{\lambda h} - r_2}{\sqrt{1 + h\lambda + \frac{9}{4}h^2\lambda^2}}$$

$$c_2 = \frac{r_1 Y_0 - Y_1}{r_1 - r_2} = \frac{r_1 - e^{\lambda h}}{\sqrt{1 + h\lambda + \frac{9}{4}h^2\lambda^2}} \tag{9.4.12}$$

If $\lambda h$ is small, then $r_1 \approx 1 + \lambda h$, $r_2 \approx (\lambda h)/2$, $c_1 \approx 1$, and $c_2 \approx 0$. Thus as $h \to 0$, $c_1 r_1^i \to e^{\lambda x_i}$ and $c_2 r_2^i \to 0$. Consider a case when $\lambda h$ is not small; for example, $\lambda = -100$ and $h = 0.05$. Then $r_1 \approx 0.36421$, $r_2 \approx -6.86421$, $c_1 \approx 0.95055$, and $c_2 \approx 0.05132$. The general solution of Equation (9.4.7) is

$$Y_i = (0.95055)(0.36421)^i + (0.05132)(-6.86421)^i \quad \text{for } i = 0, 1, \ldots \tag{9.4.13}$$

As $i$ becomes large, $c_2 r_2^i$ of Equation (9.4.13) overtakes $c_1 r_1^i$ of Equation (9.4.13) and causes the numerical solution to diverge from $e^{\lambda x_i}$. For this reason, the Adams–Bashforth method is considered to be **weakly stable**. The term $c_2 r_2^i$ in Equation (9.4.10) does not correspond to any solution of $dy/dx = \lambda y$, therefore it is called a **parasitic** or **spurious solution** of Equation (9.4.7).

When we apply Equation (9.4.2) to Equation (9.4.5), we get a linear difference equation

$$Y_{i+1} = \alpha_1 Y_i + \alpha_2 Y_{i-1} + \cdots + \alpha_k Y_{i-k+1} + h\lambda[\beta_0 Y_{i+1} + \beta_1 Y_i + \cdots + \beta_k Y_{i-k+1}]$$

or

$$(1 - h\lambda\beta_0)Y_{i+1} = (\alpha_1 + h\lambda\beta_1)Y_i + (\alpha_2 + h\lambda\beta_2)Y_{i-1} + \cdots + (\alpha_k + h\lambda\beta_k)Y_{i-k+1}$$
$$\textbf{(9.4.14)}$$

for $i = k, k + 1, \ldots$.

The characteristic polynomial of Equation (9.4.14) is

$$Q(h\lambda) = (1 - h\lambda\beta_0)r^k - (\alpha_1 + h\lambda\beta_1)r^{k-1} - \cdots - (\alpha_k + h\lambda\beta_k) = 0$$
$$\textbf{(9.4.15)}$$

The general solution of Equation (9.4.14) is given by

$$Y_i = c_1 r_1^i + c_2 r_2^i + \cdots + c_k r_k^i \quad \text{for } i = 0, 1, \ldots \qquad \textbf{(9.4.16)}$$

where $r_1, r_2, \ldots, r_k$ are the roots of Equation (9.4.15). One of these $k$ roots, say $r_1$, represents the approximate solution corresponding to the exact solution of Equation (9.4.5) while the other $k - 1$ roots $r_2, r_3, \ldots, r_k$ are parasitic or spurious and have no connection with Equation (9.4.5). For a strongly stable method all parasitic terms will approach zero as $i \to \infty$ and for sufficiently small $h$ the remaining term will approximate $y(x_i)$. The region of absolute stability of a multistep method consists of the values of $h\lambda$ in the complex plane for which the absolute value of all the roots of Equation (9.4.15) are less than or equal to one and for which $|r_k| = 1$ is simple. Different numerical methods have different regions of asbolute stability. Gear (1971) discusses the stability regions of different numerical methods. Stability is necessary for convergence. A linear multistep method is convergent if and only if it is strongly stable and consistent (for the proof, see Stoer and Bulirsch 1980).

When we apply the fourth-order Runge–Kutta method given by Equation (8.5.28) to Equation (9.4.5), then

$$k_1 = hf(x_i, Y_i) = h\lambda Y_i$$

$$k_2 = hf\left(x_i + \frac{h}{2}, Y_i + \frac{k_1}{2}\right) = h\lambda\left(Y_i + \frac{k_1}{2}\right) = h\lambda\left(1 + \frac{h\lambda}{2}\right)Y_i$$

$$k_3 = hf\left(x_i + \frac{h}{2}, Y_i + \frac{k_2}{2}\right) = h\lambda\left(Y_i + \frac{k_2}{2}\right) = h\lambda\left(1 + \frac{h\lambda}{2} + \left(\frac{h\lambda}{2}\right)^2\right)Y_i$$

$$k_4 = hf(x_i + h, Y_i + k_3) = h\lambda(Y_i + k_3) = h\lambda\left(1 + h\lambda + \frac{(h\lambda)^2}{2} + \frac{(h\lambda)^3}{4}\right)Y_i$$

and

$$Y_{i+1} = Y_i + \frac{1}{6}(k_1 + 2k_2 + 2k_3 + k_4)$$

$$= Y_i\left[1 + h\lambda + \frac{(h\lambda)^2}{2!} + \frac{(h\lambda)^3}{3!} + \frac{(h\lambda)^4}{4!}\right] \quad \text{for } i = 0, 1, \ldots \quad \textbf{(9.4.17)}$$

The solution of Equation (9.4.17) is

$$Y_i = c_1\left[1 + h\lambda + \frac{(h\lambda)^2}{2!} + \frac{(h\lambda)^3}{3!} + \frac{(h\lambda)^4}{4!}\right]^i \quad \text{for } i = 0, 1, \ldots \quad \textbf{(9.4.18)}$$

where $c_1$ is an arbitrary constant. For $i = 0$, $Y_0 = c_1 = y(x_0) = 1$. Further it can be shown (Exercise 13) that

$$\lim_{h \to 0}\left(1 + h\lambda + \frac{(h\lambda)^2}{2!} + \frac{(h\lambda)^3}{3!} + \frac{(h\lambda)^4}{4!}\right)^{x_i/h} = e^{\lambda x_i}$$

Hence as $h \to 0$, the solution of Equation (9.4.18) approaches $e^{\lambda x_i}$ which is the exact solution. One-step methods do not have parasitic solutions and therefore are strongly stable. If

$$\left|1 + h\lambda + \frac{(h\lambda)^2}{2!} + \frac{(h\lambda)^3}{3!} + \frac{(h\lambda)^4}{4!}\right| \leq 1 \quad \textbf{(9.4.19)}$$

in Equation (9.4.17), then Equation (9.4.18) converges to $e^{\lambda x_i}$. Thus Equation (9.4.19) yields a set of complex values $h\lambda$ that form a unit circle in the complex $h\lambda$-plane centered at $(-1, 0)$. The set of all $h\lambda$ for which Equation (9.4.19) is true is called the region of absolute stability of the fourth-order Runge–Kutta method [Equation (8.5.28)].

## Stiff Equations

Consider the solution $y(x) = (1/100) + (99/100)e^{-100x}$ of the differential equation $dy/dx = -100y + 1$, $y(0) = 1$. The solution decreases very rapidly from 1 to 1/100; for example $y(0.1) = (1/100) + (99/100)e^{-10} \approx 0.01004$.

In order to get an accurate solution, we expect to use a small step size $h$ initially. Once this part of the solution is computed, we are tempted to increase the step size $h$ since the solution varies slowly in the rest of the part. Although the transient part of the solution dies out, the differential equation remains the same. This causes instability with large step size $h$. Most of the methods we described in this chapter exhibit instability when applied to such problems. Differential equations with solutions of widely different scale components are known as **stiff differential equations**. Very small step size throughout a large interval is very expensive and therefore implicit methods are used for stiff problems.

**EXAMPLE 9.4.2**    Approximate the solution of the initial-value problem

$$\frac{dy}{dx} = -100y + 1 \qquad y(0) = 1 \quad \textbf{(9.4.20)}$$

at $x = 1$ using the (1) Euler forward method, (2) backward Euler method, (3) fourth-order Runge–Kutta method, and (4) fourth-order predictor-corrector method using various $h$.

The exact solution is $y(x) = (1/100) + (99/100)e^{-100x}$ and therefore $y(1) = (1/100) + (99/100)e^{-100} \approx 0.01$. The computed values of $y(x)$ at $x = 1$ are given in Table 9.4.1.

**Table 9.4.1**

|  | $h = 0.1$ | $h = 0.05$ | $h = 0.02$ | $h = 0.01$ |
|---|---|---|---|---|
| Euler method | $-0.31067 \times 10^{11}$ | $-0.43541 \times 10^{13}$ | $-0.98000$ | $0.01000$ |
| Backward Euler method | $0.01000$ | $0.01000$ | $0.01000$ | $0.01000$ |
| Runge–Kutta method | $0.12545 \times 10^{28}$ | $0.74527 \times 10^{24}$ | $0.01000$ | $0.01000$ |
| Predictor-Corrector method | $-0.21024 \times 10^{37}$ | $-0.33444 \times 10^{71}$ | $-0.91160 \times 10^{134}$ | $0.55459 \times 10^{219}$ |

The Euler forward method explodes for $h = 0.1$ and $0.05$ while the backward Euler method gives good results. We want to find explanations for these behaviors.

Since $f(x_i, Y_i) = -100Y_i + 1$, the Euler method gives

$$\begin{aligned} Y_{i+1} &= Y_i + hf(x_i, Y_i) \\ &= Y_i + h(-100Y_i + 1) \\ &= Y_i(1 - 100h) + h \quad \text{for } i = 0, 1, \ldots \end{aligned} \quad \textbf{(9.4.21)}$$

The characteristic equation of (9.4.21) is

$$r - (1 - 100h) = 0 \quad \textbf{(9.4.22)}$$

Thus for stability, $|100h - 1| < 1$. Hence $h < 0.02$. If $h > 0.02$, then $Y_{i+1}$ in Equation (9.4.21) grows rapidly at each step and leads to an unstable behavior. The solution of Equation (9.4.21)

$$Y_i = \frac{99}{100}(1 - 100h)^i + \frac{1}{100} \quad \text{for } i = 0, 1, \ldots \quad \textbf{(9.4.23)}$$

approaches $(99/100)e^{-100x_i} + (1/100)$ as $i \to \infty$ for $h < 0.02$. Euler's forward method requires that we approximate Equation (9.4.20) accurately to maintain stability, even though $(99/100)e^{-100x}$ contributes virtually nothing after $x = 0.1$.

When the backward Euler method is applied to Equation (9.4.20), we get

$$\begin{aligned} Y_{i+1} &= Y_i + hf(x_{i+1}, Y_{i+1}) \\ &= Y_i + h(-100Y_{i+1} + 1) \quad \text{for } i = 0, 1, \ldots \end{aligned} \quad \textbf{(9.4.24)}$$

Solving Equation (9.4.24) for $Y_{i+1}$ leads to

$$Y_{i+1}(1 + 100h) - Y_i = h \quad \text{for } i = 0, 1, \ldots \quad \textbf{(9.4.25)}$$

The characteristic equation of Equation (9.4.25) is

$$r(1 + 100h) - 1 = 0 \quad \textbf{(9.4.26)}$$

For stability, we need $1/|1 + 100h| < 1$. This is true for any positive value of $h$. Thus the backward Euler method is stable for all $h > 0$, therefore we call the backward Euler method an **absolutely stable method**.

We can carry out the same kind of analysis for the Runge–Kutta and predictor-corrector methods.

■ ■ ■

**EXAMPLE 9.4.3**

Approximate the solution of the initial-value problem

$$\frac{d^2y}{dx^2} + 51\frac{dy}{dx} + 50y = 0 \qquad y(0) = \frac{51}{50} \qquad \frac{dy}{dx}(0) = -2 \qquad \textbf{(9.4.27)}$$

at $x = 1$ using the (1) Euler forward method, (2) Trapezoidal method, and (3) fourth-order Runge–Kutta method.

The exact solution is given by $y(x) = e^{-x} + (e^{-50x}/50)$ and therefore $y(1) = e^{-1} + (e^{-50}/50) \approx 0.36788$. Although the second term $(e^{-50x}/50)$ contributes very little after $x = 0.2$, we will need small step size $h$ to maintain stability. We convert Equation (9.4.27) to a system of two first-order ordinary differential equations

$$\frac{dy_1}{dx} = f_1(x, y_1, y_2) = y_2 \qquad y_1(0) = \frac{51}{50}$$

$$\frac{dy_2}{dx} = f_2(x, y_1, y_2) = -51y_2 - 50y_1 \qquad y_2(0) = -2 \qquad \textbf{(9.4.28)}$$

The computed values of $y(1)$ by the Euler and Runge–Kutta methods are given in Table 9.4.2 using various $h$.

**Table 9.4.2**

|  | $h = 0.1$ | $h = 0.05$ | $h = 0.02$ | $h = 0.01$ |
|---|---|---|---|---|
| Euler method | $0.20972 \times 10^5$ | $0.66864 \times 10^2$ | 0.36417 | 0.36603 |
| Runge–Kutta method | $0.46868 \times 10^{10}$ | 0.36788 | 0.36788 | 0.36788 |
| Trapezoidal method | 0.36758 | 0.36780 | 0.36787 | 0.36788 |

The Trapezoidal method when applied to Equation (9.4.28) gives

$$Y_{1,i+1} = Y_{1,i} + \frac{h}{2}[f_1(x_i, Y_{1,i}, Y_{2,i}) + f_1(x_{i+1}, Y_{1,i+1}, Y_{2,i+1})]$$

$$= Y_{1,i} + \frac{h}{2}[Y_{2,i} + Y_{2,i+1}]$$

and

$$Y_{2,i+1} = Y_{2,i} + \frac{h}{2}[f_2(x_i, Y_{1,i}, Y_{2,i}) + f_2(x_{i+1}, Y_{1,i+1}, Y_{2,i+1})]$$

$$= Y_{2,i} + \frac{h}{2}[-51Y_{2,i} - 50Y_{1,i} - 51Y_{2,i+1} - 50Y_{1,i+1}] \quad \text{for } i = 0, 1, \ldots$$

$$\textbf{(9.4.29)}$$

Solving Equation (9.4.29) for $Y_{1,i+1}$ and $Y_{2,i+1}$, we obtain

$$Y_{1,i+1} - \frac{h}{2}Y_{2,i+1} = Y_{1,i} + \frac{h}{2}Y_{2,i}$$

and

$$25hY_{1,i+1} + \left(1 + \frac{51h}{2}\right)Y_{2,i+1} = \left(1 - \frac{51h}{2}\right)Y_{2,i} - 25hY_{1,i} \quad \text{for } i = 0, 1, \dots$$

$$(9.4.30)$$

Elimination of $Y_{2,i+1}$ and then $Y_{1,i+1}$ between equations in Equation (9.4.30) leads to

$$Y_{1,i+1} = \frac{1}{\left(1 + \frac{51h}{2} + \frac{25h^2}{2}\right)}\left[hY_{2,i} + \left(1 + \frac{51h}{2} - \frac{25h^2}{2}\right)Y_{1,i}\right]$$

and

$$Y_{2,i+1} = \frac{1}{\left(1 + \frac{51h}{2} + \frac{25h^2}{2}\right)}\left[\left(1 + \frac{51h}{2} - \frac{25h^2}{2}\right)Y_{2,i} - 50hY_{1,i}\right] \quad \text{for } i = 0, 1, \dots$$

$$(9.4.31)$$

Using $Y_{1,0} = 51/50$ and $Y_{2,0} = -2$ in Equation (9.4.31), the values of $Y_{1,i+1}$ and $Y_{2,i+1}$ are computed. The computed value of $Y_1$ at $x = 1$ is given in Table 9.4.2.

    The Runge–Kutta method of fourth order gives more accurate results once inside the region of absolute stability, while the Trapezoidal method of second order does not grow exponentially like the Runge–Kutta method for large step sizes.    ■ ■ ■

    The techniques used for stiff equations are implicit methods since some are absolutely stable. It has been proven by Dahlquist (1963) that an absolutely stable multistep method cannot have order greater than two. Explicit Runge–Kutta methods are somewhat inefficient for stiff equations. It is natural to investigate the implicit Runge–Kutta methods that are either absolutely stable or have a larger region of stability than the explicit methods. The general theory and more details about implicit Runge–Kutta methods can be found in Lapidus and Seinfeld (1971), Jain (1984), and Butcher (1987). To illustrate, we determine $w_1$, $w_2$, $\alpha$, $\beta$, $\gamma$, and $\delta$ such that

$$Y_{i+1} = y_i + w_1 k_1 + w_2 k_2 \qquad (9.4.32)$$

where

$$k_1 = hf(x_i + (\alpha + \beta)h, Y_i + \alpha k_1 + \beta k_2)$$

and

$$k_2 = hf(x_i + (\gamma + \delta)h, Y_i + \gamma k_1 + \delta k_2)$$

gives the best accuracy and the largest region of stability. A particular form is

$$Y_{i+1} = Y_i + \frac{1}{2}(k_1 + k_2) \qquad (9.4.33)$$

where

$$k_1 = hf\left(x_i + \left(\frac{1}{2} - \frac{\sqrt{3}}{6}\right)h, Y_i + \frac{1}{4}k_1 + \left(\frac{1}{4} - \frac{\sqrt{3}}{6}\right)k_2\right)$$

and

$$k_2 = hf\left(x_i + \left(\frac{1}{2} + \frac{\sqrt{3}}{6}\right)h, Y_i + \left(\frac{1}{4} + \frac{\sqrt{3}}{6}\right)k_1 + \frac{1}{4}k_2\right)$$

In the case of a stiff nonlinear equation, $f_2(x_i, Y_{1,i}, Y_{2,i})$ in Equation (9.4.28) is nonlinear and Equation (9.4.29) becomes a system of nonlinear equations. We solve a system of nonlinear equations by Newton's method or the fixed-point method. These topics will be discussed in Chapter 12.

## EXERCISES

1. Verify that the Euler formula

$$Y_{i+1} = Y_i + hf(x_i, Y_i) \quad \text{for } i = 0, 1, \ldots$$

   is consistent.

2. Investigate the stability properties of the Euler method

$$Y_{i+1} = Y_i + hf(x_i, Y_i) \quad \text{for } i = 0, 1, \ldots$$

   by applying it to the test problem $dy/dx = \lambda y$, $y(0) = 1$. Sketch the graph of $|1 + h\lambda| < 1$ where $\lambda$ may be complex.

3. Verify that the Trapezoidal formula

$$Y_{i+1} = Y_i + \frac{h}{2}[f(x_i, Y_i) + f(x_{i+1}, Y_{i+1})] \quad \text{for } i = 0, 1, \ldots$$

   is consistent.

4. Investigate the stability properties of the Trapezoidal method

$$Y_{i+1} = Y_i + \frac{h}{2}[f(x_i, Y_i) + f(x_{i+1}, Y_{i+1})] \quad \text{for } i = 0, 1, \ldots$$

   by applying it to the test problem $dy/dx = \lambda y$, $y(0) = 1$. Sketch the region of absolute stability.

5. Consider the initial-value problem $dy/dx = -10y$, $y(0) = 1$. Determine $e^{-5} - Y_5$ with $h = 0.1$ using each of the following methods: (a) Euler forward method, (b) backward Euler or implicit method, and (c) Modified Euler method.

In Exercises 6–9, approximate the solution at $x = 0.4$ using the (a) Euler forward method, and (b) backward Euler method with $h = 0.1$. Compare it with the exact value $y(0.4)$.

6. $dy/dx = -25(y - x)$, $y(0) = 0$.
7. $d^2y/dx^2 + 101(dy/dx) + 100y = 0$, $y(0) = 1$, $dy/dx(0) = 1$.
8. $d^2y/dx^2 + 1001(dy/dx) + 1000y = 0$, $y(0) = 1$, $dy/dx(0) = -2$.
9. $dy/dx + 50y = \sin x$, $y(0) = 0$.

10. Investigate the stability properties of the Milne method

$$Y_{i+1} = Y_{i-3} + \frac{4h}{3}[2f(x_i, Y_i) - f(x_{i-1}, Y_{i-1}) + 2f(x_{i-2}, Y_{i-2})] \quad \text{for } i = 3, 4, \ldots$$

by applying it to the test problem $dy/dx = \lambda y$, $y(0) = 1$.

11. Investigate the stability properties of the Adams–Moulton method

$$Y_{i+1} = Y_i + \frac{h}{24}[9f(x_{i+1}, Y_{i+1}) + 19f(x_i, Y_i)$$
$$- 5f(x_{i-1}, Y_{i-1}) + f(x_{i-2}, Y_{i-2})] \quad \text{for } i = 2, 3, \ldots$$

by applying it to the test problem $dy/dx = \lambda y$, $y(0) = 1$.

12. Investigate the stability properties of the Adams–Bashforth method

$$Y_{i+1} = Y_i + \frac{h}{24}[55f(x_i, Y_i) - 59f(x_{i-1}, Y_{i-1})$$
$$+ 37f(x_{i-2}, Y_{i-2}) - 9f(x_{i-3}, Y_{i-3})] \quad \text{for } i = 3, 4, \ldots$$

by applying it to the test problem $dy/dx = \lambda y$, $y(0) = 1$.

13. Prove that

$$\lim_{h \to 0} \left(1 + h\lambda + \frac{(h\lambda)^2}{2!} + \frac{(h\lambda)^3}{3!} + \frac{(h\lambda)^4}{4!}\right)^{x_i/h} = e^{\lambda x_i}$$

## COMPUTER EXERCISES

C.1. Write a subroutine to approximate the solution of an initial-value problem

$$\frac{dy}{dx} = f(x, y) \qquad y(x_0) = y0$$

using the (a) Euler forward method, (b) backward Euler method, (c) fourth-order Runge–Kutta method, and (d) Trapezoidal method.

C.2. Using Exercise C.1, write a program to approximate the solution of an initial-value problem

$$\frac{dy}{dx} = f(x, y) \qquad y(x_0) = y0$$

on a given interval $[x_0 = a, b]$. Read $a$, $b$, $N$, and $y0$. Use various $h$.

C.3. Using Exercise C.2, approximate the solution of the initial-value problem $dy/dx = -25(y - x)$, $y(0) = 0$ for $0 \le x \le 1$. Compare your results with the exact solution.

C.4. Using Exercise C.2, approximate the solution of the initial-value problem $dy/dx + 50y = \sin x$, $y(0) = 0$ for $0 \le x \le 1$. Compare your results with the exact solution.

**C.5.** Approximate the solution of the initial-value problem

$$\frac{d^2y}{dx^2} + 101\frac{dy}{dx} + 100y = 0 \qquad y(0) = 1 \qquad \frac{dy}{dx}(0) = 1$$

at $x = 1$ using the (a) Euler forward method, (b) Trapezoidal method, and (c) fourth-order Runge–Kutta method. Use $h = 0.1, 0.05, 0.02$, and $0.01$.

# SUGGESTED READINGS

J. C. Butcher, *The Numerical Analysis of Ordinary Differential Equations,* John Wiley & Sons, Chichester, 1987.

G. Dahlquist, *A Special Stability Problem for Linear Multistep Methods,* BIT **3** (1963), 27–43.

G. W. Gear, *Numerical Initial Value Problems in Ordinary Differential Equations,* Prentice-Hall, Inc., Englewood Cliffs, New Jersey, 1971.

P. Henrici, *Discrete Variable Methods for Ordinary Differential Equations,* John Wiley & Sons, New York, 1962.

M. K. Jain, *Numerical Solution of Differential Equations,* 2nd ed., Wiley Eastern, New Delhi, 1984.

L. Lapidus and J. H. Seinfeld, *Numerical Solution of Ordinary Differential Equations,* Academic Press, New York, 1971.

J. Stoer and R. Bulirsch, *Introduction to Numerical Analysis,* Springer–Verlag, New York, 1980.

# 10 CURVE FITTING AND APPROXIMATION OF FUNCTIONS

$I$t is known that the height $h$ of Monterey pine trees is a linear function of rainfall $r$ given by $h = a_0 + a_1 r$ where $a_0$ and $a_1$ are constants. To determine $a_0$ and $a_1$, an experimenter grew one foot Monterey pine trees in similar controlled environments where variation in rainfall was controlled by providing different amounts of irrigation. The following table presents the amount of rainfall (in inches) and the measured height (in inches) of Monterey pine trees at the end of one year.

**Table 1**

| r (inches) | 11 | 13 | 17 | 19 | 23 | 27 |
|------------|----|----|----|----|----|----|
| h (inches) | 20 | 21 | 24 | 26 | 32 | 34 |

This data contains experimental errors. How can we determine $a_0$ and $a_1$? We cannot use a random pair of points to determine $a_0$ and $a_1$ since the table contains experimental errors in the data. Therefore, it is more appropriate to find the line that fits all the data points by selecting $a_0$ and $a_1$. In this chapter, we develop methods to fit various types of functions to a given set of data points.

---

## 10.1  INTRODUCTION

The least squares solution of $A\mathbf{x} = \mathbf{b}$ where $A$ is an $m \times n$ matrix is derived. A function $g(x)$ is said to fit a given set of data points $\{(x_1, y_1), (x_2, y_2), \ldots, (x_N, y_N)\}$ if its graph comes close to the data points. The least squares method is used to fit a polynomial or a set of functions to a given set of data points. The set of exponential functions is also considered because it is used in the analysis of radioactive decay, population growth, and other applications. If $A$ has linearly dependent rows or columns, then $Ax = b$ is a singular system. The singular value decomposition of a given matrix $A$ is derived. Then the least squares solution of $Ax = b$ is expressed as $x = A^+b$ where $A^+$ is the pseudoinverse of $A$.

Another set of functions is trigonometric polynomials. The numerical Fourier analysis is extensively used for cyclic phenomena and time series and therefore Fast Fourier Transform is discussed.

Another approach is to have exact matching at given data points, but higher-degree polynomials lead to wiggles. We fit piecewise a table by lower-degree polynomials and require these fits to be smooth (first derivative continuous) at the interface. The piecewise fits are called **splines**. We also discuss cubic splines, a fairly new development of growing importance.

Finally, we will consider the rational approximations of given functions.

## 10.2    LEAST SQUARES APPROXIMATIONS

So far we discussed the techniques to compute $x$ of a given linear system $Ax = b$ where $A$ is a square matrix. If $A$ is nonsingular, then there exists a unique solution. In this section, we turn our attention to a system of $m$ equations in $n$ unknowns where $m \neq n$. Thus if $A$ has $m$ rows and $n$ columns, then $x$ is a vector with $n$ components and $b$ is a vector with $m$ components. If $m > n$, then we have more equations than unknowns. Such systems are usually **overdetermined**. Overdetermined systems do arise in practice and need to be solved. From Table 1, we have

$$a_0 + 11a_1 = 20 \qquad a_0 + 19a_1 = 26$$
$$a_0 + 13a_1 = 21 \qquad a_0 + 23a_1 = 32$$
$$a_0 + 17a_1 = 24 \qquad a_0 + 27a_1 = 34$$

in other words,

$$\begin{bmatrix} 1 & 11 \\ 1 & 13 \\ 1 & 17 \\ 1 & 19 \\ 1 & 23 \\ 1 & 27 \end{bmatrix} \begin{bmatrix} a_0 \\ a_1 \end{bmatrix} = \begin{bmatrix} 20 \\ 21 \\ 24 \\ 26 \\ 32 \\ 34 \end{bmatrix} \qquad \textbf{(10.2.1)}$$

One possibility is to determine $a_0$ and $a_1$ from a part of Equation (10.2.1) by ignoring the rest. However, since the data comes from the same source, it is difficult to know which equations contain large errors. Thus we cannot justify determining $a_0$ and $a_1$ from a part of Equation (10.2.1) by ignoring the rest. It seems reasonable to choose $a_0$ and $a_1$ such that the average error in these six equations is minimum. There are many ways to define this average error, but the most convenient and often used is the sum of squares

$$E^2 = (20 - a_0 - 11a_1)^2 + (21 - a_0 - 13a_1)^2 + (24 - a_0 - 17a_1)^2$$
$$+ (26 - a_0 - 19a_1)^2 + (32 - a_0 - 23a_1)^2 + (34 - a_0 - 27a_1)^2$$

Consider a system

$$
\begin{aligned}
a_{11}x_1 + a_{12}x_2 + \cdots + a_{1n}x_n &= b_1 \\
a_{21}x_1 + a_{22}x_2 + \cdots + a_{2n}x_n &= b_2 \\
\vdots \qquad \vdots \qquad \vdots \qquad \vdots \qquad \vdots & \\
a_{m1}x_1 + a_{m2}x_2 + \cdots + a_{mn}x_n &= b_m
\end{aligned}
$$

(10.2.2)

in other words,

$$A\mathbf{x} = \mathbf{b}$$

where $A$ is $m \times n$ and $m > n$. Define the **residual vector**

$$
\begin{aligned}
\mathbf{r} &= \mathbf{b} - A\mathbf{x} \\
&= \left[ b_1 - \sum_{i=1}^{n} a_{1i}x_i, \; b_2 - \sum_{i=1}^{n} a_{2i}x_i, \; \ldots, \; b_m - \sum_{i=1}^{n} a_{mi}x_i \right]^t \\
&= [r_1, r_2, \ldots, r_m]^t
\end{aligned}
$$

(10.2.3)

Then

$$
\begin{aligned}
E^2 &= r_1^2 + r_2^2 + \cdots + r_m^2 \\
&= \mathbf{r}'\mathbf{r} \\
&= (\mathbf{b} - A\mathbf{x})'(\mathbf{b} - A\mathbf{x}) \\
&= (b_1 - a_{11}x_1 - a_{12}x_2 - \cdots - a_{1n}x_n)^2 + (b_2 - a_{21}x_1 - a_{22}x_2 - \cdots - a_{2n}x_n)^2 \\
&\quad + \cdots + (b_m - a_{m1}x_1 - a_{m2}x_2 - \cdots - a_{mn}x_n)^2
\end{aligned}
$$

We wish to find $\mathbf{x} \in R^n$ for which $E^2$ is minimum. A vector $\mathbf{x}$ that makes $E^2$ minimum is called a **least squares solution** of $A\mathbf{x} = \mathbf{b}$. $E^2$ is minimum where

$$\frac{\partial}{\partial x_1}(\mathbf{r}'\mathbf{r}) = \frac{\partial}{\partial x_2}(\mathbf{r}'\mathbf{r}) = \cdots = \frac{\partial}{\partial x_n}(\mathbf{r}'\mathbf{r}) = 0 \qquad (10.2.4)$$

Since $\mathbf{r}'\mathbf{r} = r_1^2 + r_2^2 + \cdots + r_m^2$,

$$
\begin{aligned}
\frac{\partial}{\partial x_1}(\mathbf{r}'\mathbf{r}) &= \frac{\partial}{\partial x_1}(r_1^2 + r_2^2 + \cdots + r_m^2) \\
&= 2r_1 \frac{\partial r_1}{\partial x_1} + 2r_2 \frac{\partial r_2}{\partial x_1} + \cdots + 2r_m \frac{\partial r_m}{\partial x_1} \\
&= -2r_1 a_{11} - 2r_2 a_{21} - \cdots - 2r_m a_{m1} \\
&= -2 \sum_{i=1}^{m} r_i a_{i1}
\end{aligned}
$$

Thus, it can be shown that

$$\frac{\partial}{\partial x_j}(\mathbf{r}'\mathbf{r}) = -2 \sum_{i=1}^{m} r_i a_{ij} \qquad (10.2.5)$$

Substituting Equation (10.2.5) in Equation (10.2.4) yields

$$\sum_{i=1}^{m} r_i a_{ij} = \sum_{i=1}^{m} a_{ij} r_i = 0 \quad \text{for } j = 1, 2, \ldots, n \qquad (10.2.6)$$

in other words,

$$\begin{bmatrix} a_{11} & a_{21} & \cdots & a_{m1} \\ a_{12} & a_{22} & \cdots & a_{m2} \\ \vdots & \vdots & \ddots & \vdots \\ a_{1n} & a_{2n} & \cdots & a_{mn} \end{bmatrix} \begin{bmatrix} r_1 \\ \vdots \\ r_m \end{bmatrix} = \begin{bmatrix} 0 \\ 0 \\ \vdots \\ 0 \end{bmatrix}$$

The system of Equation (10.2.6) is

$$A'\mathbf{r} = \mathbf{0} \tag{10.2.7}$$

Substituting $\mathbf{r} = \mathbf{b} - A\mathbf{x}$ from Equation (10.2.3) in Equation (10.2.7), we get

$$A'(\mathbf{b} - A\mathbf{x}) = \mathbf{0} \tag{10.2.8}$$

Therefore

$$A'A\mathbf{x} = A'\mathbf{b} \tag{10.2.9}$$

Equation (10.2.9) is called a **normal equation**. A solution $\mathbf{x}$ of Equation (10.2.9) will be a least squares solution of Equation (10.2.2) if the columns of $A$ are linearly independent.

**EXAMPLE 10.2.1**    The surface tension $\sigma$ of a liquid is known to be a linear function of temperature $T$ given by $\sigma = a_0 + a_1 T$. Find $a_0$ and $a_1$ to fit the following measurements of surface tension at given temperatures:

| Temperature $T$ | 0°C | 20°C | 40°C | 60°C | 80°C | 100°C |
|---|---|---|---|---|---|---|
| Surface Tension $\sigma$ | 81.4 | 77.7 | 74.2 | 72.4 | 70.3 | 68.8 |

Using

$$A = \begin{bmatrix} 1 & 0 \\ 1 & 20 \\ 1 & 40 \\ 1 & 60 \\ 1 & 80 \\ 1 & 100 \end{bmatrix} \qquad \mathbf{a} = \begin{bmatrix} a_0 \\ a_1 \end{bmatrix} \qquad \mathbf{b} = \begin{bmatrix} 81.4 \\ 77.7 \\ 74.2 \\ 72.4 \\ 70.3 \\ 68.8 \end{bmatrix}$$

Equation (10.2.9) $A'A\mathbf{x} = A'\mathbf{b}$ reduces to

$$\begin{bmatrix} 1 & 1 & 1 & 1 & 1 & 1 \\ 0 & 20 & 40 & 60 & 80 & 100 \end{bmatrix} \begin{bmatrix} 1 & 0 \\ 1 & 20 \\ 1 & 40 \\ 1 & 60 \\ 1 & 80 \\ 1 & 100 \end{bmatrix} \begin{bmatrix} a_0 \\ a_1 \end{bmatrix}$$

$$= \begin{bmatrix} 1 & 1 & 1 & 1 & 1 & 1 \\ 0 & 20 & 40 & 60 & 80 & 100 \end{bmatrix} \begin{bmatrix} 81.4 \\ 77.7 \\ 74.2 \\ 72.4 \\ 70.3 \\ 68.8 \end{bmatrix}$$

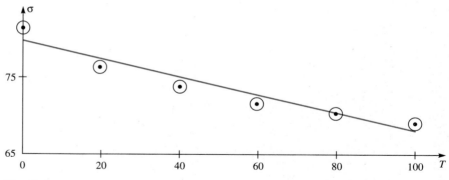

**Fig. 10.2.1**

This simplifies to

$$\begin{bmatrix} 6 & 300 \\ 300 & 22,000 \end{bmatrix} \begin{bmatrix} a_0 \\ a_1 \end{bmatrix} = \begin{bmatrix} 444.8 \\ 21,370.0 \end{bmatrix}$$

Observe that the matrix $A'A$ is symmetric. Solving, we get $a_0 \approx 80.3$ and $a_1 \approx -0.1$. Thus

$$\sigma = 80.3 - 0.1T$$

The graphs of the data points and $\sigma = 80.3 - 0.1T$ are shown in Figure 10.2.1.
Also, $e_i = P_1(x_i) - y_i$ are given in Table 10.2.1 where $P_1(x) = 80.3 - 0.1x$.

**Table 10.2.1**

| $x_i$ | $y_i$ | $P_1(x_i)$ | $|e_i|$ | $e_i^2$ |
|---|---|---|---|---|
| 0 | 81.4 | 80.3 | 1.1 | 1.21 |
| 20 | 77.7 | 77.9 | 0.2 | 0.04 |
| 40 | 74.2 | 75.4 | 1.2 | 1.44 |
| 60 | 72.4 | 72.9 | 0.5 | 0.25 |
| 80 | 70.3 | 70.4 | 0.1 | 0.01 |
| 100 | 68.8 | 67.9 | 0.9 | 0.81 |
| | | | 4.0 | 3.76 |

■ ■ ■

Many times an overdetermined system arises when we try to find $a_0$ and $a_1$ such that $y = a_0 + a_1x$ is the least squares fit to the data given in Table 10.2.2.

**Table 10.2.2**

| $x$ | $x_1$ | $x_2$ | $x_3$ | . . . | $x_N$ |
|---|---|---|---|---|---|
| $y$ | $y_1$ | $y_2$ | $y_3$ | . . . | $y_N$ |

For each pair $(x_i, y_i)$ in Table 10.2.2, the equation $y_i = a_0 + a_1x_i$ should hold. Therefore

$$\begin{bmatrix} 1 & x_1 \\ 1 & x_2 \\ \vdots & \vdots \\ 1 & x_N \end{bmatrix} \begin{bmatrix} a_0 \\ a_1 \end{bmatrix} = \begin{bmatrix} y_1 \\ y_2 \\ \vdots \\ y_N \end{bmatrix}$$

The normal equation $A'A\mathbf{x} = A'\mathbf{b}$ reduces to

$$\begin{bmatrix} 1 & 1 & \ldots & 1 \\ x_1 & x_2 & \ldots & x_N \end{bmatrix} \begin{bmatrix} 1 & x_1 \\ 1 & x_2 \\ \vdots & \vdots \\ 1 & x_N \end{bmatrix} \begin{bmatrix} a_0 \\ a_1 \end{bmatrix} = \begin{bmatrix} 1 & 1 & \ldots & 1 \\ x_1 & x_2 & \ldots & x_N \end{bmatrix} \begin{bmatrix} y_1 \\ y_2 \\ \vdots \\ y_N \end{bmatrix} \qquad \textbf{(10.2.10)}$$

Equation (10.2.10) simplifies to

$$\begin{bmatrix} N & \sum\limits_{i=1}^{N} x_i \\ \sum\limits_{i=1}^{N} x_i & \sum\limits_{i=1}^{N} x_i^2 \end{bmatrix} \begin{bmatrix} a_0 \\ a_1 \end{bmatrix} = \begin{bmatrix} \sum\limits_{i=1}^{N} y_i \\ \sum\limits_{i=1}^{N} x_i y_i \end{bmatrix} \qquad \textbf{(10.2.11)}$$

$A'A$ in Equation (10.2.11) is symmetric. Solving Equation (10.2.11) for $a_0$ and $a_1$, we get

$$a_0 = \frac{\sum\limits_{i=1}^{N} y_i \sum\limits_{i=1}^{N} x_i^2 - \left(\sum\limits_{i=1}^{N} x_i y_i\right)\left(\sum\limits_{i=1}^{N} x_i\right)}{N\left(\sum\limits_{i=1}^{N} x_i^2\right) - \left(\sum\limits_{i=1}^{N} x_i\right)^2}$$

$$a_1 = \frac{N\left(\sum\limits_{i=1}^{N} x_i y_i\right) - \left(\sum\limits_{i=1}^{N} y_i\right)\left(\sum\limits_{i=1}^{N} x_i\right)}{N\left(\sum\limits_{i=1}^{N} x_i^2\right) - \left(\sum\limits_{i=1}^{N} x_i\right)^2} \qquad \textbf{(10.2.12)}$$

We could use a polynomial of degree $M$ given by

$$P_M(x) = a_0 + a_1x + \cdots + a_Mx^M \qquad \textbf{(10.2.13)}$$

where the coefficients $a_0, a_1, \ldots, a_M$ are to be determined to fit a given set of data points $\{(x_1, y_1), (x_2, y_2), \ldots, (x_N, y_N)\}$. Since each pair $(x_i, y_i)$ satisfies Equation (10.2.13), we get

$$\begin{bmatrix} 1 & x_1 & x_1^2 & \ldots & x_1^M \\ 1 & x_2 & x_2^2 & \ldots & x_2^M \\ \vdots & \vdots & \vdots & \ddots & \vdots \\ 1 & x_N & x_N^2 & \ldots & x_N^M \end{bmatrix} \begin{bmatrix} a_0 \\ a_1 \\ \vdots \\ a_M \end{bmatrix} = \begin{bmatrix} y_1 \\ y_2 \\ \vdots \\ y_N \end{bmatrix} \qquad \textbf{(10.2.14)}$$

The least squares solution of Equation (10.2.14) is given by

$$A'A\mathbf{a} = A'\mathbf{b} \qquad \textbf{(10.2.15)}$$

where

$$A = \begin{bmatrix} 1 & x_1 & x_1^2 & \cdots & x_1^M \\ 1 & x_2 & x_2^2 & \cdots & x_2^M \\ \vdots & \vdots & \vdots & \ddots & \vdots \\ 1 & x_N & x_N^2 & \cdots & x_N^M \end{bmatrix} \quad \mathbf{a} = \begin{bmatrix} a_0 \\ a_1 \\ \vdots \\ a_M \end{bmatrix} \quad \mathbf{b} = \begin{bmatrix} y_1 \\ y_2 \\ \vdots \\ y_N \end{bmatrix} \quad \text{(10.2.16)}$$

The $N \times (M + 1)$ matrix $A$ is a Vandermonde matrix.

## Underdetermined Systems

Consider the equation

$$x_1 + x_2 = 1 \quad \text{(10.2.17)}$$

Equation (10.2.17) has an infinite number of solutions. The point $P(s_1, s_2)$ on the graph of Equation (10.2.17) closest to the origin as shown in Figure 10.2.2 is the least squares

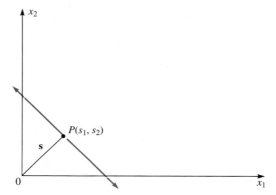

**Fig. 10.2.2**

solution of Equation (10.2.17). If $(s_1, s_2)$ is the **least squares solution** of Equation (10.2.17), then the vector **s** joining the origin to $(s_1, s_2)$ is perpendicular to the line (10.2.17).

The vector **s** is perpendicular to $x_1 + x_2 = 1$. The set of points forming the vector **s** has the form $\gamma \begin{bmatrix} 1 \\ 1 \end{bmatrix}$ where $\gamma$ is a real number. $P(s_1, s_2)$ is on the vector **s**, hence

$$\gamma \begin{bmatrix} 1 \\ 1 \end{bmatrix} = \begin{bmatrix} \gamma \\ \gamma \end{bmatrix} = \begin{bmatrix} s_1 \\ s_2 \end{bmatrix} \quad \text{(10.2.18)}$$

Comparing the elements in Equation (10.2.18), we get $s_1 = \gamma$ and $s_2 = \gamma$. Since $P(s_1, s_2)$ is on (10.2.17), $s_1 + s_2 = 2\gamma = 1$. Thus $\gamma = \frac{1}{2}$. Hence the least squares solution of Equation (10.2.17) is $(\frac{1}{2}, \frac{1}{2})$.

Consider a general underdetermined linear system

$$A\mathbf{x} = \mathbf{b} \quad \text{(10.2.19)}$$

where $A$ is $m \times n$, $\mathbf{x}$ is $n \times 1$, $\mathbf{b}$ is $m \times 1$, and $m < n$. In other words, Equation (10.2.19) contains more unknowns than the number of equations. There are an infinite number of solutions of Equation (10.2.19), but we are interested in finding the least squares solution. Suppose $\mathbf{y}$ is any solution of Equation (10.2.19) and $\mathbf{z}$ is a vector such that $A\mathbf{z} = \mathbf{0}$. Thus $A(\mathbf{y} + \mathbf{z}) = A\mathbf{y} + A\mathbf{z} = A\mathbf{y} = \mathbf{b}$. Thus $\mathbf{y} + \mathbf{z}$ is also a solution of Equation (10.2.19).

We have $A\mathbf{x} = \mathbf{0}$ is the null space of $A$ and $A\mathbf{x} = \mathbf{b}$ is the solution space of $A$. The solution vector $\mathbf{s}$ through the origin is also orthogonal to the null space of $A$. The set of vectors orthogonal to the null space of $A$ is a linear combination of the rows of $A$. In other words, the set of vectors orthogonal to the null space of $A$ has the form $A'\mathbf{\Gamma}$ where $\mathbf{\Gamma}$ is an arbitrary vector. Thus if $\mathbf{s}$ denotes the least squares solution of $A\mathbf{x} = \mathbf{b}$, then $\mathbf{s} = A'\mathbf{\Gamma}$ for arbitrary $\mathbf{\Gamma}$.

For the least squares solution $\mathbf{s}$ of $A\mathbf{x} = \mathbf{b}$,

1. solve $A\mathbf{s} = A(A'\mathbf{\Gamma}) = \mathbf{b}$ for $\mathbf{\Gamma}$ and then
2. solve $\mathbf{s} = A'\mathbf{\Gamma}$ for $\mathbf{s}$.

**EXAMPLE 10.2.2**    Find the least squares solution of the underdetermined system

$$2x_1 + 3x_2 + x_3 = 6 \quad \text{and} \quad x_1 - 2x_2 + 3x_3 = 2$$

We have

$$\begin{bmatrix} 2 & 3 & 1 \\ 1 & -2 & 3 \end{bmatrix} \begin{bmatrix} x_1 \\ x_2 \\ x_3 \end{bmatrix} = \begin{bmatrix} 6 \\ 2 \end{bmatrix}$$

First we solve $A A'\mathbf{\Gamma} = \mathbf{b}$ for $\mathbf{\Gamma}$. We have

$$\begin{bmatrix} 2 & 3 & 1 \\ 1 & -2 & 3 \end{bmatrix} \begin{bmatrix} 2 & 1 \\ 3 & -2 \\ 1 & 3 \end{bmatrix} \begin{bmatrix} \gamma_1 \\ \gamma_2 \end{bmatrix} = \begin{bmatrix} 6 \\ 2 \end{bmatrix}$$

which reduces to

$$\begin{bmatrix} 14 & -1 \\ -1 & 14 \end{bmatrix} \begin{bmatrix} \gamma_1 \\ \gamma_2 \end{bmatrix} = \begin{bmatrix} 6 \\ 2 \end{bmatrix}$$

Solving for $\mathbf{\Gamma}$, we get $\gamma_1 = 86/195$ and $\gamma_2 = 34/195$.

Further

$$\mathbf{s} = A'\mathbf{\Gamma} = \begin{bmatrix} 2 & 1 \\ 3 & -2 \\ 1 & 3 \end{bmatrix} \begin{bmatrix} 86/195 \\ 34/195 \end{bmatrix} = \begin{bmatrix} 206/195 \\ 190/195 \\ 188/195 \end{bmatrix}$$

So the least squares solution is

$$x_1 = \frac{206}{195} = 1.05641 \qquad x_2 = \frac{190}{195} = 0.97436 \qquad x_3 = \frac{188}{195} = 0.96410$$

Notice that $x_1$, $x_2$, and $x_3$ satisfy the equations exactly as expected.    ■ ■ ■

1. Using the least squares method, find the linear polynomial that fits the following data:

| $x_i$ | $-1$ | $0$ | $1$ |
|---|---|---|---|
| $y_i$ | $-3$ | $-1$ | $2$ |

2. Using the least squares method, find the constant function that fits the following data:

| $x_i$ | 2 | 3 | 4 |
|---|---|---|---|
| $y_i$ | 2 | 9/4 | 7/4 |

3. Using the least squares method, find the quadratic polynomial that fits the following data:

| $x_i$ | 1 | 2 | 3 | 4 |
|---|---|---|---|---|
| $y_i$ | 1 | 2 | 4 | 9 |

4. In the denominator of Equation (10.2.12), we have

$$\text{deno} = N \left( \sum_{i=1}^{N} x_i^2 \right) - \left( \sum_{i=1}^{N} x_i \right)^2$$

Let $\bar{x} = (\sum_{i=1}^{N} x_i)/N$. Then verify that deno $= N \sum_{i=1}^{N} (x_i - \bar{x})^2$. Also verify that

$$\text{deno} = \frac{1}{2} \sum_{i=1}^{N} \sum_{j=i}^{N} (x_i - x_j)^2$$

by expanding $\sum_{i=1}^{N} \sum_{j=i}^{N} (x_i - x_j)^2$. Therefore deno is zero only if all points $x_1$, $x_2, \ldots, x_N$ are identical.

5. Hooke's law states that when a force is applied to a spring of uniform material, the increase in length of the spring is proportional to the applied force; in other words, $F = kx$ where $F$ is the force (in pounds), $k$ is the spring constant (the constant of proportionality), and $x$ is the increase in length of the spring (in inches). In order to determine $k$, an experiment was performed. Using the least squares method, determine $k$ that fits the following data:

| $x_i$ | 0.1 | 0.3 | 0.5 | 0.7 | 0.9 |
|---|---|---|---|---|---|
| $y_i = F_i$ | 1.1 | 3.2 | 4.9 | 6.8 | 9.1 |

Plot these points and the linear least squares approximation on the same axes.

6. The gravitational constant $g$ can be determined by $d = \frac{1}{2}gt^2$ where $d$ is the distance in feet and $t$ is the time in seconds. An experiment was performed to

determine $g$. Using the least squares method, determine $g$ that fits the following data:

| $t_i$ | 0.5 | 1 | 1.5 | 2.0 |
|---|---|---|---|---|
| $d_i$ | 4.2 | 16.1 | 35.9 | 64.2 |

7.  Determine the approximate solution in the least squares sense of the overdetermined system

$$x + y = 2 \qquad x - y = 2 \qquad 3x - 2y = 2$$

8.  Show that the solution to a least squares problem is not necessarily unique by determining solutions of the system

$$x - y = 1 \qquad 2x - 2y = 3 \qquad -x + y = 1$$

9.  Prove that a matrix $A'A$ in Equation (10.2.9) is a positive definite matrix if the columns of the matrix $A$ are linearly independent.

10.  Determine the least squares solution of $2x_1 + 3x_2 = 5$.

11.  Determine the least squares solution of the underdetermined system of linear equations

$$x_1 + x_2 - 6x_3 = 1 \quad \text{and} \quad 2x_1 + 3x_2 + 5x_3 = 2$$

**COMPUTER EXERCISES**

**C.1.**  Write a subroutine that fits a straight line $y = a_0 + a_1 x$ to a given table using Equation (10.2.12).

**C.2.**  Using Exercise C.1, determine $a_0$ and $a_1$ such that $h = a_0 + a_1 r$ fits to the data points given in the beginning of this chapter in Table 1.

## 10.3  THE GENERAL LEAST SQUARES PROBLEM

Consider a mixture of three radioactive materials. We know their rates of decay but we do not know how much of each is in the mixture. Let these three unknown amounts be $a_0$, $a_1$, and $a_2$. Then our experimental data would behave like

$$y = a_0 e^{-\alpha_1 t} + a_1 e^{-\alpha_2 t} + a_2 e^{-\alpha_3 t}$$

We extend the least squares procedure to determine $a_0$, $a_1$, and $a_2$ in this section. Let $\phi_0(x)$, $\phi_1(x)$, $\ldots$, $\phi_M(x)$ be linearly independent functions and $w_1, w_2, \ldots, w_N$ be positive numbers called weights. Different values of $w_i$ allow us to attach varying degrees of importance to different data points. If all points are given equal importance, then we take $w_i = 1$ for $i = 1, 2, \ldots, N$. Let $(x_1, y_1), (x_2, y_2), \ldots, (x_N, y_N)$ be given $N$ data points and we want to fit

$$g(x) = \sum_{j=0}^{M} a_j \, \phi_j(x) \tag{10.3.1}$$

by selecting $a_0, a_1, \ldots, a_M$ in such a way that

$$E(a_0, a_1, \ldots, a_M) = \sum_{i=1}^{N} w_i \left[ y_i - \sum_{j=0}^{M} a_j \, \phi_j(x_i) \right]^2 \tag{10.3.2}$$

is minimum. The partial derivative with respect to $a_k$ is

$$\frac{\partial E}{\partial a_k} = -2 \sum_{i=1}^{N} w_i \left[ y_i - \sum_{j=0}^{M} a_j \, \phi_j(x_i) \right] \phi_k(x_i) \tag{10.3.3}$$

for $k = 0, 1, \ldots, M$. Setting the derivatives in Equation (10.3.3) to zero and simplifying, we get the normal equations for Equation (10.3.1)

$$a_0 \left[ \sum_{i=1}^{N} w_i \phi_0(x_i)\phi_k(x_i) \right] + a_1 \left[ \sum_{i=1}^{N} w_i \phi_1(x_i)\phi_k(x_i) \right] + a_2 \left[ \sum_{i=1}^{N} w_i \phi_2(x_i)\phi_k(x_i) \right]$$

$$+ \cdots + a_M \left[ \sum_{i=1}^{N} w_i \phi_M(x_i)\phi_k(x_i) \right] = \sum_{i=1}^{N} w_i y_i \phi_k(x_i) \quad \text{for } k = 0, 1, \ldots, M$$

This can be written in matrix form as

$$\begin{bmatrix} \sum_{i=1}^{N} w_i \phi_0(x_i)\phi_0(x_i) & \sum_{i=1}^{N} w_i \phi_1(x_i)\phi_0(x_i) & \cdots & \sum_{i=1}^{N} w_i \phi_M(x_i)\phi_0(x_i) \\ \sum_{i=1}^{N} w_i \phi_0(x_i)\phi_1(x_i) & \sum_{i=1}^{N} w_i \phi_1(x_i)\phi_1(x_i) & \cdots & \sum_{i=1}^{N} w_i \phi_M(x_i)\phi_1(x_i) \\ \vdots & \vdots & \ddots & \vdots \\ \sum_{i=1}^{N} w_i \phi_0(x_i)\phi_M(x_i) & \sum_{i=1}^{N} w_i \phi_1(x_i)\phi_M(x_i) & \cdots & \sum_{i=1}^{N} w_i \phi_M(x_i)\phi_M(x_i) \end{bmatrix} \begin{bmatrix} a_0 \\ a_1 \\ \vdots \\ a_M \end{bmatrix}$$

$$= \begin{bmatrix} \sum_{i=1}^{N} w_i y_i \phi_0(x_i) \\ \sum_{i=1}^{N} w_i y_i \phi_1(x_i) \\ \vdots \\ \sum_{i=1}^{N} w_i y_i \phi_M(x_i) \end{bmatrix} \tag{10.3.4}$$

Equation (10.3.4) can be written as

$$A'A\mathbf{a} = A'\mathbf{b} \tag{10.3.5}$$

where

$$A = \begin{bmatrix} \sqrt{w_1}\,\phi_0(x_1) & \sqrt{w_1}\,\phi_1(x_1) & \cdots & \sqrt{w_1}\,\phi_M(x_1) \\ \sqrt{w_2}\,\phi_0(x_2) & \sqrt{w_2}\,\phi_1(x_2) & \cdots & \sqrt{w_2}\,\phi_M(x_2) \\ \vdots & \vdots & \ddots & \vdots \\ \sqrt{w_N}\,\phi_0(x_N) & \sqrt{w_N}\,\phi_1(x_N) & \cdots & \sqrt{w_N}\,\phi_M(x_N) \end{bmatrix}$$

$$\mathbf{a} = \begin{bmatrix} a_0 \\ a_1 \\ \vdots \\ a_M \end{bmatrix} \qquad \mathbf{b} = \begin{bmatrix} \sqrt{w_1}\,y_1 \\ \sqrt{w_2}\,y_2 \\ \vdots \\ \sqrt{w_N}\,y_N \end{bmatrix} \tag{10.3.6}$$

If the functions $\phi_i(x)$ are linearly dependent, $A$ can be singular. The most reliable method for computing the coefficients of a general least squares problem is the singular value decomposition (SVD) method to be discussed in the next section.

**EXAMPLE 10.3.1**    Construct the least squares approximation of the form $g(x) = a_0 + a_1 e^x + a_2 e^{-x}$ that fits the following data:

| $x_i$ | 0 | 1 | 2 | 3 |
|---|---|---|---|---|
| $y_i$ | 6 | 7.5 | 16.2 | 41.3 |
| $w_i$ | 1 | 4 | 4 | 1 |

We have $\phi_0(x) = 1$, $\phi_1(x) = e^x$, and $\phi_2(x) = e^{-x}$ with $w_1 = 1$, $w_2 = 4$, $w_3 = 4$, and $w_4 = 1$. Then Equation (10.3.6) becomes

$$A = \begin{bmatrix} 1 & e^0 & e^0 \\ 2 & 2e^1 & 2e^{-1} \\ 2 & 2e^2 & 2e^{-2} \\ 1 & e^3 & e^{-3} \end{bmatrix} \quad \mathbf{a} = \begin{bmatrix} a_0 \\ a_1 \\ a_2 \end{bmatrix} \quad \mathbf{b} = \begin{bmatrix} 6 \\ 2(7.5) \\ 2(16.2) \\ 41.3 \end{bmatrix} = \begin{bmatrix} 6 \\ 15.0 \\ 32.4 \\ 41.3 \end{bmatrix}$$

Hence Equation (10.3.5) gives

$$\begin{bmatrix} 10 & 61.51 & 3.06 \\ 61.51 & 652.38 & 10 \\ 3.06 & 10 & 1.62 \end{bmatrix} \begin{bmatrix} a_0 \\ a_1 \\ a_2 \end{bmatrix} = \begin{bmatrix} 142.10 \\ 1395.89 \\ 27.86 \end{bmatrix}$$

Solving these equations, we get $a_0 \approx 0.99$, $a_1 \approx 2.00$, and $a_2 \approx 2.99$ and the least squares fit is $g(x) = 0.99 + 2.00e^x + 2.99e^{-x}$.    ■ ■ ■

We consider the exponential form

$$g(x) = a_0 e^{a_1 x} \tag{10.3.7}$$

to fit given $N$ data points $(x_1, y_1), (x_2, y_2), \ldots, (x_N, y_N)$ with weights $w_1, w_2, \ldots, w_N$. The least squares procedure requires that we minimize

$$E(a_0, a_1) = \sum_{i=1}^{N} w_i[y_i - a_0 e^{a_1 x_i}]^2 \tag{10.3.8}$$

Necessary conditions for a minimum to occur at $a_0$ and $a_1$ are

$$\frac{\partial E}{\partial a_0} = \frac{\partial E}{\partial a_1} = 0$$

which give

$$\frac{\partial E}{\partial a_0} = -2 \sum_{i=1}^{N} w_i[y_i - a_0 e^{a_1 x_i}]e^{a_1 x_i} = 0$$

and

$$\frac{\partial E}{\partial a_1} = -2 \sum_{i=1}^{N} w_i[y_i - a_0 e^{a_1 x_i}]x_i a_0 e^{a_1 x_i} = 0 \tag{10.3.9}$$

Simplifying Equation (10.3.9), the normal equations for the exponential fit are given by

$$a_0\left(\sum_{i=1}^{N} w_i e^{2a_1 x_i}\right) - \sum_{i=1}^{N} w_i y_i e^{a_1 x_i} = 0$$

and

$$a_0\left(\sum_{i=1}^{N} w_i x_i e^{2a_1 x_i}\right) - \sum_{i=1}^{N} w_i y_i x_i e^{a_1 x_i} = 0 \qquad \textbf{(10.3.10)}$$

The system given by Equation (10.3.10) is nonlinear in $a_0$ and $a_1$ and is solved by the methods to be discussed in Chapter 12. However, another common approach is to take the natural logarithm of both sides of Equation (10.3.7) and get

$$\ln g(x) = \ln a_0 + a_1 x \qquad \textbf{(10.3.11)}$$

Substituting

$$\ln g(x) = Y \qquad \ln a_0 = A \qquad x = X \qquad a_1 = B \qquad \textbf{(10.3.12)}$$

in Equation (10.3.11) leads to

$$Y = A + BX \qquad \textbf{(10.3.13)}$$

The values of $A$ and $B$ are obtained by minimizing

$$E(A, B) = \sum_{i=1}^{N} w_i [Y_i - (A + BX_i)]^2 \qquad \textbf{(10.3.14)}$$

where

$$X_i = x_i \quad \text{and} \quad Y_i = \ln y_i \qquad \textbf{(10.3.15)}$$

It can be verified (Exercise 2) that the normal equations for Equation (10.3.13) are given by

$$A\left(\sum_{i=1}^{N} w_i\right) + B\left(\sum_{i=1}^{N} w_i X_i\right) = \sum_{i=1}^{N} w_i Y_i$$

and

$$A\left(\sum_{i=1}^{N} w_i X_i\right) + B\left(\sum_{i=1}^{N} w_i X_i^2\right) = \sum_{i=1}^{N} w_i X_i Y_i \qquad \textbf{(10.3.16)}$$

Solving Equation (10.3.16) for $A$ and $B$, yields $a_0 = e^A$ and $a_1 = B$. However, this finds the least squares approximation of the transformed problem and not of the original problem.

**EXAMPLE 10.3.2**   Construct the least squares approximation of the form $y = a_0 e^{a_1 x}$ (using Equation (10.3.12)) that fits the following data:

| $x_i$ | 0 | 1 | 2 | 3 |
|-------|-----|-------|-------|-------|
| $y_i$ | 2.1 | 1.214 | 0.736 | 0.447 |
| $w_i$ | 1 | 1 | 1 | 1 |

**Table 10.3.1**

| $w_i$ | $X_i$ | $Y_i$ | $w_i X_i$ | $w_i X_i^2$ | $w_i Y_i$ | $w_i X_i Y_i$ |
|---|---|---|---|---|---|---|
| 1 | 0 | 0.74194 | 0 | 0 | 0.74194 | 0 |
| 1 | 1 | 0.19392 | 1 | 1 | 0.19392 | 0.19392 |
| 1 | 2 | −0.30653 | 2 | 4 | −0.30653 | −0.61306 |
| 1 | 3 | −0.80520 | 3 | 9 | −0.80520 | −2.41560 |
| | | | 6 | 14 | −0.17587 | −2.83474 |

The data corresponding to the new variables $X = x$, $Y = \ln y$ is given in Table 10.3.1.

The linear system of Equation (10.3.16) for $A$ and $B$ is

$$4A + 6B = -0.17587$$

$$6A + 14B = -2.83474$$

Solving these equations, we get $A = 0.72731$ and $B = -0.51418$ and hence $a_0 = e^A = 2.06951$ and $a_1 = B$. The resulting exponential fit is given by $g(x) = 2.06951\ e^{-0.51418x}$. ■ ■ ■

## EXERCISES

1. Using the least squares method, find the function of the form $g(x) = a_0 e^x$ that fits the following data:

   | $x_i$ | 0 | 1 | 2 | 3 |
   |---|---|---|---|---|
   | $y_i$ | 1.1 | 3.0 | 8.12 | 22 |

2. Verify that the normal equations for Equation (10.3.13) are given by Equation (10.3.16).

3. Find the normal equations for fitting the following guess functions to $N$ data points using the least squares method:
   (a) $g(x) = a_0\sqrt{x} + a_1 x$   (b) $g(x) = a_0 \sin a_1 x$   (c) $g(x) = a_0 + (a_1/x^2)$

4. In some cases by using proper transformations, the guess function can be transformed to a linear function whose normal equations are linear equations. Find proper transformations to transform the following guess functions to linear functions:
   (a) $g(x) = a_0 + a_1\sqrt{x}$   (b) $g(x) = a_0 x^{a_1}$   (c) $g(x) = 1/(a_0 x + a_1)$
   (d) $g(x) = a_0 x e^{-a_1 x}$   (e) $g(x) = 1/(a_0 x + a_1)^2$

5. The ideal gas law is known to be $pv^\gamma = c$ where $\gamma$ and $c$ are to be determined. An experiment was performed to determine $\gamma$ and $c$. Using the least squares method, determine $\gamma$ and $c$ that fit the following data:

   | $v$ (cm³) | 50 | 60 | 70 | 80 | 90 |
   |---|---|---|---|---|---|
   | $p$ (kg/cm²) | 63.9 | 52.0 | 39.9 | 22.8 | 16.7 |

▄▄▄▄▄▄▄▄▄▄ C O M P U T E R   E X E R C I S E ▄▄▄▄▄▄▄▄▄▄

**C.1.** The following data comes from a process that behaves like $y = f(x) = axe^{-bx}$, $a > 0$:

| $x$ | 0.1 | 0.2 | 0.3 | 0.4 | 0.5 | 0.6 | 0.7 | 0.8 | 0.9 | 1.0 |
|---|---|---|---|---|---|---|---|---|---|---|
| $y$ | 0.158 | 0.332 | 0.523 | 0.733 | 0.963 | 1.215 | 1.49 | 1.79 | 2.117 | 2.473 |

Take the natural log on both sides of $y = axe^{-bx}$ to obtain $\ln y = \ln a + \ln x - bx$. Set $z = \ln(y/x)$ and then fit $z = a_0 + a_1 x$ to the data points.

## 10.4 THE SINGULAR VALUE DECOMPOSITION AND THE PSEUDOINVERSE

In Sections 10.2 and 10.3, we developed the least squares solutions of the overdetermined and underdetermined systems $A\mathbf{x} = \mathbf{b}$ where the columns and rows of an $m \times n$ matrix were independent. When $A$ has linearly dependent rows or columns, then we call the system $A\mathbf{x} = \mathbf{b}$ **singular**. The least squares solution of the singular system $A\mathbf{x} = \mathbf{b}$ can be obtained using the **singular value decomposition** of $A$. Since we require the eigenvalues and eigenvectors of $A'A$ and $AA'$ for the singular decomposition of $A$, we study the properties of the eigenvalues and eigenvectors of $A'A$ and $AA'$ where $A$ is an $m \times n$ matrix.

Since $(A'A)' = A'(A')' = A'A$, $A'A$ is symmetric. Similarly we can prove that $AA'$ is symmetric. It can also be shown that the eigenvalues of symmetric matrices $A'A$ and $AA'$ are real numbers and also the associated eigenvectors are real (Theorem 13.1.4).

Let $\mathbf{x}$ be an eigenvector of $A'A$ corresponding to a nonzero eigenvalue $\lambda$. Then

$$A'A\mathbf{x} = \lambda\mathbf{x} \qquad (10.4.1)$$

Multiplying Equation (10.4.1) by $\mathbf{x}'$ on the left yields

$$\mathbf{x}'(A'A\mathbf{x}) = \mathbf{x}'(\lambda\mathbf{x}) = \lambda\mathbf{x}'\mathbf{x}$$

Also $\mathbf{x}'(A'A\mathbf{x}) = (\mathbf{x}'A')(A\mathbf{x}) = (A\mathbf{x})'(A\mathbf{x}) \geq 0$. Therefore $\lambda\mathbf{x}'\mathbf{x} \geq 0$, and the eigenvalues of $A'A$ are real and nonnegative.

Multiplication by $A$ on the left of Equation (10.4.1) leads to

$$A(A'A\mathbf{x}) = (AA')(A\mathbf{x}) = \lambda(A\mathbf{x})$$

Thus we have

$$(AA')(A\mathbf{x}) = \lambda(A\mathbf{x}) \qquad (10.4.2)$$

From Equation (10.4.2), we have $A\mathbf{x}$ as an eigenvector of $AA'$ corresponding to its nonzero eigenvalue $\lambda$ where $\mathbf{x}$ is an eigenvector of $A'A$ corresponding to its nonzero eigenvalue $\lambda$. This leads to the following theorem.

**Theorem 10.4.1**  $A'A$ and $AA'$ have the same nonzero eigenvalues and if $\mathbf{x}$ is an eigenvector of $A'A$ corresponding to its nonzero eigenvalue $\lambda$, then $A\mathbf{x}$ is an eigenvector of $AA'$ corresponding to its nonzero eigenvalue $\lambda$. ■

**EXAMPLE 10.4.1**

Find the eigenvalues of $A'A$ and $AA'$ if $A = \begin{bmatrix} 1 & 1 \\ 1 & 1 \\ 2 & 2 \end{bmatrix}$

Since

$$A'A = \begin{bmatrix} 1 & 1 & 2 \\ 1 & 1 & 2 \end{bmatrix} \begin{bmatrix} 1 & 1 \\ 1 & 1 \\ 2 & 2 \end{bmatrix} = \begin{bmatrix} 6 & 6 \\ 6 & 6 \end{bmatrix}$$

then

$$\det(A'A - \lambda I) = \lambda(\lambda - 12)$$

Also we have

$$AA' = \begin{bmatrix} 1 & 1 \\ 1 & 1 \\ 2 & 2 \end{bmatrix} \begin{bmatrix} 1 & 1 & 2 \\ 1 & 1 & 2 \end{bmatrix} = \begin{bmatrix} 2 & 2 & 4 \\ 2 & 2 & 4 \\ 4 & 4 & 8 \end{bmatrix}$$

and

$$\det(AA' - \lambda I) = -\lambda^2(\lambda - 12)$$

$A'A$ and $AA'$ are symmetric and have the same nonzero eigenvalue 12.   ■ ■ ■

The singular value decomposition method has a long history. For square matrices, it was discovered by Beltrami in 1873 and Jordan in 1874 independently. Eckart and Young extended this technique in the 1930s to rectangular matrices. Golub and his colleagues (1983) developed it as a computational tool. The singular value decomposition is another matrix factorization.

**Theorem 10.4.2**

**(Singular Value Decomposition)**   Let $A$ be an $m \times n$ matrix with real elements (assume $m \geq n$ for convenience). Then there are orthogonal matrices $U$ and $V$ of orders $m$ and $n$ respectively such that

$$A = USV' \qquad \textbf{(10.4.3)}$$

where $S$ is an $m \times n$ matrix whose off-diagonal elements are zeros and diagonal elements are

$$\mu_1 \geq \mu_2 \geq \cdots \geq \mu_n \geq 0$$

*Proof*   The eigenvalues of $A'A$ are nonnegative and can be arranged so that

$$\lambda_1 \geq \lambda_2 \geq \cdots \geq \lambda_n \geq 0$$

Since $A'A$ is symmetric, there is an orthogonal matrix $V$ (Theorem 13.1.6) such that

$$V'(A'A)V = D \qquad \textbf{(10.4.4)}$$

where $D$ is a diagonal matrix whose diagonal elements are $\lambda_1, \lambda_2, \ldots, \lambda_n$.

Assume that exactly $r$ of the $\lambda_i$'s are nonzero. Define

$$\mu_i = \sqrt{\lambda_i} \quad \text{for } i = 1, 2, \ldots, r \tag{10.4.5}$$

Let

$$S = \begin{bmatrix} \mu_1 & 0 & 0 & \cdots & 0 & \cdots & \cdots & 0 \\ 0 & \mu_2 & 0 & \cdots & 0 & \cdots & \cdots & 0 \\ \vdots & \vdots & \vdots & \ddots & \vdots & \ddots & \ddots & \vdots \\ 0 & 0 & 0 & \cdots & \mu_r & \cdots & \cdots & 0 \\ \vdots & \vdots & \vdots & \ddots & \vdots & \ddots & \ddots & \vdots \\ 0 & 0 & 0 & \cdots & 0 & \cdots & \cdots & 0 \end{bmatrix} = \begin{bmatrix} S_r & \bigcirc \\ \bigcirc & \bigcirc \end{bmatrix} \tag{10.4.6}$$

where $S$ is an $m \times n$ matrix whose diagonal elements are $\mu_1, \mu_2, \ldots, \mu_r, 0, 0, \ldots,$ 0 and $S_r$ is an $r \times r$ diagonal matrix with diagonal elements $\mu_1, \mu_2, \ldots, \mu_r$. Since $S'S = D$, Equation (10.4.4) reduces to

$$V'(A'A)V = (V'A')(AV) = (AV)'(AV) = S'S \tag{10.4.7}$$

Let $\mathbf{v}_1, \mathbf{v}_2, \ldots, \mathbf{v}_r$ be the column vectors of $V$ corresponding to the eigenvalues $\mu_1^2,$ $\mu_2^2, \ldots, \mu_r^2$ and $\mathbf{v}_{r+1}, \mathbf{v}_{r+2}, \ldots, \mathbf{v}_n$ be column vectors of $V$ corresponding to the eigenvalue 0. Thus the orthogonal matrix $V = [\mathbf{v}_1, \mathbf{v}_2, \ldots, \mathbf{v}_r, \ldots, \mathbf{v}_n]$.

Let $W = AV$. Then Equation (10.4.7) reduces to

$$W'W = S'S \tag{10.4.8}$$

The matrix $W = [\mathbf{w}_1, \mathbf{w}_2, \ldots, \mathbf{w}_r, \ldots, \mathbf{w}_n]$ where $\mathbf{w}_i \in R^m$. From Equation (10.4.8), we have

$$\mathbf{w}_i'\mathbf{w}_i = \begin{cases} \mu_i^2 & \text{if } 1 \le i \le r \\ 0 & i > r \end{cases} \tag{10.4.9}$$

and

$$\mathbf{w}_i'\mathbf{w}_j = 0 \quad \text{if } i \ne j \tag{10.4.10}$$

Since $\mathbf{w}_i = \mathbf{0}$ for $i > r$ from Equations (10.4.9) and (10.4.10), the first $r$ columns of $W$ are linearly independent. Hence $r \le m$.

Define

$$\mathbf{u}_i = \frac{1}{\mu_i} \mathbf{w}_i = \frac{1}{\mu_i} A\mathbf{v}_i \quad \text{for } i = 1, 2, \ldots, r \tag{10.4.11}$$

From Equations (10.4.9) and (10.4.10), Equation (10.4.11) is an orthonormal set. If $r < m$, then choose $\mathbf{u}_{r+1}, \mathbf{u}_{r+2}, \ldots, \mathbf{u}_m$ so that $U = [\mathbf{u}_1, \mathbf{u}_2, \ldots, \mathbf{u}_m]$ is an orthogonal matrix.

From Equation (10.4.11), it follows that $US = AV$. Hence $A = USV'$.   ■

The numbers $\mu_1, \mu_2, \ldots, \mu_r$ in Equation (10.4.6) are unique and are called the **singular values** of $A$.

*Note:* we have

$$AA^t = (USV^t)(USV^t)^t$$
$$= (USV^t)(V(US)^t)$$
$$= USS^tU^t$$

Since $U$ diagonalizes $AA^t$, $\{\mathbf{u}_{r+1}, \mathbf{u}_r, \ldots, \mathbf{u}_m\}$ are orthonormal eigenvectors of $AA^t$ corresponding to the eigenvalue 0.

---

**EXAMPLE 10.4.2**    Find the singular values and the SVD (singular value decomposition) of the matrix

$$A = \begin{bmatrix} 1 & 1 \\ 1 & 1 \\ 2 & 2 \end{bmatrix}$$

From Example 10.4.1, $A^tA = \begin{bmatrix} 6 & 6 \\ 6 & 6 \end{bmatrix}$ has eigenvalues $\lambda = 12$ and 0 and corres-

ponding orthonormal eigenvectors $\mathbf{v}_1 = \begin{bmatrix} 1/\sqrt{2} \\ 1/\sqrt{2} \end{bmatrix}$ and $\mathbf{v}_2 = \begin{bmatrix} 1/\sqrt{2} \\ -1/\sqrt{2} \end{bmatrix}$. Hence $V =$

$(1/\sqrt{2}) \begin{bmatrix} 1 & 1 \\ 1 & -1 \end{bmatrix}$. Since $\mu_1 = \sqrt{12}$, $\mu_2 = 0$,

$$S = \begin{bmatrix} \sqrt{12} & 0 \\ 0 & 0 \\ 0 & 0 \end{bmatrix}$$

From Equation (10.4.11), $\mathbf{u}_i = (A\mathbf{v}_i)/\mu_i$.
For $i = 1$,

$$\mathbf{u}_1 = \frac{A\mathbf{v}_1}{\mu_1} = \frac{1}{\sqrt{12}\sqrt{2}} \begin{bmatrix} 1 & 1 \\ 1 & 1 \\ 2 & 2 \end{bmatrix} \begin{bmatrix} 1 \\ 1 \end{bmatrix} = \frac{1}{2\sqrt{6}} \begin{bmatrix} 2 \\ 2 \\ 4 \end{bmatrix} = \frac{1}{\sqrt{6}} \begin{bmatrix} 1 \\ 1 \\ 2 \end{bmatrix}$$

The remaining column vectors $\mathbf{u}_2$ and $\mathbf{u}_3$ are the eigenvectors of $AA^t$ corresponding to the repeated eigenvalue 0. It can be verified that the eigenvectors of $AA^t$ are

$$\begin{bmatrix} -1 \\ 1 \\ 0 \end{bmatrix} \quad \text{and} \quad \begin{bmatrix} -2 \\ 0 \\ 1 \end{bmatrix}$$

corresponding to the eigenvalue 0. We find the orthonormal vectors

$$\mathbf{u}_2 = \frac{1}{\sqrt{2}} \begin{bmatrix} -1 \\ 1 \\ 0 \end{bmatrix} \quad \text{and} \quad \mathbf{u}_3 = \frac{1}{\sqrt{3}} \begin{bmatrix} -1 \\ -1 \\ 1 \end{bmatrix}$$

using the Gram–Schmidt process (Appendix E). Thus

$$U = \begin{bmatrix} 1/\sqrt{6} & -1/\sqrt{2} & -1/\sqrt{3} \\ 1/\sqrt{6} & 1/\sqrt{2} & -1/\sqrt{3} \\ 2/\sqrt{6} & 0 & 1/\sqrt{3} \end{bmatrix}$$

is orthogonal. Hence

$$A = USV' = \begin{bmatrix} 1/\sqrt{6} & -1/\sqrt{2} & -1/\sqrt{3} \\ 1/\sqrt{6} & 1/\sqrt{2} & -1/\sqrt{3} \\ 2/\sqrt{6} & 0 & 1/\sqrt{3} \end{bmatrix} \begin{bmatrix} \sqrt{12} & 0 \\ 0 & 0 \\ 0 & 0 \end{bmatrix} \begin{bmatrix} 1/\sqrt{2} & 1/\sqrt{2} \\ 1/\sqrt{2} & -1/\sqrt{2} \end{bmatrix}$$

$$= \begin{bmatrix} 1/\sqrt{6} & -1/\sqrt{2} & -1/\sqrt{3} \\ 1/\sqrt{6} & 1/\sqrt{2} & -1/\sqrt{3} \\ 2/\sqrt{6} & 0 & 1/\sqrt{3} \end{bmatrix} \begin{bmatrix} \sqrt{6} & \sqrt{6} \\ 0 & 0 \\ 0 & 0 \end{bmatrix} = \begin{bmatrix} 1 & 1 \\ 1 & 1 \\ 2 & 2 \end{bmatrix} \qquad \blacksquare\blacksquare\blacksquare$$

The **rank** of a matrix is the maximum number of independent columns in the matrix. The rank of a general matrix $A$ is equal to the rank of the diagonal matrix $S$ where $A = USV'$. Thus the rank of a matrix $A$ is the number of nonzero eigenvalues of $AA'$.

If $A$ is a nonsingular matrix of order $n$, then the solution of a linear equation $A\mathbf{x} = \mathbf{b}$ is given by $\mathbf{x} = A^{-1}\mathbf{b}$ where $A^{-1}$ is the inverse of $A$. For any arbitrary $m \times n$ matrix $A$, there is an $n \times m$ matrix $A^+$, the so-called **pseudoinverse**. The least squares solution of $A\mathbf{x} = \mathbf{b}$ where $A$ is a rectangular or a singular matrix can be expressed in terms of the pseudoinverse $A^+$ as $\mathbf{x} = A^+\mathbf{b}$. The pseudoinverse $A^+ = A^{-1}$ when $m = n$ and $A$ is nonsingular. We cannot expect $AA^+ = I$ if $m > n$ since the ranks of $A$, $A^+$, and $A^+A$ are at most $m$. $A^+$ has the following properties (Moore–Penrose properties):

1.  $AA^+A = A$
2.  $A^+AA^+ = A^+$
3.  $(AA^+)' = AA^+$
4.  $(A^+A)' = A^+A$ $\qquad\qquad\qquad\qquad\qquad\qquad\qquad\qquad$ **(10.4.12)**

Alternately, these properties can be used to define the pseudoinverse of a matrix $A$.

The **pseudoinverse** of a number $\mu$ is

$$\mu^+ = \begin{cases} 1/\mu & \text{if } \mu \neq 0 \\ 0 & \mu = 0 \end{cases}$$

and satisfies the four properties given by Equation (10.4.12).

The pseudoinverse of $3 \times 2$ diagonal matrix

$$S = \begin{bmatrix} 2 & 0 \\ 0 & 3 \\ 0 & 0 \end{bmatrix}$$

is defined as the $2 \times 3$ diagonal matrix

$$S^+ = \begin{bmatrix} \frac{1}{2} & 0 & 0 \\ 0 & \frac{1}{3} & 0 \end{bmatrix}$$

We have

$$SS^+ = \begin{bmatrix} 2 & 0 \\ 0 & 3 \\ 0 & 0 \end{bmatrix} \begin{bmatrix} \frac{1}{2} & 0 & 0 \\ 0 & \frac{1}{3} & 0 \end{bmatrix} = \begin{bmatrix} 1 & 0 & 0 \\ 0 & 1 & 0 \\ 0 & 0 & 0 \end{bmatrix}$$

It can be verified that $S^+$ satisfies Equation (10.4.12).

The pseudoinverse of an $m \times n$ diagonal matrix $S$ can be defined similarly.

If $A$ is a nonsingular $n \times n$ matrix with $A = USV'$, then $A^{-1} = (USV')^{-1} = (V')^{-1}(US)^{-1} = VS^{-1}U'$. If $A$ is an $m \times n$ matrix with $A = USV'$, we define the **pseudoinverse** of $A$ to be the $n \times m$ matrix

$$A^+ = VS^+U' \tag{10.4.13}$$

By computing the SVD of $A$, we compute the SVD of $A^+$ since we rarely need the pseudoinverse.

**EXAMPLE 10.4.3** Find $A^+$ if $A = \begin{bmatrix} 1 & 1 \\ 1 & 1 \\ 2 & 2 \end{bmatrix}$.

From Example 10.4.2, we have

$$U = \begin{bmatrix} 1/\sqrt{6} & -1/\sqrt{2} & -1/\sqrt{3} \\ 1/\sqrt{6} & 1/\sqrt{2} & -1/\sqrt{3} \\ 2/\sqrt{6} & 0 & 1/\sqrt{3} \end{bmatrix} \quad S = \begin{bmatrix} \sqrt{12} & 0 \\ 0 & 0 \\ 0 & 0 \end{bmatrix} \quad V' = \begin{bmatrix} 1/\sqrt{2} & 1/\sqrt{2} \\ 1/\sqrt{2} & -1/\sqrt{2} \end{bmatrix}$$

Therefore $S^+ = \begin{bmatrix} 1/\sqrt{12} & 0 & 0 \\ 0 & 0 & 0 \end{bmatrix}$ and

$$A^+ = VS^+U'$$

$$= \frac{1}{\sqrt{2}} \begin{bmatrix} 1 & 1 \\ 1 & -1 \end{bmatrix} \begin{bmatrix} 1/\sqrt{12} & 0 & 0 \\ 0 & 0 & 0 \end{bmatrix} \begin{bmatrix} 1/\sqrt{6} & 1/\sqrt{6} & 2/\sqrt{6} \\ -1/\sqrt{2} & 1/\sqrt{2} & 0 \\ -1/\sqrt{3} & -1/\sqrt{3} & 1/\sqrt{3} \end{bmatrix}$$

$$= \frac{1}{\sqrt{2}} \begin{bmatrix} 1 & 1 \\ 1 & -1 \end{bmatrix} \begin{bmatrix} 1/(6\sqrt{2}) & 1/(6\sqrt{2}) & \sqrt{2}/6 \\ 0 & 0 & 0 \end{bmatrix}$$

$$= \frac{1}{\sqrt{2}} \begin{bmatrix} 1/(6\sqrt{2}) & 1/(6\sqrt{2}) & \sqrt{2}/6 \\ 1/(6\sqrt{2}) & 1/(6\sqrt{2}) & \sqrt{2}/6 \end{bmatrix}$$

$$= \begin{bmatrix} 1/12 & 1/12 & 1/6 \\ 1/12 & 1/12 & 1/6 \end{bmatrix} = \frac{1}{12} \begin{bmatrix} 1 & 1 & 2 \\ 1 & 1 & 2 \end{bmatrix}$$

It can be verified that

$$AA^+ = \frac{1}{12} \begin{bmatrix} 1 & 1 \\ 1 & 1 \\ 2 & 2 \end{bmatrix} \begin{bmatrix} 1 & 1 & 2 \\ 1 & 1 & 2 \end{bmatrix} = \frac{1}{12} \begin{bmatrix} 2 & 2 & 4 \\ 2 & 2 & 4 \\ 4 & 4 & 8 \end{bmatrix}$$    ■ ■ ■

Let us express the least squares solution of $A\mathbf{x} = \mathbf{b}$ in terms of the pseudoinverse $A^+$. Let $A$ be an $m \times n$ matrix of rank $r < n$ with singular value decomposition $USV'$. Since $U$ is orthogonal,

$$\begin{aligned} E^2 &= (\mathbf{b} - A\mathbf{x})'(\mathbf{b} - A\mathbf{x}) \\ &= [U'(\mathbf{b} - A\mathbf{x})]'[U'(\mathbf{b} - A\mathbf{x})] \\ &= [U'\mathbf{b} - U'A\mathbf{x}]'[U'\mathbf{b} - U'A\mathbf{x}] \\ &= [U'\mathbf{b} - SV'\mathbf{x}]'[U'\mathbf{b} - SV'\mathbf{x}] \end{aligned}$$    (10.4.14)

Let $\mathbf{c} = U'\mathbf{b}$ and $\mathbf{y} = V'\mathbf{x}$. Then Equation (10.4.14) reduces to

$$E^2 = (\mathbf{c} - S\mathbf{y})'(\mathbf{c} - S\mathbf{y})$$    (10.4.15)

Since $A$ has rank $r$, $\mu_1 \geq \mu_2 \geq \cdots \geq \mu_r > 0$ and $\mu_{r+1} = \mu_{r+2} = \cdots = \mu_n = 0$. Thus Equation (10.4.15) gives

$$E^2 = \sum_{i=1}^{r} |c_i - \mu_i y_i|^2 + \sum_{i=r+1}^{n} |c_i|^2$$    (10.4.16)

Since $\mathbf{c} = U'\mathbf{b}$ is independent of $\mathbf{x}$, $E^2$ is minimum when

$$y_i = \frac{c_i}{\mu_i} \quad \text{for } i = 1, 2, \ldots, r$$    (10.4.17)

Since $V$ is orthogonal,

$$\mathbf{y}'\mathbf{y} = (V'\mathbf{x})'(V'\mathbf{x}) = \mathbf{x}'VV'\mathbf{x} = \mathbf{x}'\mathbf{x}$$

Thus $\mathbf{x}'\mathbf{x}$ is minimum when $\mathbf{y}'\mathbf{y}$ is minimum, and $\mathbf{y}'\mathbf{y}$ is minimum when $y_{r+1} = y_{r+2} = \cdots = y_n = 0$. For convenience, let us express

$$\mathbf{c} = \begin{bmatrix} c_1 \\ c_2 \\ \vdots \\ c_r \\ c_{r+1} \\ \vdots \\ c_n \end{bmatrix} = \begin{bmatrix} \mathbf{d}_1 \\ \mathbf{d}_2 \end{bmatrix} = U'\mathbf{b} \quad \text{where } \mathbf{d}_1 = \begin{bmatrix} c_1 \\ c_2 \\ \vdots \\ c_r \end{bmatrix} \quad \text{and} \quad \mathbf{d}_2 = \begin{bmatrix} c_{r+1} \\ c_{r+2} \\ \vdots \\ c_n \end{bmatrix}$$

and

$$\mathbf{y} = \begin{bmatrix} y_1 \\ y_2 \\ \vdots \\ y_r \\ 0 \\ \vdots \\ 0 \end{bmatrix} = \begin{bmatrix} \mathbf{z}_1 \\ \mathbf{z}_2 \end{bmatrix} = V'\mathbf{x} \quad \text{where } \mathbf{z}_1 = \begin{bmatrix} y_1 \\ y_2 \\ \vdots \\ y_r \end{bmatrix} \quad \text{and} \quad \mathbf{z}_2 = \begin{bmatrix} 0 \\ 0 \\ \vdots \\ 0 \end{bmatrix}$$    (10.4.18)

Then

$$\begin{aligned}
\mathbf{x} &= V\mathbf{y} \\
&= V\begin{bmatrix} \mathbf{z}_1 \\ \mathbf{z}_2 \end{bmatrix} = V\begin{bmatrix} S_1^{-1}\mathbf{d}_1 \\ \mathbf{0} \end{bmatrix} \\
&= V\begin{bmatrix} S_1^{-1} & O \\ O & O \end{bmatrix}\begin{bmatrix} \mathbf{d}_1 \\ \mathbf{d}_2 \end{bmatrix} \\
&= V\begin{bmatrix} S_1^{-1} & O \\ O & O \end{bmatrix}U'\mathbf{b} \\
&= VS^+U'\mathbf{b} \\
&= A^+\mathbf{b}
\end{aligned}$$
(10.4.19)

This proves the next theorem.

**Theorem 10.4.3**

If $A$ is an $m \times n$ matrix with singular value decomposition $A = USV'$ and $A\mathbf{x} = \mathbf{b}$, then the least squares solution of $A\mathbf{x} = \mathbf{b}$ is given by

$$\mathbf{x} = A^+\mathbf{b}$$
(10.4.20)

∎

If the singular value decomposition of $A$ is known, then the least squares solution of $A\mathbf{x} = \mathbf{b}$ is computed in two steps:

1. Compute $y_i = c_i/\mu_i = \mathbf{u}_i'\mathbf{b}/\mu_i$ for $i = 1, 2, \ldots, r$.
2. Compute $\mathbf{x} = V\mathbf{y} = y_1\mathbf{v}_1 + y_2\mathbf{v}_2 + \cdots + y_r\mathbf{v}_r$.

**EXAMPLE 10.4.4**

Find the least squares solution of the linear system

$$\begin{bmatrix} 1 & 1 \\ 1 & 1 \\ 2 & 2 \end{bmatrix}\begin{bmatrix} x_1 \\ x_2 \end{bmatrix} = \begin{bmatrix} 3 \\ 4 \\ 5 \end{bmatrix}$$

From Example 10.4.2, we have

$$U = \begin{bmatrix} 1/\sqrt{6} & -1/\sqrt{2} & -1/\sqrt{3} \\ 1/\sqrt{6} & 1/\sqrt{2} & -1/\sqrt{3} \\ 2/\sqrt{6} & 0 & 1/\sqrt{3} \end{bmatrix} \quad S = \begin{bmatrix} \sqrt{12} & 0 \\ 0 & 0 \\ 0 & 0 \end{bmatrix} \quad V' = \begin{bmatrix} 1/\sqrt{2} & 1/\sqrt{2} \\ 1/\sqrt{2} & -1/\sqrt{2} \end{bmatrix}$$

Then

$$y_1 = \frac{c_1}{\mu_1} = \frac{\mathbf{u}_1'\mathbf{b}}{\mu_1} = \frac{1}{\sqrt{12}}[1/\sqrt{6} \quad 1/\sqrt{6} \quad 2/\sqrt{6}]\begin{bmatrix} 3 \\ 4 \\ 5 \end{bmatrix}$$

$$= \frac{1}{\sqrt{12}}\left(\frac{3}{\sqrt{6}} + \frac{4}{\sqrt{6}} + \frac{10}{\sqrt{6}}\right) = \frac{17}{6\sqrt{2}}$$

We have $\mathbf{x} = V\mathbf{y} = y_1\mathbf{v}_1$. Thus

$$\mathbf{x} = \begin{bmatrix} x_1 \\ x_2 \end{bmatrix} = y_1\mathbf{v}_1 = \frac{17}{6\sqrt{2}} \begin{bmatrix} 1/\sqrt{2} \\ 1/\sqrt{2} \end{bmatrix} = \begin{bmatrix} 17/12 \\ 17/12 \end{bmatrix} \qquad \blacksquare\acute{\ }\blacksquare\ \blacksquare$$

Since SVD employs orthogonal matrices, it is a powerful computational tool. One option is to compute the eigenvalues and eigenvectors of $A'A$ (or $AA'$). We will study several methods of computing eigenvalues in Chapter 13. The smaller singular values are computed inaccurately by this method.

The second option is to reduce $A$ to a bidiagonal matrix using orthogonal matrices. If an $m \times n$ matrix $A$ has a singular value decomposition $A = USV'$ and $B = HAP'$ where $H$ and $P$ are orthogonal matrices of orders $m$ and $n$ respectively, then $B = (HU)S(PV)'$. Thus $A$ and $B$ have the same singular values. In order to compute the singular values of a matrix $A$, obtain another simpler matrix $B$ with the same singular values using orthogonal transformations. Golub and Kahan (1965) have shown that $A$ can be reduced to an upper bidiagonal matrix with the same singular values using the Householder transformation (Section 13.7). Then the singular value decomposition of the bidiagonal matrix is obtained using the QR algorithm (Section 13.10). Complete discussion is given in Golub and Van Loan (1989), Hager (1988), and Watkins (1991).

## EXERCISES

For Exercises 1–3, find a singular value decomposition.

1. $[1 \quad 2]$    2. $\begin{bmatrix} 1 \\ 2 \end{bmatrix}$    3. $\begin{bmatrix} 1 & 1 \\ 0 & 0 \\ 1 & 1 \end{bmatrix}$

4. Prove that if $A$ is symmetric, then $A^+$ is also symmetric.
5. If $A$ is positive definite, then what is its singular value decomposition?

For Exercises 6–8, find the pseudoinverse $A^+$.

6. $[1 \quad 2]$    7. $\begin{bmatrix} 1 \\ 2 \end{bmatrix}$    8. $\begin{bmatrix} 1 & 1 \\ 0 & 0 \\ 1 & 1 \end{bmatrix}$

9. Prove that $A^{++} = A$.
10. Prove that $(A')^+ = (A^+)'$.
11. If $\mathbf{v}_1$ and $\mathbf{v}_2$ are orthogonal eigenvectors of $A'A$, then show that $A\mathbf{v}_1$ and $A\mathbf{v}_2$ are also orthogonal.
12. Let $A$ be an $m \times n$ matrix and $A = USV'$ where $U$ and $V$ are orthogonal matrices of orders $m$ and $n$, respectively, and $S$ is an $m \times n$ diagonal matrix. Express $A'$, $A'A$, and $AA'$ in terms of $U$, $S$, and $V$.

**13.** Find the least squares solution of the following system:

$$x_1 + x_2 = 3.1$$

$$0x_1 + 0x_2 = 5.1$$

$$x_1 + x_2 = 3$$

**14.** Show that Equation (10.4.13) satisfies Equation (10.4.12).

## 10.5   FOURIER APPROXIMATION

Many natural phenomena display periodic behavior like the tide coming in and going out or a weight suspended from a spring going up and down. Such phenomena are described by periodic functions. A function $f$ is said to have **period** $p > 0$ if $f(x + p) = f(x)$ for all real numbers $x$. For convenience, assume that $f$ has a period $2\pi$. We approximate such a function $f$ by using a trigonometric polynomial of degree $M$ and period $2\pi$ given by

$$P(x) = a_0 + \sum_{k=1}^{M} (a_k \cos kx + b_k \sin kx) \tag{10.5.1}$$

where $a_0, a_1, \ldots, a_M$ and $b_1, b_2, \ldots, b_M$ are real or complex constants. We can reduce Equation (10.5.1) using trigonometric addition formulas to

$$P(x) = \alpha_0 + \sum_{k=1}^{M} (\alpha_k(\cos x)^k + \beta_k(\sin x)^k)$$

This justifies the use of the phrase trigonometric polynomial. Using Euler's formula

$$e^{Ix} = \cos x + I \sin x \quad \text{where } I = \sqrt{-1}$$

we obtain

$$\cos x = \frac{e^{Ix} + e^{-Ix}}{2} \quad \text{and} \quad \sin x = \frac{e^{Ix} - e^{-Ix}}{2I} \tag{10.5.2}$$

Substituting for $\cos kx$ and $\sin kx$ from Equation (10.5.2) in Equation (10.5.1),

$$P(x) = a_0 + \sum_{k=1}^{M} \left[ a_k \left( \frac{e^{Ikx} + e^{-Ikx}}{2} \right) + b_k \left( \frac{e^{Ikx} - e^{-Ikx}}{2I} \right) \right]$$

$$= a_0 + \sum_{k=1}^{M} \left[ e^{Ikx} \left( \frac{a_k - Ib_k}{2} \right) + e^{-Ikx} \left( \frac{a_k + Ib_k}{2} \right) \right] \tag{10.5.3}$$

Substituting

$$c_0 = a_0 \qquad c_k = \frac{1}{2}(a_k - Ib_k) \qquad c_{-k} = \frac{1}{2}(a_k + Ib_k) \tag{10.5.4}$$

Equation (10.5.3) becomes

$$P(x) = c_0 + \sum_{k=1}^{M} [c_k e^{Ikx} + c_{-k} e^{-Ikx}] = \sum_{k=-M}^{M} c_k e^{Ikx} \tag{10.5.5}$$

If Equation (10.5.5) represents a real function, then $P(x) = \overline{P(x)}$; so,

$$\sum_{k=-M}^{M} c_k e^{Ikx} = \sum_{k=-M}^{M} \overline{c_k e^{Ikx}} = \sum_{k=-M}^{M} \overline{c_k} e^{-Ikx} = \sum_{k=-M}^{M} \overline{c_{-(-k)}} e^{I(-k)x}$$

By rearranging the terms, it can be shown that

$$\sum_{k=-M}^{M} \overline{c_{-(-k)}} e^{I(-k)x} = \sum_{k=-M}^{M} \overline{c_{-k}} e^{Ikx}$$

Therefore

$$\sum_{k=-M}^{M} c_k e^{Ikx} = \sum_{k=-M}^{M} \overline{c_{-k}} e^{Ikx}$$

Hence

$$c_k = \overline{c_{-k}} \quad \text{for all } k \tag{10.5.6}$$

Taking the complex conjugate on both sides of Equation (10.5.6) yields

$$\overline{c_k} = \overline{\overline{c_{-k}}} = c_{-k} \quad \text{for all } k$$

From Equation (10.5.4),

$$a_k = c_k + c_{-k} \quad \text{and} \quad b_k = I(c_k - c_{-k}) \tag{10.5.7}$$

If Equation (10.5.5) represents a real function, then Equation (10.5.7) reduces to

$$a_k = c_k + \overline{c_k} = 2\text{Re } c_k \quad \text{and} \quad b_k = I(c_k - \overline{c_k}) = -2\text{Im } c_k \tag{10.5.8}$$

Let us consider the set of experimental data points $(x_0, y_0)$, $(x_1, y_1)$, . . . , $(x_N, y_N)$ where $x_0, x_1, \ldots, x_N$ are equally spaced as shown in Figure 10.5.1.

**Fig. 10.5.1**

Assume that $y_0, y_1, \ldots, y_N$ are real numbers. Since we assume that the set of data points are to be fitted to a periodic function, $y_N = y_0$. Since $x_0, x_1, \ldots, x_N$ are equally spaced,

$$x_i = \frac{2\pi i}{N} \quad \text{for } i = 0, 1, \ldots, N \tag{10.5.9}$$

Since we want to fit a trigonometric polynomial of degree $M$ to the set of data points $(x_0, y_0)$, $(x_1, y_1)$, $(x_2, y_2)$, . . . , $(x_{N-1}, y_{N-1})$ in the least squares sense, we need to determine $c_k$ such that

$$E(c_{-M}, c_{-M+1}, \ldots, c_{M-1}, c_M) = \sum_{i=0}^{N-1} [y_i - P(x_i)]^2$$

$$= \sum_{i=0}^{N-1} \left[ y_i - \sum_{k=-M}^{M} c_k e^{Ikx_i} \right]^2 \tag{10.5.10}$$

is minimum. Thus

$$\frac{\partial E}{\partial c_j} = \sum_{i=0}^{N-1} 2 \left[ y_i - \sum_{k=-M}^{M} c_k \, e^{Ikx_i} \right] (-e^{Ijx_i}) = 0 \qquad \textbf{(10.5.11)}$$

for $j = -M, -M + 1, \ldots, M - 1, M$. Multiplying Equation (10.5.11) by $(-e^{-2Ijx_i})/2$, we get

$$\sum_{i=0}^{N-1} \left[ y_i - \sum_{k=-M}^{M} c_k \, e^{Ikx_i} \right] e^{-Ijx_i} = 0 \qquad \textbf{(10.5.12)}$$

for $j = -M, -M + 1, \ldots, M - 1, M$. Equation (10.5.12) is equivalent to

$$\sum_{i=0}^{N-1} y_i \, e^{-Ijx_i} = \sum_{i=0}^{N-1} \left( \sum_{k=-M}^{M} c_k \, e^{I(k-j)x_i} \right)$$

$$= \sum_{k=-M}^{M} c_k \left( \sum_{i=0}^{N-1} e^{I(k-j)x_i} \right) \qquad \textbf{(10.5.13)}$$

for $j = -M, -M + 1, \ldots, M - 1, M$.

Let us find $\sum_{i=0}^{N-1} e^{I(k-j)x_i}$ in order to simplify Equation (10.5.13). Substituting for $x_i$ from Equation (10.5.9),

$$\sum_{i=0}^{N-1} e^{I(k-j)x_i} = \sum_{i=0}^{N-1} e^{i \, I(k-j)2\pi/N}$$

$$= \sum_{i=0}^{N-1} e^{Iix_{k-j}}$$

$$= \sum_{i=0}^{N-1} (e^{Ix_{k-j}})^i$$

$$= \sum_{i=0}^{N-1} z^i \qquad \textbf{(10.5.14)}$$

where $z = e^{Ix_{k-j}}$. Since $\sum_{i=0}^{N-1} z^i$ is a geometric series,

$$\sum_{i=0}^{N-1} z^i = \begin{cases} (z^N - 1)/(z - 1) & \text{if } k \neq j \\ N & \text{if } k = j \end{cases}$$

Further $z^N = e^{Ix_{k-j}N} = e^{2\pi I(k-j)} = 1$. Therefore

$$\sum_{i=0}^{N-1} z^i = \begin{cases} 0 & \text{if } k \neq j \\ N & \text{if } k = j \end{cases}$$

Hence

$$\sum_{i=0}^{N-1} e^{I(k-j)x_i} = \sum_{i=0}^{N-1} z^i = \begin{cases} 0 & \text{if } k \neq j \\ N & \text{if } k = j \end{cases} \qquad \textbf{(10.5.15)}$$

Substituting Equation (10.5.15) in Equation (10.5.13) yields

$$\sum_{i=0}^{N-1} y_i \, e^{-Ijx_i} = Nc_j \qquad \textbf{(10.5.16)}$$

for $j = -M, -M + 1, \ldots, M - 1, M$. Hence

$$
\begin{aligned}
c_j &= \frac{1}{N} \sum_{i=0}^{N-1} y_i \, e^{-ljx_i} \\
&= \frac{1}{N} \sum_{i=0}^{N-1} y_i \, e^{-(2\pi lij)/N} \\
&= \frac{1}{N} \sum_{i=0}^{N-1} y_i \, \omega^{ij}
\end{aligned}
$$

for $j = -M, -M + 1, \ldots, M - 1, M$, where $\omega = e^{-(2\pi l)/N}$. Replacing $j$ by $k$, we get

$$
c_k = \frac{1}{N} \sum_{i=0}^{N-1} y_i \, \omega^{ik} \qquad (10.5.17)
$$

for $k = -M, -M + 1, \ldots, M - 1, M$.

If $y_0, y_1, \ldots, y_{N-1}$ are real numbers, then by using Equation (10.5.17) it can be verified that $c_{-k} = \overline{c_k}$ for $k = 0, 1, \ldots, M$ (Exercise 2). Thus for real numbers $(x_i, y_i)$, we need to compute $c_k$ for $k = 0, 1, \ldots, M$.

**EXAMPLE 10.5.1**    On a certain day the tide at Trinidad, California is recorded as follows:

| Time | 6 A.M. | 8 | 10 | 12 P.M. | 2 | 4 | 6 |
|------|--------|-----|-----|---------|-----|-----|-----|
| Tide | 2 ft | 7/2 | 5 | 17/2 | 7 | 9/2 | 2 |

Find the best fit trigonometric polynomial of degree two in the least squares sense by assuming a 12 hour period and compute the tide at 11 A.M. and 5 P.M.

The data we want to fit is as follows:

| $x_i$ | 0 | 2 | 4 | 6 | 8 | 10 | 12 |
|-------|---|-----|---|------|---|-----|----|
| $y_i$ | 2 | 7/2 | 5 | 17/2 | 7 | 9/2 | 2 |

Since we have derived Equation (10.5.17) for period $2\pi$, substitute $X_i = (\pi/6)x_i$ and $Y_i = y_i$ for $i = 0, 1, \ldots, 6$. Thus we have

| $X_i$ | 0 | $\pi/3$ | $2\pi/3$ | $\pi$ | $4\pi/3$ | $5\pi/3$ | $2\pi$ |
|-------|---|---------|----------|-------|----------|----------|--------|
| $Y_i$ | 2 | 7/2 | 5 | 17/2 | 7 | 9/2 | 2 |

Therefore $N = 6$, $\omega = e^{-(2\pi l)/6} = e^{-(\pi l)/3}$, and $c_k = \left(\frac{1}{6}\right) \sum_{i=0}^{5} \omega^{ik} Y_i$ for $k = -2, -1, 0, 1, 2$. Thus

$$
\begin{bmatrix}
c_{-2} \\
c_{-1} \\
c_0 \\
c_1 \\
c_2
\end{bmatrix}
=
\frac{1}{6}
\begin{bmatrix}
\omega^0 & \omega^{-2} & \omega^{-4} & \omega^{-6} & \omega^{-8} & \omega^{-10} \\
w^0 & w^{-1} & w^{-2} & w^{-3} & w^{-4} & w^{-5} \\
w^0 & w^0 & w^0 & w^0 & w^0 & w^0 \\
w^0 & w^1 & w^2 & w^3 & w^4 & w^5 \\
w^0 & w^2 & w^4 & w^6 & w^8 & w^{10}
\end{bmatrix}
\begin{bmatrix}
2 \\
7/2 \\
5 \\
17/2 \\
7 \\
9/2
\end{bmatrix}
$$

Substituting $w = e^{-(\pi I)/3} = \cos(-\pi/3) + I\sin(-\pi/3) = (1 - I\sqrt{3})/2$, $w^2 = e^{(-2\pi I)/3} = \cos(-2\pi/3) + I\sin(-2\pi/3) = (-1 - I\sqrt{3})/2$ and so forth, we get

$$c_2 = \frac{1}{6}\left\{2 + \frac{7}{2}w^2 + 5w^4 + \frac{17}{2}w^6 + 7w^8 + \frac{9}{2}w^{10}\right\}$$

$$= \frac{1}{6}\left\{2 + \frac{7}{2}\left(\frac{-1 - I\sqrt{3}}{2}\right) + 5\left(\frac{-1 + I\sqrt{3}}{2}\right) + \frac{17}{2}\right.$$

$$\left. + 7\left(\frac{-1 - I\sqrt{3}}{2}\right) + \frac{9}{2}\left(\frac{-1 + I\sqrt{3}}{2}\right)\right\}$$

$$= \frac{1}{12}(1 - I\sqrt{3})$$

$$c_1 = \frac{1}{6}\left\{2 + \frac{7}{2}w^1 + 5w^2 + \frac{17}{2}w^3 + 7w^4 + \frac{9}{2}w^5\right\}$$

$$= \frac{1}{6}\left\{2 + \frac{7}{2}\frac{(1 - I\sqrt{3})}{2} + 5\left(\frac{-1 - I\sqrt{3}}{2}\right) + \frac{17}{2}(-1)\right.$$

$$\left. + 7\left(\frac{-1 + I\sqrt{3}}{2}\right) + \frac{9}{2}\frac{(1 + I\sqrt{3})}{2}\right\}$$

$$= \frac{1}{12}(-17 + I3\sqrt{3})$$

$$c_0 = \frac{1}{6}\left\{2 + \frac{7}{2} + 5 + \frac{17}{2} + 7 + \frac{9}{2}\right\} = \frac{61}{12}$$

From Equations (10.5.4) and (10.5.7), we have $a_0 = c_0 = 61/62$, $a_1 = c_1 + \overline{c_1} = -17/6$, $a_2 = c_2 + \overline{c_2} = \frac{1}{6}$, $b_1 = I(c_1 - \overline{c_1}) = I(6\sqrt{3}\,I/12) = -\sqrt{3}/2$, and $b_2 = I(c_2 - \overline{c_2}) = I(-2\sqrt{3}\,I/12) = \sqrt{3}/6$. Therefore

$$P(X) = \frac{1}{6}\cos 2X - \frac{17}{6}\cos X + \frac{61}{12} + \frac{\sqrt{3}}{6}\sin 2X - \frac{3\sqrt{3}}{6}\sin X$$

Substituting $X = (\pi/6)x$, we get

$$P(x) = \frac{1}{6}\cos\frac{\pi}{3}x - \frac{17}{6}\cos\frac{\pi}{6}x + \frac{61}{12} + \frac{\sqrt{3}}{6}\sin\frac{\pi}{3}x - \frac{3\sqrt{3}}{6}\sin\frac{\pi}{6}x$$

The tides at 11 A.M. and 5 P.M. are given by

$$P(5) = \frac{1}{6}\cos\frac{5\pi}{3} - \frac{17}{6}\cos\frac{5\pi}{6} + \frac{61}{12} + \frac{\sqrt{3}}{6}\sin\frac{5\pi}{3} - \frac{3\sqrt{3}}{6}\sin\frac{5\pi}{6}$$

$$= \frac{1}{6}\cdot\frac{1}{2} - \frac{17}{6}\left(-\frac{\sqrt{3}}{2}\right) + \frac{61}{12} + \frac{\sqrt{3}}{6}\left(-\frac{\sqrt{3}}{2}\right) - \frac{3\sqrt{3}}{6}\cdot\frac{1}{2}$$

$$= \frac{59 + 14\sqrt{3}}{12} \approx 6.93739 \approx 6.9 \text{ ft}$$

and

$$P(11) = \frac{1}{6} \cos \frac{11\pi}{3} - \frac{17}{6} \cos \frac{11\pi}{6} + \frac{61}{12} + \frac{\sqrt{3}}{6} \sin \frac{11\pi}{3} - \frac{3\sqrt{3}}{6} \sin \frac{11\pi}{6}$$

$$= \frac{1}{6} \cdot \frac{1}{2} - \frac{17}{6} \left( \frac{\sqrt{3}}{2} \right) + \frac{61}{12} + \frac{\sqrt{3}}{6} \left( -\frac{\sqrt{3}}{2} \right) - \frac{3\sqrt{3}}{6} \left( \frac{-1}{2} \right)$$

$$= \frac{59 - 14\sqrt{3}}{12} \approx 2.89594 \approx 2.9 \text{ ft} \qquad \blacksquare \ \blacksquare \ \blacksquare$$

Let us consider a somewhat general series of the form

$$a_0 + \sum_{k=1}^{\infty} (a_k \cos kx + \sin kx) \qquad \textbf{(10.5.18)}$$

On the set of values where the series (10.5.18) converges, it defines a function $f$. The value of $f$ at each number $x$ is the sum of the series for that value of $x$. In such a case, the series of Equation (10.5.18) is said to be the **Fourier series** of $f$. Can every function be represented by a Fourier series? The answer is given by the following theorem (for the proof, see Kaplan 1973).

**Theorem 10.5.1**    Let $f$ and $f'$ be piecewise continuous on the interval $[0, 2\pi]$ and $f$ be $2\pi$-periodic. Then $f$ has a Fourier series

$$f(x) = a_0 + \sum_{k=1}^{\infty} (a_k \cos kx + b_k \sin kx) \qquad \textbf{(10.5.19)}$$

where the coefficients $a_0$, $a_k$, and $b_k$ are given by Equations (10.5.20), (10.5.21), and (10.5.22). The Fourier series converges to $f(x)$ at all points where $f$ is continuous and to $(f(a^+) + f(a^-))/2$ at all points $x = a$ where $f$ is discontinuous. $f(a^+) = \lim_{x \to a^+} f(x)$ and $f(a^-) = \lim_{x \to a^-} f(x)$ denote the right-hand and left-hand limits respectively. $\blacksquare$

We determine the coefficients $a_0$, $a_k$, and $b_k$ using the orthogonality property (Appendix E) of $\sin kx$ and $\cos kx$. Integrating both sides of Equation (10.5.19) from 0 to $2\pi$, we have

$$a_0 = \frac{1}{2\pi} \int_0^{2\pi} f(x) \, dx$$

Multiplying both sides of Equation (10.5.19) by $\cos jx$ and integrating from 0 to $2\pi$, yields (Appendix E)

$$a_j = \frac{1}{\pi} \int_0^{2\pi} f(x) \cos jx \, dx$$

for $j = 1, 2, \ldots$ Similarly multiplying both sides of Equation (10.5.20) by $\sin jx$ and integrating from 0 to $2\pi$, we get

$$b_j = \frac{1}{\pi} \int_0^{2\pi} f(x) \sin jx \, dx$$

for $j = 1, 2, \ldots$ Writing $k$ in place of $j$ we have the so called **Euler formulas**

$$a_0 = \frac{1}{2\pi} \int_0^{2\pi} f(x) \, dx \qquad\qquad \textbf{(10.5.20)}$$

$$a_k = \frac{1}{\pi} \int_0^{2\pi} f(x) \cos kx \, dx \qquad\qquad \textbf{(10.5.21)}$$

$$b_k = \frac{1}{\pi} \int_0^{2\pi} f(x) \sin kx \, dx \qquad\qquad \textbf{(10.5.22)}$$

for $k = 1, 2, \ldots$

Using Equations (10.5.2) and (10.5.4), Equation (10.5.19) reduces to

$$f(x) = \sum_{k=-\infty}^{\infty} c_k e^{Ikx} \qquad\qquad \textbf{(10.5.23)}$$

Multiplying both sides of Equation (10.5.23) by $e^{-Ijx}$ and integrating from 0 to $2\pi$, we get

$$c_j = \frac{1}{2\pi} \int_0^{2\pi} f(x) e^{-Ijx} \, dx$$

for $j = \ldots, -2, -1, 0, 1, 2, \ldots$ Changing $j$ to $k$, we get (Exercise 14)

$$c_k = \frac{1}{2\pi} \int_0^{2\pi} f(x) e^{-Ikx} \, dx \qquad\qquad \textbf{(10.5.24)}$$

for $k = \ldots, -2, -1, 0, 1, 2, \ldots$

How do we compute Fourier coefficients $c_k$ of the series of Equation (10.5.23) when a periodic function $f$ is known empirically? Suppose $f$ is known at equidistant points $x_i = (2\pi i)/N$ for $i = 0, 1, \ldots, N$. Then we use the Trapezoidal, Simpson, or Romberg method. The Trapezoidal rule gives

$$c_k \approx \hat{c}_k = \frac{\pi}{N} \left\{ \frac{1}{2\pi} \left[ f(x_0)e^{-Ikx_0} + 2 \sum_{i=1}^{N-1} f(x_i)e^{-Ikx_i} + f(x_N)e^{-Ikx_N} \right] \right\}$$

$$= \frac{1}{N} \left\{ \frac{1}{2} f(x_0)e^{-Ikx_0} + \sum_{i=1}^{N-1} f(x_i)e^{-Ikx_i} + \frac{1}{2} f(x_N)e^{-Ikx_N} \right\}$$

$$= \frac{1}{N} \sum_{i=0}^{N-1} f(x_i)e^{-Ikx_i} = \frac{1}{N} \sum_{i=0}^{N-1} f(x_i)\omega^{ik} \qquad\qquad \textbf{(10.5.25)}$$

where $\omega = e^{-(2\pi I)/N}$ since $f(x_0) = f(x_N)$ because of the periodicity of $f$. This is identical to the coefficients given by Equation (10.5.17) of the trigonometric polynomial (10.5.5). This implies that the truncated Fourier series for $k = M$ is also the best trigonometric polynomial of degree $M$ in the least squares sense.

**EXAMPLE 10.5.2**

The $2\pi$-periodic function is given by $f(x) = x$, $0 < x < 2\pi$. Find (a) the Fourier series of $f$ and (b) the trigonometric polynomial of degree three and seven using $N = 4$ and 8.

(a) For $k = 0$, we get from Equation (10.5.24)

$$c_0 = \frac{1}{2\pi} \int_0^{2\pi} x\, dx = \frac{1}{2\pi} \left( \frac{x^2}{2} \right) \Big|_0^{2\pi} = \pi$$

$$c_k = \frac{1}{2\pi} \int_0^{2\pi} x e^{-Ikx}\, dx$$

$$= \frac{1}{2\pi} \left[ \frac{Ixe^{-Ikx}}{k} + \frac{e^{-Ikx}}{k^2} \right] \Big|_0^{2\pi} = \frac{I}{k}$$

From Equations (10.5.4) and (10.5.7),

$$a_0 = c_0 = \pi \qquad a_k = c_k + \overline{c_k} = 0 \qquad b_k = I(c_k - \overline{c_k}) = -\frac{2}{k}$$

Substituting the values of $a_0$, $a_k$, and $b_k$ in Equation (10.5.19), we get

$$x = \pi - \sum_{k=1}^{\infty} \frac{2 \sin kx}{k} \qquad\qquad \textbf{(10.5.26)}$$

The graph of $f(x)$ and the partial sum $S_4 = \pi - 2 \sin x - \sin 2x - (2 \sin 3x)/3 - (\sin 4x)/2$ are shown in Figure 10.5.2.

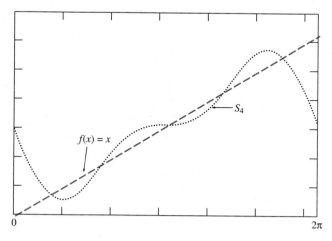

**Fig. 10.5.2**

The agreement between $f(x)$ and $S_4$ is not good near the ends of the interval $0 \le x \le 2\pi$ since $f$ is not continuous at the ends of the interval $[0, 2\pi]$. For large $N$, the graph of $S_N$ contains wiggles near the endpoints where $f$ is not continuous. This is known as **Gibbs phenomenon** and can cause considerable difficulty in applications of the Fourier series.

(b)  From Equation (10.5.25),

$$\hat{c}_k = \frac{1}{N} \sum_{i=0}^{N-1} f(x_i)\omega^{ik} \quad \text{where } \omega = e^{-(2\pi I)/N}$$

Since $f$ is not defined at 0 and $2\pi$, the Fourier series converges to $(f(0^-) + f(0^+))/2$ and $(f(2\pi^-) + f(2\pi^+))/2$ at 0 and $2\pi$ respectively. Therefore we use $f(0) = (f(0^-) + f(0^+))/2 = (2\pi + 0)/2 = \pi$.

Using $N = 4$ and $x_i = (2\pi i)/4 = (\pi i)/2$ for $i = 0, 1, 2,$ and 3, Equation (10.5.25) reduces to

$$\hat{c}_k = \frac{1}{4} \sum_{i=0}^{3} f(x_i)\omega^{ik} \quad \text{where } \omega = e^{-(2\pi I)/4} = e^{-(\pi I)/2} = \cos\frac{\pi}{2} - I \sin\frac{\pi}{2} = -I$$

$$\text{(10.5.27)}$$

We have $\omega^1 = -I$, $\omega^2 = \omega^1 \cdot \omega^1 = (-I)(-I) = -1$, $\omega^3 = \omega^2 \cdot \omega^1 = (-1)(-I) = I$, and $\omega^4 = \omega^3 \cdot \omega^1 = I(-I) = 1$.

For $k = 0$ and 1, Equation (10.5.27) reduces to

$$\hat{c}_0 = \frac{1}{4} \sum_{i=0}^{3} f(x_i) = \frac{1}{4} \{f(x_0) + f(x_1) + f(x_2) + f(x_3)\}$$

$$= \frac{1}{4} \left\{ \pi + \frac{\pi}{2} + \frac{2\pi}{2} + \frac{3\pi}{2} \right\} = \pi$$

and

$$\hat{c}_1 = \frac{1}{4} \sum_{i=0}^{3} f(x_i)\omega^i = \frac{1}{4} \{f(x_0) + f(x_1)\omega^1 + f(x_2)\omega^2 + f(x_3)\omega^3\}$$

$$= \frac{1}{4} \left\{ \pi + \frac{\pi}{2}(-I) + \frac{2\pi}{2}(-1) + \frac{3\pi}{2}(I) \right\} = \frac{\pi}{4}I \approx 0.78540\, I$$

Similarly, we compute $\hat{c}_0$, $\hat{c}_1$, and $\hat{c}_2$ using $N = 8$. The results are given in Table 10.5.1.

**Table 10.5.1**

|  | Fourier Series | N = 4 | N = 8 |
|---|---|---|---|
| $c_0$ | $\pi$ | $\pi$ | $\pi$ |
| $c_1$ | $I$ | $(\pi I)/4 \approx 0.78540\, I$ | $[\pi(2 + \sqrt{2})/(8\sqrt{2})]\, I \approx 0.94806\, I$ |
| $c_2$ | $I/2 \approx 0.5\, I$ | — | $(\pi/8)\, I \approx 0.39270\, I$ |
| $c_3$ | $I/3 \approx 0.33333\, I$ | — | $[\pi(2 - \sqrt{2})/(8\sqrt{2})]\, I \approx 0.16266\, I$ |

The coefficient $\hat{c}_1$ improves significantly when the number of interpolation points increases from $N = 4$ to $N = 8$. As the number of interpolation points increases, the values of the coefficients $\hat{c}_k$ approach those of the coefficients $c_k$ of the Fourier series of $f(x)$.

■ ■ ■

---

**EXERCISES**

---

1.  Verify that $e^{2Ix} = (e^{Ix})^2 = (\cos x + I \sin x)^2 = \cos 2x + I \sin 2x$.
2.  If $y_0, y_1, \ldots, y_{N-1}$ are real numbers, then using Equation (10.5.17) verify that $c_{-k} = \bar{c}_k$.
3.  The following temperatures have been recorded in Arcata, California:

| Time | Midnight | 4 A.M. | 8 A.M. | Noon | 4 P.M. | 8 P.M. |
|------|----------|--------|--------|------|--------|--------|
| **Temperature (F)** | 45° | 46° | 50° | 60° | 62° | 58° |

Find the best fit trigonometric polynomial of degree two in the least squares sense by assuming a 24 hour period and compute the temperature at 11 A.M. and 5 P.M.

4.  Find the Fourier series for the $2\pi$-periodic function given by

$$f(x) = \begin{cases} x & 0 < x < \pi \\ 0 & \pi < x < 2\pi \end{cases}$$

5.  Find the Fourier series for the $2\pi$-periodic function given by

$$f(x) = \begin{cases} x & 0 < x < \pi \\ 2\pi - x & \pi < x < 2\pi \end{cases}$$

6.  Find the third-degree trigonometric polynomial of the form (10.5.1) approximating the $2\pi$-periodic function given in Exercise 4 and compare its coefficients with those of Exercise 4. Use $N = 8$.
7.  Find the first-degree trigonometric polynomial of the form (10.5.1) approximating the $2\pi$-periodic function given in Exercise 5 and compare its coefficients with those of Exercise 5. Use $N = 4$.
8.  Find the Fourier series for the $2\pi$-periodic function given by

$$f(x) = \begin{cases} 1 & 0 < x < \pi \\ -1 & \pi < x < 2\pi \end{cases}$$

9.  Find the Fourier series for the $2\pi$-periodic function given by $f(x) = x^2$ on the interval $(0, 2\pi)$.
10.  Find the third-degree trigonometric polynomial of the form (10.5.1) approximating the $2\pi$-periodic function given in Exercise 8 and compare its coefficients with those of Exercise 8. Use $N = 8$.
11.  Find the first-degree trigonometric polynomial of the form (10.5.1) approximating the $2\pi$-periodic function given in Exercise 9 and compare its coefficients with those of Exercise 9. Use $N = 4$.
12.  Find the Fourier series for the $2\pi$-periodic function given by $f(x) = |\sin x|$ on the interval $(0, 2\pi)$.
13.  Find the first-degree trigonometric polynomial of the form (10.5.1) approximating the $2\pi$-periodic function given in Exercise 12 and compare its coefficients with those of Exercise 12. Use $N = 4$.

**14.** (a) Show that

$$\int_0^{2\pi} e^{Ikx} e^{-Ijx}\, dx = \begin{cases} 0 & \text{if } k \neq j \\ 2\pi & \text{if } k = j \end{cases}$$

(b) Using the result of part (a), prove Equation (10.5.24).

**15.** By separating real and imaginary parts of $\sum_{i=0}^{N-1} e^{(2\pi Iik)/N}$, prove that

$$\sum_{i=0}^{N-1} \cos\frac{2\pi ik}{N} = \begin{cases} N & \text{if } k \text{ divides } N \\ 0 & \text{otherwise} \end{cases} \quad \text{and} \quad \sum_{i=0}^{N-1} \sin\frac{2\pi ik}{N} = 0$$

## COMPUTER EXERCISES

**C.1.** Plot the graphs of the $2\pi$-periodic function $f(x) = x$, $0 < x < 2\pi$ and $S_N(x) = \pi - \sum_{k=1}^{N} (2/k) \sin kx$ for $N = 10, 20$, and $30$ on the interval $[-\pi, 3\pi]$. Notice the Gibbs phenomenon.

**C.2.** Write a subroutine to compute the coefficients $a_0$, $a_k$, and $b_k$ given by Equations (10.5.20), (10.5.21), and (10.5.22) of the Fourier series using the Trapezoidal or Simpson rule.

**C.3.** Using Exercise C.2, compute the Fourier coefficients $a_0$, $a_k$, and $b_k$ for the $2\pi$-periodic function $f(x) = x$, $0 < x < 2\pi$. Plot the graphs of the $2\pi$-periodic function $f$ and $S_N = a_0 + \sum_{k=1}^{N} (a_k \cos kx + b_k \sin kx)$ on the interval $[-\pi, 3\pi]$ using $N = 5, 10$, and $20$.

## 10.6    THE FAST FOURIER TRANSFORM

For given values of $y_i$, the central problem is to compute $c_k$ as efficiently as possible so that

$$Nc_k = b_k = \sum_{i=0}^{N-1} y_i\, \omega^{ik} \tag{10.6.1}$$

for $k = 0, 1, \ldots, N - 1$ where

$$\omega = e^{-(2\pi I)/N} \qquad I = \sqrt{-1} \tag{10.6.2}$$

Equation (10.6.1) has a wide variety of applications. Besides communications, it is widely used in spectroscopy, metallurgy, and nonlinear system analysis. It is also used as a basic tool of analysis of numerical approximations of partial differential equations. For digital image processing, it is common to have $N = 2^{10} = 1024$. The work is quite substantial and it becomes prohibitive if this operation is repeated. One of the earliest algorithms developed by Danielson and Lanczos in 1942 is called the **Fast Fourier Transform**. The term Fast Fourier Transform (FFT) was first used in 1967.

The FFT is based on the following properties of the exponents:

$$\omega^{x+y} = \omega^x \cdot \omega^y \tag{10.6.3}$$

$$\omega^N = (e^{-(2\pi I)/N})^N = e^{-2\pi I} = 1 \tag{10.6.4}$$

$$\omega^{N/2} = (e^{-(2\pi I)/N})^{N/2} = e^{-\pi I} = -1 \tag{10.6.5}$$

Danielson and Lanczos observed that the transform of size $N$ can be rewritten as the sum of the two transforms of size $N/2$. Assume that $N$ is even. Then

$$Nc_k = b_k = \sum_{i=0}^{N-1} y_i \, \omega^{ik}$$

$$= \sum_{i=0}^{(N/2)-1} y_{2i} \, \omega^{2ik} + \sum_{i=0}^{(N/2)-1} y_{(2i+1)} \, \omega^{(2i+1)k}$$

$$= \sum_{i=0}^{(N/2)-1} y_{2i} \, \omega^{2ik} + \omega^k \sum_{i=0}^{(N/2)-1} y_{(2i+1)} \, \omega^{2ik} \tag{10.6.6}$$

$$= b_k^e + \omega^k b_k^o \tag{10.6.7}$$

for $k = 0, 1, \ldots, (N/2) - 1$. Notice that $b_k^e$ denotes the transform of length $N/2$ formed from the even components of $y_i$ while $b_k^o$ denotes the corresponding transform from the odd components of $y_i$.

We compute $b_{(N/2)}, b_{(N/2)+1}, \ldots, b_{N-1}$ by computing $b_{(N/2)+k}$ for $k = 0, 1, \ldots, (N/2) - 1$. From Equation (10.6.6), we have

$$b_{(N/2)+k} = \sum_{i=0}^{(N/2)-1} y_{2i} \, \omega^{2i((N/2)+k)} + \omega^{(N/2)+k} \sum_{i=0}^{(N/2)-1} y_{(2i+1)} \, \omega^{2i((N/2)+k)}$$

$$= \sum_{i=0}^{(N/2)-1} y_{2i} \, (\omega^N)^i \, \omega^{2ik} + \omega^{N/2} \cdot \omega^k \sum_{i=0}^{(N/2)-1} y_{(2i+1)} \, \omega^{2ik}$$

$$= b_k^e - \omega^k b_k^o \tag{10.6.8}$$

for $k = 0, 1, \ldots, (N/2) - 1$. Thus $b_{(N/2)+k}$ involves precisely the same terms as $b_k$ with a different sign. Therefore, no new multiplications are required for Equation (10.6.8). We have reduced the number of multiplications by almost one-half.

If $N/2$ is divisible by 2, then we can compute $b_k^e$ by computing the transforms $b_k^{ee}$ and $b_k^{eo}$ of length $N/4$. Thus we define $b_k^{ee}$ and $b_k^{eo}$ to be the Fourier Transforms of even-even and even-odd components of $y_{2i}$. If $N/4$ is divisible by 2, then we repeat the same procedure. If $N = 2^n$, then this procedure can be used $n$ times. This leads to the subdivision of the data all the way to length one.

**EXAMPLE 10.6.1**

Using the FFT, find $c_k$ for $k = 0, 1, \ldots, 7$ using the data points $(x_k, e^{-4x_k/\pi})$ where $x_k = (k\pi)/4$ for $k = 0, 1, \ldots, 7$.

From Equation (10.6.7),

$$b_k = \sum_{i=0}^{3} y_{2i} \, \omega^{2ik} + \omega^k \sum_{i=0}^{3} y_{(2i+1)} \, \omega^{2ik}$$

$$= b_k^e + \omega^k b_k^o \quad \text{for } k = 0, 1, 2, \text{ and } 3 \tag{10.6.9}$$

From Equation (10.6.8)

$$b_{4+k} = b_k^e - \omega^k b_k^o \quad \text{for } k = 0, 1, 2, \text{ and } 3 \tag{10.6.10}$$

We have

$$b_k^e = \sum_{i=0}^{3} y_{2i} \, \omega^{2ik}$$

$$= \sum_{i=0}^{1} y_{4i} \, \omega^{4ik} + \omega^{2k} \sum_{i=0}^{1} y_{(4i+2)} \, \omega^{4ik}$$

$$= b_k^{ee} + \omega^{2k} b_k^{eo} \quad \text{for } k = 0, 1 \tag{10.6.11}$$

Similarly

$$b_{2+k}^e = b_k^{ee} - \omega^{2k} b_k^{eo} \quad \text{for } k = 0, 1 \tag{10.6.12}$$

Further we get

$$b_k^o = b_k^{oe} + \omega^{2k} b_k^{oo} \tag{10.6.13}$$

and

$$b_{2+k}^o = b_k^{oe} - \omega^{2k} b_k^{oo} \quad \text{for } k = 0, 1 \tag{10.6.14}$$

We have

$$
\begin{aligned}
b_k^{ee} &= \sum_{i=0}^{1} y_{4i}\, \omega^{4ik} \\
&= y_0\, \omega^0 + \omega^{4k} y_4 \\
&= y_0 + \omega^{4k} y_4 \\
&= b_k^{eee} + \omega^{4k} b_k^{eeo} \quad \text{for } k = 0, 1
\end{aligned}
$$

This leads to

$$b_0^{ee} = b_0^{eee} + \omega^0 b_0^{eeo} = y_0 + y_4 = 1.01832$$

$$b_1^{ee} = b_1^{eee} + \omega^4 b_1^{eeo} = y_0 - y_4 = 0.98168$$

Similarly

$$b_0^{eo} = b_0^{eoe} + \omega^0 b_0^{eoo} = y_2 + y_6 = 0.13781$$

$$b_1^{eo} = b_1^{eoe} + \omega^4 b_1^{eoo} = y_2 - y_6 = 0.13286$$

$$b_0^{oe} = b_0^{oee} + \omega^0 b_0^{oeo} = y_1 + y_5 = 0.37462$$

$$b_1^{oe} = b_1^{oee} + \omega^4 b_1^{oeo} = y_1 - y_5 = 0.36114$$

$$b_0^{oo} = b_0^{ooe} + \omega^0 b_0^{ooo} = y_3 + y_7 = 0.05070$$

$$b_1^{oo} = b_1^{ooe} + \omega^4 b_0^{ooo} = y_3 - y_7 = 0.04888 \tag{10.6.15}$$

We have $\omega = e^{-(\pi/4)I} = \cos(\pi/4) - I \sin(\pi/4) = (1 - I)/\sqrt{2}$, therefore

$$\omega^2 = e^{-(\pi/2)I} = \cos\frac{\pi}{2} - I \sin\frac{\pi}{2} = -I \qquad \omega^3 = -\frac{(1 - I)}{\sqrt{2}} \qquad \omega^4 = -1$$

and so forth. Substituting the required values from Equation (10.6.15) in Equations (10.6.11) and (10.6.12),

$$b_0^e = b_0^{ee} + \omega^0 b_0^{eo} = 1.01832 + 0.13781 = 1.15613$$

$$b_1^e = b_1^{ee} + \omega^2 b_1^{eo} = 0.98168 - I\, 0.13286$$

$$b_2^e = b_0^{ee} - \omega^0 b_0^{eo} = 1.01832 - 0.13781 = 0.88051$$

$$b_3^e = b_1^{ee} - \omega^2 b_1^{eo} = 0.98168 + I\, 0.13286 \tag{10.6.16}$$

Substituting the required values from Equation (10.6.15) in Equations (10.6.13) and (10.6.14),

$$b_0^o = b_0^{oe} + \omega^0 b_0^{oo} = 0.37462 + 0.05070 = 0.42532$$

$$b_1^o = b_1^{oe} + \omega^2 b_1^{oo} = 0.36114 - I\,0.04888$$

$$b_2^o = b_0^{oe} - \omega^0 b_0^{oo} = 0.37462 - 0.00507 = 0.32392$$

$$b_3^o = b_1^{oe} - \omega^2 b_1^{oo} = 0.36114 + I\,0.04888 \tag{10.6.17}$$

Substituting the required values from Equations (10.6.16) and (10.6.17) in Equations (10.6.9) and (10.6.10),

$$b_0 = b_0^e + \omega^0 b_0^o = 1.15613 + 0.42532 = 1.58145$$

$$b_1 = b_1^e + \omega^1 b_1^o = (0.98168 - I\,0.13286) + \frac{(1 - I)}{\sqrt{2}}(0.36114 - I\,0.04888)$$
$$= 1.20248 - I\,0.42279$$

$$b_2 = b_2^e + \omega^2 b_2^o = 0.88051 - I\,0.32392$$

$$b_3 = b_3^e + \omega^3 b_3^o = (0.98168 + I\,0.13268) - \frac{(1 - I)}{\sqrt{2}}(0.36114 + I\,0.04888)$$
$$= 0.69175 + I\,0.35906$$

$$b_4 = b_0^e - \omega^0 b_0^o = 1.15613 - 0.42532 = 0.73081$$

$$b_5 = b_1^e - \omega^1 b_1^o = (0.98168 - I\,0.13286) - \frac{(1 - I)}{\sqrt{2}}(0.36114 - I\,0.04888)$$
$$= 0.76088 - I\,0.15707$$

$$b_6 = b_2^e - \omega^2 b_2^o = 0.88051 + I\,0.32392$$

$$b_7 = b_3^e - \omega^3 b_3^o = (0.98168 + I\,0.13286) + \frac{(1 - I)}{\sqrt{2}}(0.36114 + I\,0.04888)$$
$$= 1.27161 - I\,0.08254 \tag{10.6.18}$$

$c_k = b_k/8$ for $k = 0, 1, \ldots, 7$ could be computed from Equation (10.6.18).   ■ ■ ■

The basic idea of the Fourier Transform is simple. The question is, what is the value of $b_j^{eeo}$ in terms of $y_i$? We have subdivided the data all the way down to transforms of length one as shown in Figure 10.6.1, therefore $j$ is not important.

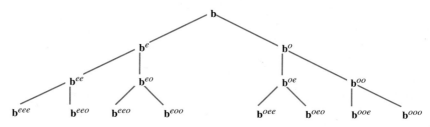

**Fig. 10.6.1**

Reverse the pattern of *eeo* (*oee*) and then substitute for $e = 0$ and $o = 1$. Thus we have in binary the value of $i = (100)_2 = 4$. This is true since we have the successive subdivision of $i$ into even and odd. Thus we have $b_0^{eeo} = b_1^{eeo} = y_4$. This is known as **bit reversal**. The bit reversal number of $(100110)_2 = 1 \times 2^5 + 0 \times 2^4 + 0 \times 2^3 + 1 \times 2^3 + 1 \times 2^2 + 1 \times 2^1 + 0 \times 2^0 = 38$ is $(011001)_2 = 0 \times 2^5 + 1 \times 2^4 + 1 \times 2^3 + 0 \times 2^2 + 0 \times 2^1 + 1 \times 2^0 = 25$. The idea of bit reversal can be exploited for bookkeeping and this makes FFT simple and powerful.

We take our original data $y_i$ and rearrange it in bit-reverse order as shown in Figure 10.6.2. Thus we store our data in the order of the number obtained by bit

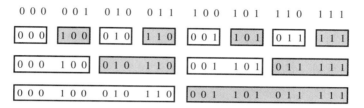

Fig. 10.6.2

reversing $i$. We combine properly adjacent pairs as shown in Figure 10.6.2 to get two-point transforms given by Equation (10.6.15). Then combine properly adjacent pairs as shown in Figure 10.6.2 to get 4-point transforms given by Equations (10.6.16) and (10.6.17).

Next, properly combine the last two halves into the final transform to get Equation (10.6.18). In this way an FFT algorithm consists of two sections. The first section sorts the data into bit-reverse order and the second section calculates the transforms of length $2, 4, \ldots, N$. There are a number of variants of the discussed FFT algorithm. In literature this algorithm is called the Cooley–Tukey FFT algorithm. For the general case of $N$ and more detailed information about FFT and its applications see Brigham (1988).

**Algorithm**

**Subroutine** FFT (**x, y, b**, $N$)
Comment: This subroutine computes $b_j$ given by Equation (10.6.1) using the FFT described in this section. $\omega = e^{-(2\pi I)/N}$ where $I = \sqrt{-1}$ is complex. Assume that $N = 2^{\gamma+1}$ where $\gamma$ is a positive integer; **x** and **y** are real numbers.
This part uses the bit reversal procedure.
  1 **Print** $N$, **x, y**, $\gamma$
  2 **Repeat** steps 3–7 **with** $j = 0, 1, \ldots, N - 1$
      3 **Call** Bitre ($i, j, \gamma$)
      4 **If** $j \geq i$, **then** return to 3
      5 temp $= y_j$
      6 $y_j = y_i$
      7 $y_i =$ temp
Comment: This part computes $b_j = b_j^{(0)}$; $\omega$, $b_j^{(\gamma)}, b_j^{(\gamma-1)}, \ldots, b_j^{(0)}$ are complex numbers.
  8 $\omega = e^{-(2\pi I)/N}$
  9 **Print** $\omega$

10 **Repeat** step 11 **with** $j = 0, 1, \ldots, N - 1$
   11  $b_j^{(\gamma+1)} = y_j$
12 **Repeat** steps 13–16 **with** $k = \gamma, \gamma - 1, \ldots, 0$
   13  $l = 2^{\gamma-k}$
   14  **Repeat** steps 15–16 **with** $j = 0, 2, \ldots, N - 2$
      15  $b_j^{(k)} = b_j^{(k+1)} + \omega^{(N*J)/(2l)} b_{j+1}^{(k+1)}$
      16  $b_{j+1}^{(k)} = b_j^{(k+1)} + \omega^{(N*(J+1))/(2l)} b_{j+1}^{(k+1)}$
17 **Print** $b_j^{(0)}$ and exit.

**Subroutine** Bitre $(I, J, \gamma)$
Comment: This subroutine converts a given positive integer $J$ into a binary number, reverses that binary number, and converts the reverse binary number back to a positive integer $I$. We have $N = 2^{\gamma+1}$.
 1 **Print** $J, \gamma$
 2 **If** $J = 0$, **then** $I = J$ and exit.
 3  $M_0 = \text{Int}(J/2)$
 4  $b_0 = J - 2M_0$
 5  $n = 0$
 6  $n = n + 1$
 7  $M_n = \text{Int}(M_{n-1}/2)$
 8  $b_n = M_{n-1} - 2M_n$
 9 **If** $M_n \neq 0$, **then** return to 6
10 **Repeat** step 11 **with** $i = n + 1, n, \ldots, \gamma$
   11  $b_i = 0$
12 **Print b**
13 Isum $= 0$
14 **Repeat** steps 15–16 **with** $i = 0, 1, \ldots, \gamma$
   15  $Ia = 2^i$
   16  Isum $=$ Isum $+ (Ia)b_{\gamma-i}$
17  $I = $ Isum
18 **Print** $J$ and $I$ and exit    ◀

EXERCISES

1. Verify Equations (10.6.4) and (10.6.5).
2. For $N = 4$, prepare figures similar to Figures 10.6.1 and 10.6.2.
3. Using the FFT procedure, find $c_k$ for $k = 0, 1, 2,$ and 3 using the data points $(x_k, e^{(-2x_k)/\pi})$ where $x_k = k(\pi/2)$ for $k = 0, 1, 2,$ and 3.
4. For $N = 4$, verify that Equation (10.6.1) reduces to

$$
\begin{bmatrix} b_0 \\ b_1 \\ b_2 \\ b_3 \end{bmatrix} = \begin{bmatrix} 1 & 1 & 1 & 1 \\ 1 & \omega & \omega^2 & \omega^3 \\ 1 & \omega^2 & \omega^4 & \omega^6 \\ 1 & \omega^3 & \omega^6 & \omega^9 \end{bmatrix} \begin{bmatrix} y_0 \\ y_1 \\ y_2 \\ y_3 \end{bmatrix}
$$

Also, verify that

$$
\begin{bmatrix} b_0 \\ b_2 \\ b_1 \\ b_3 \end{bmatrix} =
\begin{bmatrix} 1 & 1 & 0 & 0 \\ 1 & \omega^2 & 0 & 0 \\ 0 & 0 & 1 & \omega \\ 0 & 0 & 1 & \omega^3 \end{bmatrix}
\begin{bmatrix} 1 & 0 & 1 & 0 \\ 0 & 1 & 0 & 1 \\ 1 & 0 & \omega^2 & 0 \\ 0 & 1 & 0 & \omega^2 \end{bmatrix}
\begin{bmatrix} y_0 \\ y_1 \\ y_2 \\ y_3 \end{bmatrix}
$$

Relate this to our FFT algorithm.

5. Let

$$
F = \begin{bmatrix} 1 & 1 & 1 & 1 \\ 1 & \omega & \omega^2 & \omega^3 \\ 1 & \omega^2 & \omega^4 & \omega^6 \\ 1 & \omega^3 & \omega^6 & \omega^9 \end{bmatrix}
$$

Verify $F\bar{F} = \bar{F}F = 4I$ for $N = 4$.

6. For $N = 16$, prepare figures similar to Figures 10.6.1 and 10.6.2.

## COMPUTER EXERCISES

C.1. Write a subroutine to convert a given positive integer $k$ into a binary number, reverse that binary number, and convert the reverse binary number back to a positive integer $l$.

C.2. Write a subroutine to compute $b_k$ given by Equation (10.6.1) using the FFT for given $(x_0, y_0), (x_1, y_1), \ldots, (x_N, y_N)$ where $x_0, x_1, \ldots, x_N, y_0, y_1, \ldots, y_N$ are real numbers and $N = 2^{\gamma+1}$. Use Exercise C.1.

C.3. Let $f(x) = e^{-x}$, $0 \le x < 2\pi$. Compute $b_k$ directly from Equation (10.6.1) and by using the FFT for $N = 8, 16, 32$, and 64. Use Exercise C.2. Compare computing times.

C.4. Let $f(x) = x^2$, $0 \le x < 2\pi$. Compute $b_k$ given by Equation (10.6.1) using the FFT for $N = 8, 16, 32$, and 64. Use Exercise C.2.

## 10.7   PIECEWISE POLYNOMIAL APPROXIMATIONS: CUBIC SPLINES

We want to fit a smooth curve to a set of points $(x_0, y_0), (x_1, y_1), \ldots, (x_N, y_N)$. We assume that the data is not necessarily subject to error; therefore, the least squares approach is inappropriate. For large $N$, an interpolation polynomial will be of high degree. A polynomial of degree $N$ can have $N - 1$ relative maxima and minima and the graph has to pass through $(N + 1)$ points.

An alternate approach is to use a lower-degree polynomial in each subinterval. We interpolated trigonometric, logarithmic, and exponential functions using linear interpolation in each subinterval. Thus, the function is replaced by the broken line or a piecewise linear function whose values match the values of $f(x)$ at $x_0, x_1, \ldots, x_N$ as shown in Figure 10.7.1.

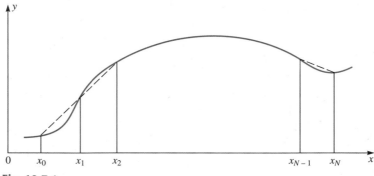

**Fig. 10.7.1**

The line in each subinterval is not smooth at the interface. However, we can overcome this by using a higher-degree polynomial in each subinterval and requiring it to be smooth at each interface. This piecewise polynomial is called a **spline**. This name is derived from a draftman's tool (a flexible strip) that passes through each point smoothly.

The simplest spline function is a polygonal path consisting of line segments that passes through $(x_0, y_0)$, $(x_1, y_1)$, . . . , $(x_N, y_N)$ as shown in Figure 10.7.1. The resulting linear spline function is given by

$$S(x) = \begin{cases} y_0 + \dfrac{y_1 - y_0}{x_1 - x_0}(x - x_0) & x \in [x_0, x_1] \\[2mm] y_1 + \dfrac{y_2 - y_1}{x_2 - x_1}(x - x_1) & x \in [x_1, x_2] \\ \vdots & \vdots \\ y_{N-1} + \dfrac{y_N - y_{N-1}}{x_N - x_{N-1}}(x - x_{N-1}) & x \in [x_{N-1}, x_N] \end{cases} \qquad \textbf{(10.7.1)}$$

The points $x_0, x_1, \ldots, x_N$ at which $S$ changes its character are known as **knots**. By the nature of its derivation, $S$ is a continuous function on $[x_0, x_N]$; however the derivative of $S$ is not continuous at the knots. A spline of degree one has "kinks" at each knot.

To remove the kinks in Figure 10.7.1, the first derivative of $S$ must be continuous at each knot. Thus we need $S$ to be a piecewise quadratic polynomial or a higher-degree polynomial. A function $S$ is a spline of degree two if $S$ is a piecewise quadratic polynomial such that $S$ and its derivative $S'$ are continuous on a given interval.

Consider a quadratic spline curve $S(x)$ passing through $(x_0, y_0)$, $(x_1, y_1)$, and $(x_2, y_2)$ as shown in Figure 10.7.2.

The equation of $S$ is given by

$$S(x) = \begin{cases} S_0(x) & x \in [x_0, x_1] \\ S_1(x) & x \in [x_1, x_2] \end{cases} \qquad \textbf{(10.7.2)}$$

The values of $S$ at $x_0$, $x_1$, and $x_2$ must match with $y_0$, $y_1$, and $y_2$ and $S'$ must be continuous at the interface $(x_1, y_1)$. Thus

**1.**  $S_0(x_0) = y_0 \qquad S_0(x_1) = y_1 = S_1(x_1) \qquad S_1(x_2) = y_2$

**2.**  $S_0'(x_1) = S_1'(x_1)$ $\qquad\qquad\qquad\qquad\qquad\qquad\qquad\qquad$ **(10.7.3)**

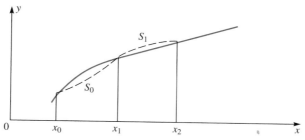

**Fig. 10.7.2**

We are looking for a quadratic spline and since it passes through $(x_0, y_0)$, $(x_1, y_1)$, and $(x_2, y_2)$, it is convenient to consider

$$S_0(x) = a_0 + b_0(x - x_0) + c_0(x - x_0)^2$$

and

$$S_1(x) = a_1 + b_1(x - x_1) + c_1(x - x_1)^2 \qquad \textbf{(10.7.4)}$$

where $a_0$, $b_0$, $c_0$, $a_1$, $b_1$, and $c_1$ are to be determined. Let $h_0 = x_1 - x_0$ and $h_1 = x_2 - x_1$. Using Equation (10.7.3), $S_0'(x) = b_0 + 2c_0(x - x_0)$, and $S_1'(x) = b_1 + 2c_1(x - x_1)$ in Equation (10.7.4), we get

$$S_0(x_0) = a_0 = y_0 \qquad S_1(x_1) = a_1 = y_1$$

$$S_0(x_1) = a_0 + b_0 h_0 + c_0 h_0^2 = S_1(x_1) = y_1$$

$$S_1(x_2) = a_1 + b_1 h_1 + c_1 h_1^2 = y_2$$

$$S_0'(x_1) = b_0 + 2c_0 h_0 = S_1'(x_1) = b_1 \qquad \textbf{(10.7.5)}$$

Equation (10.7.5) gives only five equations for six unknowns. Therefore, one additional condition is required. It can be either $y_0'$ or $y_2'$. This procedure can be extended to $N + 1$ points or $N$ intervals. Quadratic splines have a smoother appearance but could have instantaneous changes in curvature at each knot. Therefore, cubic splines are most widely used in practice.

Let us consider a cubic spline curve $S(x)$ passing through $(x_0, y_0)$, $(x_1, y_1)$, ..., $(x_N, y_N)$ as shown in Figure 10.7.3.

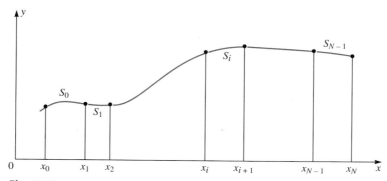

**Fig. 10.7.3**

$S(x)$ consists of $N$ portions of cubic polynomials such that

$$S(x) = \begin{cases} S_0(x) & x_0 \leq x \leq x_1 \\ S_1(x) & x_1 \leq x \leq x_2 \\ \vdots & \\ S_{N-1}(x) & x_{N-1} \leq x \leq x_N \end{cases} \tag{10.7.6}$$

where $S$, $S'$, and $S''$ are continuous on $[x_0, x_N]$. Then

$$S_i(x_i) = y_i \quad \text{and} \quad S_i(x_{i+1}) = y_{i+1} \quad \text{for } i = 0, 1, \ldots, N-1 \tag{10.7.7}$$

These conditions give rise to $2N$ equations. Moreover, we want $S'$ and $S''$ to be continuous at each knot, therefore

$$S'_{i-1}(x_i) = S'_i(x_i) \quad \text{and} \quad S''_{i-1}(x_i) = S''_i(x_i) \quad \text{for } i = 1, 2, \ldots, N-1 \tag{10.7.8}$$

Equation (10.7.8) provides $2(N-1)$ equations.

Let

$$S_i(x) = a_i + b_i(x - x_i) + c_i(x - x_i)^2 + d_i(x - x_i)^3 \tag{10.7.9}$$

be a cubic polynomial defined on $[x_i, x_{i+1}]$. Each cubic polynomial $S_i$ has $a_i$, $b_i$, $c_i$, and $d_i$ as unknowns; therefore there are $4N$ unknowns. We have $(4N - 2)$ equations given by (10.7.7) and (10.7.8) for $4N$ unknowns. We still need two more equations that can be obtained by imposing boundary conditions at the endpoints $x_0$ and $x_N$. The most commonly used boundary conditions are

$$S''_0(x_0) = S''_{N-1}(x_N) = 0 \tag{10.7.10}$$

and

$$S'_0(x_0) = y'_0 \quad \text{and} \quad S'_{N-1}(x_N) = y'_N \tag{10.7.11}$$

The boundary conditions given by Equation (10.7.10) are called **free** or **natural boundary conditions** while the boundary conditions given by Equation (10.7.11) are called **clamped boundary conditions**.

From Equation (10.7.9), we have

$$S'_i(x) = b_i + 2c_i(x - x_i) + 3d_i(x - x_i)^2 \tag{10.7.12}$$

and

$$S''_i(x) = 2c_i + 6d_i(x - x_i) \tag{10.7.13}$$

From Equation (10.7.9), we get

$$S_i(x_i) = a_i = y_i \tag{10.7.14}$$

and

$$S_i(x_{i+1}) = a_i + b_i h_i + c_i h_i^2 + d_i h_i^3 = y_{i+1} = a_{i+1} \tag{10.7.15}$$

for $i = 0, 1, \ldots, N-1$, where

$$h_i = x_{i+1} - x_i \quad \text{for } i = 0, 1, \ldots, N-1 \tag{10.7.16}$$

Substituting $a_i = y_i$ from Equation (10.7.14) in Equation (10.7.15), we get

$$b_i h_i + c_i h_i^2 + d_i h_i^3 = y_{i+1} - y_i \tag{10.7.17}$$

Since $S_i'(x_i) = b_i = S_{i-1}'(x_i)$, define (for convenience) $S_N'(x_N) = b_N = S_{N-1}'(x_N)$. Hence from Equation (10.7.12),

$$b_i = b_{i-1} + 2c_{i-1}h_{i-1} + 3d_{i-1}h_{i-1}^2 \quad \text{for } i = 1, 2, \ldots, N$$

This can be rewritten as

$$2c_{i-1}h_{i-1} + 3d_{i-1}h_{i-1}^2 = b_i - b_{i-1} \quad \text{for } i = 1, 2, \ldots, N \tag{10.7.18}$$

Since $S_i''(x_i) = 2c_i = S_{i-1}''(x_i)$, define (for convenience) $S_N''(x_N) = 2c_N = S_{N-1}''(x_N)$. Then from Equation (10.7.13),

$$2c_i = 2c_{i-1} + 6d_{i-1}h_{i-1} \quad \text{for } i = 1, 2, \ldots, N \tag{10.7.19}$$

Solving Equation (10.7.19) for $d_{i-1}$,

$$d_{i-1} = \frac{c_i - c_{i-1}}{3h_{i-1}} \quad \text{for } i = 1, 2, \ldots, N \tag{10.7.20}$$

Substituting for $d_i$ from Equation (10.7.20) in Equation (10.7.17) and then solving for $b_i$ gives

$$b_i = \frac{y_{i+1} - y_i}{h_i} - \frac{h_i}{3}(c_{i+1} + 2c_i) \quad \text{for } i = 0, 1, \ldots, N - 1 \tag{10.7.21}$$

Substituting for $b_i$ from Equation (10.7.21) and $d_i$ from Equation (10.7.20) in Equation (10.7.18) and then simplifying yields

$$h_{i-1}c_{i-1} + 2(h_i + h_{i-1})c_i + h_ic_{i+1} = 3\left[\frac{y_{i+1} - y_i}{h_i} - \frac{y_i - y_{i-1}}{h_{i-1}}\right] \tag{10.7.22}$$

for $i = 1, 2, \ldots, N - 1$.

For natural or free boundary conditions, from Equation (10.7.13), we have

$$c_0 = 0 \quad \text{and} \quad c_N = 0 \tag{10.7.23}$$

Equations (10.7.22) and (10.7.23) can be written as

$$\begin{bmatrix} 1 & 0 & 0 & \cdots & 0 & 0 & 0 \\ h_0 & 2(h_0 + h_1) & h_1 & \cdots & 0 & 0 & 0 \\ 0 & h_1 & 2(h_1 + h_2) & \cdots & 0 & 0 & 0 \\ \vdots & \vdots & \vdots & \ddots & \vdots & \vdots & \vdots \\ 0 & 0 & 0 & \cdots & h_{N-2} & 2(h_{N-2} + h_{N-1}) & h_{N-1} \\ 0 & 0 & 0 & \cdots & 0 & 0 & 1 \end{bmatrix} \begin{bmatrix} c_0 \\ c_1 \\ c_2 \\ \vdots \\ c_{N-1} \\ c_N \end{bmatrix}$$

$$= \begin{bmatrix} 0 \\ 3[(y_2 - y_1)/h_1 - (y_1 - y_0)/h_0] \\ 3[(y_3 - y_2)/h_2 - (y_2 - y_1)/h_1] \\ \vdots \\ 3[(y_N - y_{N-1})/h_{N-1} - (y_{N-1} - y_{N-2})/h_{N-2}] \\ 0 \end{bmatrix} \tag{10.7.24}$$

in other words, $A\mathbf{c} = \mathbf{u}$.

For the clamped boundary conditions, from Equations (10.7.11) and (10.7.12)

$$S_0'(x_0) = b_0 = y_0'$$

and

$$S_{N-1}'(x_N) = b_{N-1} + 2c_{N-1}h_{N-1} + 3d_{N-1}h_{N-1}^2 = y_N' \qquad \textbf{(10.7.25)}$$

Substituting the values of $b_0$, $b_{N-1}$, and $d_{N-1}$ from Equations (10.7.21) and (10.7.20) in Equation (10.7.25), we get

$$y_0' = \frac{y_1 - y_0}{h_0} - \frac{h_0}{3}(c_1 + 2c_0)$$

and

$$y_N' = \frac{y_N - y_{N-1}}{h_{N-1}} + \frac{h_{N-1}}{3}(2c_N + c_{N-1}) \qquad \textbf{(10.7.26)}$$

Equations (10.7.26) can be rewritten as

$$2h_0 c_0 + h_0 c_1 = \frac{3(y_1 - y_0)}{h_0} - 3y_0'$$

and

$$h_{N-1}c_{N-1} + 2h_{N-1}c_N = 3y_N' - \frac{3(y_N - y_{N-1})}{h_{N-1}} \qquad \textbf{(10.7.27)}$$

Equations (10.7.22) and (10.7.27) can be written as

$$
\begin{bmatrix}
2h_0 & h_0 & 0 & \cdots & 0 & 0 & 0 \\
h_0 & 2(h_0 + h_1) & h_1 & \cdots & 0 & 0 & 0 \\
0 & h_1 & 2(h_1 + h_2) & \cdots & 0 & 0 & 0 \\
\vdots & \vdots & \vdots & \ddots & \vdots & \vdots & \vdots \\
0 & 0 & 0 & \cdots & h_{N-2} & 2(h_{N-2} + h_{N-1}) & h_{N-1} \\
0 & 0 & 0 & \cdots & 0 & h_{N-1} & 2h_{N-1}
\end{bmatrix}
\begin{bmatrix}
c_0 \\ c_1 \\ c_2 \\ \vdots \\ c_{N-1} \\ c_N
\end{bmatrix}
$$

$$
=
\begin{bmatrix}
3[(y_1 - y_0)/h_0] - 3y_0' \\
3[(y_2 - y_1)/h_1 - (y_1 - y_0)/h_0] \\
3[(y_3 - y_2)/h_2 - (y_2 - y_1)/h_1] \\
\vdots \\
3[(y_N - y_{N-1})/h_{N-1} - (y_{N-1} - y_{N-2})/h_{N-2}] \\
3[y_N' - (y_N - y_{N-1})/h_{N-1}]
\end{bmatrix}
\qquad \textbf{(10.7.28)}
$$

in other words, $A_1 \mathbf{c} = \mathbf{u}$.

The systems of Equations (10.7.28) and (10.7.24) are tridiagonal systems and can be solved by the *LU* decomposition method developed in Section 7.7 for $c_i$. Once $c_i$ is known, Equations (10.7.20) and (10.7.21) can be used to compute $b_i$ and $d_i$; $a_i$ is known from Equation (10.7.14). $S_i$ is known from Equation (10.7.9) since $a_i$, $b_i$, $c_i$, and $d_i$ are already computed.

**EXAMPLE 10.7.1**  Construct the cubic spline with (a) free and (b) clamped boundary conditions to approximate $f(x) = e^x$ using the following data:

| $x_i$ | 0 | 1/4 | 1 |
|---|---|---|---|
| $y_i = e^{x_i}$ | 1 | $e^{1/4}$ | $e$ |

We have $x_0 = 0$, $x_1 = \frac{1}{4}$, and $x_2 = 1$, therefore $h_0 = x_1 - x_0 = \frac{1}{4}$ and $h_1 = x_2 - x_1 = 1 - (\frac{1}{4}) = \frac{3}{4}$.

(a) For free boundary conditions, Equation (10.7.24) reduces to

$$\begin{bmatrix} 1 & 0 & 0 \\ 1/4 & 2 & 3/4 \\ 0 & 0 & 1 \end{bmatrix} \begin{bmatrix} c_0 \\ c_1 \\ c_2 \end{bmatrix} = \begin{bmatrix} 0 \\ 4(e - 4e^{1/4} + 3) \\ 0 \end{bmatrix}$$

Solving the system of equations, we get $c_0 = 0$, $c_1 = 2(e - 4e^{1/4} + 3)$, and $c_2 = 0$. Equation (10.7.20) gives

$$d_0 = \frac{c_1 - c_0}{3h_0} = \frac{8}{3}(e - 4e^{1/4} + 3)$$

$$d_1 = \frac{c_2 - c_1}{3h_1} = -\frac{8}{9}(e - 4e^{1/4} + 3)$$

From Equation (10.7.21),

$$b_0 = \frac{y_1 - y_0}{h_0} - \frac{h_0}{3}(c_1 + 2c_0) = \frac{-e + 28e^{1/4} - 27}{6}$$

$$b_1 = \frac{y_2 - y_1}{h_1} - \frac{h_1}{3}(c_2 + 2c_1) = \frac{e + 8e^{1/4} - 9}{3}$$

We get $a_0 = y_0 = 1$ and $a_1 = y_1 = e^{1/4}$ from Equation (10.7.14). Thus

$$S(x) = \begin{cases} 1 + \frac{1}{6}(-e + 28e^{1/4} - 27)x + \frac{8}{3}(e - 4e^{1/4} + 3)x^3 & 0 \le x \le \frac{1}{4} \\ e^{1/4} + \frac{1}{3}(e + 8e^{1/4} - 9)(x - \frac{1}{4}) + 2(e - 4e^{1/4} + 3)(x - \frac{1}{4})^2 & \\ \quad - \frac{8}{9}(e - 4e^{1/4} + 3)(x - \frac{1}{4})^3 & \frac{1}{4} \le x \le 1 \end{cases}$$

$$(10.7.29)$$

The values given in Table 10.7.1 are computed at various $x_i$ using Equation (10.7.29). Also, the absolute value of the actual errors due to using Equation (10.7.29) are given in the fourth column of Table 10.7.1. The errors are due to large step size $h_i$ and the natural boundary conditions. We have $(d^2 e^x)/dx^2 = e^x$ and therefore $y''(0) = e^0 = 1$, $y''(1) = e^1 = e$; yet we use $S''(0) = 0$ and $S''(1) = 0$.

**Table 10.7.1**

| $x_i$ | $e^{x_i}$ | Natural Boundary Conditions Eq. (10.7.29) | \|Actual Error\| for Eq. (10.7.29) | Clamped Boundary Conditions Eq. (10.7.30) | \|Actual Error\| for Eq. (10.7.30) |
|---|---|---|---|---|---|
| 0.10000 | 1.10828 | 1.10546 | 0.00029 | 1.10517 | 0.00311 |
| 0.20000 | 1.19133 | 1.22023 | 0.00117 | 1.22146 | 0.03007 |
| 0.30000 | 1.34961 | 1.35338 | 0.00352 | 1.34986 | 0.00025 |
| 0.40000 | 1.49087 | 1.50800 | 0.01618 | 1.49182 | 0.00095 |
| 0.50000 | 1.64713 | 1.68125 | 0.03253 | 1.64872 | 0.00159 |
| 0.60000 | 1.82023 | 1.87003 | 0.04791 | 0.82212 | 0.00189 |
| 0.70000 | 2.01204 | 2.07122 | 0.05747 | 2.01375 | 0.00171 |
| 0.80000 | 2.22442 | 2.28174 | 0.05619 | 2.22554 | 0.00112 |
| 0.90000 | 2.45921 | 2.49846 | 0.03885 | 2.45960 | 0.00039 |
| 1.00000 | 2.71828 | 2.71828 | 0.00000 | 2.71828 | 0.00000 |

(b) For clamped boundary conditions, we have $y_0' = 1$ and $y_2' = e$, therefore Equation (10.7.28) reduces to

$$\begin{bmatrix} 1/2 & 1/4 & 0 \\ 1/4 & 2 & 3/4 \\ 0 & 3/4 & 3/2 \end{bmatrix} \begin{bmatrix} c_0 \\ c_1 \\ c_2 \end{bmatrix} = \begin{bmatrix} 12e^{1/4} - 15 \\ 4(e - 4e^{1/4} + 3) \\ -e + 4e^{1/4} \end{bmatrix}$$

Solving the system of equations, we get $c_0 = (192e^{1/4} - 9e - 219)/6$, $c_1 = (-48e^{1/4} + 9e + 39)/3$, and $c_2 = (64e^{1/4} - 13e - 39)/6$.
From Equation (10.7.20),

$$d_0 = \frac{c_1 - c_0}{3h_0} = \frac{2}{9}[-288e^{1/4} + 17e + 297]$$

$$d_1 = \frac{c_2 - c_1}{3h_1} = \frac{2}{27}[160e^{1/4} - 31e - 117]$$

Equation (10.7.21) gives

$$b_0 = \frac{y_1 - y_0}{h_0} - \frac{h_0}{3}(c_1 + 2c_0) = \frac{1}{3}[8e^{1/4} - 7]$$

$$b_1 = \frac{y_2 - y_1}{h_1} - \frac{h_1}{3}(c_2 + 2c_1) = \frac{1}{24}[96e^{1/4} + 9e - 117]$$

Thus

$$S(x) = \begin{cases} 1 + \frac{1}{3}[8e^{1/4} - 7]x + \frac{1}{6}[192e^{1/4} - 9e - 219]x^2 \\ \quad + \frac{2}{9}[-288e^{1/4} + 17e + 297]x^3 & 0 \le x \le \frac{1}{4} \\ e^{1/4} + \frac{1}{24}[96e^{1/4} + 9e - 117](x - \frac{1}{4}) \\ \quad + \frac{1}{3}[-48e^{1/4} + 9e + 39](x - \frac{1}{4})^2 \\ \quad + \frac{2}{27}[160e^{1/4} - 31e - 117](x - \frac{1}{4})^3 & \frac{1}{4} \le x \le 1 \end{cases}$$

$$\text{(10.7.30)}$$

The values computed by using Equation (10.7.30) at various $x_i$ are given in the fifth column and the absolute value of the actual errors in the sixth column of Table 10.7.1. The errors are due to large step size $h_i$.   ■ ■ ■

Since cubic splines are well behaved, they are sometimes used in computer graphics routines for curve plotting.

**Algorithm**

**Subroutine** Spline (**a**, **b**, **c**, **d**, **x**, **y**, $N$)
Comment: This subroutine computes the coefficients $a_i$, $b_i$, $c_i$, and $d_i$ (10.7.9) using Equations (10.7.22), (10.7.21), (10.7.20), and (10.7.14). Data points $(x_0, y_0)$, $(x_1, y_1)$, ..., $(x_N, y_N)$ and the clamped boundary conditions (10.7.11) are given. We have **a** = $[a_0, a_1, \ldots, a_N]$, **b** = $[b_0, b_1, \ldots, b_N]$, **c** = $[c_0, c_1, \ldots, c_N]$, **d** = $[d_0, d_1, \ldots, d_N]$, **x** = $[x_0, x_1, \ldots, x_N]$, and **y** = $[y_0, y_1, \ldots, y_N]$.
 1 **Print x, y**, $N$
 2 **Repeat** steps 3–4 **with** $i = 0, 1, \ldots, N - 1$
        3 $h_i = x_{i+1} - x_i$
        4 $a_i = y_i$
Comment: This part computes a solution of a tridiagonal system (10.7.28) using **subroutine** Tridi of Section 7.7. We use **p**, **q**, **r**, and **z** instead of **a**, **b**, **c**, and **x** in subroutine Tridi.
 5 $p_1 = h_0$
 6 $q_1 = 2h_0$
 7 $s_1 = 3[(y_1 - y_0)/h_0 - y_0']$
 8 **Repeat** steps 9–12 **with** $i = 2, 3, \ldots, N$
        9 $p_i = h_{i-2}$
       10 $q_i = 2(h_{i-2} + h_{i-1})$
       11 $r_i = h_{i-1}$
       12 $s_i = 3[(y_i - y_{i-1})/h_{i-1} - (y_{i-1} - y_{i-2})/h_{i-2}]$
13 $p_{N+1} = h_{N-1}$
14 $q_{N+1} = 2h_{N-1}$
15 $s_{N+1} = 3y_N' - 3[(y_N - y_{N-1})/h_{N-1}]$
16 **Call** Tridi (**p**, **q**, **r**, **s**, $N+1$, **z**)
17 **Repeat** step 18 **with** $i = 0, 1, \ldots, N$
       18 $c_i = z_{i+1}$
19 **Repeat** steps 20–21 **with** $i = 0, 1, \ldots, N-1$
       20 $b_i = (y_{i+1} - y_i)/h_i - h_i(c_{i+1} + 2c_i)/3$
       21 $d_i = (c_{i+1} - c_i)/(3h_i)$
22 **Print a, b, c**, and 'finished' and exit.   ◀

# EXERCISES

**1.** Determine $a_0$, $a_1$, $b_0$, and $b_1$ so that $S(x)$ is a linear spline where

$$S(x) = \begin{cases} a_0 + b_0x & 0 \le x \le 1 \\ a_1 + b_1(x - 1) & 1 \le x \le 2 \end{cases}$$

with $S(0) = 0$, $S(1) = 1$, and $S(2) = 8$.

2. Determine $a_0$, $a_1$, $b_0$, $b_1$, $c_0$, and $c_1$ so that $S(x)$ is a quadratic spline where

$$S(x) = \begin{cases} a_0 + b_0 x + c_0 x^2 & 0 \le x \le 1 \\ a_1 + b_1(x - 1) + c_1(x - 1)^2 & 1 \le x \le 2 \end{cases}$$

with $S(0) = 0$, $S(1) = 1$, $S(2) = 8$, and $S'(0) = 0$.

3. Determine $a_0$, $a_1$, $b_0$, $b_1$, $c_0$, $c_1$, $d_0$, and $d_1$ so that $S(x)$ is a natural cubic spline where

$$S(x) = \begin{cases} a_0 + b_0 x + c_0 x^2 + d_0 x^3 & 0 \le x \le 1 \\ a_1 + b_1(x - 1) + c_1(x - 1)^2 + d_1(x - 1)^3 & 1 \le x \le 2 \end{cases}$$

with $S(0) = 0$, $S(1) = 1$, and $S(2) = 8$.

4. Determine $a_0$, $a_1$, $b_0$, $b_1$, $c_0$, $c_1$, $d_0$, and $d_1$ so that $S(x)$ is a clamped cubic spline where

$$S(x) = \begin{cases} a_0 + b_0 x + c_0 x^2 + d_0 x^3 & 0 \le x \le 1 \\ a_1 + b_1(x - 1) + c_1(x - 1)^2 + d_1(x - 1)^3 & 1 \le x \le 2 \end{cases}$$

with $S(0) = 0$, $S(1) = 1$, $S(2) = 8$, $S'(0) = 0$, $S'(2) = 12$.

5. Construct a free cubic spline to approximate $f(x) = \cos x$ using the following data:

| $x_i$ | 0.0 | 0.4 | 1.0 |
|---|---|---|---|
| $y_i = \cos x_i$ | 1.00000 | 0.92106 | 0.54030 |

Using your constructed cubic spline, find $f(0.5)$ and compare it with 0.87758.

6. Construct a clamped cubic spline to approximate $f(x) = \cos x$ using the following data:

| $x_i$ | 0.0 | 0.4 | 1.0 |
|---|---|---|---|
| $y_i = \cos x_i$ | 1.00000 | 0.92106 | 0.54030 |

$$f'(0) = 0.00000 \quad \text{and} \quad f'(1) = -0.84147$$

Using your constructed cubic spline, find $f(0.5)$ and compare it with 0.87758.

7. Construct a clamped cubic spline to approximate $f(x) = x^3 - 1$ using the following data:

| $x_i$ | -2 | 0 | 1 |
|---|---|---|---|
| $y_i$ | -9 | -1 | 0 |

Using your constructed cubic spline, find $f(-1)$ and compare it with the exact value $-2$.

8. Construct a free cubic spline to approximate $f(x) = x^3 - 1$ using the following data:

| $x_i$ | -2 | 0 | 1 |
|---|---|---|---|
| $y_i$ | -9 | -1 | 0 |

Using your constructed cubic spline, find $f(-1)$ and compare it with the exact value $-2$.

**C.1.** Write a subroutine to solve a tridiagonal system $A\mathbf{x} = \mathbf{b}$ using the factorization method.

**C.2.** Write a subroutine that computes the coefficients $a_i$, $b_i$, $c_i$, and $d_i$ of Equation (10.7.9) using Equations (10.7.28), (10.7.21), (10.7.20), and (10.7.14). Use Exercise C.1 or C.1, p. 219.

**C.3.** Determine a clamped cubic spline to approximate $f(x) = 1/(1 + x^2)$ in the interval $[-5, 5]$ using 5 and 11 equidistant points. Represent the spline function graphically and find the maximum error. Use Exercise C.2.

**C.4.** Determine a clamped cubic spline to approximate $f(x) = \sqrt{1 - x^2}$ in the interval $[-1, 1]$ using 5 and 11 equidistant points. Represent the spline function graphically and find the maximum error. Use Exercise C.2.

**C.5.** Use a free cubic spline to approximate the population of the United States in 1954, 1967, and 1978 from the following table:

| Year | 1950 | 1960 | 1970 | 1980 | 1990 |
|---|---|---|---|---|---|
| **Population (in thousands)** | 151325 | 179323 | 203302 | 226542 | 248709 |

## 10.8   RATIONAL FUNCTION APPROXIMATION

A rational function $r$ of degree $N + M$ has the form

$$r_{N,M}(x) = \frac{p_N(x)}{q_M(x)} = \frac{p_0 + p_1 x + p_2 x^2 + \cdots + p_N x^N}{q_0 + q_1 x + q_2 x^2 + \cdots + q_M x^M} \qquad (10.8.1)$$

where $p_N(x)$ and $q_M(x)$ are polynomials of degree $N$ and $M$ respectively. For $r_{N,M}(x)$ to be defined at 0, we require $q_0 \neq 0$. In fact, we assume $q_0 = 1$. If $q_0 \neq 1$, then we replace $p_N(x)$ by $p_N(x)/q_0$ and $q_M(x)$ by $q_M(x)/q_0$. Hence the rational function $r_{N,M}(x)$ has $N + M + 1$ parameters $p_0, p_1, \ldots, p_N, q_1, q_2, \ldots, q_M$. The graphs of rational functions assume shapes that the graph of a polynomial cannot assume. In addition, rational functions provide a better approximation of functions that have infinite discontinuities near but outside the interval of approximation. For example, consider the function tan $x$ in the interval $0 \leq x \leq \pi/3$. Also, empirical evidence shows that for a given value of $N + M$, the most accurate results are usually given by a rational function with $N = M$ or $M + 1$. Thus rational functions can provide a useful class of approximations.

We want to approximate $f(x)$ by $r_{N,M}(x)$ on $[a, b]$. The Padé rational approximation technique requires $f^{(k)}(\alpha) = r_{N,M}^{(k)}(\alpha)$ for $k = 0, 1, \ldots, N + M$. Thus the Padé approximation is an application of Taylor's approximation to a rational function. We assume $\alpha = 0$ to ensure ease in manipulation. We can always achieve this by a suitable change of variable. To determine the coefficients $p_i$ and $q_i$, expand $f(x)$ in terms of its Maclaurin series. Therefore, assume that $f$ has the Maclaurin expansion $f(x) = \sum_{i=0}^{\infty} a_i x^i$ where $a_i = f^{(i)}(0)/i!$. Then

$$f(x) - r_{N,M}(x) = \sum_{i=0}^{\infty} a_i x^i - \frac{\sum_{i=0}^{N} p_i x^i}{\sum_{i=0}^{M} q_i x^i}$$

$$= \frac{\left(\sum_{i=0}^{\infty} a_i x^i\right)\left(\sum_{i=0}^{M} q_i x^i\right) - \sum_{i=0}^{N} p_i x^i}{\sum_{i=0}^{M} q_i x^i} \qquad \textbf{(10.8.2)}$$

We have $N + M + 1$ constants $p_i$ and $q_i$ to be determined by equating to zero $f^{(k)}(x) - r_{N,M}^{(k)}(x)$ for $k = 0, 1, \ldots, N + M$. This is achieved by equating the numerator of the right-hand side of Equation (10.8.2) to a power series with its leading term of degree $N + M + 1$; in other words,

$$\left(\sum_{i=0}^{\infty} a_i x^i\right)\left(\sum_{i=0}^{M} q_i x^i\right) - \left(\sum_{i=0}^{N} p_i x^i\right) = \sum_{i=N+M+1}^{\infty} d_i x^i \qquad \textbf{(10.8.3)}$$

We have $q_0 = 1$. Multiplying out the left-hand side of Equation (10.8.3) and equating the coefficients of the powers of $x^i$ to zero for $i = 0, 1, \ldots, N + M$, we get a system of $N + M + 1$ linear equations:

$$
\begin{aligned}
a_0 - p_0 &= 0 \\
q_1 a_0 + \qquad a_1 - p_1 &= 0 \\
q_2 a_0 + \qquad q_1 a_1 + \quad a_2 - p_2 &= 0 \\
q_3 a_0 + \qquad q_2 a_1 + q_1 a_2 + a_3 - p_3 &= 0 \\
\vdots \qquad\qquad \vdots \quad\ \vdots \quad\ \vdots \quad\ \vdots \quad\ \vdots \\
q_M a_{N-M} + q_{M-1} a_{N-M+1} + \cdots + a_N - p_N &= 0
\end{aligned} \qquad \textbf{(10.8.4)}
$$

and

$$
\begin{aligned}
q_M a_{N-M+1} + q_{M-1} a_{N-M+2} + \cdots + q_1 a_N + a_{N+1} &= 0 \\
q_M a_{N-M+2} + q_{M-1} a_{N-M+3} + \cdots + q_1 a_{N+1} + a_{N+2} &= 0 \\
\vdots \\
q_M a_N + q_{M-1} a_{N+1} + \cdots + q_1 a_{N+M-1} + a_{N+M} &= 0
\end{aligned} \qquad \textbf{(10.8.5)}
$$

*Note:* (1) The sum of the subscripts of the factors of each product is constant for each equation in (10.8.4) and (10.8.5). This constant increases consecutively from 0 to $N + M$. (2) The $M$ equations in Equation (10.8.5) can be written as

$$
\begin{bmatrix}
a_{N-M+1} & a_{N-M+2} & \cdots & a_N \\
a_{N-M+2} & a_{N-M+3} & \cdots & a_{N+1} \\
\vdots & \vdots & \ddots & \vdots \\
a_N & a_{N+1} & \cdots & a_{N+M-1}
\end{bmatrix}
\begin{bmatrix}
q_M \\
q_{M-1} \\
\vdots \\
q_1
\end{bmatrix}
=
\begin{bmatrix}
-a_{N+1} \\
-a_{N+2} \\
\vdots \\
-a_{N+M}
\end{bmatrix} \qquad \textbf{(10.8.6)}
$$

or

$$\sum_{i=0}^{M} a_{l-i} q_i = 0 \quad \text{for } l = N + 1, N + 2, \ldots, N + M \qquad \textbf{(10.8.7)}$$

where $a_{l-i} = 0$ if $l < i$ and $q_0 = 1$ must be solved for $q_1, q_2, \ldots, q_M$. Then Equation (10.8.4) can be written as

$$p_l = \sum_{i=0}^{l} a_{l-i} q_i \quad \text{for } l = 0, 1, \ldots, N \tag{10.8.8}$$

where $q_i = 0$ if $i > M$. This gives $p_l$ explicitly.

**EXAMPLE 10.8.1**   Determine the Padé approximation of degree five with various $N$ and $M$ for $f(x) = e^x$. Compare the results at $x = 0, 0.2, 0.4, 0.6,$ and $1.0$.

The Maclaurin polynomial of the fifth degree for $e^x$ is

$$r_{5,0} = 1 + x + \frac{1}{2!} x^2 + \frac{1}{3!} x^3 + \frac{1}{4!} x^4 + \frac{1}{5!} x^5$$

Thus $a_i = 1/i!$.

Further

$$r_{4,1} = \frac{p_0 + p_1 x + p_2 x^2 + p_3 x^3 + p_4 x^4}{1 + q_1 x}$$

where $p_i$ and $q_i$ from Equations (10.8.7) and (10.8.8) are given by

$$\sum_{i=0}^{l} a_{l-i} q_i = 0 \quad \text{for } l = 5 \tag{10.8.9}$$

and

$$p_l = \sum_{i=0}^{l} a_{l-i} q_i \quad \text{for } l = 0, 1, 2, 3, \text{ and } 4 \tag{10.8.10}$$

Equation (10.8.9) reduces to

$$a_5 q_0 + a_4 q_1 = 0$$

Hence $q_1 = -(a_5/a_4)q_0 = -\frac{1}{5}$. For $l = 0$, Equation (10.8.10) reduces to

$$p_0 = a_0 q_0 = 1$$

From Equation (10.8.10), for $l = 1$, we get

$$p_1 = \sum_{i=0}^{1} a_{1-i} q_i = a_1 q_0 + a_0 q_1 = 1 - \frac{1}{5} = \frac{4}{5}$$

Similarly

$$p_2 = \sum_{i=0}^{2} a_{2-i} q_i = a_2 q_0 + a_1 q_1 + a_0 q_2 = \frac{1}{2!} - \frac{1}{5} = \frac{3}{10}$$

$$p_3 = \frac{1}{15} \quad \text{and} \quad p_4 = \frac{1}{120}$$

Thus

$$r_{4,1} = \frac{1 + \frac{4}{5} x + \frac{3}{10} x^2 + \frac{1}{15} x^3 + \frac{1}{120} x^4}{1 - \frac{1}{5} x}$$

It can be verified that

$$r_{3,2} = \frac{1 + \frac{6}{10}x + \frac{3}{20}x^2 + \frac{1}{60}x^3}{1 - \frac{2}{5}x - \frac{1}{20}x^2}$$

$$r_{2,3} = \frac{1 + \frac{2}{5}x + \frac{1}{20}x^2}{1 - \frac{3}{5}x + \frac{3}{20}x^2 - \frac{1}{60}x^3}$$

$$r_{1,4} = \frac{1 + \frac{1}{5}x}{1 - \frac{4}{5}x + \frac{3}{10}x^2 - \frac{1}{15}x^3 + \frac{1}{120}x^4}$$

$$r_{0,5} = \frac{1}{1 - x + \frac{x^2}{2!} - \frac{x^3}{3!} + \frac{x^4}{4!} - \frac{x^5}{5!}}$$

The errors $|r_{N,M}(x) - e^x|$ for various values of $x$, $N$, and $M$ are given in Table 10.8.1.

### Table 10.8.1

| $x$ | $e^x$ | $\|r_{5,0}(x) - e^x\|$ | $\|r_{4,1}(x) - e^x\|$ | $\|r_{3,2}(x) - e^x\|$ | $\|r_{2,3}(x) - e^x\|$ | $\|r_{1,4}(x) - e^x\|$ | $\|r_{0,5}(x) - e^x\|$ |
|---|---|---|---|---|---|---|---|
| 0.0 | 1.00000 | 0.00000 | 0.00000 | 0.00000 | 0.00000 | 0.00000 | 0.00000 |
| 0.2 | 1.22140 | 0.00000 | 0.00001 | 0.00000 | 0.00000 | 0.00000 | 0.00000 |
| 0.4 | 1.49182 | 0.00001 | 0.00023 | 0.00000 | 0.00000 | 0.00000 | 0.00001 |
| 0.6 | 1.82212 | 0.00007 | 0.00121 | 0.00001 | 0.00001 | 0.00003 | 0.00020 |
| 0.8 | 2.22554 | 0.00041 | 0.00395 | 0.00007 | 0.00010 | 0.00025 | 0.00162 |
| 1.0 | 2.71828 | 0.00162 | 0.00995 | 0.00033 | 0.00047 | 0.00130 | 0.00899 |

■ ■ ■

We can evaluate $r_{3,2}(x)$ using nested multiplication as

$$r_{3,2}(x) = \frac{1 + x\left(\frac{6}{10} + x\left(\frac{3}{20} + \frac{1}{60}x\right)\right)}{1 - x\left(\frac{2}{5} + \frac{1}{20}x\right)} \qquad \textbf{(10.8.11)}$$

which requires five multiplications, five additions/subtractions, and one division. Another approach is to write $r_{3,2}(x)$ as continued division by performing a series of divisions and reciprocations. A **continued fraction** is an expression of the form

$$f(x) = P_0(x) + \cfrac{P_1(x)}{Q_1(x) + \cfrac{P_2(x)}{Q_2(x) + \cfrac{P_3(x)}{Q_3(x) + \cdots}}}$$

where $P_i(x)$ and $Q_i(x)$ are linear functions or constants. Thus

$$
r_{3,2}(x) = \frac{\dfrac{1}{3}x^3 + 3x^2 + 12x + 20}{x^2 - 8x + 20}
$$

$$
= \frac{1}{3}x + \frac{17}{3} + \frac{\dfrac{152}{3}x - \dfrac{280}{3}}{x^2 - 8x + 20}
$$

$$
= \frac{1}{3}x + \frac{17}{3} + \frac{\dfrac{152}{3}\left(x - \dfrac{35}{19}\right)}{x^2 - 8x + 20}
$$

$$
= \frac{1}{3}x + \frac{17}{3} + \frac{\dfrac{152}{3}}{\dfrac{x^2 - 8x + 20}{x - \dfrac{35}{19}}}
$$

$$
= \frac{1}{3}x + \frac{17}{3} + \frac{\dfrac{152}{3}}{x - \dfrac{117}{19} + \dfrac{\dfrac{3125}{361}}{x - \dfrac{35}{19}}}
\tag{10.8.12}
$$

The expression form in Equation (10.8.12) of a rational function is called **continued fraction form**. It requires one multiplication, five additions/subtractions and two divisions. If division and multiplication require the same amount of time, then Equation (10.8.12) is preferable.

We can improve rational function approximation using the set of Chebyshev polynomials. The Chebyshev rational function approximation method proceeds in a manner similar to the Padé approximation method by replacing each term $x^i$ in the Padé approximation by the $i$th degree Chebyshev polynomial $T_i$ (Exercises 10, 11, 12, and 13).

**E X E R C I S E S**

1. Determine the Padé approximation $r_{1,1}(x)$ for $f(x) = e^x$.
2. Determine the Padé approximation $r_{2,2}(x)$ for $f(x) = e^x$.
3. Compute $e^x - r_{1,1}(x)$ for $x = 0.1, 0.5,$ and $1.0$ and comment on the differences.
4. Compute $e^x - r_{2,2}(x)$ for $x = -0.1, -0.5,$ and $1.0$ and comment on the differences.
5. Determine the Padé approximation $r_{2,2}(x)$ for $f(x) = \tan^{-1} x$.
6. Express 34/5 in continued fraction form.
7. Express the following rational functions in continued fraction form:

    (a) $\dfrac{x^2 + x + 1}{x + 1}$    (b) $\dfrac{x^2 + x + 1}{x^2 - x - 1}$

8. Determine the Padé approximation $r_{3,3}(x)$ for $f(x) = \sin x$. Express the Padé approximation as a continued fraction.

9. Determine the Padé approximation $r_{2,2}(x)$ for $f(x) = \ln(1 + x)$. Express the Padé approximation as a continued fraction.

10. The expansion of $f(x)$ in a series of Chebyshev polynomials is given by

$$f(x) = \frac{a_0}{2} + \sum_{i=0}^{\infty} a_i T_i(x)$$

where

$$a_i = \frac{2}{\pi} \int_{-1}^{1} \frac{f(x)T_i(x)}{\sqrt{1 - x^2}} dx$$

(a) Find $a_0$, $a_1$, and $a_2$ if $f(x) = x^2$.

(b) Find $a_0$, $a_1$, $a_2$, and $a_3$ if $f(x) = e^x$.

11. In order to approximate $f(x)$ by an $(N + M)$th-degree rational function $r_{N,M}(x)$ given by

$$r_{N,M}(x) = \frac{\displaystyle\sum_{i=0}^{N} p_i T_i(x)}{\displaystyle\sum_{i=0}^{M} q_i T_i(x)} \quad \text{with } q_0 = 1$$

determine $p_i$ and $q_i$ so that the coefficients of $T_i(x)$ for $i = 0, 1, \ldots, N + M$ in $f(x) - r_{N,M}(x)$ vanishes. Verify that this leads to

$$\frac{a_0}{2} \sum_{i=0}^{M} q_i T_i(x) + \sum_{i=0}^{\infty} \sum_{j=0}^{M} a_i q_j T_i(x)T_j(x) - \sum_{i=0}^{N} p_i T_i(x) = \sum_{i=N+M+1}^{\infty} d_i T_i(x)$$

**(10.8.13)**

12. Verify that

$$T_{i+j}(x) + T_{|i-j|}(x) = 2T_i(x)T_j(x) \qquad \textbf{(10.8.14)}$$

13. Using Equation (10.8.14) in Equation (10.8.13), verify that the set of $N + M + 1$ equations in $N + M + 1$ constants $p_i$ and $q_i$ is

$$p_0 = \frac{1}{2} \sum_{i=0}^{M} q_i a_i$$

$$p_l = \frac{1}{2} \sum_{i=0}^{M} q_i(a_{|l-i|} + a_{l+i}) \quad \text{for } l = 1, 2, \ldots, N + M$$

where $p_l = 0$ if $l > N$.

14. Find the Chebyshev rational approximation $r_{2,2}(x)$ for $f(x) = e^x$ using Exercises 10(b) and 13. Compare it with the Padé approximation $r_{2,2}(x)$ given in Exercise 2.

## COMPUTER EXERCISES

**C.1.** Write a subroutine Padé to approximate a polynomial $P_{N+M}(x)$ by a rational function $r_{N,M}(x)$ by computing the coefficients $\mathbf{p} = [p_0, p_1, \ldots, p_N]$ of the numerator and $\mathbf{q} = [q_0, q_1, \ldots, q_M]$ of the denominator using Equations (10.8.7) and (10.8.8).

**C.2.**   Determine the Padé approximation of degree nine with various $N$ and $M$ for $f(x) = \arctan x$. Compare the results at $x = 0.0, 0.2, 0.4, 0.6, 0.8$, and $1.0$. Use Exercise C.1.

**C.3.**   Determine the Padé approximation of degree nine with various $N$ and $M$ for $f(x) = \sin x$. Compare the results at $x = 0.0, 0.2, 0.4, 0.6, 0.8$, and $1.0$. Use Exercise C.1.

**C.4.**   Write a subroutine to convert a rational function into a continued fraction.

**C.5.**   Write a subroutine to evaluate a continued fraction.

**C.6.**   Evaluate the following at $x = -0.4, -0.1, 0.1$, and $0.4$ using Exercises C.4 and C.5:

(a) $\dfrac{30x + 21x^2 + x^3}{30 + 36x + 9x^2}$   (b) $\dfrac{945x + 735x^3 + 64x^5}{945 + 1050x^2 + 225x^4}$

## SUGGESTED READINGS

E. O. Brigham, *The Fast Fourier Transform and its Applications,* Prentice Hall, Englewood Cliffs, New Jersey, 1988.

G. H. Golub and W. Kahan, *Calculating the singular values and pseudo-inverse of a matrix,* SIAM J. Numer. Anal. Ser. B **2**(1965), 205–224.

G. H. Golub and C. F. Van Loan, *Matrix Computations,* 2nd ed., Johns Hopkins University Press, Baltimore, Maryland, 1989.

W. Kaplan, *Advanced Calculus,* 2nd ed., Addison–Wesley, Reading, Massachusetts, 1973.

W. W. Hager, *Applied Numerical Linear Algebra,* Prentice Hall, Englewood Cliffs, New Jersey, 1988.

D. S. Watkins, *Fundamentals of Matrix Computations,* John Wiley & Sons, New York, 1991.

# Chapter

# 11 ITERATIVE METHODS FOR SYSTEMS OF EQUATIONS

## 11.1 INTRODUCTION

In this chapter we discuss iterative methods to solve $A\mathbf{x} = \mathbf{b}$ where $A$ is a square, sparse matrix of large order. A **sparse matrix** is a matrix in which most coefficients are zero. Such systems arise a great deal in the numerical solution of boundary-value problems and partial differential equations. In order to discuss the convergence of iterative methods, norms are also discussed.

Before we discuss the concept of an iterative method, we explain how large systems of linear equations can arise in practice and the special properties their coefficient matrices have.

Consider the steady-state temperature distribution in a unit length cylindrical rod with the steady energy source $q(x) = 1$ and an insulated lateral surface whose ends are held at constant temperatures $T_0$ and $T_1$. The temperature function $v(x)$ satisfies the heat equation

$$\frac{d^2v}{dx^2} + 1 = 0 \qquad 0 < x < 1 \tag{11.1.1}$$

with boundary conditions

$$v(0) = T_0 \quad \text{and} \quad v(1) = T_1 \tag{11.1.2}$$

Our goal is to reduce this problem to a system of linear equations.

Divide the interval $[0, 1]$ into a number of subintervals of equal length $h$ by introducing grid points

$$0 = x_0 < x_1 < x_2 \cdots < x_N = 1$$

as shown in Figure 11.1.1

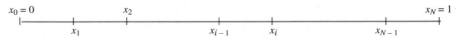

**Fig. 11.1.1**

The number $h = (1 - 0)/N$ is called the **grid spacing**. Then $x_i = ih$ for $i = 0, 1, \ldots, N$; $x_0$ and $x_N$ are called the **boundary grid points**, and $x_1, x_2, \ldots, x_{N-1}$ are called the **interior grid points**.

At each interior grid point $x_i$ for $i = 1, 2, \ldots, N - 1$, Equation (11.1.1) is satisfied and we have

$$\frac{d^2v}{dx^2}(x_i) + 1 = 0 \tag{11.1.3}$$

The second derivative $\dfrac{d^2v}{dx^2}$ at $x_i$ from Table 5.3.1 is given by

$$\frac{d^2v}{dx^2}(x_i) = \frac{v(x_i - h) - 2v(x_i) + v(x_i + h)}{h^2} - \frac{h^2}{12}v^{(4)}(\xi_i) \tag{11.1.4}$$

where $x_i - h < \xi_i < x_i + h$. Substituting Equation (11.1.4) into Equation (11.1.3), we get at each interior grid point

$$v(x_i - h) - 2v(x_i) + v(x_i + h) + h^2 - \frac{h^4}{12}v^{(4)}(\xi_i) = 0 \tag{11.1.5}$$

If we choose $h$ small enough, we can ignore the last term in Equation (11.1.5). Denoting the approximate value of $v$ at $x_i$ by $V_i$ [i.e., $V_i \cong v(x_i)$], the approximate value of $v$ at $x_i + h$ by $V_{i+1}$ [i.e., $V_{i+1} \cong v(x_i + h)$], and the approximate value of $v$ at $x_i - h$ by $V_{i-1}$ [i.e., $V_{i-1} \cong v(x_i - h)$], Equation (11.1.5) reduces to

$$V_{i-1} - 2V_i + V_{i+1} = -h^2 \quad \text{for } i = 1, 2, \ldots, N - 1 \tag{11.1.6}$$

with

$$V_0 = v(0) = T_0 \qquad V_N = v(1) = T_1 \tag{11.1.7}$$

The first and last equations of (11.1.6) contain $V_0$ and $V_N$ whose values are given by Equation (11.1.7). We write Equation (11.1.6) in the form $A\mathbf{v} = \mathbf{b}$ to see the form of the matrix $A$. Let $N = 7$. Then $h = \frac{1}{7}$ and

$$
\begin{aligned}
V_0 - 2V_1 + V_2 & & & & & = -h^2 \\
V_1 - 2V_2 + V_3 & & & & & = -h^2 \\
V_2 - 2V_3 + V_4 & & & & & = -h^2 \\
V_3 - 2V_4 + V_5 & & & & & = -h^2 \\
V_4 - 2V_5 + V_6 & & & & & = -h^2 \\
V_5 - 2V_6 + V_7 & & & & & = -h^2
\end{aligned} \tag{11.1.8}
$$

in other words,

$$A\mathbf{v} = \begin{bmatrix} -2 & 1 & 0 & 0 & 0 & 0 \\ 1 & -2 & 1 & 0 & 0 & 0 \\ 0 & 1 & -2 & 1 & 0 & 0 \\ 0 & 0 & 1 & -2 & 1 & 0 \\ 0 & 0 & 0 & 1 & -2 & 1 \\ 0 & 0 & 0 & 0 & 1 & -2 \end{bmatrix} \begin{bmatrix} V_1 \\ V_2 \\ V_3 \\ V_4 \\ V_5 \\ V_6 \end{bmatrix} = \begin{bmatrix} -T_0 - h^2 \\ -h^2 \\ -h^2 \\ -h^2 \\ -h^2 \\ -T_1 - h^2 \end{bmatrix} = \mathbf{b} \tag{11.1.9}$$

We have reduced a given physical problem to a system of linear equations. The matrix $A$ in Equation (11.1.9) is sparse; to be precise it is tridiagonal and symmetric.

The Gaussian elimination and $LU$ decomposition methods are efficient for such sparse systems; however, a system of order $10^4$ or $10^5$ arising in the numerical solution of partial differential equations is not uncommon. Therefore, from the viewpoint of storage, it becomes difficult to use Gaussian elimination or its variants. The nonzero elements of $A$ are known by a simple formula and can be generated as needed. The iterative methods solve $Ax = b$ by storing a few vectors of length $N$ ($\mathbf{b}$ and $\mathbf{x}$ are usually stored).

We solved $f(x) = 0$ by the fixed-point iteration method by first writing it in the form $x = g(x)$ and then solving $x^{(k+1)} = g(x^{(k)})$ for $k = 0, 1, \ldots$ by initially guessing $x^{(0)}$. Similarly we rewrite $Ax = b$ in the form $\mathbf{x} = B\mathbf{x} + \mathbf{c}$, where $B$ is an $N \times N$ matrix and $\mathbf{c}$ is a vector and then compute $\mathbf{x}^{(k+1)} = B\mathbf{x}^{(k)} + \mathbf{c}$ for $k = 0, 1, \ldots$ after selecting $\mathbf{x}^{(0)}$ until it converges.

**EXAMPLE 11.1.1**  Solve Equation (11.1.8) for $V_0 = T_0 = 2$ and $V_7 = T_1 = 3$.
Equation (11.1.8) can be written as

$$V_1 = \frac{1}{2}(V_0 + V_2 + h^2) = \frac{1}{2}V_2 + 1 + (h^2/2) \qquad V_4 = \frac{1}{2}(V_3 + V_5 + h^2)$$

$$V_2 = \frac{1}{2}(V_1 + V_3 + h^2) \qquad\qquad\qquad V_5 = \frac{1}{2}(V_4 + V_6 + h^2)$$

$$V_3 = \frac{1}{2}(V_2 + V_4 + h^2) \qquad\qquad\qquad V_6 = \frac{1}{2}(V_5 + V_7 + h^2)$$

$$\qquad\qquad\qquad\qquad\qquad\qquad\qquad = \frac{1}{2}V_5 + \frac{3}{2} + (h^2/2)$$

This can be written as

$$\begin{bmatrix} V_1 \\ V_2 \\ V_3 \\ V_4 \\ V_5 \\ V_6 \end{bmatrix}^{(k+1)} = \begin{bmatrix} 0 & 1/2 & 0 & 0 & 0 & 0 \\ 1/2 & 0 & 1/2 & 0 & 0 & 0 \\ 0 & 1/2 & 0 & 1/2 & 0 & 0 \\ 0 & 0 & 1/2 & 0 & 1/2 & 0 \\ 0 & 0 & 0 & 1/2 & 0 & 1/2 \\ 0 & 0 & 0 & 0 & 1/2 & 0 \end{bmatrix} \begin{bmatrix} V_1 \\ V_2 \\ V_3 \\ V_4 \\ V_5 \\ V_6 \end{bmatrix}^{(k)} + \begin{bmatrix} 1 + (h^2/2) \\ h^2/2 \\ h^2/2 \\ h^2/2 \\ h^2/2 \\ 3/2 + (h^2/2) \end{bmatrix}$$

$$\textbf{(11.1.10)}$$

in other words,

$$\mathbf{v}^{(k+1)} = B\mathbf{v}^{(k)} + \mathbf{c}$$

Let $\mathbf{v}^{(0)} = [0, 0, 0, 0, 0, 0]'$ (this is our initial selection). Then for $k = 0$, Equation (11.1.10) reduces to

$$V_1^{(1)} = \frac{1}{2}(V_0^{(0)} + V_2^{(0)} + h^2) = 1.01020 \qquad V_4^{(1)} = \frac{1}{2}(V_3^{(0)} + V_5^{(0)} + h^2) = 0.01020$$

$$V_2^{(1)} = \frac{1}{2}(V_1^{(0)} + V_3^{(0)} + h^2) = 0.01020 \qquad V_5^{(1)} = \frac{1}{2}(V_4^{(0)} + V_6^{(0)} + h^2) = 0.01020$$

$$V_3^{(1)} = \frac{1}{2}(V_2^{(0)} + V_4^{(0)} + h^2) = 0.01020 \qquad V_6^{(1)} = \frac{1}{2}(V_5^{(0)} + V_7^{(0)} + h^2)$$

$$\qquad\qquad\qquad\qquad\qquad\qquad\qquad = 1.51020 \qquad \textbf{(11.1.11)}$$

**Table 11.1.1**

| $k$ | $V_0^{(k)}$ | $V_1^{(k)}$ | $V_2^{(k)}$ | $V_3^{(k)}$ | $V_4^{(k)}$ | $V_5^{(k)}$ | $V_6^{(k)}$ | $V_7^{(k)}$ |
|---|---|---|---|---|---|---|---|---|
| 0 | 2.00000 | 0.00000 | 0.00000 | 0.00000 | 0.00000 | 0.00000 | 0.00000 | 3.00000 |
| 1 | 2.00000 | 1.01020 | 0.01020 | 0.01020 | 0.01020 | 0.01020 | 1.51020 | 3.00000 |
| 2 | 2.00000 | 1.01531 | 0.52041 | 0.02041 | 0.02041 | 0.77041 | 1.51531 | 3.00000 |
| 3 | 2.00000 | 1.27041 | 0.52806 | 0.28061 | 0.40561 | 0.77806 | 1.89541 | 3.00000 |
| 4 | 2.00000 | 1.27423 | 0.78571 | 0.47704 | 0.53954 | 1.16071 | 1.89923 | 3.00000 |
| 5 | 2.00000 | 1.40306 | 0.88584 | 0.67283 | 0.82908 | 1.22959 | 2.09056 | 3.00000 |
| 10 | 2.00000 | 1.71044 | 1.48643 | 1.44238 | 1.56055 | 1.92783 | 2.41063 | 3.00000 |
| 20 | 2.00000 | 2.03021 | 2.06817 | 2.16034 | 2.29528 | 2.50303 | 2.74095 | 3.00000 |
| 40 | 2.00000 | 2.18248 | 2.34805 | 2.50249 | 2.64437 | 2.77741 | 2.89633 | 3.00000 |
| 60 | 2.00000 | 2.20140 | 2.38282 | 2.54499 | 2.68773 | 2.81149 | 2.91563 | 3.00000 |
| 80 | 2.00000 | 2.20375 | 2.38714 | 2.55027 | 2.69311 | 2.81573 | 2.91803 | 3.00000 |
| 91 | 2.00000 | 2.20397 | 2.38756 | 2.55078 | 2.69364 | 2.81613 | 2.91826 | 3.00000 |
| 92 | 2.00000 | 2.20399 | 2.38758 | 2.55081 | 2.69366 | 2.81615 | 2.91827 | 3.00000 |

The additional iterates are computed from Equation (11.1.10) in a similar way and are given in Table 11.1.1. The method used in this example is called the **Jacobi method**.

■ ■ ■

## 11.2 THE JACOBI AND GAUSS–SEIDEL METHODS

Consider a linear system $A\mathbf{x} = \mathbf{b}$ given by

$$
\begin{aligned}
a_{11} x_1 + a_{12} x_2 + \cdots + a_{1N} x_N &= b_1 \\
a_{21} x_1 + a_{22} x_2 + \cdots + a_{2N} x_N &= b_2 \\
\vdots \qquad\qquad \vdots \qquad\qquad \vdots \qquad \vdots & \\
a_{N1} x_1 + a_{N2} x_2 + \cdots + a_{NN} x_N &= b_N
\end{aligned} \qquad \textbf{(11.2.1)}
$$

Solve the first equation for $x_1$, the second equation for $x_2$, and so forth. Then

$$
\begin{aligned}
x_1 &= \frac{1}{a_{11}} (b_1 - a_{12} x_2 - \cdots - a_{1N} x_N) \\
x_2 &= \frac{1}{a_{22}} (b_2 - a_{21} x_1 - \cdots - a_{2N} x_N) \\
\vdots \qquad \vdots \qquad\qquad\qquad\qquad \vdots & \\
x_N &= \frac{1}{a_{NN}} (b_N - a_{N1} x_1 - \cdots - a_{N,N-1} x_{N-1})
\end{aligned} \qquad \textbf{(11.2.2)}
$$

The system given by Equation (11.2.2) can be written compactly as

$$
x_i = \frac{1}{a_{ii}} \left( b_i - \sum_{\substack{j=1 \\ j \neq i}}^{N} a_{ij} x_j \right) \quad \text{for } i = 1, 2, \ldots, N \qquad \textbf{(11.2.3)}
$$

provided $a_{ii} \neq 0$.

The entire sequence of **Jacobi iterates** is defined from Equation (11.2.3) as

$$x_i^{(k+1)} = \frac{1}{a_{ii}} \left( b_i - \sum_{\substack{j=1 \\ j \neq i}}^{N} a_{ij} x_j^{(k)} \right) \quad \text{for } i = 1, 2, \ldots, N \quad \text{and} \quad k = 0, 1, \ldots \qquad \textbf{(11.2.4)}$$

Equation (11.2.4) is easy to program for computation.

**▶ Algorithm**

**Subroutine** Jacobi $(N, A, \mathbf{b}, \mathbf{x}, \text{eps}, \text{maxit})$
Comment: This program solves $A\mathbf{x} = \mathbf{b}$ using the Jacobi method given by Equation (11.2.4). $N$ is the number of equations and unknowns; $a_{ij}$ for $i, j = 1, 2, \ldots, N$ are the elements of a matrix $A$; $b_1, b_2, \ldots, b_N$ are the elements of $\mathbf{b}$; $x_1, x_2, \ldots, x_N$ are the initial guesses of $\mathbf{x}$; eps is the tolerance and maxit is the maximum number of iterations.
  1 **Print** $N$, $A$, eps
  2 **Repeat** step 3 **with** $i = 1, 2, \ldots, N$
      3 **If** $|a_{ii}| <$ eps, **then print** 'Algorithm fails because $|a_{ii}| <$ eps' and exit.
  4 **Print** $\mathbf{b}$, $\mathbf{x}$, maxit
  5 $k = 0$
Comment: $\text{xold}_i$ for $i = 1, 2, \ldots, N$ are the old values of $\mathbf{x}$.
  6 **Repeat** steps 7–17 **while** $k \leq$ maxit
      7 $k = k + 1$
      8 **Repeat** step 9 **with** $i = 1, 2, \ldots, N$
          9 $\text{xold}_i = x_i$
     10 **Repeat** steps 11–14 **with** $i = 1, 2, \ldots, N$
         11 sum $= 0$
         12 **Repeat** step 13 **with** $j = 1, 2, \ldots, N$
             13 **If** $j \neq i$, **then** sum $=$ sum $+ a_{ij} * \text{xold}_j$
         14 $x_i = (b_i - \text{sum})/a_{ii}$
     15 **Repeat** step 16 **with** $i = 1, 2, \ldots, N$
         16 **If** $|x_i - \text{xold}_i| \geq$ eps $|x_i|$, **then** return to 6
     17 **Print** $k$, $\mathbf{x}$, and 'finished' and exit.
  18 **Print** 'algorithm fails: no convergence' and exit.   ◀

It is useful to write Equation (11.2.4) in matrix-vector notation to study the convergence of the Jacobi method. For this purpose let

$$A = \begin{bmatrix} a_{11} & a_{12} & a_{13} & \cdots & a_{1N} \\ a_{21} & a_{22} & a_{23} & \cdots & a_{2N} \\ a_{31} & a_{32} & a_{33} & \cdots & a_{3N} \\ \vdots & \vdots & \vdots & \ddots & \vdots \\ a_{N1} & a_{N2} & a_{N3} & \cdots & a_{NN} \end{bmatrix}$$

$$= \begin{bmatrix} 0 & 0 & 0 & \cdots & 0 \\ a_{21} & 0 & 0 & \cdots & 0 \\ a_{31} & a_{32} & 0 & \cdots & 0 \\ \vdots & \vdots & \vdots & \ddots & \vdots \\ a_{N1} & a_{N2} & a_{N3} & \cdots & 0 \end{bmatrix} + \begin{bmatrix} a_{11} & 0 & 0 & \cdots & 0 \\ 0 & a_{22} & 0 & \cdots & 0 \\ 0 & 0 & a_{33} & \cdots & 0 \\ \vdots & \vdots & \vdots & \ddots & \vdots \\ 0 & 0 & 0 & \cdots & a_{NN} \end{bmatrix}$$

$$+ \begin{bmatrix} 0 & a_{12} & a_{13} & \cdots & a_{1N} \\ 0 & 0 & a_{23} & \cdots & a_{2N} \\ 0 & 0 & 0 & \cdots & a_{3N} \\ \vdots & \vdots & \vdots & \ddots & \vdots \\ 0 & 0 & 0 & \cdots & 0 \end{bmatrix}$$

$$= L + D + U \qquad \textbf{(11.2.5)}$$

where $L$, $D$, and $U$ are the strictly lower, diagonal, and strictly upper triangular parts of $A$. The equation $A\mathbf{x} = \mathbf{b}$, which is $(L + D + U)\mathbf{x} = \mathbf{b}$, can be written as

$$D\mathbf{x} = -(L + U)\mathbf{x} + \mathbf{b}$$

This reduces to

$$\mathbf{x} = -D^{-1}(L + U)\mathbf{x} + D^{-1}\mathbf{b} \qquad \textbf{(11.2.6)}$$

and the sequence of iterates is given by

$$\mathbf{x}^{(k+1)} = -D^{-1}(L + U)\mathbf{x}^{(k)} + D^{-1}\mathbf{b} \quad \text{for } k = 0, 1, \dots \qquad \textbf{(11.2.7)}$$

In Equation (11.1.11), we have $V_2^{(1)} = (V_1^{(0)} + V_3^{(0)} + h^2)/2$, although $V_1^{(1)} = 1.01020$ is available. It would seem natural to use the updated value $V_1^{(1)}$ rather than $V_1^{(0)}$. If we use the updated values as soon as they are available, then Equation (11.1.11) becomes

$$V_1^{(1)} = \frac{1}{2}(V_0^{(0)} + V_2^{(0)} + h^2) = 1.01020 \qquad V_4^{(1)} = \frac{1}{2}(V_3^{(1)} + V_5^{(0)} + h^2) = 0.14413$$

$$V_2^{(1)} = \frac{1}{2}(V_1^{(1)} + V_3^{(0)} + h^2) = 0.51531 \qquad V_5^{(1)} = \frac{1}{2}(V_4^{(1)} + V_6^{(0)} + h^2) = 0.08227$$

$$V_3^{(1)} = \frac{1}{2}(V_2^{(1)} + V_4^{(0)} + h^2) = 0.26786 \qquad V_6^{(1)} = \frac{1}{2}(V_5^{(1)} + V_7^{(0)} + h^2) = 1.55134$$

$$\textbf{(11.2.8)}$$

In most cases use of the updated values speed up the convergence. The additional iterates given in Table 11.2.1 are generated in a similar manner. This is called the **Gauss–Seidel method**.

**Table 11.2.1**

| $k$ | $V_0^{(k)}$ | $V_1^{(k)}$ | $V_2^{(k)}$ | $V_3^{(k)}$ | $V_4^{(k)}$ | $V_5^{(k)}$ | $V_6^{(k)}$ | $V_7^{(k)}$ |
|---|---|---|---|---|---|---|---|---|
| 0 | 2.00000 | 0.00000 | 0.00000 | 0.00000 | 0.00000 | 0.00000 | 0.00000 | 3.00000 |
| 1 | 2.00000 | 1.01020 | 0.51531 | 0.26786 | 0.14413 | 0.08227 | 1.55134 | 3.00000 |
| 2 | 2.00000 | 1.26786 | 0.77806 | 0.47130 | 0.28699 | 0.92937 | 1.97489 | 3.00000 |
| 3 | 2.00000 | 1.39923 | 0.94547 | 0.62643 | 0.78811 | 1.39170 | 2.20605 | 3.00000 |
| 4 | 2.00000 | 1.48294 | 1.06489 | 0.93670 | 1.17441 | 1.70043 | 2.36042 | 3.00000 |
| 5 | 2.00000 | 1.54265 | 1.24988 | 1.22235 | 1.47159 | 1.92621 | 2.47331 | 3.00000 |
| 10 | 2.00000 | 1.94482 | 1.96628 | 2.07705 | 2.26662 | 2.50756 | 2.76399 | 3.00000 |
| 20 | 2.00000 | 2.17179 | 2.33533 | 2.49212 | 2.64081 | 2.77799 | 2.89920 | 3.00000 |
| 30 | 2.00000 | 2.20007 | 2.38124 | 2.54370 | 2.68729 | 2.81156 | 2.91599 | 3.00000 |
| 40 | 2.00000 | 2.20358 | 2.38695 | 2.55011 | 2.69306 | 2.81573 | 2.91807 | 3.00000 |
| 49 | 2.00000 | 2.20401 | 2.38763 | 2.55088 | 2.69375 | 2.81624 | 2.91832 | 3.00000 |
| 50 | 2.00000 | 2.20402 | 2.38765 | 2.55091 | 2.69378 | 2.81625 | 2.91833 | 3.00000 |

In order to solve Equation (11.2.1) by the Gauss–Seidel method, we rewrite Equation (11.2.3) as

$$x_i = \frac{1}{a_{ii}}\left[ b_i - \sum_{j=1}^{i-1} a_{ij}x_j - \sum_{j=i+1}^{N} a_{ij}x_j \right] \quad \text{for } i = 1, 2, \dots, N \qquad \textbf{(11.2.9)}$$

provided $a_{ii} \neq 0$.

Since the updated values of $x_1, x_2, \ldots, x_{i-1}$ for the $i$th equation are available, the Gauss–Seidel iteration sequence can be defined from Equation (11.2.9) as

$$x_i^{(k+1)} = \frac{1}{a_{ii}}\left[ b_i - \sum_{j=1}^{i-1} a_{ij} x_j^{(k+1)} - \sum_{j=i+1}^{N} a_{ij} x_j^{(k)} \right] \tag{11.2.10}$$

for $i = 1, 2, \ldots, N$, $k = 0, 1, \ldots$, and given $\mathbf{x}^{(0)}$.

Each updated component $x_i^{(k+1)}$ is used in the calculation of the next component and therefore, for computer calculations, the new value can be immediately stored at the location where the old value was stored. This reduces the number of necessary locations (although you may need old values for stopping criteria).

Equation (11.2.10) can be written as

$$a_{11} x_1^{(k+1)} = b_1 - a_{12} x_2^{(k)} - \cdots \\ - a_{1N} x_N^{(k)}$$

$$a_{21} x_1^{(k+1)} + a_{22} x_2^{(k+1)} = b_2 - a_{23} x_3^{(k)} - \cdots \\ - a_{2N} x_N^{(k)}$$

$$\vdots$$

$$a_{N1} x_1^{(k+1)} + a_{N2} x_2^{(k+1)} + a_{N3} x_3^{(k+1)} + \cdots + a_{NN} x_N^{(k+1)} = b_N$$

In matrix notation

$$\begin{bmatrix} a_{11} & 0 & \cdots & 0 \\ a_{21} & a_{22} & \cdots & 0 \\ \vdots & \vdots & \ddots & \vdots \\ a_{N1} & a_{N2} & \cdots & a_{NN} \end{bmatrix} \begin{bmatrix} x_1^{(k+1)} \\ x_2^{(k+1)} \\ \vdots \\ x_N^{(k+1)} \end{bmatrix} = \begin{bmatrix} b_1 \\ b_2 \\ \vdots \\ b_N \end{bmatrix} - \begin{bmatrix} 0 & a_{12} & \cdots & a_{1N} \\ 0 & 0 & \cdots & a_{2N} \\ \vdots & \vdots & \ddots & \vdots \\ 0 & 0 & \cdots & 0 \end{bmatrix} \begin{bmatrix} x_1^{(k)} \\ x_2^{(k)} \\ \vdots \\ x_N^{(k)} \end{bmatrix} \tag{11.2.11}$$

Equation (11.2.11) can be written by using Equation (11.2.5) as

$$(D + L)\,\mathbf{x}^{(k+1)} = \mathbf{b} - U\mathbf{x}^{(k)}$$

Solving for $\mathbf{x}^{(k+1)}$ yields

$$\mathbf{x}^{(k+1)} = -(D + L)^{-1} U\mathbf{x}^{(k)} + (D + L)^{-1} \mathbf{b} \tag{11.2.12}$$

**Algorithm**

**Subroutine** Seidel ($N$, $A$, $\mathbf{b}$, $\mathbf{x}$, eps, maxit)
Comment: This program solves $A\mathbf{x} = \mathbf{b}$ using the Gauss–Seidel method given by Equation (11.2.10). $N$ is the number of equations and unknowns; $a_{ij}$ for $i, j = 1, 2, \ldots, N$ are the elements of a matrix $A$; $b_1, b_2, \ldots, b_N$ are the elements of $\mathbf{b}$; $x_1, x_2, \ldots, x_N$ are the initial guesses of $\mathbf{x}$; eps is the tolerance and maxit is the maximum number of iterations.

  1 **Print** $N$, $A$, eps
  2 **Repeat** step 3 **with** $i = 1, 2, \ldots, N$
       3 **If** $|a_{ii}| <$ eps, **then print** 'Algorithm fails because $|a_{ii}| <$ eps' and exit.
  4 **Print** $\mathbf{b}$, $\mathbf{x}$, maxit
  5 $k = 0$
Comment: xold$_i$ for $i = 1, 2, \ldots, N$ are the old values of $\mathbf{x}$.
  6 **Repeat** steps 7–17 while $k \leq$ maxit
       7 $k = k + 1$

    8 **Repeat** step 9 **with** $i = 1, 2, \ldots, N$
      9 $\text{xold}_i = x_i$
   10 **Repeat** steps 11–14 **with** $i = 1, 2, \ldots, N$
    11 sum $= 0$
    12 **Repeat** step 13 **with** $j = 1, 2, \ldots, N$
      13 **If** $j \neq i$, **then** sum $=$ sum $+ a_{ij} * x_j$
    14 $x_i = (b_i - \text{sum})/a_{ii}$
   15 **Repeat** step 16 **with** $i = 1, 2, \ldots, N$
    16 **If** $|x_i - \text{xold}_i| \geq \text{eps}\,|x_i|$, **then** return to 6
  17 **Print** $k$, **x**, and 'finished' and exit.
18 **Print** 'algorithm fails: no convergence' and exit.    ◀

**EXAMPLE 11.2.1**    Solve

$$x_1 + 10x_2 = 11$$
$$8x_1 + \phantom{10}x_2 = 9$$

using (a) the Jacobi method, (b) the Gauss–Seidel method, and (c) express the system in the forms of Equations (11.2.7) and (11.2.12).

    Solving the first equation for $x_1$ and the second equation for $x_2$ yields

$$x_1 = 11 - 10x_2 \qquad x_2 = 9 - 8x_1 \tag{11.2.13}$$

(a)   The Jacobi iterates are given by

$$x_1^{(k+1)} = 11 - 10x_2^{(k)} \quad \text{and} \quad x_2^{(k+1)} = 9 - 8x_1^{(k)} \quad \text{for } k = 0, 1, \ldots \tag{11.2.14}$$

Let $x_1^{(0)} = 0$ and $x_2^{(0)} = 0$. Then we have from Equation (11.2.14) for $k = 0$,

$$x_1^{(1)} = 11 - 10x_2^{(0)} = 11 \qquad x_2^{(1)} = 9 - 8x_1^{(0)} = 9$$

in other words, $x_1^{(1)} = 11$ and $x_2^{(1)} = 9$. For $k = 1$, we have from Equation (11.2.14)

$$x_1^{(2)} = 11 - 10x_2^{(1)} = 11 - 90 = -79 \qquad x_2^{(2)} = 9 - 8x_1^{(1)} = 9 - 88 = -79$$

Continuing similarly, the additional iterates given in Table 11.2.2 are computed. It is clear that it is not converging to the exact values $x_1 = 1$ and $x_2 = 1$.

**Table 11.2.2**

| | Jacobi Method | | Gauss–Seidel Method | |
|---|---|---|---|---|
| $k$ | $x_1^{(k)}$ | $x_2^{(k)}$ | $x_1^{(k)}$ | $x_2^{(k)}$ |
| 0 | 0.0 | 0.0 | 0.0 | 0.0 |
| 1 | 11.0 | 9.0 | 11.0 | $-79.0$ |
| 2 | $-79.0$ | $-79.0$ | 801.0 | $-6399.0$ |
| 3 | 801.0 | 641.0 | 64,001.0 | $-511,999.0$ |
| 4 | $-6399.0$ | $-6399.0$ | 5,120,001.0 | $-40,959,999.0$ |
| 5 | 64,001.0 | 51,201.0 | 409,600,001.0 | $-3,276,799,999.0$ |

(b)   From Equation (11.2.13), the Gauss–Seidel iterates are given by

$$x_1^{(k+1)} = 11 - 10x_2^{(k)} \quad \text{and} \quad x_2^{(k+1)} = 9 - 8x_1^{(k+1)} \quad \text{for } k = 0, 1, \ldots \quad \textbf{(11.2.15)}$$

Let $x_1^{(0)} = 0$ and $x_2^{(0)} = 0$. Then from Equation (11.2.15) we get for $k = 0$,

$$x_1^{(1)} = 11 - 10x_2^{(0)} = 11 \qquad x_2^{(1)} = 9 - 8x_1^{(1)} = 9 - 88 = -79$$

For $k = 1$, we have from Equation (11.2.15)

$$x_1^{(2)} = 11 - 10x_2^{(1)} = 11 + 790 = 801$$
$$x_2^{(2)} = 9 - 8x_1^{(2)} = 9 - 8(801) = -6399$$

The additional iterates given in Table 11.2.2 are generated in a similar manner. It is diverging.

Thus in Example 11.2.1, the iterates are not converging.

(c)   We have

$$A = \begin{bmatrix} 1 & 10 \\ 8 & 1 \end{bmatrix} = \begin{bmatrix} 0 & 0 \\ 8 & 0 \end{bmatrix} + \begin{bmatrix} 1 & 0 \\ 0 & 1 \end{bmatrix} + \begin{bmatrix} 0 & 10 \\ 0 & 0 \end{bmatrix} = L + D + U$$

$$D^{-1} = \begin{bmatrix} 1 & 0 \\ 0 & 1 \end{bmatrix} \quad \text{and} \quad (D + L)^{-1} = \begin{bmatrix} 1 & 0 \\ -8 & 1 \end{bmatrix}$$

Therefore $\mathbf{x}^{(k+1)} = -D^{-1}(L + U)\mathbf{x}^{(k)} + D^{-1}\mathbf{b}$ reduces to

$$\begin{bmatrix} x_1 \\ x_2 \end{bmatrix}^{(k+1)} = -\begin{bmatrix} 1 & 0 \\ 0 & 1 \end{bmatrix}\begin{bmatrix} 0 & 10 \\ 8 & 0 \end{bmatrix}\begin{bmatrix} x_1 \\ x_2 \end{bmatrix}^{(k)} + \begin{bmatrix} 1 & 0 \\ 0 & 1 \end{bmatrix}\begin{bmatrix} 11 \\ 9 \end{bmatrix}$$

$$= -\begin{bmatrix} 0 & 10 \\ 8 & 0 \end{bmatrix}\begin{bmatrix} x_1 \\ x_2 \end{bmatrix}^{(k)} + \begin{bmatrix} 11 \\ 9 \end{bmatrix}$$

and $\mathbf{x}^{(k+1)} = -(D + L)^{-1}U\mathbf{x}^{(k)} + (D + L)^{-1}\mathbf{b}$ reduces to

$$\begin{bmatrix} x_1 \\ x_2 \end{bmatrix}^{(k+1)} = -\begin{bmatrix} 1 & 0 \\ -8 & 1 \end{bmatrix}\begin{bmatrix} 0 & 10 \\ 0 & 0 \end{bmatrix}\begin{bmatrix} x_1 \\ x_2 \end{bmatrix}^{(k)} + \begin{bmatrix} 1 & 0 \\ -8 & 1 \end{bmatrix}\begin{bmatrix} 11 \\ 9 \end{bmatrix}$$

$$= -\begin{bmatrix} 0 & 10 \\ 0 & -80 \end{bmatrix}\begin{bmatrix} x_1 \\ x_2 \end{bmatrix}^{(k)} + \begin{bmatrix} 11 \\ -79 \end{bmatrix}$$

■ ■ ■

Our next question is to determine the conditions under which the sequence of iterates generated by a given iterative method converges to the solution of a system. For this purpose we need vector and matrix norms, which are discussed in the next two sections.

## EXERCISES

For Exercises 1–3, carry out the first four iterations using (a) the Jacobi method and (b) the Gauss–Seidel method using $\mathbf{x}^{(0)} = \mathbf{0}$. Compare it with the exact solution.

**1.** $4x_1 + x_2 = 5$
  $x_1 - 5x_2 = -4$
Exact solution: $x_1 = 1$, $x_2 = 1$.

**2.** $3x_1 + x_2 - x_3 = 8$
  $x_1 + 5x_2 + x_3 = -2$
  $x_1 - x_2 + 8x_3 = 4$
Exact solution: $x_1 = 3$, $x_2 = -1$, $x_3 = 0$.

**3.** $x_1 - 5x_2 = -4$
  $4x_1 + x_2 = 5$
Exact solution: $x_1 = 1$, $x_2 = 1$.

For Exercises 4–6, find (a) the Jacobi matrix $-D^{-1}(L + U)$ and its eigenvalues and (b) the Gauss–Seidel matrix $-(D + L)^{-1} U$ and its eigenvalues.

**4.** $\begin{bmatrix} 4 & 1 \\ 1 & -5 \end{bmatrix}$

**5.** $\begin{bmatrix} 3 & 1 & -1 \\ 1 & 5 & 1 \\ 1 & -1 & 8 \end{bmatrix}$

**6.** $\begin{bmatrix} 1 & -5 \\ 4 & 1 \end{bmatrix}$

## COMPUTER EXERCISES

**C.1.** Write a subroutine to solve a system of linear equations $Ax = b$ by the Jacobi method using Equation (11.2.4). Your subroutine should terminate iterations when $|x_i^{(k+1)} - x_i^{(k)}| \le \text{eps}\, |x_i^{(k+1)}|$ for $i = 1, 2, \dots, N$ or $k$ reaches the maximum number of iterations maxit. The parameters are $N$, $A$, $\mathbf{b}$, initial guess $\mathbf{x}^{(0)}$, the maximum number of iterations maxit, and the tolerance eps. Print the values of $\mathbf{x}$ at various values of $k$.

**C.2.** Write a subroutine to solve a system of linear equations $Ax = b$ by the Gauss–Seidel method using Equation (11.2.10). Use the instructions given in Exercise C.1.

**C.3.** Write a computer program to solve a system of linear equations $Ax = b$ using Exercises C.1 and C.2. Read $N$, initial guess $\mathbf{x}^{(0)}$, the maximum number of iterations maxit, and tolerance eps. Also, read $A$ and $\mathbf{b}$ or generate $A$ and $\mathbf{b}$.

For Exercises C.4–C.7, use Exercise (C.3) to solve the system of equations by using $\text{eps} = 10^{-5}$ and $\text{maxit} = 100$. Use various $\mathbf{x}^{(0)}$ and compare the required number of iterations by the Jacobi and Gauss–Seidel methods.

**C.4.** $4x_1 + x_2 + x_3 = 6$
  $x_1 + 6x_2 + 2x_3 = 9$
  $2x_1 + x_2 + 8x_3 = 11$

**C.5.** The system of equations given by Equation (11.1.6) for $N = 10$ and $v_0 = v(0) = T_0 = 2$, $v_N = v(N) = T_1 = 3$.

**C.6.**
$$
\begin{aligned}
x_1 \phantom{+ x_2} + x_3 &= 2 \\
-x_1 + x_2 \phantom{+ x_3} &= 0 \\
x_1 + 2x_2 - 3x_3 &= 0
\end{aligned}
$$

Note that the Jacobi method converges but the Gauss–Seidel method diverges.

**C.7.**
$$
\begin{aligned}
x_1 + 6x_2 + 2x_3 &= 9 \\
2x_1 + x_2 + 8x_3 &= 11 \\
4x_1 + x_2 + x_3 &= 6
\end{aligned}
$$

## 11.3  VECTOR NORMS

When we computed the solution of a nonlinear equation $f(x) = 0$, we used $|r_i - r|$ to compare the computed solution $r_i$ with the exact solution $r$. We computed $|r_{i+1} - r_i|/|r_{i+1}|$ to terminate iterations. Also, we had $\lim_{i \to \infty} r_i = r$. When we solve $A\mathbf{x} = \mathbf{b}$ by an iterative scheme, how should we define $\lim_{k \to \infty} \mathbf{x}^{(k)} = \mathbf{x}$? For relative error, the division of a vector by another vector is undefined. Thus we need some meaningful way to measure the ''size'' of a vector and the distance between two vectors. We extend the concept of absolute value or magnitude from real numbers to vectors. For a vector $\mathbf{x} = [x_1, x_2, \ldots, x_N]'$, the Euclidean length of $\mathbf{x}$, denoted by $\|\mathbf{x}\|_E$, is defined by $\|\mathbf{x}\|_E = \sqrt{x_1^2 + x_2^2 + \cdots + x_N^2}$. The Euclidean length $\|\mathbf{x}\|_E$ has the following geometrical length properties:

**(i)**   $\|\mathbf{x}\|_E \geq 0$ and $\|\mathbf{x}\|_E = 0$ if and only if $\mathbf{x} = \mathbf{0}$

**(ii)**  $\|\alpha\mathbf{x}\|_E = |\alpha| \, \|\mathbf{x}\|_E$ for all $\alpha \in R$

**(iii)** $\|\mathbf{x}\|_E + \|\mathbf{y}\|_E \geq \|\mathbf{x} + \mathbf{y}\|_E$ where $\mathbf{y}$ is any vector which originates from either end of $\mathbf{x}$ in $R^N$. In $R^2$ and $R^3$, (iii) states that the sum of the lengths of two sides of a triangle is at least as large as the length of the third side. Therefore (iii) is called the **triangle inequality**.

$\|\mathbf{x}\|_E$ is a single number that gives an estimate of the "size" of vector $\mathbf{x}$. There are various other numbers that have these properties; also, in many situations it is convenient to measure the size of a vector in other ways. We shall use the word "norm" for any number associated with $\mathbf{x}$ that has these properties of the length of a physical vector. More precisely, let $R^N = \{\mathbf{x} | \mathbf{x} = [x_1, x_2, \ldots, x_N]' \text{ where } x_1, x_2, \ldots, x_N \in R\}$. Then a **vector norm** on $R^N$ is a real valued function $\| \ \|$ from $R^N$ to $R$ with the following distance-like properties:

**(i)**   $\|\mathbf{x}\| \geq 0$ for all $\mathbf{x} \in R^N$ and $\|\mathbf{x}\| = 0$ if and only if $\mathbf{x} = \mathbf{0}$

**(ii)**  $\|\alpha\mathbf{x}\| = |\alpha| \, \|\mathbf{x}\|$ for all $\alpha \in R$ and $\mathbf{x} \in R^N$

**(iii)** $\|\mathbf{x} + \mathbf{y}\| \leq \|\mathbf{x}\| + \|\mathbf{y}\|$ for all $\mathbf{x}, \mathbf{y} \in R^N$.          **(11.3.1)**

A useful class of norms are the $l_p$ **norms** defined by

$$
\|\mathbf{x}\|_p = \left( \sum_{i=1}^{N} |x_i|^p \right)^{1/p} \quad \text{for } p \geq 1 \tag{11.3.2}
$$

Three particular cases

$$(11.3.3a) \qquad \|\mathbf{x}\|_1 = |x_1| + |x_2| + \cdots + |x_N| = \sum_{i=1}^{N} |x_i|$$

$$(11.3.3b) \qquad \|\mathbf{x}\|_2 = \sqrt{x_1^2 + x_2^2 + \cdots + x_N^2} = \sqrt{\sum_{i=1}^{N} |x_i|^2}$$

$$(11.3.3c) \qquad \|\mathbf{x}\|_\infty = \max\{|x_1|, |x_2|, \ldots, |x_N|\} = \max_i |x_i| \qquad \textbf{(11.3.3)}$$

are the most important for us.

It can be verified that $\lim_{p \to \infty} \|\mathbf{x}\|_p = \max_i |x_i| = \|\mathbf{x}\|_\infty$ (Exercise 3). The $l_\infty$ norm is often called the **maximum norm** or **uniform norm**.

**EXAMPLE 11.3.1**   Find $\|\mathbf{x}\|_1$, $\|\mathbf{x}\|_2$, and $\|\mathbf{x}\|_\infty$ if $\mathbf{x} = [1, 2, -1, 0]'$.

We get from Equation (11.3.3), $\|\mathbf{x}\|_1 = |1| + |2| + |-1| + |0| = 4$, $\|\mathbf{x}\|_2 = \sqrt{(1)^2 + (2)^2 + (-1)^2 + (0)^2} = \sqrt{6}$, and $\|\mathbf{x}\|_\infty = \max\{|1|, |2|, |-1|, |0|\} = 2$.   ∎ ∎ ∎

To verify that Equation (11.3.3a) actually defines norm, we verify that (i), (ii), and (iii) of Equation (11.3.1) are satisfied.

**(i)**   Clearly $\|\mathbf{x}\|_1 = |x_1| + |x_2| + \cdots |x_N| \geq 0$. If $\mathbf{x} = [0, 0, \ldots, 0]'$, then $\|\mathbf{x}\|_1 = |0| + |0| + \cdots + |0| = 0$. Conversely, if $\mathbf{x} = [x_1, x_2, \ldots, x_N]'$ and $\|\mathbf{x}\|_1 = 0$, then $|x_1| + |x_2| + \cdots + |x_N| = 0$ and therefore $|x_1| = |x_2| = \cdots = |x_N| = 0$. Hence $\mathbf{x} = \mathbf{0}$.

**(ii)**   $\alpha\mathbf{x} = [\alpha x_1, \alpha x_2, \ldots, \alpha x_N]'$, so $\|\alpha\mathbf{x}\|_1 = |\alpha x_1| + |\alpha x_2| + \cdots + |\alpha x_N| = |\alpha|(|x_1| + |x_2| + \cdots + |x_N|) = |\alpha| \|\mathbf{x}\|_1$.

**(iii)**   Let $\mathbf{y} = [y_1, y_2, \ldots, y_N]'$. Then $\mathbf{x} + \mathbf{y} = [x_1 + y_1, x_2 + y_2, \ldots, x_N + y_N]'$ and therefore

$$\begin{aligned}
\|\mathbf{x} + \mathbf{y}\|_1 &= |x_1 + y_1| + |x_2 + y_2| + \cdots + |x_N + y_N| \\
&\leq (|x_1| + |y_1|) + (|x_2| + |y_2|) + \cdots + (|x_N| + |y_N|) \\
&= (|x_1| + |x_2| + \cdots + |x_N|) + (|y_1| + |y_2| + \cdots + |y_N|) \\
&= \|\mathbf{x}\|_1 + \|\mathbf{y}\|_1.
\end{aligned}$$

Hence $\|\mathbf{x} + \mathbf{y}\|_1 \leq \|\mathbf{x}\|_1 + \|\mathbf{y}\|_1$.

Similarly, it can be verified that Equation (11.3.3b) satisfies (i) and (ii) of Equation (11.3.1). The proof of part (iii) of Equation (11.3.1) for Equation (11.3.3b) depends on the following well-known inequality.

**Theorem 11.3.1**   **(Cauchy–Schwarz Inequality)**   For each $\mathbf{x} = [x_1, x_2, \ldots, x_N]'$ and $\mathbf{y} = [y_1, y_2, \ldots, y_N]'$ in $R^N$,

$$\sum_{i=1}^{N} |x_i \, y_i| \leq \left[ \left( \sum_{i=1}^{N} x_i^2 \right) \left( \sum_{i=1}^{N} y_i^2 \right) \right]^{1/2} \qquad \textbf{(11.3.4)}$$

*Proof*  Let $\mathbf{x} \neq \mathbf{0}$ and $\mathbf{y} \neq \mathbf{0}$. For any real number $\alpha$,

$$
\begin{aligned}
0 \leq \|\mathbf{x} - \alpha\mathbf{y}\|_2^2 &= (x_1 - \alpha y_1)^2 + (x_2 - \alpha y_2)^2 + \cdots + (x_N - \alpha y_N)^2 \\
&= x_1^2 + x_2^2 + \cdots + x_N^2 - 2\alpha\,(x_1\,y_1 + x_2\,y_2 + \cdots + x_N\,y_N) \\
&\quad + \alpha^2\,(y_1^2 + y_2^2 + \cdots + y_N^2) \\
&= \|\mathbf{x}\|_2^2 - 2\alpha \sum_{i=1}^{N} x_i\,y_i + \alpha^2\,\|\mathbf{y}\|_2^2
\end{aligned}
$$

Thus

$$
2\alpha \sum_{i=1}^{N} x_i\,y_i \leq \|\mathbf{x}\|_2^2 + \alpha^2\,\|\mathbf{y}\|_2^2
$$

Since this is true for any $\alpha$, letting $\alpha = \|\mathbf{x}\|_2 / \|\mathbf{y}\|_2$ yields

$$
2\,\frac{\|\mathbf{x}\|_2}{\|\mathbf{y}\|_2} \sum_{i=1}^{N} x_i\,y_i \leq 2\,\|\mathbf{x}\|_2^2
$$

Thus

$$
\begin{aligned}
&\leq \sum_{i=1}^{N} x_i^2 + 2 \sum_{i=1}^{N} |x_i\,y_i| + \sum_{i=1}^{N} y_i^2 \\
&\leq \|\mathbf{x}\|_2^2 + 2\,\|\mathbf{x}\|_2\,\|\mathbf{y}\|_2 + \|\mathbf{y}\|_2^2 \\
&\sum_{i=1}^{N} x_i\,y_i \leq \|\mathbf{x}\|_2\,\|\mathbf{y}\|_2
\end{aligned}
$$

If $x_i\,y_i < 0$, then replace $x_i$ by $-x_i$ and call the new vector $\tilde{\mathbf{x}}$. Thus $\sum_{i=1}^{N} |x_i\,y_i| \leq \|\tilde{\mathbf{x}}\|_2\,\|\mathbf{y}\|_2 = \|\mathbf{x}\|_2\,\|\mathbf{y}\|_2$. Hence Equation (11.3.4) is established.  ∎

$$
\begin{aligned}
\|\mathbf{x} + \mathbf{y}\|_2^2 &= \sum_{i=1}^{N} (x_i + y_i)^2 = \sum_{i=1}^{N} (x_i^2 + 2x_i\,y_i + y_i^2) \\
&= (\|\mathbf{x}\|_2 + \|\mathbf{y}\|_2)^2
\end{aligned}
$$

Hence

$$
\|\mathbf{x} + \mathbf{y}\|_2 \leq \|\mathbf{x}\|_2 + \|\mathbf{y}\|_2
$$

It can be verified that Equation (11.3.3c) defines a vector norm (Exercise 4).

For any norm, the set $\{\mathbf{x}\mid \|\mathbf{x}\| = 1\}$ is called the **unit sphere**, although this set resembles a sphere only in the Euclidean case. To visualize these norms, consider $R^2$. For $l_1$ norm, $\|\mathbf{x}\|_1 = |x_1| + |x_2| = 1$ describes a square whose diagonals coincide with the coordinate axes. For example, when $x_1$ and $x_2$ are positive, $x_1 + x_2 = 1$ describes the line joining the points $(1, 0)$ and $(0, 1)$. For the $l_2$ norm, $\|\mathbf{x}\|_2 = \sqrt{x_1^2 + x_2^2} = 1$ describes a unit circle with center at the origin. For the $l_\infty$ norm, $\|\mathbf{x}\|_\infty = \max\{|x_1|, |x_2|\} = 1$ describes a square with sides parallel to the coordinate axes. For example, $|x_1| = 1$ describes the left and right sides of the square. In $R^2$, $l_1$, $l_2$, and $l_\infty$ are sketched in Figure 11.3.1.

As seen in Figure 11.3.1, the unit sphere in the $p$-norm deforms from the square inscribed in the circle to the square circumscribing the circle as $p$ increases.

Two vector norms are said to be **equivalent** if the ratio of a vector's length in one norm to a vector's length in another norm is bounded from above and below by

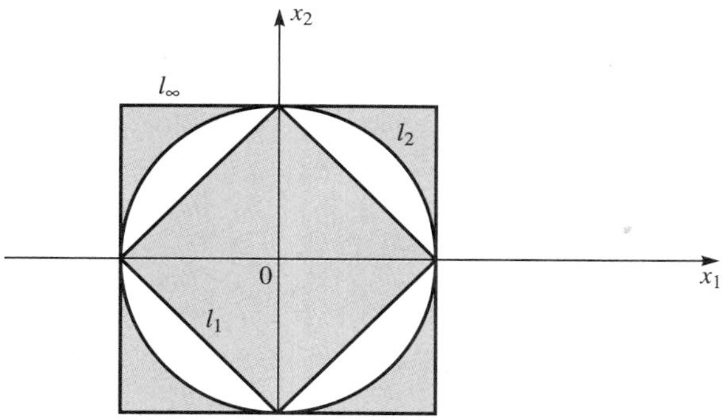

**Fig. 11.3.1**

constants that are independent of the vector. For example, since (Exercise 8)

$$\|\mathbf{x}\|_{\infty} \le \|\mathbf{x}\|_1 \le N \|\mathbf{x}\|_{\infty} \tag{11.3.5}$$

$1 \le \|\mathbf{x}\|_1/\|\mathbf{x}\|_{\infty} \le N$. Thus $l_1$ and $l_{\infty}$ are equivalent.

In general,

$$\|\mathbf{x}\|_r \le \|\mathbf{x}\|_q \le N^{(r-q)/(rq)} \|\mathbf{x}\|_r \tag{11.3.6}$$

if $r \ge q$. Since $1 \le \|\mathbf{x}\|_q/\|\mathbf{x}\|_r \le N^{(r-q)/(rq)}$, the $r$-norm and $q$-norm are equivalent.

In $R^2$, the sequence of vectors $\begin{bmatrix} 1/2 \\ 1.1 \end{bmatrix}, \begin{bmatrix} 1/2^2 \\ 1.01 \end{bmatrix}, \begin{bmatrix} 1/2^3 \\ 1.001 \end{bmatrix}, \dots, \begin{bmatrix} 1/2^k \\ 1 + 10^{-k} \end{bmatrix}, \dots$

converges to $\begin{bmatrix} 0 \\ 1 \end{bmatrix}$ since $\lim_{k \to \infty} 1/2^k = 0$ and $\lim_{k \to \infty} (1 + 10^{-k}) = 1$.

In $R^N$, we say that $\lim_{k \to \infty} \mathbf{x}^{(k)} = \mathbf{x}$ if and only if $\lim_{k \to \infty} x_i^{(k)} = x_i$ for each $i = 1, 2, \dots, N$.

Since $\lim_{k \to \infty} x_i^{(k)} = x_i$, $\lim_{k \to \infty} |x_i^{(k)} - x_i| = 0$ for $i = 1, 2, \dots, N$. Hence $\lim_{k \to \infty} \max_i \{|x_i^{(k)} - x_i|\} = 0$ and therefore $\lim_{k \to \infty} \|\mathbf{x}^{(k)} - \mathbf{x}\|_{\infty} = 0$. We can reverse all these steps and show that if $\lim_{k \to \infty} \|\mathbf{x}^{(k)} - \mathbf{x}\|_{\infty} = 0$, then $\lim_{k \to \infty} \mathbf{x}^{(k)} = \mathbf{x}$. This leads to the next theorem.

**Theorem 11.3.2**    $\lim_{k \to \infty} \mathbf{x}^{(k)} = \mathbf{x}$ if and only if $\lim_{k \to \infty} \|\mathbf{x}^{(k)} - \mathbf{x}\|_{\infty} = 0$.    ∎

Since

$$\|\mathbf{x}^{(k)} - \mathbf{x}\|_{\infty} \le \|\mathbf{x}^{(k)} - \mathbf{x}\|_1 \le N \|\mathbf{x}^{(k)} - \mathbf{x}\|_{\infty}$$

from Equation (11.3.5), if $\lim_{k \to \infty} \|\mathbf{x}^{(k)} - \mathbf{x}\|_{\infty} = 0$, then also $\lim_{k \to \infty} \|\mathbf{x}^{(k)} - \mathbf{x}\|_1 = 0$ (why?).

Similarly using Equation (11.3.6), we can show that if $\lim_{k \to \infty} \|\mathbf{x}^{(k)} - \mathbf{x}\|_{\infty} = 0$, then $\lim_{k \to \infty} \|\mathbf{x}^{(k)} - \mathbf{x}\|_2 = 0$.

Thus in Theorem 11.3.2, we can replace the $l_{\infty}$ norm by the $l_1$ norm or the $l_2$ norm. Can we replace the $l_{\infty}$ norm by any norm? In order to answer this question, let us look at the norm function.

In $R^1$, $f(x) = \|x\| = |x|$ is continuous at every $x \in R$. In $R^N$, $f(\mathbf{x}) = \|\mathbf{x}\|$ is continuous at every $\mathbf{x} \in R^N$ (Exercise 7).

Let $S \subset R^N$ defined by $S = \{\mathbf{x}| \|\mathbf{x}\|_2 = 1, \mathbf{x} \in R^N\}$. Since $f(\mathbf{x}) = \|\mathbf{x}\|$ is continuous on $S$ which is closed and bounded, by the Theorem of Weierstrass (a generalization of the Extreme Value Theorem) $\|\mathbf{x}\|$ attains its minimum and maximum at some points of $S$. That is, for some $\mathbf{x}_1$, $\mathbf{x}_2$ in $S$, $\|\mathbf{x}_1\| = \min_{\mathbf{x} \in S} \|\mathbf{x}\| = m$ and $\|\mathbf{x}_2\| = \max_{\mathbf{x} \in S} \|\mathbf{x}\| = M$. Thus we have

$$m \leq \|\mathbf{x}\| \leq M \quad \text{for all } \mathbf{x} \in S \tag{11.3.7}$$

Let $\mathbf{y} \in R^N$ and $\mathbf{y} \neq \mathbf{0}$. Then $\mathbf{x} = \mathbf{y}/\|\mathbf{y}\|_2 \in S$ and Equation (11.3.7) becomes

$$m \|\mathbf{y}\|_2 \leq \|\mathbf{y}\| \leq M \|\mathbf{y}\|_2$$

Thus in general,

$$m \|\mathbf{x}\|_2 \leq \|\mathbf{x}\| \leq M \|\mathbf{x}\|_2 \tag{11.3.8}$$

We have not specified any particular norm. Therefore, for another norm $\| . \|'$,

$$m' \|\mathbf{x}\|_2 \leq \|\mathbf{x}\|' \leq M' \|\mathbf{x}\|_2 \tag{11.3.9}$$

From Equations (11.3.8) and (11.3.9), the following theorem is established.

**Theorem 11.3.3**     **(Norm Equivalence Theorem)**   Let $\| . \|$ and $\| . \|'$ be any two norms on $R^N$ (or $C^N$). Then there exist constants $c_1 > 0$ and $c_2 > 0$ such that for all $\mathbf{x} \in R^N$,

$$c_1 \|\mathbf{x}\|' \leq \|\mathbf{x}\| \leq c_2 \|\mathbf{x}\|' \tag{11.3.10}$$

∎

Note that this theorem applies only to finite dimensional spaces.

Because of the equivalence of the norms, convergence in some norm implies convergence in any other norm for a finite dimensional space. Hence Theorem 11.3.2 is generalized as follows.

**Theorem 11.3.4**     $\lim_{k \to \infty} \mathbf{x}^{(k)} = \mathbf{x}$ if and only if $\lim_{k \to \infty} \|\mathbf{x}^{(k)} - \mathbf{x}\| = 0$ where $\| \|$ represents any norm on $V$.

∎

**EXERCISES**

1.  Find $\|\mathbf{x}\|_1$, $\|\mathbf{x}\|_2$, and $\|\mathbf{x}\|_\infty$ for the following vectors:
    (a)  $\mathbf{x} = [1, 2, 3, 4]^t$
    (b)  $\mathbf{x} = [\cos i, \sin i, 10^{-i}]^t$, where $i$ is an integer
    (c)  $\mathbf{x} = [-1, -2, -\frac{1}{2}, -\frac{1}{4}]^t$

**2.** Let

$$A = \begin{bmatrix} 1 & 2 & 3 & 0 \\ 2 & 3 & -1 & 2 \\ -3 & 1 & 0 & 0 \\ 1 & 0 & 0 & 2 \end{bmatrix} \quad \text{and} \quad \mathbf{x} = [1, -1, 1, 1]'.$$

Find $\|A\mathbf{x}\|_1$, $\|A\mathbf{x}\|_2$, and $\|A\mathbf{x}\|_\infty$.

**3.** Verify that $\|\mathbf{x}\|_\infty \le \|\mathbf{x}\|_p \le N^{1/p} \|\mathbf{x}\|_\infty$ for $p \ge 1$. Then show that $\lim_{p \to \infty} \|\mathbf{x}\|_p = \|\mathbf{x}\|_\infty = \max_i |x_i|$.

**4.** Verify that Equation (11.3.3c) is a norm on $R^N$.

**5.** Check whether the following sequences are convergent and if they are convergent, find their limits:
 (a) $\mathbf{x}^{(k)} = [1/k, \sin(1/k), e^{-k}]'$
 (b) $\mathbf{x}^{(k)} = [(\cos k)/k, e^{1/k}, 1]'$
 (c) $\mathbf{x}^{(k)} = [k, 0, (k+1)/(k-1), \sin k]'$

**6.** Let $\|\cdot\|$ be any norm on $R^N$. Show that for all $\mathbf{x}, \mathbf{y} \in R^N$, $|\, \|\mathbf{x}\| - \|\mathbf{y}\| \,| \le \|\mathbf{x} - \mathbf{y}\|$.

**7.** Using Exercise 6, prove that $f(x) = \|\mathbf{x}\|$ is continuous for all $\mathbf{x} \in R^N$.

**8.** Prove Equation (11.3.5).

**9.** Derive Equation (11.3.10) from Equations (11.3.8) and (11.3.9).

**10.** Prove that $\|\mathbf{x}\|_2 \le \|\mathbf{x}\|_1 \le \sqrt{N} \|\mathbf{x}\|_2$.

## 11.4 MATRIX NORMS

Let $M_N$ be the set of all $N \times N$ matrices. Then $M_N$ is a vector space of dimension $N^2$ and so we can measure the "size" of a matrix by using any vector norm of $R^{N^2}$. However, $M_N$ is not just a high-dimensional vector space, it has a multiplication operation. It is useful in making estimates to relate the "size" of $AB$ to the "sizes" of $A$ and $B$. Thus we impose an additional condition on a matrix norm

$$\|AB\| \le \|A\| \|B\| \quad \text{for all } A, B \in M_N$$

With this additional requirement, the vector norms of Section 11.3 do not all become matrix norms. Thus a **matrix norm** on $M_N$ is a real valued function $\|\ \|$ from $M_N$ to $R$ with the following properties:

 **(i)** $\|A\| \ge 0$ for all $A \in M_N$ and $\|A\| = 0$ if and only if $A = 0$
 **(ii)** $\|\alpha A\| = |\alpha| \|A\|$ for all $\alpha \in R$ and $A \in M_N$
 **(iii)** $\|A + B\| \le \|A\| + \|B\|$ for all $A, B \in M_N$
 **(iv)** $\|AB\| \le \|A\| \|B\|$ for all $A, B \in M_N$. **(11.4.1)**

There are many ways to construct matrix norms that satisfy the preceeding conditions. However, since matrices and vectors appear jointly in our case, we define a matrix norm $\|A\|$ in such a way that it is compatible with a vector norm by requiring

$$\|A\mathbf{x}\| \le \|A\| \|\mathbf{x}\|$$

for all $A \in M_N$ and for all $\mathbf{x} \in R^N$ (note that $\|A\mathbf{x}\|$ and $\|\mathbf{x}\|$ are vector norms). Let us derive a matrix norm from a given vector norm. The matrix $A$ transforms a vector $\mathbf{x}$

into another vector $A\mathbf{x}$. If the elements of $A$ are large, then for some choices of $\mathbf{x}$, the components of $A\mathbf{x}$ are large. To measure the amplification of $A\mathbf{x}$ to $\mathbf{x}$, we form the ratio of $\|A\mathbf{x}\|/\|\mathbf{x}\|$. The maximum amplification over all possible $\mathbf{x}$ can be taken as the definition of the norm $A$ and therefore for each vector norm, we define a corresponding matrix norm

$$\|A\| = \max_{\|\mathbf{x}\| \neq 0} \frac{\|A\mathbf{x}\|}{\|\mathbf{x}\|} \tag{11.4.2}$$

This matrix norm is often called a **compatible matrix norm** or a **natural norm**. Since we are using the same notation $\|\ \|$ to denote the matrix norm and vector norm, the distinction will be clear from the usage. For example, $A\mathbf{x}$ is a vector so $\|A\mathbf{x}\|$ denotes a vector norm; $A$ is a matrix, therefore, $\|A\|$ denotes a matrix norm.

Let $\mathbf{z} = \mathbf{x}/\|\mathbf{x}\|$ for any $\mathbf{x} \neq \mathbf{0}$. Then $\|\mathbf{z}\| = 1$, $\|A\mathbf{z}\| = \|A\mathbf{x}\|/\|\mathbf{x}\|$ and so the definition of Equation (11.4.2) is equivalent to

$$\|A\| = \max_{\|\mathbf{z}\|=1} \|A\mathbf{z}\| \tag{11.4.3}$$

We want to show that Equation (11.4.3) indeed defines a matrix norm by verifying the properties given by Equation (11.4.1) by noting that $\|A\mathbf{z}\|$ is a vector norm.

**(i)**   Since $\|\mathbf{z}\| = 1$, $\mathbf{z} \neq \mathbf{0}$ and $A \neq 0$, then $\|A\| = \max_{\|\mathbf{z}\|=1} \|A\mathbf{z}\| \geq 0$. If $A = 0$, then $A\mathbf{z} = \mathbf{0}$ and $\|A\mathbf{z}\| = 0$ for all $\mathbf{z}$ such that $\|\mathbf{z}\| = 1$. Thus if $A = 0$, then $\|A\| = \max_{\|\mathbf{z}\|=1} \|A\mathbf{z}\| = 0$. If $\mathbf{z} \neq \mathbf{0}$ and $\|A\| = 0$, then $\max_{\|\mathbf{z}\|=1} \|A\mathbf{z}\| = 0$. Since $\max_{\|\mathbf{z}\|=1} \|A\mathbf{z}\| = 0$, then $A\mathbf{z} = \mathbf{0}$ and hence $A = 0$.

**(ii)**   $\|\alpha A\| = \max_{\|\mathbf{z}\|=1} \|\alpha A\mathbf{z}\| = \max_{\|\mathbf{z}\|=1} |\alpha| \|A\mathbf{z}\|$
$= |\alpha| \max_{\|\mathbf{z}\|=1} \|A\mathbf{z}\| = |\alpha| \|A\|$

**(iii)**   $\|A + B\| = \max_{\|\mathbf{z}\|=1} \|(A + B)\mathbf{z}\| = \max_{\|\mathbf{z}\|=1} \|A\mathbf{z} + B\mathbf{z}\|$
$\leq \max_{\|\mathbf{z}\|=1} (\|A\mathbf{z}\| + \|B\mathbf{z}\|) \leq \max_{\|\mathbf{z}\|=1} \|A\mathbf{z}\| + \max_{\|\mathbf{z}\|=1} \|B\mathbf{z}\|$
$= \|A\| + \|B\|$
This implies $\|A + B\| \leq \|A\| + \|B\|$.

**(iv)**   $\|AB\| = \max_{\|\mathbf{z}\|=1} \|(AB)\mathbf{z}\| = \max_{\|\mathbf{z}\|=1} \|A(B\mathbf{z})\| = \max_{\|\mathbf{z}\|=1} \frac{\|A(B\mathbf{z})\|}{\|B\mathbf{z}\|} \|B\mathbf{z}\|$
$\leq \max_{\|B\mathbf{z}\| \neq 0} \frac{\|(AB)\mathbf{z}\|}{\|B\mathbf{z}\|} \max_{\|\mathbf{z}\|=1} \|B\mathbf{z}\| = \|A\| \|B\|$

Hence $\|AB\| \leq \|A\| \|B\|$. Thus Equation (11.4.3) defines a matrix norm.

Let us determine the matrix norms that are induced by the $l_1$, $l_2$, and $l_\infty$ norms defined by Equation (11.3.3). Since $\|\mathbf{z}\|_1 = 1$, $\sum_{j=1}^{N} |z_j| = 1$. Then

$$\|A\mathbf{z}\|_1 = \sum_{i=1}^{N} \left| \sum_{j=1}^{N} a_{ij} z_j \right| \leq \sum_{i=1}^{N} \sum_{j=1}^{N} |a_{ij}| |z_j|$$

$$= \sum_{j=1}^{N} |z_j| \left( \sum_{i=1}^{N} |a_{ij}| \right) \leq \left( \max_{j} \sum_{i=1}^{N} |a_{ij}| \right) \left( \sum_{j=1}^{N} |z_j| \right)$$

$$= \max_{j} \sum_{i=1}^{N} |a_{ij}|$$

Hence

$$\|A\|_1 = \max_{\|\mathbf{z}\|_1 = 1} \|A\mathbf{z}\|_1 \le \max_j \sum_{i=1}^N |a_{ij}| \qquad (11.4.4)$$

If the maximum is attained for $j = k$, then select $\mathbf{z} = \mathbf{e}_k = [0, 0, \ldots, 1, 0, \ldots, 0]'$. Then

$$\|A\mathbf{z}\|_1 = \sum_{i=1}^N \left| \sum_{j=1}^N a_{ij} \mathbf{e}_k \right| = \sum_{i=1}^N |a_{ik}|$$

Hence, for $\mathbf{z} = \mathbf{e}_k$, equality is attained in Equation (11.4.4), and thus

$$\|A\|_1 = \max_j \sum_{i=1}^N |a_{ij}| \text{ (maximum absolute column sum)} \qquad (11.4.5)$$

Next, let us consider the $l_2$ norm. If $\|\mathbf{z}\|_2 = 1$, then $\|\mathbf{z}\|_2^2 = \mathbf{z}'\mathbf{z} = z_1^2 + z_2^2 + \cdots + z_N^2 = 1$. Also

$$\|A\mathbf{z}\|_2^2 = (A\mathbf{z})'(A\mathbf{z}) = \mathbf{z}'A'A\mathbf{z} \ge 0 \qquad (11.4.6)$$

Since $A'A$ is symmetric (why?), it has orthonormal eigenvectors (Theorem 13.1.6) $\mathbf{x}_1, \mathbf{x}_2, \ldots, \mathbf{x}_N$ such that

$$\mathbf{x}_i'\mathbf{x}_j = \delta_{ij} \quad \text{and} \quad A'A\mathbf{x}_k = \lambda_k \mathbf{x}_k \qquad (11.4.7)$$

where $\lambda_k$ is an eigenvalue of $A'A$ corresponding to an eigenvector $\mathbf{x}_k$. Multiplying both sides of $A'A\mathbf{x}_k = \lambda_k \mathbf{x}_k$ by $\mathbf{x}_k'$ on the left, we get

$$\mathbf{x}_k'A'A\mathbf{x}_k = \lambda_k \mathbf{x}_k'\mathbf{x}_k = \lambda_k$$

Since $\mathbf{x}_k'A'A\mathbf{x}_k = (A\mathbf{x}_k)'(A\mathbf{x}_k) \ge 0$, $\lambda_k \ge 0$. Let us express $\mathbf{z}$ in terms of the linearly independent eigenvectors $\mathbf{x}_k$; in other words,

$$\mathbf{z} = \sum_{k=1}^N \alpha_k \mathbf{x}_k \quad \text{where } \alpha_k \text{ are constants} \qquad (11.4.8)$$

Then substituting Equation (11.4.8) in Equation (11.4.6), we get

$$\begin{aligned}\|A\mathbf{z}\|_2^2 = \mathbf{z}'A'A\mathbf{z} &= \sum_{i=1}^N \alpha_i \mathbf{x}_i' A'A \sum_{k=1}^N \alpha_k \mathbf{x}_k \\ &= \sum_{i=1}^N \alpha_i \mathbf{x}_i' \sum_{k=1}^N \alpha_k A'A\mathbf{x}_k \\ &= \sum_{i=1}^N \alpha_i \mathbf{x}_i' \sum_{k=1}^N \alpha_k \lambda_k \mathbf{x}_k \quad \text{from Equation (11.4.7)} \\ &= \sum_{k=1}^N \lambda_k \alpha_k^2 \le \max_k |\lambda_k| \sum_{k=1}^N \alpha_k^2 = \max_k |\lambda_k|\end{aligned}$$

since $\|\mathbf{z}\|_2^2 = 1$.
    Thus

$$\|A\|_2 = \max_{\|\mathbf{z}\|=1} \|A\mathbf{z}\|_2 \le \sqrt{\max_k |\lambda_k|} \qquad (11.4.9)$$

If $\mathbf{z} = \mathbf{x}_k$ corresponds to $\max_k \lambda_k$, then

$$\|A\mathbf{z}\|_2^2 = \mathbf{x}_k'A'A\mathbf{x}_k = \mathbf{x}_k'\lambda_k \mathbf{x}_k = \max_k \lambda_k$$

Thus equality is attained in Equation (11.4.9) and we get

$$\|A\|_2 = \sqrt{\max_k \lambda_k} = \{\text{maximum eigenvalue of } A'A\}^{1/2}$$

Let $B$ be an $N \times N$ matrix with eigenvalues $\lambda_1, \lambda_2, \ldots, \lambda_N$. Then $\sigma(B) = \{\lambda_1, \lambda_2, \ldots, \lambda_N\}$ is called the **spectrum** of $B$ and $\rho(B) = \max \{|\lambda_1|, |\lambda_2|, \ldots, |\lambda_N|\}$ is called the **spectral radius** of $B$. Thus

$$\|A\|_2 = \sqrt{\rho(A'A)} \qquad \qquad \textbf{(11.4.10)}$$

Note that if $\lambda = \alpha + I\beta$ where $\alpha$ and $\beta$ are real numbers and $I = \sqrt{-1}$, then $|\lambda| = \sqrt{\alpha^2 + \beta^2}$.

It can be proven (Exercise 2) that the matrix norm induced by the $l_\infty$ norm is

$$\|A\|_\infty = \max_i \sum_{j=1}^{N} |a_{ij}| \text{ (maximum absolute row sum)} \qquad \qquad \textbf{(11.4.11)}$$

For convenience, the vector norms and associated matrix norms are given in Table 11.4.1.

**Table 11.4.1**

| Vector Norm | Associated Matrix Norm |
|---|---|
| $\|\mathbf{x}\|_1 = \sum_{i=1}^{N} \|x_i\|$ | $\|A\|_1 = \max_j \sum_{i=1}^{N} \|a_{ij}\|$ (maximum absolute column sum) |
| $\|\mathbf{x}\|_2 = \left( \sum_{i=1}^{N} x_i^2 \right)^{1/2}$ | $\|A\|_2 = \sqrt{\rho(A'A)}$ (maximum absolute eigenvalue of $A'A)^{1/2}$ |
| $\|\mathbf{x}\|_\infty = \max_i \|x_i\|$ | $\|A\|_\infty = \max_i \sum_{j=1}^{N} \|a_{ij}\|$ (maximum absolute row sum) |

EXAMPLE 11.4.1    Find $\|A\|_1$, $\|A\|_2$, and $\|A\|_\infty$ if

$$A = \begin{bmatrix} 1 & 2 & 3 \\ 3 & 4 & 5 \\ 5 & 6 & 7 \end{bmatrix}$$

$\|A\|_1 = \max \{|1| + |3| + |5|, |2| + |4| + |6|, |3| + |5| + |7|\} = \max \{9, 12, 15\} = 15$.

$$A'A = \begin{bmatrix} 1 & 3 & 5 \\ 2 & 4 & 6 \\ 3 & 5 & 7 \end{bmatrix} \begin{bmatrix} 1 & 2 & 3 \\ 3 & 4 & 5 \\ 5 & 6 & 7 \end{bmatrix} = \begin{bmatrix} 35 & 44 & 53 \\ 44 & 56 & 68 \\ 53 & 68 & 83 \end{bmatrix}$$

We can verify that the eigenvalues of $A'A$ are 0.0, 0.083, and 173.17 and therefore

$$\sigma(A'A) = \{0, 0.083, 173.17\}$$

The spectral radius of $A' A$

$$\rho (A' A) = \max \{0, 0.083, 173.17\} = 173.17$$

Hence

$$\|A\|_2 = \sqrt{\rho (A' A)} = \sqrt{173.17} = 13.16$$

$\|A\|_\infty = \max \{|1| + |2| + |3|, |3| + |4| + |5|, |5| + |6| + |7|\} = \max \{6, 12, 18\} = 18.$

∎∎∎

Since every matrix norm satisfies the conditions (11.3.1), the function $f(A) = \|A\|$ is continuous and Theorem 11.3.3 applies.

**Theorem 11.4.1**    Let $\| . \|$ and $\| . \|'$ be any two matrix norms on $M_N$. Then there exist constants $c_1 > 0$ and $c_2 > 0$ such that for all $A \in M_N$,

$$c_1 \|A\|' \le \|A\| \le c_2 \|A\|' \qquad \qquad \textbf{(11.4.12)}$$

∎

For convenience, we omit the subscripts for the norms whenever we use any compatible matrix and vector norms. Let $\lambda$ be an eigenvalue of $A$ such that $\rho (A) = |\lambda|$ and $\mathbf{x}$ be an eigenvector of $A$ corresponding to $\lambda$ such that $\|\mathbf{x}\| = 1$. Then $\rho (A) = |\lambda| = |\lambda| \|\mathbf{x}\| = \|\lambda\mathbf{x}\| = \|A\mathbf{x}\| \le \|A\| \|\mathbf{x}\| = \|A\|$ and therefore $\rho (A) \le \|A\|$.

On the other hand, for each $N \times N$ matrix and a given $\epsilon > 0$, there is a compatible matrix norm such that

$$\|A\| \le \rho (A) + \epsilon$$

The proof can be found in Isaacson and Keller (1966). The results are combined in the following theorem.

**Theorem 11.4.2**    Let $A$ be any $N \times N$ matrix and $\epsilon > 0$ be given. Then there exists a compatible norm $\|A\|$ such that

$$\rho (A) \le \|A\| \le \rho (A) + \epsilon \qquad \qquad ∎$$

This is a very helpful theorem to find bounds for the spectral radius since it holds for any norm and $\|A\|_1$ and $\|A\|_\infty$ norms can be easily calculated.

**EXAMPLE 11.4.2**    Find a bound on $\rho (A)$ if

$$A = \begin{bmatrix} 1 & 2 & 3 \\ 3 & 4 & 5 \\ 5 & 6 & 7 \end{bmatrix}$$

Since $\|A\|_1 = 15$ and $\|A\|_\infty = 18$, then $\rho (A) \le 15$. In fact $\lambda_1 = 0$, $\lambda_2 = -0.9282$ and $\lambda_3 = 12.9282$. Thus $\rho (A) = 12.9282 < 15$.

∎∎∎

**EXAMPLE 11.4.3**     Let $A = \begin{bmatrix} 1/2 & 0 \\ 0 & 1/4 \end{bmatrix}$. Then $A^2 = \begin{bmatrix} 1/2^2 & 0 \\ 0 & 1/4^2 \end{bmatrix}$, $A^3 = \begin{bmatrix} 1/2^3 & 0 \\ 0 & 1/4^3 \end{bmatrix}$, ..., $A^m = \begin{bmatrix} 1/2^m & 0 \\ 0 & 1/4^m \end{bmatrix}$. Then $\lim_{m \to \infty} A^m = \begin{bmatrix} \lim_{m \to \infty} 1/2^m & \lim_{m \to \infty} 0 \\ \lim_{m \to \infty} 0 & \lim_{m \to \infty} 1/4^m \end{bmatrix} = \begin{bmatrix} 0 & 0 \\ 0 & 0 \end{bmatrix}$. ∎∎∎

A square matrix $A$ is said to be **convergent** if $\lim_{m \to \infty} A^m = 0$.

In Example 11.4.3, $A$ is a convergent matrix.

We need to investigate the conditions under which a given matrix $A$ is convergent for various iterative procedures. We can verify (Exercise 8) the following theorem.

**Theorem 11.4.3**     $\lim_{m \to \infty} A^m = 0$ if and only if $\lim_{m \to \infty} \|A^m\| = 0$.     ∎

Now $\|A^m\| = \|AA^{m-1}\| \le \|A\| \|A^{m-1}\| = \|A\| \|AA^{m-2}\| \le \|A\| \|A\| \|AA^{m-2}\| \le \cdots \le \|A\|^m$. If $\|A\| < 1$, then $\lim_{m \to \infty} \|A^m\| = 0$. Therefore $A$ is convergent. This proves the next theorem.

**Theorem 11.4.4**     If $\|A\| < 1$, then $A$ is convergent.     ∎

Let $\rho (A) < 1$. Then from Theorem 11.4.2 we find an $\epsilon > 0$ and a compatible matrix norm such that

$$\|A\| \le \rho (A) + \epsilon < 1$$

Suppose that $\rho (A) \ge 1$ and let $\lambda$ be an eigenvalue of $A$ such that $|\lambda| \ge 1$. If $\mathbf{x}$ is its corresponding eigenvector, then

$$\|A^m \mathbf{x}\| = \|A^{m-1} A\mathbf{x}\| = \|A^{m-1} \lambda\mathbf{x}\| = \|\lambda (A^{m-2} A\mathbf{x})\|$$
$$= \cdots = \|\lambda^k \mathbf{x}\| = |\lambda^k| \|\mathbf{x}\| \ge \|\mathbf{x}\|$$

Thus $\|\mathbf{x}\| \le \|A^m \mathbf{x}\| \le \|A^m\| \|\mathbf{x}\|$. Since $\|A^m\| \|\mathbf{x}\| \ge \|\mathbf{x}\|$, $\|A^m\| \ge 1$ for all $m$. This leads to the next theorem.

**Theorem 11.4.5**     Let $A$ be an $N \times N$ matrix. Then $A$ is convergent if and only if $\rho (A) < 1$.     ∎

Just as $1/(1 - x) = 1 + x + x^2 + \cdots$ if $|x| < 1$ is a very useful geometric series, its matrix analogue given by the following theorem is very useful.

**Theorem 11.4.6**     Let $A$ be an $N \times N$ matrix such that $\rho (A) < 1$. Then $(I - A)^{-1}$ exists, and $(I - A)^{-1} = I + A + A^2 + A^3 + \cdots$.

*Proof*    Let $\lambda$ be an eigenvalue of $A$ such that $\rho (A) = |\lambda|$ and $\mathbf{x}$ be an eigenvector of $A$ corresponding to $\lambda$. Since $A\mathbf{x} = \lambda\mathbf{x}$, then $(I - A) \mathbf{x} = (I - \lambda I) \mathbf{x} = (1 - \lambda) \mathbf{x}$. Hence $(1 - \lambda)$ is an eigenvalue of $(I - A)$. We know that $\rho (A) < 1$, therefore $|\lambda| < 1$ and so $1 - \lambda \ne 0$. Hence $I - A$ is invertible. Let

$$S_m = I + A + A^2 + \cdots + A^m$$

Then

$$(I - A) S_m = (I - A)(I + A + A^2 + \cdots + A^m)$$
$$= I + A + A^2 + \cdots + A^m - A - A^2 - \cdots - A^{m+1}$$
$$= I - A^{m+1}$$

Since $\rho(A) < 1$, $\lim_{m \to \infty} A^m = 0$. Hence

$$\lim_{m \to \infty} (I - A) S_m = \lim_{m \to \infty} (I - A^{m+1}) = I$$

Thus

$$\lim_{m \to \infty} (I - A) S_m = (I - A) \lim_{m \to \infty} S_m = I$$

and hence

$$\lim_{m \to \infty} S_m = (I - A)^{-1} = I + A + A^2 + \cdots$$

∎

### EXERCISES

1. Find $\|A\|_1$, $\|A\|_2$, and $\|A\|_\infty$ for the following matrices:

(a) $A = \begin{bmatrix} 1 & 0 \\ 0 & 1 \end{bmatrix}$ (b) $A = \begin{bmatrix} 5 & 3 \\ 0 & -2 \end{bmatrix}$ (c) $A = \begin{bmatrix} -1 & 0 & 0 \\ 0 & 3 & 1 \\ 0 & 1 & -2 \end{bmatrix}$

2. Verify that the matrix norm induced by the $l_\infty$ norm is $\|A\|_\infty = \max_i \sum_{j=1}^{N} |a_{ij}|$ (maximum absolute row sum).

3. Determine whether $\|A\|_@$ defined by
   (a) $\|A\|_@ = \sqrt{\sum_{j=1}^{n} \sum_{i=1}^{n} a_{ij}^2}$ (b) $\|A\|_@ = \sum_{j=1}^{n} \sum_{i=1}^{n} |a_{ij}|$
   (c) $\|A\|_@ = \max_{1 \le i, j \le n} |a_{ij}|$ define matrix norms.

4. Let $A$ be an $N \times N$ matrix. Then prove that $\rho(A) < 1$ if and only if $\|A\| < 1$ for some compatible norm.

5. Show that $\|A\|_2 = \rho(A)$ if $A$ is symmetric.

6. Show that $\|A\|_2^2 \le \|A\|_1 \|A\|_\infty$ (*Hint*: $\rho(A^t A) \le \|A^t A\|_1$ and $\|A^t\|_1 = \|A\|_\infty$).

7. Show that $1/\|A^{-1}\| \le |\lambda| \le \|A\|$ where $\lambda$ is any eigenvalue of a nonsingular matrix $A$.

8. Prove Theorem 11.4.3.

9. Determine whether the following matrices are convergent:

(a) $\begin{bmatrix} 1 & \frac{1}{4} \\ 0 & \frac{1}{2} \end{bmatrix}$

(b) $\begin{bmatrix} \frac{1}{2} & 0 \\ 8 & \frac{1}{3} \end{bmatrix}$

10. Prove: $\lim_{m \to \infty} A^m \mathbf{x} = \mathbf{0}$ for all $\mathbf{x} \neq \mathbf{0}$ if and only if $\lim_{m \to \infty} A^m = 0$.

11. Use Theorem 11.4.6 to compute the inverse of

$$\begin{bmatrix} 3/4 & 0 & 0 \\ 0 & 3/4 & 0 \\ 0 & 0 & 3/4 \end{bmatrix}$$

## 11.5   CONVERGENCE OF THE JACOBI AND GAUSS–SEIDEL METHODS

The Jacobi and Gauss–Seidel iterates can be written in the form

$$\mathbf{x}^{(k+1)} = B\mathbf{x}^{(k)} + \mathbf{c} \quad \text{for } k = 0, 1, \ldots \tag{11.5.1}$$

where $B = -D^{-1}(L + U)$ and $\mathbf{c} = D^{-1}\mathbf{b}$ in the Jacobi method, and $B = -(D + L)^{-1}U$ and $\mathbf{c} = (D + L)^{-1}\mathbf{b}$ in the Gauss–Seidel method.

Equation (11.5.1) was derived by rearranging the original equation $A\mathbf{x} = \mathbf{b}$ in the form

$$\mathbf{x} = B\mathbf{x} + \mathbf{c} \tag{11.5.2}$$

The solution of Equation (11.5.2) is also the solution of $A\mathbf{x} = \mathbf{b}$.

The Jacobi and Gauss–Seidel methods are a form of the fixed-point iteration and the general form of both iterations is just an "infinite" Horner's rule.

We want to determine conditions under which the sequence of iterates generated by Equation (11.5.1) converges to the solution of Equation (11.5.2). The error in the $n$th approximation is given by $\mathbf{e}^{(n)} = \mathbf{x}^{(n)} - \mathbf{x}$. Subtracting Equation (11.5.2) from the $n$th approximation given by Equation (11.5.1), we get

$$\mathbf{e}^{(n)} = \mathbf{x}^{(n)} - \mathbf{x} = B\mathbf{x}^{(n-1)} + \mathbf{c} - (B\mathbf{x} + \mathbf{c}) = B(\mathbf{x}^{(n-1)} - \mathbf{x}) = B\mathbf{e}^{(n-1)}$$

Hence $\mathbf{e}^{(n)} = B\mathbf{e}^{(n-1)} = B(B\mathbf{e}^{(n-2)}) = \cdots = B^n\mathbf{e}^{(0)}$.

If the sequence $\mathbf{x}^{(1)}, \mathbf{x}^{(2)}, \ldots, \mathbf{x}^{(n)}, \ldots$ converges to the solution of Equation (11.5.2), then $\lim_{n \to \infty} \mathbf{e}^{(n)} = \mathbf{0}$. Since $\mathbf{x}^{(0)}$ is arbitrary, $\mathbf{e}^{(0)}$ is arbitrary. Hence if $\lim_{n \to \infty} \mathbf{e}^{(n)} = \mathbf{0}$, then $\lim_{n \to \infty} B^n = 0$ (Exercise 10, p. 383). Thus if (11.5.1) converges, then $B$ must be convergent. From Theorem 11.4.5, $B$ is convergent if and only if $\rho(B) < 1$. Hence if (11.5.1) converges, then $\rho(B) < 1$.

We will also prove that if $\rho(B) < 1$, then the iterates given by Equation (11.5.1) converge to the solution of Equation (11.5.2) and thus we will prove that $\rho(B) < 1$ is necessary and sufficient for the convergence of Equation (11.5.1) to Equation (11.5.2).

Let $\rho(B) < 1$. Then from Equation (11.5.1),

$$
\begin{aligned}
\mathbf{x}^{(k+1)} &= B\mathbf{x}^{(k)} + \mathbf{c} \\
&= B(B\mathbf{x}^{(k-1)} + \mathbf{c}) + \mathbf{c} \\
&= B^2\mathbf{x}^{(k-1)} + (B + I)\mathbf{c} \\
&= B^2(B\mathbf{x}^{(k-2)} + \mathbf{c}) + (B + I)\mathbf{c} \\
&= B^3\mathbf{x}^{(k-2)} + (B^2 + B + I)\mathbf{c} \\
&\ \ \vdots \\
&= B^{k+1}\mathbf{x}^{(0)} + (B^k + B^{k-1} + \cdots + B + I)\mathbf{c}
\end{aligned}
$$

Hence

$$
\begin{aligned}
\lim_{k \to \infty} \mathbf{x}^{(k+1)} &= \lim_{k \to \infty} B^{k+1}\mathbf{x}^{(0)} + \lim_{k \to \infty} \left( \sum_{i=0}^{k} B^i \right) \mathbf{c} \\
&= 0 + (I - B)^{-1}\mathbf{c} \quad \text{from Theorems 11.4.5 and 11.4.6} \\
&= (I - B)^{-1}\mathbf{c}
\end{aligned}
$$

Further let $\lim_{k \to \infty} \mathbf{x}^{(k+1)} = \mathbf{x}$. Then $\mathbf{x} = (I - B)^{-1}\mathbf{c}$ which is $\mathbf{x} = B\mathbf{x} + \mathbf{c}$. This proves the following important theorem.

**Theorem 11.5.1**

The sequence generated by

$$\mathbf{x}^{(k+1)} = B\mathbf{x}^{(k)} + \mathbf{c} \quad \text{for } k = 0, 1, \ldots$$

converges to the unique solution $\mathbf{x}$ which is given by

$$\mathbf{x} = B\mathbf{x} + \mathbf{c}$$

for any starting vector $\mathbf{x}^{(0)} \in R^N$ if and only if $\rho(B) < 1$.   ■

Since it is extremely difficult to find $\rho(B)$, we often accept the sufficient conditions under which the sequence of iterations defined by Equation (11.5.1) will converge. Such a sufficient condition follows from Theorem 11.4.2 which states that $\rho(B) \le \|B\|$. This leads to the following theorem.

**Theorem 11.5.2**

If $\|B\| < 1$, then the sequence defined by

$$\mathbf{x}^{(k+1)} = B\mathbf{x}^{(k)} + \mathbf{c} \quad \text{for } k = 0, 1, \ldots$$

converges to the unique solution $\mathbf{x}$ which is given by

$$\mathbf{x} = B\mathbf{x} + \mathbf{c}$$

for any $\mathbf{x}^{(0)} \in R^N$.   ■

We would like to give general conditions which would guarantee the convergence of the Jacobi and Gauss–Seidel methods. Theorems 11.5.1 and 11.5.2 apply to any iterative process of the form of Equation (11.5.1) and are the basic results for such iterative methods.

For the Jacobi method, $B = -D^{-1}(L + U)$. From Equation (11.2.5),

$$D^{-1}(L + U) = \begin{bmatrix} 1/a_{11} & 0 & 0 & \cdots & 0 \\ 0 & 1/a_{22} & 0 & \cdots & 0 \\ 0 & 0 & 1/a_{33} & \cdots & 0 \\ \vdots & \vdots & \vdots & \ddots & \vdots \\ 0 & 0 & 0 & \cdots & 1/a_{NN} \end{bmatrix} \begin{bmatrix} 0 & a_{12} & a_{13} & \cdots & a_{1N} \\ a_{21} & 0 & a_{23} & \cdots & a_{2N} \\ a_{31} & a_{32} & 0 & \cdots & a_{3N} \\ \vdots & \vdots & \vdots & \ddots & \vdots \\ a_{N1} & a_{N2} & a_{N3} & \cdots & 0 \end{bmatrix}$$

$$= \begin{bmatrix} 0 & a_{12}/a_{11} & a_{13}/a_{11} & \cdots & a_{1N}/a_{11} \\ a_{21}/a_{22} & 0 & a_{23}/a_{22} & \cdots & a_{2N}/a_{22} \\ a_{31}/a_{33} & a_{32}/a_{33} & 0 & \cdots & a_{3N}/a_{33} \\ \vdots & \vdots & \vdots & \ddots & \vdots \\ a_{N1}/a_{NN} & a_{N2}/a_{NN} & a_{N3}/a_{NN} & \cdots & 0 \end{bmatrix}$$

Therefore

$$\|B\|_\infty = \|-D^{-1}(L + U)\|_\infty = \max_i \left\{ \sum_{\substack{j=1 \\ j \ne i}}^{N} \left| \frac{a_{ij}}{a_{ii}} \right| \right\} < 1$$

Thus if $\sum_{j \ne i, j=1}^{N} |a_{ij}| < |a_{ii}|$ for $i = 1, 2, \ldots, N$, then the Jacobi iterates defined by Equation (11.2.4) converge for any initial guess $\mathbf{x}^{(0)}$. The matrices in which $|a_{ii}| > \sum_{j \ne i, j=1}^{N} |a_{ij}|$ for $i = 1, 2, \ldots, N$ are important and are called **strictly diagonally dominant** since each main diagonal entry is larger in absolute value than the sum of the absolute values of all the off-diagonal entries in that row.

**EXAMPLE 11.5.1**    Consider the matrix

$$P = \begin{bmatrix} 3 & 1 & 1 \\ -1 & 5 & 2 \\ 0 & -2 & 3 \end{bmatrix}$$

Since $|3| > |1| + |1|$, $|5| > |-1| + |2|$, and $|3| > |0| + |-2|$, $P$ is a strictly diagonally dominant matrix. ■ ■ ■

Let $A$ be a strictly diagonally dominant $N \times N$ matrix. We would also like to show that the sequence generated by Equation (11.5.1) converges for the Gauss–Seidel method. In this case, we have $B = -(D + L)^{-1} U$.

Let $B\mathbf{x} = \lambda \mathbf{x}$. Then $-(D + L)^{-1} U\mathbf{x} = \lambda \mathbf{x}$. Multiplying both sides by $-(D + L)$, we get

$$U\mathbf{x} = -\lambda (D + L) \mathbf{x}$$

This can be rewritten as

$$\lambda D\mathbf{x} = -\lambda L\mathbf{x} - U\mathbf{x} \tag{11.5.3}$$

Let $x_i^{(k+1)}$ be maximal; that is, $|x_i^{(k+1)}| \geq |x_j^{(k)}|$ or $|x_j^{(k+1)}|$ for all $j$ for any $k$. Then Equation (11.5.3) can be written as

$$\lambda \, a_{ii} x_i^{(k+1)} = -\lambda \sum_{j=1}^{i-1} a_{ij} x_j^{(k+1)} - \sum_{j=i+1}^{N} a_{ij} x_j^{(k)} \tag{11.5.4}$$

Dividing Equation (11.5.4) by $x_i^{(k+1)}$, we get

$$\lambda \, a_{ii} = -\lambda \sum_{j=1}^{i-1} a_{ij} \frac{x_j^{(k+1)}}{x_i^{(k+1)}} - \sum_{j=i+1}^{N} a_{ij} \frac{x_j^{(k)}}{x_i^{(k+1)}}$$

Therefore

$$|\lambda \, a_{ii}| \leq |\lambda| \sum_{j=1}^{i-1} |a_{ij}| + \sum_{j=i+1}^{N} |a_{ij}| \tag{11.5.5}$$

Let $|\lambda| \geq 1$. Then Equation (11.5.5) can be written as

$$|\lambda| \, |a_{ii}| \leq |\lambda| \sum_{j=1}^{i-1} |a_{ij}| + |\lambda| \sum_{j=i+1}^{N} |a_{ij}| \tag{11.5.6}$$

Dividing Equation (11.5.6) by $|\lambda|$, we get

$$|a_{ii}| \leq \sum_{j=1}^{i-1} |a_{ij}| + \sum_{j=i+1}^{N} |a_{ij}| \tag{11.5.7}$$

This contradicts the fact that $A$ is strictly diagonally dominant and therefore $|\lambda| < 1$. Hence $\rho(B) < 1$; therefore Equation (11.2.10) converges whenever $A$ is a strictly diagonally dominant matrix. This proves the following theorem.

**Theorem 11.5.3**    Let $A$ be a strictly diagonally dominant matrix. Then the Jacobi and Gauss–Seidel iterations converge to the unique solution of $A\mathbf{x} = \mathbf{b}$ for any $\mathbf{x}^{(0)}$. ■

Strict diagonal dominance of the coefficient matrix of a system is sufficient to guarantee convergence of the Jacobi and Gauss–Seidel methods, but it is not necessary as can be seen in the following example.

**EXAMPLE 11.5.2**   Consider the linear system

$$
\begin{array}{rrrcr}
-2x_1 & + & x_2 & & = -2 \\
x_1 & - & 2x_2 & + & x_3 = 0 \\
& & x_2 & - & 2x_3 = -3
\end{array}
$$

In this case

$$
A = \begin{bmatrix} -2 & 1 & 0 \\ 1 & -2 & 1 \\ 0 & 1 & -2 \end{bmatrix} = \begin{bmatrix} 0 & 0 & 0 \\ 1 & 0 & 0 \\ 0 & 1 & 0 \end{bmatrix} + \begin{bmatrix} -2 & 0 & 0 \\ 0 & -2 & 0 \\ 0 & 0 & -2 \end{bmatrix} + \begin{bmatrix} 0 & 1 & 0 \\ 0 & 0 & 1 \\ 0 & 0 & 0 \end{bmatrix}
$$

The Jacobi matrix is

$$
B_J = -D^{-1}(L + U) = \begin{bmatrix} 0 & 1/2 & 0 \\ 1/2 & 0 & 1/2 \\ 0 & 1/2 & 0 \end{bmatrix}
$$

The eigenvalues are $\lambda_i = \cos(i\pi/4)$ for $i = 1, 2,$ and 3. Thus $1/\sqrt{2}$, 0, and $-1/\sqrt{2}$ are the eigenvalues. Therefore $\rho(B_J) = 1/\sqrt{2} = 0.70711$.
  The Gauss–Seidel matrix is

$$
B_G = -(D + L)^{-1}U = \begin{bmatrix} -1/2 & 0 & 0 \\ -1/4 & -1/2 & 0 \\ -1/8 & -1/4 & -1/2 \end{bmatrix} \begin{bmatrix} 0 & 1 & 0 \\ 0 & 0 & 1 \\ 0 & 0 & 0 \end{bmatrix} = \begin{bmatrix} 0 & 1/2 & 0 \\ 0 & 1/4 & 1/2 \\ 0 & 1/8 & 1/4 \end{bmatrix}
$$

The eigenvalues are 0, 0, and $\frac{1}{2}$. Therefore $\rho(B_G) = \frac{1}{2} = 0.5 < \rho(B_J)$.
  Although the matrix $A$ is not strictly diagonally dominant, the Jacobi and Gauss–Seidel methods converge. Starting with $\mathbf{x}^{(0)} = [0, 0, 0]'$, we obtain the values that are given in Table 11.5.1. In both cases iterations were terminated when $\|\mathbf{x}^{(k+1)} - \mathbf{x}^{(k)}\|_\infty \le 10^{-5} \|\mathbf{x}^{(k+1)}\|_\infty$.
  The exact solution is $x_1 = 9/4$, $x_2 = 5/2$, and $x_3 = 11/4$.

**Table 11.5.1**

| | Jacobi Method | | | Gauss–Seidel Method | | |
|---|---|---|---|---|---|---|
| $k$ | $x_1^{(k)}$ | $x_2^{(k)}$ | $x_3^{(k)}$ | $x_1^{(k)}$ | $x_2^{(k)}$ | $x_3^{(k)}$ |
| 0 | 0.00000 | 0.00000 | 0.00000 | 0.00000 | 0.00000 | 0.00000 |
| 1 | 1.00000 | 0.00000 | 1.50000 | 1.00000 | 0.50000 | 1.75000 |
| 2 | 1.00000 | 1.25000 | 1.50000 | 1.25000 | 1.50000 | 2.25000 |
| 5 | 1.93750 | 1.87500 | 2.43750 | 2.12500 | 2.37500 | 2.68750 |
| 10 | 2.17188 | 2.42188 | 2.67188 | 2.24609 | 2.49609 | 2.74805 |
| 15 | 2.24023 | 2.48047 | 2.74023 | 2.24988 | 2.49988 | 2.74994 |
| 17 | 2.24512 | 2.49023 | 2.74512 | 2.24997 | 2.49997 | 2.74998 |
| 20 | 2.24756 | 2.49756 | 2.74756 | | | |
| 25 | 2.24969 | 2.49939 | 2.74969 | | | |
| 30 | 2.24992 | 2.49992 | 2.74992 | | | |
| 32 | 2.24996 | 2.49996 | 2.74996 | | | |

From Table 11.5.1, it is clear that both methods are converging, but the number of iterations required for the Jacobi method to converge is almost double the number of iterations required for the Gauss–Seidel method.

For $\|B\| < 1$, the sequence generated by $\mathbf{x}^{(k+1)} = B\mathbf{x}^{(k)} + \mathbf{c}$ for $k = 0, 1, \ldots$ converges to the unique solution $\mathbf{x}$ which is given by $\mathbf{x} = B\mathbf{x} + \mathbf{c}$. We can show (Exercise 7(a)) that $\|\mathbf{x}^{(k+1)} - \mathbf{x}\| \leq \|B\|^{k+1} \|\mathbf{x}^{(0)} - \mathbf{x}\|$. From Theorem 11.4.2 we know that $\|B\| \leq \rho(B) + \epsilon$; therefore $\|\mathbf{x}^{(k+1)} - \mathbf{x}\| \leq (\rho(B) + \epsilon)^{(k+1)} \|\mathbf{x}^{(0)} - \mathbf{x}\|$. Thus the smaller the spectral radius of $B$, the more rapidly iterations converge. In the last example, since $\rho(B_G) = \frac{1}{2} = 0.5 < \rho(B_J) = 1/\sqrt{2} \approx 0.70711$, the Gauss–Seidel method converges faster than the Jacobi method.

If $\mathbf{x}^{(0)} = \mathbf{0}$, then $\|\mathbf{x}^{(k)} - \mathbf{x}\|/\|\mathbf{x}\| \leq \|B\|^k$. Here $\|\mathbf{x}^{(k)} - \mathbf{x}\|/\|\mathbf{x}\|$ represents the relative error. In some norm, we approximate $\|B\|$ by $\rho(B)$, and thus the relative error after $k$ iterations is given by $[\rho(B)]^k$. Let us say we want $10^{-t}$ accuracy after $k$ iterations. Then

$$[\rho(B)]^k \leq 10^{-t}$$

By taking the log on both sides and since $0 < \rho(B) < 1$, we get

$$k \geq \frac{t}{-\log \rho(B)}$$

The convergence speed of the iterative method $\mathbf{x}^{(k+1)} = B\mathbf{x}^{(k)} + \mathbf{c}$ is defined by $R = -\log \rho(B) = \log (1/\rho(B))$. Thus $k \geq t/R$. We select the iterative method with maximum $R$; in other words, with minimum $\rho(B)$. In our example, the convergence speed of the Jacobi method is $R = -\log (1/\sqrt{2}) = 0.15051$, and the convergence speed of the Gauss–Seidel method is $R = -\log (1/2) = 0.30103$.

**EXERCISES**

For Exercises 1–4, determine whether the Jacobi method, the Gauss–Seidel method, or both converge when we use them. Also find the convergence speed in case of convergence.

1.  $4x_1 + x_2 = 5$
    $x_1 - 5x_2 = -4$

2.  $x_1 - 5x_2 = -4$
    $4x_1 + x_2 = 5$

3.  $x_1 \qquad + x_3 = 2$
    $-x_1 + x_2 \qquad = 0$
    $x_1 + 2x_2 - 3x_3 = 0$

4.  $x_1 + \frac{1}{2}x_2 + \frac{1}{2}x_3 = 2$
    $\frac{1}{2}x_1 + x_2 + \frac{1}{2}x_3 = 2$
    $\frac{1}{2}x_1 + \frac{1}{2}x_2 + x_3 = 2$

For Exercises 5–6, find the values of $\alpha$ and $\beta$ (if they exist) such that (a) the Jacobi method converges but the Gauss–Seidel method does not converge, (b) the Gauss–Seidel method converges but the Jacobi method does not converge, and (c) both methods converge.

5. $\begin{bmatrix} 1 & \alpha \\ \beta & 1 \end{bmatrix}$

6. $\begin{bmatrix} 1 & \alpha & \beta \\ 1 & 1 & 1 \\ -\beta & \alpha & 1 \end{bmatrix}$

7. Let $\|B\| < 1$ and the sequence generated by $\mathbf{x}^{(k+1)} = B\mathbf{x}^{(k)} + \mathbf{c}$ for $k = 0, 1, \ldots$ converges to the unique solution $\mathbf{x}$ which is given by $\mathbf{x} = B\mathbf{x} + \mathbf{c}$. Then prove
   (a) $\|\mathbf{x}^{(k+1)} - \mathbf{x}\| \le \|B\|^{(k+1)} \|\mathbf{x}^{(0)} - \mathbf{x}\|$
   (b) $\|\mathbf{x}^{(k+1)} - \mathbf{x}\| \le \dfrac{\|B\|^{(k+1)}}{1 - \|B\|} \|\mathbf{x}^{(1)} - \mathbf{x}^{(0)}\|$ (*Hint:* $\mathbf{x}^{(k+1)} - \mathbf{x} = \mathbf{x}^{(k+1)} - \mathbf{x}^{(k+2)} + \mathbf{x}^{(k+2)} - \mathbf{x}$).

## 11.6   SUCCESSIVE OVERRELAXATION (SOR) METHOD

The Jacobi and Gauss–Seidel methods in Example 11.1.1 converge. It is often possible to substantially improve the rate of convergence of the Gauss–Seidel method. In order to solve $A\mathbf{x} = \mathbf{b}$, we compute the Gauss–Seidel iterate

$$\hat{x}_i^{(k+1)} = \frac{1}{a_{ii}} \left[ b_i - \sum_{j=1}^{i-1} a_{ij} x_j^{(k+1)} - \sum_{j=i+1}^{N} a_{ij} x_j^{(k)} \right] \qquad \textbf{(11.6.1)}$$

for $i = 1, 2, \ldots, N$ as an intermediate step and then calculate the final value from

$$x_i^{(k+1)} = x_i^{(k)} + \omega \, (\hat{x}_i^{(k+1)} - x_i^{(k)}) \qquad \textbf{(11.6.2)}$$

where $\omega$, the relaxation parameter, is to be determined so that the convergence speed is maximized. If $\omega = 1$, then Equation (11.6.2) reduces to the Gauss–Seidel iteration (11.6.1).

In order to write this procedure in matrix form, substitution of Equation (11.6.1) into Equation (11.6.2) gives

$$x_i^{(k+1)} = (1 - \omega) \, x_i^{(k)} + \frac{\omega}{a_{ii}} \left[ b_i - \sum_{j=1}^{i-1} a_{ij} x_j^{(k+1)} - \sum_{j=i+1}^{N} a_{ij} x_j^{(k)} \right] \qquad \textbf{(11.6.3)}$$

for $i = 1, 2, \ldots, N$. Equation (11.6.3) can be rearranged in the form

$$a_{ii} x_i^{(k+1)} + \omega \sum_{j=1}^{i-1} a_{ij} x_j^{(k+1)} = (1 - \omega) \, a_{ii} x_i^{(k)} - \omega \sum_{j=i+1}^{N} a_{ij} x_j^{(k)} + \omega b_i \qquad \textbf{(11.6.4)}$$

for $i = 1, 2, \ldots, N$. This relationship between the old iterates $x_i^{(k)}$ and the new $x_i^{(k+1)}$ can be written in matrix-vector form as

$$(D + \omega L) \, \mathbf{x}^{(k+1)} = (1 - \omega) \, D\mathbf{x}^{(k)} - \omega U\mathbf{x}^{(k)} + \omega \mathbf{b} \qquad \textbf{(11.6.5)}$$

by using the decomposition $A = L + D + U$. Let $a_{ii} \neq 0$ for each $i$. Then $(D + \omega L)$ is nonsingular and therefore Equation (11.6.5) can be written as

$$\mathbf{x}^{(k+1)} = (D + \omega L)^{-1} [(1 - \omega) D - \omega U] \mathbf{x}^{(k)} + \omega (D + \omega L)^{-1} \mathbf{b} \quad \textbf{(11.6.6)}$$

This defines the **successive relaxation method** and is called the **underrelaxation method** if $\omega < 1$ and the **overrelaxation method** if $\omega > 1$. However, for the actual computations Equations (11.6.1) and (11.6.2) are used. If $\omega = 1$, then Equation (11.6.6) reduces to the Gauss–Seidel matrix form (11.2.12).

Let

$$B = (D + \omega L)^{-1} [(1 - \omega) D - \omega U] \quad \textbf{(11.6.7)}$$

and

$$\mathbf{c} = \omega (D + \omega L)^{-1} \mathbf{b} \quad \textbf{(11.6.8)}$$

Then Equation (11.6.6) reduces to our well-known form

$$\mathbf{x}^{(k+1)} = B\mathbf{x}^{(k)} + \mathbf{c} \quad \textbf{(11.6.9)}$$

The sequence of iterates in Equation (11.6.9) converges by Theorem 11.5.1 if $\rho(B) < 1$.

Since $L$ and $U$ are strictly lower and upper triangular matrices, $D + \omega L$ and $(1 - \omega) D - \omega U$ are lower and upper matrices. Hence

$$\det (D + \omega L)^{-1} = \det D^{-1} \quad \text{and} \quad \det [(1 - \omega) D - \omega U] = \det (1 - \omega)D$$
$$\textbf{(11.6.10)}$$

Hence

$$\begin{aligned}
\det B &= \det (D + \omega L)^{-1} \det [(1 - \omega) D - \omega U] \\
&= \det D^{-1} \det (1 - \omega) D \\
&= \det D^{-1} \cdot \det D \cdot \det (1 - \omega) I \\
&= (1 - \omega)^N
\end{aligned}$$

Let $\lambda_1, \lambda_2, \ldots, \lambda_N$ be the eigenvalues of $B$. Then it can be verified that $\det B = \lambda_1 \lambda_2 \ldots \lambda_N$ (Exercise 14, p. 427). Thus $\max_i |\lambda_i| \geq |1 - \omega|$. For $\rho(B) < 1$, we must have $|1 - \omega| < 1$. Hence (11.6.6) converges if $0 < \omega < 2$ where $\omega$ is real. Further from Exercise 7(a), p. 389,

$$\|\mathbf{x}^{(k+1)} - \mathbf{x}\| \leq \|B\|^{k+1} \|\mathbf{x}^{(0)} - \mathbf{x}\|$$

Since $\rho(B) \leq \|B\|$, the smaller $\rho(B)$, the more rapidly (11.6.9) converges. The convergence speed of (11.6.9) is maximized by selecting $\omega$ in Equation (11.6.7) to make $\rho(B)$ as small as possible.

For an arbitrary set of linear equations, we do not yet have a formula to find the optimal value $\omega_o$ of $\omega$ which minimizes $\rho(B)$; however, for many of the difference equations approximating the first- and second-order partial differential equations, $\omega_o$ can be calculated from the following theorem (for the proof, refer to Young 1972).

**Theorem 11.6.1**    Let $A$ be a positive definite and tridiagonal $N \times N$ matrix. Then $\rho(B_G) = [\rho(B_J)]^2 < 1$ and the optimal $\omega_o$ for Equation (11.6.3) is given by

$$\omega_o = \frac{2[1 - \sqrt{1 - [\rho(B_J)]^2}]}{[\rho(B_J)]^2} \qquad (11.6.11)$$

∎

Furthermore, the SOR method applied to the equation $A\mathbf{x} = \mathbf{b}$ converges for all $\omega$ in the range $0 < \omega < 2$.

**EXAMPLE 11.6.1**    Solve Example 11.1.1 using the SOR method.
We want to solve the system

$$
\begin{aligned}
-2V_1 + V_2 & & & & & & = -V_0 - h^2 \\
V_1 - 2V_2 + V_3 & & & & & = -h^2 \\
V_2 - 2V_3 + V_4 & & & & = -h^2 \\
V_3 - 2V_4 + V_5 & & & = -h^2 \\
V_4 - 2V_5 + V_6 & & = -h^2 \\
V_5 - 2V_6 & = -V_7 - h^2
\end{aligned}
$$

The matrix $A$ is given by

$$
A = \begin{bmatrix}
-2 & 1 & 0 & 0 & 0 & 0 \\
1 & -2 & 1 & 0 & 0 & 0 \\
0 & 1 & -2 & 1 & 0 & 0 \\
0 & 0 & 1 & -2 & 1 & 0 \\
0 & 0 & 0 & 1 & -2 & 1 \\
0 & 0 & 0 & 0 & 1 & -2
\end{bmatrix} = \begin{bmatrix}
-2 & 0 & 0 & 0 & 0 & 0 \\
0 & -2 & 0 & 0 & 0 & 0 \\
0 & 0 & -2 & 0 & 0 & 0 \\
0 & 0 & 0 & -2 & 0 & 0 \\
0 & 0 & 0 & 0 & -2 & 0 \\
0 & 0 & 0 & 0 & 0 & -2
\end{bmatrix}
$$

$$
+ \begin{bmatrix}
0 & 1 & 0 & 0 & 0 & 0 \\
1 & 0 & 1 & 0 & 0 & 0 \\
0 & 1 & 0 & 1 & 0 & 0 \\
0 & 0 & 1 & 0 & 1 & 0 \\
0 & 0 & 0 & 1 & 0 & 1 \\
0 & 0 & 0 & 0 & 1 & 0
\end{bmatrix}
$$

$$= D + (L + U)$$

We can verify that

$$
-D^{-1}(L + U) = \begin{bmatrix}
0 & 1/2 & 0 & 0 & 0 & 0 \\
1/2 & 0 & 1/2 & 0 & 0 & 0 \\
0 & 1/2 & 0 & 1/2 & 0 & 0 \\
0 & 0 & 1/2 & 0 & 1/2 & 0 \\
0 & 0 & 0 & 1/2 & 0 & 1/2 \\
0 & 0 & 0 & 0 & 1/2 & 0
\end{bmatrix}
$$

and its eigenvalues are given by $\mu_i = \cos i\pi h = \cos(i\pi/(N + 1))$ for $i = 1, 2, \ldots,$ 6 and $N = 6$. Also it can be seen that $\mu_1 = -\mu_6$, $\mu_2 = -\mu_5$, $\mu_3 = -\mu_4$, and $\rho(B_J) = \mu_{max} = \cos \pi/7$. Therefore

$$\omega_0 = \frac{2}{\mu_{max}^2}\left[1 - \sqrt{1 - \mu_{max}^2}\right] = \frac{2(1 - \sin \pi/7)}{\cos^2 \pi/7} = \frac{2}{1 + \sin \pi/7} \approx 1.39481$$

From the boundary conditions we have $V_0 = 2$ and $V_7 = 3$. Select $V_1^{(0)} = V_2^{(0)} = V_3^{(0)} = V_4^{(0)} = V_5^{(0)} = V_6^{(0)} = 0$. We have

$$\hat{V}_i^{(k+1)} = \frac{1}{2}(V_{i+1}^{(k)} + V_{i-1}^{(k+1)} + h^2) \qquad \textbf{(11.6.12)}$$

and

$$V_i^{(k+1)} = V_i^{(k)} + \omega_0(\hat{V}_i^{(k+1)} - V_i^{(k)}) \qquad \textbf{(11.6.13)}$$

for $i = 1, 2, \ldots, 6$ and $k = 0, 1, 2, \ldots$ from Equations (11.6.1) and (11.6.2).

For $k = 0$ and $i = 1$, we get

$$\hat{V}_1^{(1)} = \frac{1}{2}(V_2^{(0)} + V_0^{(1)} + h^2) = \frac{1}{2}(0.0 + 2.0 + 0.02041) = 1.01020$$

$$V_1^{(1)} = V_1^{(0)} + \omega_0(\hat{V}_1^{(1)} - V_1^{(0)}) = 0 + 1.39481(1.01020 - 0) = 1.41155$$

For $k = 0$ and $i = 2$, Equations (11.6.12) and (11.6.13) become

$$\hat{V}_2^{(1)} = \frac{1}{2}(V_3^{(0)} + V_1^{(1)} + h^2) = \frac{1}{2}(0.0 + 1.41155 + 0.02041) = 0.71598$$

$$V_2^{(1)} = V_2^{(0)} + \omega_0(\hat{V}_2^{(1)} - V_2^{(0)}) = 0 + 1.39481(0.71598 - 0) = 0.99866$$

Similarly for $k = 0$ and $i = 3$,

$$\hat{V}_3^{(1)} = \frac{1}{2}(V_4^{(0)} + V_2^{(1)} + h^2) = \frac{1}{2}(0.0 + 0.99866 + 0.02041) = 0.50953$$

$$V_3^{(1)} = V_3^{(0)} + \omega_0(\hat{V}_3^{(1)} - V_3^{(0)}) = 0 + 1.39481(0.50953 - 0) = 0.71070$$

We can continue the SOR method.

The computations were terminated once we had $\|\mathbf{V}^{(k)} - \mathbf{V}^{(k+1)}\| < 10^{-5} \|\mathbf{V}^{(k+1)}\|$. In Table 11.6.1 the number of required iterations corresponding to $\omega$ is given for this problem.

From Table 11.6.1, it can be seen that the Gauss–Seidel method required 51 iterations while the SOR method with $\omega_0$ required only 18 iterations. The SOR method

**Table 11.6.1**

| $\omega$ | Number of Iterations | $\omega$ | Number of Iterations |
|---|---|---|---|
| 1.0 (Gauss–Seidel) | 51 | 1.4 | 17 |
| 1.1 | 42 | 1.5 | 20 |
| 1.2 | 34 | 1.6 | 27 |
| 1.3 | 26 | 1.7 | 35 |
| 1.39481 | 18 | 1.8 | 56 |

**Table 11.6.2**

| N(= 1/h) | ω | Number of Iterations | N(= 1/h) | ω | Number of Iterations |
|---|---|---|---|---|---|
| 10 | 1.0 (Gauss–Seidel) | 96 | 20 | 1.0 (Gauss–Seidel) | 328 |
| 10 | 1.1 | 80 | 20 | 1.1 | 275 |
| 10 | 1.2 | 66 | 20 | 1.2 | 229 |
| 10 | 1.3 | 53 | 20 | 1.3 | 190 |
| 10 | 1.4 | 42 | 20 | 1.4 | 154 |
| 10 | 1.5 | 30 | 20 | 1.5 | 122 |
| 10 | 1.52786 | 25 | 20 | 1.6 | 91 |
| 10 | 1.6 | 27 | 20 | 1.7 | 60 |
| 10 | 1.7 | 38 | 20 | 1.72945 | 47 |
| 10 | 1.8 | 53 | 20 | 1.8 | 61 |

requires far fewer iterations for the larger system than the Gauss–Seidel method does. Let us pick up various values of $h$ and solve the system of Equation (11.1.6) with $v_0 = V_0 = 2$ and $v_N = V_N = 3$. The results are given in Table 11.6.2. ■ ■ ■

**Algorithm**

**Subroutine** SOR ($N$, $A$, **b**, **x**, eps, maxit, ω)
Comment: This program solves $A\mathbf{x} = \mathbf{b}$ using the SOR method given by Equations (11.6.1) and (11.6.2). $N$ is the number of equations and unknowns; $a_{ij}$ for $i, j = 1, 2, \ldots, N$ are the elements of a matrix $A$; $b_1, b_2, \ldots, b_N$ are the elements of **b**; $x_1, x_2, \ldots, x_N$ are the initial guesses of **x**; eps is the tolerance; maxit is the maximum number of iterations and ω is the relaxation parameter.
1 **Print** $N$, $A$, eps
2 **Repeat** step 3 **with** $i = 1, 2, \ldots, N$
   3 **If** $|a_{ii}| <$ eps, **then print** 'Algorithm fails because $|a_{ii}| <$ eps' and exit.
4 **Print b, x,** maxit, ω
5 $k = 0$
Comment: xold$_i$, $1 \leq i \leq N$ are the old values of **x**.
6 **Repeat** steps 7–18 **while** $k \leq$ maxit
   7 $k = k + 1$
   8 **Repeat** step 9 **with** $i = 1, 2, \ldots, N$
      9 xold$_i = x_i$
   10 **Repeat** steps 11–15 **with** $i = 1, 2, \ldots, N$
      11 sum $= 0$
      12 **Repeat** step 13 **with** $j = 1, 2, \ldots, N$
         13 **If** $j \neq i$, **then** sum $=$ sum $+ a_{ij} * x_j$
      14 $x_i = (b_i -$ sum$)/a_{ii}$
      15 $x_i =$ xold$_i + ω (x_i -$ xold$_i)$
   16 **Repeat** step 17 **with** $i = 1, 2, \ldots, N$
      17 **If** $|x_i -$ xold$_i| \geq$ eps $|x_i|$, **then** return to 6
   18 **Print** $k$, **x**, and 'finished' and exit
19 **Print** 'Algorithm fails: no convergence' and exit.

The optimal value of $\omega$ is not always predictable in advance because $\rho(B_J)$ is not always known. If we want to solve many cases with the same system but with different boundary conditions, it may be appropriate to solve the system with different $\omega$ and pick the value of $\omega$ that gives the fastest convergence. This inefficient procedure can be improved by an iterative procedure. For the iterative procedure, refer to Young and Gregory (1972).

---

## EXERCISES

1. For the matrix

$$A = \begin{bmatrix} -2 & 1 & 0 \\ 1 & -2 & 1 \\ 0 & 1 & -2 \end{bmatrix}$$

find (a) $B_J = -D^{-1}(L + U)$, (b) the eigenvalues of $B_J$, and (c) $\omega_o$ for the SOR method.

2. For the matrix

$$A = \begin{bmatrix} -2 & 1 & 0 & 0 \\ 1 & -2 & 1 & 0 \\ 0 & 1 & -2 & 1 \\ 0 & 0 & 1 & -2 \end{bmatrix}$$

find (a) $B_J = -D^{-1}(L + U)$, (b) the eigenvalues of $B_J$, and (c) $\omega_o$ for the SOR method.

3. For the matrix

$$A = \begin{bmatrix} -4 & 1 & 0 \\ 1 & -4 & 1 \\ 0 & 1 & -4 \end{bmatrix}$$

find (a) $B_J = -D^{-1}(L + U)$, (b) the eigenvalues of $B_J$, and (c) $\omega_o$ for the SOR method.

4. Carry out the first three iterations of the SOR method with $\omega_o$ from Exercise 1 for the system

$$\begin{aligned} -2x_1 + x_2 \phantom{+ x_3} &= -1 \\ x_1 - 2x_2 + x_3 &= 0 \\ x_2 - 2x_3 &= -1 \end{aligned}$$

5. Carry out the first three iterations of the SOR method with $\omega_o$ from Exercise 3 for the system

$$\begin{aligned} -4x_1 + x_2 \phantom{+ x_3} &= -3 \\ x_1 - 4x_2 + x_3 &= -2 \\ x_2 - 4x_3 &= -3 \end{aligned}$$

**COMPUTER EXERCISES**

**C.1.** Write a subroutine to solve the system of linear $A\mathbf{x} = \mathbf{b}$ by the SOR method using Equations (11.6.1) and (11.6.2). Your subroutine should terminate iterations when $\|\mathbf{x}^{(k+1)} - \mathbf{x}^{(k)}\|_\infty \leq$ eps $\|\mathbf{x}^{(k+1)}\|_\infty$ or $k$ reaches the maximum number of iterations maxit. The parameters are $A$, $N$, $\mathbf{b}$, initial guess $\mathbf{x}^{(0)}$, $\omega$, the maximum number of iterations maxit and the tolerance eps. Print the values of $\mathbf{x}$ at various values of $k$.

**C.2.** Write a computer program to solve the system of linear equations

$$-2x_1 + x_2 = -1$$
$$x_{i-1} - 2x_i + x_{i+1} = 0 \quad \text{for } i = 2, 3, \ldots, N-1$$
$$x_{N-1} - 2x_N = -1$$

for $N = 10$ and $20$. Use Exercise C.1 and eps $= 10^{-5}$. Select various $\omega$ and study the required number of iterations and $\omega$.

## SUGGESTED READINGS

E. Isaacson and H. B. Keller, *Analysis of Numerical Methods*, John Wiley & Sons, New York, 1966.

D. M. Young, *Iterative Solution of Large Linear Systems*, Academic Press, New York, 1971.

D. M. Young and R. T. Gregory, *A Survey of Numerical Mathematics Vols. I & II*, Addison-Wesley, Reading, Massachusetts, 1972.

# 12 NUMERICAL SOLUTIONS OF SYSTEMS OF NONLINEAR EQUATIONS

$\mathbf{I}$n Chapter 6 we solved the system of four nonlinear equations

$$a_0 + a_1 = 2$$

$$a_0 x_0 + a_1 x_1 = 0$$

$$a_0 x_0^2 + a_1 x_1^2 = \frac{2}{3}$$

$$a_0 x_0^3 + a_1 x_1^3 = 0 \qquad \textbf{(1)}$$

in four unknowns $a_0$, $a_1$, $x_0$, and $x_1$ to derive the two-point Gaussian formula. In this chapter, we develop methods to approximate a solution of a system of nonlinear equations.

## 12.1 INTRODUCTION

We discuss methods to find the numerical solution of a system of nonlinear equations in this chapter. We discussed in Chapter 2 methods to find the numerical solution of a single nonlinear equation $f(x) = 0$. We generalize the fixed-point method and Newton's method to find the numerical solution of a system of nonlinear equations in this chapter. Also, the steepest descent method is discussed.

In this chapter we are concerned with the solution of $N$ nonlinear equations in $N$ unknowns $x_1, x_2, \ldots, x_N$ given by

$$f_1(x_1, x_2, \ldots, x_N) = 0$$
$$f_2(x_1, x_2, \ldots, x_N) = 0$$
$$\vdots$$
$$f_N(x_1, x_2, \ldots, x_N) = 0 \qquad \textbf{(12.1.1)}$$

If the values $r_1, r_2, \ldots, r_N$ exist such that

$$f_1(r_1, r_2, \ldots, r_N) = 0$$
$$f_2(r_1, r_2, \ldots, r_N) = 0$$
$$\vdots$$
$$f_N(r_1, r_2, \ldots, r_N) = 0$$

then $\mathbf{r} = (r_1, r_2, \ldots, r_N)$ is called a **solution** of Equation (12.1.1).

To simplify the writing of Equation (12.1.1) and to aid in the understanding of analysis, it is convenient to use vector notation. Let $\mathbf{x} = (x_1, x_2, \ldots, x_N)$. Then Equation (12.1.1) can be written as $f_i(\mathbf{x}) = 0$. Furthermore, define

$$\mathbf{f}(\mathbf{x}) = [f_1(\mathbf{x}), f_2(\mathbf{x}), \ldots, f_N(\mathbf{x})]^t$$

Then Equation (12.1.1) reduces to

$$\mathbf{f}(\mathbf{x}) = \mathbf{0} \tag{12.1.2}$$

**EXAMPLE 12.1.1**   Consider the system of four equations in four unknowns given by

$$e^{x_1} \sin x_2 + \quad x_4 = 0$$
$$x_1 x_3 + x_2 x_4 = 0$$
$$\ln \cos x_1 + \quad x_1^2 = 0$$
$$x_1^2 + x_2^2 - x_1 x_2 - x_3 x_4 = 0$$

Express the system in the form of Equation (12.1.2).

We have $f_1(x_1, x_2, x_3, x_4) = e^{x_1} \sin x_2 + x_4$, $f_2(x_1, x_2, x_3, x_4) = x_1 x_3 + x_2 x_4$, $f_3(x_1, x_2, x_3, x_4) = \ln \cos x_1 + x_1^2$, and $f_4(x_1, x_2, x_3, x_4) = x_1^2 + x_2^2 - x_1 x_2 - x_3 x_4$. Then

$$\mathbf{f}(\mathbf{x}) = \mathbf{f}(x_1, x_2, x_3, x_4) = \begin{bmatrix} f_1(x_1, x_2, x_3, x_4) \\ f_2(x_1, x_2, x_3, x_4) \\ f_3(x_1, x_2, x_3, x_4) \\ f_4(x_1, x_2, x_3, x_4) \end{bmatrix} = \begin{bmatrix} e^{x_1} \sin x_2 + x_4 \\ x_1 x_3 + x_2 x_4 \\ \ln \cos x_1 + x_1^2 \\ x_1^2 + x_2^2 - x_1 x_2 - x_3 x_4 \end{bmatrix} = \begin{bmatrix} 0 \\ 0 \\ 0 \\ 0 \end{bmatrix}$$

Since $\mathbf{f}(\mathbf{0}) = \mathbf{0}$, $\mathbf{0} = (0, 0, 0, 0)$ is a solution of our system $\mathbf{f}(\mathbf{x}) = \mathbf{0}$.   ■ ■ ■

For simplicity we first consider the case for $N = 2$, then a generalization for any $N$ will be stated. Consider

$$f_1(x_1, x_2) = 0 \quad \text{and} \quad f_2(x_1, x_2) = 0 \tag{12.1.3}$$

Each of these equations defines, in general, a **curve** in the $x_1$, $x_2$ plane, therefore the problem of solving a system of equations is considered to be equivalent to the problem of finding the point or points of intersection of these two curves.

The solution of Equation (12.1.3) can also be looked upon as the point or points of intersection of the zero curves in the $x_1$, $x_2$ plane of the surfaces $z = f_1(x_1, x_2)$ and $z = f_2(x_1, x_2)$. This approach is used to generalize Newton's method.

Consider the systems

1. $x_1^2 - x_2 = 0$, $-2x_1 + x_2 + 1 = 0$
2. $\sin x_1 - x_2 = 0$, $\cos x_1 - x_2 = 0$

Figure 12.1.1 shows the graphs of systems 1 and 2. For system 1, there is only one solution, while for system 2 there are infinite solutions.

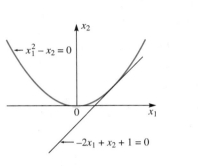
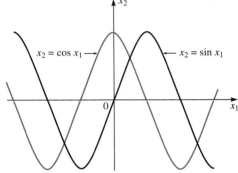

**Fig. 12.1.1**

Solving the first equation of system 1 for $x_2$, we get $x_2 = x_1^2$. Substituting for $x_2$ in the second equation of system 1, we get $x_1^2 - 2x_1 + 1 = 0$. Thus we are able to reduce the problem to a single equation in a single variable. In this chapter, we assume that such a reduction to a single equation is difficult or impossible.

We are able to sketch the curves for Equation (12.1.3). In general, it is very difficult to sketch the multidimensional surfaces defined by Equation (12.1.2). Thus, it is difficult to find an approximation of a root of Equation (12.1.2). The bisection and regula falsi methods which always converged in the one-dimensional case cannot be used because approximation of a root for Equation (12.1.2) is not known. Also, this makes it difficult to choose a method that converges. Because of the nonlinearity of Equation (12.1.2), it is generally difficult to know when Equation (12.1.2) has solutions and how many. This is when the fixed-point iteration method becomes an important tool.

## 12.2   FIXED-POINT ITERATION METHOD

An iterative method for solving $f(x) = 0$ was developed in Chapter 2 by first writing $f(x) = 0$ into an equation of the form $x = g(x)$ such that $r = g(r)$ whenever $f(r) = 0$. A root $r$ of $x = g(x)$ is called a fixed point of $g$. Also, we developed iterative methods for solving a linear system of equations $A\mathbf{x} = \mathbf{b}$ by writing $A\mathbf{x} = \mathbf{b}$ as $\mathbf{x} = B\mathbf{x} + \mathbf{c}$ in Chapter 11.

We generalize the fixed-point method by writing $\mathbf{f}(\mathbf{x}) = \mathbf{0}$ as

$$\mathbf{x} = \mathbf{g}(\mathbf{x}) \tag{12.2.1}$$

such that $\mathbf{r} = \mathbf{g}(\mathbf{r})$ whenever $\mathbf{f}(\mathbf{r}) = \mathbf{0}$. For example, one choice could be $\mathbf{x} = \mathbf{g}(\mathbf{x}) = \mathbf{x} + A\mathbf{f}(\mathbf{x})$ where $A$ is a nonsingular $N \times N$ matrix. A root $\mathbf{r}$ of $\mathbf{x} = \mathbf{g}(\mathbf{x})$ is called a **fixed point** of $\mathbf{g}$. To find $\mathbf{r}$, we start with $\mathbf{x}^{(0)}$ and use the algorithm

$$\mathbf{x}^{(k+1)} = \mathbf{g}(\mathbf{x}^{(k)}) \quad \text{for } k = 0, 1, \ldots \qquad (12.2.2)$$

For simplicity, consider $N = 2$. Then we have

$$f_1(x_1, x_2) = 0 \quad \text{and} \quad f_2(x_1, x_2) = 0 \qquad (12.2.3)$$

We solve the first equation of (12.2.3) for $x_1$ and the second equation for $x_2$ as

$$x_1 = g_1(x_1, x_2) \quad \text{and} \quad x_2 = g_2(x_1, x_2) \qquad (12.2.4)$$

Also, we can solve the second equation of (12.2.3) for $x_1$ and the first equation for $x_2$ and represent them in the form of Equation (12.2.4).

Starting with $\mathbf{x}^{(0)} = (x_1^{(0)}, x_2^{(0)})$, the iterations are given by

$$x_1^{(k+1)} = g_1(x_1^{(k)}, x_2^{(k)}) \quad \text{and} \quad x_2^{(k+1)} = g_2(x_1^{(k)}, x_2^{(k)}) \qquad (12.2.5)$$

for $k = 0, 1, \ldots$. We have a choice of using or not using the latest values of iterations. When we use the latest value of $x_1^{(k+1)}$, we get the modified iterative equations

$$x_1^{*(k+1)} = g_1(x_1^{*(k)}, x_2^{*(k)}) \quad \text{and} \quad x_2^{*(k+1)} = g_2(x_1^{*(k+1)}, x_2^{*(k)}) \qquad (12.2.6)$$

This is similar to the Gauss–Seidel method for a system of linear equations.

**EXAMPLE 12.2.1**  The system of equations

$$f_1(x_1, x_2) = x_1 + \cos x_2 = 0 \quad \text{and} \quad f_2(x_1, x_2) = x_2 + \sin x_1 = 0$$

can be written in the form of Equation (12.2.4) in any one of the following ways:

**1.**  $x_1 = g_1(x_1, x_2) = -\cos x_2 \quad \text{and} \quad x_2 = g_2(x_1, x_2) = -\sin x_1$

**2.**  $x_1 = g_1(x_1, x_2) = x_1 - \dfrac{2}{3}\left[(x_1 + \cos x_2) + \dfrac{1}{\sqrt{2}}(x_2 + \sin x_1)\right]$

$x_2 = g_2(x_1, x_2) = x_2 - \dfrac{2}{3}\left[-\dfrac{1}{\sqrt{2}}(x_1 + \cos x_2) + (x_2 + \sin x_1)\right]$

**3.**  $x_1 = g_1(x_1, x_2) = \arcsin(-x_2)$, and $x_2 = g_2(x_1, x_2) = \arccos(-x_1)$. This is obtained by solving the second equation $x_2 + \sin x_1 = 0$ for $x_1$ and the first equation $x_1 + \cos x_2 = 0$ for $x_2$.

In case 1, Equation (12.2.5) becomes

$$\begin{bmatrix} x_1^{(k+1)} \\ x_2^{(k+1)} \end{bmatrix} = \begin{bmatrix} -\cos x_2^{(k)} \\ -\sin x_1^{(k)} \end{bmatrix} \qquad (12.2.7)$$

Starting with $k = 0$, $x_1^{(0)} = 2$ and $x_2^{(0)} = 2$, we get from Equation (12.2.7)

$$\begin{bmatrix} x_1^{(1)} \\ x_2^{(1)} \end{bmatrix} = \begin{bmatrix} -\cos x_2^{(0)} \\ -\sin x_1^{(0)} \end{bmatrix} = \begin{bmatrix} -\cos 2 \\ -\sin 2 \end{bmatrix} = \begin{bmatrix} 0.41615 \\ -0.90930 \end{bmatrix}$$

For $k = 1$, Equation (12.2.7) gives

$$\begin{bmatrix} x_1^{(2)} \\ x_2^{(2)} \end{bmatrix} = \begin{bmatrix} -\cos x_2^{(1)} \\ -\sin x_1^{(1)} \end{bmatrix} = \begin{bmatrix} -\cos(-0.90930) \\ -\sin(0.41615) \end{bmatrix} = \begin{bmatrix} -0.61430 \\ -0.40424 \end{bmatrix}$$

We continue and the computed values are given in Table 12.2.1.

**Table 12.2.1**

| $k$ | $x_1^{(k)}$ | $x_2^{(k)}$ | $x_1^{*(k)}$ | $x_2^{*(k)}$ |
|---|---|---|---|---|
| 0  | 2.00000  | 2.00000  | 2.00000  | 2.00000  |
| 1  | 0.41615  | -0.90930 | 0.41615  | -0.40424 |
| 2  | -0.61430 | -0.40424 | -0.91940 | 0.79524  |
| 5  | -0.70011 | 0.74360  | -0.75377 | 0.68439  |
| 10 | -0.76136 | 0.68439  | -0.76847 | 0.69503  |
| 15 | -0.76958 | 0.69378  | -0.76816 | 0.69482  |
| 16 | -0.76883 | 0.69583  | -0.76817 | 0.69482  |
| 20 | -0.76831 | 0.69503  |          |          |
| 25 | -0.76814 | 0.69484  |          |          |
| 30 | -0.76817 | 0.69482  |          |          |

The modified equations of Equation (12.2.6) become

$$\begin{bmatrix} x_1^{*(k+1)} \\ x_2^{*(k+1)} \end{bmatrix} = \begin{bmatrix} -\cos x_2^{*(k)} \\ -\sin x_1^{*(k+1)} \end{bmatrix} \qquad (12.2.8)$$

Starting with $k = 0$, $x_1^{*(0)} = 2$, and $x_2^{*(0)} = 2$, Equation (12.2.8) becomes

$$\begin{bmatrix} x_1^{*(1)} \\ x_2^{*(1)} \end{bmatrix} = \begin{bmatrix} -\cos x_2^{*(0)} \\ -\sin x_1^{*(1)} \end{bmatrix} = \begin{bmatrix} -\cos 2 \\ -\sin(0.41615) \end{bmatrix} = \begin{bmatrix} 0.41615 \\ -0.40424 \end{bmatrix}$$

For $k = 1$, Equation (12.2.8) gives

$$\begin{bmatrix} x_1^{*(2)} \\ x_2^{*(2)} \end{bmatrix} = \begin{bmatrix} -\cos x_2^{*(1)} \\ -\sin x_1^{*(2)} \end{bmatrix} = \begin{bmatrix} -\cos(-0.40424) \\ -\sin(-0.91940) \end{bmatrix} = \begin{bmatrix} -0.91940 \\ 0.79524 \end{bmatrix}$$

The results obtained by using Equation (12.2.8) are also given in Table 12.2.1 and are indicated by $x_1^*$ and $x_2^*$ in the fourth and fifth columns. This clearly accelerates the convergence as in the Gauss–Seidel method for linear systems. However, keep in mind that the Gauss–Seidel method does not always accelerate convergence. The sequences in both cases were terminated at the $k$th iteration once we had $\|\mathbf{x}^{(k+1)} - \mathbf{x}^{(k)}\| < 10^{-5} \|\mathbf{x}^{(k+1)}\|$.

For case 2, Equation (12.2.5) becomes

$$\begin{bmatrix} x_1^{(k+1)} \\ x_2^{(k+1)} \end{bmatrix} = \begin{bmatrix} x_1^{(k)} \\ x_2^{(k)} \end{bmatrix} - \frac{2}{3} \begin{bmatrix} 1 & 1/\sqrt{2} \\ -1/\sqrt{2} & 1 \end{bmatrix} \begin{bmatrix} x_1^{(k)} + \cos x_2^{(k)} \\ x_2^{(k)} + \sin x_1^{(k)} \end{bmatrix} \qquad (12.2.9)$$

For $k = 0$, $x_1^{(0)} = 2$, and $x_2^{(0)} = 2$, we get from Equation (12.2.9)

$$
\begin{bmatrix} x_1^{(1)} \\ x_2^{(1)} \end{bmatrix} = \begin{bmatrix} 2 \\ 2 \end{bmatrix} - \frac{2}{3} \begin{bmatrix} 1 & 1/\sqrt{2} \\ -1/\sqrt{2} & 1 \end{bmatrix} \begin{bmatrix} 2 + \cos 2 \\ 2 + \sin 2 \end{bmatrix}
$$

$$
= \begin{bmatrix} 2 - \dfrac{2}{3}(1.58385 + 2.05718) \\ 2 - \dfrac{2}{3}(-1.11995 + 2.90930) \end{bmatrix}
$$

$$
= \begin{bmatrix} -0.42736 \\ 0.80710 \end{bmatrix}
$$

The computed values are given in Table 12.2.2. Also, the values computed by using the latest values $x_1^{(k+1)}$ instead of $x_1^{(k)}$ are given in the fourth and fifth columns of Table 12.2.2. In this case, we need just five iterations.

**Table 12.2.2**

| $k$ | $x_1^{(k)}$ | $x_2^{(k)}$ | $x_1^{*(k)}$ | $x_2^{*(k)}$ |
|---|---|---|---|---|
| 0 | 2.00000 | 2.00000 | 2.00000 | 2.00000 |
| 1 | −0.42736 | 0.80710 | −0.42736 | 0.54535 |
| 2 | −0.78861 | 0.66991 | −0.77411 | 0.68594 |
| 3 | −0.76685 | 0.69399 | −0.76773 | 0.69452 |
| 4 | −0.76814 | 0.69478 | −0.76816 | 0.69481 |
| 5 | −0.76817 | 0.69482 | −0.76817 | 0.69482 |

In case 3, Equation (12.2.5) becomes

$$
\begin{bmatrix} x_1^{(k+1)} \\ x_2^{(k+1)} \end{bmatrix} = \begin{bmatrix} \arcsin(-x_2^{(k)}) \\ \arccos(-x_1^{(k)}) \end{bmatrix} \tag{12.2.10}
$$

Starting with $k = 0$, $x_1^{(0)} = 2$ and $x_2^{(0)} = 2$, we get from Equation (12.2.10), $x_1^{(1)} = \arcsin(-2)$ which is undefined. Let us start with $x_1^{(0)} = 0.1$ and $x_2^{(0)} = 0.1$. The results are given in Table 12.2.3.

**Table 12.2.3**

| $k$ | $x_1^{(k)}$ | $x_2^{(k)}$ | $x_1^{*(k)}$ | $x_2^{*(k)}$ |
|---|---|---|---|---|
| 0 | 0.1 | 0.1 | 0.1 | 0.1 |
| 1 | −0.10017 | 1.67096 | −0.10017 | 1.47046 |
| 2 | undefined | | undefined | |

■ ■ ■

It is natural to ask what makes the various iteration forms behave the way they do as shown in Tables 12.2.1, 12.2.2, and 12.2.3. The answer is given by an extension of the Fixed-point Theorem to the two-dimensional case.

**Theorem 12.2.1**

Let $D = \{(x_1, x_2) \mid a_1 \le x_1 \le b_1, a_2 \le x_2 \le b_2$ where $a_1, a_2, b_1$, and $b_2$ are constants$\}$ be a rectangle in $R^2$. Assume that $\mathbf{g}$ is continuous at all points in $D$ such that $\mathbf{g}(\mathbf{x}) \in D$ whenever $\mathbf{x} \in D$. Then $\mathbf{g}$ has a fixed point in $D$.

Further, let $G$ be the **Jacobian matrix** of the functions $g_1$ and $g_2$ defined at $\mathbf{x}$ by

$$G(\mathbf{x}) = \begin{bmatrix} \dfrac{\partial g_1}{\partial x_1}(\mathbf{x}) & \dfrac{\partial g_1}{\partial x_2}(\mathbf{x}) \\[2mm] \dfrac{\partial g_2}{\partial x_1}(\mathbf{x}) & \dfrac{\partial g_2}{\partial x_2}(\mathbf{x}) \end{bmatrix}$$

such that $\|G(\mathbf{x})\| < 1$ for every element $\mathbf{x}$ in $D$. Then for any $\mathbf{x}^{(0)}$ in $D$, the sequence of iterates $\mathbf{x}^{(k+1)} = \mathbf{g}(\mathbf{x}^{(k)})$ for $k = 0, 1, \ldots$ converges to a unique fixed point $\mathbf{r}$ in $D$.

The proof of the first part is the extension of the proof of the Fixed-point Theorem 2.9.1, therefore it is left for the reader.

The error at the $(k + 1)$th step is given by

$$
\begin{aligned}
\mathbf{e}^{(k+1)} &= \begin{bmatrix} r_1 - x_1^{(k+1)} \\ r_2 - x_2^{(k+1)} \end{bmatrix} = \begin{bmatrix} g_1(r_1, r_2) - g_1(x_1^{(k)}, x_2^{(k)}) \\ g_2(r_1, r_2) - g_2(x_1^{(k)}, x_2^{(k)}) \end{bmatrix} \\[2mm]
&= \begin{bmatrix} g_1(x_1^{(k)} + r_1 - x_1^{(k)}, x_2^{(k)} + r_2 - x_2^{(k)}) - g_1(x_1^{(k)}, x_2^{(k)}) \\ g_2(x_1^{(k)} + r_1 - x_1^{(k)}, x_2^{(k)} + r_2 - x_2^{(k)}) - g_2(x_1^{(k)}, x_2^{(k)}) \end{bmatrix} \\[2mm]
&= \begin{bmatrix} (r_1 - x_1^{(k)}) \dfrac{\partial g_1}{\partial x_1}(\hat{x}_1^{(k)}, \hat{x}_2^{(k)}) + (r_2 - x_2^{(k)}) \dfrac{\partial g_1}{\partial x_2}(\hat{x}_1^{(k)}, \hat{x}_2^{(k)}) \\[2mm] (r_1 - x_1^{(k)}) \dfrac{\partial g_2}{\partial x_1}(\tilde{x}_1^{(k)}, \tilde{x}_2^{(k)}) + (r_2 - x_2^{(k)}) \dfrac{\partial g_2}{\partial x_2}(\tilde{x}_1^{(k)}, \tilde{x}_2^{(k)}) \end{bmatrix} \\[2mm]
&= \begin{bmatrix} \dfrac{\partial g_1}{\partial x_1}(\hat{x}_1^{(k)}, \hat{x}_2^{(k)}) & \dfrac{\partial g_1}{\partial x_2}(\hat{x}_1^{(k)}, \hat{x}_2^{(k)}) \\[2mm] \dfrac{\partial g_2}{\partial x_1}(\tilde{x}_1^{(k)}, \tilde{x}_2^{(k)}) & \dfrac{\partial g_2}{\partial x_2}(\tilde{x}_1^{(k)}, \tilde{x}_2^{(k)}) \end{bmatrix} \begin{bmatrix} r_1 - x_1^{(k)} \\ r_2 - x_2^{(k)} \end{bmatrix}
\end{aligned}
\tag{12.2.11}
$$

where $(\hat{x}_1^{(k)}, \hat{x}_2^{(k)})$ and $(\tilde{x}_1^{(k)}, \tilde{x}_2^{(k)})$ are the points on the line segment joining $(x_1^{(k)}, x_2^{(k)})$ and $(r_1, r_2)$. Then Equation (12.2.11) can be written as

$$\mathbf{e}^{(k+1)} = G_k \mathbf{e}^{(k)} \tag{12.2.12}$$

where

$$G_k = \begin{bmatrix} \dfrac{\partial g_1}{\partial x_1}(\hat{x}_1^{(k)}, \hat{x}_2^{(k)}) & \dfrac{\partial g_1}{\partial x_2}(\hat{x}_1^{(k)}, \hat{x}_2^{(k)}) \\[2mm] \dfrac{\partial g_2}{\partial x_1}(\tilde{x}_1^{(k)}, \tilde{x}_2^{(k)}) & \dfrac{\partial g_2}{\partial x_2}(\tilde{x}_1^{(k)}, \tilde{x}_2^{(k)}) \end{bmatrix}$$

Thus

$$
\begin{aligned}
\mathbf{e}^{(k+1)} = G_k \mathbf{e}^{(k)} &= G_k (G_{k-1} \mathbf{e}^{(k-1)}) = G_k G_{k-1}(\mathbf{e}^{(k-1)}) \\
&= G_k G_{k-1} (G_{k-2} \mathbf{e}^{(k-2)}) = \cdots = G_k G_{k-1} \cdots G_1 G_0 \mathbf{e}^{(0)}
\end{aligned}
\tag{12.2.13}
$$

Taking the norm on both sides of Equation (12.2.13) yields

$$
\begin{aligned}
\|\mathbf{e}^{(k+1)}\| &= \|G_k G_{k-1} \cdots G_1 G_0 \mathbf{e}^{(0)}\| \\
&\le \|G_k\| \, \|G_{k-1} \cdots G_1 G_0 \mathbf{e}^{(0)}\| \\
&\;\;\vdots \\
&\le \|G_k\| \, \|G_{k-1}\| \cdots \|G_0\| \, \|\mathbf{e}^{(0)}\|
\end{aligned}
\tag{12.2.14}
$$

Let $\max_k \|G_k\| = \|G\| \leq M < 1$. Then Equation (12.2.14) reduces to

$$0 \leq \|\mathbf{e}^{(k+1)}\| \leq M^{k+1}\|\mathbf{e}^{(0)}\|$$

Since $M < 1$, $\lim_{k\to\infty} M^{k+1}\|\mathbf{e}^{(0)}\| = 0$. By the Sandwich Theorem A.0, $\lim_{k\to\infty} \|\mathbf{e}^{(k+1)}\| = 0$. Hence if $\|G\| < 1$, then the sequence $\mathbf{x}^{(k+1)}$ converges to $\mathbf{r}$.

For the uniqueness part, use the norm for absolute value of $G$ for $g'$ in the uniqueness part of the Fixed-point Theorem. Hence the theorem. ∎

1. $\|G(\mathbf{x})\|_\infty = \max \left\{ \left| \dfrac{\partial g_1}{\partial x_1}(\mathbf{x}) \right| + \left| \dfrac{\partial g_1}{\partial x_2}(\mathbf{x}) \right|, \left| \dfrac{\partial g_2}{\partial x_1}(\mathbf{x}) \right| + \left| \dfrac{\partial g_2}{\partial x_2}(\mathbf{x}) \right| \right\}$ and therefore if $\|G(\mathbf{x})\|_\infty < 1$, then also the second part of Theorem 12.2.1 will be satisfied. Thus it is very convenient to check whether

$$\left| \frac{\partial g_1}{\partial x_1}(\mathbf{x}) \right| + \left| \frac{\partial g_1}{\partial x_2}(\mathbf{x}) \right| < 1 \quad \text{and} \quad \left| \frac{\partial g_2}{\partial x_1}(\mathbf{x}) \right| + \left| \frac{\partial g_2}{\partial x_2}(\mathbf{x}) \right| < 1$$

2. Theorem 12.2.1 can be generalized for a system of $N$ nonlinear equations in $N$ unknowns.

**EXAMPLE 12.2.2**

Analyze the three cases of Example 12.2.1.
*Case 1* Since $g_1(x_1, x_2) = -\cos x_2$, $g_2(x_1, x_2) = -\sin x_1$,

$$G(\mathbf{x}) = \begin{bmatrix} \dfrac{\partial g_1}{\partial x_1}(\mathbf{x}) & \dfrac{\partial g_1}{\partial x_2}(\mathbf{x}) \\ \dfrac{\partial g_2}{\partial x_1}(\mathbf{x}) & \dfrac{\partial g_2}{\partial x_2}(\mathbf{x}) \end{bmatrix} = \begin{bmatrix} 0 & \sin x_2 \\ -\cos x_1 & 0 \end{bmatrix}$$

Hence $\rho(G(\mathbf{x})) \leq 1$ for all $\mathbf{x} \in R^2$. For any $\mathbf{x}^{(0)} \in R^2$, $g(\mathbf{x}^{(0)}) \in D$ where $D = \{(x_1, x_2) \mid -1 \leq x_1 \leq 1, -1 \leq x_2 \leq 1\}$. Then $\mathbf{x}^{(1)} \in D$ and $g(\mathbf{x}^{(1)}) \in D$. Starting with any $\mathbf{x}^{(0)}$ in $R^2$, $g(\mathbf{x}) \in D$ and Equation (12.2.7) converges, though it can be verified that $\|G(\mathbf{x})\|_\infty \leq 1$.

*Case 2* Since $g_1(x_1, x_2) = x_2 - \dfrac{2}{3}\left[ (x_1 + \cos x_2) + \dfrac{1}{\sqrt{2}}(x_2 + \sin x_1) \right]$, $g_2(x_1, x_2) = x_2 - \dfrac{2}{3}\left[ -\dfrac{1}{\sqrt{2}}(x_1 + \cos x_2) + (x_2 + \sin x_1) \right]$,

$$G(\mathbf{x}) = \begin{bmatrix} \dfrac{\partial g_1}{\partial x_1}(\mathbf{x}) & \dfrac{\partial g_1}{\partial x_2}(\mathbf{x}) \\ \dfrac{\partial g_2}{\partial x_1}(\mathbf{x}) & \dfrac{\partial g_2}{\partial x_2}(\mathbf{x}) \end{bmatrix} = \frac{1}{3}\begin{bmatrix} 1 - \sqrt{2}\cos x_1 & -\sqrt{2}(1 - \sqrt{2}\sin x_2) \\ \sqrt{2}(1 - \sqrt{2}\cos x_1) & 1 - \sqrt{2}\sin x_2 \end{bmatrix}$$

So $\|G(\mathbf{x})\|_\infty < 1$ if $\mathbf{x} \in D$ where $D = \left\{ (x_1, x_2) \mid -\dfrac{\pi}{2} \leq x_1 \leq \dfrac{\pi}{2}, 0 \leq x_2 \leq \dfrac{\pi}{2} \right\}$. Further it can be verified that $g(\mathbf{x}) \in D$ whenever $\mathbf{x} \in D$. In fact, starting with any $\mathbf{x}^{(0)}$ in $R^2$, the sequence of iterates converges to the solution. Keep in mind that our theorem does not give necessary conditions.

*Case 3*   The domain of $g_1(x_1, x_2) = \arcsin(-x_2)$ and $g_2(x_1, x_2) = \arccos(-x_1)$ is given by $\left\{(x_1, x_2) \mid -\pi \leq x_1 \leq 0, -\dfrac{\pi}{2} \leq x_2 \leq \dfrac{\pi}{2}\right\}$. However we cannot find a set $D$ such that $\mathbf{g}(\mathbf{x}) \in D$ whenever $\mathbf{x} \in D$. Thus there is no sequence of iterates to converge to a fixed point in the domain of $g_1$ and $g_2$. Further

$$G(\mathbf{x}) = \begin{bmatrix} \dfrac{\partial g_1}{\partial x_1}(\mathbf{x}) & \dfrac{\partial g_1}{\partial x_2}(\mathbf{x}) \\[2mm] \dfrac{\partial g_2}{\partial x_1}(\mathbf{x}) & \dfrac{\partial g_2}{\partial x_2}(\mathbf{x}) \end{bmatrix} = \begin{bmatrix} 0 & -1/\sqrt{1 - x_2^2} \\[2mm] 1/\sqrt{1 - x_1^2} & 0 \end{bmatrix}$$

and $\|G(\mathbf{x})\|_\infty = \max\left\{1/\sqrt{1 - x_2^2},\ 1/\sqrt{1 - x_1^2}\right\} \not< 1$ for any $\mathbf{x}^{(0)}$ in $R^2$. Hence Equation (12.2.10) does not converge.    ■ ■ ■

---

## 12.3   THE NEWTON METHOD

As with the fixed-point method, for simplicity we will consider a system of two equations in two unknowns:

$$f_1(x_1, x_2) = 0 \quad \text{and} \quad f_2(x_1, x_2) = 0 \tag{12.3.1}$$

Let $r_1$ and $r_2$ be the roots of Equation (12.3.1), that is,

$$f_1(r_1, r_2) = 0 \quad \text{and} \quad f_2(r_1, r_2) = 0 \tag{12.3.2}$$

As with Newton's method for a single variable, we expand $f_1(r_1, r_2)$ and $f_2(r_1, r_2)$ by Taylor's Theorem A.9 about the point $(x_1^{(0)}, x_2^{(0)})$ which is close to the point $(r_1, r_2)$ and get

$$
\begin{aligned}
f_1(r_1, r_2) &= f_1(x_1^{(0)} + (r_1 - x_1^{(0)}), x_2^{(0)} + (r_2 - x_2^{(0)})) \\
&= f_1(x_1^{(0)}, x_2^{(0)}) + \left[(r_1 - x_1^{(0)})\frac{\partial f_1}{\partial x_1}(x_1^{(0)}, x_2^{(0)}) + (r_2 - x_2^{(0)})\frac{\partial f_1}{\partial x_2}(x_1^{(0)}, x_2^{(0)})\right] \\
&\quad + \frac{1}{2!}\left[(r_1 - x_1^{(0)})^2\frac{\partial^2 f_1}{\partial x_1^2}(\xi_1, \eta_1) + 2(r_1 - x_1^{(0)})(r_2 - x_2^{(0)})\frac{\partial^2 f_1}{\partial x_1 \partial x_2}(\xi_1, \eta_1)\right. \\
&\quad \left. + (r_2 - x_2^{(0)})^2\frac{\partial^2 f_1}{\partial x_2^2}(\xi_1, \eta_1)\right]
\end{aligned}
$$

$$
\begin{aligned}
f_2(r_1, r_2) &= f_2(x_1^{(0)} + (r_1 - x_1^{(0)}), x_2^{(0)} + (r_2 - x_2^{(0)})) \\
&= f_2(x_1^{(0)}, x_2^{(0)}) + \left[(r_1 - x_1^{(0)})\frac{\partial f_2}{\partial x_1}(x_1^{(0)}, x_2^{(0)}) + (r_2 - x_2^{(0)})\frac{\partial f_2}{\partial x_2}(x_1^{(0)}, x_2^{(0)})\right] \\
&\quad + \frac{1}{2!}\left[(r_1 - x_1^{(0)})^2\frac{\partial^2 f_2}{\partial x_1^2}(\xi_2, \eta_2) + 2(r_1 - x_1^{(0)})(r_2 - x_2^{(0)})\frac{\partial^2 f_2}{\partial x_1 \partial x_2}(\xi_2, \eta_2)\right. \\
&\quad \left. + (r_2 - x_2^{(0)})^2\frac{\partial^2 f_2}{\partial x_2^2}(\xi_2, \eta_2)\right]
\end{aligned}
\tag{12.3.3}
$$

where $(\xi_1, \eta_1)$ and $(\xi_2, \eta_2)$ are the points on the line joining the points $(x_1^{(0)}, x_2^{(0)})$ and $(r_1, r_2)$. Let us assume that $(x_1^{(0)}, x_2^{(0)})$ is close to $(r_1, r_2)$. Then the factors $(r_1 - x_1^{(0)})^2$,

$(r_1 - x_1^{(0)})(r_2 - x_2^{(0)})$, and $(r_2 - x_2^{(0)})^2$ in Equation (12.3.3) are small. Neglecting the second-order terms in Equation (12.3.3) and using Equation (12.3.2) in Equation (12.3.3), we get

$$0 \approx f_1(x_1^{(0)}, x_2^{(0)}) + (r_1 - x_1^{(0)}) \frac{\partial f_1}{\partial x_1}(x_1^{(0)}, x_2^{(0)}) + (r_2 - x_2^{(0)}) \frac{\partial f_1}{\partial x_2}(x_1^{(0)}, x_2^{(0)})$$

$$0 \approx f_2(x_1^{(0)}, x_2^{(0)}) + (r_1 - x_1^{(0)}) \frac{\partial f_2}{\partial x_1}(x_1^{(0)}, x_2^{(0)}) + (r_2 - x_2^{(0)}) \frac{\partial f_2}{\partial x_2}(x_1^{(0)}, x_2^{(0)}) \quad \textbf{(12.3.4)}$$

In matrix notation, Equation (12.3.4) becomes

$$\begin{bmatrix} f_1(x_1^{(0)}, x_2^{(0)}) \\ f_2(x_1^{(0)}, x_2^{(0)}) \end{bmatrix} + \begin{bmatrix} \frac{\partial f_1}{\partial x_1}(x_1^{(0)}, x_2^{(0)}) & \frac{\partial f_1}{\partial x_2}(x_1^{(0)}, x_2^{(0)}) \\ \frac{\partial f_2}{\partial x_1}(x_1^{(0)}, x_2^{(0)}) & \frac{\partial f_2}{\partial x_2}(x_1^{(0)}, x_2^{(0)}) \end{bmatrix} \begin{bmatrix} r_1 - x_1^{(0)} \\ r_2 - x_2^{(0)} \end{bmatrix} \approx \begin{bmatrix} 0 \\ 0 \end{bmatrix} \quad \textbf{(12.3.5)}$$

in other words,

$$\mathbf{f}(\mathbf{x}^{(0)}) + J(\mathbf{x}^{(0)})(\mathbf{r} - \mathbf{x}^{(0)}) \approx \mathbf{0} \quad \textbf{(12.3.6)}$$

where

$$\mathbf{r} = \begin{bmatrix} r_1 \\ r_2 \end{bmatrix} \qquad \mathbf{x}^{(0)} = \begin{bmatrix} x_1^{(0)} \\ x_2^{(0)} \end{bmatrix} \qquad \mathbf{f}(\mathbf{x}^{(0)}) = \begin{bmatrix} f_1(x_1^{(0)}, x_2^{(0)}) \\ f_2(x_1^{(0)}, x_2^{(0)}) \end{bmatrix}$$

$$J(\mathbf{x}^{(0)}) = \begin{bmatrix} \frac{\partial f_1}{\partial x_1}(x_1^{(0)}, x_2^{(0)}) & \frac{\partial f_1}{\partial x_2}(x_1^{(0)}, x_2^{(0)}) \\ \frac{\partial f_2}{\partial x_1}(x_1^{(0)}, x_2^{(0)}) & \frac{\partial f_2}{\partial x_2}(x_1^{(0)}, x_2^{(0)}) \end{bmatrix}$$

Solving Equation (12.3.6) for $\mathbf{r}$, we get

$$\mathbf{r} \approx \mathbf{x}^{(0)} - J^{-1}(\mathbf{x}^{(0)}) \mathbf{f}(\mathbf{x}^{(0)}) \quad \textbf{(12.3.7)}$$

provided $J^{-1}(\mathbf{x}^{(0)})$ exists. The solution $\mathbf{r}$ of Equation (12.3.7) will not be precisely the root $\mathbf{r}$ since Equation (12.3.7) is not the same as Equation (12.3.3). This $\mathbf{r}$ is merely an estimate and writing it as $\mathbf{x}^{(1)}$, Equation (12.3.7) becomes

$$\mathbf{x}^{(1)} = \mathbf{x}^{(0)} - J^{-1}(\mathbf{x}^{(0)}) \mathbf{f}(\mathbf{x}^{(0)})$$

Having this new estimate $\mathbf{x}^{(1)}$ of $\mathbf{r}$, we repeat this process. This suggests the iterative method

$$\mathbf{x}^{(k+1)} = \mathbf{x}^{(k)} - J^{-1}(\mathbf{x}^{(k)}) \mathbf{f}(\mathbf{x}^{(k)}) \quad \text{for } k = 1, 2, \ldots \quad \textbf{(12.3.8)}$$

We derived Equation (12.3.8) using a system of two equations in two unknowns. We can use exactly the same procedure for a system of $N$ equations in $N$ unknowns and verify that Equation (12.3.8) is also true for that system. The matrix $J(\mathbf{x}^{(k)})$ of partial derivatives at $\mathbf{x}^{(k)}$ is called the **Jacobian matrix** and we are assuming that the Jacobian matrices $J(\mathbf{x}^{(k)})$ are nonsingular. The procedure described by Equation (12.3.8) is called the **Newton method for several variables** and is similar to Newton's method for one variable

$$x_{k+1} = x_k - \frac{f(x_k)}{f'(x_k)} \quad \text{for } k = 1, 2, \ldots$$

There is a geometric interpretation of Newton's method. In $(x, y, z)$ space the equation of the tangent plane to a surface $z = f_1(x, y)$ at the point $(x^{(0)}, y^{(0)}, f_1(x^{(0)}, y^{(0)}))$ is given by

$$z = f_1(x^{(0)}, y^{(0)}) + (x - x^{(0)}) \frac{\partial f_1}{\partial x}(x^{(0)}, y^{(0)}) + (y - y^{(0)}) \frac{\partial f_1}{\partial y}(x^{(0)}, y^{(0)}) \quad \textbf{(12.3.9)}$$

Similarly

$$z = f_2(x^{(0)}, y^{(0)}) + (x - x^{(0)}) \frac{\partial f_2}{\partial x}(x^{(0)}, y^{(0)}) + (y - y^{(0)}) \frac{\partial f_2}{\partial y}(x^{(0)}, y^{(0)}) \quad \textbf{(12.3.10)}$$

represents the tangent plane to the surface $z = f_2(x, y)$ at the point $(x^{(0)}, y^{(0)}, f_2(x^{(0)}, y^{(0)}))$. The intersection point of the planes (12.3.9) and (12.3.10) with the plane $z = 0$ (i.e., $xy$-plane) is the same as the point determined from the solution of Equation (12.3.4). Denoting the coordinates of this point in the $xy$-plane by $(x^{(1)}, y^{(1)})$, the equation of the tangent plane to the surface $z = f_1(x, y)$ at the point $(x^{(1)}, y^{(1)}, f_1(x^{(1)}, y^{(1)}))$ is given by

$$z = f_1(x^{(1)}, y^{(1)}) + (x - x^{(1)}) \frac{\partial f_1}{\partial x}(x^{(1)}, y^{(1)}) + (y - y^{(1)}) \frac{\partial f_1}{\partial y}(x^{(1)}, y^{(1)})$$

Similarly, we have

$$z = f_2(x^{(1)}, y^{(1)}) + (x - x^{(1)}) \frac{\partial f_2}{\partial x}(x^{(1)}, y^{(1)}) + (y - y^{(1)}) \frac{\partial f_2}{\partial y}(x^{(1)}, y^{(1)})$$

The intersection point $(x^{(2)}, y^{(2)})$ of these planes with $z = 0$ is the same point as was determined by the solution of Equation (12.3.8). Thus in passing from one dimension to two dimensions, tangent lines are replaced by tangent planes. For higher dimensions, tangent hyperplanes are used.

It is usually computationally inefficient to compute inverses of matrices. Therefore, we carry out Equation (12.3.8) in the following manner for higher values of $N$:

1.  Define $\mathbf{z}^{(k+1)} = \mathbf{x}^{(k+1)} - \mathbf{x}^{(k)}$.
2.  Solve $J(\mathbf{x}^{(k)}) \, \mathbf{z}^{(k+1)} = -\mathbf{f}(\mathbf{x}^{(k)})$ by one of the methods discussed in Chapter 7 or 11 for $\mathbf{z}^{(k+1)}$ and then find $\mathbf{x}^{(k+1)}$ by $\mathbf{x}^{(k+1)} = \mathbf{z}^{(k+1)} + \mathbf{x}^{(k)}$.

**EXAMPLE 12.3.1**    Solve

$$x_1 + \cos x_2 = 0$$

$$x_2 + \sin x_1 = 0$$

using Newton's method with starting value $(x_1^{(0)}, x_2^{(0)}) = (2, 2)$.

We have

$$\mathbf{f}(\mathbf{x}) = \begin{bmatrix} f_1(x_1, x_2) \\ f_2(x_1, x_2) \end{bmatrix} = \begin{bmatrix} x_1 + \cos x_2 \\ x_2 + \sin x_1 \end{bmatrix}$$

$$J(\mathbf{x}) = \begin{bmatrix} \dfrac{\partial f_1}{\partial x_1} & \dfrac{\partial f_1}{\partial x_2} \\ \dfrac{\partial f_2}{\partial x_1} & \dfrac{\partial f_2}{\partial x_2} \end{bmatrix} = \begin{bmatrix} 1 & -\sin x_2 \\ \cos x_1 & 1 \end{bmatrix}$$

For $k = 0$, we have

$$J(\mathbf{x}^{(0)})\, \mathbf{z}^{(1)} = -\mathbf{f}(\mathbf{x}^{(0)})$$

Therefore

$$\begin{bmatrix} 1 & -\sin x_2^{(0)} \\ \cos x_1^{(0)} & 1 \end{bmatrix} \begin{bmatrix} z_1^{(1)} \\ z_2^{(1)} \end{bmatrix} = - \begin{bmatrix} x_1^{(0)} + \cos x_2^{(0)} \\ x_2^{(0)} + \sin x_1^{(0)} \end{bmatrix}$$

This gives

$$\begin{bmatrix} 1 & -0.90930 \\ -0.41615 & 1 \end{bmatrix} \begin{bmatrix} z_1^{(1)} \\ z_2^{(1)} \end{bmatrix} = - \begin{bmatrix} 1.58385 \\ 2.90930 \end{bmatrix}$$

Since $z_1^{(1)} = -6.80386$ and $z_2^{(1)} = -5.74070$ by using $LU$ decomposition,

$$x_1^{(1)} = z_1^{(1)} + x_1^{(0)} = -6.80386 + 2.0 = -4.80386$$

and

$$x_2^{(1)} = z_2^{(1)} + x_2^{(0)} = -5.74070 + 2.0 = -3.74070$$

The results of this iterative procedure are shown in Table 12.3.1.

**Table 12.3.1**

| $k$ | $x_1^{(k)}$ | $x_2^{(k)}$ | $f_1(x_1^{(k)}, x_2^{(k)})$ | $f_2(x_1^{(k)}, x_2^{(k)})$ |
|---|---|---|---|---|
| 0 | 2.00000 | 2.00000 | 1.58385 | 2.90930 |
| 1 | -4.80386 | -3.74070 | -5.62970 | -2.74488 |
| 2 | 2.02210 | -1.61932 | 1.97360 | -0.71944 |
| 3 | 0.14683 | -1.71776 | 0.00039 | -1.57146 |
| 4 | -72.43031 | 71.64989 | -73.25182 | 71.82269 |
| 5 | 1.24697 | 72.39606 | 0.25668 | 73.34409 |
| 6 | 11.64343 | -4.25609 | 11.20280 | -5.05348 |
| 7 | 7.31950 | 3.40676 | 6.35445 | 4.26729 |
| 8 | 1.27668 | 2.21765 | 0.67400 | 3.17471 |
| 9 | -1.32812 | -0.20194 | -0.34844 | -1.17264 |
| 10 | -1.20914 | 0.94211 | -0.62106 | 0.00680 |
| 11 | -0.73054 | 0.76597 | -0.00983 | 0.09870 |
| 12 | -0.76918 | 0.69605 | -0.00180 | 0.00051 |
| 13 | -0.76817 | 0.69482 | 0.00000 | 0.00000 |

The sequence was terminated once we had $\|\mathbf{x}^{(k+1)} - \mathbf{x}^{(k)}\|_\infty < 10^{-5}\|\mathbf{x}^{(k+1)}\|_\infty$.
Once we are in the neighborhood of a root, the Newton method converges very fast, as can be seen from our computations in Table 12.3.1. ■ ■ ■

**Algorithm**

**Subroutine** nonnew $(N, A, \mathbf{b}, \mathbf{x}, \mathbf{z}, \text{eps}, \text{maxit})$
Comment: $N$ is the number of equations and unknowns; $x_1, x_2, \ldots, x_N$ are the initial guesses of $\mathbf{x}$; eps is tolerance and maxit is the maximum number of iterations.
1 **Print** $N$, $\mathbf{x}$, eps, maxit
2 $k = 0$

    3 **Repeat** steps 4–9, **while** $k \leq$ maxit
      4 **Call** Jac($A$, **b**, **x**, $N$)
Comment: Subroutine Jac evaluates $J(\mathbf{x})$ and $-\mathbf{f}(\mathbf{x})$. We denote $a(1,1) = \partial f_1 / \partial x_1(\mathbf{x})$, $a(1,2) = \partial f_1 / \partial x_2(\mathbf{x}), \ldots, a(1,N) = \partial f_1 / \partial x_N(\mathbf{x})$, $a(2,1) = \partial f_2 / \partial x_1(\mathbf{x})$, $a(2,2) = \partial f_2 / \partial x_2(\mathbf{x})$, and so forth. Also, $b(1) = -f_1(\mathbf{x})$, $b(2) = -f_2(\mathbf{x}), \ldots, b(N) = -f_N(\mathbf{x})$. The output is the matrix $A$ and **b**.
      5 **Call** Solver($N$, $A$, **b**, **z**, eps)
Comment: **Subroutine** Solver solves $A\mathbf{z} = \mathbf{b}$ by using Gaussian elimination. We can use the $LU$ decomposition method or any iterative method. The output is **z**.
      6 $\mathbf{x} = \mathbf{z} + \mathbf{x}$
      7 Print **x**
      8 **If** $\|\mathbf{z}\| <$ eps $\|\mathbf{x}\|$, **then print** 'finished' and exit.
      9 $k = k + 1$
  10 **Print** 'Algorithm fails: no convergence' and exit. ◀

---

**EXERCISES**

1. Sketch the graphs of the curves given by $x_1^2 + x_2^2 - 1 = 0$ and $e^{x_1} - x_2 = 0$. Find the intersection points of these curves.

2. (a) Transform the nonlinear system

$$x_1^2 + x_2^2 - 1 = 0$$

$$e^{x_1} - x_2 \quad = 0$$

    into a fixed-point problem $x_1 = g_1(x_1, x_2)$ and $x_2 = g_2(x_1, x_2)$ in at least two different ways.
  (b) Starting with $x_1^{(0)} = x_2^{(0)} = -0.2$, compute $x_1^{(k+1)} = g_1(x_1^{(k)}, x_2^{(k)})$ and $x_2^{(k+1)} = g_2(x_1^{(k)}, x_2^{(k)})$ for $k = 0, 1, 2$, and 3 for any form of your choice in (a).
  (c) Starting with $x_1^{*(0)} = x_2^{*(0)} = -0.2$, compute $x_1^{*(k+1)} = g_1(x_1^{*(k)}, x_2^{*(k)})$ and $x_2^{*(k+1)} = g_2(x_1^{*(k+1)}, x_2^{*(k)})$ for $k = 0, 1, 2$, and 3 for the form you choose in (b).

3. Use Theorem 12.2.1 to find a function **g** and a set $S \subset R^2$ such that **g** has a fixed point in $S$ for the nonlinear system

$$x_1^2 + x_2^2 - 1 = 0$$

$$e^{x_1} - x_2 \quad = 0$$

4. Sketch the graphs of the curves given by $x_1^2 + x_2^2 - x_1 = 0$ and $x_1^2 - x_2^2 - x_2 = 0$. Find the intersection points of these curves exactly.

5. The nonlinear system

$$x_1^2 + x_2^2 - x_1 = 0$$

$$x_1^2 - x_2^2 - x_2 = 0$$

    can be transformed into the fixed-point problem

$$x_1 = g_1(x_1, x_2) = x_1^2 + x_2^2$$

$$x_2 = g_2(x_1, x_2) = x_1^2 - x_2^2$$

(a) Starting with $x_1^{(0)} = x_2^{(0)} = 0.2$, compute $x_1^{(k+1)} = (x_1^{(k)})^2 + (x_2^{(k)})^2$ and $x_2^{(k+1)} = (x_1^{(k)})^2 - (x_2^{(k)})^2$ for $k = 0, 1, 2$, and 3.

(b) Starting with $x_1^{*(0)} = x_2^{*(0)} = 0.2$, compute $x_1^{*(k+1)} = (x_1^{*(k)})^2 + (x_2^{*(k)})^2$ and $x_2^{*(k+1)} = (x_1^{*(k+1)})^2 - (x_2^{*(k)})^2$ for $k = 0, 1, 2$, and 3.

(c) Starting with $x_1^{(0)} = x_2^{(0)} = 1$, compute $x_1^{(k+1)} = (x_1^{(k)})^2 + (x_2^{(k)})^2$ and $x_2^{(k+1)} = (x_1^{(k)})^2 - (x_2^{(k)})^2$ for $k = 0, 1, 2$, and 3.

6.  Use Theorem 12.2.1 to find a function **g** and a set $S \subset R^2$ such that **g** has a fixed point in $S$ for the nonlinear system

$$x_1^2 + x_2^2 - 4 = 0$$

$$2x_1 - x_2^2 \quad = 0$$

7.  Starting with $x_1^{(0)} = x_2^{(0)} = 1$, carry out the first three iterations of Newton's method for the equations

$$f_1(x_1, x_2) = x_1^2 + x_2^2 - 4 = 0$$

$$f_2(x_1, x_2) = 2x_1 - x_2^2 = 0$$

8.  Starting with $x_1^{(0)} = x_2^{(0)} = 0.2$, carry out the first three iterations of Newton's method for the equations

$$f_1(x_1, x_2) = x_1^2 + x_2^2 - x_1 = 0$$

$$f_2(x_1, x_2) = x_1^2 - x_2^2 - x_2 = 0$$

9.  Show that Newton's method converges in exactly one iteration when applied to a system of linear equations

$$a_{11}x_1 + a_{12}x_2 + \cdots + a_{1N}x_N - b_1 = 0$$
$$a_{21}x_1 + a_{22}x_2 + \cdots + a_{2N}x_N - b_2 = 0$$
$$\vdots$$
$$a_{N1}x_1 + a_{N2}x_2 + \cdots + a_{NN}x_N - b_N = 0$$

10. If $r_1, r_2$, and $r_3$ are the roots of a third-degree polynomial equation $a_3x^3 + a_2x^2 + a_1x + a_0 = 0$, then (Exercise 4, Section 3.1)

$$r_1 + r_2 + r_3 = -\frac{a_2}{a_3} \qquad r_1r_2 + r_2r_3 + r_3r_1 = \frac{a_1}{a_3} \qquad r_1r_2r_3 = -\frac{a_0}{a_3}$$

We can solve these three nonlinear equations and find the roots of a polynomial equation. Consider $x^3 + 2x^2 - x - 2 = 0$. We can find the roots by solving the nonlinear equations

$$x_1 + x_2 + x_3 = -2 \qquad x_1x_2 + x_2x_3 + x_3x_1 = -1 \qquad x_1x_2x_3 = 2$$

Solve these equations using the fixed-point method accurate to within $10^{-2}$.

11. Find the solution of the following nonlinear system using Newton's method accurate to within $10^{-2}$:

$$x_1 + x_2 + x_3 = -2$$

$$x_1x_2 + x_2x_3 + x_3x_1 = -1$$

$$x_1x_2x_3 = 2$$

Use $x_1^{(0)} = -1, x_2^{(0)} = 2, x_3^{(0)} = -2$.

12. We can modify Newton's method

$$\mathbf{x}^{(k+1)} = \mathbf{x}^{(k)} - J^{-1}(\mathbf{x}^{(k)})\, \mathbf{f}(\mathbf{x}^{(k)})$$

by replacing $\mathbf{x}^{(k)}$ by $\mathbf{x}^{(0)}$. Then we have

$$\mathbf{x}^{(k+1)} = \mathbf{x}^{(k)} - J^{-1}(\mathbf{x}^{(0)})\, \mathbf{f}(\mathbf{x}^{(k)})$$

This reduces some work but slows its convergence. Apply this algorithm to Exercise 7.

13. In order to find the extrema of $f(x, y)$, we need to solve $\partial f/\partial x = 0$ and $\partial f/\partial y = 0$. These equations are usually nonlinear and we can use Newton's method to solve them. Use Newton's method with $x^{(0)} = y^{(0)} = 0$ to find the extrema of $f$ if $f(x, y) = x^2 - 4xy + y^3 + 4y$.

## COMPUTER EXERCISES

In order to solve a system of nonlinear equations by Newton's method, we need to solve

$$J(\mathbf{x}^{(k)})\mathbf{z}^{(k+1)} = -\mathbf{f}(\mathbf{x}^{(k)})$$

where $J(\mathbf{x}^{(k)})$ depends on the given equations.

C.1. Write a subroutine to solve a system of linear equations given by $J(\mathbf{x}^{(k)})\mathbf{z}^{(k+1)} = -\mathbf{f}(\mathbf{x}^{(k)})$ by $LU$ decomposition or by the Gaussian elimination method.

C.2. Write a computer program to solve the system

$$x_1 + x_2 + x_3 = -2$$
$$x_1 x_2 + x_2 x_3 + x_3 x_1 = -1$$
$$x_1 x_2 x_3 = 2$$

accurate to within $10^{-5}$ using Exercise C.1. Also, use a subroutine to compute $J(\mathbf{x}^{(k)})$ and $-\mathbf{f}(\mathbf{x}^{(k)})$.

C.3. Write a computer program to solve the system

$$x_1 + x_2 + x_3 = -2$$
$$x_1 x_2 + x_2 x_3 + x_3 x_1 = -1$$
$$x_1 x_2 x_3 = 2$$

accurate to within $10^{-5}$ using the fixed-point iteration method.

C.4. The growth rate is proportional to the size of a population and the death rate is proportional to the number of two-party interactions. This leads to a logistic model

$$\frac{dp}{dt} = ap - bp^2 \qquad p(0) = p_0$$

where $a$ and $b$ are parameters. The solution of this differential equation is given by

$$p(t) = \frac{ap_0}{bp_0 + (a - bp_0)e^{-at}} \qquad \textbf{(C.4.1)}$$

To test this model, consider the United States population. The United States population as given by the U.S. Census Bureau is given in the following table.

| Year | 1930 | 1940 | 1950 | 1960 | 1970 | 1980 | 1990 |
|------|------|------|------|------|------|------|------|
| **Population (in millions)** | 123.20 | 132.16 | 151.33 | 179.32 | 203.30 | 226.54 | 248.71 |

In Equation (C.4.1), $t = 0$ corresponds to 1930, therefore $p(0) = p_0 = 123.20$. To determine $a$ and $b$, we use $p(20) = 151.33$ and $p(60) = 248.71$. Thus

$$151.33 = \frac{123.20a}{123.20b + (a - 123.20\,b)e^{-20a}} \qquad \textbf{(C.4.2)}$$

$$248.71 = \frac{123.20a}{123.20b + (a - 123.20b)e^{-60a}} \qquad \textbf{(C.4.3)}$$

Equations (C.4.2) and (C.4.3) are two nonlinear equations in $a$ and $b$.

Solve the system given by Equations (C.4.2) and (C.4.3) for $a$ and $b$. Using $a$ and $b$ in Equation (C.4.1), compute the population for 1940, 1960, and 1970. Compare the predicted values with the values in the table.

**C.5.** Solve Equation (1) given at the beginning of this chapter by using any method of this chapter.

**C.6.** The finite difference method reduces the boundary-value problem $d^2y/dx^2 = (3/2)\, y^2$, $y(0) = 4$, $y(1) = 1$ to the following system of nonlinear equations:

$$-2x_1 - \frac{3}{32}x_1^2 + x_2 + 4 = 0$$

$$x_1 - 2x_2 - \frac{3}{32}x_2^2 + x_3 = 0$$

$$x_2 - 2x_3 - \frac{3}{32}x_3^2 + 1 = 0$$

Solve the system by using any method of this chapter.

---

## 12.4   THE STEEPEST DESCENT METHOD

The Newton and fixed-point methods converge if an accurate initial approximation is provided. The steepest descent method considered in this section provides an initial approximation of a solution.

We observe that the roots of a system of equations

$$f_1(x_1, x_2) = 0 \quad \text{and} \quad f_2(x_1, x_2) = 0 \qquad \textbf{(12.4.1)}$$

reduce

$$Q(x_1, x_2) = \frac{1}{2} [f_1(x_1, x_2)]^2 + \frac{1}{2} [f_2(x_1, x_2)]^2 \qquad \textbf{(12.4.2)}$$

to zero; otherwise $Q(x_1, x_2) > 0$. Also, the numbers $x_1, x_2$ for which $Q(x_1, x_2)$ is zero are the roots of Equation (12.4.2). Thus, we have converted the problem of finding the roots of Equation (12.4.1) to that of minimizing $Q(x_1, x_2)$ given by Equation (12.4.2) in some region $R$.

The local minimization of a function is one of the oldest problems. The method of steepest descent was proposed by Cauchy in 1845. The graph of a function is similar to a landscape of valleys and hills. The bottom of the valleys represents the local minima of a function. The idea is to reach the lowest point of the valley. In order to reach the lowest point, we follow the example of water descending from a hill. Water moves in the direction of steepest descent.

Let the graph of $Q$ be bowl-shaped with its lowest point at $(r_1, r_2, Q(r_1, r_2))$. At that lowest point $\partial Q/\partial x_1(r_1, r_2) = 0$ and $\partial Q/\partial x_2(r_1, r_2) = 0$ if they exist. The gradient of $Q(x_1, x_2)$ is defined to be

$$\nabla Q(x_1, x_2) = \frac{\partial Q}{\partial x_1}(x_1, x_2)\mathbf{i_1} + \frac{\partial Q}{\partial x_2}(x_1, x_2)\mathbf{i_2} \qquad \textbf{(12.4.3)}$$

where $\mathbf{i_1}$ and $\mathbf{i_2}$ are unit vectors in the directions of the $x_1$ and $x_2$ axes respectively. It is a standard result in multidimensional calculus that $Q(x_1, x_2)$ increases most rapidly in the direction of $\nabla Q(x_1, x_2)$ and $Q(x_1, x_2)$ decreases most rapidly in the direction of $-\nabla Q(x_1, x_2)$.

We start with $(x_1^{(0)}, x_2^{(0)}, Q(x_1^{(0)}, x_2^{(0)}))$. We want to reach $(r_1, r_2, Q(r_1, r_2))$. We move on the surface to $(x_1^{(1)}, x_2^{(1)}, Q(x_1^{(1)}, x_2^{(1)}))$ such that $Q(x_1^{(1)}, x_2^{(1)}) < Q(x_1^{(0)}, x_2^{(0)})$ in the direction of $-\nabla Q(x_1^{(0)}, x_2^{(0)})$ where the surface is steepest and $Q(x_1, x_2)$ decreases most rapidly. The equation of a straight line $L$ through $(x_1^{(0)}, x_2^{(0)}, Q(x_1^{(0)}, x_2^{(0)}))$ and in the direction of $-\nabla Q(x_1^{(0)}, x_2^{(0)})$ is given by

$$\frac{x_1 - x_1^{(0)}}{-\dfrac{\partial Q}{\partial x_1}(x_1^{(0)}, x_2^{(0)})} = \frac{x_2 - x_2^{(0)}}{-\dfrac{\partial Q}{\partial x_2}(x_1^{(0)}, x_2^{(0)})} = \frac{x_3 - Q(x_1^{(0)}, x_2^{(0)})}{1} = t \qquad \textbf{(12.4.4)}$$

If we move a small distance along the straight line $L$, then

$$Q(x_1^{(1)}, x_2^{(1)}) = Q(x_1^{(0)} - t\frac{\partial Q}{\partial x_1}(x_1^{(0)}, x_2^{(0)}), x_2^{(0)} - t\frac{\partial Q}{\partial x_2}(x_1^{(0)}, x_2^{(0)})) < Q(x_1^{(0)}, x_2^{(0)})$$

Denote

$$x_1^{(1)} = x_1^{(0)} - t\frac{\partial Q}{\partial x_1}(x_1^{(0)}, x_2^{(0)}) \quad \text{and} \quad x_2^{(1)} = x_2^{(0)} - t\frac{\partial Q}{\partial x_2}(x_1^{(0)}, x_2^{(0)})$$

where $t \geq 0$ that controls how far we move. It remains to be determined. Continuing, we get

$$x_1^{(k+1)} = x_1^{(k)} - t\frac{\partial Q}{\partial x_1}(x_1^{(k)}, x_2^{(k)}) \quad \text{and} \quad x_2^{(k+1)} = x_2^{(k)} - t\frac{\partial Q}{\partial x_2}(x_1^{(k)}, x_2^{(k)}) \qquad \textbf{(12.4.5)}$$

for $k = 0, 1, \ldots$. In order to determine $t$, consider

$$\phi(t) = Q(x_1^{(k+1)}, x_2^{(k+1)})$$

$$= Q(x_1^{(k)} - t\frac{\partial Q}{\partial x_1}(x_1^{(k)}, x_2^{(k)}), x_2^{(k)} - t\frac{\partial Q}{\partial x_2}(x_1^{(k)}, x_2^{(k)}))$$

$$= \frac{1}{2}\left\{\left[f_1(x_1^{(k)} - t\frac{\partial Q}{\partial x_1}(x_1^{(k)} x_2^{(k)}), x_2^{(k)} - t\frac{\partial Q}{\partial x_2}(x_1^{(k)}, x_2^{(k)}))\right]^2\right.$$

$$\left. + \left[f_2(x_1^{(k)} - t\frac{\partial Q}{\partial x_1}(x_1^{(k)}, x_2^{(k)}), x_2^{(k)} - t\frac{\partial Q}{\partial x_2}(x_1^{(k)}, x_2^{(k)}))\right]^2\right\} \qquad \textbf{(12.4.6)}$$

We select $t$ such that $\phi(t)$ is minimum. Thus

$$\phi'(t) = 0 \qquad \textbf{(12.4.7)}$$

The least positive real root of Equation (12.4.7) is the required value of $t$. Equation (12.4.7) can be solved numerically. However, we give a formula to approximate $t$. Assume that $t$ is small, so the square and higher powers of $t$ can be neglected in Equation (12.4.6). Then using Taylor's Theorem A.9, Equation (12.4.6) reduces to

$$\phi(t) = \frac{1}{2}\sum_{i=1}^{2}\left[f_i(x_1^{(k)}, x_2^{(k)}) - t\frac{\partial Q}{\partial x_1}(x_1^{(k)}, x_2^{(k)})\frac{\partial f_i}{\partial x_1}(x_1^{(k)}, x_2^{(k)}) - t\frac{\partial Q}{\partial x_2}(x_1^{(k)}, x_2^{(k)})\frac{\partial f_i}{\partial x_2}(x_1^{(k)}, x_2^{(k)})\right]^2$$
$$\textbf{(12.4.8)}$$

Differentiating Equation (12.4.8) with respect to $t$ on both sides yields

$$\phi'(t) = \sum_{i=1}^{2}\left[f_i(x_1^{(k)}, x_2^{(k)}) - t\frac{\partial Q}{\partial x_1}(x_1^{(k)}, x_2^{(k)})\frac{\partial f_i}{\partial x_1}(x_1^{(k)}, x_2^{(k)}) - t\frac{\partial Q}{\partial x_2}(x_1^{(k)}, x_2^{(k)})\frac{\partial f_i}{\partial x_2}(x_1^{(k)}, x_2^{(k)})\right]$$

$$\left\{-\frac{\partial Q}{\partial x_1}(x_1^{(k)}, x_2^{(k)})\frac{\partial f_i}{\partial x_1}(x_1^{(k)}, x_2^{(k)}) - \frac{\partial Q}{\partial x_2}(x_1^{(k)}, x_2^{(k)})\frac{\partial f_i}{\partial x_2}(x_1^{(k)}, x_2^{(k)})\right\} = 0 \qquad \textbf{(12.4.9)}$$

Solving Equation (12.4.9) for $t$ gives

$$t = \frac{\sum_{i=1}^{2}f_i(x_1^{(k)}, x_2^{(k)})\left[\frac{\partial Q}{\partial x_1}(x_1^{(k)}, x_2^{(k)})\frac{\partial f_i}{\partial x_1}(x_1^{(k)}, x_2^{(k)}) + \frac{\partial Q}{\partial x_2}(x_1^{(k)}, x_2^{(k)})\frac{\partial f_i}{\partial x_2}(x_1^{(k)}, x_2^{(k)})\right]}{\sum_{i=1}^{2}\left[\left(\frac{\partial Q}{\partial x_1}\frac{\partial f_i}{\partial x_1}\right) + \left(\frac{\partial Q}{\partial x_2}\frac{\partial f_i}{\partial x_2}\right)\right]^2} \qquad \textbf{(12.4.10)}$$

This method can be generalized for a system of $N$ nonlinear equations in $N$ unknowns.

**EXAMPLE 12.4.1**

Solve $x_1 + \cos x_2 = 0$ and $x_2 + \sin x_1 = 0$ using the steepest descent method starting with $(x_1^{(0)}, x_2^{(0)}) = (2, 2)$.

We have $f_1(x_1, x_2) = x_1 + \cos x_2$ and $f_2(x_1, x_2) = x_2 + \sin x_1$, therefore

$$Q(x_1, x_2) = \frac{1}{2}(x_1 + \cos x_2)^2 + \frac{1}{2}(x_2 + \sin x_1)^2$$

Therefore

$$\frac{\partial f_1}{\partial x_1} = 1 \qquad \frac{\partial f_1}{\partial x_2} = -\sin x_2 \qquad \frac{\partial f_2}{\partial x_1} = \cos x_1 \qquad \frac{\partial f_2}{\partial x_2} = 1$$

$$\frac{\partial Q}{\partial x_1} = (x_1 + \cos x_2) + \cos x_1 \, (x_2 + \sin x_1)$$

$$\frac{\partial Q}{\partial x_2} = -\sin x_2 \, (x_1 + \cos x_2) + (x_2 + \sin x_1)$$

For $(x_1^{(0)}, x_2^{(0)}) = (2, 2)$, we have $f_1(2, 2) = 2 + \cos 2 = 1.58385$, $f_2(2, 2) = 2 + \sin 2 = 2.90930$, $\partial f_1/\partial x_1(2, 2) = 1$, $\partial f_1/\partial x_2 = -\sin 2 = -0.90930$, $\partial f_2/\partial x_1 = \cos 2 = -0.41615$, $\partial f_2/\partial x_2(2, 2) = 1$, $\partial Q/\partial x_1(2, 2) = (2 + \cos 2) + \cos 2 \, (2 + \sin 2) = (2 - 0.41615) + (-0.41615)(2 + 0.90930) = 0.37316$, and $\partial Q/\partial x_2(2, 2) = -\sin 2 \, (2 + \cos 2) + (2 + \sin 2) = 1.46910$.

From Equation (12.4.10),

$$t = \frac{2.29751}{2.65287} = 0.86605$$

For $k = 0$ and $t = 0.86605$, Equation (12.4.5) reduces to

$$x_1^{(1)} = x_1^{(0)} - t[(x_1^{(0)} + \cos x_2^{(0)}) + (\cos x_1^{(0)})(x_2^{(0)} + \sin x_1^{(0)})]$$

$$= 2 - 0.86605 \, [(2 + \cos 2) + (\cos 2)(2 + \sin 2)] = 1.67683$$

$$x_2^{(1)} = x_2^{(0)} - t[-(\sin x_2^{(0)})(x_1^{(0)} + \cos x_2^{(0)}) + (x_2^{(0)} + \sin x_1^{(0)})]$$

$$= 2 - 0.86605 \, [-(\sin 2)(2 + \cos 2) + (2 + \sin 2)] = 0.72770$$

The computed results are given in Table 12.4.1

**Table 12.4.1**

| $k$ | $x_1^{(k)}$ | $x_2^{(k)}$ | $Q(x_1^{(k)}, x_2^{(k)})$ | $t$ |
|---|---|---|---|---|
| 0 | 2 | 2 | 5.48630 | 0.86605 |
| 1 | 1.67683 | 0.72770 | 4.41955 | 1.06758 |
| 2 | −0.71591 | 0.61020 | 0.00643 | 0.81024 |
| 3 | −0.77168 | 0.69567 | 0.00001 | 0.67353 |
| 4 | −0.76814 | 0.69504 | 0.00000 | 0.69356 |
| 5 | −0.76818 | 0.69482 | 0.00000 | 0.67322 |
| 6 | −0.76817 | 0.69482 | 0.00000 | |

■ ■ ■

Note that $t$ is not as small as we assumed in the derivation of Equation (12.4.10). Alternatively, we can keep the value of $t$ the same unless $Q(x_1^{(k)}, x_2^{(k)})$ increases from one iteration to the next. If $Q(x_1^{(k)}, x_2^{(k)})$ increases, then decrease the value of $t$. This procedure reduces the computational work at each step but may require more iterations.

**Algorithm**

**Subroutine** Stdes($N$, **x**, **b**, **z**, eps, maxit)
Comment: $N$ is the number of equations and unknowns; $x_1, x_2, \ldots, x_N$ are the initial guesses of **x**; eps is the tolerance, and maxit is the maximum number of iterations.
  1 **Print** $N$, **x**, eps, maxit
  2 $k = 0$
  3 **Repeat** steps 4–9, **while** $k \leq$ maxit
      4 Call Para ($t$, **b**, $Q$)
Comment: Subroutine Para computes the parameter $t$ using Equation (12.4.10). We denote $c(1) = f_1(\mathbf{x})$, $c(2) = f_2(\mathbf{x})$, $\ldots$, $c(N) = f_N(\mathbf{x})$, $a(1,1) = \partial f_1/\partial x_1(\mathbf{x})$, $a(1,2) = \partial f_1/\partial x_2(\mathbf{x})$, $\ldots$, $a(1,N) = \partial f_1/\partial x_N(\mathbf{x})$, $a(2,1) = \partial f_2/\partial x_1(\mathbf{x})$ and so forth, $b(1) = \partial Q/\partial x_1(\mathbf{x})$, $b(2) = \partial Q/\partial x_2(\mathbf{x})$, $\ldots$, $b(N) = \partial Q/\partial x_N(\mathbf{x})$. Compute $Q$ and $t$.
      5 $\mathbf{z} = \mathbf{x} - t\,\mathbf{b}$
      6 **Print z**
      7 **If** $\|\mathbf{z} - \mathbf{x}\| < \text{eps}\|\mathbf{z}\|$, **then print** 'finished' and exit.
      8 $\mathbf{x} = \mathbf{z}$
      9 $k = k + 1$
 10 **Print** 'Algorithm fails: no convergence' and exit.

---

## E X E R C I S E S

1.  Starting with $(x_1^{(0)}, x_2^{(0)}) = (0, 0)$, carry out the first two iterations of the steepest descent method for the system of equations

$$x_1 + x_2 - 2 = 0 \quad \text{and} \quad x_1 - x_2 = 0$$

2.  Starting with $(x_1^{(0)}, x_2^{(0)}) = (0, 0)$, carry out the first two iterations of the steepest descent method for the system of equations

$$x_1^2 + x_2^2 = 1 \quad \text{and} \quad x_1 x_2 = 1$$

3.  Starting with $(x_1^{(0)}, x_2^{(0)}, x_3^{(0)}) = (0, 0, 0)$, carry out the first two iterations of the steepest descent method for the system of equations

$$x_1 + x_2 + x_3 = 3 \quad x_1 - x_2 + x_3 = 1 \quad x_1 + x_2 - x_3 = 1$$

4.  Starting with $(x_1^{(0)}, x_2^{(0)}, x_3^{(0)}) = (0, 0, 0)$, carry out the first two iterations of the steepest descent method for the system of equations

$$\frac{x_1^2}{4} + \frac{x_2^2}{4} + \frac{x_3^2}{4} - 1 = 0 \qquad x_1^2 + x_2^2 - x_3^2 - 1 = 0 \qquad (x_1 - 1)^2 + x_2^2 - 1 = 0$$

5.  Starting with $(x_1^{(0)}, x_2^{(0)}) = (0, 0)$, carry out the first three iterations of the steepest descent method to find the minimum value of the function

$$F(x_1, x_2) = x_1^2 - 4x_1 x_2 + x_2^3 + 4x_2$$

**COMPUTER EXERCISES**

**C.1.**   Write a subroutine to compute $t$ using Equation (12.4.10) for Equation (12.4.5).

**C.2.**   Write a computer program to solve the system

$$x_1 - \sin(x_1 + x_2) = 0 \qquad x_2 - \cos(x_1 - x_2) = 0$$

accurate to within $10^{-5}$ using Exercise C.1.

**C.3.**   Use the steepest descent method to find the minimum of

$$f(x_1, x_2) = 25(x_1^2 - x_2)^8 + (1 - x_1)^8$$

Use Exercise C.1.

**C.4.**   Write a computer program to solve the system

$$x_1 + x_2 + x_3 = -2$$

$$x_1 x_2 + x_2 x_3 + x_3 x_1 = -1$$

$$x_1 x_2 x_3 = 2$$

accurate to within $10^{-5}$ using the steepest descent method.

## SUGGESTED READINGS

J. Ortega and W. Rheinboldt, *Iterative Solution of Nonlinear Equations in Several Variables*, Academic Press, New York, 1970.

A. L. Peressini, F. E. Sullivan, and J. J. Uhl, Jr., *The Mathematics of Nonlinear Programming*, Springer–Verlag, New York, 1988.

# 13 EIGENVALUES AND EIGENVECTORS

The Lorenz equations (1963) given by

$$\frac{dx}{dt} = -Px + Py$$

$$\frac{dy}{dt} = -y + rx - xz$$

$$\frac{dz}{dt} = -bz + xy$$

where $P$, $r$, and $b$ are constants, played a significant role in the development of ideas about chaos. The linear stability analysis of the Lorenz equations leads to the eigenvalues of the matrix

$$\begin{bmatrix} -P & P & 0 \\ r - z_s & -1 & -x_s \\ y_s & x_s & -1 \end{bmatrix}$$

where $x_s$, $y_s$, and $z_s$ are the steady solutions of the Lorenz equations. In this chapter, we develop methods to approximate the eigenvalues and the corresponding eigenvectors of a given matrix.

## 13.1  INTRODUCTION

The problem of calculating the eigenvalues and eigenvectors of a given square matrix occurs in many fields ranging from economics to physics. The overwhelming majority of matrices in applications have real elements and often are symmetric; therefore, we develop methods for real symmetric matrices in this chapter.

Methods for real nonsymmetric matrices are also discussed. The general problem of finding all eigenvalues and eigenvectors of an arbitrary nonsymmetric matrix is quite difficult. For the general situation, the reader is referred to Wilkinson (1965), Stewart (1973), and Golub and Van Loan (1989).

Usually the eigenvalues of a given matrix are computed in two stages. First, by using a similarity transformation, a given matrix is reduced to a simpler form (e.g., tridiagonal or Hessenberg). Then the eigenvalues of this simpler form are calculated along with eigenvectors if they are desired. The main exception to this procedure is the power method, which calculates a single dominant eigenvalue and an associated eigenvector of a given matrix.

## Preliminaries

Let $A$ be a real $N \times N$ matrix. The problem is to determine all values of $\lambda$ for which the system of $N$ linear equations in $N$ unknowns

$$A\mathbf{x} = \lambda\mathbf{x} \tag{13.1.1}$$

has a nontrivial solution $\mathbf{x}$. Obviously $\mathbf{x} = \mathbf{0}$ is a solution of Equation (13.1.1) and so we ignore it by requiring $\mathbf{x} \neq \mathbf{0}$. We do allow $\lambda = 0$ and in that case $A$ is singular (why?). Rewriting $\lambda\mathbf{x}$ as $\lambda I\mathbf{x}$ where $I$ is an identity matrix, and moving this term to the left side, Equation (13.1.1) becomes

$$(A - \lambda I)\mathbf{x} = \mathbf{0} \tag{13.1.2}$$

Since $\mathbf{x} \neq \mathbf{0}$, $\det(A - \lambda I) = 0$ in Equation (13.1.2) and in return,

$$p(\lambda) = \det(A - \lambda I) = 0 \tag{13.1.3}$$

becomes an equation to determine $\lambda$. By a cofactor expansion of $\det(A - \lambda I)$ it can be seen that

$$p(\lambda) = (-1)^N \lambda^N + c_{N-1} \lambda^{N-1} + c_{N-2} \lambda^{N-2} + \ldots + c_1 \lambda + c_0 \tag{13.1.4}$$

where $c_0, c_1, \ldots, c_{N-1}$ are to be determined from the elements of $A$. The polynomial of Equation (13.1.4) is called the **characteristic polynomial** of $A$. The $N$ roots of Equation (13.1.3) are called the **characteristics roots** or **eigenvalues** of $A$. Corresponding to each eigenvalue $\lambda$, a nontrivial $\mathbf{x}$ satisfying Equation (13.1.2) is called an **eigenvector**. Such eigenvectors are arbitrary up to a constant multiplier.

For a given $N \times N$ matrix $A$, theoretically the eigenvalues of $A$ can be computed by finding the zeros of $p(\lambda)$ given by Equation (13.1.4). Then we can determine the associated eigenvectors by solving the associated linear system (13.1.2) or (13.1.1). In practice, it is difficult to obtain $p(\lambda)$ and it is equally difficult to find the roots of an $N$th degree polynomial equation except for small values of $N$, due in part to the problem of ill-conditioning of higher-degree polynomials.

Since $A$ has real elements, Equation (13.1.4) is a polynomial with real coefficients. Therefore, if the complex number $\lambda$ is a root of Equation (13.1.4), then the complex conjugate $\overline{\lambda}$ is also a root of Equation (13.1.4). Let $\lambda$ be complex. Since $A$ is real in $A\mathbf{x} = \lambda\mathbf{x}$, then $\mathbf{x}$ has to be complex. Let $\lambda$ be real and its associated eigenvector of $A$ be $\mathbf{x} = \mathbf{u} + i\mathbf{v}$. Then $A\mathbf{x} = A(\mathbf{u} + i\mathbf{v}) = \lambda(\mathbf{u} + i\mathbf{v})$. Thus comparing real and imaginary parts, $A\mathbf{u} = \lambda\mathbf{u}$ and $A\mathbf{v} = \lambda\mathbf{v}$. We have real associated eigenvectors $\mathbf{u}$ and $\mathbf{v}$ of $A$ corresponding to a real eigenvalue $\lambda$. The results are combined in the following theorem.

**Theorem 13.1.1**    Let $A$ be a real $N \times N$ matrix. Then

1.  If $\lambda$ is a complex eigenvalue of $A$, then $\overline{\lambda}$ is also an eigenvlue of $A$.
2.  If $\lambda$ is a complex eigenvalue of $A$, then its associated eigenvector $\mathbf{x}$ is also complex.
3.  If $\lambda$ is a real eigenvalue of $A$, then there exist real associated eigenvectors. ∎

Two $N \times N$ matrices $A$ and $B$ are called **similar** if there is a nonsingular matrix $S$ such that $B = S^{-1} AS$. Let us find the relationship between the eigenvalues and eigenvectors of similar matrices. Let $\lambda$ be an eigenvalue of $B$. Then

$$\begin{aligned} \det(B - \lambda I) &= \det(S^{-1} AS - \lambda S^{-1} IS) \\ &= \det S^{-1} (A - \lambda I)S \\ &= \det S^{-1} \det(A - \lambda I) \det S \\ &= \det(A - \lambda I) \end{aligned}$$

Thus $\lambda$ is also an eigenvalue of $A$. In other words, similar matrices have the same eigenvalues. This also shows their multiplicities are the same.

Let $\mathbf{x}$ be an eigenvector of $B$ corresponding to $\lambda$. Then

$$B\mathbf{x} = S^{-1} AS\mathbf{x} = \lambda \mathbf{x} \tag{13.1.5}$$

Multiplying Equation (13.1.5) by $S$ on the left, we get

$$AS\mathbf{x} = \lambda S\mathbf{x}$$

Therefore $S\mathbf{x}$ is an eigenvector of $A$ corresponding to $\lambda$. This leads to the following theorem.

**Theorem 13.1.2**    Let $A$ and $B$ be two $N \times N$ similar matrices where $B = S^{-1} AS$. Then $A$ and $B$ have the same eigenvalues. Further, let $\mathbf{x}$ be an eigenvector of $B$ corresponding to an eigenvalue $\lambda$. Then $S\mathbf{x}$ is an eigenvector of $A$ corresponding to an eigenvalue $\lambda$ of $A$.

∎

This theorem suggests a method to find the eigenvalues of a given matrix $A$. Apply a sequence of similarity transformations to $A$ that transforms $A$ into a special matrix whose eigenvalues can be computed easily. We can attempt to reduce $A$ to a triangular or diagonal matrix by using similarity transformations, then find the eigenvalues by reading the values on the diagonal of the reduced matrix.

It should be noted that the elementary row operations may change the eigenvalues. Also the eigenvalues and eigenvectors are not preserved when the matrix is premultiplied or postmultiplied by another matrix.

Now we consider the relationship between eigenvectors.

**Theorem 13.1.3**    (a)    Let $A$ be an $N \times N$ matrix and $\mathbf{x}_1, \mathbf{x}_2, \ldots, \mathbf{x}_N$ be eigenvectors of $A$ corresponding to distinct eigenvalues $\lambda_1, \lambda_2, \ldots, \lambda_N$. Then $\mathbf{x}_1, \mathbf{x}_2, \ldots, x_N$ are linearly independent.

(b)    Let $P$ be a matrix whose columns consist of $\mathbf{x}_1, \mathbf{x}_2, \ldots, \mathbf{x}_N$ of part (a). Then $P$ is nonsingular and $P^{-1} AP = D$ where $D$ is a diagonal matrix whose elements are $\lambda_1, \lambda_2, \ldots, \lambda_N$.

***Proof*** (a) This part will be proven by mathematical induction. For $n = 1$, $\alpha_1 \mathbf{x}_1 = \mathbf{0}$ and since $\mathbf{x}_1 \neq \mathbf{0}$, $\alpha_1 = 0$. The theorem is trivially true for $n = 1$.

Assume that the theorem is true for $n = k$ eigenvectors. Eigenvectors $\mathbf{x}_1, \mathbf{x}_2, \ldots, \mathbf{x}_k$ corresponding to eigenvalues $\lambda_1, \lambda_2, \ldots, \lambda_k$ are linearly independent. Now we want to prove that the eigenvectors $\mathbf{x}_1, \mathbf{x}_2, \ldots, \mathbf{x}_{k+1}$ corresponding to eigenvalues $\lambda_1, \lambda_2, \ldots, \lambda_{k+1}$ are linearly independent. Let

$$\beta_1 \mathbf{x}_1 + \beta_2 \mathbf{x}_2 + \cdots + \beta_k \mathbf{x}_k + \beta_{k+1} \mathbf{x}_{k+1} = \mathbf{0} \tag{13.1.6}$$

We want to prove $\beta_1 = \beta_2 = \cdots = \beta_k = \beta_{k+1} = 0$. Multiply Equation (13.1.6) by $A - \lambda_{k+1} I$ on the left, then Equation (13.1.6) becomes

$$\begin{aligned} \beta_1 (A\mathbf{x}_1 - \lambda_{k+1} \mathbf{x}_1) + \beta_2 (A\mathbf{x}_2 - \lambda_{k+1} \mathbf{x}_2) + \cdots \\ + \beta_k (A\mathbf{x}_k - \lambda_{k+1} \mathbf{x}_k) \\ + \beta_{k+1} (A\mathbf{x}_{k+1} - \lambda_{k+1} \mathbf{x}_{k+1}) = \mathbf{0} \end{aligned} \tag{13.1.7}$$

Since $A\mathbf{x}_i = \lambda_i \mathbf{x}_i$ for $i = 1, 2, \ldots, k + 1$, Equation (13.1.7) becomes

$$\beta_1(\lambda_1 - \lambda_{k+1})\mathbf{x}_1 + \beta_2(\lambda_2 - \lambda_{k+1})\mathbf{x}_2 + \cdots + \beta_k(\lambda_k - \lambda_{k+1})\mathbf{x}_k = \mathbf{0}$$

Further $\lambda_1, \lambda_2, \ldots, \lambda_{k+1}$ are distinct and $\mathbf{x}_1, \mathbf{x}_2, \ldots, \mathbf{x}_k$ are linearly independent, therefore $\beta_1 = \beta_2 = \cdots = \beta_k = 0$. We have $\beta_1 = \beta_2 = \cdots = \beta_k = 0$, hence Equation (13.1.6) becomes

$$\beta_{k+1} \mathbf{x}_{k+1} = \mathbf{0}$$

Since $\mathbf{x}_{k+1} \neq \mathbf{0}$, then $\beta_{k+1} = 0$. Thus in Equation (13.1.6) we have $\beta_1 = \beta_2 = \cdots = \beta_{k+1} = 0$ and therefore $\mathbf{x}_1, \mathbf{x}_2, \ldots, \mathbf{x}_{k+1}$ are linearly independent. This proves part (a).

(b) Let us write $P = [\mathbf{x}_1, \mathbf{x}_2, \ldots, \mathbf{x}_N]$. Since the column vectors $\mathbf{x}_1, \mathbf{x}_2, \ldots, \mathbf{x}_N$ are linearly independent, $P$ is nonsingular. Further

$$\begin{aligned} AP &= A[\mathbf{x}_1, \mathbf{x}_2, \ldots, \mathbf{x}_N] \\ &= [A\mathbf{x}_1, A\mathbf{x}_2, \ldots, A\mathbf{x}_N] \\ &= [\lambda_1\mathbf{x}_1, \lambda_2 \mathbf{x}_2, \ldots, \lambda_N \mathbf{x}_N] \end{aligned} \tag{13.1.8}$$

and

$$PD = [\mathbf{x}_1, \mathbf{x}_2, \ldots, \mathbf{x}_N] \begin{bmatrix} \lambda_1 & 0 & \ldots & 0 \\ 0 & \lambda_2 & \ldots & 0 \\ \vdots & \vdots & \ddots & \vdots \\ 0 & 0 & \ldots & \lambda_N \end{bmatrix}$$
$$= [\lambda_1 \mathbf{x}_1, \lambda_2 \mathbf{x}_2, \ldots, \lambda_N \mathbf{x}_N] \tag{13.1.9}$$

From Equations (13.1.8) and (13.1.9), we get

$$AP = PD \tag{13.1.10}$$

Multiplying Equation (13.1.10) by $P^{-1}$ on the left, we get

$$P^{-1} AP = D \tag{13.1.11}$$

Hence part (b) is proven. ∎

Since we are interested in developing the methods to find the eigenvalues and the corresponding eigenvectors of a given real symmetric matrix, let us look at the eigenvalues and eigenvectors of a real symmetric matrix.

Let $A$ be a real symmetric matrix of order $N$ and $\mathbf{x}$ be an eigenvector corresponding to its eigenvalue $\lambda$. Then

$$A\mathbf{x} = \lambda\mathbf{x} \tag{13.1.1}$$

Taking the complex conjugate on both sides of Equation (13.1.1) and noting that $A$ is real, we get

$$A\bar{\mathbf{x}} = \bar{\lambda}\bar{\mathbf{x}} \tag{13.1.12}$$

where $\bar{\mathbf{x}}$ is the complex congugate of the vector $\mathbf{x}$. Taking the transpose on both sides of Equation (13.1.1) and then multiplying both sides on the right by $\bar{\mathbf{x}}$ gives

$$(A\mathbf{x})'\,\bar{\mathbf{x}} = (\lambda\mathbf{x})'\,\bar{\mathbf{x}} = \lambda\mathbf{x}'\,\bar{\mathbf{x}} \tag{13.1.13}$$

Simplifying the left-hand side of Equation (13.1.13) and using Equation (13.1.12) and noting $A' = A$, we get

$$(A\mathbf{x})'\,\bar{\mathbf{x}} = \mathbf{x}'A'\,\bar{\mathbf{x}} = \mathbf{x}'A\,\bar{\mathbf{x}} = \mathbf{x}'(\bar{\lambda}\bar{\mathbf{x}}) = \bar{\lambda}\mathbf{x}'\,\bar{\mathbf{x}} \tag{13.1.14}$$

Comparison of Equations (13.1.13) and (13.1.14) gives

$$\lambda\mathbf{x}'\,\bar{\mathbf{x}} = \bar{\lambda}\mathbf{x}'\,\bar{\mathbf{x}}$$

Thus $(\lambda - \bar{\lambda})\mathbf{x}'\,\bar{\mathbf{x}} = 0$. Since $\mathbf{x}'\,\bar{\mathbf{x}} \neq 0$, $\lambda = \bar{\lambda}$. This is possible only if $\lambda$ is real.

Since the eigenvalues and elements of $A$ are real, then by Theorem 13.1.1, there exist real corresponding eigenvectors. This proves the following theorem.

**Theorem 13.1.4**    The eigenvalues of a real symmetric matrix are all real numbers and also there exist real associated eigenvectors.    ■

Further, let $\lambda_j$ and $\lambda_k$ be distinct eigenvalues and $\mathbf{x}_j$ and $\mathbf{x}_k$ be associated eigenvectors of $A$. Then

$$A\mathbf{x}_j = \lambda_j\,\mathbf{x}_j \tag{13.1.15}$$

and

$$A\mathbf{x}_k = \lambda_k\,\mathbf{x}_k \tag{13.1.16}$$

Multiplying Equation (13.1.15) by $\mathbf{x}_k'$ on the left side, we get

$$\mathbf{x}_k'\,(A\mathbf{x}_j) = \lambda_j\,\mathbf{x}_k'\,\mathbf{x}_j \tag{13.1.17}$$

Taking the transpose on both sides of Equation (13.1.17) yields

$$(\mathbf{x}_k'\,(A\mathbf{x}_j))' = (A\mathbf{x}_j)'\,\mathbf{x}_k = \mathbf{x}_j'A'\,\mathbf{x}_k = \mathbf{x}_j'\,(A\mathbf{x}_k)$$
$$= \mathbf{x}_j'\,(\lambda_k\,\mathbf{x}_k) = \lambda_k\,\mathbf{x}_j'\,\mathbf{x}_k$$

Also

$$(\mathbf{x}_k'\,(A\mathbf{x}_j))' = (\lambda_j\,\mathbf{x}_k'\,\mathbf{x}_j)' = \lambda_j\,\mathbf{x}_j'\,\mathbf{x}_k$$

Therefore $(\mathbf{x}_k^t(A\mathbf{x}_j))^t = \lambda_k \mathbf{x}_j^t \mathbf{x}_k = \lambda_j \mathbf{x}_j^t \mathbf{x}_k$. Hence $(\lambda_k - \lambda_j) \mathbf{x}_j^t \mathbf{x}_k = 0$. Since $\lambda_k \neq \lambda_j$, $\mathbf{x}_j^t \mathbf{x}_k = 0$. Thus $\mathbf{x}_j$ and $\mathbf{x}_k$ are orthogonal. If a real symmetric matrix has $N$ distinct eigenvalues, then the associated eigenvectors $\mathbf{x}_1, \mathbf{x}_2, \ldots, \mathbf{x}_N$ are mutually orthogonal. Also it can be proven that if $\mathbf{x}_1, \mathbf{x}_2, \ldots, \mathbf{x}_N$ are mutually orthogonal nonzero vectors, then they are also linearly independent (Exercise 9). The following theorem combines these results.

**Theorem 13.1.5**

Let $A$ be a real symmetric matrix of order $N$. Then

(a)  The eigenvalues $\lambda_1, \lambda_2, \ldots, \lambda_N$ of $A$ are real numbers.
(b)  The eigenvectors $\mathbf{x}_1, \mathbf{x}_2, \ldots, \mathbf{x}_N$ corresponding to distinct eigenvalues $\lambda_1, \lambda_2, \ldots, \lambda_N$ are orthogonal.
(c)  If $\lambda_1, \lambda_2, \ldots, \lambda_N$ are distinct eigenvalues and $P = [\mathbf{x}_1, \mathbf{x}_2, \ldots, \mathbf{x}_N]$, then $P^t AP = D$ where $D$ is a diagonal matrix whose diagonal elements are $\lambda_1, \lambda_2, \ldots, \lambda_N$. ∎

Part (c) follows from Theorem 13.1.3(b). Part (c) can be generalized as:

**Theorem 13.1.6**

Let $A$ be a real symmetric matrix with $\lambda_1, \lambda_2, \ldots, \lambda_N$ not necessarily distinct eigenvalues. Then there exists an orthogonal matrix $Q$ such that $Q^t AQ = D$ where $D$ is a diagonal matrix whose diagonal elements are $\lambda_1, \lambda_2, \ldots, \lambda_N$. ∎

For the proof, see Steinberg (1974).

**EXAMPLE 13.1.1**

Find an orthogonal matrix $P$ which diagonalizes the matrix $A = \begin{bmatrix} 1 & 2 \\ 2 & 1 \end{bmatrix}$.

This is a symmetric matrix and therefore by Theorem 13.1.4 the eigenvalues are real numbers. From Example 10 of Appendix B, we have $\lambda_1 = -1$ with corresponding eigenvector $\mathbf{x}_1 = \begin{bmatrix} 1/\sqrt{2} \\ -1/\sqrt{2} \end{bmatrix}$ and $\lambda_2 = 3$ with corresponding eigenvector $\mathbf{x}_2 = \begin{bmatrix} 1/\sqrt{2} \\ 1/\sqrt{2} \end{bmatrix}$. Since $\lambda_1$ and $\lambda_2$ are distinct, $\mathbf{x}_1$ and $\mathbf{x}_2$ are orthogonal and $P = [\mathbf{x}_1, \mathbf{x}_2] = \begin{bmatrix} 1/\sqrt{2} & 1/\sqrt{2} \\ -1/\sqrt{2} & 1/\sqrt{2} \end{bmatrix}$. Hence

$$P^t AP = \begin{bmatrix} 1/\sqrt{2} & -1/\sqrt{2} \\ 1/\sqrt{2} & 1/\sqrt{2} \end{bmatrix} \begin{bmatrix} 1 & 2 \\ 2 & 1 \end{bmatrix} \begin{bmatrix} 1/\sqrt{2} & 1/\sqrt{2} \\ -1/\sqrt{2} & 1/\sqrt{2} \end{bmatrix}$$

$$= \begin{bmatrix} -1 & 0 \\ 0 & 3 \end{bmatrix}$$

■ ■ ■

A matrix $A$ is said to be **diagonalizable** if there exists a diagonal matrix $D$ such that $A$ is similar to $D$. By Theorem 13.1.6 all symmetric matrices are diagonalizable.

All matrices are not diagonalizable. To see this, consider $A = \begin{bmatrix} 0 & 1 \\ 0 & 0 \end{bmatrix}$. The eigenvalues of $A$ are $\lambda_1 = \lambda_2 = 0$. Thus if $A$ is similar to $D$, then there exists a nonsingular matrix $S$ such that $S^{-1}AS = \begin{bmatrix} 0 & 0 \\ 0 & 0 \end{bmatrix}$. Then $A = S \begin{bmatrix} 0 & 0 \\ 0 & 0 \end{bmatrix} S^{-1} = \begin{bmatrix} 0 & 0 \\ 0 & 0 \end{bmatrix}$ which is a contradiction. Therefore $\begin{bmatrix} 0 & 1 \\ 0 & 0 \end{bmatrix}$ is not diagonalizable. Often a matrix with repeated eigenvalues is not similar to a diagonal matrix. What is the most compact form to which a given matrix $A$ is similar? Matrix $A$ is similar to a Jordan matrix.

We define a **Jordan block matrix** to be a square matrix whose elements are zero except along its diagonal, which are all equal, and one above the diagonal; thus

$$J_i = \begin{bmatrix} \lambda_i & 1 & 0 & \dots & 0 & 0 \\ 0 & \lambda_i & 1 & \dots & 0 & 0 \\ \vdots & \vdots & \vdots & \ddots & \vdots & \\ 0 & 0 & 0 & \dots & \lambda_i & 1 \\ 0 & 0 & 0 & \dots & 0 & \lambda_i \end{bmatrix}$$

A **Jordan matrix** has the form

$$J = \begin{bmatrix} J_1 & & & \\ & J_2 & & \\ & & \ddots & \\ & & & J_k \end{bmatrix}$$

**EXAMPLE 13.1.2**

$$\begin{bmatrix} 1 & 1 & 0 \\ 0 & 1 & 0 \\ 0 & 0 & 2 \end{bmatrix}, \begin{bmatrix} 1 & 1 & 0 & 0 & 0 & 0 \\ 0 & 1 & 0 & 0 & 0 & 0 \\ 0 & 0 & 2 & 1 & 0 & 0 \\ 0 & 0 & 0 & 2 & 1 & 0 \\ 0 & 0 & 0 & 0 & 2 & 0 \\ 0 & 0 & 0 & 0 & 0 & 1 \end{bmatrix}, \text{ and } \begin{bmatrix} 5 & 0 & 0 & 0 & 0 \\ 0 & 0 & 1 & 0 & 0 \\ 0 & 0 & 0 & 0 & 0 \\ 0 & 0 & 0 & 1 & 1 \\ 0 & 0 & 0 & 0 & 1 \end{bmatrix}$$

are Jordan matrices.    ■ ■ ■

**Theorem 13.1.7**

**(The Jordan Canonical Form)**    Let $A$ be an $N \times N$ matrix. Then there exists an invertible $N \times N$ matrix $M$ such that

$$M^{-1}AM = J$$

where $J$ is a Jordan matrix whose diagonal elements are the eigenvalues of $A$.    ■

For a proof and general discussion see G. Strang (1976).

**EXAMPLE 13.1.3** Find an invertible matrix $M$ that transforms $\begin{bmatrix} 10 & -9 \\ 4 & -2 \end{bmatrix}$ to its Jordan canonical form.

The characteristic polynomial is given by $|A - \lambda I| = \begin{vmatrix} 10 - \lambda & -9 \\ 4 & -2 - \lambda \end{vmatrix} = -(10 - \lambda)(2 + \lambda) + 36 = \lambda^2 - 8\lambda + 16 = (\lambda - 4)^2 = 0$. Thus $\lambda = 4$ is an eigenvalue of multiplicity two.

For $\lambda = 4$,

$$(A - 4I)\mathbf{x} = \begin{bmatrix} 6 & -9 \\ 4 & -6 \end{bmatrix} \begin{bmatrix} x_1 \\ x_2 \end{bmatrix} = \begin{bmatrix} 0 \\ 0 \end{bmatrix}$$

This gives $\mathbf{x}_1 = \begin{bmatrix} 3 \\ 2 \end{bmatrix}$. To find another vector $\mathbf{x}_2$, we compute $(A - 4I)\mathbf{x}_2 = \mathbf{x}_1$. This gives

$$6x_1 - 9x_2 = 3$$
$$4x_1 - 6x_2 = 2$$

Thus $2x_1 - 3x_2 = 1$. Hence $x_1 = (1 + 3x_2)/2$. Thus $\mathbf{x}_2 = \begin{bmatrix} \frac{1}{2} \\ 0 \end{bmatrix}$.

Let $M = \begin{bmatrix} 3 & \frac{1}{2} \\ 2 & 0 \end{bmatrix}$. Then $M^{-1} = \begin{bmatrix} 0 & \frac{1}{2} \\ 2 & -3 \end{bmatrix}$ and

$$M^{-1}AM = \begin{bmatrix} 0 & \frac{1}{2} \\ 2 & -3 \end{bmatrix} \begin{bmatrix} 10 & -9 \\ 4 & -2 \end{bmatrix} \begin{bmatrix} 3 & \frac{1}{2} \\ 2 & 0 \end{bmatrix}$$

$$= \begin{bmatrix} 0 & \frac{1}{2} \\ 2 & -3 \end{bmatrix} \begin{bmatrix} 12 & 5 \\ 8 & 2 \end{bmatrix} = \begin{bmatrix} 4 & 1 \\ 0 & 4 \end{bmatrix} = J$$ ∎

This method can be generalized to obtain a Jordan canonical form for any $N \times N$ matrix $A$.

If $A$ is not symmetric, then its eigenvalues and its corresponding eigenvectors could be complex. For diagonalization or reduction to Jordan canonical form, we need its eigenvectors. Since eigenvectors are complex, the matrix constructed from these complex eigenvectors has elements that are complex numbers. Let $A$ be a matrix whose elements are complex numbers. We have discussed real symmetric matrices and their important properties are given by Theorems 13.1.5 and 13.1.6. We extend the idea of symmetry for a complex matrix in such a way that Theorems 13.1.5 and 13.1.6 hold true. First let us define the **conjugate transpose** of $A$. We denote the conjugate transpose of $A$ by $A^H$ and define it by

$$A^H = \bar{A}^t$$  (13.1.18)

**EXAMPLE 13.1.4** Consider

$$A = \begin{bmatrix} 1 & 2 + i \\ 2 - i & 2 \end{bmatrix}$$

Then

$$\bar{A} = \begin{bmatrix} \overline{1} & \overline{2+i} \\ \overline{2-i} & \overline{2} \end{bmatrix} = \begin{bmatrix} 1 & 2-i \\ 2+i & 2 \end{bmatrix}$$

and so

$$\bar{A}' = \begin{bmatrix} 1 & 2+i \\ 2-i & 2 \end{bmatrix} = A^H$$

If $A = A^H$, then $A$ is called a **Hermitian matrix**. Thus $A = \begin{bmatrix} 1 & 2+i \\ 2-i & 2 \end{bmatrix}$ is a Hermitian matrix. Two real vectors $\mathbf{x}$ and $\mathbf{y}$ are orthogonal if $\mathbf{x}'\mathbf{y} = 0$. Two complex vectors $\mathbf{x}$ and $\mathbf{y}$ are orthogonal if $\mathbf{x}^H \mathbf{y} = 0$. Note that if $\mathbf{x}$ is real, then $\mathbf{x}^H = \mathbf{x}'$.

$P$ is orthogonal if $PP^t = P^t P = I$. We extend this concept to matrices with complex elements. $U$ is called a **unitary matrix** if

$$U^H U = UU^H = I \tag{13.1.19}$$

We can prove (Exercise 17) the following theorem.

**Theorem 13.1.8**

Let $A$ be a Hermitian matrix of order $N$. Then

(a)   The eigenvalues $\lambda_1, \lambda_2, \ldots, \lambda_N$ are real numbers.
(b)   The eigenvectors $\mathbf{x}_1, \mathbf{x}_2, \ldots, \mathbf{x}_N$ corresponding to distinct eigenvalues $\lambda_1, \lambda_2, \ldots, \lambda_N$ are orthogonal.
(c)   If $\lambda_1, \lambda_2, \ldots, \lambda_N$ are distinct eigenvalues and $U = [\mathbf{x}_1, \mathbf{x}_2, \ldots, \mathbf{x}_N]$, then $U^H AU = D$ where $D$ is a diagonal matrix whose diagonal elements are $\lambda_1, \lambda_2, \ldots, \lambda_N$. ∎

We can generalize Theorem 13.1.6 for Hermitian matrices. Since an orthogonal matrix diagonalizes a symmetric real matrix, what is the simplest standard form to which a general matrix can be reduced by a unitary matrix? The Schur Theorem gives that answer.

**Theorem 13.1.9**

(**Schur Theorem**)   Let $A$ be an $N \times N$ matrix. Then there is a unitary matrix $U$ such that $U^H AU = T$ where $T$ is an upper triangular matrix with the eigenvalues of $A$ on the diagonal. ∎

For the proof, see G. Strang (1976).

Just as symmetric matrices are important, the subset of the set of symmetric matrices, called the set of positive definite matrices (Section 7.6) is very important. An $N \times N$ symmetric matrix $A$ is called **positive definite** if $\mathbf{x}' A\mathbf{x} > 0$ for all nonzero $\mathbf{x} \in R^N$.

**EXAMPLE 13.1.5**    Let $A = \begin{bmatrix} 2 & 1 \\ 1 & 4 \end{bmatrix}$. Then

$$\mathbf{x}' A \mathbf{x} = [x_1, x_2] \begin{bmatrix} 2 & 1 \\ 1 & 4 \end{bmatrix} \begin{bmatrix} x_1 \\ x_2 \end{bmatrix} = [x_1, x_2] \begin{bmatrix} 2x_1 + x_2 \\ x_1 + 4x_2 \end{bmatrix} = 2x_1^2 + 2x_1 x_2 + 4x_2^2$$
$$= x_1^2 + 2x_1 x_2 + x_2^2 + x_1^2 + 3x_2^2 = (x_1 + x_2)^2 + x_1^2 + 3x_2^2 > 0$$

whenever $\mathbf{x} \neq \mathbf{0}$.    ■ ■ ■

It is tedious to use the definition to check whether a given matrix is positive definite. A common test for positive definite is provided by the following theorem.

**Theorem 13.1.10**    A real symmetric matrix $A$ is positive definite if and only if all of its eigenvalues are real and positive.    ■

First, let us prove that if $A$ is positive definite, then all its eigenvalues are real and positive. Since $A$ is real symmetric, its eigenvalues are real numbers. Let $\lambda_i$ be any real eigenvalue and $\mathbf{x}_i$ be a real eigenvector corresponding to $\lambda_i$ such that $\mathbf{x}_i' \mathbf{x}_i = 1$. Then $A\mathbf{x}_i = \lambda_i \mathbf{x}_i$. Therefore $\mathbf{x}_i' A \mathbf{x}_i = \lambda_i \mathbf{x}_i' \mathbf{x}_i = \lambda_i > 0$, since $\mathbf{x}_i' A \mathbf{x}_i > 0$ for any nonzero $\mathbf{x} \in R^N$ because $A$ is a positive definite matrix. Thus all eigenvalues are positive when $A$ is positive definite. Now let us prove that if all eigenvalues of $A$ are real and positive, then $A$ is positive definite. Since $A$ is symmetric, then because of Theorem 13.1.6, $A$ has $\mathbf{x}_1, \mathbf{x}_2, \ldots, \mathbf{x}_N$ orthogonal eigenvectors. Let $\mathbf{x} \in R^N$ be any vector. Then $\mathbf{x}$ can be expressed as

$$\mathbf{x} = c_1 \mathbf{x}_1 + c_2 \mathbf{x}_2 + \cdots + c_N \mathbf{x}_N$$

where $c_1, c_2, \ldots, c_N$ are constants. Further $\mathbf{x}' = c_1 \mathbf{x}_1' + c_2 \mathbf{x}_2' + \cdots + c_N \mathbf{x}_N'$ and

$$A\mathbf{x} = A(c_1 \mathbf{x}_1 + c_2 \mathbf{x}_2 + \cdots + c_N \mathbf{x}_N) = c_1 \lambda_1 \mathbf{x}_1 + c_2 \lambda_2 \mathbf{x}_2 + \cdots + c_N \lambda_N \mathbf{x}_N$$

Therefore, since $\mathbf{x}_i' \mathbf{x}_j = 0$, if $i \neq j$,

$$\mathbf{x}' A \mathbf{x} = (c_1 \mathbf{x}_1' + c_2 \mathbf{x}_2' + \cdots + c_N \mathbf{x}_N')(c_1 \lambda_1 \mathbf{x}_1 + c_2 \lambda_2 \mathbf{x}_2 + \cdots + c_N \lambda_N \mathbf{x}_N)$$
$$= c_1^2 \lambda_1 + c_2^2 \lambda_2 + \cdots + c_N^2 \lambda_N > 0$$

because $\lambda_i > 0$. Thus $A$ is positive definite.
Other tests are considered in Exercises 20 and 22.

**EXERCISES**

In Exercises 1–3, find a matrix $P$ that diagonalizes $A$ and determine $P^{-1} A P$.

**1.**  $A = \begin{bmatrix} 1 & 0 \\ 5 & 2 \end{bmatrix}$

**2.**  $A = \begin{bmatrix} 2 & 0 & -2 \\ 0 & 3 & 0 \\ 0 & 0 & 1 \end{bmatrix}$

**3.**  $A = \begin{bmatrix} 3 & 0 & 0 \\ 0 & 2 & 1 \\ 0 & 1 & 1 \end{bmatrix}$

In Exercises 4–6, diagonalize $A$ when possible.

**4.**  $A = \begin{bmatrix} 1 & 0 & 0 \\ 0 & 1 & 0 \\ 0 & 2 & 1 \end{bmatrix}$

**5.**  $A = \begin{bmatrix} 1 & 0 \\ 1 & 1 \end{bmatrix}$

**6.**  $A = \begin{bmatrix} 2 & -3 \\ 1 & -1 \end{bmatrix}$

**7.**  Prove that the characteristic polynomials of $A$ and $A^t$ are identical, and therefore $A$ and $A^t$ have the same eigenvalues.

**8.**  Prove that if $\lambda_i$ is an eigenvalue of $A$ with corresponding eigenvector $\mathbf{x}_i$, then $(\lambda_i + \mu)$ is an eigenvalue of $(A + \mu I)$ with corresponding eigenvector $\mathbf{x}_i$.

**9.**  If $\mathbf{x}_1, \mathbf{x}_2, \ldots, \mathbf{x}_N$ are mutually orthogonal nonzero vectors, then prove that they are linearly independent.

**10.**  Write the possible Jordan canonical forms for $A$ whose characteristic polynomial is $(\lambda - 1)^3$.

Transform the matrices in Exercises 11–13 into Jordan canonical form.

**11.**  $\begin{bmatrix} 2 & -9/4 \\ 1 & -1 \end{bmatrix}$

**12.**  $\begin{bmatrix} 2 & 0 & 0 \\ 0 & 1 & 0 \\ 0 & 2 & 1 \end{bmatrix}$

**13.**  $\begin{bmatrix} 0 & 1 & 3 \\ 0 & 0 & 1 \\ 0 & 0 & 0 \end{bmatrix}$

**14.**  Let $A$ be any $N \times N$ matrix. Then by using the Jordan canonical form, prove that $\det A = \lambda_1 \lambda_2 \ldots \lambda_N$ where $\lambda_1, \lambda_2, \ldots, \lambda_N$ are the eigenvalues of $A$.

**15.**  If $A$ and $B$ are similar, show that (a) $A^t$ and $B^t$ are similar and (b) $A^{-1}$ and $B^{-1}$ are similar.

**16.**  If $A$ is invertible, show that $AB$ is similar to $BA$ for all $B$.

**17.**  Prove Theorem 13.1.8.

**18.**  Given $p(x) = a_0 + a_1 x + a_2 x^2 + x^3$, show that

$$A = \begin{bmatrix} 0 & 1 & 0 \\ 0 & 0 & 1 \\ -a_0 & -a_1 & -a_2 \end{bmatrix}$$

is a matrix whose characteristic polynomial equals $p(x)$. Generalize to polynomials of degree $N$ with the coefficient of $x^N$ being one. This matrix is called the companion matrix of $p(x)$.

**19.** Are the following matrices positive definite?

(a) $\begin{bmatrix} 4 & 3 \\ 3 & 5 \end{bmatrix}$    (b) $\begin{bmatrix} 5 & -1 & -1 \\ -1 & 3 & 1 \\ -1 & 1 & 3 \end{bmatrix}$    (c) $\begin{bmatrix} 2 & -1 & 1 \\ -1 & 3 & 6 \\ 0 & 6 & 7 \end{bmatrix}$

**20.** Prove that a real symmetric matrix $A$ is positive definite if and only if each of the leading principal submatrices $A_i$ has a positive determinant.

**21.** Prove that if $A$ is positive definite, then $A^2$ and $A^{-1}$ are also positive definite.

**22.** Prove that if any diagonal entry $a_{ii}$ of the symmetric matrix $A$ is nonpositive, then $A$ cannot be positive definite. (*Hint*: Let $a_{11} < 0$. Calculate $\mathbf{e}_1^t A \mathbf{e}_1$.)

## 13.2  THE GERSCHGORIN THEOREM

In some instances it helps to have crude approximations of the eigenvalues of a matrix. For this purpose there are remarkable results that are extremely easy to use.

Let

$$A = \begin{bmatrix} a_{11} & a_{12} & \cdots & a_{1N} \\ a_{21} & a_{22} & \cdots & a_{2N} \\ \vdots & \vdots & \ddots & \vdots \\ a_{N1} & a_{N2} & \cdots & a_{NN} \end{bmatrix} \tag{13.2.1}$$

and $\mathbf{x} = [x_1, x_2, \ldots, x_N]^t$ be an eigenvector of $A$ corresponding to its eigenvalue $\lambda$. Let $|x_i|$ be the largest absolute component of $\mathbf{x}$; in other words, $|x_i| \geq |x_j|$ for $j = 1, 2, \ldots, N$. Since $A\mathbf{x} = \lambda \mathbf{x}$, its $i$th component is given by

$$a_{i1} x_1 + a_{i2} x_2 + \cdots + a_{ii} x_i + \cdots + a_{iN} x_N = \lambda x_i \tag{13.2.2}$$

Equation (13.2.2) can be written as

$$\sum_{\substack{k=1 \\ k \neq i}}^{N} a_{ik} x_k = (\lambda - a_{ii}) x_i$$

Thus we have

$$|(\lambda - a_{ii}) x_i| = \left| \sum_{\substack{k=1 \\ k \neq i}}^{N} a_{ik} x_k \right|$$

Since $|x_i| \neq 0$,

$$|(\lambda - a_{ii})| \leq \sum_{\substack{k=1 \\ k \neq i}}^{N} |a_{ik}| \left| \frac{x_k}{x_i} \right|$$

Since $|x_k/x_i| \leq 1$, we get

$$|(\lambda - a_{ii})| \leq \sum_{\substack{k=1 \\ k \neq i}}^{N} |a_{ik}| \qquad \text{(13.2.3)}$$

Let

$$r_i = \sum_{\substack{k=1 \\ k \neq i}}^{N} |a_{ik}| \quad \text{for } i = 1, 2, \ldots, N \qquad \text{(13.2.4)}$$

that is, $r_i$ is the sum of the absolute values of the off-diagonal elements of the $i$th row of $A$. Then Equation (13.2.3) becomes

$$|\lambda - a_{ii}| \leq r_i \quad \text{for } i = 1, 2, \ldots, N \qquad \text{(13.2.5)}$$

Further, let the discs with center $a_{ii}$ and radius $r_i$ be denoted by

$$D_i = \left\{ x \,\middle|\, |x - a_{ii}| \leq r_i \right\} \qquad \text{(13.2.6)}$$

Then $\lambda$ lies in the $i$th disc $D_i$; however, we do not know which $i$ to use since $\mathbf{x}$ is not known. Since Equation (13.2.5) holds for each eigenvalue of $A$, then each eigenvalue must lie in the union $D$ of these discs given by

$$D = \bigcup_{i=1}^{N} D_i \qquad \text{(13.2.7)}$$

This proves the following theorem.

**Theorem 13.2.1**    **(Gerschgorin's Theorem)**   Let $A$ be an $N \times N$ matrix. Then all the eigenvalues of $A$ lie in the set $D$ given by Equation (13.2.7).    ∎

Since $\det(A - \lambda I) = \det(A - \lambda I)' = \det(A' - \lambda I)$, then $A$ and $A'$ have the same eigenvalues. Thus applying Gerschgorin's Theorem to $A'$, all the eigenvalues of $A'$ lie in the set

$$\hat{D} = \bigcup_{i=1}^{N} \hat{D}_i \quad \text{where } \hat{D}_i = \left\{ x \,\middle|\, |x - a_{ii}| \leq c_i \right\}$$

and

$$c_i = \sum_{\substack{k=1 \\ k \neq i}}^{N} |a'_{ik}| = \sum_{\substack{k=1 \\ k \neq i}}^{N} |a_{ki}|$$

That is, $c_i$ is the sum of the absolute values of the off-diagonal elements of the $i$th column of $A$. Hence the eigenvalues of $A$ must lie in the intersection of $D$ and $\hat{D}$.

**EXAMPLE 13.2.1**    Use Gerschgorin's Theorem to obtain bounds on the magnitude of the eigenvalues for $\begin{bmatrix} 1 & -2 \\ 3 & 2 \end{bmatrix}$ and graph the discs.

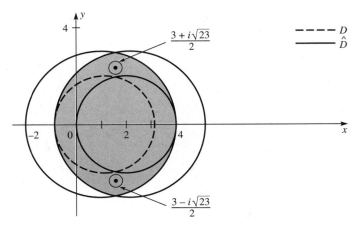

**Fig. 13.2.1**

$$D_1 = \left\{ x \,\middle|\, |x - 1| \le 2 \right\} \qquad D_2 = \left\{ x \,\middle|\, |x - 2| \le 3 \right\}$$

$$\hat{D}_1 = \left\{ x \,\middle|\, |x - 1| \le 3 \right\} \qquad \hat{D}_2 = \left\{ x \,\middle|\, |x - 2| \le 2 \right\}$$

These discs and the eigenvalues $(3 + i\sqrt{23})/2$ and $(3 - i\sqrt{23})/2$ are shown in Figure 13.2.1.    ■ ■ ■

---

## 13.3  THE POWER METHOD

Many times we are interested only in the dominant eigenvalue (the largest eigenvalue in magnitude). Let $A$ be an $N \times N$ matrix with eigenvalues $\lambda_1, \lambda_2, \ldots, \lambda_N$ such that $|\lambda_1| > |\lambda_2| \ge |\lambda_3| \cdots \ge |\lambda_N|$. Assume that $A$ has $N$ linearly independent eigenvectors $\mathbf{v}_1, \mathbf{v}_2, \ldots, \mathbf{v}_N$ associated with each of these eigenvalues. Since $\{\mathbf{v}_1, \mathbf{v}_2, \ldots, \mathbf{v}_N\}$ forms a basis of $R^N$, we can express any given vector $\mathbf{x}^{(0)}$ as

$$\mathbf{x}^{(0)} = \alpha_1 \mathbf{v}_1 + \alpha_2 \mathbf{v}_2 + \cdots + \alpha_N \mathbf{v}_N \tag{13.3.1}$$

where $\alpha_1, \alpha_2, \ldots, \alpha_N$ are constants. Multiplying both sides of Equation (13.3.1) by $A$ gives

$$\begin{aligned} A\mathbf{x}^{(0)} &= \alpha_1 A\mathbf{v}_1 + \alpha_2 A\mathbf{v}_2 + \cdots + \alpha_N A\mathbf{v}_N \\ &= \alpha_1 \lambda_1 \mathbf{v}_1 + \alpha_2 \lambda_2 \mathbf{v}_2 + \cdots + \alpha_N \lambda_N \mathbf{v}_N \end{aligned}$$

Inductively, for any positive integer $k$,

$$\begin{aligned} A^k \mathbf{x}^{(0)} &= \alpha_1 \lambda_1^k \mathbf{v}_1 + \alpha_2 \lambda_2^k \mathbf{v}_2 + \cdots + \alpha_N \lambda_N^k \mathbf{v}_N \\ &= \lambda_1^k \left\{ \alpha_1 \mathbf{v}_1 + \alpha_2 \left( \frac{\lambda_2}{\lambda_1} \right)^k \mathbf{v}_2 + \cdots + \alpha_N \left( \frac{\lambda_N}{\lambda_1} \right)^k \mathbf{v}_N \right\} \end{aligned}$$

Since $|\lambda_i/\lambda_1| < 1$ for $i \geq 2$, then

$$A^k \mathbf{x}^{(0)} \to \lambda_1^k \alpha_1 \mathbf{v}_1 \quad \text{as } k \to \infty \tag{13.3.2}$$

For any given vector $\mathbf{x}^{(0)}$ we generate the sequence given by

$$\mathbf{x}^{(k)} = A\mathbf{x}^{(k-1)} \quad \text{for } k = 1, 2, \ldots \tag{13.3.3}$$

We can verify that

$$\mathbf{x}^{(k)} = A^k \mathbf{x}^{(0)} \quad \text{for } k = 1, 2, \ldots \tag{13.3.4}$$

Thus

$$\mathbf{x}^{(k)} \to \lambda_1^k \alpha_1 \mathbf{v}_1 \quad \text{as } k \to \infty \tag{13.3.5}$$

Since the sequence of Equation (13.3.4) converges to zero if $|\lambda_1| < 1$ and diverges if $|\lambda_1| > 1$, Equation (13.3.4) may not be a practical sequence to compute the dominant eigenvalue. It is desirable to keep $\mathbf{x}^{(k)}$ within computational limits by scaling $\mathbf{x}^{(k)}$. This can be done by dividing $\mathbf{x}^{(k)}$ by its absolute largest component which is denoted by $\|\mathbf{x}^{(k)}\|_\infty$ at each step. Let $\mathbf{x}^{(0)}$ be an initial guess. Then define $\mathbf{z}^{(0)} = \mathbf{x}^{(0)}/\|\mathbf{x}^{(0)}\|_\infty$. Compute $\mathbf{x}^{(1)} = A\mathbf{z}^{(0)}$. Then define $\mathbf{z}^{(1)} = \mathbf{x}^{(1)}/\|\mathbf{x}^{(1)}\|_\infty$ and continue. We obtain

$$\begin{cases} \mathbf{z}^{(k)} = \mathbf{x}^{(k)}/\|\mathbf{x}^{(k)}\|_\infty \\ \mathbf{x}^{(k+1)} = A\mathbf{z}^{(k)} \end{cases} \quad \text{for } k = 0, 1, \ldots \tag{13.3.6}$$

Since $\mathbf{v}_1, \mathbf{v}_2, \ldots, \mathbf{v}_N$ are linearly independent eigenvectors, we express $\mathbf{z}^{(0)}$ as

$$\mathbf{z}^{(0)} = \beta_1 \mathbf{v}_1 + \beta_2 \mathbf{v}_2 + \cdots + \beta_N \mathbf{v}_N$$

where $\beta_1, \beta_2, \ldots, \beta_N$ are constants. We compute

$$\mathbf{x}^{(1)} = A\mathbf{z}^{(0)} \quad \text{and} \quad \mathbf{z}^{(1)} = \mathbf{x}^{(1)}/\|\mathbf{x}^{(1)}\|_\infty = A\mathbf{z}^{(0)}/\|A\mathbf{z}^{(0)}\|_\infty$$

Similarly

$$\mathbf{z}^{(k)} = \frac{A^k \mathbf{z}^{(0)}}{\|A^k \mathbf{z}_0\|_\infty} = \frac{\lambda_1^k}{\|A^k \mathbf{z}_0\|_\infty} \left\{ \beta_1 \mathbf{v}_1 + \beta_2 \left(\frac{\lambda_2}{\lambda_1}\right)^k \mathbf{v}_2 + \cdots + \beta_N \left(\frac{\lambda_N}{\lambda_1}\right)^k \mathbf{v}_N \right\} \tag{13.3.7}$$

As $k \to \infty$,

$$\mathbf{z}^{(k)} \to \frac{\beta_1 \mathbf{v}_1}{\|A^k \mathbf{z}_0\|_\infty} \lambda_1^k \tag{13.3.8}$$

Multiplying Equation (13.3.7) on the left by $A$ yields

$$\mathbf{x}^{(k+1)} = A\mathbf{z}^{(k)}$$
$$= \frac{\lambda_1^{k+1}}{\|A^k \mathbf{z}_0\|_\infty} \left\{ \beta_1 \mathbf{v}_1 + \beta_2 \left(\frac{\lambda_2}{\lambda_1}\right)^{k+1} \mathbf{v}_2 + \cdots + \beta_N \left(\frac{\lambda_N}{\lambda_1}\right)^{k+1} \mathbf{v}_N \right\} \tag{13.3.9}$$

As $k \to \infty$,

$$\mathbf{x}^{(k+1)} \to \frac{\beta_1 \mathbf{v}_1}{\|A^k \mathbf{z}_0\|_\infty} \lambda_1^{k+1} = \lambda_1 \left(\frac{\beta_1 \mathbf{v}_1}{\|A^k \mathbf{z}_0\|_\infty}\right) \lambda_1^k \tag{13.3.10}$$

Using Equation (13.3.8) in Equation (13.3.10), we get $\mathbf{x}^{(k+1)} \rightarrow \lambda_1 \mathbf{z}^{(k)}$. For large $k$, we have

$$\mathbf{x}^{(k+1)} = A\mathbf{z}^{(k)} \approx \lambda_1 \mathbf{z}^{(k)} \tag{13.3.11}$$

Multiplying Equation (13.3.11) on the left by $(\mathbf{z}^{(k)})'$, we get

$$\lambda_1 \approx \frac{(\mathbf{z}^{(k)})' \, (A\mathbf{z}^{(k)})}{(\mathbf{z}^{(k)})' \, \mathbf{z}^{(k)}} \tag{13.3.12}$$

Denoting

$$\lambda_1^{(k)} = \frac{(\mathbf{z}^{(k)})' \, (A\mathbf{z}^{(k)})}{(\mathbf{z}^{(k)})' \, \mathbf{z}^{(k)}} \quad \text{for } k = 0, 1, \ldots \tag{13.3.13}$$

we have $\lambda_1^{(k)} \rightarrow \lambda_1$ as $k \rightarrow \infty$.

It follows from Equation (13.3.8) that $\mathbf{z}^{(k)} \rightarrow c\mathbf{v}_1$ as $k \rightarrow \infty$ where $c$ is a constant. Thus $\mathbf{z}^{(k)}$ tends to some multiple of the eigenvector $\mathbf{v}_1$. Further, it follows from Equation (13.3.7) that the convergence of $\mathbf{z}^{(k)}$ to a scalar multiple of $\mathbf{v}_1$ depends upon how fast the ratios $(\lambda_i/\lambda_1)^k$ for $i = 2, 3, \ldots, N$ go to zero.

We combine Equations (13.3.6) and (13.3.13) to get

1.  $\mathbf{z}^{(k)} = \mathbf{x}^{(k)}/\|\mathbf{x}^{(k)}\|_\infty$
2.  $\mathbf{x}^{(k+1)} = A\mathbf{z}^{(k)}$
3.  $\lambda_1^{(k+1)} = (\mathbf{z}^{(k)})' \, \mathbf{x}^{(k+1)}/((\mathbf{z}^{(k)})' \, \mathbf{z}^{(k)})$ for $k = 0, 1, \ldots$ $\tag{13.3.14}$

Then as $k \rightarrow \infty$, $\lambda_1^{(k+1)} \rightarrow \lambda_1$ and $\mathbf{z}^{(k)} \rightarrow \mathbf{v}_1$.

**EXAMPLE 13.3.1**    Find the dominant eigenvalue and the corresponding eigenvector of the matrix

$$A = \begin{bmatrix} 1 & -1 & 4 \\ 3 & 2 & -1 \\ 2 & 1 & -1 \end{bmatrix} \quad \text{using } \mathbf{x}^{(0)} = \begin{bmatrix} 1 \\ 1 \\ 1 \end{bmatrix}$$

Since $\|\mathbf{x}^{(0)}\|_\infty = 1$, $\mathbf{z}^{(0)} = \mathbf{x}^{(0)}/\|\mathbf{x}^{(0)}\|_\infty = \begin{bmatrix} 1 \\ 1 \\ 1 \end{bmatrix}$,

$$\mathbf{x}^{(1)} = A\mathbf{z}^{(0)} = \begin{bmatrix} 1 & -1 & 4 \\ 3 & 2 & -1 \\ 2 & 1 & -1 \end{bmatrix} \begin{bmatrix} 1 \\ 1 \\ 1 \end{bmatrix} = \begin{bmatrix} 4 \\ 4 \\ 2 \end{bmatrix}$$

and

$$\lambda_1^{(1)} = \frac{(\mathbf{z}^{(0)})' \, \mathbf{x}^{(1)}}{(\mathbf{z}^{(0)})' \, \mathbf{z}^{(0)}} = \frac{[1, 1, 1] \begin{bmatrix} 4 \\ 4 \\ 2 \end{bmatrix}}{[1, 1, 1] \begin{bmatrix} 1 \\ 1 \\ 1 \end{bmatrix}} = \frac{4 + 4 + 2}{1 + 1 + 1} = \frac{10}{3} = 3.33333$$

For $k = 1$, we get from Equation (13.3.14)

$$\mathbf{z}^{(1)} = \frac{\mathbf{x}^{(1)}}{\|\mathbf{x}^{(1)}\|_\infty} = \frac{\begin{bmatrix} 4 \\ 4 \\ 2 \end{bmatrix}}{4} = \begin{bmatrix} 1 \\ 1 \\ 0.5 \end{bmatrix}$$

$$\mathbf{x}^{(2)} = A\mathbf{z}^{(1)} = \begin{bmatrix} 1 & -1 & 4 \\ 3 & 2 & -1 \\ 2 & 1 & -1 \end{bmatrix} \begin{bmatrix} 1 \\ 1 \\ 0.5 \end{bmatrix} = \begin{bmatrix} 2 \\ 4.5 \\ 2.5 \end{bmatrix}$$

and

$$\lambda_2^{(1)} = \frac{(\mathbf{z}^{(1)})^t \mathbf{x}^{(2)}}{(\mathbf{z}^{(1)})^t \mathbf{z}^{(1)}} = \frac{[1, 1, 0.5]\begin{bmatrix} 2 \\ 4.5 \\ 2.5 \end{bmatrix}}{[1, 1, 0.5]\begin{bmatrix} 1 \\ 1 \\ 0.5 \end{bmatrix}} = \frac{2 + 4.5 + 1.25}{1 + 1 + 0.25} = \frac{7.75}{2.25} = 3.44444$$

Continuing in this fashion, the values of $\mathbf{z}^{(k)}$, $\mathbf{x}^{(k+1)}$, and $\lambda_1^{(k+1)}$ are computed. The values of $\lambda_1^{(k+1)}$ and $\mathbf{z}^{(k)}$ are given in Table 13.3.1. The calculations were continued until $|\lambda^{(k)} - \lambda^{(k+1)}| < 10^{-5} |\lambda^{(k+1)}|$.

**Table 13.3.1**

| $k$ | $\lambda_1^{(k+1)}$ | $(\mathbf{z}^{(k)})^t$ | | |
|---|---|---|---|---|
| 0 | 3.33333 | (1.00000, | 1.00000, | 1.00000) |
| 1 | 3.44444 | (1.00000, | 1.00000, | 0.50000) |
| 2 | 2.82787 | (0.44444, | 1.00000, | 0.55556) |
| 3 | 3.18008 | (0.60000, | 1.00000, | 0.48000) |
| 4 | 2.89580 | (0.45783, | 1.00000, | 0.51807) |
| 5 | 3.07392 | (0.53586, | 1.00000, | 0.48945) |
| 10 | 2.99033 | (0.49569, | 1.00000, | 0.50144) |
| 15 | 3.00127 | (0.50057, | 1.00000, | 0.49981) |
| 20 | 2.99983 | (0.44992, | 1.00000, | 0.50003) |
| 25 | 3.00002 | (0.50001, | 1.00000, | 0.50000) |
| 26 | 2.99999 | (0.49999, | 1.00000, | 0.50000) |
| 27 | 3.00001 | (0.50000, | 1.00000, | 0.50000) |

Thus we have the dominant eigenvalue $\lambda = 3.00001$ and the corresponding eigenvector [0.500000, 1.00000, 0.50000]. The exact values of the eigenvalues are 3, $-2$, and 1. Convergence is slow, but can be accelerated by using Aitken's extrapolation technique.

■ ■ ■

The main advantage of the power method is that only matrix-vector multiplications and scalings are required. There are no operations on matrix $A$ itself.

**Algorithm**

**Subroutine** Power ($N$, $A$, eigen, $\mathbf{x}$, eps, maxit)

Comment: This program computes the largest eigenvalue in magnitude (dominant eigenvalue) and its corresponding eigenvector $\mathbf{x}$ of a given $N \times N$ matrix $A$ using Equation (13.3.14), where $a_{ij}$ for $i, j = 1, 1, \ldots, N$ are the elements of the matrix $A$; $x_1, x_2, \ldots, x_N$ are the initial guesses of $\mathbf{x}$; eps is the tolerance and maxit is the maximum number of iterations.

1  **Print** $N$, $A$, $\mathbf{x}$, eps, maxit
2  $k = 0$
3  eigen = 0
4  $k = k + 1$
5  **If** $k >$ maxit, **then print** 'algorithm fails: no convergence' and exit.
6  eigen1 = eigen
7  temp = $|x_1|$
8  **Repeat** step 9 **with** $i = 2, 3, \ldots, N$
    9  **If** $|x_i| >$ temp, **then** temp = $|x_i|$
10  **Repeat** step 11 **with** $i = 1, 2, \ldots, N$
    11  $z_i = x_i/$temp
12  **Repeat** steps 13–16 **with** $i = 1, 2, \ldots, N$
    13  temp = 0
    14  **Repeat** step 15 **with** $j = 1, 2, \ldots, N$
        15  temp = temp + $a_{ij} * z_j$
    16  $x_i =$ temp
17  sum = 0
18  sum1 = 0
19  **Repeat** steps 20–21 **with** $i = 1, 2, \ldots, N$
    20  sum = sum + $z_i * x_i$
    21  sum1 = sum1 + $z_i * z_i$
22  eigen = sum/sum1
23  **If** $|$eigen $-$ eigen1$| >$ eps $|$eigen$|$, **then** return to 4
24  **Print** $k$, eigen, $\mathbf{x}$, and 'finished' and exit.

◀

## 13.4  DEFLATION

The power method finds the dominant eigenvalue. There are many ways to compute other eigenvalues. Let $A$ be real, having real and distinct eigenvalues. Assume that the eigenvector $\mathbf{v}_1$ corresponding to the dominant eigenvlue $\lambda_1$ of $A$ is known. Then the eigenvector $\mathbf{v}_1$ can be normalized so that the first component of $\mathbf{v}_1$ is one; that is, $\mathbf{v}_1 = [1, v_{21}, v_{31}, \ldots, v_{N1}]^t$. Let $\mathbf{a}^t = [a_{11}, a_{12}, \ldots, a_{1N}]$ be the first row of $A$.

Define

$$A_1 = A - \mathbf{v}_1 \, \mathbf{a}' \tag{13.4.1}$$

$$
= \begin{bmatrix}
a_{11} & a_{12} & \cdots & a_{1N} \\
a_{21} & a_{22} & \cdots & a_{2N} \\
\vdots & \vdots & \ddots & \vdots \\
a_{N1} & a_{N2} & \cdots & a_{NN}
\end{bmatrix}
- \begin{bmatrix}
1 \\
v_{21} \\
\vdots \\
v_{N1}
\end{bmatrix}
[a_{11}, a_{12}, \ldots, a_{1N}]
$$

$$
= \begin{bmatrix}
0 & 0 & \cdots & 0 \\
a_{21} - v_{21} a_{11} & a_{22} - v_{21} a_{12} & \cdots & a_{2N} - v_{21} a_{1N} \\
\vdots & \vdots & \ddots & \vdots \\
a_{N1} - v_{N1} a_{11} & a_{N2} - v_{N1} a_{12} & \cdots & a_{NN} - v_{N1} a_{1N}
\end{bmatrix} \tag{13.4.2}
$$

Let $|\lambda_2|$ be the second dominant eigenvalue and $\mathbf{v}_2$ be the corresponding eigenvector of $A$ such that $\mathbf{v}_2 = [1, v_{22}, v_{32}, \ldots, v_{N2}]'$. Then

$$
\begin{aligned}
A_1 (\mathbf{v}_1 - \mathbf{v}_2) &= (A - \mathbf{v}_1 \, \mathbf{a}')(\mathbf{v}_1 - \mathbf{v}_2) \\
&= A\mathbf{v}_1 - A\mathbf{v}_2 - (\mathbf{v}_1 \, \mathbf{a}' \, \mathbf{v}_1) + \mathbf{v}_1 (\mathbf{a}' \, \mathbf{v}_2) \\
&= \lambda_1 \, \mathbf{v}_1 - \lambda_2 \, \mathbf{v}_2 - \lambda_1 \, \mathbf{v}_1 + \lambda_2 \, \mathbf{v}_1 \\
&= \lambda_2 \, (\mathbf{v}_1 - \mathbf{v}_2)
\end{aligned}
$$

Since $\mathbf{a}' \, \mathbf{v}_1 = \lambda_1$, $\mathbf{a}' \, \mathbf{v}_2 = \lambda_2$ (note that the first component of $\mathbf{v}_1$ as well as $\mathbf{v}_2$ is one). Thus $\lambda_2$ is an eigenvalue of $A_1$ and $\mathbf{v}_1 - \mathbf{v}_2$ is an eigenvector of $A_1$ corresponding to $\lambda_2$. Since the first component of $\mathbf{v}_1 - \mathbf{v}_2$ is zero, the first column of $A_1$ is irrelevant for $A_1 (\mathbf{v}_1 - \mathbf{v}_2)$. In particular, define $B$ from Equation (13.4.2) by deleting the first row and first column of $A_1$ as

$$
B = \begin{bmatrix}
a_{22} - v_{21} a_{12} & a_{23} - v_{21} a_{13} & \cdots & a_{2N} - v_{21} a_{1N} \\
a_{32} - v_{31} a_{12} & a_{33} - v_{31} a_{13} & \cdots & a_{3N} - v_{31} a_{1N} \\
\vdots & \vdots & \ddots & \vdots \\
a_{N2} - v_{N1} a_{12} & a_{N3} - v_{N1} a_{13} & \cdots & a_{NN} - v_{N1} a_{1N}
\end{bmatrix} \tag{13.4.3}
$$

Thus $B$ is the lower $(N - 1) \times (N - 1)$ block of $A_1$ and has the eigenvalues $\lambda_2$, $\lambda_3, \ldots, \lambda_N$, which are the same as $A$ except $\lambda_1$. The process of finding an $(N - 1) \times (N - 1)$ matrix $B$ whose eigenvalues are identical to those of $A$, except $\lambda_1$, is called **deflation**. By applying the power method to $B$, we will get $\lambda_2$ and the eigenvector $\mathbf{v}_2^*$ of $B$ corresponding to $\lambda_2$. Then adding a zero as a first component, we get a vector $\mathbf{z}$. Consider the relationship

$$\mathbf{v}_1 - \mathbf{v}_2 = c\mathbf{z} \tag{13.4.4}$$

where $c$ is to be determined. Multiplying Equation (13.4.4) on the left by $\mathbf{a}'$, we get

$$\mathbf{a}' \, \mathbf{v}_1 - \mathbf{a}' \, \mathbf{v}_2 = c\mathbf{a}' \, \mathbf{z} \tag{13.4.5}$$

Since $\mathbf{a}' \, \mathbf{v}_1 = \lambda_1$ and $\mathbf{a}' \, \mathbf{v}_2 = \lambda_2$, we get from Equation (13.4.5)

$$c = \frac{\lambda_1 - \lambda_2}{\mathbf{a}' \, \mathbf{z}} \tag{13.4.6}$$

**EXAMPLE 13.4.1**    Use deflation to approximate the remaining eigenvalues and corresponding eigenvectors of

$$A = \begin{bmatrix} 1 & -1 & 4 \\ 3 & 2 & -1 \\ 2 & 1 & -1 \end{bmatrix}$$

We previously (Example 13.3.1) found the dominant eigenvalue three of $A$ and its corresponding eigenvector $[\frac{1}{2}, 1, \frac{1}{2}]'$. Since the first component of $\mathbf{v}_1$ must be one, we use $\mathbf{v}_1 = 2[\frac{1}{2}, 1, \frac{1}{2}]' = [1, 2, 1]'$. Then

$$A_1 = A - \mathbf{v}_1 \mathbf{a}'$$

$$= \begin{bmatrix} 1 & -1 & 4 \\ 3 & 2 & -1 \\ 2 & 1 & -1 \end{bmatrix} - \begin{bmatrix} 1 \\ 2 \\ 1 \end{bmatrix} [1 \quad -1 \quad 4]$$

$$= \begin{bmatrix} 1 & -1 & 4 \\ 3 & 2 & -1 \\ 2 & 1 & -1 \end{bmatrix} - \begin{bmatrix} 1 & -1 & 4 \\ 2 & -2 & 8 \\ 1 & -1 & 4 \end{bmatrix} = \begin{bmatrix} 0 & 0 & 0 \\ 1 & 4 & -9 \\ 1 & 2 & -5 \end{bmatrix}$$

Deleting the first row and first column of $A_1$, the deflated matrix $B$ is given by

$$B = \begin{bmatrix} 4 & -9 \\ 2 & -5 \end{bmatrix}$$

By using the power method, we find the eigenvalue $\lambda = -2$ and the corresponding eigenvector $\begin{bmatrix} 1 \\ \frac{2}{3} \end{bmatrix}$. Adding the zero component to $\begin{bmatrix} 1 \\ \frac{2}{3} \end{bmatrix}$, we get

$$\mathbf{z} = \begin{bmatrix} 0 \\ 1 \\ \frac{2}{3} \end{bmatrix}$$

To find the corresponding eigenvector of $A$, we find $c$ which is given by Equation (13.4.6) as

$$c = \frac{\lambda_1 - \lambda_2}{\mathbf{a}' \mathbf{z}} = \frac{3 + 2}{5/3} = 3$$

Thus from Equation (13.4.4), we get

$$\mathbf{v}_2 = \mathbf{v}_1 - c\mathbf{z} = \begin{bmatrix} 1 \\ 2 \\ 1 \end{bmatrix} - 3 \begin{bmatrix} 0 \\ 1 \\ \frac{2}{3} \end{bmatrix} = \begin{bmatrix} 1 \\ -1 \\ -1 \end{bmatrix}$$

Similarly using the deflation method, we find the eigenvalue $\lambda = 1$ of $B$. Therefore $\lambda = 1$ is the eigenvalue of $A$ and the corresponding eigenvector

$$\mathbf{v}_3 = \begin{bmatrix} 1 \\ -4 \\ -1 \end{bmatrix}$$

■ ■ ■

In most cases, we do not know exactly the value $\lambda_1$ and so the deflated matrix $B$ will not have precisely the eigenvalues $\lambda_2, \lambda_3, \ldots, \lambda_N$. When we deflate $B$, we introduce additional errors. We are not sure that the eigenvalues obtained after several deflations are those of the original matrix; therefore, the power method along with deflation is generally not recommended to find all eigenvalues. The power method is quite useful to find the first few dominant eigenvalues.

## The Inverse Power Method

Often we need the eigenvalue with the smallest magnitude. The power method can be easily modified to find the smallest magnitude eigenvalue provided $A^{-1}$ exists.

Suppose $A\mathbf{x} = \lambda\mathbf{x}$ where $A$ is nonsingular. Then multiplying both sides by $A^{-1}$ on the left, we get

$$A^{-1}A\mathbf{x} = I\mathbf{x} = \lambda A^{-1}\mathbf{x}$$

Thus we have $A^{-1}\mathbf{x} = (1/\lambda)\mathbf{x}$. Hence $(1/\lambda)$ is an eigenvalue of $A^{-1}$. Consequently, if $\lambda_1, \lambda_2, \ldots, \lambda_N$ are the eigenvalues of $A$, then $1/\lambda_1, 1/\lambda_2, \ldots, 1/\lambda_N$ are the corresponding eigenvalues of $A^{-1}$.

It follows that the dominant eigenvalue of $A^{-1}$ will give the smallest magnitude eigenvalue of $A$. To find the smallest magnitude eigenvalue of $A$, we find the dominant eigenvalue of $A^{-1}$ using the power method. Let $\mathbf{x}^{(0)}$ be an initial guess. Then define $\mathbf{z}^{(0)} = \mathbf{x}^{(0)}/\|\mathbf{x}^{(0)}\|_\infty$. We compute $\mathbf{x}^{(1)} = A^{-1}\mathbf{z}^{(0)}$ and continue. However, it is not necessary to find $A^{-1}$ to use this algorithm. Rewrite $\mathbf{x}^{(1)} = A^{-1}\mathbf{z}^{(0)}$ as

$$A\mathbf{x}^{(1)} = \mathbf{z}^{(0)} \tag{13.4.7}$$

and solve for $\mathbf{x}^{(1)}$ using the $LU$ decomposition method. Store these factors to compute other vectors. Thus we have

1.  $\mathbf{z}^{(k)} = \mathbf{x}^{(k)}/\|\mathbf{x}^{(k)}\|_\infty$
2.  $LU\,\mathbf{x}^{(k+1)} = \mathbf{z}^{(k)}$
3.  $\lambda_1^{(k+1)} = ((\mathbf{z}^{(k)})^t\,\mathbf{x}^{(k+1)})/((\mathbf{z}^{(k)})^t\,(\mathbf{z}^{(k)}))$ for $k = 0, 1, \ldots$ $\tag{13.4.8}$

In step 2, we solve the system for $x^{(k+1)}$. Thus as $k \to \infty$, $\lambda^{(k+1)} \to \lambda$ and $\mathbf{z}^{(k)} \to \mathbf{v}_1$. The smallest magnitude eigenvalue $\lambda$ and its associated eigenvector $\mathbf{v}_1$ are computed using the inverse power method.

## EXERCISES

For Exercises 1–4, use the Gerschgorin Theorem to obtain bounds on the magnitude of the eigenvalues and graph the Gerschgorin discs. Compare your results with the exact eigenvalues by computing the eigenvalues directly.

1.  $\begin{bmatrix} 4 & -1 \\ 2 & 2 \end{bmatrix}$

2.  $\begin{bmatrix} 4 & -9 \\ 2 & -5 \end{bmatrix}$

**3.** $\begin{bmatrix} 1 & -1 & 4 \\ 3 & 2 & -1 \\ 2 & 1 & -1 \end{bmatrix}$

**4.** $\begin{bmatrix} 2 & 1 & 1 \\ 1 & 1 & 0 \\ 1 & 0 & 2 \end{bmatrix}$

For Exercises 5–7, perform three iterations of the power method using $\mathbf{x}^{(0)} = [1, 1]'$ or $[1, 1, 1]'$. Compare your results with the exact dominant eigenvalue.

**5.** $\begin{bmatrix} 4 & -9 \\ 2 & -5 \end{bmatrix}$

**6.** $\begin{bmatrix} 4 & -1 \\ 2 & 2 \end{bmatrix}$

**7.** $\begin{bmatrix} 0 & 1 & 1 \\ 1 & 0 & 1 \\ 1 & 1 & 0 \end{bmatrix}$

**8.** Using $\lambda = -2$ and $\mathbf{v}_1 = \begin{bmatrix} 1 \\ \frac{2}{3} \end{bmatrix}$, deflate the matrix $B = \begin{bmatrix} 4 & -9 \\ 2 & -5 \end{bmatrix}$ to find the remaining eigenvalue and a corresponding eigenvector of $A$ in Example 13.4.1.

**9.** Prove that the matrix $A_1$ given by Equation (13.4.2) has the eigenvalues $\lambda_2, \lambda_3, \ldots, \lambda_N, 0$ where $\lambda_2, \lambda_3, \ldots, \lambda_N$ are the eigenvalues of $A$ given by Equation (13.2.1).

**10.** Given $p(x) = x^2 - x - 2$,
 (a) Show that the characteristic polynomial of the companion matrix
$$A = \begin{bmatrix} 0 & 1 \\ 2 & 1 \end{bmatrix} \text{ equals } p(x).$$
 (b) Apply the Gerschgorin Theorem to obtain bounds on the eigenvalues of $A$. Determine your results by computing the eigenvalues directly.
 (c) Apply the power method to find the dominant eigenvalue of $A$.
 (d) Apply the inverse power method to find the smallest magnitude eigenvalue of $A$.

**11.** Perform three iterations to determine the smallest magnitude eigenvalue and the corresponding eigenvector of the matrix in Exercise 5 using the inverse power method. Use $\mathbf{x}^{(0)} = [1, 1]$.

---

## COMPUTER EXERCISES

**C.1.** Write a subroutine using Equation (13.3.14) with $\mathbf{x}^{(0)} = [1, 1, \ldots, 1]'$ to generate a sequence to determine the dominant eigenvalue and a corresponding eigenvector of a given matrix $A$. Terminate iterative procedure when $|\lambda^{(k+1)} - \lambda^{(k)}| < 10^{-5} |\lambda^{(k+1)}|$ or the number of iterations exceeds 100.

**C.2.** Write a program to compute the dominant eigenvalue and the corresponding eigenvector of a given $N \times N$ matrix $A$ using Exercise C.1.

**C.3.** Use Exercise C.2 to compute the dominant eigenvalue and the corresponding eigenvector of

$$A = \begin{bmatrix} 3 & 0 & 1 \\ 0 & -3 & 0 \\ 1 & 0 & 3 \end{bmatrix}$$

**C.4.** Write a subroutine using the inverse power method [Equation (13.4.8)] with $\mathbf{x} = [1, 1, \ldots, 1]'$ to generate a sequence to determine the smallest magnitude eigenvalue and the corresponding eigenvector of a given matrix $A$. Terminate iterative procedure when $|\lambda^{(k+1)} - \lambda^{(k)}| < 10^{-5} |\lambda^{(k+1)}|$ or the number of iterations exceeds 100.

**C.5.** Write a program to compute the smallest magnitude eigenvalue and the corresponding eigenvector of a given $N \times N$ matrix $A$ using Exercise C.4.

**C.6.** Use Exercise C.5 to compute the smallest magnitude eigenvalue and the corresponding eigenvector of the matrix $A$ given in Exercise C.3.

**C.7.** The displacement $y(x)$ from the axis of rotation of a tightly stretched flexible string of length one and linear density $\rho$ rotating with uniform angular velocity $\omega$ about its equilibrium position along the $x$-axis is given by

$$\frac{d^2y}{dx^2} + \lambda y = 0; \qquad y(0) = 0, \qquad y(1) = 0 \qquad \textbf{(C.7.1)}$$

where $\lambda = (\rho\omega^2)/T$ and $T$ is the uniform tensile force. This problem can be transformed into a matrix eigenvalue problem by dividing the interval $[0, 1]$ into $N$ subintervals and replacing the second derivative at each grid point $x_i$ by

$$\frac{d^2y}{dx^2}(x_i) \approx \frac{Y_{i-1} - 2Y_i + Y_{i+1}}{h^2}$$

This substitution reduces Equation (C.7.1) to $A\mathbf{Y} = \lambda\mathbf{Y}$. As $N$ increases the smallest magnitude eigenvalue of matrix $A$ approaches the smallest magnitude eigenvalue of Equation (C.7.1). Use Exercise C.5 to compute the smallest value of angular velocity $\omega$ for which the string starts to display from its undeformed position. Use various values of $N$ and compare its corresponding computed smallest value of angular velocity $\omega$.

**C.8.** In order to solve the system

$$\begin{array}{rcl} -2V_1 + V_2 & = & -V_0 - h^2 \\ V_1 - 2V_2 + V_3 & = & -h^2 \\ V_2 - 2V_3 + V_4 & = & -h^2 \\ V_3 - 2V_4 + V_5 & = & -h^2 \\ V_4 - 2V_5 + V_6 & = & -h^2 \\ V_5 - 2V_6 & = & -V_7 - h^2 \end{array}$$

using the SOR in Example 11.6.1, we need $\omega_o$ given by Equation (11.6.11). To find $\omega_o$, we need $\rho(B_J)$ where

$$B_J = -D^{-1}(L + U) = \begin{bmatrix} 0 & 1/2 & 0 & 0 & 0 & 0 \\ 1/2 & 0 & 1/2 & 0 & 0 & 0 \\ 0 & 1/2 & 0 & 1/2 & 0 & 0 \\ 0 & 0 & 1/2 & 0 & 1/2 & 0 \\ 0 & 0 & 0 & 1/2 & 0 & 1/2 \\ 0 & 0 & 0 & 0 & 1/2 & 0 \end{bmatrix}$$

Verify that the dominant eigenvalue of $B_J$ is $\cos(\pi/7) \approx 0.90097$ using Exercise C.2.

# The Symmetric Eigenvalue Problem

The symmetric matrices are extremely important in applications and the eigen problem for symmetric matrices is simpler than the eigen problem for matrices in general. Our object in the next four sections is to develop methods that compute all the eigenvalues and eigenvectors when $A$ is a real symmetric matrix. We shall assume throughout the next four sections that $A$ is a real symmetric matrix. Theorem 13.1.6 assures us that there is an orthogonal matrix $Q$ such that $Q^t AQ = D$ where $D$ is a diagonal matrix whose diagonal elements are the eigenvalues of $A$. How do we go from $A$ to $D$? We can produce one zero at each step (as in Gaussian elimination) or we can work with a whole column at once. Let us first try to produce one zero at each step. This is called the **Jacobi method**.

## 13.5   THE JACOBI METHOD

Let us first consider the simple case of a $2 \times 2$ matrix. Let

$$A = \begin{bmatrix} a_{11} & a_{12} \\ a_{21} & a_{22} \end{bmatrix} \quad \text{where } a_{12} = a_{21} \tag{13.5.1}$$

How do we find $Q$?

If we are given a general equation of a conic, then the basic problem in analytic geometry is to "rotate the axes" through an angle $\theta$ to make the new axes coincide with the major and minor axes of the conic. This has the effect of removing the cross-product term. This suggests that $Q$ must coincide with the plane rotation matrix; therefore, let

$$Q = \begin{bmatrix} \cos\theta & -\sin\theta \\ \sin\theta & \cos\theta \end{bmatrix} \tag{13.5.2}$$

It can be verified that $Q'Q = I$; in other words, $Q$ is an orthogonal matrix. Our goal is to find $\theta$ such that

$$Q'AQ = \begin{bmatrix} \lambda_1 & 0 \\ 0 & \lambda_2 \end{bmatrix}$$

where $\lambda_1$ and $\lambda_2$ are the eigenvalues of $A$. We have

$$Q'AQ = \begin{bmatrix} \cos\theta & \sin\theta \\ -\sin\theta & \cos\theta \end{bmatrix} \begin{bmatrix} a_{11} & a_{12} \\ a_{21} & a_{22} \end{bmatrix} \begin{bmatrix} \cos\theta & -\sin\theta \\ \sin\theta & \cos\theta \end{bmatrix}$$

$$= \begin{bmatrix} \cos\theta & \sin\theta \\ -\sin\theta & \cos\theta \end{bmatrix} \begin{bmatrix} a_{11}\cos\theta + a_{12}\sin\theta & -a_{11}\sin\theta + a_{12}\cos\theta \\ a_{21}\cos\theta + a_{22}\sin\theta & -a_{21}\sin\theta + a_{22}\cos\theta \end{bmatrix}$$

$$= \begin{bmatrix} a_{11}\cos^2\theta + a_{12}\sin\theta\cos\theta + a_{22}\sin^2\theta & (a_{22}-a_{11})\sin\theta\cos\theta + a_{12}(\cos^2\theta - \sin^2\theta) \\ (a_{22}-a_{11})\sin\theta\cos\theta + a_{12}(\cos^2\theta - \sin^2\theta) & a_{11}\sin^2\theta - 2a_{12}\sin\theta\cos\theta + a_{22}\cos^2\theta \end{bmatrix}$$

Since we want $Q'AQ$ to be a diagonal matrix,

$$(a_{22} - a_{11})\sin\theta\cos\theta + a_{12}(\cos^2\theta - \sin^2\theta) = 0$$

or

$$\frac{(a_{22} - a_{11})}{2}\sin 2\theta + a_{12}\cos 2\theta = 0 \tag{13.5.3}$$

Therefore

$$\tan 2\theta = \frac{2a_{12}}{a_{11} - a_{22}} \quad \text{if } a_{22} \neq a_{11} \tag{13.5.4}$$

If $a_{11} = a_{22}$, then Equation (13.5.3) becomes $a_{12}\cos 2\theta = 0$ and since $a_{12} \neq 0$, we get $\cos 2\theta = 0$. Therefore we choose $\theta = \pi/4$ if $a_{11} = a_{22}$. In order to get a rotation as small as possible, we shall always take $\theta$ to lie in the range

$$|\theta| \leq \pi/4$$

More generally, we define the $Q(i, j)$ matrix which can be considered to be a generalization of Equation (13.5.2) in the sense that this orthogonal similarity transformation produces zero at the $(i, j)$th entry. Thus

$$Q(i, j) = \begin{pmatrix} 1 & 0 & \cdots & 0 & \cdots & 0 & \cdots & 0 \\ 0 & 1 & \cdots & 0 & \cdots & 0 & \cdots & 0 \\ \cdots & \cdots & \cdots & & \cdots & & \cdots & \cdots \\ 0 & 0 & \cdots & \cos\theta & \cdots & -\sin\theta & \cdots & 0 \\ \cdots & \cdots & \cdots & \cdots & & \cdots & \cdots & \cdots \\ 0 & 0 & \cdots & \sin\theta & \cdots & \cos\theta & \cdots & 0 \\ \cdots & \cdots & \cdots & \cdots & & \cdots & \cdots & \cdots \\ 0 & 0 & \cdots & 0 & \cdots & 0 & \cdots & 1 \end{pmatrix} \begin{matrix} \\ \\ \\ \leftarrow i\text{th row} \\ \\ \leftarrow j\text{th row} \\ \\ \\ \end{matrix} \tag{13.5.5}$$

with $i$th column and $j$th column indicated above.

For example, for $N = 3$ we have

$$Q(1, 2) = \begin{bmatrix} \cos\theta & -\sin\theta & 0 \\ \sin\theta & \cos\theta & 0 \\ 0 & 0 & 1 \end{bmatrix} \quad \text{and} \quad Q(2, 3) = \begin{bmatrix} 1 & 0 & 0 \\ 0 & \cos\theta & -\sin\theta \\ 0 & \sin\theta & \cos\theta \end{bmatrix}$$

Now we want $Q'(i, j)\,AQ(i, j)$ to have zeros at the $(i, j)$ and $(j, i)$ entries. Let

$$A^{(1)} = Q'(i, j)\,AQ(i, j) \tag{13.5.6}$$

Then it can be verified (see Exercise 7) that $A^{(1)}$ is symmetric. Further $AQ(i, j)$ changes only the $i$ and $j$ columns of $A$ and $Q'(i, j)A$ changes only the $i$ and $j$ rows of $A$. Thus the elements of $A$ and $A^{(1)}$ are the same except for the elements in the $i$th and $j$th rows and columns as shown below:

$$A^{(1)} = \begin{bmatrix} a_{11} & a_{12} & \dots & a_{1i}^{(1)} & \dots & a_{1j}^{(1)} & \dots & a_{1N} \\ a_{21} & a_{22} & \dots & a_{2i}^{(1)} & \dots & a_{2j}^{(1)} & \dots & a_{2N} \\ \dots & \dots & \dots & \dots & \dots & \dots & \dots \\ a_{i1}^{(1)} & a_{i2}^{(1)} & \dots & a_{ii}^{(1)} & \dots & a_{ij}^{(1)} & \dots & a_{iN}^{(1)} \\ \dots & \dots & \dots & \dots & \dots & \dots & \dots \\ a_{j1}^{(1)} & a_{j2}^{(1)} & \dots & a_{ji}^{(1)} & \dots & a_{jj}^{(1)} & \dots & a_{jN}^{(1)} \\ \dots & \dots & \dots & \dots & \dots & \dots & \dots \\ a_{N1} & a_{N2} & \dots & a_{Ni}^{(1)} & \dots & a_{Nj}^{(1)} & \dots & a_{NN} \end{bmatrix} \tag{13.5.7}$$

Multiplying $Q'(i, j)\,AQ(i, j)$ given in Equation (13.5.6) and using the symmetry of $A$, we get

$$a_{ki}^{(1)} = a_{ik}^{(1)} = a_{ik}\cos\theta + a_{jk}\sin\theta \quad k \neq i \quad \text{and} \quad k \neq j \tag{13.5.8}$$

$$a_{kj}^{(1)} = a_{jk}^{(1)} = -a_{ik}\sin\theta + a_{jk}\cos\theta \quad k \neq i \quad \text{and} \quad k \neq j \tag{13.5.9}$$

$$a_{ii}^{(1)} = a_{ii}\cos^2\theta + 2a_{ij}\sin\theta\cos\theta + a_{jj}\sin^2\theta \tag{13.5.10}$$

$$a_{jj}^{(1)} = a_{ii}\sin^2\theta - 2a_{ij}\sin\theta\cos\theta + a_{jj}\cos^2\theta \tag{13.5.11}$$

$$a_{ij}^{(1)} = a_{ji}^{(1)} = a_{ij}\cos 2\theta + \frac{1}{2}(a_{jj} - a_{ii})\sin 2\theta \tag{13.5.12}$$

The idea of the Jacobi method is to have the $(i, j)$ and $(j, i)$ entries in $A^{(1)}$ be zero; that is, from Equation (13.5.12) we have for $a_{ii} \neq a_{jj}$

$$\tan 2\theta = \frac{2a_{ij}}{a_{ii} - a_{jj}} \tag{13.5.13}$$

We shall select $\theta$ such that $|\theta| \leq \pi/4$. When $a_{ii} = a_{jj}$, let $\theta = \pi/4$ if $a_{ij} > 0$ and $\theta = -\pi/4$ if $a_{ij} < 0$.

Next, we select another nonzero off-diagonal element. The procedure is repeated. The $(i, j)$ and $(j, i)$ entries which were reduced to zeros in $A^{(1)}$ will no longer be zeros in $A^{(2)}$. In the method suggested by Jacobi in 1846, the largest off-diagonal element $a_{ij}$ was searched for and then was set to zero. If this procedure is repeated indefinitely, we would like to show that the limit of the sequence $A^{(k)}$ as $k \to \infty$ will be a diagonal matrix $D$ with the eigenvalues of $A$ as its diagonal elements.

Consider the sum of the squares of the off-diagonal elements of $A$, $S = \sum_{i=1}^{N} \sum_{\substack{j=1 \\ j \neq i}}^{N} a_{ij}^2$. The sum of the squares of the off-diagonal elements of $A^{(1)}$, $S^{(1)} = \sum_{i=1}^{N} \sum_{\substack{j=1 \\ j \neq i}}^{N} (a_{ij}^{(1)})^2$. From Equations (13.5.8) and (13.5.9), we have $(a_{ik}^{(1)})^2 + (a_{jk}^{(1)})^2 = a_{ik}^2 + a_{jk}^2$. Since $\theta$ was chosen so that $a_{ij}^{(1)} = a_{ji}^{(1)} = 0$, then

$$S^{(1)} = S - 2a_{ij}^2 \tag{13.5.14}$$

The question is how to relate $S$ and $a_{ij}$? Let $|a_{ij}| = c$ be the largest off-diagonal element. Then

$$S = \sum_{i=1}^{N} \sum_{\substack{j=1 \\ j \neq i}}^{N} a_{ij}^2 \leq \sum_{i=1}^{N} \sum_{\substack{j=1 \\ j \neq i}}^{N} c^2 = \sum_{i=1}^{N} (N-1)\, c^2 = N(N-1)\, c^2$$

Hence

$$c^2 = a_{ij}^2 \geq \frac{S}{N(N-1)} \tag{13.5.15}$$

Using Equation (13.5.15) in Equation (13.5.14), we get

$$S^{(1)} \leq \left(1 - \frac{2}{N(N-1)}\right) S$$

Applying these results to the sequence $A^{(k)}$ yields

$$S^{(k)} \leq \left(1 - \frac{2}{N(N-1)}\right)^k S \tag{13.5.16}$$

As $k \to \infty$, $S^{(k)} \to 0$ since the right-hand side of Equation (13.5.16) tends to zero as $k \to \infty$. Since $A^{(k)}$ is similar to $A$ and $A^{(k)}$ has off-diagonal elements which tend to zero, then $\lim_{k \to \infty} A^{(k)} = D$.

Since a search to find the largest element is time consuming on a computer, a better strategy is the **cyclic Jacobi method** where we select the elements to be eliminated in a strict order. Perhaps we can simply proceed down the rows: $(1, 2)$, $(1, 3)$, ..., $(1, N)$; $(2, 3)$, $(2, 4)$, ..., $(2, N)$; ...; $(N - 1, N)$. Henrici (1958) has shown that the cyclic Jacobi method converges if the angles are suitably restricted.

Since $S^{(1)} = S - 2a_{ij}^2$ indicates that a small $a_{ij}$ would be wasteful; Pope and Tompkins (1957) suggested a refinement. They suggested associating a threshold value with each sweep. If the magnitude of the off-diagonal element is below the threshold value, then ignore that rotation. Of course, we can lower the threshold value during each sweep.

EXAMPLE 13.5.1    Find the eigenvalues of

$$A = \begin{bmatrix} 1 & 2 & 3 \\ 2 & 3 & 4 \\ 3 & 4 & 5 \end{bmatrix}$$

using the cyclic Jacobi method.

We proceed down the rows. First we get $a_{12} = a_{21} = 0$. Then $a_{13}$, $a_{31}$, $a_{23}$, and $a_{32}$ are set to zero. We have

$$Q(1, 2) = \begin{bmatrix} \cos\theta & -\sin\theta & 0 \\ \sin\theta & \cos\theta & 0 \\ 0 & 0 & 1 \end{bmatrix}$$

where $\tan 2\theta = 2a_{12}/(a_{11} - a_{22}) = 4/(1 - 3) = -2$. Therefore $\theta = \frac{1}{2}\tan^{-1}(-2) = -1.10715/2 = -0.55357$. This gives

$$Q(1, 2) = \begin{bmatrix} 0.85065 & 0.52573 & 0 \\ -0.52573 & 0.85065 & 0 \\ 0 & 0 & 1.0 \end{bmatrix}$$

Therefore

$$\begin{aligned} A^{(1)} &= Q'(1, 2)\, AQ(1, 2) \\ &= \begin{bmatrix} 0.85065 & -0.52573 & 0 \\ 0.52573 & 0.85065 & 0 \\ 0 & 0 & 1.0 \end{bmatrix} \begin{bmatrix} 1 & 2 & 3 \\ 2 & 3 & 4 \\ 3 & 4 & 5 \end{bmatrix} \begin{bmatrix} 0.85065 & 0.52573 & 0 \\ -0.52573 & 0.85065 & 0 \\ 0 & 0 & 1.0 \end{bmatrix} \\ &= \begin{bmatrix} -0.23607 & 0.0 & 0.44903 \\ 0.0 & 4.23607 & 4.97980 \\ 0.44903 & 4.97980 & 5.00000 \end{bmatrix} \end{aligned}$$

Now

$$Q(1, 3) = \begin{bmatrix} \cos\theta & 0 & -\sin\theta \\ 0 & 1 & 0 \\ \sin\theta & 0 & \cos\theta \end{bmatrix}$$

where $\tan 2\theta = 2a_{13}^{(1)}/(a_{11}^{(1)} - a_{33}^{(1)}) = 2(0.44903)/(-0.23607 - 5) = -0.17152$. Hence $\theta = -0.084931$. Then

$$A^{(2)} = Q'(1, 3)\, A^{(1)}\, Q(1, 3) = \begin{bmatrix} -0.27430 & -0.42243 & 0.0 \\ -0.42243 & 4.23607 & 4.96185 \\ 0.0 & 4.96185 & 5.03823 \end{bmatrix}$$

Observe that $a_{12}^{(2)} \neq 0$, while $a_{12}^{(1)} = 0$. Continuing, we get

$$A^{(5)} = \begin{bmatrix} -0.00003 & -0.00424 & 0.0 \\ -0.00424 & -0.61987 & -0.19138 \\ 0.0 & -0.19138 & 9.61990 \end{bmatrix} \quad \text{and}$$

$$A^{(8)} = \begin{bmatrix} 0.00000 & 0.00000 & 0.00000 \\ 0.00000 & -0.62348 & 0.00000 \\ 0.00000 & 0.00000 & 9.62348 \end{bmatrix}$$

Thus the eigenvalues are $0$, $-0.62348$, and $9.62348$.    ■ ■ ■

**Algorithm**

**Subroutine** Jacobi ($N$, $A$, eps, maxit)
Comment: This program reduces a given real and symmetric $N \times N$ matrix to a diagonal matrix using the cyclic Jacobi method given by Equations (13.5.8)–(13.5.13); eps is the tolerance and maxit is the maximum number of iterations.

1 **Print** $N$, $A$, eps, maxit
2 $m = 0$
3 **If** $m >$ maxit, **then print** 'algorithm fails: no convergence' and exit.
4 **Repeat** steps 5–26 **with** $i = 1, 2, \ldots, N$
    5 **Repeat** steps 6–26 **with** $j = i + 1, i + 2, \ldots, N$
      6 **If** $|a_{ii} - a_{jj}| >$ eps, **then** $\theta = (\frac{1}{2})(\tan^{-1}(2a_{ij})/(a_{ii} - a_{jj}))$
      7 **If** $|a_{ii} - a_{jj}| <$ eps, **then** $\theta = (\pi/4) * (a_{ij}/|a_{ij}|)$
      8 **If** $|\theta| > \pi/4$, **then** $\theta = \theta - (\pi/2) * (\theta/|\theta|)$
      9 $c = \cos\theta$
      10 $s = \sin\theta$
      11 $cc = c * c$
      12 $ss = s * s$
      13 $cs = c * s$
      14 $aij1 = a_{ij}$
      15 $aii1 = a_{ii}$
      16 $ajj1 = a_{jj}$
      17 $a_{ii} = aii1 * cc + 2 * aij1 * cs + ajj1 * ss$
      18 $a_{jj} = aii1 * ss - 2 * aij1 * cs + ajj1 * cc$
      19 **Repeat** steps 20–26 **with** $k = 1, 2, \ldots, N$
        20 **If** $k = i$ or $k = j$, **then** return to 19
        21 $aki1 = a_{ki}$
        22 $akj1 = a_{kj}$
        23 $a_{ki} = aki1 * c + akj1 * s$
        24 $a_{ik} = a_{ki}$
        25 $a_{kj} = -akj1 * s + akj1 * c$
        26 $a_{jk} = a_{kj}$
27 $m = m + 1$
28 **Repeat** steps 29–31 **with** $i = 1, 2, \ldots, N$
    29 **Repeat** steps 30–31 **with** $j = 1, 2, \ldots, N$
      30 **If** $j = i$, **then** return to 29
      31 **If** $|a_{ij}| >$ eps, **then** return to 3
32 **Print** $m$, $A$, and 'finished' and exit.

## 13.6 THE GIVENS METHOD

In the Jacobi method, elements in the $(i, j)$ and $(j, i)$ positions were reduced to zero by using the orthogonal matrix $Q(i, j)$; these entries do not necessarily remain zero during subsequent transformations. Givens (1958) proposed a method that preserves the zeros on the off-diagonal positions once they are created.

In the Jacobi method the element $a_{ij}^{(1)}$ was annihilated in Equation (13.5.6) by using the orthogonal matrix $Q(i, j)$ given by Equation (13.5.5) where $\theta$ was determined by equating Equation (13.5.12) to zero. In fact, we have Equations (13.5.8) through (13.5.12) to determine $\theta$. Rather than equating Equation (13.5.12) to zero (annihilating $a_{ij}^{(1)}$), let us annihilate $a_{kj}^{(1)}$; that is, equate Equation (13.5.9) to zero. Thus

$$a_{kj}^{(1)} = -a_{ik}\sin\theta + a_{jk}\cos\theta = 0 \tag{13.6.1}$$

This equation determines $\theta$ and gives

$$\tan \theta = \frac{a_{jk}}{a_{ik}} \tag{13.6.2}$$

We can choose for $k$ any number between one and $N$ except $i$ and $j$. For simplicity choose $k = i - 1$. Hence

$$\tan \theta = \frac{a_{j,i-1}}{a_{i,i-1}} \tag{13.6.3}$$

If we use $\theta$ given by Equation (13.6.3) in the orthogonal matrix $Q(i, j)$ given by Equation (13.5.5), then $a^{(1)}_{i-1,j}$ and $a^{(1)}_{j,i-1}$ are annihilated in the matrix $A^{(1)}$ given by Equation (13.5.6).

Let us try to annihilate $a^{(1)}_{12} = a^{(1)}_{i-1,j} = a^{(1)}_{j,i-1}$. We will need the orthogonal matrix $Q(2, 2)$ which is not defined. Since we cannot diagonalize a matrix through this procedure, we will try to reduce a given matrix to a tridiagonal matrix. We annihilate $a^{(1)}_{13}$ and $a^{(1)}_{31}$ by using the orthogonal matrix $Q(2, 3)$. Further we reduce the entries in the $(1, 4)$ and $(4, 1)$ positions to zeros by using the orthogonal matrix $Q(2, 4)$. The orthogonal matrix $Q(2, 4)$ alters the second and fourth rows and columns in $A^{(1)}$. Thus the entries in the $(1, 3)$ and $(3, 1)$ positions of $A^{(1)}$ that are zero remain zero. By repeating this procedure, we can transform $A$ into a matrix in which the $(1, 3)$, $(1, 4)$, ..., $(1, N)$ positions in the first row and $(3, 1)$, $(4, 1)$, ..., $(N, 1)$ positions in the first column are zero. We proceed to reduce the elements in the $(2, 4)$ and $(4, 2)$ positions in the resulting matrix to zero by using $Q(3, 4)$. The orthogonal matrix $Q(3, 4)$ alters the third and fourth rows and columns in the resulting matrix and therefore the zero elements in the first row remain zero. Further, the resulting matrix is symmetric so the zero elements in the first column remain zero. We continue with $(2, 5)$, $(2, 6)$, ..., $(2, N)$ and $(5, 2)$, $(6, 2)$, ..., $(N, 2)$ positions. This procedure continues until $A$ is reduced to a tridiagonal matrix

$$T = \begin{bmatrix} b_1 & c_1 & 0 & 0 & \dots & 0 \\ c_1 & b_2 & c_2 & 0 & \dots & 0 \\ 0 & c_2 & b_3 & c_3 & \dots & 0 \\ \vdots & \vdots & \vdots & \vdots & \ddots & \vdots \\ 0 & 0 & 0 & 0 & \dots & b_N \end{bmatrix} \tag{13.6.4}$$

**EXAMPLE 13.6.1**  Reduce

$$A = \begin{bmatrix} 1 & 2 & 3 \\ 2 & 3 & 4 \\ 3 & 4 & 5 \end{bmatrix}$$

to a tridiagonal form using the Givens method.

We reduce the entry at the position $(1, 3)$ to zero. Therefore we have $i - 1 = 1$, so $i = 2$ and $j = 3$. From Equation (13.6.3), we get

$$\tan \theta = \frac{a_{31}}{a_{21}} = \frac{3}{2} = 1.5$$

Therefore

$$Q(2, 3) = \begin{bmatrix} 1 & 0 & 0 \\ 0 & \cos\theta & -\sin\theta \\ 0 & \sin\theta & \cos\theta \end{bmatrix} = \begin{bmatrix} 1 & 0 & 0 \\ 0 & 2/\sqrt{13} & -3/\sqrt{13} \\ 0 & 3/\sqrt{13} & 2/\sqrt{13} \end{bmatrix}$$

Thus

$$Q'(2, 3)\, AQ(2, 3) = \begin{bmatrix} 1 & 0 & 0 \\ 0 & 2/\sqrt{13} & 3/\sqrt{13} \\ 0 & -3/\sqrt{13} & 2/\sqrt{13} \end{bmatrix} \begin{bmatrix} 1 & 2 & 3 \\ 2 & 3 & 4 \\ 3 & 4 & 5 \end{bmatrix} \begin{bmatrix} 1 & 0 & 0 \\ 0 & 2/\sqrt{13} & -3/\sqrt{13} \\ 0 & 3/\sqrt{13} & 2/\sqrt{13} \end{bmatrix}$$

$$= \begin{bmatrix} 1 & 0 & 0 \\ 0 & 2/\sqrt{13} & 3/\sqrt{13} \\ 0 & -3/\sqrt{13} & 2/\sqrt{13} \end{bmatrix} \begin{bmatrix} 1 & 13/\sqrt{13} & 0 \\ 2 & 18/\sqrt{13} & -1/\sqrt{13} \\ 3 & 23/\sqrt{13} & -2/\sqrt{13} \end{bmatrix}$$

$$= \begin{bmatrix} 1 & \sqrt{13} & 0 \\ \sqrt{13} & 105/\sqrt{13} & -8/\sqrt{13} \\ 0 & -8/\sqrt{13} & -1/\sqrt{13} \end{bmatrix}$$

■ ■ ■

In the Jacobi method, the eigenvalues are exhibited on the diagonal of a diagonalized matrix. We are not so fortunate when we use the Givens method. However, the tridiagonalization requires a finite number of steps, unlike the Jacobi method that requires an unknown number of iterations to converge.

The next task is to find the eigenvalues of a tridiagonal matrix $T$ given by Equation (13.6.4). We will outline the classical method of Sturm sequences along with the Householder method. Also, we will discuss a method based on $QR$ Transformation. The Householder method, to be discussed next, is more efficient than the Givens method, so the Givens method is generally not used.

EXERCISES

For each of the matrices in Exercises 1–3, use Jacobi's method to find a rotation matrix $Q$ and a diagonal matrix $D = \text{diag}(\lambda_1, \lambda_2)$ such that $Q'AQ = D$.

**1.** $A = \begin{bmatrix} 2 & 1 \\ 1 & 2 \end{bmatrix}$

**2.** $A = \begin{bmatrix} 2 & 3 \\ 3 & 1 \end{bmatrix}$

**3.** $A = \begin{bmatrix} 5 & 1 \\ 1 & 3 \end{bmatrix}$

Perform two iterations of cyclic Jacobi's method for each of the matrices in Exercises 4–6.

**4.** $\begin{bmatrix} 1 & 1 & 2 \\ 1 & 2 & 1 \\ 2 & 1 & 2 \end{bmatrix}$

**5.** $\begin{bmatrix} 1 & -1 & 2 \\ -1 & 1 & 0 \\ 2 & 0 & 2 \end{bmatrix}$

**6.** $\begin{bmatrix} 2 & 0 & 1 \\ 0 & 1 & -2 \\ 1 & -2 & 3 \end{bmatrix}$

**7.** Verify that the matrix $A^{(1)}$ given by Equation (13.5.6) is symmetric.

**8.** Show that if $A = \begin{bmatrix} \cos\theta & -\sin\theta \\ \sin\theta & \cos\theta \end{bmatrix}$, then $A^i = \begin{bmatrix} \cos i\theta & -\sin i\theta \\ \sin i\theta & \cos i\theta \end{bmatrix}$ for $i = \pm 1, \pm 2, \ldots$.

For Exercises 9–11, reduce each of the matrices to a tridiagonal form using the Givens method.

**9.** $\begin{bmatrix} 0 & 3 & 2 \\ 3 & 0 & 1 \\ 2 & 1 & 0 \end{bmatrix}$

**10.** $\begin{bmatrix} 2 & -1 & 1 & 4 \\ -1 & 3 & 1 & 2 \\ 1 & 1 & 5 & -3 \\ 4 & 2 & -3 & 6 \end{bmatrix}$

**11.** $\begin{bmatrix} 1 & 1 & 3 \\ 1 & -2 & 1 \\ 3 & 1 & 3 \end{bmatrix}$

**12.** Construct the Givens rotation matrix $Q$ that reduces the entry at the position (2, 1) of the matrix

$$\begin{bmatrix} 1 & 2 & 3 \\ 2 & 3 & 4 \\ 3 & 4 & 5 \end{bmatrix}$$

to zero.

## COMPUTER EXERCISES

**C.1.** Write a subroutine to reduce a given symmetric real matrix to a diagonal matrix using cyclic Jacobi's method. Find $\theta$ by using Equation (13.5.13) and then compute all elements of $A^{(1)}$ by using Equations (13.5.8)–(13.5.12). Continue this procedure until $(\sum_{i=1}^{N} \sum_{\substack{j=1 \\ j \neq i}}^{N} |a_{ij}^{(k)}|) < 10^{-5}$.

**C.2.** Write a computer program to find the eigenvalues of a given real symmetric matrix using Exercise C.1.

**C.3.** Using Exercise C.2, find the eigenvalues of

$$A = \begin{bmatrix} 1 & 2 & 3 & 4 & 5 \\ 2 & 3 & 4 & 5 & 6 \\ 3 & 4 & 5 & 6 & 7 \\ 4 & 5 & 6 & 7 & 8 \\ 5 & 6 & 7 & 8 & 9 \end{bmatrix}$$

## 13.7   THE HOUSEHOLDER METHOD

In the Givens method we annihilate each pair of symmetric elements by orthogonal transformations. Householder suggested that a matrix of the form

$$P = I - \frac{2\mathbf{v}\,\mathbf{v}'}{\|\mathbf{v}\|_2^2} \tag{13.7.1}$$

annihilates the required part of a whole column if the components of $\mathbf{v}$ are properly chosen.

Let $\mathbf{x} = [x_1, x_2, \ldots, x_N]'$ be the vector composed of the first column of $A$. We would like to construct $\mathbf{v}$ in Equation (13.7.1) such that $P\mathbf{x}$ is a multiple of $\mathbf{e}_1 = [1, 0, \ldots, 0]'$. Since $P\mathbf{x}$ is a multiple of $\mathbf{e}_1$,

$$\mathbf{v} = \mathbf{x} + \alpha \mathbf{e}_1 \tag{13.7.2}$$

where $\alpha$ is to be determined. Then

$$P\mathbf{x} = \left(I - \frac{2\mathbf{v}\,\mathbf{v}'}{\|\mathbf{v}\|_2^2}\right)\mathbf{x} = \mathbf{x} - \frac{2\mathbf{v}\,(\mathbf{v}'\,\mathbf{x})}{\|\mathbf{v}\|_2^2} \tag{13.7.3}$$

Further

$$\begin{aligned} \mathbf{v}'\,\mathbf{x} = (\mathbf{x}' + \alpha\,\mathbf{e}_1')\,\mathbf{x} &= \mathbf{x}'\,\mathbf{x} + \alpha\mathbf{e}_1'\,\mathbf{x} \\ &= \|\mathbf{x}\|_2^2 + \alpha\,x_1 \end{aligned} \tag{13.7.4}$$

Substituting $\mathbf{v}$ from Equation (13.7.2) and $\mathbf{v}'\,\mathbf{x}$ from Equation (13.7.4) into Equation (13.7.3), we get

$$\begin{aligned} P\mathbf{x} &= \mathbf{x} - \frac{2(\mathbf{x} + \alpha\mathbf{e}_1)(\|\mathbf{x}\|_2^2 + \alpha x_1)}{\|\mathbf{v}\|_2^2} \\ &= \frac{\mathbf{x}\,(\|\mathbf{v}\|_2^2 - 2(\|\mathbf{x}\|_2^2 + \alpha x_1)) - 2\alpha\mathbf{e}_1\,(\|\mathbf{x}\|_2^2 + \alpha x_1)}{\|\mathbf{v}\|_2^2} \end{aligned} \tag{13.7.5}$$

We want the coefficient of $\mathbf{x}$ to be zero in Equation (13.7.5); therefore,

$$\|\mathbf{v}\|_2^2 - 2(\|\mathbf{x}\|_2^2 + \alpha x_1) = 0 \tag{13.7.6}$$

From this equation we determine $\alpha$. We want to find $\|\mathbf{v}\|_2^2$ since we need it in Equation (13.7.6). Hence

$$
\begin{aligned}
\|\mathbf{v}\|_2^2 = \mathbf{v}'\,\mathbf{v} &= (\mathbf{x} + \alpha\mathbf{e}_1)'\,(\mathbf{x} + \alpha\mathbf{e}_1) \\
&= (\mathbf{x}' + \alpha\mathbf{e}_1')(\mathbf{x} + \alpha\mathbf{e}_1) \\
&= \mathbf{x}'\,\mathbf{x} + \alpha\mathbf{e}_1'\,\mathbf{x} + \alpha\mathbf{x}'\,\mathbf{e}_1 + \alpha^2\,\mathbf{e}_1'\,\mathbf{e}_1 \\
&= \|\mathbf{x}\|_2^2 + 2\alpha\,x_1 + \alpha^2
\end{aligned}
\tag{13.7.7}
$$

Substituting Equation (13.7.7) in Equation (13.7.6) gives

$$
\alpha^2 = \|\mathbf{x}\|_2^2 \quad \text{so } \alpha = \pm\sqrt{\|\mathbf{x}\|_2^2}
\tag{13.7.8}
$$

Substituting Equations (13.7.7) and (13.7.8) in Equation (13.7.5), we get

$$
P\mathbf{x} = \frac{-2\alpha\,\mathbf{e}_1\,(\|\mathbf{x}\|_2^2 + \alpha\,x_1)}{2(\|\mathbf{x}\|_2^2 + \alpha\,x_1)} = -\alpha\,\mathbf{e}_1
\tag{13.7.9}
$$

Thus $\mathbf{v} = \mathbf{x} \pm \|\mathbf{x}\|_2\,\mathbf{e}_1$. To avoid cancellation we choose the sign to agree with the sign of $x_1$. Thus

$$
\mathbf{v} = \mathbf{x} + (\operatorname{sign} x_1)\,\|\mathbf{x}\|_2\,\mathbf{e}_1
\tag{13.7.10}
$$

This simple determination of $\mathbf{v}$ makes the $P$ matrix a powerful tool.

**EXAMPLE 13.7.1**    Let $\mathbf{x} = [0, 4, 3]'$. Find $P$ such that $P\mathbf{x}$ is a multiple of $\mathbf{e}_1 = [1, 0, 0]'$.
From Equation (13.7.10) we have

$$
\mathbf{v} = \mathbf{x} + (\operatorname{sign} x_1)\,\|\mathbf{x}\|_2\,\mathbf{e}_1
$$

$$
= \begin{bmatrix} 0 \\ 4 \\ 3 \end{bmatrix} + \sqrt{0 + 16 + 9}\begin{bmatrix} 1 \\ 0 \\ 0 \end{bmatrix} = \begin{bmatrix} 5 \\ 4 \\ 3 \end{bmatrix}
$$

Since $\|\mathbf{v}\|_2^2 = \mathbf{v}'\,\mathbf{v} = 25 + 16 + 9 = 50$,

$$
P = I - 2\frac{\mathbf{v}\,\mathbf{v}'}{\|\mathbf{v}\|_2^2}
$$

$$
= \begin{bmatrix} 1 & 0 & 0 \\ 0 & 1 & 0 \\ 0 & 0 & 1 \end{bmatrix} - \frac{2}{50}\begin{bmatrix} 5 \\ 4 \\ 3 \end{bmatrix}\begin{bmatrix} 5 & 4 & 3 \end{bmatrix}
$$

$$
= \begin{bmatrix} 1 & 0 & 0 \\ 0 & 1 & 0 \\ 0 & 0 & 1 \end{bmatrix} - \frac{1}{25}\begin{bmatrix} 25 & 20 & 15 \\ 20 & 16 & 12 \\ 15 & 12 & 9 \end{bmatrix}
$$

$$
= \frac{1}{25}\begin{bmatrix} 0 & -20 & -15 \\ -20 & 9 & -12 \\ -15 & -12 & 16 \end{bmatrix}
$$

It can be verified that

$$
Px = \frac{1}{25}\begin{bmatrix} 0 & -20 & -15 \\ -20 & 9 & -12 \\ -15 & -12 & 16 \end{bmatrix}\begin{bmatrix} 0 \\ 4 \\ 3 \end{bmatrix} = \begin{bmatrix} -5 \\ 0 \\ 0 \end{bmatrix} = -5\begin{bmatrix} 1 \\ 0 \\ 0 \end{bmatrix}
$$

$$
= -\alpha\, \mathbf{e}_1
$$

■ ■ ■

Since $P$ is an important matrix, it is essential to know the properties of matrix $P$. We can verify (Exercise 4) that (1) $P' = P$ and (2) $P'P = I$. Thus $P$ is symmetric and orthogonal.

The matrix $P$ may be used in a similarity transformation with our matrix $A$. We consider the first column of $A$ to be $\mathbf{x}$ and then construct $P$ accordingly. Then $PA$ gives a new matrix having $-\alpha$ in the (1, 1) position and zeros in the rest of the first column.

Consider the matrix

$$
A = \begin{bmatrix} 0 & 4 & 3 \\ 4 & 2 & 1 \\ 3 & 1 & 2 \end{bmatrix}
$$

We want to transform its first column $\mathbf{x} = [0, 4, 3]'$ into a scalar multiple of $\mathbf{e}_1 = [1, 0, 0]'$. From Example 13.7.1, we have

$$
P = \frac{1}{25}\begin{bmatrix} 0 & -20 & -15 \\ -20 & 9 & -12 \\ -15 & -12 & 16 \end{bmatrix}
$$

Then

$$
PA = \frac{1}{25}\begin{bmatrix} 0 & -20 & -15 \\ -20 & 9 & -12 \\ -15 & -12 & 16 \end{bmatrix}\begin{bmatrix} 0 & 4 & 3 \\ 4 & 2 & 1 \\ 3 & 1 & 2 \end{bmatrix} = \frac{1}{25}\begin{bmatrix} -125 & -55 & -50 \\ 0 & -74 & -75 \\ 0 & -68 & -25 \end{bmatrix}
$$

Thus $PA$ gives a new matrix having $-5$ in the (1, 1) position and zeros in the rest of the first column. Since we are looking for a similar matrix, consider

$$
P^{-1}AP = PAP = \frac{1}{625}\begin{bmatrix} -125 & -55 & -50 \\ 0 & -74 & -75 \\ 0 & -68 & -25 \end{bmatrix}\begin{bmatrix} 0 & -20 & -15 \\ -20 & 9 & -12 \\ -15 & -12 & 16 \end{bmatrix}
$$

$$
= \frac{1}{625}\begin{bmatrix} 1850 & 2605 & 1735 \\ 2605 & 234 & -312 \\ 1735 & -312 & 416 \end{bmatrix}
$$

$PAP$ is symmetric, but we lost all the zeros in the first column and we do not have zeros in the first row either. We should be less ambitious and reduce matrix $A$ to a triadiagonal matrix. We attempt to reduce $\mathbf{x} = [4, 3]'$ to a scalar multiple of $\mathbf{e}_1 = [1, 0]'$. Since $\|\mathbf{x}\|_2 = 5$,

$$
\mathbf{v} = \mathbf{x} + (\operatorname{sign} x_1)\,\|\mathbf{x}\|_2\, \mathbf{e}_1
$$

$$
= \begin{bmatrix} 4 \\ 3 \end{bmatrix} + 5\begin{bmatrix} 1 \\ 0 \end{bmatrix} = \begin{bmatrix} 9 \\ 3 \end{bmatrix}
$$

Further $\|\mathbf{v}\|_2^2 = \mathbf{v}' \mathbf{v} = 81 + 9 = 90$. Hence

$$P = I - 2\frac{\mathbf{v}\,\mathbf{v}'}{\|\mathbf{v}\|_2^2} = \begin{bmatrix} 1 & 0 \\ 0 & 0 \end{bmatrix} - \frac{2}{90}\begin{bmatrix} 9 \\ 3 \end{bmatrix}[9 \quad 3]$$

$$= \begin{bmatrix} 1 & 0 \\ 0 & 1 \end{bmatrix} - \frac{1}{45}\begin{bmatrix} 81 & 27 \\ 27 & 9 \end{bmatrix} = -\frac{1}{5}\begin{bmatrix} 4 & 3 \\ 3 & -4 \end{bmatrix}$$

At this point $P$ is only of order two, so it can be embedded into the lower right corner of a $3 \times 3$ matrix given by

$$U = \begin{bmatrix} 1 & 0 & 0 \\ 0 & -4/5 & -3/5 \\ 0 & -3/5 & 4/5 \end{bmatrix}$$

It can be verified that

$$U'AU = (UA)\,U = \begin{bmatrix} 0 & 4 & 3 \\ -5 & -11/5 & -2 \\ 0 & -2/5 & 1 \end{bmatrix}\begin{bmatrix} 1 & 0 & 0 \\ 0 & -4/5 & -3/5 \\ 0 & -3/5 & 4/5 \end{bmatrix}$$

$$= \begin{bmatrix} 0 & -5 & 0 \\ -5 & 74/25 & -7/25 \\ 0 & -7/25 & 26/25 \end{bmatrix}$$

Thus $UA$ produces appropriate zeros and the post multiplication by $U$ keeps those zeros because of the first column and first row of $U$. Thus $(UA)U$ is a tridiagonal matrix which is similar to $A$.

Let $A$ be a real symmetric $N \times N$ matrix. In order to reduce $A$ to a tridiagonal form, we select the lower $N - 1$ elements of the first column of $A$ as $\mathbf{x}$ for the first Householder matrix $P_{N-1}$. Then

$$U_1 A = \begin{bmatrix} 1 & 0 & & 0 & \cdots & 0 \\ 0 & & (N-1) \times (N-1) & & & \\ 0 & & P_{N-1} & & & \\ \vdots & \vdots & & \vdots & \ddots & \vdots \\ 0 & & & & & \end{bmatrix}\begin{bmatrix} a_{11} & a_{12} & a_{13} & \cdots & a_{1N} \\ a_{21} & a_{22} & a_{23} & \cdots & a_{2N} \\ a_{31} & a_{32} & a_{33} & \cdots & a_{3N} \\ \vdots & \vdots & \vdots & \ddots & \vdots \\ a_{N1} & a_{N2} & a_{N3} & \cdots & a_{NN} \end{bmatrix}$$

$$= \begin{bmatrix} a_{11} & a_{12} & a_{13} & \cdots & a_{1N} \\ \alpha_1 & a'_{22} & a'_{23} & \cdots & a'_{2N} \\ 0 & a'_{32} & a'_{33} & \cdots & a'_{3N} \\ \vdots & \vdots & \vdots & \ddots & \vdots \\ 0 & a'_{N2} & a'_{N3} & \cdots & a'_{NN} \end{bmatrix}$$

Further $U'_1 = U_1$, therefore

$$A^{(1)} = U'_1 A U_1 = (U_1 A) U_1 = \begin{bmatrix} a_{11} & a_{12} & a_{13} & \dots & a_{1N} \\ \alpha_1 & a'_{22} & a'_{23} & \dots & a'_{2N} \\ 0 & a'_{32} & a'_{33} & \dots & a'_{3N} \\ \vdots & \vdots & \vdots & \ddots & \vdots \\ 0 & a'_{N2} & a'_{N3} & \dots & a'_{NN} \end{bmatrix}$$

$$\begin{bmatrix} 1 & 0 & & 0 & & \dots & 0 \\ 0 & & (N-1) \times (N-1) & & & & \\ 0 & & & P_{N-1} & & & \\ \vdots & \vdots & & \vdots & & \ddots & \vdots \\ 0 & & & & & & \end{bmatrix}$$

$$= \begin{bmatrix} a_{11} & \alpha_1 & 0 & \dots & 0 \\ \alpha_1 & a_{22}^{(1)} & a_{23}^{(1)} & \dots & a_{2N}^{(1)} \\ 0 & a_{32}^{(1)} & a_{33}^{(1)} & \dots & a_{3N}^{(1)} \\ \vdots & \vdots & \vdots & \ddots & \vdots \\ 0 & a_{N2}^{(1)} & a_{N3}^{(1)} & \dots & a_{NN}^{(1)} \end{bmatrix}$$

Now we select the bottom $(N-2)$ elements of the second column of $A^{(1)}$ as $\mathbf{x}$ for the second Householder matrix $P_{N-2}$ and from it construct

$$U_2 = \begin{bmatrix} 1 & 0 & & 0 & & 0 & \dots & 0 \\ 0 & 1 & & 0 & & 0 & \dots & 0 \\ 0 & 0 & (N-2) \times (N-2) & & & & & \\ \vdots & \vdots & & P_{N-2} & & & & \\ 0 & 0 & & & & & & \end{bmatrix}$$

Then

$$A^{(2)} = U_2 A^{(1)} U_2 = \begin{bmatrix} a_{11} & \alpha_1 & 0 & \dots & 0 \\ \alpha_1 & a_{22}^{(1)} & \alpha_2 & \dots & 0 \\ 0 & \alpha_2 & a_{33}^{(2)} & \dots & a_{3N}^{(2)} \\ \vdots & \vdots & \vdots & \ddots & \vdots \\ 0 & 0 & a_{3N}^{(2)} & \dots & a_{NN}^{(2)} \end{bmatrix}$$

The $(N-2) \times (N-2)$ matrix $P_{N-2}$ creates one additional row and column of the required tridiagonal matrix and the upper identify block of $U_2$ keeps the tridiagonalization achieved in the first step. Continuing, we reduce $A$ by using $(N-1)$ transformations to a tridiagonal form $T$ given by

$$T = \begin{bmatrix} b_1 & \alpha_1 & 0 & \dots & 0 \\ \alpha_1 & b_2 & \alpha_2 & \dots & 0 \\ 0 & \alpha_2 & b_3 & \dots & 0 \\ \vdots & \vdots & \vdots & \ddots & \vdots \\ 0 & 0 & 0 & \dots & b_N \end{bmatrix} \qquad \textbf{(13.7.11)}$$

**Algorithm**

**Subroutine** Houder ($N$, $A$, $C$)

Comment: This program reduces a given real and symmetric $N \times N$ matrix to a tridiagonal matrix like Equation (13.7.11) using the Householder transformation given by Equation (13.7.1).

  1 **Print** $N$, $A$
  2 **Repeat** steps 3–33 **with** $i = 1, 2, \ldots, N - 2$
      3 **Repeat** steps 4–6 **with** $l = 1, 2, \ldots, N$
         4 **Repeat** steps 5–6 **with** $m = 1, 2, \ldots, N$
            5 **If** $m = l$, **then** $u_{lm} = 1$
            6 **If** $m \neq l$, **then** $u_{lm} = 0$

Comment: Steps 7–18 generate **v** given by Equation (13.7.10).

  7 xnorm $= 0$
  8 **Repeat** steps 9–11 **with** $j = i + 1, i + 2, \ldots, N$
      9 $e_j = 0$
    10 $x_j = a_{ij}$
    11 xnorm $=$ xnorm $+ x_j * x_j$
 12 $e_{i+1} = 1$
 13 xnorm $= \sqrt{\text{xnorm}}$
 14 vnorm $= 0$
 15 **Repeat** steps 16–17 **with** $j = i + 1, i + 2, \ldots, N$
    16 $v_j = x_j + \text{sign}(x_{i+1}) * \text{xnorm} * e_j$
    17 vnorm $=$ vnorm $+ v_j * v_j$

Comment: Steps 3–6 and 18–20 generate the required matrix $U$.

 18 **Repeat** steps 19–20 **with** $j = i + 1, i + 2, \ldots, N$
    19 **Repeat** step 20 **with** $k = i + 1, i + 2, \ldots, N$
       20 $u_{jk} = u_{jk} - (2 * v_j * v_k / \text{vnorm})$

Comment: This part multiplies $U(AU)$. We have $B = AU$ and $C = UB = U(AU)$.

 21 **Repeat** steps 22–26 **with** $m = 1, 2, \ldots, N$
    22 **Repeat** steps 23–26 **with** $l = 1, 2, \ldots, N$
      23 sum $= 0$
      24 **Repeat** step 25 **with** $k = 1, 2, \ldots, N$
         25 sum $=$ sum $+ a_{mk} * u_{kl}$
      26 $b_{ml} =$ sum
 27 **Repeat** steps 28–32 **with** $m = 1, 2, \ldots, N$
    28 **Repeat** steps 29–32 **with** $l = 1, 2, \ldots, N$
      29 sum $= 0$
      30 **Repeat** step 31 **with** $k = 1, 2, \ldots, N$
         31 sum $=$ sum $+ u_{mk} * b_{kl}$
      32 $c_{ml} =$ sum
33 **Print** $C$ and 'finished' and exit. ◀

---

## 13.8   THE EIGENVALUES OF A SYMMETRIC TRIDIAGONAL MATRIX

We will compute the characteristic polynomial of a matrix $T$ given by Equation (13.7.11) and then use it to find the eigenvalues of $T$. We use this approach because $T$ is a special matrix and it is easy to find the characteristic polynomial of $T$. Let $\alpha_i \neq 0$ for $i = 1$,

2, ..., $N - 1$. There is no loss of generality because the characteristic polynomial trivially factors into two polynomials if some $\alpha_i = 0$.

To compute $P_N(\lambda) = \det(T - \lambda I)$, define the **sequence**

$$p_i(\lambda) = \begin{vmatrix} b_1 - \lambda & \alpha_1 & 0 & \dots & 0 \\ \alpha_1 & b_2 - \lambda & \alpha_2 & \dots & 0 \\ 0 & \alpha_2 & b_3 - \lambda & \dots & 0 \\ \vdots & \vdots & \vdots & \ddots & \vdots \\ 0 & 0 & 0 & \dots & b_i - \lambda \end{vmatrix} \qquad \textbf{(13.8.1)}$$

for $i = 1, 2, \ldots, N$ and

$$p_0(\lambda) = 1$$

By direct evaluation

$$\begin{aligned} p_1(\lambda) &= b_1 - \lambda \\ p_2(\lambda) &= (b_2 - \lambda)(b_1 - \lambda) - \alpha_1^2 \\ &= (b_2 - \lambda)p_1(\lambda) - \alpha_1^2 p_0(\lambda) \end{aligned}$$

By expanding the determinant of Equation (13.8.1) in the last row using minors, we get

$$p_i(\lambda) = (b_i - \lambda)p_{i-1}(\lambda) - \alpha_{i-1}^2 p_{i-2}(\lambda) \qquad \textbf{(13.8.2)}$$

for $i = 2, 3, \ldots, N$. The sequence $\{p_i(\lambda)\}$ is called a **Sturm sequence**.

**EXAMPLE 13.8.1**   Find the characteristic polynomial of

$$T = \begin{bmatrix} -2 & 1 & 0 & 0 & 0 & 0 \\ 1 & -2 & 1 & 0 & 0 & 0 \\ 0 & 1 & -2 & 1 & 0 & 0 \\ 0 & 0 & 1 & -2 & 1 & 0 \\ 0 & 0 & 0 & 1 & -2 & 1 \\ 0 & 0 & 0 & 0 & 1 & -2 \end{bmatrix}$$

We have $p_0(\lambda) = 1$, $p_1(\lambda) = -2 - \lambda$,

$$\begin{aligned} p_2(\lambda) &= (-2 - \lambda)\, p_1(\lambda) - 1\, p_0(\lambda) = (-2 - \lambda)^2 - 1 \\ p_3(\lambda) &= (-2 - \lambda)\, p_2(\lambda) - 1\, p_1(\lambda) = (-2 - \lambda)\, [(-2 - \lambda)^2 - 2] \\ p_4(\lambda) &= (-2 - \lambda)\, p_3(\lambda) - 1\, p_2(\lambda) = (-2 - \lambda)^4 - 3\, (-2 - \lambda)^2 + 1 \\ p_5(\lambda) &= (-2 - \lambda)\, p_4(\lambda) - 1\, p_3(\lambda) = (-2 - \lambda)\, [(-2 - \lambda)^4 - 4\, (-2 - \lambda)^2 + 3] \\ p_6(\lambda) &= (-2 - \lambda)\, p_5(\lambda) - 1\, p_4(\lambda) = (-2 - \lambda)^6 - 5\, (-2 - \lambda)^4 \\ &\quad + 6\, (-2 - \lambda)^2 - 1 \end{aligned}$$

■ ■ ■

The triple recursion relation of (13.8.2) is very useful to evaluate $p_6(\lambda)$. At this point, we can use many polynomial root finding methods since $p_N(\lambda)$ is a polynomial. But the sequence $\{p_i(\lambda)\}$ has special properties and these properties make it easy to isolate the eigenvalues of $T$. Let us study those properties.

We first examine the roots of equations $p_i(\lambda) = 0$ for $i = 2, 3, \ldots, N$. As $\lambda \to -\infty$, it follows from Equation (13.8.2) that $p_i(\lambda) > 0$ for $i = 0, 1, \ldots, N$.

For $i = 1$, we have only one root $b_1$.

For $i = 2$, $p_2(\lambda) = (b_2 - \lambda)(b_1 - \lambda) - \alpha_1^2$. Therefore, we have $p_2(-\infty) > 0$, $p_2(b_1) < 0$, and $p_2(\infty) > 0$. Let $r_1$ and $r_2$ be the roots of $p_2(\lambda) = 0$ such that $r_1 < r_2$. Then

$$-\infty < r_1 < b_1 < r_2 < \infty \qquad \text{(why?)}$$

Hence the roots of $p_2(\lambda) = 0$ are separated by the root of $p_1(\lambda) = 0$. Let $p_2(\lambda) = (r_1 - \lambda)(r_2 - \lambda)$. Then

$$p_3(\lambda) = (b_3 - \lambda)(r_1 - \lambda)(r_2 - \lambda) - \alpha_2^2(b_1 - \lambda)$$

Let us examine the sign of $p_3(\lambda)$ at suitable points:

| $\lambda$ | $-\infty$ | $r_1$ | $r_2$ | $\infty$ |
|---|---|---|---|---|
| **Sign of $p_3(\lambda)$** | + | − | + | − |

Then $p_3(\lambda)$ has three roots $\beta_1$, $\beta_2$, and $\beta_3$ such that

$$-\infty < \beta_1 < r_1 < \beta_2 < r_2 < \beta_3 < \infty$$

In general, it can be shown that $p_i(\lambda) = 0$ has $i$ real roots, separated by the roots of $p_{i-1}(\lambda) = 0$. Also, no two consecutive Sturm polynomials have a common zero.

We have $p_1(b_1) = 0$. Then $p_2(b_1) = -\alpha_1^2$. Thus $p_2(b_1)\, p_0(b_1) < 0$. Let $p_i(\lambda_0) = 0$. Then

$$p_{i+1}(\lambda_0) = (b_{i+1} - \lambda_0)\, p_i(\lambda_0) - \alpha_i^2\, p_{i-1}(\lambda_0)$$
$$= -\alpha_i^2\, p_{i-1}(\lambda_0)$$

Multiplying both sides by $p_{i-1}(\lambda_0)$ yields

$$p_{i+1}(\lambda_0)\, p_{i-1}(\lambda_0) = -\alpha_i^2\, p_{i-1}^2(\lambda_0) < 0$$

Thus if $p_i(\lambda_0) = 0$, then $p_{i+1}(\lambda_0)\, p_{i-1}(\lambda_0) < 0$.

Define $C(\lambda)$ to be the **number of agreements in sign** between consecutive terms in the sequence

$$1, p_1(\lambda), p_2(\lambda), \dots, p_N(\lambda)$$

where if some $p_i(\lambda) = 0$, then we take its sign opposite to the sign of $p_{i-1}(\lambda)$.

It is clear that $C(-\infty) = N$ and $C(\infty) = 0$. Let $a$ and $b$ be two real numbers such that $p_i(\lambda) \neq 0$ for $i = 1, 2, \dots, N$ in the closed interval $[a, b]$. Then $C(a) = C(b)$.

Let $p_i(\lambda) = 0$ for $1 < i < N$ have a root $\rho$ in the interval $[a, b]$. Then $p_{i+1}(\rho)\, p_{i-1}(\rho) < 0$. This is also true in the interval $[\rho - \epsilon, \rho + \epsilon]$ since $p_{i+1}(\rho) \neq 0$ and $p_{i-1}(\rho) \neq 0$.

Let $p_{i-1}(\rho) < 0$. Then we have the following combination of signs:

| $\lambda$ | $p_{i-1}$ | $p_i$ | $p_{i+1}$ |
|---|---|---|---|
| $\rho - \epsilon$ | − | − | + |
| $\rho + \epsilon$ | − | + | + |

The number of sign changes does not change when we pass through a root of $p_i(\lambda) = 0$ if $i < N$.

It is a different story when $i = N$. Let $\rho_1, \rho_2, \ldots, \rho_N$ be the roots of $p_N(\lambda) = 0$ and $\sigma_1, \sigma_2, \ldots, \sigma_{N-1}$ be the roots of $p_{N-1}(\lambda) = 0$. Then $-\infty < \rho_1 < \sigma_1 < \rho_2 < \sigma_2$ $\cdots \rho_{N-1} < \sigma_{N-1} < \rho_N < \infty$ and therefore we have

| $\lambda$ | $p_0$ | $p_1$ | $\cdots$ | $p_{N-1}$ | $p_N$ |
|---|---|---|---|---|---|
| $\rho_1 - \epsilon$ | (same signs | | | $+$ | $+$ |
| $\rho_1 + \epsilon$ | in both cases) | | | $+$ | $-$ |

Thus $C(\rho_1 - \epsilon) - C(\rho_1 + \epsilon) = 1$. Let $\lambda$ be in the open interval which contains $\sigma_1$. Then we have

| $\lambda$ | $p_0$ | $p_1$ | $\cdots$ | $p_{N-2}$ | $p_{N-1}$ | $p_N$ |
|---|---|---|---|---|---|---|
| $\sigma_1 - \epsilon$ | (same signs | | | $+$ | $+$ | $-$ |
| $\sigma_1 + \epsilon$ | in both cases) | | | $+$ | $-$ | $-$ |

and so $C(\sigma_1 - \epsilon) - C(\sigma_1 + \epsilon) = 0$. Further let $\lambda$ be in the open interval which contains $\rho_2$. Then we have

| $\lambda$ | $p_0$ | $p_1$ | $\cdots$ | $p_{N-2}$ | $p_{N-1}$ | $p_N$ |
|---|---|---|---|---|---|---|
| $\rho_2 - \epsilon$ | (same signs | | | | $-$ | $-$ |
| $\rho_2 + \epsilon$ | in both cases) | | | | $-$ | $+$ |

Hence $C(\rho_2 - \epsilon) - C(\rho_2 + \epsilon) = 1$. Continuing the same way through all the roots $\rho_i$, we conclude that the number of sign changes decreases by one each time we pass a root $\rho_i$.

**Theorem 13.8.1**   For $a < b$, the number of roots of $p_N(\lambda) = \det(T - \lambda I)$ in the interval $a < \lambda \leq b$ is given by $C(a) - C(b)$. ∎

**EXAMPLE 13.8.2**   Discuss intervals containing the eigenvalues of $T$ given in Example 13.8.1.
   $P_i(\lambda)$ for $i = 1, 2, \ldots, 6$ are given in Example 13.8.1. We record the signs of $p_i(\lambda)$ for various $\lambda$ in Table 13.8.1.

**Table 13.8.1**

| $\lambda$ | $-5$ | $-4$ | $-3$ | $-2$ | $-1$ | $0$ |
|---|---|---|---|---|---|---|
| $p_0(\lambda)$ | $+$ | $+$ | $+$ | $+$ | $+$ | $+$ |
| $p_1(\lambda)$ | $+$ | $+$ | $+$ | $0(-)$ | $-$ | $-$ |
| $p_2(\lambda)$ | $+$ | $+$ | $0(-)$ | $-$ | $0(+)$ | $+$ |
| $p_3(\lambda)$ | $+$ | $+$ | $-$ | $0(+)$ | $+$ | $-$ |
| $p_4(\lambda)$ | $+$ | $+$ | $-$ | $+$ | $-$ | $+$ |
| $p_5(\lambda)$ | $+$ | $+$ | $0(+)$ | $0(-)$ | $0(+)$ | $-$ |
| $p_6(\lambda)$ | $+$ | $+$ | $+$ | $-$ | $+$ | $+$ |
| $C(\lambda)$ | $6$ | $6$ | $4$ | $3$ | $2$ | $0$ |

Note that $p_2(-3) = 0$ and therefore we take the opposite sign of $p_1(-3)$. The signs are shown in parentheses near 0. There are two eigenvalues in each of the intervals $[-4, -3]$, $[-1, 0]$ and one eigenvalue in each of the intervals $[-3, -2]$, $[-2, -1]$. The Gerschgorin Theorem yields that all eigenvalues are in $[-4, 0]$. ■ ■ ■

**Algorithm**

**Subroutine** Sturm $(N, \mathbf{b}, \boldsymbol{\alpha}, \boldsymbol{\Gamma}, a, bb, \text{eps}, \text{maxit})$
Comment: This program computes the eigenvalues of a symmetric tridiagonal $N \times N$ matrix $T$ given by Equation (13.7.11), where $\mathbf{b} = [b_1, b_2, \ldots, b_N]$ and $\boldsymbol{\alpha} = [\alpha_1, \alpha_2, \ldots, \alpha_{N-1}]$ are the elements of the matrix $T$. We use the Strum sequence $\mathbf{p} = [p_0, p_1, \ldots, p_N]$ given by Equation (13.8.2) to evaluate the characteristic polynomial of the matrix $T$. The interval $[a, bb]$ contains all the eigenvalues of matrix $T$ and is determined by the Gerschgorin Theorem. We divide the interval $[a, bb]$ into $N$ equal subintervals. In each subinterval, we determine whether there are eigenvalues. If there is more than one eigenvalue in a particular subinterval, then we bisect that particular subinterval. We continue to bisect the subinterval until the interval contains only one eigenvalue. Once we determine the interval containing only one eigenvalue, then use the bisection method (Subroutine bisect) to find the eigenvalue within the given tolerance eps. $\boldsymbol{\Gamma} = [\gamma_1, \gamma_2, \ldots, \gamma_N]$ is the set of the eigenvalues of $T$, and maxit is the maximum number of iterations for the bisection method.

1  **Print** $N$, $\mathbf{b}$, $\boldsymbol{\alpha}$, $a$, $bb$, eps, maxit
2  $h = (bb - a)/N$
3  $l = 0$ (This is the counter for the eigenvalues)
4  $k = 0$
5  $k1 = N$
6  $p_0 = 1$
7  $x = a$
8  $y = a + h$
9  $\lambda = y$
10  $p_1 = b_1 - \lambda$
11  **If** $p_1 = 0$, **then** sign $p_1 = -$sign $p_0$
12  **If** (sign $p_0$) $*$ (sign $p_1$) $> 0$, **then** $k = k + 1$
13  **Repeat** steps 14–16 **with** $i = 2, 3, \ldots, N$
    14  $p_i = (b_i - \lambda) * p_{i-1} - \alpha_{i-1}^2 * p_{i-2}$
    15  **If** $p_i = 0$, **then** sign $p_i = -$sign $p_{i-1}$
    16  **If** (sign $p_i$) $*$ (sign $p_{i-1}$) $> 0$, **then** $k = k + 1$
17  **If** $k1 - k = 0$, **then**
    18  $x = y$
    19  $y = y + h$
    20  return to 9
21  **If** $k1 - k > 1$, **then**
    22  $y = (x + y)/2$
    23  return to 9
24  **If** $k1 - k = 1$, **then**
    25  $l = l + 1$
    26  **Call** bisect $(N, \mathbf{b}, \boldsymbol{\alpha}, x, y, \text{eigen}, \text{eps}, \text{maxit})$
    27  $\gamma_l = \text{eigen}$

28 **If** $l = N$, **then print** $\Gamma$ and 'finished' and exit.
29 $k1 = k1 - 1$
30 $x = y$
31 $y = y + h$
32 return to 9 ◀

**Algorithm**

**Subroutine** bisect($N$, **b**, **α**, $x$, $y$, eigen, eps, maxit)
Comment: The interval $[x, y]$ contains the eigenvalue of the matrix $T$ whose elements are given by **b** and **α**; eps is the tolerance; maxit is the maximum number of iterations; $p_0, p_1, \ldots, p_N$ are the elements of the Sturm sequence.
1 **Print** $N$, **b**, **α**, $x$, $y$, eps, maxit
2 $m = 0$
3 $xm = x$
4 $ym = y$
5 $p_0 = 1$
6 $p_1 = b_1 - ym$
7 **Repeat** step 8 **with** $i = 2, 3, \ldots, N$
    8 $p_i = (b_i - ym) * p_{i-1} - \alpha_{i-1}^2 * p_{i-2}$
9 sp1 $= \text{sign}(p_N)$
10 **Repeat** steps 11–19 **while** $m \le$ maxit
    11 eigen $= (xm + ym)/2$
    12 **If** $|ym - xm| \le$ eps $|ym|$, **then print** $m$, eigen, and 'finished' and exit
    13 $p_1 = b_1 -$ eigen
    14 **Repeat** step 15 **with** $i = 2, 3, \ldots, N$
        15 $p_i = (b_i -$ eigen$) * p_{i-1} - \alpha_{i-1}^2 * p_{i-2}$
    16 sp2 $= \text{sign}(p_N)$
    17 **If** (sp1)(sp2) $< 0$, **then** $xm =$ eigen; **otherwise** $ym =$ eigen
    18 sp1 $=$ sp2
    19 $m = m + 1$
20 **Print** 'algorithm fails: no convergence' and exit. ◀

E X E R C I S E S

1. Let $\mathbf{x} = [1, 1, 1]'$. Find a matrix $P$ such that $P\mathbf{x}$ is a multiple of $\mathbf{e}_1 = [1, 0, 0]'$.
2. Let $\mathbf{x} = [1, 1, 1]'$. Find a matrix $P$ such that $P\mathbf{x}$ is a multiple of $\mathbf{e}_3 = [0, 0, 1]'$.
3. Let $\mathbf{x} = [3, 2, 1, 1]'$. Find a matrix $P$ such that $P\mathbf{x}$ is a multiple of $\mathbf{e}_1 = [1, 0, 0, 0]'$.
4. Prove (a) $P' = P$ and (b) $P'P = I$ where $P = I - (2\mathbf{v}\mathbf{v}'/(\mathbf{v}'\mathbf{v}))$.

Reduce each of the following matrices in Exercises 5–7 to a tridiagonal form using the Householder method:

5. $\begin{bmatrix} 0 & 3 & 2 \\ 3 & 0 & 1 \\ 2 & 1 & 0 \end{bmatrix}$

6. $\begin{bmatrix} 2 & -1 & 1 & 4 \\ -1 & 3 & 1 & 2 \\ 1 & 1 & 5 & -3 \\ 4 & 2 & -3 & 6 \end{bmatrix}$

7. $\begin{bmatrix} 1 & 1 & 3 \\ 1 & -2 & 1 \\ 3 & 1 & 3 \end{bmatrix}$

8. Find the number of eigenvalues of

$$\begin{bmatrix} 0 & 3 & 2 \\ 3 & 0 & 1 \\ 2 & 1 & 0 \end{bmatrix}$$

that lie in the interval $[0, 2]$.

9. Find the number of eigenvalues of

$$\begin{bmatrix} 2 & 4 & 0 & 0 \\ 4 & 1 & 3 & 0 \\ 0 & 3 & 2 & -1 \\ 0 & 0 & -1 & 3 \end{bmatrix}$$

that lie in the interval $[0, 3]$.

10. Prove that no two consecutive Sturm polynomials have a common zero.

11. Prove that $C(\lambda)$ is the number of roots of $p_N(\lambda) = 0$ greater than or equal to $\lambda$. (*Hint*: use Theorem 13.8.1).

12. A matrix occurring in the numerical solution of partial differential equations is the $N \times N$ tridiagonal matrix

$$T = \begin{bmatrix} b & c & 0 & \cdots & 0 \\ c & b & c & \cdots & 0 \\ 0 & c & b & \cdots & 0 \\ \vdots & \vdots & \vdots & \ddots & \vdots \\ 0 & 0 & 0 & \cdots & b \end{bmatrix}$$

where $b$ and $c$ are real numbers. By direct substitution and using trigonometric identities, show that the eigenvalues and corresponding eigenvectors are given by

$$\lambda_i = b + 2c \cos \frac{i\pi}{N+1} \quad \text{and} \quad \mathbf{x}_i = \left[ \sin \frac{i\pi}{N+1}, \sin \frac{2i\pi}{N+1}, \ldots, \sin \frac{Ni\pi}{N+1} \right]^t$$

for $i = 1, 2, \ldots, N$.

**COMPUTER EXERCISES**

**C.1.**  Write a subroutine that accepts $\mathbf{x}$ as input with $N$ elements and computes $\mathbf{v}$ using Equation (13.7.10).

**C.2.**  Write a subroutine that calls the subroutine developed in Exercise C.1 and computes the $N \times N$ matrix $P$ using Equation (13.7.1).

**C.3.**  Write a computer program that reduces a given $N \times N$ matrix using Exercises C.1 and C.2 to tridiagonal form.

**C.4.**  Write a computer program to reduce a given $N \times N$ matrix to tridiagonal form using the Givens method.

**C.5.**  Write a subroutine that counts the agreements in sign between consecutive terms of the $p_i$ for $i = 1, 2, \ldots, N$ given by Equation (13.8.2) at various real numbers. Then locate the intervals that contain the eigenvalues. Use the bisection method to find these eigenvalues.

**C.6.**  Reduce

$$\begin{bmatrix} 1 & 2 & 3 & 4 & 5 \\ 2 & 3 & 4 & 5 & 6 \\ 3 & 4 & 5 & 6 & 7 \\ 4 & 5 & 6 & 7 & 8 \\ 5 & 6 & 7 & 8 & 9 \end{bmatrix}$$

to a tridiagonal matrix by using Exercises C.1, C.2, and C.3. Then find its eigenvalues using Exercise C.5.

**C.7.**  Reduce

$$\begin{bmatrix} 1 & 2 & 3 & 4 & 5 \\ 2 & 3 & 4 & 5 & 6 \\ 3 & 4 & 5 & 6 & 7 \\ 4 & 5 & 6 & 7 & 8 \\ 5 & 6 & 7 & 8 & 9 \end{bmatrix}$$

to a tridiagonal matrix by using Exercise C.4. Then find its eigenvalues using Exercise C.5.

**C.8.**  Using Exercises C.1, C.2, C.3, and C.5, find all eigenvalues of the matrix given in Exercise 6.

**C.9.**  Using Exercises C.4 and C.5, find all eigenvalues of the matrix given in Exercise 6.

**C.10.**  A matrix occurring often in the numerical solution of partial differential equations is the $N \times N$ tridiagonal matrix

$$T_N = \begin{bmatrix} -2 & 1 & 0 & \ldots & 0 \\ 1 & -2 & 1 & \ldots & 0 \\ 0 & 1 & -2 & \ldots & 0 \\ \vdots & \vdots & \vdots & \ddots & \vdots \\ 0 & 0 & 0 & \ldots & -2 \end{bmatrix}$$

Find the eigenvalues of $T_{10}$.

# The Nonsymmetric Eigenvalue Problem

In the next two sections, we deal with the problem of computing the eigenvalues of nonsymmetric real matrices. Of course, any method for computing the eigenvalues of nonsymmetric matrices will work for symmetric matrices, although not necessarily as efficiently as the methods of the previous sections. We discuss in these sections two closely-related factorization methods, the *LR* and *QR* methods. The *LR* method is easier to apply but may be numerically unstable while the *QR* method is stable.

## 13.9  THE *LR* ALGORITHM

The *LR* algorithm was developed by Rutishauser (1958) for finding the eigenvalues of a given real $N \times N$ matrix $A$. We split $A$ into two triangular matrices $L$ and $R$ where $L$ is a lower triangular matrix with $l_{ii} = 1$ for $i = 1, 2, \ldots, N$ and $R$ is an upper triangular matrix such that $A = LR$. We use $R$ instead of $U$ to facilitate comparison with Rutishauser's work. ($L$ and $R$ are the first letters of the words "left" and "right".) Also, we use $A_1$ for the given original matrix $A$.

The key to this method is the observation that if $A_1 = L_1 R_1$, then, if we multiply $L_1$ and $R_1$ in the reverse order, the matrix $A_2 = R_1 L_1$ has the same eigenvalues as $A_1$. This is because

$$A_2 = R_1 L_1 = L_1^{-1} A_1 L_1 \qquad \textbf{(13.9.1)}$$

Thus $A_2$ and $A_1$ are **similar**.

Now $A_2$ may be decomposed into two triangular matrices $L_2$ and $R_2$ such that $A_2 = L_2 R_2 = R_1 L_1$ and then continuing $R_2 L_2 = A_3 = L_3 R_3$. Thus we have

$$A_i = L_i R_i \qquad \textbf{(13.9.2)}$$
$$A_{i+1} = R_i L_i \quad \text{for } i = 1, 2, \ldots$$

**EXAMPLE 13.9.1**   Find the eigenvalues of

$$\begin{bmatrix} 1 & 2 & 3 \\ 3 & 4 & 5 \\ 5 & 6 & 7 \end{bmatrix}$$

using the *LR* algorithm.

The decomposition of $A_1 = A$ gives

$$L_1 = \begin{bmatrix} 1 & 0 & 0 \\ 3 & 1 & 0 \\ 5 & 2 & 1 \end{bmatrix} \quad \text{and} \quad R_1 = \begin{bmatrix} 1 & 2 & 3 \\ 0 & -2 & -4 \\ 0 & 0 & 0 \end{bmatrix}$$

Then

$$A_2 = R_1 L_1 = \begin{bmatrix} 1 & 2 & 3 \\ 0 & -2 & -4 \\ 0 & 0 & 0 \end{bmatrix} \begin{bmatrix} 1 & 0 & 0 \\ 3 & 1 & 0 \\ 5 & 2 & 1 \end{bmatrix} = \begin{bmatrix} 22 & 8 & 3 \\ -26 & -10 & -4 \\ 0 & 0 & 0 \end{bmatrix}$$

Further $A_2$ decomposes as

$$L_2 = \begin{bmatrix} 1 & 0 & 0 \\ -1.18182 & 1 & 0 \\ 0 & 0 & 1 \end{bmatrix} \quad \text{and} \quad R_2 = \begin{bmatrix} 22 & 8 & 3 \\ 0 & -0.54545 & -0.45455 \\ 0 & 0 & 0 \end{bmatrix}$$

such that $A_2 = L_2 R_2$. We define

$$A_3 = R_2 L_2 = \begin{bmatrix} 12.54545 & 8 & 3 \\ 0.64463 & -0.54545 & -0.45455 \\ 0 & 0 & 0 \end{bmatrix}$$

The decomposition of $A_3$ gives

$$L_3 = \begin{bmatrix} 1 & 0 & 0 \\ 0.05138 & 1 & 0 \\ 0 & 0 & 1 \end{bmatrix} \quad \text{and} \quad R_3 = \begin{bmatrix} 12.54545 & 8 & 3 \\ 0 & -0.95652 & -0.6087 \\ 0 & 0 & 0 \end{bmatrix}$$

Continuing, we get the sequence of $A_i$ which is given in Table 13.9.1 along with $L_i$ and $R_i$.

**Table 13.9.1**

| $i$ | $A_i$ | $L_i$ | $R_i$ |
|---|---|---|---|
| 4 | $\begin{bmatrix} 12.95652 & 8.0 & 3.0 \\ -0.04915 & -0.95652 & -0.60870 \\ 0.0 & 0.0 & 0.0 \end{bmatrix}$ | $\begin{bmatrix} 1.0 & 0.0 & 0.0 \\ -0.00379 & 1.0 & 0.0 \\ 0.0 & 0.0 & 1.0 \end{bmatrix}$ | $\begin{bmatrix} 12.95652 & 8.0 & 3.0 \\ 0.0 & -0.92617 & -0.59732 \\ 0.0 & 0.0 & 0.0 \end{bmatrix}$ |
| 5 | $\begin{bmatrix} 12.95652 & 8.0 & 3.0 \\ 0.00351 & -0.92617 & -0.59732 \\ 0.0 & 0.0 & 0.0 \end{bmatrix}$ | $\begin{bmatrix} 1.0 & 0.0 & 0.0 \\ 0.00027 & 1.0 & 0.0 \\ 0.0 & 0.0 & 1.0 \end{bmatrix}$ | $\begin{bmatrix} 12.92617 & 8.0 & 3.0 \\ 0.0 & -0.92835 & -0.59813 \\ 0.0 & 0.0 & 0.0 \end{bmatrix}$ |
| 6 | $\begin{bmatrix} 12.92835 & 8.3 & 3.0 \\ -0.00025 & -0.92835 & -0.59813 \\ 0.0 & 0.0 & 0.0 \end{bmatrix}$ | $\begin{bmatrix} 1.0 & 0.0 & 0.0 \\ -0.00002 & 1.0 & 0.0 \\ 0.0 & 0.0 & 1.0 \end{bmatrix}$ | $\begin{bmatrix} 12.92835 & 8.3 & 3.0 \\ 0.0 & -0.92819 & -0.59807 \\ 0.0 & 0.0 & 0.0 \end{bmatrix}$ |
| 7 | $\begin{bmatrix} 12.92819 & 8.3 & 3.0 \\ 0.00002 & -0.92819 & -0.59807 \\ 0.0 & 0.0 & 0.0 \end{bmatrix}$ | $\begin{bmatrix} 1.0 & 0.0 & 0.0 \\ 0 & 1.0 & 0.0 \\ 0.0 & 0.0 & 1.0 \end{bmatrix}$ | $\begin{bmatrix} 12.92819 & 8.0 & 3.0 \\ 0.0 & -0.92820 & -0.59808 \\ 0.0 & 0.0 & 0.0 \end{bmatrix}$ |

We can verify that the eigenvalues are $\lambda_1 = 6 + 4\sqrt{3} \approx 12.92820$, $\lambda_2 = 6 - 4\sqrt{3} \approx -0.92820$, and $\lambda_3 = 0.0$. ■ ■ ■

Observe from Table 13.9.1 that $L_i$ converges to $I$ and $A_i$ converges to $R_i$. We read the eigenvalues on the diagonal of $R_i$ or $A_i$. We will show that

$$\lim_{i \to \infty} L_i = I \quad \text{and} \quad \lim_{i \to \infty} A_i = \lim_{i \to \infty} R_i = \begin{bmatrix} \lambda_1 & * & * & \cdots & * \\ 0 & \lambda_2 & * & \cdots & * \\ 0 & 0 & \lambda_3 & \cdots & * \\ \vdots & \vdots & \vdots & \ddots & \cdots \\ 0 & 0 & 0 & \cdots & \lambda_N \end{bmatrix}$$

Unfortunately, there are several difficulties associated with this method. First, the triangular decomposition may not exist at every stage and second, even if it exists, the numerical calculations used to obtain it may not be stable.

Let us assume that the triangular decomposition exists at every stage of a real matrix $A$ and for convenience consider $|\lambda_1| > |\lambda_2| > \cdots > |\lambda_N|$. From Equation (13.9.2),

$$A_{i+1} = R_i L_i = R_i A_i R_i^{-1} = R_i R_{i-1} L_{i-1} R_i^{-1} = R_i R_{i-1} A_{i-1} R_{i-1}^{-1} R_i^{-1}$$

Repeating, we get

$$A_{i+1} = R_i R_{i-1} \cdots R_2 R_1 A_1 R_1^{-1} R_2^{-1} \cdots R_{i-1}^{-1} R_i^{-1} \tag{13.9.3}$$

Similarly

$$A_{i+1} = L_i^{-1} L_{i-1}^{-1} \cdots L_2^{-1} L_1^{-1} A_1 L_1 L_2 \cdots L_{i-1} L_i \tag{13.9.4}$$

For convenience let

$$P_i = R_i R_{i-1} \cdots R_2 R_1 \quad \text{and} \quad N_i = L_1 L_2 \cdots L_{i-1} L_i$$

Then,

$$A_{i+1} = P_i A_1 P_i^{-1} = N_i^{-1} A_1 N_i \tag{13.9.5}$$

Further, observe that

$$P_i = R_i P_{i-1} \quad \text{and} \quad N_i = N_{i-1} L_i$$

Therefore, $L_i = N_{i-1}^{-1} N_i$. In order to show $\lim_{i \to \infty} L_i = I$, we want to show $N_i$ converges to a nonsingular matrix.

From Equation (13.9.5), we get

$$N_i P_i = N_{i-1} L_i R_i P_{i-1} = N_{i-1} A_i P_{i-1} = N_{i-1} P_{i-1} A_1$$

Repeated application of this result leads to

$$N_i P_i = (A_1)^i \tag{13.9.6}$$

In order to study the convergence of $N_i$, let us look at $(A_1)^i$ decomposed into the lower unit triangular matrix $N_i$ and upper triangular matrix $P_i$.

Assume that $A_1$ is diagonalizable. Then there exists a nonsingular matrix $V$ such that $A_1 = VDV^{-1}$. Let $U = V^{-1} = L_U R_U$ and $V = L_V R_V$ where $L_U$ and $L_V$ are lower triangular matrices with diagonal elements one and $R_U$ and $R_V$ are upper triangular matrices. Then

$$A_1^2 = A_1 A_1 = (VDV^{-1})(VDV^{-1}) = VD (V^{-1} V) DV^{-1} = VD^2 V^{-1}$$

Continuing,

$$\begin{aligned} A_1^i &= VD^i V^{-1} = L_V R_V D^i L_U R_U \\ &= L_V R_V D^i L_U (D^{-i} D^i) R_U \\ &= L_V R_V (D^i L_U D^{-i}) D^i R_U \end{aligned} \tag{13.9.7}$$

Let us look at $D^i L_U D^{-i}$. For convenience, let

$$D = \begin{bmatrix} \lambda_1 & 0 & 0 \\ 0 & \lambda_2 & 0 \\ 0 & 0 & \lambda_3 \end{bmatrix} \quad \text{and} \quad L_U = \begin{bmatrix} 1 & 0 & 0 \\ l_{21} & 1 & 0 \\ l_{31} & l_{32} & 1 \end{bmatrix}$$

Then it can be verified that

$$D^i L_U D^{-i} = \begin{bmatrix} 1 & 0 & 0 \\ l_{21}\left(\dfrac{\lambda_2}{\lambda_1}\right)^i & 1 & 0 \\ l_{31}\left(\dfrac{\lambda_3}{\lambda_1}\right)^i & l_{32}\left(\dfrac{\lambda_3}{\lambda_2}\right)^i & 1 \end{bmatrix}$$

Since $|\lambda_1| > |\lambda_2| > |\lambda_3|$, $\lim_{i \to \infty} D^i L_U D^{-i} = I$. Similarly for an $N \times N$ matrix, we can prove that $\lim_{i \to \infty} D^i L_U D^{-i} = I$. Comparing Equation (13.9.6) with Equation (13.9.7), we get $N_i \to L_V$ as $i \to \infty$ since $R_V D^i R_U$ is an upper triangular matrix. Thus $N_i$ converges to a nonsingular matrix.

Therefore $\lim_{i \to \infty} L_i = \lim_{i \to \infty} N_{i-1}^{-1} N_i = I$ (since $l_{ii} = 1$ for $i = 1, 2, \ldots, N$). Further

$$\begin{aligned} \lim_{i \to \infty} R_i &= \lim_{i \to \infty} L_i^{-1} A_i = \lim_{i \to \infty} L_i^{-1} N_{i-1}^{-1} A_1 N_{i-1} \\ &= \lim_{i \to \infty} N_i^{-1} N_{i-1} N_{i-1}^{-1} A_1 N_{i-1} = \lim_{i \to \infty} N_i^{-1} A_1 N_i \end{aligned}$$

Since $\lim_{i \to \infty} N_i = L_V$, $\lim_{i \to \infty} R_i$ exists. Hence $\lim_{i \to \infty} A_i = \lim_{i \to \infty} L_i R_i = \lim_{i \to \infty} R_i$. This completes the proof.

## 13.10 THE *QR* ALGORITHM

Another factorization method is the *QR* algorithm of Francis (1961). This method decomposes an arbitrary matrix $A$ into a product $QR$ where $Q$ is an orthogonal matrix and $R$ is an upper triangular matrix. Since $Q$ is orthogonal, $Q^t = Q^{-1}$. Let $A_1 = A$. Then

$$A_1 = Q_1 R_1 \tag{13.10.1}$$

Now, as in the *LR* algorithm, consider the matrix formed by writing the factors in Equation (13.10.1) in reverse order; that is,

$$A_2 = R_1 Q_1 = Q_1^{-1} A_1 Q_1$$

Since $A_1$ and $A_2$ are similar, they have the same eigenvalues. We decompose $A_2$ as a product $Q_2 R_2$; that is, $A_2 = R_1 Q_1 = Q_2 R_2$. We continue this procedure which is given by

$$\begin{aligned} A_i &= Q_i R_i \\ A_{i+1} &= R_i Q_i \quad \text{for } i = 1, 2, \ldots \end{aligned} \tag{13.10.2}$$

where $Q_i$ is orthogonal and $R_i$ is upper triangular. Note that a decomposition of $A_i$ as $Q_i R_i$ exists provided $A$ is nonsingular and can be computed by "plane rotation" matrices, Householder matrices, or a Gram–Schmidt orthogonalization procedure (Appendix E). Let $P_1^{(i)}, P_2^{(i)}, \ldots, P_{N-1}^{(i)}$ be Householder matrices such that

$$P_{N-1}^{(i)} P_{N-2}^{(i)} \cdots P_1^{(i)} A_i = R_i \qquad (13.10.3)$$

where $R_i$ is an upper triangular matrix. Since $(P_1^{(i)})(P_1^{(i)})' = (P_1^{(i)})' (P_1^{(i)}) = I$, then Equation (13.10.3) becomes

$$A_i = (P_{N-1}^{(i)} P_{N-2}^{(i)} \cdots P_1^{(i)})^{-1} R_i \qquad (13.10.4)$$

Let $Q_i' = P_{N-1}^{(i)} P_{N-2}^{(i)} \cdots P_1^{(i)}$. Then Equation (13.10.4) becomes $A_i = (Q_i')^{-1} R_i$. Since $P_1^{(i)}, P_2^{(i)}, \ldots, P_{N-1}^{(i)}$ are orthogonal, then $Q_i$ is orthogonal, so $Q_i' = Q_i^{-1}$. Thus

$$A_i = Q_i R_i \qquad (13.10.5)$$

and

$$A_{i+1} = R_i Q_i = R_i P_{N-1}^{(i)} P_{N-2}^{(i)} \cdots P_1^{(i)} \qquad (13.10.6)$$

It can be proven (Exercise 8) in much the same way as the *LR* algorithm that as $i \to \infty$, the sequence $\{A_i\}$ converges to a triangular matrix if $|\lambda_1| > |\lambda_2| \cdots > |\lambda_N|$.

**EXAMPLE 13.10.1**   Find the eigenvalues of

$$A_1 = \begin{bmatrix} 1 & 3 & 3 \\ 2 & 3 & 0 \\ 2 & 0 & 3 \end{bmatrix}$$

using the *QR* algorithm.

Since Householder's matrices are orthogonal, we will use them to decompose $A_1$. We want to transform $\mathbf{x} = \begin{bmatrix} 1 \\ 2 \\ 2 \end{bmatrix}$ as a scalar multiple of $\begin{bmatrix} 1 \\ 0 \\ 0 \end{bmatrix}$. Therefore

$$\mathbf{v} = \mathbf{x} + (\text{sign } x_1) \|\mathbf{x}\|_2^2 \mathbf{e}_1 = \begin{bmatrix} 1 \\ 2 \\ 2 \end{bmatrix} + 3 \begin{bmatrix} 1 \\ 0 \\ 0 \end{bmatrix} = 2 \begin{bmatrix} 2 \\ 1 \\ 1 \end{bmatrix}$$

Hence

$$P_1 = I - 2 \frac{\mathbf{v} \mathbf{v}'}{\|\mathbf{v}\|_2^2} = \begin{bmatrix} 1 & 0 & 0 \\ 0 & 1 & 0 \\ 0 & 0 & 1 \end{bmatrix} - \frac{1}{3} \begin{bmatrix} 4 & 2 & 2 \\ 2 & 1 & 1 \\ 2 & 1 & 1 \end{bmatrix} = -\frac{1}{3} \begin{bmatrix} 1 & 2 & 2 \\ 2 & -2 & 1 \\ 2 & 1 & -2 \end{bmatrix}$$

Therefore

$$P_1 A_1 = -\frac{1}{3} \begin{bmatrix} 1 & 2 & 2 \\ 2 & -2 & 1 \\ 2 & 1 & -2 \end{bmatrix} \begin{bmatrix} 1 & 3 & 3 \\ 2 & 3 & 0 \\ 2 & 0 & 3 \end{bmatrix} = \begin{bmatrix} -3 & -3 & -3 \\ 0 & 0 & -3 \\ 0 & -3 & 0 \end{bmatrix}$$

Now we want to transform $\begin{bmatrix} 0 \\ -3 \end{bmatrix}$ to $\begin{bmatrix} 1 \\ 0 \end{bmatrix}$. Therefore

$$\mathbf{v} = \begin{bmatrix} 0 \\ -3 \end{bmatrix} + 3 \begin{bmatrix} 1 \\ 0 \end{bmatrix} = \begin{bmatrix} 3 \\ -3 \end{bmatrix}$$

Therefore

$$H = I - 2\frac{\mathbf{v}\,\mathbf{v}^t}{\|\mathbf{v}\|_2^2} = \begin{bmatrix} 1 & 0 \\ 0 & 1 \end{bmatrix} - \frac{1}{9}\begin{bmatrix} 9 & -9 \\ -9 & 9 \end{bmatrix} = \begin{bmatrix} 0 & 1 \\ 1 & 0 \end{bmatrix}$$

Hence

$$P_2 = \begin{bmatrix} 1 & 0 & 0 \\ 0 & 0 & 1 \\ 0 & 1 & 0 \end{bmatrix}$$

$$P_2\,(P_1 A_1) = \begin{bmatrix} 1 & 0 & 0 \\ 0 & 0 & 1 \\ 0 & 1 & 0 \end{bmatrix}\begin{bmatrix} -3 & -3 & -3 \\ 0 & -3 & 0 \\ 0 & 0 & -3 \end{bmatrix} = \begin{bmatrix} -3 & -3 & -3 \\ 0 & -3 & 0 \\ 0 & 0 & -3 \end{bmatrix} = R_1$$

Therefore

$$A_1 = P_1^{-1} P_2^{-1} R_1 = (P_2 P_1)^{-1} R_1 = Q_1^{-1} R_1 = Q_1 R_1$$

Thus

$$Q_1 = P_2 P_1 = P_1 P_2 = -\frac{1}{3}\begin{bmatrix} 1 & 2 & 2 \\ 2 & -2 & 1 \\ 2 & 1 & -2 \end{bmatrix}\begin{bmatrix} 1 & 0 & 0 \\ 0 & 0 & 1 \\ 0 & 1 & 0 \end{bmatrix} = -\frac{1}{3}\begin{bmatrix} 1 & 2 & 2 \\ 2 & 1 & -2 \\ 2 & -2 & 1 \end{bmatrix}$$

Therefore we get

$$A_2 = R_1 Q_1 = -\frac{1}{3}\begin{bmatrix} -3 & -3 & -3 \\ 0 & -3 & 0 \\ 0 & 0 & -3 \end{bmatrix}\begin{bmatrix} 1 & 2 & 2 \\ 2 & 1 & -2 \\ 2 & -2 & 1 \end{bmatrix} = \begin{bmatrix} 5 & 1 & 1 \\ 2 & 1 & -2 \\ 2 & -2 & 1 \end{bmatrix}$$

Continuing this procedure, we decompose $A_2$ as

$$R_2 = \begin{bmatrix} -5.74456 & -0.52223 & -0.52223 \\ 0.0 & -2.39317 & 1.36752 \\ 0.0 & 0.0 & -1.96396 \end{bmatrix} \quad \text{and}$$

$$Q_2 = \begin{bmatrix} -0.87038 & -0.22792 & -0.43643 \\ -0.34815 & -0.34188 & 0.87287 \\ -0.34815 & 0.91168 & 0.21821 \end{bmatrix}$$

Thus

$$A_3 = R_2 Q_2 = \begin{bmatrix} 5.36363 & 1.01173 & 1.93732 \\ 0.35708 & 2.06493 & -1.79051 \\ 0.68376 & -1.79051 & -0.42857 \end{bmatrix}$$

Similarly

$$A_5 = \begin{bmatrix} 5.59183 & 0.22475 & 1.46374 \\ 0.01012 & 2.89423 & -0.68881 \\ 0.06591 & -0.68881 & -1.48606 \end{bmatrix}$$

$$A_{10} = \begin{bmatrix} 5.60557 & -0.00955 & -1.41405 \\ 0.0 & 2.99978 & -0.03110 \\ 0.00012 & -0.03110 & -1.60536 \end{bmatrix}$$

$$A_{15} = \begin{bmatrix} 5.60555 & 0.00041 & 1.41421 \\ 0.0 & 2.99999 & -0.00136 \\ 0.0 & -0.00136 & -1.60555 \end{bmatrix}$$

The exact values are $2 + \sqrt{13} \approx 5.60555$, 3, and $2 - \sqrt{13} \approx -1.60555$.   ■ ■ ■

**Algorithm**

**Subroutine** $QR(A, N, \text{eps}, \text{maxit})$
Comment: This program reduces a given real matrix $A$ with real eigenvalues to a triangular matrix using the $QR$ algorithm. By calling Subroutine ortho$(A, Q, R)$, $A$ is decomposed into an orthogonal matrix $Q$ and an upper triangular matrix $R$. eps is the tolerance and maxit is the maximum number of iterations.
  1 **Print** $N$, $A$, eps, maxit
  2 $k = 0$
  3 **Repeat** steps 4–10 **while** $k \le$ maxit
       4 $k = k + 1$
       5 **Call** ortho$(N, A, Q, R)$
  Comment: This part calls Subroutine prod$(N, A, R, Q)$. Subroutine prod multiplies two $N \times N$ matrices $R$ and $Q$ and stores it as $N \times N$ matrix $A$. We have $A = RQ$.
       6 **Call** prod $(N, A, R, Q)$
       7 **Repeat** steps 8–9 **with** $i = 1, 2, \ldots, N$
            8 **Repeat** 9 **with** $j = 1, 2, \ldots, i - 1$
                 9 **If** $|a_{ij}| >$ eps, **then** return to 3
       10 **Print** $k$, $A$, and 'finished' and exit
  11 **Print** 'algorithm fails: no convergence' and exit.    ◀

**Algorithm**

**Subroutine** ortho$(N, A, Q, R)$
Comment: This subroutine decomposes a given matrix $A$ into two factors $Q$ and $R$ using the Householder transformation (13.7.1). We get $P_{N-1} P_{N-2} \ldots P_1 A = R$ and $Q = P_1 P_2 \ldots P_{N-1}$. Initially $Q = I$ and $R = A$.
  1 **Print** $N$, $A$
  2 **Repeat** steps 3–6 **with** $i = 1, 2, \ldots, N$
       3 **Repeat** steps 4–6 **with** $j = 1, 2, \ldots, N$
            4 **If** $j = i$, **then** $q_{ij} = 1$
            5 **If** $j \ne i$, **then** $q_{ij} = 0$
            6 $r_{ij} = a_{ij}$
  7 **Repeat** steps 8–39 **with** $i = 1, 2, \ldots, N - 1$
       8 **Repeat** steps 9–11 **with** $l = 1, 2, \ldots, N$
            9 **Repeat** steps 10–11 **with** $m = 1, 2, \ldots, N$

10 **If** $m = l$, **then** $p_{lm} = 1$
11 **If** $m \neq l$, **then** $p_{lm} = 0$
Comment: Steps 12–23 generate **v** given by Equation (13.7.10).
12 xnorm = 0
13 **Repeat** steps 14–16 **with** $j = i, i + 1, \ldots, N$
14 $e_j = 0$
15 $x_j = r_{ji}$
16 xnorm = xnorm + $x_j * x_j$
17 $e_i = 1$
18 xnorm = $\sqrt{\text{xnorm}}$
19 vnorm = 0
20 **Repeat** steps 21–22 **with** $j = i, i + 1, \ldots, N$
21 $v_j = x_j + \text{sign}(x_i) * \text{xnorm} * e_j$
22 vnorm = vnorm + $v_j * v_j$
23 **Repeat** steps 24–25 **with** $j = i, i + 1, \ldots, N$
24 **Repeat** step 25 **with** $k = i, i + 1, \ldots, N$
25 $p_{jk} = p_{jk} - (2 * v_j * v_k/\text{vnorm})$
Comment: This part multiplies *PR* and stores it temporarily as *A*. Also *QP* is multiplied and stored temporarily as *H*.
26 **Repeat** steps 27–34 **with** $m = 1, 2, \ldots, N$
27 **Repeat** steps 28–34 **with** $l = 1, 2, \ldots, N$
28 sum = 0
29 sum1 = 0
30 **Repeat** steps 31–32 **with** $k = 1, 2, \ldots, N$
31 sum = sum + $p_{mk} * r_{kl}$
32 sum1 = sum1 + $q_{mk} * p_{kl}$
33 $a_{ml}$ = sum
34 $h_{ml}$ = sum1
Comment: This part transfers *A* to *R* which is $P_{N-1} P_{N-2} \ldots P_1 A$. Also it transfers *H* to *Q* which is $P_1 P_2 \ldots P_{N-1} = Q$.
35 **Repeat** steps 36–38 **with** $m = 1, 2, \ldots, N$
36 **Repeat** steps 37–38 **with** $l = 1, 2, \ldots, N$
37 $r_{ml} = a_{ml}$
38 $q_{ml} = h_{ml}$
39 **Print** *Q* and *R* and exit.   ◀

There is an important restriction for the convergence in both algorithms; namely, $|\lambda_1| > |\lambda_2| > \cdots > |\lambda_N|$. If *A* is real, then all its factors $Q_i$ and $R_i$ are also real and therefore there is no hope that $A_i$ could converge to a triangular matrix with complex eigenvalues. However, in such cases $A_i$ will converge to an almost-triangular form as shown for a $4 \times 4$ matrix

$$A_i = \begin{bmatrix} \lambda_1 & * & * & * \\ 0 & a_{22}^{(i)} & a_{23}^{(i)} & * \\ 0 & a_{32}^{(i)} & a_{33}^{(i)} & * \\ 0 & 0 & 0 & \lambda_4 \end{bmatrix}$$

For this 4 × 4 matrix, we have two real eigenvalues $\lambda_1$ and $\lambda_4$ with distinct absolute values and two complex conjugate pairs of eigenvalues. The elements of the 2 × 2 matrix

$$\begin{bmatrix} a_{22}^{(i)} & a_{23}^{(i)} \\ a_{32}^{(i)} & a_{33}^{(i)} \end{bmatrix}$$

do not converge, but their eigenvalues converge to a complex conjugate pair of eigenvalues of $A$. These eigenvalues can be computed as the roots of the characteristic polynomial of this 2 × 2 submatrix. This analysis also applies to the $LR$ algorithm.

The $LR$ and $QR$ algorithms can be effectively used by first reducing the matrix $A$ to a form in which the decomposition can be computed fast. If $A$ is symmetric, then it is reduced to a similar symmetric tridiagonal form and if $A$ is not symmetric, then it is reduced to a similar Hessenberg matrix. A matrix $H$ is **Hessenberg** if

$$h_{ij} = 0 \quad \text{whenever } i > j + 1$$

In the case of a 4 × 4 matrix, we have the Hessenberg matrix $H$ given by

$$H = \begin{bmatrix} h_{11} & h_{12} & h_{13} & h_{14} \\ h_{21} & h_{22} & h_{23} & h_{24} \\ 0 & h_{32} & h_{33} & h_{34} \\ 0 & 0 & h_{43} & h_{44} \end{bmatrix}$$

The original matrix $A$ can be reduced to Hessenberg form by using the Householder transformations.

## EXERCISES

1. The first few iterations of the $QR$ method are $A = A_1 = Q_1 R_1$, $A_2 = R_1 Q_1 = Q_2 R_2$, $A_3 = R_2 Q_2 = Q_3 R_3$. Verify (a) $Q_1 Q_2 R_2 R_1 = Q_1 R_1 Q_1 R_1 = A^2$ and (b) $(Q_1 Q_2)^t A (Q_1 Q_2) = A_3$.

2. Find $L$ and $R$ factors (if they exist) for each of the following matrices such that $A = LR$:

   (a) $A = \begin{bmatrix} 0 & 2 \\ 1 & 1 \end{bmatrix}$    (b) $A = \begin{bmatrix} \cos\theta & -\sin\theta \\ \sin\theta & \cos\theta \end{bmatrix}$

   where $0 < \theta < \pi/2$.

3. Find $Q$ and $R$ factors for each of the matrices given in Exercise 2 such that $A = QR$ using the Householder transformation.

4. Find $Q$ and $R$ factors for each of the matrices given in Exercise 2 such that $A = QR$ using the Givens transformation.

5. Perform three iterations (if possible) of the $LR$ algorithm for each of the matrices given in Exercise 2.

6. Perform three iterations of the $QR$ algorithm for each of the matrices given in Exercise 2.

7. Prove that the Hessenberg form of a matrix is preserved under $QR$ transformations.

8. Prove that if $|\lambda_1| > |\lambda_2| > \cdots > |\lambda_N|$, then $\lim_{i \to \infty} A_i$ converges to a triangular matrix in the $QR$ algorithm.
9. For the $LR$ algorithm if $L_1 R_1 = L_2 R_2$, then show that there exists a diagonal matrix $D$ such that $R_1 = DR_2$ and $L_2 = L_1 D$.

## COMPUTER EXERCISES

**C.1.** Write a subroutine to factor a given $N \times N$ matrix $A$ as $LR$ using the methods of Chapter 7.

**C.2.** Write a computer program to find the eigenvalues using Exercise C.1 and the $LR$ algorithm. For stopping the iteration, use $\|R_{i-1} - R_i\| < 10^{-5} \|R_i\|$.

**C.3.** Write a subroutine to factor a given $N \times N$ matrix $A$ as $QR$ by using Householder's transformations given by Equation (13.7.1).

**C.4.** Write a computer program to find the eigenvalues by using Exercise C.3 and the $QR$ algorithm. For stopping the iteration, use $\|R_{i-1} - R_i\| < 10^{-5} \|R_i\|$.

**C.5.** Write a subroutine to reduce a given $N \times N$ matrix $A$ to a similar Hessenberg matrix by using Householder's transformations given by Equation (13.7.1).

**C.6.** Write a computer program to find the eigenvalues of a given $N \times N$ matrix $A$ by using Exercise C.5 and the $QR$ algorithm.

Find the eigenvalues of each of the following matrices using Exercises C.6, C.2, or C.4.

**C.7.**
$$\begin{bmatrix} 1 & 3 & 3 \\ 2 & 3 & 0 \\ 2 & 0 & 3 \end{bmatrix}$$

**C.8.**
$$\begin{bmatrix} 12 & 3 & 1 \\ -9 & -2 & -3 \\ 14 & 6 & 2 \end{bmatrix}$$

**C.9.**
$$\begin{bmatrix} 2 & 1 & 3 & 1 \\ -1 & 2 & 2 & 1 \\ 1 & 0 & 1 & 0 \\ -1 & -1 & -1 & 1 \end{bmatrix}$$

**C.10.**
$$\begin{bmatrix} 2 & -1 & 1 & 4 \\ -1 & 3 & 1 & 2 \\ 1 & 1 & 5 & -3 \\ 4 & 2 & -3 & 6 \end{bmatrix}$$

**C.11.** Show that the characteristic polynomial of the companion matrix

$$\begin{bmatrix} 0 & 1 & 0 & 0 & 0 \\ 0 & 0 & 1 & 0 & 0 \\ 0 & 0 & 0 & 1 & 0 \\ 0 & 0 & 0 & 0 & 1 \\ -a_0 & -a_1 & -a_2 & -a_3 & -a_4 \end{bmatrix}$$

is $\lambda^5 + a_4 \lambda^4 + a_3 \lambda^3 + a_2 \lambda^2 + a_1 \lambda + a_0 = 0$. Using the $QR$ algorithm, compute the roots of the polynomial equation $x^5 - 3x^4 + 6x^3 - 2x + 2 = 0$.

**C.12.** Find the eigenvalues of the $10 \times 10$ Frank matrix

$$\begin{bmatrix} 10 & 9 & 8 & 7 & 6 & 5 & 4 & 3 & 2 & 1 \\ 9 & 9 & 8 & 7 & 6 & 5 & 4 & 3 & 2 & 1 \\ 0 & 8 & 8 & 7 & 6 & 5 & 4 & 3 & 2 & 1 \\ 0 & 0 & 7 & 7 & 6 & 5 & 4 & 3 & 2 & 1 \\ 0 & 0 & 0 & 6 & 6 & 5 & 4 & 3 & 2 & 1 \\ 0 & 0 & 0 & 0 & 5 & 5 & 4 & 3 & 2 & 1 \\ 0 & 0 & 0 & 0 & 0 & 4 & 4 & 3 & 2 & 1 \\ 0 & 0 & 0 & 0 & 0 & 0 & 3 & 3 & 2 & 1 \\ 0 & 0 & 0 & 0 & 0 & 0 & 0 & 2 & 2 & 1 \\ 0 & 0 & 0 & 0 & 0 & 0 & 0 & 0 & 1 & 1 \end{bmatrix}$$

## SUGGESTED READINGS

J. G. F. Francis, *The QR Transformation: A Unitary Analogue to the LR Transformation, Parts I and II*, Comp. J. **4** (1961), 265–72, 332–45.

W. Givins, *Computation of Plane Unitary Rotations Transforming a General Matrix to Triangular Form*, SIAM J. Applied Math. **6** (1958), 26–50.

G. H. Golub and C. F. Van Loan, *Matrix Computations*, 2nd ed., Johns Hopkins Univ. Press, Baltimore, 1989.

P. Henrici, *On the Speed of Convergence of Cyclic and Quasicyclic Jacobi Methods for Computing the Eigenvalues of Hermitian Matrices*, SIAM J. Applied Math. **6** (1958), 144–162.

D. A. Pope and C. Tompkins, *Maximizing Functions of Rotations: Experiments Concerning Speed of Diagonalization of Symmetric Matrices Using Jacobi's Method*, J. Assoc. Comp. Mach. **4** (1957), 459–466.

H. Rutishauser, *Solution of Eigenvalue Problems with the LR Transformation*, Nat. Bur. Stand. App. Math. ser. **49** (1958), 47–81.

D. I. Steinberg, *Computational Matrix Algebra*, McGraw-Hill Book Company, New York, 1974.

G. W. Stewart, *Introduction to Matrix Computations*, Academic Press, New York, 1973.

G. Strang, *Linear Algebra and its Applications*, 3rd ed., Harcourt Brace Jovanovich, Publishers, San Diego, 1988.

J. H. Wilkinson, *The Algebraic Eigenvalue Problem*, Oxford Univ. Press, London, 1965.

# 14 NUMERICAL SOLUTION OF BOUNDARY-VALUE PROBLEMS

There exists a very thin viscous layer near the surface of a body when fluid flows over a surface such as an airplane wing. This thin layer is known as the boundary layer. It is well known that the boundary layer separates from the surface before reaching the trailing edge if pressure increases in the flow direction, causing a considerable increase in drag. Therefore, it is desirable to prevent separation. A boundary layer is readily observed on a thin flat plate set up parallel to the free stream. The governing equations of motion and continuity reduce to a single ordinary differential equation in nondimensional stream function $f(\eta)$ (Tritton 1988) given by

$$f\frac{d^2f}{d\eta^2} + 2\frac{d^3f}{d\eta^3} = 0$$

with boundary conditions $f(0) = df/d\eta(0) = 0$ and $df/d\eta \to 1$ as $\eta \to \infty$. In this chapter, we develop methods to approximate the solution of a given boundary-value problem.

## 14.1 INTRODUCTION

In Chapters 8 and 9 we considered methods for solving initial-value problems. In this chapter we consider ordinary differential equations with conditions imposed at more than one point. Consider a second-order ordinary differential equation

$$y'' + y = 0$$

with the boundary conditions

$$y(0) = 1 \quad \text{and} \quad y(\pi/2) = 1$$

**473**

We denote the first $n$ derivatives of $y$ with respect to $x$ by $y', y'', \ldots, y^{(n)}$.

An example of a fourth-order boundary-value problem is

$$y^{(4)} - y = 1 \qquad y(0) = y'(0) = 0 \qquad y(1) = y'(1) = 0$$

We shall develop numerical methods to solve a general second-order ordinary differential equation

$$y'' = f(x, y, y') \qquad a \le x \le b \tag{14.1.1}$$

subject to the boundary conditions

$$\gamma_1 y(a) + \gamma_2 y'(a) = \alpha \quad \text{and} \quad \gamma_3 y(b) + \gamma_4 y'(b) = \beta \tag{14.1.2}$$

Before we attempt to develop numerical methods, we would like to know the conditions that ensure that the solution to Equations (14.1.1) and (14.1.2) exists and is unique. The following theorem gives the required condition.

**Theorem 14.1.1**     Let $f$, $\partial f/\partial y$, and $\partial f/\partial y'$ be continuous on $R = \{(x, y, y') | a \le x \le b, y^2 + y'^2 < \infty\}$ and $\partial f/\partial y > 0$, $|\partial f/\partial y'| \le M$ for some positive constant $M$ on $R$. Further, let the coefficients in Equation (14.1.2) satisfy

$$\gamma_1\gamma_2 \le 0 \qquad \gamma_3\gamma_4 \ge 0 \qquad |\gamma_1| + |\gamma_3| \ne 0$$

Then the boundary-value problem (14.1.1) and (14.1.2) has a unique solution (Keller 1968). ∎

**EXAMPLE 14.1.1**     Consider the boundary-value problem

$$y'' - y = x; \qquad y(0) = 0, \quad y(1) = -1$$

We have $f(x, y, y') = x + y$, $\partial f/\partial y = 1 > 0$ for all $(x, y, y')$ and $|\partial f/\partial y'| = |0| \le M$ for any positive constant $M$. Further $\gamma_1 = 1$, $\gamma_2 = 0$, $\gamma_3 = 1$, $\gamma_4 = 0$ so $\gamma_1\gamma_2 = 0$, $\gamma_3\gamma_4 = 0$ and $|\gamma_1| + |\gamma_3| = 2$. Thus this boundary-value problem has a unique solution. We can verify that $y(x) = -x$ is the solution.  ■ ■ ■

**EXAMPLE 14.1.2**     Consider the boundary-value problem

$$y'' + y = 0; \qquad y'(0) = 0, \quad y'(\pi/2) = -1$$

Since $f(x, y, y') = -y$, $\partial f/\partial y = -1 < 0$. Also $|\gamma_1| + |\gamma_3| = 0$, so we cannot apply the uniqueness theorem. From this we cannot conclude that uniqueness is impossible, we simply cannot draw any conclusion about uniqueness in this problem. In fact, $y(x) = \cos x$ is the solution.  ■ ■ ■

## 14.2   THE SHOOTING METHOD

Let us assume that the solution to the second-order ordinary differential equation

$$y'' = f(x, y, y') \qquad a \le x \le b \tag{14.2.1a}$$

subject to the boundary conditions

$$y(a) = \alpha \quad \text{and} \quad y(b) = \beta \tag{14.2.1b}$$

for given numbers $\alpha$ and $\beta$ exists and is unique.

In Chapters 8 and 9, we considered several methods for the numerical solution of second-order ordinary differential equations with the given initial conditions $y(a)$ and $y'(a)$. For our problem, $y(a) = \alpha$ is known but $y'(a)$ is unknown. One way of finding $y'(a)$ is to guess. Let us therefore guess $y'(a) = t^{(0)}$ and then solve the initial-value problem

$$y'' = f(x, y, y'); \qquad y(a) = \alpha, \quad y'(a) = t^{(0)} \tag{14.2.2}$$

from $x = a$ to $x = b$. If the corresponding solution happens to satisfy $y(b) = \beta$, then we are done. In general $y(b) \ne \beta$, so guess another value $y'(a) = t^{(1)}$ and solve the corresponding initial-value problem. The process is repeated until the solution of the initial-value problem agrees with the specified tolerance with $y(b) = \beta$.

The method based on this idea is known as the **shooting method** because it resembles an artilleryman trying to fire a cannonball at a fixed target. The artilleryman sets the elevation of the gun and fires preliminary rounds at the target. After successive shots, he zeroes in on the target.

If linear second-order ordinary differential equations and linear boundary conditions are involved, then only one adjustment of the initial condition is required.

In order to illustrate this idea, consider a boundary-value problem whose unique solution can be found analytically; for example,

$$y'' + y = \frac{2x}{\pi}; \qquad y(0) = 0, \quad y\left(\frac{\pi}{2}\right) = 0 \tag{14.2.3}$$

We can verify that

$$y(x) = -\sin x + \frac{2x}{\pi} \tag{14.2.4}$$

is the solution of Equation (14.2.3). Then $y'(x) = -\cos x + (2/\pi)$ and

$$y'(0) = -1 + \frac{2}{\pi} \tag{14.2.5}$$

In order to solve Equation (14.2.3) by the method of an initial-value problem, we supply the initial condition $y'(0) = t$. Thus we solve the initial-value problem

$$y'' + y = \frac{2x}{\pi}; \qquad y(0) = 0, \quad y'(0) = t \tag{14.2.6}$$

We can verify that

$$y(x, t) = \left(t - \frac{2}{\pi}\right) \sin x + \frac{2x}{\pi} \qquad (14.2.7)$$

is the required solution of Equation (14.2.6).

Let $y'(0) = t^{(0)} = 0$. Then from Equation (14.2.7), we get

$$y\left(\frac{\pi}{2}, t^{(0)}\right) = -\frac{2}{\pi} + 1 = 0.36338$$

This is far from the required boundary condition $y(\pi/2) = 0$. Let us consider another initial condition $y'(0) = t^{(1)} = -1$. From Equation (14.2.7) we get

$$y\left(\frac{\pi}{2}, t^{(1)}\right) = -\frac{2}{\pi} = -0.63661$$

Now $y'(0) = t = 0$ gives $y(\pi/2, t^{(0)}) = -2/\pi + 1 = 0.36338$ and $y'(0) = t = -1$ gives $y(\pi/2, t^{(1)}) = -2/\pi = -0.63661$, while we want $y(\pi/2) = 0$. By using linear interpolation, $t = -1 + (2/\pi)$ gives $y(\pi/2) = 0$. Thus $y'(0) = t = -1 + (2/\pi)$ is the required initial condition. The same condition is given by Equation (14.2.5). Since this is a linear second-order ordinary differential equation, we came to the conclusion that $y'(0) = -1 + (2/\pi)$ by using linear interpolation.

We are, in the process, trying to solve the equation $y(\pi/2, t) = 0$. From Equation (14.2.7) we have $y(\pi/2, t) = t - (2/\pi) + 1 = 0$. Thus $t = -1 + (2/\pi)$.

In general, to solve (14.2.1a) and (14.2.1b), we replace them by the initial-value problem

$$y'' = f(x, y, y'); \qquad y(a) = \alpha, \quad y'(a) = t \qquad (14.2.8)$$

and then solve the nonlinear equation

$$y(b, t) - \beta = F(t) = 0 \qquad (14.2.9)$$

where $y(b, t)$ is given by the solution $y(x, t)$ of Equation (14.2.8). We can solve Equation (14.2.9) conceptually by a number of methods discussed in Chapter 2. The analytical form of $y(b, t)$ is unknown. We know the value of $y(b, t)$ for a given value of $t$ at $x = b$. Let us solve Equation (14.2.9) by Newton's method. The iteration formula is given by

$$t^{(k+1)} = t^{(k)} - \frac{F(t^{(k)})}{F'(t^{(k)})}$$

$$= t^{(k)} - \frac{y(b, t^{(k)}) - \beta}{F'(t^{(k)})} \quad \text{for } k = 0, 1, \dots \qquad (14.2.10)$$

In this equation, we know $y(b, t^{(k)})$ and $\beta$. In order to use Equation (14.2.10), we need $F'(t) = d/dt(F(t)) = \partial/\partial t\ (y(b, t) - \beta) = \partial/\partial t(y(b, t))$. Let $z = \partial y/\partial t$. There is no explicit representation of $z(x, t)$, but it is possible to compute $z(b, t)$ by solving the initial-value problem in $z$.

We have $\partial/\partial t\ (y') = \partial/\partial t\ (\partial y/\partial x) = \partial/\partial x\ (\partial y/\partial t) = \partial z/\partial x = z'$ assuming the order of differentiation with respect to $x$ and $t$ can be interchanged. Similarly, we can show that

$$\frac{\partial}{\partial t}\ (y'') = \frac{\partial^2}{\partial x^2}\left(\frac{\partial y}{\partial t}\right) = \frac{\partial^2 z}{\partial x^2} = z''$$

and therefore

$$z'' = \frac{\partial^2 z}{\partial x^2} = \frac{\partial}{\partial t}(y'') = \frac{\partial}{\partial t}(f(x, y(x, t), y'(x, t)))$$

$$= \frac{\partial f}{\partial y}\frac{\partial y}{\partial t} + \frac{\partial f}{\partial y'}\frac{\partial y'}{\partial t}$$

$$= \frac{\partial f}{\partial y}z + \frac{\partial f}{\partial y'}z' \qquad (14.2.11)$$

If the conditions in Equation (14.2.8) are also differentiated with respect to $t$, then we get

$$z(a) = 0 \quad \text{and} \quad z'(a) = 1 \qquad (14.2.12)$$

Hence $z$ satisfies the initial-value problem

$$z'' = \frac{\partial f}{\partial y}z + \frac{\partial f}{\partial y'}z'; \qquad z(a) = 0, \quad z'(a) = 1 \qquad (14.2.13)$$

In this equation $\partial f/\partial y$ and $\partial f/\partial y'$ are functions of $x$, $y$, and $y'$ since $f$ is a function of $x$, $y$, and $y'$. Since $y$ and $y'$ are not known exactly, Equation (14.2.13) cannot be solved independent of Equation (14.2.8). Let $y_1 = y$, $y_2 = y_1'$, $y_3 = z$, and $y_4 = z'$. Then Equations (14.2.8) and (14.2.13) reduce to

$$y_1' = y_2$$
$$y_2' = f(x, y_1, y_1')$$
$$y_3' = y_4$$
$$y_4' = z'' = \frac{\partial f}{\partial y_1}y_3 + \frac{\partial f}{\partial y_2}y_4 \qquad (14.2.14a)$$

with initial conditions

$$y_1(a) = \alpha \qquad y_2(a) = t^{(k)} \qquad y_3(a) = 0 \qquad y_4(a) = 1 \qquad (14.2.14b)$$

This system of four first-order equations with given initial conditions can be solved by a number of methods discussed in Chapters 8 and 9. Then $y(b, t^{(k)}) = y_1(b)$ and $z(b, t^{(k)}) = y_3(b)$ are known from the numerical solution of Equations (14.2.14a) and (14.2.14b). Substituting $F'(t) = \partial/\partial t(y(b, t)) = z(b, t) = y_3(b)$ in Equation (14.2.10), we get

$$t^{(k+1)} = t^{(k)} - \frac{(y(b, t^{(k)}) - \beta)}{z(b, t^{(k)})} = t^{(k)} - \frac{(y_1(b) - \beta)}{y_3(b)} \qquad (14.2.15)$$

Substituting the values of $y_1(b)$, $y_3(b)$, and $\beta$ in Equation (14.2.15), we calculate $t^{(k+1)}$.

The shooting method replaces Equations (14.2.1a) and (14.2.1b) by a system of four first-order equations [Equation (14.2.14a)] with initial conditions [Equation (14.2.14b)] and an iterative equation [Equation (14.2.15)] for $t^{(k+1)}$. Guess $t^{(0)}$, then solve Equations (14.2.14a) and (14.2.14b) and find $y(b, t^{(0)}) = y_1(b)$ and $z(b, t^{(0)}) = y_3(b)$. Find $t^{(1)}$ by using Equation (14.2.15). Repeat these steps until $t$ or $|y_1(b) - \beta|$ is reached within the specified degree of accuracy. If $t^{(0)}$ is sufficiently close to the exact value, then Newton's method converges rapidly. If nothing is known about the exact solution, then use $t^{(0)} = (\beta - \alpha)/(b - a)$.

**EXAMPLE 14.2.1**     Use the shooting method to solve the boundary-value problem

$$y'' = -yy' + y^3; \qquad y(0) = 1, \quad y(1) = \frac{1}{2}$$

We can verify that the exact solution is given by $y(x) = 1/(x + 1)$. Since $f(x, y, y') = -yy' + y^3$,

$$\frac{\partial f}{\partial y} = -y' + 3y^2 \quad \text{and} \quad \frac{\partial f}{\partial y'} = -y$$

Thus Equation (14.2.14a) reduces to

$$y_1' = y_2$$
$$y_2' = -y_1 y_2 + y_1^3$$
$$y_3' = y_4$$
$$y_4' = (-y_2 + 3y_1^2)y_3 + (-y_1)y_4$$

and the initial conditions (14.2.14b) reduce to

$$y_1(0) = 1 \qquad y_2(0) = t^{(0)} = \frac{\frac{1}{2} - 1}{1 - 0} = -\frac{1}{2} \qquad y_3(0) = 0 \qquad y_4(0) = 1$$

The fourth-order Runge–Kutta method is used to solve the system with $h = 0.1$. We get $y(1, t^{(0)}) = y_1(1, -\frac{1}{2}) = 0.97124$ and $z(1, t^{(0)}) = y_3(1, -\frac{1}{2}) = 0.95930$. From Equation (14.2.15), we get

$$
\begin{aligned}
t^{(1)} &= t^{(0)} - \frac{(y(1, t^{(0)}) - \frac{1}{2})}{z(1, t^{(0)})} \\
&= -\frac{1}{2} - \frac{(0.97124 - 0.5)}{0.95930} = -0.99124
\end{aligned}
$$

We continue and stop when

$$\left| t^{(k+1)} - t^{(k)} \right| < 10^{-5} \left| t^{(k+1)} \right|$$

The computed values of $t$, $y(1, t)$, and $z(1, t)$ are given in Table 14.2.1.

**Table 14.2.1**

| $t$ | $y(1, t)$ | $z(1, t)$ |
|---|---|---|
| $-0.50000$ | 0.97124 | 0.95930 |
| $-0.99124$ | 0.50823 | 0.94523 |
| $-0.99994$ | 0.50007 | 0.94535 |
| $-1.00001$ | 0.50000 | 0.94535 |

The values obtained by using $t = -1.00001$ are given in Table 14.2.2 along with the exact value $1/(x + 1)$.

**Table 14.2.2**

| $x_i$ | $y(x_i)$ | $1/(x_i + 1)$ | \|Actual Error\| |
|-------|----------|---------------|------------------|
| 0.0 | 1.00000 | 1.00000 | 0.00000 |
| 0.1 | 0.90909 | 0.90909 | 0.00000 |
| 0.2 | 0.83334 | 0.83333 | 0.00001 |
| 0.3 | 0.76923 | 0.76923 | 0.00000 |
| 0.4 | 0.71429 | 0.71429 | 0.00000 |
| 0.5 | 0.66667 | 0.66667 | 0.00000 |
| 0.6 | 0.62500 | 0.62500 | 0.00000 |
| 0.7 | 0.58824 | 0.58824 | 0.00000 |
| 0.8 | 0.55556 | 0.55556 | 0.00000 |
| 0.9 | 0.52632 | 0.52632 | 0.00000 |
| 1.0 | 0.50000 | 0.50000 | 0.00000 |

■ ■ ■

**Algorithm**

**Subroutine** Shoot ($a$, $b$, $N$, $\alpha$, $\beta$, eps, maxit)

Comment: This subroutine approximates the solution of the boundary-value problem $y'' = f(x, y, y')$; $y(a) = \alpha$, $y(b) = \beta$ using the shooting method. We have

$$\frac{dy_1}{dx} = f_1(x, y_1, y_2, y_3, y_4) = y_2 = y'$$

$$\frac{dy_2}{dx} = f_2(x, y_1, y_2, y_3, y_4) = f(x, y, y') = f(x, y_1, y_2)$$

$$\frac{dy_3}{dx} = f_3(x, y_1, y_2, y_3, y_4) = y_4$$

$$\frac{dy_4}{dx} = f_4(x, y_1, y_2, y_3, y_4) = \frac{\partial f}{\partial y_1} y_3 + \frac{\partial f}{\partial y_2} y_4$$

with initial conditions $y_1(a) = \alpha_1 = \alpha$, $y_2(a) = \alpha_2 = t = (\beta - \alpha)/(b - a)$, $y_3(a) = \alpha_3 = 0$, and $y_4(a) = \alpha_4 = 1$. We solve this system by using Subroutine System which uses the classical Runge–Kutta method of order four from Section 8.7. The tolerance is eps and maxit is the maximum number of iterations. We have $\mathbf{y} = [y_1, y_2, y_3, y_4]'$, $\mathbf{f} = [f_1, f_2, f_3, f_4]'$ and $\boldsymbol{\alpha} = [\alpha_1, \alpha_2, \alpha_3, \alpha_4]'$.

1 **Print** $a$, $b$, $N$, $\alpha$, $\beta$, eps, maxit
2 $h = (b - a)/N$
3 $x = a$
4 $\alpha_1 = \alpha$
5 $\alpha_2 = (\beta - \alpha)/(b - a)$
6 $\alpha_3 = 0$
7 $\alpha_4 = 1$
8 $k = 0$
9 **Repeat** steps 10–14 **while** $k \leq$ maxit
   10 **Call** System ($a$, $b$, $N$, 4, $\boldsymbol{\alpha}$, $\mathbf{y}$, $\mathbf{f}$)
   11 **If** $|y_1(b) - \beta| <$ eps, **then print** 'finished,' $x$, and $y_1$ and exit
   12 $t = t - [(y_1(b) - \beta)/y_3(b)]$
   13 $\alpha_2 = t$

14 $k = k + 1$
15 **Print** 'Algorithm fails: no convergence' and exit.   ◀

Consider the linear second-order ordinary differential equation

$$y'' = p(x)y' + q(x)y + r(x) \qquad \textbf{(14.2.16)}$$

subject to the boundary conditions

$$y(a) = \alpha \quad \text{and} \quad y(b) = \beta \qquad \textbf{(14.2.17)}$$

In this case Equation (14.2.14a) reduces to

$$
\begin{aligned}
y_1' &= y_2 \\
y_2' &= p(x)y_2 + q(x)y_1 + r(x) \qquad &\textbf{(14.2.18a)} \\
y_3' &= y_4 \\
y_4' &= q(x)y_3 + p(x)y_4 \qquad &\textbf{(14.2.19a)}
\end{aligned}
$$

and the initial conditions (14.2.14b) reduce to

$$y_1(a) = \alpha \qquad y_2(a) = t \qquad \textbf{(14.2.18b)}$$

$$y_3(a) = 0 \qquad y_4(a) = 1 \qquad \textbf{(14.2.19b)}$$

Since Equation (14.2.19a) is independent of $y_1$ and $y_2$, we can solve Equation (14.2.18a) with Equation (14.2.18b) and Equation (14.2.19a) with Equation (14.2.19b) independently by using the fourth-order Runge–Kutta method.

Alternatively, we can replace the linear boundary-value problem (14.2.16) and (14.2.17) by the following initial-value problems:

$$y'' = p(x)y' + q(x)y + r(x); \qquad y(a) = \alpha, \quad y'(a) = 0 \qquad \textbf{(14.2.20)}$$

$$y'' = p(x)y' + q(x)y + r(x); \qquad y(a) = \alpha, \quad y'(a) = 1 \qquad \textbf{(14.2.21)}$$

Let $y_1$ and $y_2$ be the solutions of Equations (14.2.20) and (14.2.21) respectively. We want to find $c_1$ and $c_2$ so that

$$y(x) = c_1 y_1(x) + c_2 y_2(x) \qquad \textbf{(14.2.22)}$$

is the solution of the boundary-value problem (14.2.16) and (14.2.17). From Equation (14.2.22), we have

$$
\begin{aligned}
y'' &= c_1 y_1'' + c_2 y_2'' \\
&= c_1(p(x)y_1' + q(x)y_1 + r(x)) + c_2(p(x)y_2' + q(x)y_2 + r(x)) \\
&\qquad \text{using Equations (14.2.20) and (14.2.21)} \\
&= p(x)[c_1 y_1' + c_2 y_2'] + q(x)[c_1 y_1 + c_2 y_2] + r(x)[c_1 + c_2] \\
&= p(x)y' + q(x)y + r(x)[c_1 + c_2] \quad \text{using Equation (14.2.22)} \\
&= p(x)y' + q(x)y + r(x) \quad \text{using Equation (14.2.20)}
\end{aligned}
$$

Hence $c_1 + c_2 = 1$. From Equation (14.2.22), we have

$$y(a) = c_1 y_1(a) + c_2 y_2(a) = c_1\alpha + c_2\alpha = \alpha(c_1 + c_2) = \alpha$$

Therefore again $c_1 + c_2 = 1$. Further, from Equation (14.2.22), $y(b) = c_1 y_1(b) + c_2 y_2(b) = \beta$. Thus we have two equations

$$c_1 + c_2 = 1$$
$$c_1 y_1(b) + c_2 y_2(b) = \beta \qquad \textbf{(14.2.23)}$$

Solving Equation (14.2.23) for $c_1$ and $c_2$, we get

$$c_1 = \frac{\beta - y_2(b)}{y_1(b) - y_2(b)} \quad \text{and} \quad c_2 = \frac{y_1(b) - \beta}{y_1(b) - y_2(b)} \qquad \textbf{(14.2.24)}$$

Therefore the solution (14.2.22) of the boundary-value problem (14.2.16) and (14.2.17) is

$$y(x) = \frac{\beta - y_2(b)}{y_1(b) - y_2(b)} y_1(x) + \frac{y_1(b) - \beta}{y_1(b) - y_2(b)} y_2(x) \qquad \textbf{(14.2.25)}$$

Hence Equation (14.2.25) is the solution of (14.2.16) with (14.2.17). In order to get Equation (14.2.25), we have to solve Equations (14.2.20) and (14.2.21) using the fourth-order Runge–Kutta method.

Also we can treat Equations (14.2.18a) and (14.2.19a) as a system of four first-order equations with the initial conditions (14.2.18b) and (14.2.19b) and solve the system using the fourth-order Runge–Kutta method with $t = t^{(0)}$. Then use Equation (14.2.15) to find $t^{(1)}$. Use $t^{(1)}$ to solve the system of Equations (14.2.18a) and (14.2.19a), with Equations (14.2.18b) and (14.2.19b). Repeat this procedure until $t$ or $|y_1(b) - \beta|$ is reached within the specified degree of accuracy. It is probable that $t^{(1)}$ will provide the required accuracy.

## EXERCISES

1. Let $y_1(x)$ and $y_2(x)$ be two solutions of the linear second-order ordinary differential equation

$$y'' = p(x)y' + q(x)y + r(x)$$

   Verify that $(c_1 y_1(x) + c_2 y_2(x))/(c_1 + c_2)$ where $c_1$ and $c_2$ are constants, is also its solution.

2. Find the conditions under which the linear boundary value

$$y'' = p(x)y' + q(x)y + r(x)$$
$$\gamma_1 y(a) + \gamma_2 y'(a) = \alpha \quad \text{and} \quad \gamma_3 y(b) + \gamma_4 y'(b) = \beta$$

   has a unique solution.

3. Consider the boundary-value problem

$$y'' + \alpha y = 0; \qquad y(0) = 0, \quad y(1) = B$$

   (a) Find $\alpha$ such that there are an infinite number of solutions.
   (b) Find $\alpha$ such that there is a unique solution.
   (c) Find $\alpha$ such that there is no solution.

4. Solve the boundary-value problem

$$y'' - y = x; \qquad y(0) = 0, \quad y(1) = -1$$

Find $y'(0)$.

5. (a) Solve the initial-value problem

$$y'' - y = x; \qquad y(0) = 0, \quad y'(0) = t$$

(b) For $t = 0$, find $y(1)$.
(c) For $t = -2$, find $y(1)$.
(d) Using linear interpolation, (b), and (c), find the value of $t$ such that $y(1) = -1$. Compare this value of $t$ with $y'(0)$ of Exercise 4.

6. Verify that $y(x) = y_1(x) + [(\beta - y_1(b))/y_2(b)] y_2(x)$ is the solution of the linear boundary-value problem (14.2.16) and (14.2.17) where $y_1(x)$ is the solution of the initial-value problem

$$y'' = p(x)y' + q(x)y + r(x); \qquad y(a) = \alpha, \quad y'(a) = 0$$

and $y_2(x)$ is the solution of the initial-value problem

$$y'' = p(x)y' + q(x)y; \qquad y(a) = 0, \quad y'(a) = 1$$

## COMPUTER EXERCISES

C.1. Write a subroutine to approximate the solution of a system of four first-order ordinary differential equations given by (14.2.14a) with the initial conditions (14.2.14b) using the fourth-order Runge–Kutta method. Also, find $t$ using Newton's method such that $y(b, t) - \beta = 0$ using Equation (14.2.15).

C.2. Write a computer program to approximate the solution of a boundary-value problem

$$y'' = f(x, y, y'); \qquad y(a) = \alpha \quad \text{and} \quad y(b) = \beta$$

using Exercise C.1.

Using Exercise C.2, approximate the solution of each of the boundary-value problems of Exercises C.3–C.5. Use $N = 10$, 20, and 100. Compare each of your approximate solutions with the exact solution.

C.3. $y'' - 2y' + y = 1$; $y(0) = 1$, $y(1) = 0$.
C.4. $y'' - (1/x)\, y' + y = x^2$; $y(1) = 1$, $y(2) = 3$.
C.5. $y'' = (3/2)\, y^2$; $y(0) = 4$, $y(1) = 1$.
C.6. Write a subroutine to approximate the solution of a system of two first-order ordinary differential equations

$$\frac{dy_1}{dx} = y_2$$

$$\frac{dy_2}{dx} = f(x, y_1, y_2)$$

with boundary conditions $y_1(a) = \alpha$ and $y_2(a) = t$. Also, find $t$ using the bisection method such that $y_1(b, t) - \beta = 0$.

**C.7.** Write a computer program to approximate the solution of a linear boundary-value problem

$$y'' = p(x)y' + q(x)y + r(x); \qquad y(a) = \alpha \quad \text{and} \quad y(b) = \beta$$

using Exercise C.6. Select $t^{(0)}$ and $t^{(1)}$ such that $y(b, t^{(0)})y(b, t^{(1)}) \leq 0$.

**C.8.** Consider the linear boundary-value problem
(a) $y'' + y = 0$; $y(0) = 0$, $y(\pi/2) = 1$.
(b) Approximate the solution of (a) using Exercise C.7 with $h = \pi/20$.
(c) Find $a$ and $b$ such that $u = ax + b$ transforms the interval $[0, \pi/2]$ to the interval $[0, 1]$.
(d) Transform the linear boundary problem in (a) in terms of $u$ using (c).
(e) Approximate the solution of (d) using Exercise C.7 with $h = 0.1$.
(f) Find the exact solution of (a) and compare it with (b) and (e).

**C.9.** Using Exercise C.7, approximate the solution of the boundary-value problem of Exercise C.3.

**C.10.** A boundary layer on a thin plate is given by

$$f\frac{d^2f}{d\eta^2} + 2\frac{d^3f}{d\eta^3} = 0$$

with boundary conditions

$$f(0) = \frac{df}{d\eta}(0) = 0 \quad \text{and} \quad \frac{df}{d\eta} \to 1 \text{ as } \eta \to \infty \qquad \textbf{(C.10.1)}$$

We replace Equation (C.10.1) by a system of three first-order differential equations

$$\frac{dy_1}{d\eta} = y_2$$

$$\frac{dy_2}{d\eta} = y_3$$

$$\frac{dy_3}{d\eta} = -\frac{1}{2}y_1 y_3$$

with boundary conditions

$$y_1(0) = y_2(0) = 0 \quad \text{and} \quad y_2 \to 1 \text{ as } \eta \to \infty \qquad \textbf{(C.10.2)}$$

In practice, we replace $y_2 \to 1$ as $\eta \to \infty$ by selecting $\eta_{max}$ so that the solution shows little further change for $\eta > \eta_{max}$. Approximate the solution of Equation (C.10.1) by computing the solution of Equation (C.10.2). We start with $y_3(0) = \frac{1}{4}$ and $y_3(0) = \frac{1}{2}$ between which $y_3(0)$ is thought to lie and $\eta_{max} = 10$. Also, compute the velocity components $u = df/d\eta$ and $v = \eta(df/d\eta) - f$ at various points on the plate.

## 14.3    THE FINITE DIFFERENCE METHOD

In this section, we consider the most common numerical method for solving a boundary-value problem, the finite difference method. The basic idea underlying the finite difference method is to replace the derivatives in a differential equation by suitable difference quotients and then solve the resulting system of equations. We used this method in Section 11.1.

We illustrate this method with the following linear second-order ordinary differential equation

$$y'' = p(x)y' + q(x)y + r(x) \tag{14.3.1a}$$

subject to the boundary conditions

$$y(a) = \alpha \quad \text{and} \quad y(b) = \beta \tag{14.3.1b}$$

Let us assume that the boundary-value problem (14.3.1a) and (14.3.1b) has a unique solution. In order to approximate the solution, we replace the derivatives in Equation (14.3.1a) by finite differences. This reduces Equation (14.3.1a) to a system of linear equations. Then we solve the system. In order to accomplish this, divide the interval $[a, b]$ into $N + 1$ equal intervals whose endpoints are the mesh points $x_i = a + ih$ for $i = 0, 1, \ldots, N + 1$ as shown in Figure 14.3.1 with $x_0 = a$, $x_{N+1} = b$, and $h = (b - a)/(N + 1)$.

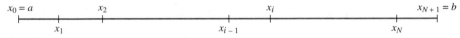

**Fig. 14.3.1**

At the interior mesh point $x = x_i$, Equation (14.3.1a) becomes

$$y''(x_i) = p(x_i)y'(x_i) + q(x_i)y(x_i) + r(x_i) \tag{14.3.2}$$

The simplest way to approximate Equation (14.3.2) is to replace the derivatives $y'(x_i)$ and $y''(x_i)$ from Table 5.3.1 by their central differences

$$y'(x_i) = \frac{y(x_i + h) - y(x_i - h)}{2h} - \frac{h^2}{6} y^{(3)}(\xi_i) \tag{14.3.3}$$

and

$$y''(x_i) = \frac{y(x_i - h) - 2y(x_i) + y(x_i + h)}{h^2} - \frac{h^2}{12} y^{(4)}(\eta_i) \tag{14.3.4}$$

Substituting Equations (14.3.3) and (14.3.4) in Equation (14.3.2), we get

$$\frac{y(x_i - h) - 2y(x_i) + y(x_i + h)}{h^2} = p(x_i) \frac{y(x_i + h) - y(x_i - h)}{2h}$$

$$+ q(x_i)y(x_i) + r(x_i) + \frac{h^2}{12} y^{(4)}(\eta_i) - \frac{h^2}{6} y^{(3)}(\xi_i) \tag{14.3.5}$$

where $\xi_i$ and $\eta_i$ are in $(x_i - h, x_i + h)$.

Since $\xi_i$ and $\eta_i$ are not known and $h^2$ is small, we ignore the last two terms of Equation (14.3.5). Denoting the approximate value of $y$ at $x_i$ by $Y_i$ (i.e., $Y_i \approx y(x_i)$), the approximate value of $y$ at $x_i + h$ by $Y_{i+1}$ (i.e., $Y_{i+1} \approx y(x_i + h)$), and the approximate value of $y$ at $x_i - h$ by $Y_{i-1}$ (i.e., $Y_{i-1} \approx y(x_i - h)$), we get from Equation (14.3.5)

$$\frac{Y_{i-1} - 2Y_i + Y_{i+1}}{h^2} = p(x_i)\frac{Y_{i+1} - Y_{i-1}}{2h} + q(x_i)Y_i + r(x_i) \qquad \textbf{(14.3.6)}$$

for $i = 1, 2, \ldots, N$. Equation (14.3.6) forms a system of $N$ equations in $N$ unknowns. Collecting like terms, Equation (14.3.6) is rewritten as

$$Y_{i-1}\left(1 + \frac{h}{2}p(x_i)\right) - Y_i(2 + h^2q(x_i)) + Y_{i+1}\left(1 - \frac{h}{2}p(x_i)\right) = h^2r(x_i) \qquad \textbf{(14.3.7)}$$

for $i = 1, 2, \ldots, N$. Let

$$a_i = 1 + \frac{h}{2}p(x_i) \qquad b_i = -(2 + h^2q(x_i)) \qquad c_i = 1 - \frac{h}{2}p(x_i) \quad \textbf{(14.3.7a)}$$

Then Equation (14.3.7) becomes

$$a_i Y_{i-1} + b_i Y_i + c_i Y_{i+1} = h^2r(x_i) \quad \text{for } i = 1, 2, \ldots, N \qquad \textbf{(14.3.8)}$$

The boundary conditions (14.3.1b) are written as

$$y(a) = y(x_0) = Y_0 = \alpha \quad \text{and} \quad y(b) = y(x_{N+1}) = Y_{N+1} = \beta \qquad \textbf{(14.3.9)}$$

In matrix form, Equations (14.3.8) and (14.3.9) become

$$A\mathbf{y} = \mathbf{s} \qquad \textbf{(14.3.10)}$$

where

$$A = \begin{bmatrix} b_1 & c_1 & 0 & \cdots & 0 & 0 \\ a_2 & b_2 & c_2 & \cdots & 0 & 0 \\ 0 & a_3 & b_3 & \cdots & 0 & 0 \\ \vdots & \vdots & \vdots & \ddots & \vdots & \vdots \\ 0 & 0 & 0 & \cdots & b_{N-1} & c_{N-1} \\ 0 & 0 & 0 & \cdots & a_N & b_N \end{bmatrix} \qquad \mathbf{y} = \begin{bmatrix} Y_1 \\ Y_2 \\ Y_3 \\ \vdots \\ Y_{N-1} \\ Y_N \end{bmatrix} \qquad \mathbf{s} = \begin{bmatrix} h^2r(x_1) - a_1\alpha \\ h^2r(x_2) \\ h^2r(x_3) \\ \vdots \\ h^2r(x_{N-1}) \\ h^2r(x_N) - c_N\beta \end{bmatrix}$$

By using any method discussed in Chapters 7 and 11, Equation (14.3.10) can be solved. This tridiagonal system can be solved very efficiently by the factorization method discussed in Section 7.7. In summary, the following steps are involved for solving Equation (14.3.1a) and (14.3.1b):

1.  Reduce Equation (14.3.1a) to its corresponding difference form [Equation (14.3.7)].
2.  Compare Equation (14.3.7) with Equation (14.3.8) to identify $a_i$, $b_i$, and $c_i$.
3.  Calculate $\alpha_i$, $\beta_i$, and $\gamma_i$ from Equation (7.7.4).
4.  Calculate $z_i$ from Equation (7.7.6).
5.  Calculate the required solutions $Y_i$ by replacing $x_i$ with $Y_i$ in Equation (7.7.7).

**EXAMPLE 14.3.1**     Solve $y'' + y = 0$; $y(0) = 1$, $y(\pi/2) = 0$ using the finite difference method.

Divide $[0, \pi/2]$ into $N + 1 = 10$ equal intervals, then $N = 9$ and $h = \pi/20 = 0.15708$. Using the central difference for $y''$, $y'' + y = 0$ reduces to

$$\frac{Y_{i-1} - 2Y_i + Y_{i+1}}{h^2} + Y_i = 0$$

which can be rearranged to

$$Y_{i-1} + (h^2 - 2)Y_i + Y_{i+1} = 0 \quad \text{for } i = 1, 2, \ldots, N \qquad \textbf{(14.3.11)}$$

along with $Y_0 = y(0) = 1$ and $Y_{10} = y(\pi/2) = 0$.

This tridiagonal system is solved by $LU$ decomposition given in Section 7.7. The computed values of $Y_i$ at various $x_i$ are given in Table 14.3.1, along with the values of the exact solution $y = \cos x$ for comparison.

**Table 14.3.1**

| $i$ | $x_i$ | $Y_i$ Using Eq. (14.3.11) | $y_i = \cos x_i$ Exact Solution | \|Actual Error\| |
|---|---|---|---|---|
| 0 | 0.0 | 1.00000 | 1.00000 | 0.00000 |
| 1 | $\pi/20$ | 0.98792 | 0.98769 | 0.00023 |
| 2 | $\pi/10$ | 0.95146 | 0.95106 | 0.00040 |
| 3 | $3\pi/20$ | 0.89152 | 0.89101 | 0.00051 |
| 4 | $4\pi/20$ | 0.80959 | 0.80902 | 0.00057 |
| 5 | $5\pi/20$ | 0.70768 | 0.70711 | 0.00057 |
| 6 | $6\pi/20$ | 0.58831 | 0.58779 | 0.00052 |
| 7 | $7\pi/20$ | 0.45442 | 0.45399 | 0.00043 |
| 8 | $8\pi/20$ | 0.30933 | 0.30902 | 0.00031 |
| 9 | $9\pi/20$ | 0.15659 | 0.15643 | 0.00016 |
| 10 | $\pi/2$ | 0.00000 | 0.00000 | 0.00000 |

■ ■ ■

In order to increase accuracy, we need to reduce the step size or use the finite difference formula with higher-order truncation error. If a formula uses more than three mesh points, then the structure of the resulting matrix is multidiagonal. The finite difference method can be used to approximate a solution of a differential equation of higher order; the structure of the resulting matrix is multidiagonal.

In the next section, we consider a method of higher order without resorting to a finite difference formula of higher order for $y'' = f(x, y)$.

**Algorithm**

**Subroutine** Finite $(ba, bb, N, b\alpha, b\beta, \mathbf{y})$
Comment: This subroutine approximates the solution of the boundary-value problem

$$y'' = p(x)y' + q(x)y + r(x); \qquad y(ba) = b\alpha, \quad y(bb) = b\beta$$

using Equations (14.3.7), (14.3.10), (7.7.4), (7.7.6), and (7.7.7). We divide $[ba, bb]$ into $N + 1$ equal intervals such that $x_0 = ba$ and $x_{N+1} = bb$. We have $\mathbf{y} = [Y_1, Y_2, \ldots, Y_N]$

  1  **Print** $ba, bb, N, b\alpha, b\beta$
  2  $h = (bb - ba)/(N + 1)$
  3  **Repeat** steps 4–8 **with** $i = 1, 2, \ldots, N$
      4  $xi = ba + ih$
      5  $a_i = 1 + (hp(xi)/2)$
      6  $b_i = -(2 + h^2 q(xi))$
      7  $c_i = 1 - (hp(xi)/2)$
      8  $s_i = h^2 r(xi)$
  9  $s_1 = s_1 - (b\alpha)a_1$
10  $s_N = s_N - (b\beta)c_N$
11  **Call** Tridi $(\mathbf{a}, \mathbf{b}, \mathbf{c}, \mathbf{s}, N, \mathbf{Y})$
12  **Repeat** steps 13–14 **with** $i = 1, 2, \ldots, N$
     13  $xi = ba + ih$
     14  **Print** $xi, Y_i$
15  **Print** 'finished' and exit.

## Mixed Boundary Conditions

Let us consider a linear second-order boundary-value problem

$$y'' = p(x)y' + q(x)y + r(x) \tag{14.3.12a}$$

with boundary conditions

$$\gamma_1 \, y(a) + \gamma_2 \, y'(a) = \alpha \quad \text{and} \quad \gamma_3 \, y(b) + \gamma_4 \, y'(b) = \beta \tag{14.3.12b}$$

Using Equations (14.3.3) and (14.3.4), Equation (14.3.12a) reduces to Equation (14.3.8) which is given by

$$a_i Y_{i-1} + b_i Y_i + c_i Y_{i+1} = h^2 r(x_i) \tag{14.3.8}$$

for $i = 1, 2, \ldots, N$. When we replace $y'(a)$ by the forward difference formula and $y'(b)$ by the backward difference formula from Table 5.3.1, Equation (14.3.12b) becomes

$$\gamma_1 \, Y_0 + \gamma_2 \, \frac{Y_1 - Y_0}{h} = \alpha \quad \text{and} \quad \gamma_3 \, Y_{N+1} + \gamma_4 \, \frac{Y_{N+1} - Y_N}{h} = \beta \tag{14.3.13}$$

Solving the first equation for $Y_0$ and the second equation for $Y_{N+1}$ and then substituting in Equation (14.3.8), Equation (14.3.8) reduces to a tridiagonal system. Unfortunately, the first derivative approximation is of the first order only. To overcome this drawback, we can use higher-order approximations in Equation (14.3.12b) for $y'(a)$ and $y'(b)$ from Table 5.3.1. Using the asymmetrical formulas from Table 5.3.1 in Equation (14.3.12b), we get

$$\gamma_1 Y_0 + \gamma_2 \, \frac{-3Y_0 + 4Y_1 - Y_2}{2h} = \alpha \quad \text{and} \quad \gamma_3 Y_{N+1} + \gamma_4 \, \frac{3Y_{N+1} - 4Y_N + Y_{N-1}}{2h} = \beta$$

$$\tag{14.3.14}$$

Solving the first equation for $Y_0$ and the second equation for $Y_{N+1}$ yields

$$Y_0 = \frac{2h\alpha - \gamma_2(4Y_1 - Y_2)}{2h\gamma_1 - 3\gamma_2} \quad \text{and} \quad Y_{N+1} = \frac{2h\beta - \gamma_4(Y_{N-1} - 4Y_N)}{2h\gamma_3 + 3\gamma_4}$$

**(14.3.15)**

For $i = 1$, Equation (14.3.8) reduces to

$$a_1 Y_0 + b_1 Y_1 + c_1 Y_2 = h^2 r(x_1)$$

Substituting the value of $Y_0$ from Equation (14.3.15), this equation reduces to

$$\left(b_1 - \frac{4a_1\gamma_2}{2h\gamma_1 - 3\gamma_2}\right) Y_1 + \left(c_1 + \frac{a_1\gamma_2}{2h\gamma_1 - 3\gamma_2}\right) Y_2 = h^2 r(x_1) - \frac{2h\alpha a_1}{2h\gamma_1 - 3\gamma_2}$$

**(14.3.16)**

For $i = N$, we get from Equation (14.3.8)

$$a_N Y_{N-1} + b_N Y_N + c_N Y_{N+1} = h^2 r(x_N)$$

Substituting the value of $Y_{N+1}$ from Equation (14.3.15) into this equation, we get

$$\left(a_N - \frac{c_N\gamma_4}{2h\gamma_3 + 3\gamma_4}\right) Y_{N-1} + \left(b_N + \frac{4c_N\gamma_4}{2h\gamma_3 + 3\gamma_4}\right) Y_N = h^2 r(x_N) - \frac{2h\beta c_N}{2h\gamma_3 + 3\gamma_4}$$

**(14.3.17)**

Thus we can write Equations (14.3.8), (14.3.16), and (14.3.17) as

$$A_1 \mathbf{y} = \mathbf{s}_1$$

**(14.3.18)**

where

$$A_1 =$$

$$
\begin{bmatrix}
b_1 - \dfrac{4a_1\gamma_2}{2h\gamma_1 - 3\gamma_2} & c_1 + \dfrac{a_1\gamma_2}{2h\gamma_1 - 3\gamma_2} & 0 & \cdots & 0 & 0 \\
a_2 & b_2 & c_2 & \cdots & 0 & 0 \\
0 & a_3 & b_3 & \cdots & 0 & 0 \\
\vdots & \vdots & \vdots & \ddots & \vdots & \vdots \\
0 & 0 & 0 & \cdots & b_{N-1} & c_{N-1} \\
0 & 0 & 0 & \cdots & a_N - \dfrac{c_N\gamma_4}{2h\gamma_3 + 3\gamma_4} & b_N + \dfrac{4c_N\gamma_4}{2h\gamma_3 + 3\gamma_4}
\end{bmatrix}
$$

$$
\mathbf{y} = \begin{bmatrix} Y_1 \\ Y_2 \\ Y_3 \\ \vdots \\ Y_{N-1} \\ Y_N \end{bmatrix}
\qquad
\mathbf{s}_1 = \begin{bmatrix} h^2 r(x_1) - \dfrac{2h\alpha a_1}{2h\gamma_1 - 3\gamma_2} \\ h^2 r(x_2) \\ h^2 r(x_3) \\ \vdots \\ h^2 r(x_{N-1}) \\ h^2 r(x_N) - \dfrac{2h\beta c_N}{2h\gamma_3 + 3\gamma_4} \end{bmatrix}
$$

**EXAMPLE 14.3.2**

Solve $y'' + y = 0$; $y'(0) = 0$, $y'(\pi/2) = -1$ using the finite difference method. Divide $[0, \pi/2]$ into $N + 1 = 10$ equal intervals. Then $h = \pi/20 = 0.15708$.

Substituting the asymmetrical formulas from Table 5.3.1 into the boundary conditions, we get

$$\frac{-3Y_0 + 4Y_1 - Y_2}{2h} = 0 \quad \text{and} \quad \frac{3Y_{N+1} - 4Y_N + Y_{N-1}}{2h} = -1$$

Therefore

$$Y_0 = \frac{1}{3}(4Y_1 - Y_2) \quad \text{and} \quad Y_{N+1} = \frac{1}{3}(4Y_N - Y_{N-1} - 2h) \qquad \textbf{(14.3.19)}$$

From Example 14.3.1, we have the difference equations

$$Y_{i-1} + (h^2 - 2)Y_i + Y_{i+1} = 0 \quad \text{for } i = 1, 2, \ldots, N \qquad \textbf{(14.3.20)}$$

For $i = 1$, $Y_0 + (h^2 - 2)Y_1 + Y_2 = 0$. Substituting $Y_0$ from Equation (14.3.19) in this equation, we get $Y_1(h^2 - (\frac{2}{3})) + (2Y_2/3) = 0$.

Substituting $N = 9$, we get from Equation (14.3.19), $Y_{10} = (4Y_9 - Y_8 - 2h)/3$. For $N = 9$, Equation (14.3.20) reduces to $Y_8 + (h^2 - 2)Y_9 + Y_{10} = 0$. Substituting for $Y_{10}$, we get $(2Y_8/3) + (h^2 - (\frac{2}{3}))Y_9 = (2h/3)$. The system of equations

$$Y_1\left(h^2 - \frac{2}{3}\right) + \frac{2}{3}Y_2 = 0$$

$$Y_{i-1} + (h^2 - 2)Y_i + Y_{i+1} = 0 \quad \text{for } i = 2, 3, \ldots, 8$$

$$\frac{2}{3}Y_8 + \left(h^2 - \frac{2}{3}\right)Y_9 = \frac{2h}{3}$$

is solved using *LU* decomposition given in Section 7.7. In Table 14.3.2, the computed values of $Y_i$ and the exact solution $y = \cos x$ at various $x_i$ are given.

**Table 14.3.2**

| $i$ | $x_i$ | $U_i$ | $y_i = \cos x_i$ | $\lvert$Actual Error$\rvert$ |
|---|---|---|---|---|
| 0 | 0.0 | 0.99087 | 1.00000 | 0.00913 |
| 1 | $\pi/20$ | 0.97880 | 0.98769 | 0.00889 |
| 2 | $2\pi/20$ | 0.94257 | 0.95106 | 0.00849 |
| 3 | $3\pi/20$ | 0.88309 | 0.89101 | 0.00792 |
| 4 | $4\pi/20$ | 0.80182 | 0.80902 | 0.00702 |
| 5 | $5\pi/20$ | 0.70076 | 0.70711 | 0.00635 |
| 6 | $6\pi/20$ | 0.58241 | 0.58779 | 0.00538 |
| 7 | $7\pi/20$ | 0.44969 | 0.45399 | 0.00430 |
| 8 | $8\pi/20$ | 0.30588 | 0.30902 | 0.00314 |
| 9 | $9\pi/20$ | 0.15452 | 0.15643 | 0.00191 |
| 10 | $\pi/2$ | −0.00065 | 0.00000 | 0.00065 |

Certainly, we need small $h$ for better accuracy.   ■ ■ ■

Also, we may use central differences in Equation (14.3.12b) with pseudonodes. We replace $y'(a)$ and $y'(b)$ by the central difference formulas from Table 5.3.1, then Equation (14.3.12b) becomes

$$\gamma_1 Y_0 + \gamma_2 \frac{Y_1 - Y_{-1}}{2h} = \alpha \quad \text{and} \quad \gamma_3 Y_{N+1} + \gamma_4 \frac{Y_{N+2} - Y_N}{2h} = \beta \quad \textbf{(14.3.21)}$$

where $Y_{-1}$ and $Y_{N+2}$ are the approximate values of $y$ at $x_{-1}$ and $x_{N+2}$ respectively as shown in Figure 14.3.2.

**Fig. 14.3.2**

Note that $x_{-1}$ and $x_{N+2}$ are not real nodes since we do not know what lies beyond the boundary points $x_0 = a$ and $x_{N+1} = b$. Solving Equation (14.3.21) for $Y_{-1}$ and $Y_{N+2}$, we get

$$Y_{-1} = \frac{2h\gamma_1}{\gamma_2} Y_0 + Y_1 - \frac{2\alpha}{\gamma_2} h \qquad Y_{N+2} = Y_N - \frac{2h\gamma_3}{\gamma_4} Y_{N+1} + \frac{2h\beta}{\gamma_4} \quad \textbf{(14.3.22)}$$

For $i = 0$, Equation (14.3.8) reduces to

$$a_0 Y_{-1} + b_0 Y_0 + c_0 Y_1 = h^2 r(x_0)$$

Substituting the value of $Y_{-1}$ from Equation (14.3.22), this equation reduces to

$$\left( b_0 + \frac{2ha_0\gamma_1}{\gamma_2} \right) Y_0 + (a_0 + c_0)Y_1 = h^2 r(x_0) + \frac{2\alpha a_0}{\gamma_2} h \quad \textbf{(14.3.23)}$$

For $i = N + 1$, Equation (14.3.8) reduces to

$$a_{N+1} Y_N + b_{N+1} Y_{N+1} + c_{N+1} Y_{N+2} = h^2 r(x_{N+1})$$

Substituting the value of $Y_{N+2}$ from Equation (14.3.22) into this equation, we get

$$(a_{N+1} + c_{N+1})Y_N + \left( b_{N+1} - \frac{c_{N+1}\gamma_3}{\gamma_4} \right) Y_{N+1} = h^2 r(x_{N+1}) - \frac{2h\beta}{\gamma_4} c_{N+1} \quad \textbf{(14.3.24)}$$

Thus Equations (14.3.23), (14.3.8), and (14.3.24) form a system of $N + 2$ equations for $Y_0, Y_1, \ldots, Y_{N+1}$.

### EXERCISES

For Exercises 1–3, use the finite difference method to approximate the solution of each of the boundary-value problems. Use $h = \frac{1}{3}$.

1.  $y'' + (\pi^2/4)y = 0;\ y(0) = 0,\ y(1) = 1.$
2.  $y'' + y = x;\ y(0) = 0,\ y(1) = 1.$
3.  $y'' - y = -x;\ y(0) = 0,\ y(1) = 1.$

For Exercises 4–7, find an approximate solution of each of the boundary-value problems (a) using the finite difference method and Equation (14.3.13) and (b) using the finite difference method and Equation (14.3.15). Use $h = \frac{1}{3}$.

**4.** $y'' - y = 0; \; y'(0) = 0, \; y(1) = 1.$
**5.** $y'' + (\pi^2/4)y = 0; \; y'(0) = 1, \; y'(1) = 0.$
**6.** $y'' + y = x; \; y'(0) = 1, \; y(1) = 0.$
**7.** $y'' - y = 2x; \; y(0) = 0, \; y(1) + y'(1) = 1.$
**8.** Consider a boundary-value problem

$$y'' = p(x)y' + q(x)y + r(x); \qquad y(a) = \alpha, \quad y(b) = \beta$$

where $q(x) \geq 0$ and $|p(x)| \leq M$ on $[a, b]$. Show that the coefficient matrix of the approximating finite difference equation is diagonally dominant if $h < 2/M$.

For Exercises 9 and 10, find an approximate solution of each of the boundary-value problems using the finite difference method and an appropriate equation from Equation (14.3.22). Use $h = \frac{1}{3}$.

**9.** $y'' - y = 0; \; y'(0) = 0, \; y(1) = 1.$
**10.** $y'' - y = 2x; \; y(0) = 0, \; y(1) + y'(1) = 1.$
**11.** Use the finite difference method to approximate the solution of the boundary-value problem

$$y^{(4)} = x; \qquad y(0) = y'(0) = 0, \quad y(1) = y'(1) = 0$$

Use $h = \frac{1}{3}$.

---

## COMPUTER EXERCISES

**C.1.** Write a subroutine to generate difference equations [Equation (14.3.7)] for a given second-order linear differential equation

$$y'' = p(x)y' + q(x)y + r(x)$$

**C.2.** Write a subroutine to solve Equation (14.3.10) by the $LU$ decomposition using Equations (7.7.4), (7.7.6), and (7.7.7).

**C.3.** Using Exercises C.1 and C.2, write a computer program to approximate the solution of a boundary-value problem

$$y'' = p(x)y' + q(x)y + r(x); \qquad y(a) = \alpha, \quad y(b) = \beta$$

Using Exercise C.3, approximate the solution of each of the boundary-value problems in Exercises C.4–C.6. Use $N = 9, 19$, and 99. Compare each of your approximate solutions with the exact solution.

**C.4.** $y'' = y' + 2y - e^x; \; y(0) = 0, \; y(1) = 1.$
**C.5.** $y'' = 25y; \; y(0) = 1, \; y(1) = e^{-5}.$
**C.6.** $y'' = 3xy' - 9y + x; \; y(0) = 1, \; y(1) = 2.$

**C.7.** Write a computer program to approximate the solution of a boundary-value problem

$$y'' = p(x)y' + q(x)y + r(x);$$
$$\gamma_1\, y(a) + \gamma_2\, y'(a) = \alpha, \quad \gamma_3\, y(b) + \gamma_4\, y'(b) = \beta$$

using Exercises C.1 and C.2, and Equation (14.3.18) or Equations (14.3.23), (14.3.8) and (14.3.24).

Using Exercise C.7, approximate the solution of each of the boundary-value problems in Exercises C.8–C.9. Use $N = 9, 19$, and $99$. Compare each of your approximate solutions with the exact solution.

**C.8.** $y'' = -16y + 32x$; $y(0) = 1$, $y'(\pi) = 2$.

**C.9.** $y'' = -4y$; $y(0) + y'(0) = 3$, $y(\pi) - y'(\pi) = -1$.

**C.10.** Bird et al. (1960) showed that the nondimensional temperature $T$ of incompressible fluid between two coaxial cylinders, the outer one rotating at a steady angular velocity, is given by

$$\frac{1}{\xi}\frac{d}{d\xi}\left(\xi\frac{dT}{d\xi}\right) + \frac{4N}{\xi^4} = 0$$

with the boundary conditions $T(k) = 0$ and $T(1) = 1$. Approximate $T(\xi)$ when $k = 0.8$ and $N = 0.4$ using Exercise C.3. Compare your approximate solution with the exact solution

$$T(\xi) = \left[(N + 1) - \frac{N}{\xi^2}\right] - \left[(N + 1) - \frac{N}{k^2}\right]\frac{\ln \xi}{\ln k}$$

**C.11.** The velocity $u$ of a Newtonian fluid between two concentric rotating cylinders (Schlichting 1960) is given by

$$\frac{d^2u}{dr^2} + \frac{d}{dr}\left(\frac{u}{r}\right) = 0$$

Let the inner and outer radii of the cylinders be $r_1$ and $r_2$ respectively, and their angular velocities be $\omega_1$ and $\omega_2$. Then the boundary conditions are given by $u(r_1) = r_1\omega_1$ and $u(r_2) = r_2\omega_2$. Compute $u(r)$ when $r_1 = \frac{1}{2}$, $\omega_1 = 5$, $r_2 = 1$, and $\omega_2 = 10$ using Exercise C.3. Compare your approximate solution with the exact solution

$$u(r) = \frac{1}{r_2^2 - r_1^2}\left[r(\omega_2\, r_2^2 - \omega_1\, r_1^2) - \frac{r_1^2\, r_2^2}{r}(\omega_2 - \omega_1)\right]$$

## 14.4  THE HERMITIAN METHOD

In this section, we consider another method for setting finite difference equations. This method gives greater accuracy, not by using more pivotal points, but by satisfying the differential equation at several points. The involved formulas contain the values of the derivatives as well as values of the function. Collatz (1966) called these methods **Hermitian methods**.

Consider the difference scheme

$$a_{-1} y(x_i - h) + a_0 y(x_i) + a_1 y(x_i + h) + B_{-1} y'(x_i - h)$$
$$+ B_0 y'(x_i) + B_1 y'(x_i + h) = 0 \qquad \textbf{(14.4.1)}$$

where constants $a_{-1}$, $a_0$, $B_{-1}$, $B_0$, and $B_1$ are to be determined. Expanding each term of Equation (14.4.1) by Taylor's Theorem A.8 about $x_i$, we get

$$a_{-1}\left\{ y(x_i) - hy'(x_i) + \frac{h^2}{2!} y''(x_i) - \frac{h^3}{3!} y^{(3)}(x_i) + \frac{h^4}{4!} y^{(4)}(x_i) - \frac{h^5}{5!} y^{(5)}(\eta_1)\right\}$$

$$+ a_0 y(x_i) + a_1\left\{ y(x_i) + hy'(x_i) + \frac{h^2}{2!} y''(x_i) + \frac{h^3}{3!} y^{(3)}(x_i) + \frac{h^4}{4!} y^{(4)}(x_i) + \frac{h^5}{5!} y^{(5)}(\eta_2)\right\}$$

$$+ B_{-1}\left\{ y'(x_i) - hy''(x_i) + \frac{h^2}{2!} y^{(3)}(x_i) - \frac{h^3}{3!} y^{(4)}(x_i) + \frac{h^4}{4!} y^{(5)}(\eta_3)\right\} + B_0 y'(x_i)$$

$$+ B_1\left\{ y'(x_i) + hy''(x_i) + \frac{h^2}{2!} y^{(3)}(x_i) + \frac{h^3}{3!} y^{(4)}(x_i) + \frac{h^4}{4!} y^{(5)}(\eta_4)\right\} = 0$$

This can be written as

$$y(x_i)[a_{-1} + a_0 + a_1] + hy'(x_i)\left[ -a_{-1} + a_1 + \frac{1}{h}(B_{-1} + B_0 + B_1)\right]$$

$$+ \frac{h^2 y''(x_i)}{2!}\left[ a_{-1} + a_1 + \frac{2}{h}(-B_{-1} + B_1)\right]$$

$$+ \frac{h^3 y'''(x_i)}{3!}\left[ -a_{-1} + a_1 + \frac{3}{h}(B_{-1} + B_1)\right]$$

$$+ \frac{h^4 y^{(4)}(x_i)}{4!}\left[ a_{-1} + a_1 + \frac{4}{h}(-B_{-1} + B_1)\right]$$

$$+ \frac{h^5}{5!}[-a_{-1} y^{(5)}(\eta_1) + a_1 y^{(5)}(\eta_2)$$

$$+ \frac{5}{h}(B_{-1} y^{(5)}(\eta_3) + B_1 y^{(5)}(\eta_4))] = 0 \qquad \textbf{(14.4.2)}$$

where $\eta_1$, $\eta_2$, $\eta_3$, and $\eta_4$ are in $(x_1 - h, x_i + h)$.

Equating the coefficients of $h^k y^{(k)}/k!$ for $k = 0, 1, \ldots, 4$ to zero of Equation (14.4.2), we get

$$a_{-1} + a_0 + a_1 = 0$$

$$-a_{-1} + a_1 + \frac{1}{h}(B_{-1} + B_0 + B_1) = 0$$

$$a_{-1} + a_1 + \frac{2}{h}(-B_{-1} + B_1) = 0$$

$$-a_{-1} + a_1 + \frac{3}{h}(B_{-1} + B_1) = 0$$

$$a_{-1} + a_1 + \frac{4}{h}(-B_{-1} + B_1) = 0$$

Since these equations are homogeneous, we solve in terms of one of the unknowns; say, $a_1$. Then

$$a_{-1} = -a_1 \qquad a_0 = 0 \qquad B_{-1} = -\frac{ha_1}{3} \qquad B_0 = -\frac{4ha_1}{3}$$

Using these values and Corollary A.2.2 in Equation (14.4.2), Equation (14.4.1) becomes

$$y(x_i + h) - y(x_i - h) - \frac{h}{3}[y'(x_i + h) + 4y'(x_i) + y'(x_i - h)] + \frac{h^5}{90} y^{(5)}(\xi) = 0$$

**(14.4.3)**

where $\xi$ is in $(x_i - h, x_i + h)$.

Since we are interested in solving $y'' = f(x, y)$, consider a difference scheme

$$a_{-1} y(x_i - h) + a_0 y(x_i) + a_1 y(x_i + h) + B_{-1} y''(x_i - h)$$
$$+ B_0 y''(x_i) + B_1 y''(x_i + h) = 0 \qquad \textbf{(14.4.4)}$$

where constants $a_{-1}$, $a_0$, $B_{-1}$, $B_0$, and $B_1$ are to be determined. We can verify that by expanding each term of Equation (14.4.4) by Taylor's Theorem A.8 about $x_i$ and equating the coefficients of $h^k y^{(k)}/k!$ for $k = 0, 1, \ldots, 4$ to zero, we get

$$a_{-1} + a_0 + a_1 = 0$$
$$-a_{-1} + a_1 = 0$$
$$a_{-1} + a_1 + \frac{2}{h^2}(B_{-1} + B_0 + B_1) = 0$$
$$-a_{-1} + a_1 + \frac{6}{h^2}(-B_{-1} + B_1) = 0$$
$$a_{-1} + a_1 + \frac{12}{h^2}(B_{-1} + B_1) = 0$$
$$-a_{-1} + a_1 + \frac{20}{h^2}(-B_{-1} + B_1) = 0 \qquad \textbf{(14.4.5)}$$

Since these equations are homogeneous, solving in terms of $a_1$ gives

$$a_{-1} = a_1 \qquad a_0 = -2a_1 \qquad B_{-1} = B_1 = -\frac{h^2 a_1}{12} \qquad B_0 = -\frac{10 a_1 h^2}{12}$$

Using these values in Equation (14.4.4), we can verify that

$$y(x_i - h) - 2y(x_i) + y(x_i + h) - \frac{h^2}{12}[y''(x_i - h)$$
$$+ 10y''(x_i) + y''(x_i + h)] - \frac{h^6}{240} y^{(6)}(\xi) = 0 \qquad \textbf{(14.4.6)}$$

where $\xi$ is in $(x_i - h, x_i + h)$.

Let us consider

$$y'' = Q(x)y + R(x) \qquad \textbf{(14.4.7)}$$

subject to the boundary conditions

$$y(a) = \alpha \quad \text{and} \quad y(b) = \beta \qquad \textbf{(14.4.8)}$$

Equation (14.4.7) does not contain the first derivative. Since Equation (14.4.7) holds true at each interior point,

$$y''(x_i - h) + 10y''(x_i) + y''(x_i + h) = Q(x_i - h)y(x_i - h) + R(x_i - h)$$
$$+ 10Q(x_i)y(x_i) + 10R(x_i) + Q(x_i + h)y(x_i + h) + R(x_i + h)$$

Approximating this equation using (14.4.6), we get

$$Y_{i-1} - 2Y_i + Y_{i+1} - \frac{h^2}{12}[Q(x_i - h)Y_{i-1} + R(x_i - h) + 10Q(x_i)Y_i$$
$$+ 10R(x_i) + Q(x_i + h)Y_{i+1} + R(x_i + h)] = 0 \quad \text{for } i = 1, 2, \dots, N$$

This simplifies to

$$Y_{i-1}\left(1 - \frac{h^2}{12}Q(x_i - h)\right) - Y_i\left(2 + \frac{10h^2}{12}Q(x_i)\right) + Y_{i+1}\left(1 - \frac{h^2}{12}Q(x_i + h)\right)$$
$$= \frac{h^2}{12}(R(x_i - h) + 10R(x_i) + R(x_i + h)) \quad \text{for } i = 1, 2, \dots, N \quad \textbf{(14.4.9)}$$

The finite difference approximation given by Equation (14.4.9) is of the fourth order while the finite difference approximation given by Equation (14.3.7) is of the second order. From the computational point of view, both methods require almost the same time. Equation (14.4.9) can be used only with the form of Equation (14.4.7). The linear ordinary differential equation

$$y'' = p(x)y' + q(x)y + r(x) \quad \textbf{(14.3.1a)}$$

can be transformed to the form (14.4.7) by selecting $z(x)$ properly in the transformation

$$y(x) = u(x)z(x) \quad \textbf{(14.4.10)}$$

Substituting Equation (14.4.10) in Equation (14.3.1a), we get

$$zu'' + (2z' - p(x)z)u' + (z'' - p(x)z' - q(x)z)u - r(x) = 0 \quad \textbf{(14.4.11)}$$

Select $z$ in Equation (14.4.11) such that $2z' - p(x)z = 0$. This gives $z(x) = e^{(1/2)\int p(x)dx}$. Thus

$$y(x) = u(x)e^{(1/2)\int p(x)dx} \quad \textbf{(14.4.12)}$$

transforms Equation (14.3.1a) into the form of Equation (14.4.7).

**EXAMPLE 14.4.1**    Transform $y'' - 3y' + 2y = 0$; $y(0) = 0$, $y(1) = 2$ into the form of Equations (14.4.7) and (14.4.8).

From Equation (14.4.12), we have $y(x) = u(x)e^{(1/2)\int 3dx}$. Substituting $y(x) = u(x)e^{3x/2}$ in $y'' - 3y' + 2y = 0$, we get $u'' - (\frac{1}{4})u = 0$. This equation does not contain the first derivative. Since $y(x) = u(x)e^{3x/2}$, we get $y(0) = e^0u(0) = 0$. Therefore $u(0) = 0$. Similarly we get $u(1) = 2e^{-3/2}$. Thus we transformed $y'' - 3y' + 2y = 0$; $y(0) = 0$, $y(1) = 2$ by $y(x) = u(x)e^{3x/2}$ to

$$u'' - \frac{1}{4}u = 0; \qquad u(0) = 0, \quad u(1) = 2e^{-3/2}.$$

∎ ∎ ∎

**EXAMPLE 14.4.2**    Solve $y'' + y = 0$; $y(0) = 1$, $y(\pi/2) = 0$ using the Hermitian method.

Since $y''(x_i) + y(x_i) = 0$, substituting $y''(x_i) = -y(x_i)$ in Equation (14.4.6) and using the approximation notations gives

$$Y_{i-1} - 2Y_i + Y_{i+1} + \frac{h^2}{12}(Y_{i-1} + 10Y_i + Y_{i+1}) = 0$$

Rewriting this equation, we get

$$\left(1 + \frac{h^2}{12}\right)Y_{i-1} + \left(\frac{10h^2}{12} - 2\right)Y_i + \left(1 + \frac{h^2}{12}\right)Y_{i+1} = 0 \qquad \textbf{(14.4.13)}$$

for $i = 1, 2, \ldots, N$.

Divide $[0, \pi/2]$ into $N + 1 = 10$ equal intervals. Then $h = \pi/20$ and $N = 9$. The system (14.4.13) is solved with $Y_0 = 1$ and $Y_{10} = 0$ by using $LU$ decomposition given in Section 7.7. In Table 14.4.1, the computed values of $Y_i$ at various $x_i$ along with the values of the exact solution $y = \cos x$ are given.

**Table 14.4.1**

| $i$ | $x_i$ | $Y_i$ Using Eq. (14.3.11) | $y_i = \cos x_i$ Exact Solution | $Y_i$ Using Eq. (14.4.13) | \|Actual Error\| |
|---|---|---|---|---|---|
| 0 | 0.0 | 1.00000 | 1.00000 | 1.00000 | 0.00000 |
| 1 | $\pi/20$ | 0.98792 | 0.98769 | 0.98769 | 0.00000 |
| 2 | $\pi/10$ | 0.95146 | 0.95106 | 0.95106 | 0.00000 |
| 3 | $3\pi/20$ | 0.89152 | 0.89101 | 0.89101 | 0.00000 |
| 4 | $4\pi/20$ | 0.80959 | 0.80902 | 0.80902 | 0.00000 |
| 5 | $5\pi/20$ | 0.70768 | 0.70711 | 0.70711 | 0.00000 |
| 6 | $6\pi/20$ | 0.58831 | 0.58779 | 0.58779 | 0.00000 |
| 7 | $7\pi/20$ | 0.45442 | 0.45399 | 0.45399 | 0.00000 |
| 8 | $8\pi/20$ | 0.30933 | 0.30902 | 0.30902 | 0.00000 |
| 9 | $9\pi/20$ | 0.15659 | 0.15643 | 0.15643 | 0.00000 |
| 10 | $\pi/2$ | 0.00000 | 0.00000 | 0.00000 | 0.00000 |

In the last column the values of $|\text{error}| = |Y_i - y(x_i)|$ where $Y_i$ are the values given by Equation (14.4.13) are given. It is clear that the values given by the Hermitian method are more accurate than those given by the second-order finite difference method with little extra work. For comparison, the values of $Y_i$ computed using the second-order finite difference method [Equation (14.3.11)] are also given in Table 14.4.1.

■ ■ ■

## 14.5   THE NONLINEAR ORDINARY DIFFERENTIAL EQUATIONS

The general nonlinear second-order ordinary differential equation

$$y'' = f(x, y, y') \qquad a < x < b \tag{14.2.1a}$$

with the boundary conditions

$$y(a) = \alpha \quad \text{and} \quad y(b) = \beta \tag{14.2.1b}$$

can be solved in an analogous way. Equation (14.2.1a) reduces to

$$Y_{i+1} - 2Y_i + Y_{i-1} = h^2 f\left(x_i, Y_i, \frac{Y_{i+1} - Y_{i-1}}{2h}\right) \tag{14.5.1}$$

for $i = 1, 2, \ldots, N$.

However, Equation (14.5.1) forms a system of $N$ nonlinear equations in $N$ unknowns. The system of nonlinear equations can be solved by using one of the iterative methods developed in Sections 12.2 and 12.3. The Newton method can be easily used because each equation involves at most three unknowns and the corresponding Jacobian matrix is tridiagonal. Let us illustrate the method by solving a nonlinear second-order boundary-value problem.

**E X A M P L E  14.5.1**   Use the finite difference method to solve the boundary-value problem

$$y'' = -yy' + y^3; \qquad y(0) = 1, \quad y(1) = \frac{1}{2}$$

We solved this example by using the shooting method. Divide $[0, 1]$ into $N + 1 = 4$ equal intervals. Then the nodal points are given by $x_i = ih$ for $i = 0, 1, 2, 3,$ and 4 and $h = \frac{1}{4}$. Applying Equation (14.5.1) to our problem, we get

$$Y_{i-1} - 2Y_i + Y_{i+1} = h^2\left[-Y_i \frac{(Y_{i+1} - Y_{i-1})}{2h} + Y_i^3\right]$$

for $i = 1, 2,$ and 3.

For $i = 1, 2, 3$ and substituting $h = \frac{1}{4}$, $y_0 = Y_0 = 1$ and $y_4 = Y_4 = \frac{1}{2}$, we get

$$f_1(Y_1, Y_2, Y_3) = -\frac{17}{8}Y_1 + \frac{1}{8}Y_1Y_2 - \frac{1}{16}Y_1^3 + Y_2 + 1 = 0$$

$$f_2(Y_1, Y_2, Y_3) = Y_1 - 2Y_2 + \frac{Y_2}{8}(Y_3 - Y_1) - \frac{Y_2^3}{16} + Y_3 = 0$$

$$f_3(Y_1, Y_2, Y_3) = Y_2 - \frac{31}{16}Y_3 - \frac{Y_2Y_3}{8} - \frac{Y_3^3}{16} + \frac{1}{2} = 0$$

Then

$$J[\mathbf{Y}] = \begin{bmatrix} \dfrac{\partial f_1}{\partial Y_1} & \dfrac{\partial f_1}{\partial Y_2} & \dfrac{\partial f_1}{\partial Y_3} \\ \dfrac{\partial f_2}{\partial Y_1} & \dfrac{\partial f_2}{\partial Y_2} & \dfrac{\partial f_2}{\partial Y_3} \\ \dfrac{\partial f_3}{\partial Y_1} & \dfrac{\partial f_3}{\partial Y_2} & \dfrac{\partial f_3}{\partial Y_3} \end{bmatrix}$$

$$= \begin{bmatrix} \dfrac{(-17 + Y_2)}{8} - \dfrac{3Y_1^2}{16} & \dfrac{Y_1}{8} + 1 & 0 \\ 1 - \dfrac{Y_2}{8} & -2 + \dfrac{(Y_3 - Y_1)}{8} - \dfrac{3Y_2^2}{16} & \dfrac{Y_2}{8} + 1 \\ 0 & 1 - \dfrac{Y_3}{8} & -\dfrac{31}{16} - \dfrac{Y_2}{8} - \dfrac{3Y_3^2}{16} \end{bmatrix}$$

Newton's method gives

$$J[\mathbf{Y}^{(k)}]\mathbf{Z}^{(k+1)} = -\mathbf{F}(\mathbf{Y}^{(k)}) \quad \text{for } k = 0, 1, \ldots$$

where

$$\mathbf{Z}^{(k+1)} = \mathbf{Y}^{(k+1)} - \mathbf{Y}^{(k)}$$

For $k = 0$ and $Y_1^{(0)} = Y_2^{(0)} = Y_3^{(0)} = 1$, we find $Z_1 = 0.18681$, $Z_2 = 0.30769$, and $Z_3 = 0.39744$. The computed values of $Y_i$ along with the exact values of $y = 1/(x + 1)$ are given in Table 14.5.1.

**Table 14.5.1**

| $i$ | $x_i$ | $Y_i$ Using Eq. (14.5.1) | $y_i = 1/(x_i + 1)$ Exact Solution | \|Actual Error\| |
|---|---|---|---|---|
| 0 | 0.00 | 1.00000 | 1.00000 | 1.00000 |
| 1 | 0.25 | 0.80115 | 0.80000 | 0.00115 |
| 2 | 0.50 | 0.66772 | 0.66667 | 0.00105 |
| 3 | 0.75 | 0.57201 | 0.57143 | 0.00058 |
| 4 | 1.00 | 0.50000 | 0.50000 | 0.00000 |

The values of |Actual Error| are reasonable since $h = \frac{1}{4}$. The results can be improved by using small $h$, but we have to solve a system of nonlinear equations with more unknowns. It is much more difficult to apply the finite difference method than the shooting method to nonlinear boundary-value problems. ■ ■ ■

## EXERCISES

**1.** Expand $y(x_i - h)$ and $y(x_i + h)$ by Taylor's Theorem about $x_i$ in $a_{-1} y(x_i - h) + a_0 y'(x_i) + a_1 y(x_i + h) = 0$ and equate the coefficients of $y(x_i)$ and $y'(x_i)$ to zero. Solve these two equations and verify that

$$-y(x_i - h) - 2hy'(x_i) + y(x_i + h) - \frac{h^3}{3}y^{(3)}(\xi_i) = 0$$

**2.** Derive Equation (14.4.5) from Equation (14.4.4).

**3.** Solve Equation (14.4.5) in terms of $a_1$.

**4.** Transform $y'' = 3xy' - 9y + x$; $y(0) = 1$, $y(1) = 2$ into a form that does not contain the first derivative.

**5.** Transform $[(1 + x^2)y']' - y = x^2 + 1$; $y(-1) = 1$, $y(1) = 0$ into a form that does not contain the first derivative.

Use the Hermitian method to approximate the solution of each of the boundary-value problems in Exercises 6–8. Use $h = \frac{1}{3}$.

**6.** $y'' + (\pi^2/4)y = 0$; $y(0) = 0$, $y(1) = 1$.

**7.** $y'' + y = x$; $y(0) = 0$, $y(1) = 1$.

**8.** $y'' - y = x$; $y(0) = 0$, $y(1) = 1$.

Approximate the solution of each of the boundary-value problems in Exercises 9–11 using the Hermitian method and Equation (14.3.15). Use $h = \frac{1}{3}$.

**9.** $y'' - y = 0$; $y'(0) = 0$, $y(1) = 1$.

**10.** $y'' + (\pi^2/4)y = 0$; $y'(0) = 1$, $y'(1) = 0$.

**11.** $y'' + y = x$; $y'(0) = 1$, $y'(1) = 1$.

Use the finite difference method to approximate the solution of each of the nonlinear boundary-value problems in Exercises 12–14. Use $h = \frac{1}{3}$.

**12.** $y'' = -2yy'$; $y(0) = 0$, $y(1) = 1$.

**13.** $y'' = (3/2)y^2$; $y(0) = 4$, $y(1) = 1$.

**14.** Develop higher-order approximations for $y'(a)$ and $y'(b)$ (of at least fourth order) to solve the boundary-value problem

$$y'' = Q(x)y + R(x); \qquad \gamma_1\, y(a) + \gamma_2\, y'(a) = \alpha, \quad \gamma_3\, y(b) + \gamma_4\, y'(b) = \beta$$

Substitute the values of $y'(a)$ and $y'(b)$ in the boundary conditions $\gamma_1\, y(a) + \gamma_2\, y'(a) = \alpha$ and $\gamma_3\, y(b) + \gamma_4\, y'(b) = \beta$. Using these two equations and Equation (14.4.9), write the system in the form $A\mathbf{y} = \mathbf{s}$.

## COMPUTER EXERCISES

**C.1.** Write a subroutine to generate difference equations [Equation (14.4.9)] for a given second-order differential equation of the form $y'' = Q(x)y + R(x)$.

**C.2.** Write a subroutine to solve Equation (14.4.9) by $LU$ decomposition using Equations (7.7.4), (7.7.6), and (7.7.7).

**C.3.** Using Exercises C.1 and C.2, write a computer program to approximate the solution of a boundary-value problem

$$y'' = Q(x)y + R(x) \qquad y(a) = \alpha \qquad y(b) = \beta$$

Using Exercise C.3, approximate the solution of each of the boundary-value problems in Exercises C.4–C.6. Use $N = 9$, 19, and 99. Compare your approximate solution with the exact solution and with the approximate solution obtained by the method of Section 14.3.

**C.4.** $y'' = y' + 2y - e^x$; $y(0) = 0$, $y(1) = 1$. (First transform it to a proper form.)

**C.5.** $y'' = 25y$; $y(0) = 1$, $y(1) = e^{-5}$.

**C.6.**  $y'' = 3xy' - 9y + x$; $y(0) = 1$, $y(1) = 2$. (First transform it to a proper form.)

**C.7.**  Using Exercises C.1, C.2, and Equation (14.3.15), write a computer program to approximate the solution of a boundary-value problem

$$y'' = Q(x)y + R(x) \qquad \gamma_1 \, y(a) + \gamma_2 \, y'(a) = \alpha \qquad \gamma_3 \, y(b) + \gamma_4 \, y'(b) = \beta$$

Using Exercise C.7, approximate the solution of each of the boundary-value problems in Exercises C.8–C.9. Use $N = 9$, 19, and 99. Compare your approximate solution with the exact solution and with the approximate solution obtained by the method of Section 14.3.

**C.8.**  $y'' = -16y + 32x$; $y(0) = 1$, $y'(\pi) = 2$.

**C.9.**  $y'' = -4y$; $y(0) + y'(0) = 3$, $y(\pi) - y'(\pi) = -1$.

**C.10.**  Write a subroutine to solve a system of linear equations given by $J[\mathbf{Y}^{(k)}]\mathbf{Z}^{(k+1)} = -\mathbf{F}(\mathbf{Y}^{(k)})$ by an $LU$ decomposition.

Using Exercise C.10, approximate the solution of each of the boundary-value problems in Exercises C.11–C.12. Compare your approximate solution with the exact solution.

**C.11.**  $y'' = \frac{1}{2}(1 + x + y)^3$; $y(0) = 0$, $y(1) = 0$. (The exact solution is $y(x) = (2/(2 - x)) - x - 1$.)

**C.12.**  $y'' = -2yy'$; $y(0) = 0$, $y(1) = 1$.

**C.13.**  The steady-state temperature distribution in a unit length cylindrical rod with an insulated lateral surface whose ends are held at zero temperature and with the steady energy source that generates energy at a rate of $\alpha e^T$ is given by

$$\frac{d^2T}{dx^2} = \alpha e^T \qquad T(0) = T(1) = 0$$

Compute $T(x)$ when $\alpha = 1$, 2, and 3.

**C.14.**  The small oscillation of a mass which is attached to two springs is given by

$$\frac{d^2y}{dx^2} = -y^3; \qquad y(0) = 0.20000, \quad y(1) = 0.18462$$

Compute $y(x)$.

**C.15.**  In a biological floc, diffusion is given by

$$\frac{d^2y}{dx^2} - \frac{\alpha^2 y}{1 + \beta y} = 0; \qquad y(0) = 1, \quad \frac{dy}{dx}(0) = 0$$

Compute $y(x)$ when $\alpha = 1$ and $\beta = 2$.

**C.16.**  The deflection angle of a cantilever beam is given by

$$\frac{d^2y}{dx^2} + \alpha \cos y = 0; \qquad y(0) = 0, \quad y'(1) = 0$$

Compute $y(x)$ when $\alpha = 0.1$, 0.2, and 0.3.

## 14.6  EIGENVALUE PROBLEMS

Consider the homogeneous second-order boundary-value problem

$$y'' + \lambda y = 0 \qquad (14.6.1a)$$

with the boundary conditions

$$y(0) = 0 \qquad y(1) = 0 \qquad (14.6.1b)$$

The general solution of Equation (14.6.1a) is

$$y(x) = A \cos \sqrt{\lambda} x + B \sin \sqrt{\lambda} x$$

Since $y(0) = A = 0$, $y(x) = B \sin \sqrt{\lambda} x$. We want $y(1) = B \sin \sqrt{\lambda} = 0$. Hence $B = 0$ or $\sin \sqrt{\lambda} = 0$. Since $B = 0$ leads to the trivial solution, and our interest is to find the nontrivial solution, then $\sin \sqrt{\lambda} = 0 = \sin n\pi$ for each positive integer $n$. Hence $\sqrt{\lambda} = n\pi$. Thus for $\lambda = n^2\pi^2 = \lambda_n$ for each positive integer $n$, we have a solution that is given by

$$y(x) = y_n = B \sin n\pi x$$

for each positive integer $n$.

The numbers $\lambda_n = n^2\pi^2$ for which Equations (14.6.1a) and (14.6.1b) have nontrivial solutions are called the **eigenvalues** and each nontrivial solution corresponding to an eigenvalue $\lambda_n$ is called an **eigenvector** or **eigenfunction** of Equations (14.6.1a) and (14.6.1b).

The eigenvalues can be obtained using the finite difference method. Let us choose $h = (b - a)/(N + 1)$. Then using Equation (14.3.4), Equation (14.6.1a) reduces to

$$Y_{i-1} - (2 - \lambda h^2)Y_i + Y_{i+1} = 0 \qquad (14.6.2)$$

for $i = 1, 2, \ldots, N$.

Let $N = 3$ and so $h = \frac{1}{4}$. Then Equation (14.6.2) reduces to

$$
\begin{aligned}
-(2 - \lambda h^2)Y_1 + \quad & Y_2 \qquad\qquad\qquad = 0 \\
Y_1 - (2 - \lambda h^2)Y_2 + \quad & Y_3 = 0 \\
Y_2 - (2 - \lambda h^2)Y_3 & = 0
\end{aligned}
\qquad (14.6.3)
$$

using $Y_0 = y(0) = 0$ and $Y_4 = y(1) = 0$. This is written as

$$
\begin{bmatrix}
-2 + \lambda h^2 & 1 & 0 \\
1 & -2 + \lambda h^2 & 1 \\
0 & 1 & -2 + \lambda h^2
\end{bmatrix}
\begin{bmatrix}
Y_1 \\
Y_2 \\
Y_3
\end{bmatrix}
=
\begin{bmatrix}
0 \\
0 \\
0
\end{bmatrix}
$$

This can also be written as the matrix equation

$$(A - \beta I)\mathbf{y} = 0 \quad \text{or} \quad A\mathbf{y} = \beta \mathbf{y} \qquad (14.6.4)$$

where

$$
A =
\begin{bmatrix}
-2 & 1 & 0 \\
1 & -2 & 1 \\
0 & 1 & -2
\end{bmatrix}
\quad \text{and} \quad \beta = -\lambda h^2 \qquad (14.6.5)
$$

Thus our eigenvalue problem of Equations (14.6.1a) and (14.6.1b) has been converted to an algebraic eigenvalue problem. We find the eigenvalues of a symmetric matrix $A$ in Equation (14.6.5) by any method developed in Chapter 13. The eigenvalues of $A$ are given by $-0.58579$, $-2.00000$, and $-3.41421$. Thus the values of $\lambda$ are given by $0.58579/h^2 = 9.37264$, $2.000/h^2 = 32.0$, and $3.41421/h^2 = 16(3.41421) = 54.62736$ to five significant digits. The first three eigenvalues are $\lambda_1 = \pi^2 = 9.86960$, $\lambda_2 = 4\pi^2 = 39.47841$, and $\lambda_3 = 9\pi^2 = 88.82643$ to five significant digits.

We can use the Hermitian method to obtain the finite difference form. Using Equation (14.4.6) in Equation (14.6.1a), we get

$$Y_{i-1} - 2Y_i + Y_{i+1} + \frac{h^2\lambda}{12}(Y_{i-1} + 10Y_i + Y_{i+1}) = 0 \qquad \textbf{(14.6.6)}$$

for $i = 1, 2, \ldots, N$ with $Y_0 = y(0) = 0$ and $Y_{N+1} = y(1) = 0$. Equation (14.6.6) is written as

$$Y_{i-1}\left(1 + \frac{h^2\lambda}{12}\right) + Y_i\left(-2 + \frac{10h^2\lambda}{12}\right) + Y_{i+1}\left(1 + \frac{h^2\lambda}{12}\right) = 0$$

For $N = 3$ and $h = \frac{1}{4}$, we get

$$Y_1\left(-2 + \frac{10h^2\lambda}{12}\right) + Y_2\left(1 + \frac{h^2\lambda}{12}\right) = 0$$

$$Y_1\left(1 + \frac{h^2\lambda}{12}\right) + Y_2\left(-2 + \frac{10h^2\lambda}{12}\right) + Y_3\left(1 + \frac{h^2\lambda}{12}\right) = 0$$

$$Y_2\left(1 + \frac{h^2\lambda}{12}\right) + Y_3\left(-2 + \frac{10h^2\lambda}{12}\right) = 0$$

This can be written as

$$\begin{bmatrix} -2 & 1 & 0 \\ 1 & -2 & 1 \\ 0 & 1 & -2 \end{bmatrix}\begin{bmatrix} Y_1 \\ Y_2 \\ Y_3 \end{bmatrix} + \frac{\lambda h^2}{12}\begin{bmatrix} 10 & 1 & 0 \\ 1 & 10 & 1 \\ 0 & 1 & 10 \end{bmatrix}\begin{bmatrix} Y_1 \\ Y_2 \\ Y_3 \end{bmatrix} = \begin{bmatrix} 0 \\ 0 \\ 0 \end{bmatrix} \qquad \textbf{(14.6.7)}$$

We verify that

$$B = \frac{1}{98}\begin{bmatrix} 9801/(990) & -1 & 1/10 \\ -1 & 10 & -1 \\ -1/10 & -1 & 99/10 \end{bmatrix}$$

is the inverse of

$$\begin{bmatrix} 10 & 1 & 0 \\ 1 & 10 & 1 \\ 0 & 1 & 10 \end{bmatrix}$$

Multiplying Equation (14.6.7) by $B$, we get

$$\begin{bmatrix} -0.21224 & 0.12244 & -0.01224 \\ 0.12244 & -0.22448 & 0.12244 \\ -0.01224 & 0.12244 & -0.21224 \end{bmatrix}\begin{bmatrix} Y_1 \\ Y_2 \\ Y_3 \end{bmatrix} + \frac{\lambda h^2}{12}\begin{bmatrix} 1 & 0 & 0 \\ 0 & 1 & 0 \\ 0 & 0 & 1 \end{bmatrix}\begin{bmatrix} Y_1 \\ Y_2 \\ Y_3 \end{bmatrix} = \begin{bmatrix} 0 \\ 0 \\ 0 \end{bmatrix}$$

This is written as

$$(A - \beta I)Y = 0 \quad \text{or} \quad AY = \beta Y$$

where

$$A = \begin{bmatrix} -0.21224 & 0.12244 & -0.01224 \\ 0.12244 & -0.22448 & 0.12244 \\ -0.01224 & 0.12244 & -0.21224 \end{bmatrix} \quad \text{and} \quad \beta = -\frac{\lambda h^2}{12}$$

We find the eigenvalues of $A$ by any method of Chapter 13. The eigenvalues of $A$ are given by $-0.05132$, $-0.20000$ and $-0.39764$ and therefore the values of $\lambda$ are given by $[12(0.05132)]/h^2 = 9.85344$, $[(12)(0.2)]/h^2 = 38.4$, and $[(12)(0.39764)]/h^2 = 76.34688$ while the corresponding exact eigenvalues are $9.86960$, $39.47841$, and $88.82643$ to five significant digits.

Our approximations of the eigenvalues got progressively worse. The smallest values of $\lambda$ are of interest in many problems. Using smaller $h$, we can improve accuracy.

A fairly general class of eigen problems in applied mathematics is given by a Sturm–Liouville system

$$(p(x)y')' + q(x)y + \lambda r(x)y = 0; \quad a_1\, y(a) - a_2\, y'(a) = 0, \quad b_1\, y(b) - b_2\, y'(b) = 0$$

where $p(x) > 0$, $q(x) \le 0$, and $r(x) > 0$ while $p'(x)$, $q(x)$, and $r(x)$ are continuous on $[a, b]$. Also $a_1^2 + a_2^2 > 0$ and $b_1^2 + b_2^2 > 0$.

We can use any numerical method we discussed to find the eigenvalues of a Sturm–Liouville system.

## EXERCISES

Find the eigenvalues and corresponding eigenfunctions for each of the systems in Exercises 1–3.

1.  $y'' + \lambda y = 0$; $y'(0) = 0$, $y(\pi) = 0$.
2.  $y'' + (1 + \lambda)y = 0$; $y(0) = 0$, $y(\pi) = 0$.
3.  $y'' + 2y' + (1 - \lambda)y = 0$; $y(0) = 0$, $y(1) = 0$.

Replace the derivatives by central differences and then find the eigenvalues of each of the Sturm–Liouville systems of Exercises 4–5. Use $N = 2$.

4.  $y'' + (1 + \lambda)y = 0$; $y(0) = 0$, $y(\pi) = 0$. Compare the eigenvalues with the eigenvalues of Exercise 2.
5.  $y'' + 2y' + (1 - \lambda)y = 0$; $y(0) = 0$, $y(1) = 0$. Compare the eigenvalues with the eigenvalues of Exercise 3.

Find the difference equations by using the Hermitian method. Then find the eigenvalues of each of the Sturm–Liouville systems in Exercises 6–7. Use $N = 2$.

6.  $y'' + (1 + \lambda)y = 0$; $y(0) = 0$, $y(\pi) = 0$. Compare the eigenvalues with the eigenvalues of Exercises 2 and 4.
7.  $y'' + 2y' + (1 - \lambda)y = 0$; $y(0) = 0$, $y(1) = 0$. Compare the eigenvalues with the eigenvalues of Exercises 3 and 5.

## 14.7    METHODS OF MINIMIZING THE RESIDUAL FUNCTION

We have discussed numerical methods that give a solution at a discrete set of points in the last six sections. In this section, we will consider methods that provide a solution of a differential equation by a linear combination of a finite number of linearly independent functions. These functions are relatively simple and are called **basis functions** or **trial functions**.

For simplicity consider a linear boundary-value problem

$$Ly = y'' + p(x)y' + q(x)y = f(x) \tag{14.7.1}$$

with the boundary conditions

$$\gamma_1 \, y(a) + \gamma_2 \, y'(a) = \alpha \quad \text{and} \quad \gamma_3 \, y(b) + \gamma_4 \, y'(b) = \beta \tag{14.7.2}$$

To find an approximate solution of Equations (14.7.1) and (14.7.2), consider a set $\{\phi_0, \phi_1, \ldots, \phi_N\}$ of twice continuously differentiable and linearly independent functions such that

$$\gamma_1 \, \phi_0(a) + \gamma_2 \, \phi_0'(a) = \alpha \quad \text{and} \quad \gamma_3 \, \phi_0(b) + \gamma_4 \, \phi_0'(b) = \beta \tag{14.7.3}$$

and

$$\gamma_1 \, \phi_i(a) + \gamma_2 \, \phi_i'(a) = 0 \quad \text{and} \quad \gamma_3 \, \phi_i(b) + \gamma_4 \, \phi_i'(b) = 0 \tag{14.7.4}$$

for $i = 1, 2, \ldots, N$. Examples of a set of basis functions are $\phi_i(x) = \sin ix$ and $\phi_i(x) = x^i(1 - x)$ for $i = 0, 1, 2, \ldots, N$.

We are looking for an approximate solution of Equation (14.7.1) in the form

$$Y_N(x) = \phi_0(x) + \sum_{i=1}^{N} c_i \, \phi_i(x) \tag{14.7.5}$$

where $c_i$ is to be determined. We need to specify the criteria for determining $c_i$ in our linear combination. There are several techniques to determine $c_i$. Since $\phi_0(x)$ satisfies Equation (14.7.3) and $\phi_i(x)$ for $i = 1, 2, \ldots, N$ satisfies Equation (14.7.4), then $Y_N(x)$ satisfies boundary conditions (14.7.2). Substituting Equation (14.7.5) in Equation (14.7.1), we get the residual function

$$R(x, c_1, c_2, c_3, \ldots, c_N) = LY_N(x) - f(x) \tag{14.7.6}$$

$$= \phi_0''(x) + \sum_{i=1}^{N} c_i \, \phi_i''(x) + p(x) \left[ \phi_0'(x) + \sum_{i=1}^{N} c_i \, \phi_i'(x) \right]$$

$$+ q(x) \left[ \phi_0(x) + \sum_{i=1}^{N} c_i \, \phi_i(x) \right] - f(x)$$

$$= \phi_0''(x) + p(x) \, \phi_0'(x) + q(x) \, \phi_0(x)$$

$$+ \sum_{i=1}^{N} c_i [\phi_i''(x) + p(x) \, \phi_i'(x) + q(x) \, \phi_i(x)] - f(x)$$

$$\tag{14.7.7}$$

This gives the measure of the extent to which $Y_N(x)$ satisfies Equation (14.7.1). If $Y_N(x)$ were an exact solution of Equation (14.7.1), then the residual function $R(x, c_1, c_2, \ldots, c_N) = 0$. Normally, we cannot expect $Y_N(x)$ to be an exact solution. As the number $N$

of the functions $\phi_i$ is increased, we hope that $R$ will become small. Since it is difficult to make $R(x, c_1, c_2, \ldots, c_N)$ identically zero, we will try to make $R(x, c_1, c_2, \ldots, c_N)$ as small as possible by selecting $c_i$ in some sense.

## The Collocation Method

In this method we require $R(x, c_1, c_2, \ldots, c_N)$ to be zero at points $x_1, x_2, \ldots, x_N$ in $[a, b]$. Thus from Equation (14.7.6) we have

$$
\begin{aligned}
R(x_k, c_1, c_2, \ldots, c_N) &= LY_N(x_k) - f(x_k) \\
&= \phi_0''(x_k) + p(x_k)\,\phi_0'(x_k) + q(x_k)\,\phi_0(x_k) \\
&\quad + \sum_{i=1}^{N} c_i\big[\phi_i''(x_k) + p(x_k)\,\phi_i'(x_k) + q(x_k)\,\phi_i(x_k)\big] - f(x_k) \\
&= 0 \quad \text{for } k = 1, 2, \ldots, N
\end{aligned}
$$

This reduces to

$$
\sum_{i=1}^{N} c_i\big[\phi_i''(x_k) + p(x_k)\phi_i'(x_k) + q(x_k)\phi_i(x_k)\big]
$$
$$
= f(x_k) - \phi_0''(x_k) - p(x_k)\phi_0'(x_k) - q(x_k)\phi_0(x_k) \quad \textbf{(14.7.8)}
$$

for $k = 1, 2, \ldots, N$. The system (14.7.8) of $N$ linear equations in $N$ unknowns $c_1, c_2, \ldots, c_N$ may be written as

$$
\mathbf{Ac} = \mathbf{b} \qquad \textbf{(14.7.9)}
$$

Solve this equation for $\mathbf{c} = [c_1, c_2, \ldots, c_N]'$ and use $\mathbf{c}$ in Equation (14.7.5) to get a solution of Equations (14.7.1) and (14.7.2).

## The Integral Method of Least Squares

This method requires that the integral

$$
I = \int_a^b R^2(x, c_1, c_2, \ldots, c_N)\,dx
$$

takes the minimum value. At the minimum point,

$$
\frac{\partial I}{\partial c_i} = 2 \int_a^b R\,\frac{\partial R}{\partial c_i}\,dx = 0 \quad \text{for } i = 1, 2, \ldots, N \qquad \textbf{(14.7.10)}
$$

if it exists. From Equation (14.7.7), we get

$$
\frac{\partial R}{\partial c_i} = \phi_i''(x) + p(x)\phi_i'(x) + q(x)\phi_i(x) \quad \text{for } i = 1, 2, \ldots, N \qquad \textbf{(14.7.11)}
$$

Substituting $R$ and $\partial R/\partial c_i$ from Equations (14.7.7) and (14.7.11) in Equation (14.7.10) yields

$$
\int_a^b \left[\phi_0''(x) + p(x)\phi_0'(x) + q(x)\phi_0(x) + \sum_{j=1}^{N} c_j(\phi_j''(x) + p(x)\phi_j'(x) + q(x)\phi_j(x)) - f(x)\right]
$$
$$
\big[\phi_i''(x) + p(x)\phi_i'(x) + q(x)\phi_i(x)\big]\,dx = 0
$$

for $i = 1, 2, \ldots, N$. This reduces to

$$\sum_{j=1}^{N} c_j \int_a^b \left[\phi_j''(x) + p(x)\phi_j'(x) + q(x)\phi_j(x)\right]\left[\phi_i''(x) + p(x)\phi_i'(x) + q(x)\phi_i(x)\right]dx$$

$$= -\int_a^b \left[\phi_0''(x) + p(x)\phi_0'(x) + q(x)\phi_0(x) - f(x)\right]\left[\phi_i''(x) + p(x)\phi_i'(x) + q(x)\phi_i(x)\right]dx$$

**(14.7.12)**

for $i = 1, 2, \ldots, N$. Equation (14.7.12) is also a system of $N$ linear equations in $N$ unknowns $c_1, c_2, \ldots, c_N$ as was Equation (14.7.8) and can be written as

$$\mathbf{Ac} = \mathbf{b}$$

Once $\mathbf{c}$ is known, the approximate solution $Y_N(x)$ can be computed from Equation (14.7.5).

## The Galerkin Method

This method is based on the concept of orthogonality of functions. We require that the basis functions $\phi_1, \phi_2, \ldots, \phi_N$ be orthogonal to the residual function $R(x, c_1, c_2, \ldots, c_N)$. This is given by

$$\int_a^b R(x, c_1, c_2, \ldots, c_N)\phi_k(x)dx = 0 \quad \text{for } k = 1, 2, \ldots, N \qquad \textbf{(14.7.13)}$$

Substituting Equation (14.7.7) in Equation (14.7.13), we get

$$\sum_{i=1}^{N} c_i \int_a^b \left[\phi_i''(x) + p(x)\phi_i'(x) + q(x)\phi_i(x)\right]\phi_k(x)dx$$

$$= \int_a^b \left[f(x) - \phi_0''(x) - p(x)\phi_0'(x) - q(x)\phi_0(x)\right]\phi_k(x)dx \qquad \textbf{(14.7.14)}$$

for $k = 1, 2, \ldots, N$. Again Equation (14.7.14) provides a system of $N$ equations in $N$ unknowns $c_1, c_2, \ldots, c_N$. This can be written as

$$\mathbf{Ac} = \mathbf{b}$$

Once this is solved for $\mathbf{c}$, use $\mathbf{c}$ to determine the appropriate solution $Y_N(x)$ in Equation (14.7.5).

## The Method of Subdomains

Let $x_0 = a < x_1 < x_2 < \cdots < x_N = b$ be $N$ points in $[a, b]$. Then the coefficients $c_i$ are found from the system of equations

$$\int_{x_{j-1}}^{x_j} R(x, c_1, c_2, \ldots, c_N)dx = 0 \quad \text{for } j = 1, 2, \ldots, N \qquad \textbf{(14.7.15)}$$

**EXAMPLE 14.7.1**     Solve $y'' + y = x$; $y(0) = 0$, $y(1) = 0$ using the (1) collocation method, (2) integral method of least squares, and (3) Galerkin method.

The basis function $\phi_0(x) = 0$ satisfies the boundary conditions. Further $\phi_i(x) = \sin i\pi x$ satisfy $\phi_i(0) = 0$ and $\phi_i(1) = 0$ for $i = 1, 2, \ldots, N$. Therefore

$$Y_N(x) = \sum_{i=1}^{N} c_i \phi_i(x) = \sum_{i=1}^{N} c_i \sin i\pi x \qquad (14.7.16)$$

satisfies the boundary conditions and gives an approximate solution of our boundary-value problem. Consider $N = 3$.

1.   Substituting Equation (14.7.16) in $y'' + y = x$, we get

$$\sum_{i=1}^{N} c_i(-i^2\pi^2 + 1) \sin i\pi x = x \qquad (14.7.17)$$

This equation is true at $x_1 = \frac{1}{4}$, $x_2 = \frac{1}{2}$, and $x_3 = \frac{3}{4}$. Substituting these values in Equation (14.7.17), we get

$$c_1 \frac{(1 - \pi^2)}{\sqrt{2}} + c_2(1 - 4\pi^2) + c_3 \frac{(1 - 9\pi^2)}{\sqrt{2}} = \frac{1}{4}$$

$$c_1(1 - \pi^2) \qquad\qquad - c_3(1 - 9\pi^2) = \frac{1}{2}$$

$$c_1 \frac{(1 - \pi^2)}{\sqrt{2}} - c_2(1 - 4\pi^2) + c_3 \frac{(1 - 9\pi^2)}{\sqrt{2}} = \frac{3}{4}$$

Solving these three equations for $c_1$, $c_2$, and $c_3$, we get

$$c_1 = \frac{\sqrt{2} + 1}{4(1 - \pi^2)} \qquad c_2 = -\frac{1}{4(1 - 4\pi^2)} \qquad c_3 = \frac{\sqrt{2} - 1}{4(1 - 9\pi^2)}$$

Substituting these values in Equation (14.7.16), we get

$$Y_3(x) = \frac{\sqrt{2} + 1}{4(1 - \pi^2)} \sin \pi x - \frac{1}{4(1 - 4\pi^2)} \sin 2\pi x + \frac{\sqrt{2} - 1}{4(1 - 9\pi^2)} \sin 3\pi x$$

$$(14.7.18)$$

The values of $Y_3(x)$ are computed at various $x$ and are given in Table 14.7.1 along with the values of the exact solution $y(x) = x - (\sin x/\sin 1)$.

**Table 14.7.1**

| $x$ | Exact | Collocation | \|Actual Error\| | Galerkin and Least Squares | \|Actual Error\| |
|-----|-------|-------------|------------------|----------------------------|------------------|
| 0.00 | 0.00000 | 0.00000 | 0.00000 | 0.00000 | 0.00000 |
| 0.1 | −0.01864 | −0.01816 | 0.00048 | −0.01536 | 0.00328 |
| 0.2 | −0.03610 | −0.03494 | 0.00116 | −0.03203 | 0.00408 |
| 0.3 | −0.05119 | −0.04924 | 0.00195 | −0.04945 | 0.00174 |
| 0.4 | −0.06278 | −0.06020 | 0.00258 | −0.06482 | 0.00204 |
| 0.5 | −0.06975 | −0.06687 | 0.00288 | −0.07419 | 0.00444 |
| 0.6 | −0.07102 | −0.06784 | 0.00318 | −0.07455 | 0.00353 |
| 0.7 | −0.06559 | −0.06159 | 0.00400 | −0.06519 | 0.00040 |
| 0.8 | −0.05250 | −0.04730 | 0.00520 | −0.04776 | 0.00474 |
| 0.9 | −0.03090 | −0.02580 | 0.00510 | −0.02509 | 0.00581 |
| 1.0 | 0.00000 | 0.00000 | 0.00000 | 0.00000 | 0.00000 |

2. Substituting $\phi_0(x) = 0$, $\phi_1(x) = \sin \pi x$, $\phi_2(x) = \sin 2\pi x$, $\phi_3(x) = \sin 3\pi x$, and $i = 1$ in Equation (14.7.12), we get

$$c_1 \int_0^1 (-\pi^2 + 1)^2 \sin^2 \pi x \, dx + c_2 \int_0^1 (-4\pi^2 + 1)(-\pi^2 + 1) \sin 2\pi x \sin \pi x \, dx$$

$$+ c_3 \int_0^1 (-9\pi^2 + 1)(-\pi^2 + 1) \sin 3\pi x \sin \pi x \, dx = \int_0^1 x(-\pi^2 + 1) \sin \pi x \, dx$$

Evaluating these integrals, we can verify that $c_1 = 2/(\pi(1 - \pi^2))$. Similarly for $i = 2$ and 3, we get from Equation (14.7.12), $c_2 = -1/(\pi(1 - 4\pi^2))$ and $c_3 = -2/(3\pi(1 - 9\pi^2))$. We get from Equation (14.7.16)

$$Y_3(x) = \frac{2}{\pi(1 - \pi^2)} \sin \pi x - \frac{1}{\pi(1 - 4\pi^2)} \sin 2\pi x - \frac{2}{3\pi(1 - 9\pi^2)} \sin 3\pi x$$

$$\textbf{(14.7.19)}$$

The computed values of $Y_3$ from Equation (14.7.19) are given with the other values in Table 14.7.1.

3. Substituting $\phi_0(x) = 0$, $\phi_1(x) = \sin \pi x$, $\phi_2(x) = \sin 2\pi x$, $\phi_3(x) = \sin 3\pi x$, and $k = 1$ in Equation (14.7.14), we get

$$c_1 \int_0^1 (-\pi^2 + 1) \sin^2 \pi x \, dx + c_2 \int_0^1 (-4\pi^2 + 1) \sin 2\pi x \sin \pi x \, dx$$

$$+ c_3 \int_0^1 (-9\pi^2 + 1) \sin 3\pi x \sin \pi x \, dx = \int_0^1 x \sin \pi x \, dx$$

Evaluating these integrals, we get $c_1 = 2/(\pi(1 - \pi^2))$. Similarly for $k = 2$ and 3, we get from Equation (14.7.14), $c_2 = -1/(\pi(1 - 4\pi^2))$ and $c_3 = -2/(3\pi(1 - 9\pi^2))$. These values are the same as in part 2 (why?), therefore $Y_3(x)$ is given by Equation (14.7.19).

We have agreement in the first two digits. Considering the small number of basis functions used in $Y_N(x)$, the results are not bad. In order to obtain more accurate results, we need to increase the number of basis functions.  ■ ■ ■

It is difficult to state which method is preferable since error depends on the equation and the approximating functions. That the solution of a differential equation can be formulated in terms of an integral is a basic idea. The finite element method that we discuss in the next section is based on the integral formulation.

## EXERCISES

1. Solve $y'' + y = x$; $y(0) = 0$, $y(1) = 0$ using (a) the collocation method, and (b) the subdomain method with the basis functions $\phi_0(x) = 0$, $\phi_1(x) = x(1 - x)$, and $\phi_2(x) = x^2(1 - x)$.

2. Solve the boundary-value problem $y'' + y = x$; $y(0) = 0$, $y(1) = 1$ using (a) the collocation method and (b) the subdomain method with basis functions $\phi_0(x) = x$, $\phi_1(x) = \sin \pi x$, and $\phi_2(x) = \sin 2\pi x$.

3. Solve the boundary-value problem in Exercise 1 using (a) the integral method of least squares and (b) the Galerkin method with the basis functions $\phi_0(x) = 0$, $\phi_1(x) = x(1 - x)$, and $\phi_2(x) = x^2(1 - x)$.

4. Solve the boundary-value problem $y'' + y = x$; $y(0) = 0$, $y(1) = 1$ using (a) the integral method of least squares and (b) the Galerkin method using the basis functions given in Exercise 2.

5. Solve the boundary-value problem $y'' + (1 + x^2)y + 1 = 0$; $y(-1) = -1$, $y(1) = 1$ using (a) the collocation method and (b) the subdomain method with the basis functions $\phi_0(x) = x$ and $\phi_1(x) = (1 - x^2)$.

6. Solve the boundary-value problem $y'' + (1 + x^2)y + 1 = 0$; $y(-1) = -1$, $y(1) = 1$ using (a) the integral method of least squares and (b) the Galerkin method with the basis functions $\phi_0(x) = x$ and $\phi_1(x) = (1 - x^2)$.

## 14.8   AN INTRODUCTION TO THE FINITE ELEMENT METHOD

We approximated the solution of a boundary-value problem by replacing derivatives at selected points by finite differences. The other way is to approximate the solution by a combination of a finite number of functions which are called the basis or trial functions. We used this approach in Section 14.7 to approximate the solution of a boundary-value problem. If the basis functions are piecewise polynomials, then these piecewise polynomials can be selected to fit the geometry of a two- or higher-dimensional problem. This is a clear advantage, and this method is called the **finite element method**. It has been used with remarkable success in almost all areas of mathematical physics; therefore, we give a brief introduction to the finite element method in this section. For more detailed discussions see Becker, Carey, and Oden (1981), Strang and Fix (1973), and Reddy (1993).

In order to sketch the basic idea, let us consider a boundary-value problem in one dimension

$$\frac{d^2u}{dx^2} + p\frac{du}{dx} + qu = f(x) \qquad 0 < x < 1 \tag{14.8.1}$$

$$u(0) = u(1) = 0 \tag{14.8.2}$$

where $p$ and $q$ are constants and $f(x)$ is a given function.

The domain $\Omega = (0, 1)$ is partitioned into $N - 1$ elements $\Omega^i$ for $i = 1, 2, \ldots, N - 1$ as shown in Figure 14.8.1. The length of each element $\Omega^i$ is denoted by $h_i$.

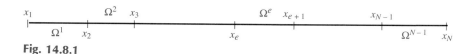

**Fig. 14.8.1**

The endpoints of each element are called **nodes** or **nodal points**. The set $\{\Omega^1, \Omega^2, \ldots, \Omega^{N-1}\}$ is called the **finite element mesh**. We assume that $\Omega^e = (x_e, x_{e+1})$ is a typical element of the finite element mesh.

Multiplying Equation (14.8.1) by a test function $v(x)$ and integrating over the domain $\Omega^e$ yields

$$\int_{x_e}^{x_{e+1}} \left( \frac{d^2u}{dx^2} + p\frac{du}{dx} + qu \right) v\, dx = \int_{x_e}^{x_{e+1}} f v\, dx \qquad (14.8.3)$$

If $u$ and $v$ are sufficiently smooth, then integrating the first term of the left side of Equation (14.8.3) by parts gives

$$\int_{x_e}^{x_{e+1}} \frac{d^2u}{dx^2} v\, dx = \frac{du}{dx} v \bigg|_{x_e}^{x_{e+1}} - \int_{x_e}^{x_{e+1}} \frac{du}{dx}\frac{dv}{dx}\, dx$$

$$= Q_2^{(e)} v(x_{e+1}) - Q_1^{(e)} v(x_e) - \int_{x_e}^{x_{e+1}} \frac{du}{dx}\frac{dv}{dx}\, dx \qquad (14.8.4)$$

where $Q_2^{(e)} = du/dx\ (x_{e+1})$ and $Q_1^{(e)} = du/dx\ (x_e)$. Substitution of Equation (14.8.4) into Equation (14.8.3) yields

$$\int_{x_e}^{x_{e+1}} \left[ -\frac{du}{dx}\frac{dv}{dx} + \left( p\frac{du}{dx} + qu \right)v \right] dx = -Q_2^{(e)} v(x_{e+1}) + Q_1^{(e)} v(x_e) + \int_{x_e}^{x_{e+1}} f v\, dx$$

$$(14.8.5)$$

Let us assume that $u$ in $\Omega^e$ can be represented as a linear combination of $N$ linearly independent functions $\phi_1^{(e)}(x), \phi_2^{(e)}(x), \ldots, \phi_N^{(e)}(x)$. Then

$$u(x) = u^{(e)}(x) = \sum_{i=1}^{N} u_i^{(e)} \phi_i^{(e)}(x) \qquad (14.8.6)$$

where $u_i^{(e)} = u(x_i)$ and $\phi_i^{(e)}(x_j) = \delta_{ij}$ are to be determined. Substituting Equation (14.8.6) for $u$ and $\phi_j^{(e)}$ for $v$ in Equation (14.8.5), we get

$$\sum_{i=1}^{N} u_i^{(e)} \int_{x_e}^{x_{e+1}} \left[ -\frac{d\phi_i^{(e)}}{dx}\frac{d\phi_j^{(e)}}{dx} + \left( p\frac{d\phi_i^{(e)}}{dx} + q\phi_i^{(e)} \right)\phi_j^{(e)} \right] dx$$

$$= -Q_2^{(e)} \phi_j^{(e)}(x_{e+1}) + Q_1^{(e)} \phi_j^{(e)}(x_e) + \int_{x_e}^{x_{e+1}} f\phi_j^{(e)}\, dx \qquad (14.8.7)$$

for $j = 1, 2, \ldots, N$. Denote

$$k_{ij}^{(e)} = \int_{x_e}^{x_{e+1}} \left[ -\frac{d\phi_i^{(e)}}{dx}\frac{d\phi_j^{(e)}}{dx} + \left( p\frac{d\phi_i^{(e)}}{dx} + q\phi_i^{(e)} \right)\phi_j^{(e)} \right] dx \qquad (14.8.8)$$

and

$$F_j^{(e)} = -Q_2^{(e)} \phi_j^{(e)}(x_{e+1}) + Q_1^{(e)} \phi_j^{(e)}(x_e) + \int_{x_e}^{x_{e+1}} f\phi_j^{(e)}\, dx \qquad (14.8.9)$$

then Equation (14.8.7) becomes a system of $N$ linear equations

$$\sum_{i=1}^{N} k_{ij}^{(e)} u_i^{(e)} = F_j^{(e)} \quad \text{for } j = 1, 2, \ldots, N \qquad (14.8.10)$$

The integrations described in Equations (14.8.8) and (14.8.9) are carried out element by element and subsequently summed up to form a global matrix. Equation (14.8.10) represents a **finite element model** for Equations (14.8.1) and (14.8.2). The matrix $K = [k_{ij}]$ is called the **stiffness matrix** for the basis functions $\phi_i$ and the column vector $\mathbf{F} = [f_j]$ is called the **load vector** for the basis functions $\phi_i$.

Equation (14.8.8) shows that $\phi_i^{(e)}$ should be at least a linear function. We express the approximate solution in $\Omega^e$ as

$$u(x) = u_{\Omega^e}(x) = u^{(e)}(x) = u_1^{(e)} \phi_1^{(e)}(x) + u_2^{(e)} \phi_2^{(e)}(x) \qquad \textbf{(14.8.11)}$$

where $u_1^{(e)} = u(x_e)$ and $u_2^{(e)} = u(x_{e+1})$. Consider linear functions

$$\phi_1^{(e)}(x) = ax + b$$
$$\phi_2^{(e)}(x) = cx + d \qquad \textbf{(14.8.12)}$$

where $a$, $b$, $c$, and $d$ are to be determined. We want at $x_e$,

$$\begin{aligned} u(x_e) = u_{\Omega^e}(x_e) &= u_1^{(e)} \phi_1^{(e)}(x_e) + u_2^{(e)} \phi_2^{(e)}(x_e) \\ &= u_1^{(e)}(ax_e + b) + u_2^{(e)}(cx_e + d) \\ &= u_1^{(e)} \end{aligned}$$

This is true if

$$ax_e + b = 1 \qquad cx_e + d = 0$$

Similarly at $x_{e+1}$,

$$\begin{aligned} u(x_{e+1}) = u_{\Omega^e}(x_{e+1}) &= u_1^{(e)} \phi_1^{(e)}(x_{e+1}) + u_2^{(e)} \phi_2^{(e)}(x_{e+1}) \\ &= u_1^{(e)}(ax_{e+1} + b) + u_2^{(e)}(cx_{e+1} + d) \\ &= u_2^{(e)} \end{aligned}$$

Again this is true whenever

$$ax_{e+1} + b = 0 \qquad cx_{e+1} + d = 1$$

Thus we have four equations in four unknowns

$$\begin{aligned} ax_e + b &= 1 \\ ax_{e+1} + b &= 0 \\ cx_e + d &= 0 \\ cx_{e+1} + d &= 1 \end{aligned} \qquad \textbf{(14.8.13)}$$

Solving the first two equations for $a$ and $b$ and the last two equations for $c$ and $d$, we get

$$a = -\frac{1}{x_{e+1} - x_e} = -\frac{1}{h_e} \qquad b = \frac{x_{e+1}}{h_e} \qquad c = \frac{1}{h_e} \qquad d = -\frac{x_e}{h_e} \quad \textbf{(14.8.14)}$$

where $h_e = x_{e+1} - x_e$. Substituting the values of $a$, $b$, $c$, and $d$ in Equation (14.8.12), we get

$$\phi_1^{(e)}(x) = \frac{x_{e+1} - x}{h_e} \qquad \phi_2^{(e)}(x) = \frac{x - x_e}{h_e} \qquad \textbf{(14.8.15)}$$

We verify that

$$\phi_1^{(e)}(x) + \phi_2^{(e)}(x) = 1 \quad \text{and} \quad \phi_1'^{(e)}(x) = -\frac{1}{h_e} \qquad \phi_2'^{(e)}(x) = \frac{1}{h_e} \quad \textbf{(14.8.16)}$$

The graphs of $\phi_1^{(e)}(x)$ and $\phi_2^{(e)}(x)$ are shown in Figure 14.8.2.

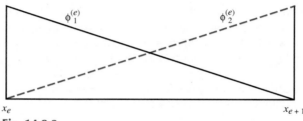

$x_e$                                                                $x_{e+1}$

**Fig. 14.8.2**

Outside the element $\Omega^e$, $u^{(e)}(x) = 0 = \phi_1^{(e)}(x) = \phi_2^{(e)}(x) = 0$. In other words, $u^{(e)}(x) = 0 = \phi_1^{(e)}(x) = \phi_2^{(e)}(x)$ for $x > x_{e+1}$ and $x < x_e$.

**EXAMPLE 14.8.1**     Using four elements of equal length and piecewise linear basis functions defined by Equation (14.8.15), calculate the finite element approximation of the boundary-value problem

$$\frac{d^2u}{dx^2} = -1 \qquad 0 < x < 1$$

$$u(0) = u(1) = 0$$

Using $p = q = 0$ and $f = -1$, Equations (14.8.8) and (14.8.9) reduce to

$$k_{ij}^{(e)} = -\int_{x_e}^{x_{e+1}} \frac{d\phi_i^{(e)}}{dx} \frac{d\phi_j^{(e)}}{dx}\, dx$$

and

$$F_j^{(e)} = -Q_2^{(e)}\, \phi_j^{(e)}(x_{e+1}) + Q_1^{(e)}\, \phi_j^{(e)}(x_e) - \int_{x_e}^{x_{e+1}} \phi_j^{(e)}\, dx \qquad \textbf{(14.8.17)}$$

We express Equation (14.8.17) for each element shown in Figure 14.8.3.

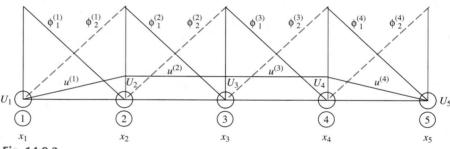

**Fig. 14.8.3**

***Element 1*** For $e = 1$, we have $\Omega^1$, $x_1 = 0$, $x_2 = \frac{1}{4}$, $h_1 = x_2 - x_1 = \frac{1}{4} = h$, $\phi_1^{(1)}(x) = 1 - 4x$, and $\phi_2^{(1)}(x) = 4x$. Then Equation (14.8.17) yields

$$k_{11}^{(1)} = -\int_0^{1/4} \frac{d\phi_1^{(1)}}{dx} \frac{d\phi_1^{(1)}}{dx} dx = -\int_0^{1/4} (-4)(-4)dx = -4$$

$$k_{12}^{(1)} = -\int_0^{1/4} \frac{d\phi_1^{(1)}}{dx} \frac{d\phi_2^{(1)}}{dx} dx = -\int_0^{1/4} (-4)(4)dx = 4 = k_{21}^{(1)}$$

$$k_{22}^{(1)} = -\int_0^{1/4} \frac{d\phi_2^{(1)}}{dx} \frac{d\phi_2^{(1)}}{dx} dx = -\int_0^{1/4} (4)(4)dx = -4$$

$$F_1^{(1)} = -Q_2^{(1)} \phi_1^{(1)}(x_1) + Q_1^{(1)} \phi_1^{(1)}(x_0) - \int_0^{1/4} (1 - 4x)dx = Q_1^{(1)} - \frac{1}{8}$$

$$F_2^{(1)} = -Q_2^{(1)} \phi_2^{(1)}(x_1) + Q_1^{(1)} \phi_2^{(1)}(x_0) - 4\int_0^{1/4} x\, dx = -Q_2^{(1)} - \frac{1}{8}$$

Hence

$$K^{(1)} = -4\begin{bmatrix} 1 & -1 \\ -1 & 1 \end{bmatrix} \quad \text{and} \quad \mathbf{F}^{(1)} = \begin{bmatrix} Q_1^{(1)} \\ -Q_2^{(1)} \end{bmatrix} - \frac{1}{8}\begin{bmatrix} 1 \\ 1 \end{bmatrix}$$

***Element 2*** For $e = 2$, we have $\Omega^2$, $x_2 = \frac{1}{4}$, $x_3 = \frac{1}{2}$, $h_2 = x_3 - x_2 = \frac{1}{4} = h$, $\phi_1^{(2)}(x) = 2 - 4x$, $\phi_2^{(2)}(x) = 4x - 1$,

$$K^{(2)} = -4\begin{bmatrix} 1 & -1 \\ -1 & 1 \end{bmatrix} \quad \text{and} \quad \mathbf{F}^{(2)} = \begin{bmatrix} Q_1^{(2)} \\ -Q_2^{(2)} \end{bmatrix} - \frac{1}{8}\begin{bmatrix} 1 \\ 1 \end{bmatrix}$$

***Element 3*** For $e = 3$, we have $\Omega^3$, $x_3 = \frac{1}{2}$, $x_4 = \frac{3}{4}$, $h_3 = x_4 - x_3 = \frac{1}{4} = h$, $\phi_1^{(3)}(x) = 3 - 4x$, $\phi_2^{(3)}(x) = 4x - 2$

$$K^{(3)} = -4\begin{bmatrix} 1 & -1 \\ -1 & 1 \end{bmatrix} \quad \text{and} \quad \mathbf{F}^{(3)} = \begin{bmatrix} Q_1^{(3)} \\ -Q_2^{(3)} \end{bmatrix} - \frac{1}{8}\begin{bmatrix} 1 \\ 1 \end{bmatrix}$$

***Element 4*** For $e = 4$, we have $\Omega^4$, $x_4 = \frac{3}{4}$, $x_5 = 1$, $h_4 = x_5 - x_4 = \frac{1}{4} = h$, $\phi_1^{(4)}(x) = 4(1 - x)$, $\phi_2^{(4)}(x) = 4x - 3$,

$$K^{(4)} = -4\begin{bmatrix} 1 & -1 \\ -1 & 1 \end{bmatrix} \quad \text{and} \quad \mathbf{F}^{(4)} = \begin{bmatrix} Q_1^{(4)} \\ -Q_2^{(4)} \end{bmatrix} - \frac{1}{8}\begin{bmatrix} 1 \\ 1 \end{bmatrix}$$

To assemble all the elements into a single equation, we note the correspondence between the global nodal values and the local nodal values:

$$U_1 = u_1^{(1)} \quad U_2 = u_2^{(1)} = u_1^{(2)} \quad U_3 = u_2^{(2)} = u_1^{(3)} \quad U_4 = u_2^{(3)} = u_1^{(4)} \quad U_5 = u_2^{(4)}$$

Then for $e = 1$, $\sum_{i=1}^5 k_{ij}^{(e)} u_i^{(e)} = F_j^{(e)}$ for $j = 1, 2, \ldots, 5$ satisfies

$$-4\begin{bmatrix} 1 & -1 & 0 & 0 & 0 \\ -1 & 1 & 0 & 0 & 0 \\ 0 & 0 & 0 & 0 & 0 \\ 0 & 0 & 0 & 0 & 0 \\ 0 & 0 & 0 & 0 & 0 \end{bmatrix}\begin{bmatrix} U_1 \\ U_2 \\ U_3 \\ U_4 \\ U_5 \end{bmatrix} = \begin{bmatrix} Q_1^{(1)} \\ -Q_2^{(1)} \\ 0 \\ 0 \\ 0 \end{bmatrix} - \frac{1}{8}\begin{bmatrix} 1 \\ 1 \\ 0 \\ 0 \\ 0 \end{bmatrix}$$

$$\text{(14.8.18)}$$

Equation (14.8.18) is not solvable since the element matrix is singular and part of the right-hand side is unknown. For $e = 2$, $\sum_{i=1}^{5} k_{ij}^{(e)} u_i^{(e)} = F_j^{(e)}$ for $j = 1, 2, \ldots, 5$ satisfies

$$-4\begin{bmatrix} 0 & 0 & 0 & 0 & 0 \\ 0 & 1 & -1 & 0 & 0 \\ 0 & -1 & 1 & 0 & 0 \\ 0 & 0 & 0 & 0 & 0 \\ 0 & 0 & 0 & 0 & 0 \end{bmatrix}\begin{bmatrix} U_1 \\ U_2 \\ U_3 \\ U_4 \\ U_5 \end{bmatrix} = \begin{bmatrix} 0 \\ Q_1^{(2)} \\ -Q_2^{(2)} \\ 0 \\ 0 \end{bmatrix} - \frac{1}{8}\begin{bmatrix} 0 \\ 1 \\ 1 \\ 0 \\ 0 \end{bmatrix}$$

Similarly expressing for $e = 3$, 4, and 5 and adding all equations, we get

$$-4\begin{bmatrix} 1 & -1 & 0 & 0 & 0 \\ -1 & 2 & -1 & 0 & 0 \\ 0 & -1 & 2 & -1 & 0 \\ 0 & 0 & -1 & 2 & -1 \\ 0 & 0 & 0 & -1 & 1 \end{bmatrix}\begin{bmatrix} U_1 \\ U_2 \\ U_3 \\ U_4 \\ U_5 \end{bmatrix} = \begin{bmatrix} Q_1^{(1)} \\ Q_1^{(2)} - Q_2^{(1)} \\ Q_1^{(3)} - Q_2^{(2)} \\ Q_1^{(4)} - Q_2^{(3)} \\ -Q_2^{(4)} \end{bmatrix} - \frac{1}{8}\begin{bmatrix} 1 \\ 2 \\ 2 \\ 2 \\ 1 \end{bmatrix} \qquad \textbf{(14.8.19)}$$

Further $Q_1^{(e+1)} - Q_2^{(e)} = du/dx(x_{e+1}) - du/dx(x_{e+1}) = 0$, hence

$$Q_1^{(2)} - Q_2^{(1)} = Q_1^{(3)} - Q_2^{(2)} = Q_1^{(4)} - Q_2^{(3)} = 0 \qquad \textbf{(14.8.20)}$$

From the boundary conditions,

$$U_1 = u_1^{(1)} = 0 \qquad U_5 = u_2^{(4)} = 0 \qquad \textbf{(14.8.21)}$$

Substitution of Equations (14.8.20) and (14.8.21) into Equation (14.8.19) yields

$$\begin{bmatrix} 2 & -1 & 0 \\ -1 & 2 & -1 \\ 0 & -1 & 2 \end{bmatrix}\begin{bmatrix} U_2 \\ U_3 \\ U_4 \end{bmatrix} = \frac{1}{32}\begin{bmatrix} 2 \\ 2 \\ 2 \end{bmatrix} = \frac{1}{16}\begin{bmatrix} 1 \\ 1 \\ 1 \end{bmatrix} \qquad \textbf{(14.8.22)}$$

and

$$4U_2 = -\frac{1}{8} + Q_1^{(1)} \qquad 4U_4 = -\frac{1}{8} - Q_2^{(4)} \qquad \textbf{(14.8.23)}$$

Solving Equation (14.8.22), we get $U_2 = U_4 = 3/32$, $U_3 = \frac{1}{8}$. Substituting $U_2 = U_4 = 3/32$ into Equation (14.8.23), we get $Q_1^{(1)} = -Q_2^{(4)} = \frac{1}{2}$.

The finite element solution for the entire domain is given by

$$u(x) = \sum_{e=1}^{4}\left(\sum_{i=1}^{2} u_i^{(e)} \phi_i^{(e)}(x)\right)$$

$$= \begin{cases} U_1\phi_1^{(1)}(x) + U_2\phi_2^{(1)}(x) & \text{if } 0 \le x \le \frac{1}{4} \\ U_2\phi_1^{(2)}(x) + U_3\phi_2^{(2)}(x) & \text{if } \frac{1}{4} \le x \le \frac{1}{2} \\ U_3\phi_1^{(3)}(x) + U_4\phi_2^{(3)}(x) & \text{if } \frac{1}{2} \le x \le \frac{3}{4} \\ U_4\phi_1^{(4)}(x) + U_5\phi_2^{(4)}(x) & \text{if } \frac{3}{4} \le x \le 1 \end{cases}$$

$$= \begin{cases} (3x)/8 & \text{if } 0 \le x \le \frac{1}{4} \\ (2x + 1)/16 & \text{if } \frac{1}{4} \le x \le \frac{1}{2} \\ (3 - 2x)/16 & \text{if } \frac{1}{2} \le x \le \frac{3}{4} \\ 3(1 - x)/8 & \text{if } \frac{3}{4} \le x \le 1 \end{cases}$$

**Table 14.8.1**

| $x$ | Finite Element Solution | Exact Solution | \|Actual Error\| |
|---|---|---|---|
| 0.00 | 0.00000 | 0.00000 | 0.00000 |
| 0.10 | 0.03750 | 0.04500 | 0.00750 |
| 0.20 | 0.07500 | 0.08000 | 0.00500 |
| 0.25 | 0.09375 | 0.09375 | 0.00000 |
| 0.30 | 0.10000 | 0.10500 | 0.00500 |
| 0.40 | 0.11250 | 0.12000 | 0.00750 |
| 0.50 | 0.12500 | 0.12500 | 0.00000 |
| 0.60 | 0.11250 | 0.12000 | 0.00750 |
| 0.70 | 0.10000 | 0.10500 | 0.00500 |
| 0.75 | 0.09375 | 0.09375 | 0.00000 |
| 0.80 | 0.07500 | 0.08000 | 0.00500 |
| 0.90 | 0.03750 | 0.04500 | 0.00750 |
| 1.00 | 0.00000 | 0.00000 | 0.00000 |

The exact solution of our problem is given by $u(x) = x(1 - x)/2$. The calculated values of the finite element solution and exact solution are given in Table 14.8.1. The finite element solution and exact solution agree with each other at the finite element nodes.

Since the exact solution varies quadratically whereas the finite element solution varies linearly between the nodes, the exact solution and finite solution differ at any point between the nodes. However, as the number of elements is increased, the finite element solution converges to the exact solution.  ■ ■ ■

**EXERCISES**

1.  Using two elements of equal length and piecewise linear basis functions defined by Equation (14.8.15), calculate the finite element approximation of the boundary-value problem

$$\frac{d^2u}{dx^2} + u = x, \qquad 0 < x < 1; \qquad u(0) = u(1) = 0$$

2.  Using three elements, calculate the finite element approximation of the boundary-value problem given in Exercise 1.

3.  Using two elements of equal length and piecewise linear basis functions defined by Equation (14.8.15), calculate the finite element approximation of the boundary-value problem

$$\frac{d^2u}{dx^2} + u = x, \qquad 0 < x < 1; \qquad u(0) = \frac{du}{dx}(1) = 0$$

4. Using three elements, calculate the finite element approximation of the boundary-value problem given in Exercise 3.

5. Using two elements of equal length and piecewise linear basis functions defined by Equation (14.8.15), calculate the finite element approximation of the boundary-value problem

$$(x + 1)\frac{d^2u}{dx^2} + \frac{du}{dx} = 0, \qquad 1 < x < 2; \qquad u(1) = 1, \qquad u(2) = 2$$

6. Find the first two eigenvalues associated with

$$\frac{d^2u}{dx^2} + \lambda u = 0, \qquad 0 < x < 1; \qquad u(0) = u(1) = 0$$

using the finite element method.

# SUGGESTED READINGS

E. B. Becker, G. F. Carey, and J. T. Oden, *Finite Elements An Introduction Volume I,* Prentice Hall, Inc., Englewood Cliffs, New Jersey, 1981.

R. B. Bird, W. E. Stewart, and E. N. Lightfoot, *Transport Phenomena,* Wiley, New York, 1960.

L. Collatz, *The Numerical Treatment of Differential Equations,* Springer–Verlag, New York, 1966.

H. B. Keller, *Numerical Methods for Two-Point Boundary Value Problems,* Blaisdell, Waltham, Massachusetts, 1968.

J. N. Reddy, *An Introduction to the Finite Element Method,* 2nd ed., McGraw–Hill Book Company, New York, 1993.

H. Schlichting, *Boundary Layer Theory,* 4th ed., McGraw–Hill Book Company, New York, 1960.

G. Strang and G. Fix, *An Analysis of the Finite Element Method,* Prentice Hall, Englewood Cliffs, N. J., 1973.

D. J. Tritton, *Physical Fluid Dynamics,* Oxford University Press, New York, 1988.

# 15 NUMERICAL SOLUTION OF PARTIAL DIFFERENTIAL EQUATIONS

The problem of viscous flow around a circular cylinder is a classical one with extensive literature. A sequence of patterns in a cylinder wake is given by Van Dyke (1982). We can also study patterns in a cylinder wake by computing numerical solutions of the nondimensional governing equation of motion (Patel 1976, 1978)

$$\frac{\partial \zeta}{\partial t} + \frac{1}{r}\left(\frac{\partial \Psi}{\partial \theta}\frac{\partial \zeta}{\partial r} - \frac{\partial \Psi}{\partial r}\frac{\partial \zeta}{\partial \theta}\right) = \frac{2}{Re}\nabla^2 \zeta$$

where the Reynolds number Re is defined by $Re = 2aU/\nu$, where $a$ is the radius of the cylinder, $U$ is the free stream velocity, and $\nu$ is the kinematic viscosity. The stream function $\Psi$ and the vorticity $\zeta$ are connected by the relation

$$\zeta = -\nabla^2 \Psi$$

where

$$\nabla^2 = \frac{\partial^2}{\partial r^2} + \frac{1}{r}\frac{\partial}{\partial r} + \frac{1}{r^2}\frac{\partial^2}{\partial \theta^2}$$

In this chapter, we develop methods to approximate the solution of a partial differential equation.

## 15.1  INTRODUCTION

We consider finite difference methods for solving partial differential equations in this chapter. Partial differential equations arise in the description of many physical processes such as heat transfer, vibrations, and fluid flows. Since numerical methods of partial differential equations is a vast subject, we do not present comprehensive coverage of the numerical methods used in solving partial differential equations in this text. In this chapter, we present some simple and general ideas for insight. More can be found in Ames (1992), Forsyth and Wasow (1960), M. K. Jain (1984), and Lapidus and Pinder (1982).

We concentrate on second-order linear partial differential equations that dominate the applications.

## 15.2  CLASSIFICATION

Let us consider some features of second-order linear partial differential equations that will be useful in studying the numerical solutions. Consider a linear second-order partial differential equation

$$A(x, y) \frac{\partial^2 u}{\partial x^2} + B(x, y) \frac{\partial^2 u}{\partial x \partial y} + C(x, y) \frac{\partial^2 u}{\partial y^2} + D(x, y) \frac{\partial u}{\partial x} + E(x, y) \frac{\partial u}{\partial y}$$
$$+ F(x, y) u = 0 \qquad (15.2.1)$$

where the coefficients $A(x, y)$, $B(x, y)$, and $C(x, y)$ are not all zero. To simplify Equation (15.2.1), we introduce new independent variables $\xi$ and $\eta$ by means of the transformations

$$\xi = \xi(x, y) \quad \text{and} \quad \eta = \eta(x, y) \qquad (15.2.2)$$

Then

$$\frac{\partial u}{\partial x} = \frac{\partial u}{\partial \xi} \frac{\partial \xi}{\partial x} + \frac{\partial u}{\partial \eta} \frac{\partial \eta}{\partial x} \qquad (15.2.3a)$$

$$\frac{\partial u}{\partial y} = \frac{\partial u}{\partial \xi} \frac{\partial \xi}{\partial y} + \frac{\partial u}{\partial \eta} \frac{\partial \eta}{\partial y} \qquad (15.2.3b)$$

$$\frac{\partial^2 u}{\partial x^2} = \frac{\partial^2 u}{\partial \xi^2} \left( \frac{\partial \xi}{\partial x} \right)^2 + 2 \frac{\partial^2 u}{\partial \xi \partial \eta} \frac{\partial \xi}{\partial x} \frac{\partial \eta}{\partial x} + \frac{\partial^2 u}{\partial \eta^2} \left( \frac{\partial \eta}{\partial x} \right)^2 + \frac{\partial u}{\partial \xi} \frac{\partial^2 \xi}{\partial x^2} + \frac{\partial u}{\partial \eta} \frac{\partial^2 \eta}{\partial x^2} \qquad (15.2.3c)$$

$$\frac{\partial^2 u}{\partial x \partial y} = \frac{\partial^2 u}{\partial \xi^2} \frac{\partial \xi}{\partial x} \frac{\partial \xi}{\partial y} + \frac{\partial^2 u}{\partial \xi \partial \eta} \left[ \frac{\partial \xi}{\partial x} \frac{\partial \eta}{\partial y} + \frac{\partial \xi}{\partial y} \frac{\partial \eta}{\partial x} \right] + \frac{\partial^2 u}{\partial \eta^2} \frac{\partial \eta}{\partial x} \frac{\partial \eta}{\partial y} + \frac{\partial u}{\partial \xi} \frac{\partial^2 \xi}{\partial x \partial y} + \frac{\partial u}{\partial \eta} \frac{\partial^2 \eta}{\partial x \partial y}$$
$$(15.2.3d)$$

and

$$\frac{\partial^2 u}{\partial y^2} = \frac{\partial^2 u}{\partial \xi^2} \left( \frac{\partial \xi}{\partial y} \right)^2 + 2 \frac{\partial^2 u}{\partial \xi \partial \eta} \frac{\partial \xi}{\partial y} \frac{\partial \eta}{\partial y} + \frac{\partial^2 u}{\partial \eta^2} \left( \frac{\partial \eta}{\partial y} \right)^2 + \frac{\partial u}{\partial \xi} \frac{\partial^2 \xi}{\partial y^2} + \frac{\partial u}{\partial \eta} \frac{\partial^2 \eta}{\partial y^2} \qquad (15.2.3e)$$

Substituting Equations (15.2.3a), (15.2.3b), (15.2.3c), (15.2.3d), and (15.2.3e) into Equation (15.2.1) and collecting similar terms, we have

$$a \frac{\partial^2 u}{\partial \xi^2} + b \frac{\partial^2 u}{\partial \xi \partial \eta} + c \frac{\partial^2 u}{\partial \eta^2} + d \frac{\partial u}{\partial \xi} + e \frac{\partial u}{\partial \eta} + fu = 0 \qquad (15.2.4)$$

where

$$a = A \left( \frac{\partial \xi}{\partial x} \right)^2 + B \frac{\partial \xi}{\partial x} \frac{\partial \xi}{\partial y} + C \left( \frac{\partial \xi}{\partial y} \right)^2 \tag{15.2.5a}$$

$$b = 2A \frac{\partial \xi}{\partial x} \frac{\partial \eta}{\partial x} + B \left( \frac{\partial \xi}{\partial x} \frac{\partial \eta}{\partial y} + \frac{\partial \xi}{\partial y} \frac{\partial \eta}{\partial x} \right) + 2C \frac{\partial \xi}{\partial y} \frac{\partial \eta}{\partial y} \tag{15.2.5b}$$

$$c = A \left( \frac{\partial \eta}{\partial x} \right)^2 + B \frac{\partial \eta}{\partial x} \frac{\partial \eta}{\partial y} + C \left( \frac{\partial \eta}{\partial y} \right)^2 \tag{15.2.5c}$$

$$d = A \frac{\partial^2 \xi}{\partial x^2} + B \frac{\partial^2 \xi}{\partial x \partial y} + C \frac{\partial^2 \xi}{\partial y^2} + D \frac{\partial \xi}{\partial x} + E \frac{\partial \xi}{\partial y} \tag{15.2.5d}$$

$$e = A \frac{\partial^2 \eta}{\partial x^2} + B \frac{\partial^2 \eta}{\partial x \partial y} + C \frac{\partial^2 \eta}{\partial y^2} + D \frac{\partial \eta}{\partial x} + E \frac{\partial \eta}{\partial y} \tag{15.2.5e}$$

$$f = F \tag{15.2.5f}$$

We can verify that

$$b^2 - 4ac = (B^2 - 4AC) \left( \frac{\partial \xi}{\partial x} \frac{\partial \eta}{\partial y} - \frac{\partial \xi}{\partial y} \frac{\partial \eta}{\partial x} \right)^2 \tag{15.2.6}$$

The right-hand side of Equation (15.2.6) contains the Jacobian

$$J = \frac{\partial(\xi, \eta)}{\partial(x, y)} = \det \begin{bmatrix} \dfrac{\partial \xi}{\partial x} & \dfrac{\partial \xi}{\partial y} \\ \dfrac{\partial \eta}{\partial x} & \dfrac{\partial \eta}{\partial y} \end{bmatrix} \tag{15.2.7}$$

which we assume is nonzero so that the transformation is locally one-to-one.

We select the transformations (15.2.2) in such a way that at least one of the second-order terms in Equation (15.2.4) drops out. Under such transformations the sign of the discriminant $b^2 - 4ac$ remains the same as $B^2 - 4AC$, since the squared factor on the right-hand side of Equation (15.2.6) is positive. If $B^2 - 4AC > 0$, then we select either $a = c = 0$ or $b = 0$ and $c = -a$. Then Equation (15.2.4) reduces to its canonical form

$$b \frac{\partial^2 u}{\partial \xi \partial \eta} + d \frac{\partial u}{\partial \xi} + e \frac{\partial u}{\partial \eta} + fu = 0 \tag{15.2.8a}$$

or

$$a \left( \frac{\partial^2 u}{\partial \xi^2} - \frac{\partial^2 u}{\partial \eta^2} \right) + d \frac{\partial u}{\partial \xi} + e \frac{\partial u}{\partial \eta} + fu = 0 \tag{15.2.8b}$$

The partial differential equation given by Equation (15.2.1) is said to be **hyperbolic** if $B^2 - 4AC > 0$.

If $B^2 - 4AC = 0$, then (15.2.1) is said to be **parabolic**. In this case we select $b = 0$ and $c = 0$; thus Equation (15.2.4) reduces to its canonical form

$$a \frac{\partial^2 u}{\partial \xi^2} + d \frac{\partial u}{\partial \xi} + e \frac{\partial u}{\partial \eta} + fu = 0 \tag{15.2.8c}$$

When $B^2 - 4AC < 0$, we select $b = 0$ and $c = a$. Then Equation (15.2.4) reduces to its canonical form

$$a\left(\frac{\partial^2 u}{\partial \xi^2} + \frac{\partial^2 u}{\partial \eta^2}\right) + d\frac{\partial u}{\partial \xi} + e\frac{\partial u}{\partial \eta} + fu = 0 \qquad \textbf{(15.2.8d)}$$

If $B^2 - 4AC < 0$, then Equation (15.2.1) is said to be **elliptic**. Any second-order linear partial differential equation like Equation (15.2.1) is either hyperbolic, parabolic, or elliptic and can be reduced to Equations (15.2.8a), (15.2.8b), (15.2.8c), or (15.2.8d).

In order to achieve Equation (15.2.8a), we select $a = 0$ and $c = 0$. Thus, from Equations (15.2.5a) and (15.2.5c), we get

$$A\left(\frac{\partial \xi}{\partial x}\right)^2 + B\frac{\partial \xi}{\partial x}\frac{\partial \xi}{\partial y} + C\left(\frac{\partial \xi}{\partial y}\right)^2 = 0 \qquad \textbf{(15.2.9)}$$

and

$$A\left(\frac{\partial \eta}{\partial x}\right)^2 + B\frac{\partial \eta}{\partial x}\frac{\partial \eta}{\partial y} + C\left(\frac{\partial \eta}{\partial y}\right)^2 = 0 \qquad \textbf{(15.2.10)}$$

We analyze the solutions of Equations (15.2.9) and (15.2.10) by considering their level curves

$$\xi(x, y) = \text{constant} \quad \text{and} \quad \eta(x, y) = \text{constant} \qquad \textbf{(15.2.11)}$$

Along each curve $\xi(x, y) = \text{constant}$; therefore,

$$\frac{\partial \xi}{\partial x} dx + \frac{\partial \xi}{\partial y} dy = 0$$

Rearranging, we get

$$\frac{dy}{dx} = \frac{-\dfrac{\partial \xi}{\partial x}}{\dfrac{\partial \xi}{\partial y}} \qquad \textbf{(15.2.12)}$$

Dividing Equation (15.2.9) by $(\partial \xi/\partial y)^2$ and then using Equation (15.2.12) yields

$$A\left(\frac{dy}{dx}\right)^2 - B\frac{dy}{dx} + C = 0 \qquad \textbf{(15.2.13)}$$

Solving Equation (15.2.13) for $dy/dx$ yields

$$\frac{dy}{dx} = \frac{B \pm \sqrt{B^2 - 4AC}}{2A} \qquad \textbf{(15.2.14)}$$

Actually Equation (15.2.14) represents two differential equations for the level curves of Equation (15.2.11) in the $(x, y)$ plane. We express the integrals of Equation (15.2.14) in the form

$$\phi(x, y) = \text{constant} \quad \text{and} \quad \psi(x, y) = \text{constant} \qquad \textbf{(15.2.15)}$$

Thus $\xi = \phi(x, y)$ and $\eta = \psi(x, y)$ are the desired solutions of Equations (15.2.9) and (15.2.10). We can verify that the Jacobian [Equation (15.2.7)] differs from zero and therefore, in the hyperbolic case, there exists at least in the small region coordinates $\xi$ and $\eta$ in terms of which Equation (15.2.1) reduces to

$$b \frac{\partial^2 u}{\partial \xi \partial \eta} + d \frac{\partial u}{\partial \xi} + e \frac{\partial u}{\partial \eta} + fu = 0 \tag{15.2.8a}$$

Dividing Equation (15.2.8a) by $b$ reduces it to the canonical form

$$\frac{\partial^2 u}{\partial \xi \partial \eta} + \frac{d}{b} \frac{\partial u}{\partial \xi} + \frac{e}{b} \frac{\partial u}{\partial \eta} + \frac{f}{b} u = 0 \tag{15.2.16}$$

In $(\xi, \eta)$ space, $\xi = \phi(x, y) = $ constant and $\eta = \psi(x, y) = $ constant are no longer curved but correspond to horizontal and vertical lines. These two families of curves are called the **characteristic curves** of Equation (15.2.1).

For a parabolic equation, set $c = 0$ and $b = 0$ to select $\xi$ and $\eta$ in Equation (15.2.2). From Equation (15.2.5c), we get

$$A \left( \frac{\partial \eta}{\partial x} \right)^2 + B \frac{\partial \eta}{\partial x} \frac{\partial \eta}{\partial y} + C \left( \frac{\partial \eta}{\partial y} \right)^2 = 0$$

This is the same as Equation (15.2.10); therefore, the level curve $\eta(x, y) = $ constant leads to Equation (15.2.14). For a parabolic equation $B^2 - 4AC = 0$; therefore, Equation (15.2.14) reduces to

$$\frac{dy}{dx} = \frac{B}{2A} \tag{15.2.17}$$

Consequently, we get just a single characteristic curve $\eta(x, y) = $ constant from Equation (15.2.17). From $B^2 - 4AC = 0$ and $c = 0$, Equation (15.2.6) yields $b = 0$. Thus $b = 0$ and $c = 0$ give $\eta(x, y) = $ constant. Hence we select for $\xi$ any function of $x$ and $y$, independent of $\eta$. Thus Equation (15.2.4) reduces to

$$a \frac{\partial^2 u}{\partial \xi^2} + d \frac{\partial u}{\partial \xi} + e \frac{\partial u}{\partial \eta} + fu = 0 \tag{15.2.18}$$

Since $a \neq 0$, we divide by $a$ and Equation (15.2.18) becomes

$$\frac{\partial^2 u}{\partial \xi^2} + \frac{d}{a} \frac{\partial u}{\partial \xi} + \frac{e}{a} \frac{\partial u}{\partial \eta} + \frac{f}{a} u = 0 \tag{15.2.19}$$

For an elliptic equation, we set $b = 0$ and $a = c$. It can be proven (Garabedian 1966) that the complex analytic solution of

$$\frac{dy}{dx} = \frac{-B \pm I \sqrt{4AC - B^2}}{2A}$$

where $I = \sqrt{-1}$ provides a pair of real functions $\xi$ and $\eta$ satisfying $b = 0$ and $a = c$. Thus Equation (15.2.4) reduces to

$$a \left( \frac{\partial^2 u}{\partial \xi^2} + \frac{\partial^2 u}{\partial \eta^2} \right) + d \frac{\partial u}{\partial \xi} + e \frac{\partial u}{\partial \eta} + fu = 0$$

Since $a \neq 0$, dividing the equation by $a$, we get

$$\frac{\partial^2 u}{\partial \xi^2} + \frac{\partial^2 u}{\partial \eta^2} + \frac{d}{a} \frac{\partial u}{\partial \xi} + \frac{e}{a} \frac{\partial u}{\partial \eta} + \frac{f}{a} u = 0 \tag{15.2.20}$$

**EXAMPLE 15.2.1**   Reduce the partial differential equation

$$\frac{\partial^2 u}{\partial x^2} + 2x \frac{\partial^2 u}{\partial x \partial y} + x^2 \frac{\partial^2 u}{\partial y^2} = 0$$

to its canonical form.

Since $A = 1$, $B = 2x$, and $C = x^2$, $B^2 - 4AC = 4x^2 - 4x^2 = 0$. Hence, the given equation is of the parabolic type. By Equation (15.2.17), the characteristic curve is given by

$$\frac{dy}{dx} = \frac{B}{2A} = x$$

Therefore $y = (x^2/2) + \text{constant}$. Hence $\eta(x, y) = y - (x^2/2)$. We select for $\xi$ any function of $x$ and $y$ such that the Jacobian [Equation (15.2.7)] is not zero. We can verify that $\xi = x$ does that. Thus $\xi = x$ and $\eta = y - (x^2/2)$. Hence Equations (15.2.5a), (15.2.5b), (15.2.5c), (15.2.5d), (15.2.5e), and (15.2.5f) give $a = A = 1$, $b = 0$, $c = 0$, $d = 0$, $e = -1$, and $f = 0$. Therefore Equation (15.2.4) reduces to

$$\frac{\partial^2 u}{\partial \xi^2} - \frac{\partial u}{\partial \eta} = 0$$

■ ■ ■

Any given linear second-order partial differential equation can be converted to one of the three canonical forms in new variables $\xi$ and $\eta$ as

1. hyperbolic form

$$\frac{\partial^2 u}{\partial \xi^2} - \frac{\partial^2 u}{\partial \eta^2} + d_1 \frac{\partial u}{\partial \xi} + e_1 \frac{\partial u}{\partial \eta} + f_1 u = 0 \qquad \textbf{(15.2.21)}$$

   or

$$\frac{\partial^2 u}{\partial \xi \partial \eta} + d_1 \frac{\partial u}{\partial \xi} + e_1 \frac{\partial u}{\partial \eta} + f_1 u = 0 \qquad \textbf{(15.2.22)}$$

2. parabolic form

$$\frac{\partial^2 u}{\partial \xi^2} + d_1 \frac{\partial u}{\partial \xi} + e_1 \frac{\partial u}{\partial \eta} + f_1 u = 0 \qquad \textbf{((15.2.23)}$$

3. elliptic form

$$\frac{\partial^2 u}{\partial \xi^2} + \frac{\partial^2 u}{\partial \eta^2} + d_1 \frac{\partial u}{\partial \xi} + e_1 \frac{\partial u}{\partial \eta} + f_1 u = 0 \qquad \textbf{(15.2.24)}$$

We need to study these three cases and therefore we develop numerical methods for each case by selecting the following well-known equations:

$$\frac{\partial^2 u}{\partial x^2} = \frac{\partial^2 u}{\partial t^2} \qquad \text{the wave equation (hyperbolic)} \qquad \textbf{(15.2.25a)}$$

$$\frac{\partial^2 u}{\partial x^2} = \frac{\partial u}{\partial t} \qquad \text{the heat equation (parabolic)} \qquad \textbf{(15.2.25b)}$$

$$\frac{\partial^2 u}{\partial x^2} + \frac{\partial^2 u}{\partial y^2} = f(x, y) \qquad \text{Poisson's equation (elliptic)} \qquad \textbf{(15.2.25c)}$$

Of course, there are equations that are of mixed type such as

$$y \frac{\partial^2 u}{\partial x^2} + \frac{\partial^2 u}{\partial y^2} = 0$$

which is elliptic for $y > 0$, hyperbolic for $y < 0$, and parabolic for $y = 0$.

We need appropriate initial or boundary conditions with Equations (15.2.25a)–(15.2.25c) to have a complete formulation for a meaningful problem. These conditions are related to the domain $D$ in which Equations (15.2.25a)–(15.2.25c) are to be solved. In general, the boundary $\alpha u(x, t) + \beta (\partial u/\partial n) (x, t) = \gamma(x, t)$, where $\partial/\partial n$ is a normal derivative to a boundary, must be specified.

Such specifications lead to a **well-posed problem**. We can also consider a well-posed problem to be a problem for which small perturbations in the auxiliary (initial or boundary conditions) conditions lead to small changes in the solution. Almost all reasonable problems are well-posed.

For the wave equation (15.2.25a), Figure 15.2.1 gives the region with appropriate conditions.

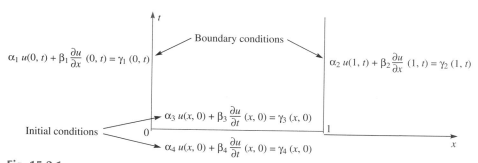

**Fig. 15.2.1**

Boundary conditions:

$$\alpha_1 u(0, t) + \beta_1 \frac{\partial u}{\partial x}(0, t) = \gamma_1(0, t) \qquad \alpha_2 u(1, t) + \beta_2 \frac{\partial u}{\partial x}(1, t) = \gamma_2(1, t) \quad \text{for } t > 0$$

Initial conditions:

$$\alpha_3 u(x, 0) + \beta_3 \frac{\partial u}{\partial t}(x, 0) = \gamma_3(x, 0) \quad 0 < x < 1$$

$$\alpha_4 u(x, 0) + \beta_4 \frac{\partial u}{\partial t}(x, 0) = \gamma_4(x, 0) \quad 0 < x < 1$$

Figure 15.2.2 shows the region with appropriate conditions for the heat equation (15.2.25b).

Boundary conditions:

$$\alpha_5 u(0, t) + \beta_5 \frac{\partial u}{\partial x}(0, t) = \gamma_5(0, t) \qquad \alpha_6 u(1, t) + \beta_6 \frac{\partial u}{\partial x}(1, t) = \gamma_6(1, t) \quad \text{for } t > 0$$

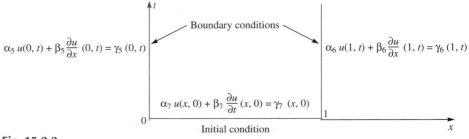

$$\alpha_5 u(0, t) + \beta_5 \frac{\partial u}{\partial x} (0, t) = \gamma_5 (0, t)$$

$$\alpha_6 u(1, t) + \beta_6 \frac{\partial u}{\partial x} (1, t) = \gamma_6 (1, t)$$

$$\alpha_7 u(x, 0) + \beta_7 \frac{\partial u}{\partial t} (x, 0) = \gamma_7 (x, 0)$$

**Fig. 15.2.2**

Initial condition:

$$\alpha_7 u(x, 0) + \beta_7 \frac{\partial u}{\partial t} (x, 0) = \gamma_7(x, 0) \quad 0 < x < 1$$

The region with appropriate conditions for the Poisson equation (15.2.25c) is shown in Figure 15.2.3.

Boundary condition:

$$\alpha_8 u(x, y) + \beta_8 \frac{\partial u}{\partial n} (x, y) = \gamma_8(x, y) \quad \text{over the curve } \partial D$$

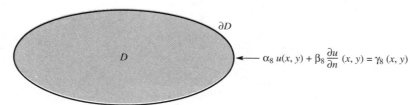

$$\alpha_8 u(x, y) + \beta_8 \frac{\partial u}{\partial n} (x, y) = \gamma_8 (x, y)$$

**Fig. 15.2.3**

## EXERCISES

For Exercises 1–6, determine the type of each of the following partial differential equations and reduce it to an appropriate canonical form:

**1.** $\dfrac{\partial^2 u}{\partial x^2} + 6 \dfrac{\partial^2 u}{\partial x \partial y} - 16 \dfrac{\partial^2 u}{\partial y^2} = 0$

**2.** $3 \dfrac{\partial^2 u}{\partial x^2} - 8 \dfrac{\partial^2 u}{\partial x \partial y} + 6 \dfrac{\partial^2 u}{\partial y^2} - u = 0$

**3.** $\dfrac{\partial^2 u}{\partial x^2} - 6 \dfrac{\partial^2 u}{\partial x \partial y} + 9 \dfrac{\partial^2 u}{\partial y^2} + \dfrac{\partial u}{\partial y} = xy$

**4.** $\dfrac{\partial^2 u}{\partial x^2} + 2 \dfrac{\partial^2 u}{\partial x \partial y} + 5 \dfrac{\partial^2 u}{\partial y^2} + \dfrac{\partial u}{\partial x} - 2 \dfrac{\partial u}{\partial y} - 3u = 0$

**5.** $\dfrac{\partial^2 u}{\partial x^2} + 4 \dfrac{\partial^2 u}{\partial x \partial y} + 5 \dfrac{\partial^2 u}{\partial y^2} - e^x u = \sin y$

**6.** $\dfrac{\partial^2 u}{\partial x^2} + (2x + 3) \dfrac{\partial^2 u}{\partial x \partial y} + 6x \dfrac{\partial^2 u}{\partial y^2} = 0$

**7.** Verify Equations (15.2.3a), (15.2.3b), (15.2.3c), (15.2.3d), and (15.2.3e).
**8.** Verify Equation (15.2.4).
**9.** Verify the identity of Equation (15.2.6).

## 15.3  PARABOLIC EQUATIONS

The typical parabolic partial differential equation has the form

$$\frac{\partial^2 u}{\partial x^2} - \frac{\partial u}{\partial t} = F(x, t) \tag{15.3.1}$$

in two independent variables $x$ and $t$. This equation, known as the one-dimensional heat equation, arises in the analysis of the conduction of heat in solids.

Consider a heat-conducting homogeneous rod of length one along the $x$-axis as shown in Figure 15.3.1. Assume that the rod has uniform cross section and that the rod is insulated laterally so that the heat flows only in the $x$ direction. Also, assume that the rod is sufficiently thin. This allows us to assume that the temperature at all points of the cross section is constant.

**Fig. 15.3.1**

Let $u(x, t)$ denote the temperature of the cross section at a point $x$ at time $t$. The temperature of the rod at time $t = 0$ is described by $u(x, 0) = f(x)$, $0 \le x \le 1$ and the two ends are maintained at constant temperatures at all times. Then the nondimensional temperature distribution $u$ in the rod for $t > 0$ is found by solving

$$\frac{\partial^2 u}{\partial x^2} = \frac{\partial u}{\partial t} \qquad 0 < x < 1 \quad 0 < t \le T \tag{15.3.2}$$

subject to

$$u(0, t) = c_0 \quad \text{and} \quad u(1, t) = c_1 \qquad 0 < t \le T \tag{15.3.3a}$$

and

$$u(x, 0) = f(x) \qquad 0 \le x \le 1 \tag{15.3.3b}$$

The auxiliary conditions given by Equations (15.3.3a) and (15.3.3b) are called the **boundary conditions** and **initial condition** respectively.

In order to approximate the solution of Equations (15.3.2), (15.3.3a), and (15.3.3b), a network is first established as shown in Figure 15.3.2, with grid spacing $\Delta x = h =$

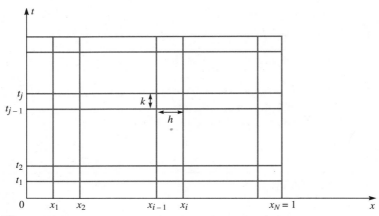

**Fig. 15.3.2**

$1/N$ and $\Delta t = k = T/M$ where $M$ and $N$ are positive integers. The lines $x = x_i$ and $t = t_j$ are called **grid lines** and their intersections are called the **mesh points** of the grid. At each interior point $(x_i, t_j)$ for $i = 1, 2, \ldots, N - 1$ and $j = 1, 2, \ldots, M$, Equation (15.3.2) is satisfied and we have

$$\frac{\partial^2 u}{\partial x^2}(x_i, t_j) = \frac{\partial u}{\partial t}(x_i, t_j) \tag{15.3.4}$$

## The Forward Difference Method

The simplest replacement of Equation (15.3.4) consists of approximating the space derivative by the centered second order differences from Table 5.3.1 given by

$$\frac{\partial^2 u}{\partial x^2}(x_i, t_j) = \frac{u(x_i - h, t_j) - 2u(x_i, t_j) + u(x_i + h, t_j)}{h^2} - \frac{h^2}{12}\frac{\partial^4 u}{\partial x^4}(\xi_i, t_j) \tag{15.3.5}$$

where $\xi_i \in (x_i - h, x_i + h)$ and the time derivative by the forward difference from Table 5.3.1 is given by

$$\frac{\partial u}{\partial t}(x_i, t_j) = \frac{u(x_i, t_j + k) - u(x_i, t_j)}{k} - \frac{k}{2}\frac{\partial^2 u}{\partial t^2}(x_i, \eta_j) \tag{15.3.6}$$

where $\eta_j \in (t_j, t_j + k)$. Substituting Equations (15.3.5) and (15.3.6) in Equation (15.3.4), we get at each interior grid point

$$\begin{aligned}
&\frac{u(x_i - h, t_j) - 2u(x_i, t_j) + u(x_i + h, t_j)}{h^2} - \frac{h^2}{12}\frac{\partial^4 u}{\partial x^4}(\xi_i, t_j) \\
&= \frac{u(x_i, t_j + k) - u(x_i, t_j)}{k} - \frac{k}{2}\frac{\partial^2 u}{\partial t^2}(x_i, \eta_j)
\end{aligned} \tag{15.3.7}$$

Assuming that $h$ and $k$ are sufficiently small allows us to ignore the last term on the left side and right side of Equation (15.3.7). Denoting the approximate value of $u$ at $(x_i, t_j)$ by $U_{i,j}$ (i.e., $U_{i,j} \approx u(x_i, t_j)$), the approximate value of $u$ at $(x_i, t_j + k)$ by $U_{i,j+1}$

(i.e., $U_{i,j+1} \approx u(x_i, t_j + k)$), and so forth, Equation (15.3.7) reduces to

$$\frac{U_{i-1,j} - 2U_{i,j} + U_{i+1,j}}{h^2} = \frac{U_{i,j+1} - U_{i,j}}{k} \qquad \textbf{(15.3.8)}$$

Solving Equation (15.3.8) for $U_{i,j+1}$, we get

$$U_{i,j+1} = r\,U_{i-1,j} + (1 - 2r)\,U_{i,j} + r\,U_{i+1,j} \qquad \textbf{(15.3.9)}$$

for $i = 1, 2, \ldots, N - 1$ and $j = 0, 1, \ldots, M - 1$ where

$$r = \frac{k}{h^2} = \frac{\Delta t}{(\Delta x)^2} \qquad \textbf{(15.3.10)}$$

The initial condition $u(x, 0) = f(x)$ reduces to

$$U_{i,0} = f(x_i) \quad \text{for } i = 0, 1, \ldots, N \qquad \textbf{(15.3.11a)}$$

The boundary conditions $u(0, t) = c_0$ and $u(1, t) = c_1$ reduce to

$$U_{0,j} = c_0 \quad \text{and} \quad U_{N,j} = c_1 \quad \text{for } j = 1, 2, \ldots, M \qquad \textbf{(15.3.11b)}$$

Since $U_{0,0}, U_{1,0}, \ldots, U_{N,0}$ are known from Equation (15.3.11a) and $U_{0,j}$ and $U_{N,j}$ are known from Equation (15.3.11b), we can compute $U_{1,1}, U_{2,1}, \ldots, U_{N-1,1}$ from Equation (15.3.9) and then $U_{1,2}, U_{2,2}$, and so on. This method is also known as an **explicit method** since $U_{i,j+1}$ is given explicitly in Equation (15.3.9). It has a local truncation error

$$-\frac{k}{2}\frac{\partial^2 u}{\partial t^2}(x_i, \eta_j) + \frac{h^2}{12}\frac{\partial^4 u}{\partial x^4}(\xi_i, t_j) \qquad \textbf{(15.3.12)}$$

The net points in the time and space differences used in Equation (15.3.9) are shown in Figure 15.3.3.

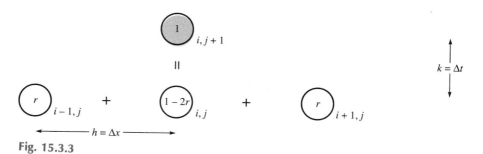

**Fig. 15.3.3**

**Algorithm**

**Subroutine** Forward($a$, $b$, $N$, $f$, $c_0$, $c_1$, **z**, $M$, $T$)
Comment: This subroutine approximates the solution of

$$\frac{\partial^2 u}{\partial x^2} = \frac{\partial u}{\partial t} \qquad u(a, t) = c_0 \quad u(b, t) = c_1 \qquad u(x, 0) = f(x) \qquad a \le x \le b$$

using the forward difference formula (15.3.9). We use $z_i = U_{i,j}$ and $v_i = U_{i,j+1}$. The output is the values of $v_i$ at time $T$. We have $\Delta x = h = (b - a)/N$ and $\Delta t = k = ak = T/M$. We have $\mathbf{z} = [z_0, z_1, \ldots, z_N]$ and $\mathbf{v} = [v_0, v_1, \ldots, v_N]$.

1 **Print** $a$, $b$, $N$, $c_0$, $c_1$, $M$, $T$
2 $h = (b - a)/N$
3 $ak = T/M$
4 $r = ak/h^2$
5 time $= 0$
6 **Repeat** steps 7–9 **with** $i = 0, 1, \ldots, N$
    7 $ai = i$
    8 $x = a + (ai)h$
    9 $z_i = f(x)$
10 **Print** time and **z**
11 $z_0 = c_0$
12 $z_N = c_1$
13 time $=$ time $+ ak$
14 **Repeat** step 15 **with** $i = 1, 2, \ldots, N - 1$
    15 $v_i = rz_{i-1} + (1 - 2r) z_i + rz_{i+1}$
16 **Repeat** step 17 **with** $i = 1, 2, \ldots, N - 1$
    17 $z_i = v_i$
18 **If** time $< T$, **then** return to 13
19 **Print** time and **v** and exit.   ◀

## The Richardson Method

We can replace the time derivative $\partial u/\partial t$ $(x_i, t_j)$ in Equation (15.3.4) by the central difference from Table 5.3.1 given by

$$\frac{\partial u}{\partial t}(x_i, t_j) = \frac{u(x_i, t_j + k) - u(x_i, t_j - k)}{2k} - \frac{k^2}{6}\frac{\partial^3 u}{\partial t^3}(x_i, \eta_j) \tag{15.3.13}$$

where $\eta_j \in (t_j - k, t_j + k)$. Substituting Equations (15.3.13) and (15.3.5) in Equation (15.3.4), we get

$$\frac{u(x_i - h, t_j) - 2u(x_i, t_j) + u(x_i + h, t_j)}{h^2} - \frac{h^2}{12}\frac{\partial^4 u}{\partial x^4}(\xi_i, t_j)$$
$$= \frac{u(x_i, t_j + k) - u(x_i, t_j - k)}{2k} - \frac{k^2}{6}\frac{\partial^3 u}{\partial t^3}(x_i, \eta_j) \tag{15.3.14}$$

Ignoring the terms in $k^2$ and $h^2$ in Equation (15.3.14) and denoting the approximate value of $u$ at $(x_i, t_j)$ by $U_{i,j}$ and so forth, Equation (15.3.14) reduces to

$$\frac{U_{i-1,j} - 2U_{i,j} + U_{i+1,j}}{h^2} = \frac{U_{i,j+1} - U_{i,j-1}}{2k} \tag{15.3.15}$$

Solving Equation (15.3.15) for $U_{i,j+1}$, we get

$$U_{i,j+1} = U_{i,j-1} + 2r(U_{i-1,j} - 2U_{i,j} + U_{i+1,j}) \tag{15.3.16}$$

for $i = 1, 2, \ldots, N - 1$ and $j = 1, 2, \ldots, M - 1$ where $r = k/h^2 = \Delta t/(\Delta x)^2$. This method is known as the **Richardson method** and has a local truncation error

$$-\frac{k^2}{6}\frac{\partial^3 u}{\partial t^3}(x_i, \eta_j) + \frac{h^2}{12}\frac{\partial^4 u}{\partial x^4}(\xi_i, t_j) \tag{15.3.17}$$

This explicit method is more accurate than the forward difference method because of higher-order truncation error; however, it has a serious stability problem as we shall see later.

In Figure 15.3.4, the net points in the time and space differences used in Equation (15.3.16) are shown.

From Figure 15.3.4 and Equation (15.3.16), the Richardson method has three levels of time. In order to start up, we need to provide $U_{i,0}$ and $U_{i,1}$ for $i = 1, 2, \ldots, N - 1$. The initial condition of Equation (15.3.11a) and boundary conditions of Equation (15.3.11b) provide $U_{i,0}$ for $i = 0, 1, \ldots, N$, $U_{0,1}$, and $U_{N,1}$ respectively. The values $U_{1,1}$, $U_{2,1}, \ldots, U_{N-1,1}$ can be obtained by another method (e.g. the forward difference method).

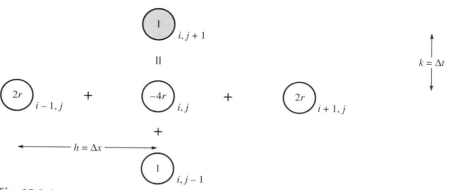

**Fig. 15.3.4**

## The Backward Difference Method

The time derivative $\partial u / \partial t\,(x_i, t_j)$ is also given in terms of the backward difference from Table 5.3.1 by

$$\frac{\partial u}{\partial t}(x_i, t_j) = \frac{u(x_i, t_j) - u(x_i, t_j - k)}{k} + \frac{k}{2}\frac{\partial^2 u}{\partial t^2}(x_i, \eta_j) \qquad \textbf{(15.3.18)}$$

where $\eta_j \in (t_j - k, t_j)$. Substituting Equations (15.3.18) and (15.3.5) in Equation (15.3.4), we get

$$\frac{u(x_i - h, t_j) - 2u(x_i, t_j) + u(x_i + h, t_j)}{h^2} - \frac{h^2}{12}\frac{\partial^4 u}{\partial x^4}(\xi_i, t_j)$$
$$= \frac{u(x_i, t_j) - u(x_i, t_j - k)}{k} + \frac{k}{2}\frac{\partial^2 u}{\partial t^2}(x_i, \eta_j) \qquad \textbf{(15.3.19)}$$

Ignoring the terms in $k$ and $h^2$ in Equation (15.3.19) and denoting the approximate value of $u$ at $(x_i, t_j)$ by $U_{i,j}$ and so forth, Equation (15.3.19) reduces to

$$\frac{U_{i-1,j} - 2U_{i,j} + U_{i+1,j}}{h^2} = \frac{U_{i,j} - U_{i,j-1}}{k} \qquad \textbf{(15.3.20)}$$

This equation can be written as

$$rU_{i-1,j} - (1 + 2r) U_{i,j} + r U_{i+1,j} = - U_{i,j-1} \qquad \textbf{(15.3.21)}$$

for $i = 1, 2, \dots, N - 1$ and $j = 1, 2, \dots, M$ where $r = k/h^2 = \Delta t/(\Delta x)^2$. This method is known as the **backward difference method** with local truncation error

$$\frac{k}{2} \frac{\partial^2 u}{\partial t^2} (x_i, \eta_j) + \frac{h^2}{12} \frac{\partial^4 u}{\partial x^4} (\xi_i, t_j) \qquad \textbf{(15.3.22)}$$

Since $U_{0,0}, U_{1,0}, \dots, U_{N,0}$ and $U_{0,j}$ and $U_{N,j}$ are known from Equations (15.3.11a) and (15.3.11b) respectively, we proceed to find $U_{1,1}, U_{2,1}, \dots, U_{N-1,1}$ from Equation (15.3.21) by solving the system of equations; then $U_{1,2}, U_{2,2}$, and so forth are found. The mesh points involved at a typical step are shown in Figure 15.3.5.

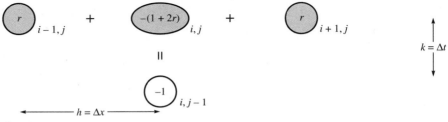

**Fig. 15.3.5**

In order to find the values of $U$ at time level $t_j$, we have to solve the system of equations. Since all unknowns are given implicitly, this method is known as an **implicit method**. There is computationally an important difference between an explicit and implicit method.

We have so far constructed three methods to approximate the values of $u(x, t)$. Before we construct any new method, we must consider the important question of whether the computed values actually represent a good approximation of the solution of Equations (15.3.2), (15.3.3a), and (15.3.3b).

**EXAMPLE 15.3.1**     Compute an approximate solution of

$$\frac{\partial^2 u}{\partial x^2} = \frac{\partial u}{\partial t} \quad 0 < x < 1 \quad t > 0$$

with the boundary conditions $u(0, t) = u(1, t) = 0, t > 0$ and initial condition $u(x, 0) = \sin \pi x, 0 \le x \le 1$ using (1) the forward difference method, (2) the Richardson method, and (3) the backward difference method. We can verify that the exact solution is given by $u(x, t) = e^{-\pi^2 t} \sin \pi x$.

Let $h = \frac{1}{4}$ and $r = \frac{1}{2}$. Then $k = rh^2 = \frac{1}{2} (\frac{1}{4})^2 = 1/32$. The nodal points are shown in Figure 15.3.6.

1.   Since the initial condition is given by $u(x_i, 0) = \sin \pi x_i$, $U_{i,0} = \sin \pi x_i$. Hence $U_{0,0} = 0$, $U_{1,0} = \sin \pi x_1 = \sin (\pi/4) = 1/\sqrt{2}$, $U_{2,0} = \sin \pi x_2 = \sin (\pi/2) = 1$, $U_{3,0} = \sin \pi x_3 = \sin (3\pi/4) = 1/\sqrt{2}$, and $U_{4,0} = \sin \pi x_4 = \sin \pi = 0$. The

**Fig. 15.3.6**

forward difference formula [Equation (15.3.9)] becomes

$$U_{i,j+1} = \frac{1}{2}(U_{i-1,j} + U_{i+1,j}) \tag{15.3.23}$$

For $j = 0$, we get from Equation (15.3.23), $U_{i,1} = \frac{1}{2}(U_{i-1,0} + U_{i+1,0})$. Hence for $i = 1, 2, 3$, we get

$$U_{1,1} = \frac{1}{2}(U_{0,0} + U_{2,0}) = \frac{1}{2}(0 + 1) = \frac{1}{2}$$

$$U_{2,1} = \frac{1}{2}(U_{1,0} + U_{3,0}) = \frac{1}{2}\left(\frac{1}{\sqrt{2}} + \frac{1}{\sqrt{2}}\right) = \frac{1}{\sqrt{2}} = 0.70711$$

$$U_{3,1} = \frac{1}{2}(U_{2,0} + U_{4,0}) = \frac{1}{2}(1 + 0) = \frac{1}{2}$$

Substituting $j = 1$ in Equation (15.3.23), we get $U_{i,2} = \frac{1}{2}(U_{i-1,1} + U_{i+1,1})$. Hence for $i = 1, 2, 3$, we get

$$U_{1,2} = \frac{1}{2}(U_{0,1} + U_{2,1}) = \frac{1}{2}(0 + 0.70711) = 0.35355$$

$$U_{2,2} = \frac{1}{2}(U_{1,1} + U_{3,1}) = \frac{1}{2}\left(\frac{1}{2} + \frac{1}{2}\right) = \frac{1}{2} = 0.5$$

$$U_{3,2} = \frac{1}{2}(U_{2,1} + U_{4,1}) = \frac{1}{2}(0.70711 + 0.) = 0.35355$$

The computed values of $U_{i,j}$ at $t = 0.5$ are given in Table 15.3.1.

2. Substituting $r = \frac{1}{2}$ in Equation (15.3.16), we get

$$U_{i,j+1} = U_{i,j-1} + (U_{i-1,j} - 2U_{i,j} + U_{i+1,j}) \tag{15.3.24}$$

Three levels of time are involved in Equation (15.3.24). For $j = 0$, we get $U_{i,1} = U_{i,-1} + (U_{i-1,0} - 2U_{i,0} + U_{i+1,0})$. Since we do not know $U_{i,-1}$, we cannot use

**Table 15.3.1**

| $x_i$ | Exact Solution | Forward Difference Method | Richardson Method | Backward Difference Method |
|---|---|---|---|---|
| 0.00 | 0.00000 | 0.00000 | 0.00000 | 0.00000 |
| 0.25 | 0.00509 | 0.00276 | $0.23036 \times 10^9$ | 0.01160 |
| 0.50 | 0.00719 | 0.00391 | $-0.32578 \times 10^9$ | 0.01641 |
| 0.75 | 0.00509 | 0.00276 | $0.23036 \times 10^9$ | 0.01160 |
| 1.00 | 0.00000 | 0.00000 | 0.00000 | 0.00000 |

Equation (15.3.24) for $j = 0$. We use the computed values $U_{i,1}$ by the forward difference method.

For $j = 1$, Equation (15.3.24) reduces to

$$U_{i,2} = U_{i,0} + (U_{i-1,1} - 2U_{i,1} + U_{i+1,1})$$

For $i = 1, 2, 3$, we get

$$U_{1,2} = U_{1,0} + (U_{0,1} - 2U_{1,1} + U_{2,1})$$

$$= \frac{1}{\sqrt{2}} + \left(0 - 1 + \frac{1}{\sqrt{2}}\right) = \frac{2}{\sqrt{2}} - 1 = 0.41421$$

$$U_{2,2} = U_{2,0} + (U_{1,1} - 2U_{2,1} + U_{3,1})$$

$$= 1 + \left(\frac{1}{2} - 2 \cdot \frac{1}{\sqrt{2}} + \frac{1}{2}\right) = 2 - \frac{2}{\sqrt{2}} = 0.58579$$

$$U_{3,2} = U_{3,0} + (U_{2,1} - 2U_{3,1} + U_{4,1}) = \frac{1}{\sqrt{2}} + \left(\frac{1}{\sqrt{2}} - 2 \cdot \frac{1}{2} + 0\right)$$

$$= \frac{2}{\sqrt{2}} - 1 = 0.41421$$

We continue to compute $U_{i,j}$. In Table 15.3.1, the computed values at $t = 0.5$ are given.

3.   Substituting $r = \frac{1}{2}$ in Equation (15.3.21), we get

$$\frac{1}{2} U_{i-1,j} - 2U_{i,j} + \frac{1}{2} U_{i+1,j} = -U_{i,j-1} \qquad \textbf{(15.3.25)}$$

For $j = 1$ and $i = 1, 2$, and 3, we get from Equation (15.3.25)

$$\frac{1}{2} U_{0,1} - 2U_{1,1} + \frac{1}{2} U_{2,1} = -U_{1,0}$$

$$\frac{1}{2} U_{1,1} - 2U_{2,1} + \frac{1}{2} U_{3,1} = -U_{2,0}$$

$$\frac{1}{2} U_{2,1} - 2U_{3,1} + \frac{1}{2} U_{4,1} = -U_{3,0}$$

This is written as

$$\begin{bmatrix} -2 & 1/2 & 0 \\ 1/2 & -2 & 1/2 \\ 0 & 1/2 & -2 \end{bmatrix} \begin{bmatrix} U_{1,1} \\ U_{2,1} \\ U_{3,1} \end{bmatrix} = - \begin{bmatrix} U_{1,0} + (U_{0,1}/2) \\ U_{2,0} \\ U_{3,0} + (U_{4,1}/2) \end{bmatrix} = - \begin{bmatrix} 1/\sqrt{2} \\ 1 \\ 1/\sqrt{2} \end{bmatrix}$$

Solving this linear system, we get

$$U_{1,1} = 0.54692 \qquad U_{2,1} = 0.77346 \qquad U_{3,1} = 0.54692$$

For $j = 2$ and $i = 1, 2$, and 3, Equation (15.3.25) reduces to

$$\begin{bmatrix} -2 & 1/2 & 0 \\ 1/2 & -2 & 1/2 \\ 0 & 1/2 & -2 \end{bmatrix} \begin{bmatrix} U_{1,2} \\ U_{2,2} \\ U_{3,2} \end{bmatrix} = - \begin{bmatrix} U_{1,1} + (U_{0,2}/2) \\ U_{2,1} \\ U_{3,1} + (U_{4,2}/2) \end{bmatrix} = - \begin{bmatrix} 0.54692 \\ 0.77346 \\ 0.54692 \end{bmatrix}$$

Solving these equations yields

$$U_{1,2} = 0.42302 \qquad U_{2,2} = 0.59824 \qquad U_{3,2} = 0.42302$$

The computed values at $t = 0.5$ are given in Table 15.3.1.

We expect the truncation error of order $k + h^2$, $k^2 + h^2$ and $k + h^2$ for the forward difference method, Richardson method, and backward difference method respectively. Since $h^2 = 0.06250$, the results for the forward difference and backward difference are not accurate. The Richardson method is supposed to give more accurate results than the other two methods, but it gives absurd results.

Let us select $k = 0.1$. Then $r = k/h^2 = 1.6$. The computed values of $U_{i,j}$ at $t = 0.5$ are given in Table 15.3.2.

From Table 15.3.2, we can again see that the Richardson method is giving results far from the exact results. The forward difference method is also getting inaccurate.

**Table 15.3.2**

| $x_i$ | Exact Solution | Forward Difference Method | Richardson Method | Backward Difference Method |
|-------|---------------|--------------------------|-------------------|----------------------------|
| 0.00 | 0.00000 | 0.00000 | 0.00000 | 0.00000 |
| 0.25 | 0.00509 | 0.00000 | $-0.42906 \times 10^5$ | 0.02591 |
| 0.50 | 0.00719 | 0.00000 | $0.60528 \times 10^5$ | 0.03665 |
| 0.75 | 0.00509 | 0.00000 | $-0.42906 \times 10^5$ | 0.02591 |
| 1.00 | 0.00000 | 0.00000 | 0.00000 | 0.00000 |

■ ■ ■

We examine the convergence of these methods in the next section.

## EXERCISES

1. Compute an approximate solution of

$$\frac{\partial^2 u}{\partial x^2} = \frac{\partial u}{\partial t} \quad 0 < x < 1 \quad t > 0$$

$$u(0, t) = u(1, t) = 0 \quad t > 0$$

$$u(x, 0) = \frac{1}{2} \sin \pi x \quad 0 \le x \le 1$$

using the forward difference method. Use $h = \frac{1}{3}$, $r = 1$, and $M = 3$ and $h = \frac{1}{3}$, $r = \frac{1}{2}$, and $M = 3$.

2. Compute an approximate solution of

$$\frac{\partial^2 u}{\partial x^2} = \frac{\partial u}{\partial t} \quad 0 < x < 1 \quad t > 0$$

$$u(0, t) = u(1, t) = 0 \quad t > 0$$

$$u(x, 0) = \frac{1}{2} \sin \pi x \quad 0 \le x \le 1$$

using the Richardson method. Use $h = \frac{1}{3}$, $r = 1$, and $M = 4$ and $h = \frac{1}{3}$, $r = \frac{1}{2}$, and $M = 4$. (*Hint:* use $U_{i,1}$ from Exercise 1 or use $U_{i,1}$ found by using any other method.)

3.  Compute an approximate solution of

$$\frac{\partial^2 u}{\partial x^2} = \frac{\partial u}{\partial t} \quad 0 < x < 1 \quad t > 0$$

$$u(0, t) = u(1, t) = 0 \quad t > 0$$

$$u(x, 0) = \frac{1}{2} \sin \pi x \quad 0 \le x \le 1$$

using the backward difference method. Use $h = \frac{1}{3}$, $r = 1$, and $M = 3$ and $h = \frac{1}{3}$, $r = \frac{1}{2}$, and $M = 3$.

4.  Compute an approximate solution of

$$\frac{\partial^2 u}{\partial x^2} = \frac{\partial u}{\partial t} - 1 \quad 0 < x < 1 \quad t > 0$$

$$u(0, t) = u(1, t) = 0 \quad t > 0$$

$$u(x, 0) = \sin \pi x + \frac{1}{2} x(x - 1) \quad 0 \le x \le 1$$

using the forward difference method. Use $h = \frac{1}{3}$, $r = 1$, and $M = 3$.

5.  Compute an approximate solution of

$$(1 - t)\frac{\partial^2 u}{\partial x^2} = \frac{\partial u}{\partial t} \quad 0 < x < 1 \quad t > 0$$

$$\frac{\partial u}{\partial x}(0, t) = \frac{1}{2} \quad u(1, t) = 1 \quad t > 0$$

$$u(x, 0) = 1 \quad 0 \le x \le 1$$

using the method of your choice. Use $h = \frac{1}{3}$, $r = \frac{1}{2}$, and $M = 3$.

---

## COMPUTER EXERCISES

**C.1.** Write a subroutine to generate difference equations given by Equation (15.3.9) for the use of the forward difference method.

**C.2.** Write a subroutine to generate difference equations given by Equation (15.3.16) for the use of the Richardson method.

**C.3.** Write a computer program using Exercises C.1 and C.2 to approximate a solution of

$$\frac{\partial^2 u}{\partial x^2} = \frac{\partial u}{\partial t} \quad a < x < b \quad t > 0$$

$$u(a, t) = c_1 \quad u(b, t) = c_2 \quad t > 0$$

$$u(x, 0) = f(x) \quad a \le x \le b$$

For Exercises C.4–C.6, use your computer program of Exercise C.3 to compute the solution of the parabolic partial differential equations. Use $h = 0.1, 0.01$, and $0.001$ and $r = 0.1, 0.5$, and $2.0$. Compare your solutions to the exact solutions at $t = 0.5$ and $1.0$.

**C.4.**

$$\frac{\partial^2 u}{\partial x^2} = \frac{\partial u}{\partial t} \quad 0 < x < 1 \quad t > 0$$

$$u(0, t) = u(1, t) = 0 \quad t > 0$$

$$u(x, 0) = \frac{1}{2} \sin \pi x \quad 0 \le x \le 1$$

The exact solution is given by $u(x, t) = (e^{-\pi^2 t} \sin \pi x)/2$.

**C.5.**

$$\frac{\partial^2 u}{\partial x^2} = \frac{\partial u}{\partial t} \quad -1 < x < 1 \quad t > 0$$

$$u(-1, t) = u(1, t) = 0 \quad t > 0$$

$$u(x, 0) = \cos \frac{\pi x}{2} \quad -1 \le x \le 1$$

The exact solution is given by $u(x, t) = e^{-(\pi^2/4)t} \cos(\pi x/2)$.

**C.6.**

$$\frac{\partial^2 u}{\partial x^2} = \frac{\partial u}{\partial t} \quad 0 < x < \pi \quad t > 0$$

$$u(0, t) = u(\pi, t) \quad t > 0$$

$$u(x, 0) = \cos\left(x - \frac{\pi}{2}\right) \quad 0 \le x \le \pi$$

The exact solution is given by $u(x, t) = e^{-t} \cos(x - (\pi/2))$.

**C.7.** Write a subroutine to solve Equation (15.3.21) by *LU* decomposition using Equations (7.7.4), (7.7.6) and (7.7.7).

**C.8.** Write a computer program using Exercise C.7 to approximate a solution of

$$\frac{\partial^2 u}{\partial x^2} = \frac{\partial u}{\partial t} \quad a < x < b \quad t > 0$$

$$u(a, t) = c_1 \quad u(b, t) = c_2 \quad t > 0$$
$$u(x, 0) = f(x) \quad a \le x \le b$$

using the backward difference method.

**C.9.** Using your computer program of Exercise C.8, compute the solution of the parabolic differential equation of Exercise C.6. Use $h = 0.1, 0.01$, and $0.001$ and $r = 0.1, 0.5$, and $2.0$. Compare your solutions to the exact solutions at $t = 0.5$ and $1.0$.

---

## 15.4    CONVERGENCE, STABILITY, AND CONSISTENCY OF DIFFERENCE SCHEMES

Let us denote the exact solution of Equations (15.3.2), (15.3.3a), and (15.3.3b) by $u(x, t)$ and the exact finite difference approximation of Equations (15.3.2), (15.3.3a), and (15.3.3b) by $U(x, t)$. Further, let the departure of the exact finite difference approximation from the exact solution be

$$w(x, t) = u(x, t) - U(x, t) \tag{15.4.1}$$

Then the finite difference method is said to be **convergent** if $w \to 0$ as $h, k \to 0$ at the fixed point $(x, t)$ everywhere in the given region. The convergence of the method assures us that the finite difference approximation can be made into an arbitrarily accurate approximation of the exact solution by proper selection of $h$ and $k$.

Consider the forward difference method. Solving Equation (15.3.7) for $u(x_i, t_j + k)$, we get

$$u(x_i, t_j + k) = ru(x_i - h, t_j) + (1 - 2r)\, u(x_i, t_j) + ru(x_i + h, t_j)$$
$$+ \frac{k^2}{2} \frac{\partial^2 u}{\partial t^2}(x_i, \eta_j) - \frac{kh^2}{12} \frac{\partial^4 u}{\partial t^4}(\xi_i, t_j) \qquad \textbf{(15.4.2)}$$

where $r = k/h^2$. Subtracting Equation (15.3.9) from Equation (15.4.2) and using the notations

$$w_{i,j+1} = w(x_i, t_j + k) = u(x_i, t_j + k) - U(x_i, t_j + k) = u(x_i, t_j + k) - U_{i,j+1}$$
$$w_{i,j} = w(x_i, t_j) = u(x_i, t_j) - U(x_i, t_j) = u(x_i, t_j) - U_{i,j}$$
$$w_{i+1,j} = w(x_i + h, t_j) = u(x_i + h, t_j) - U(x_i + h, t_j) = u(x_i + h, t_j) - U_{i+1,j}$$

we get

$$w_{i,j+1} = rw_{i-1,j} + (1 - 2r)\, w_{i,j} + rw_{i+1,j} + \frac{k^2}{2} \frac{\partial^2 u}{\partial t^2}(x_i, \eta_j) - \frac{kh^2}{12} \frac{\partial^4 u}{\partial t^4}(\xi_i, t_j) \qquad \textbf{(15.4.3)}$$

If $0 < r \le \frac{1}{2}$, then the coefficients $r$ and $1 - 2r$ of Equation (15.4.3) are positive and

$$|w_{i,j+1}| \le r\, |w_{i-1,j}| + (1 - 2r)\, |w_{i,j}| + r\, |w_{i+1,j}| + |z_{i,j}| \qquad \textbf{(15.4.4)}$$

where

$$z_{i,j} = \frac{k^2}{2} \frac{\partial^2 u}{\partial t^2}(x_i, \eta_j) - \frac{kh^2}{12} \frac{\partial^4 u}{\partial t^4}(\xi_i, t_j)$$

Let $\|W_j\|_\infty = \max_i |w_{i,j}|$ and $\|Z\|_\infty = \max_{i,j} |z_{i,j}|$. Then from Equation (15.4.4), we get

$$\|W_{j+1}\|_\infty \le \|W_j\|_\infty + \|Z\|_\infty \qquad \textbf{(15.4.5)}$$

Since $\|W_0\|_\infty = 0$, we get $\|W_1\|_\infty \le \|Z\|_\infty$, $\|W_2\|_\infty \le \|W_1\|_\infty + \|Z\|_\infty \le 2\, \|Z\|_\infty$. Continuing

$$\|W_j\|_\infty \le j\, \|Z\|_\infty = jk \frac{\|Z\|_\infty}{k}$$

$$\le T \left\| \frac{k}{2} \frac{\partial^2 u}{\partial t^2}(x_i, \eta_j) - \frac{h^2}{12} \frac{\partial^4 u}{\partial t^4}(\xi_i, t_j) \right\|_\infty \qquad \textbf{(15.4.6)}$$

where $jk = j\, \Delta t = T$. Thus from Equation (15.4.6), $\|W_j\|_\infty \to 0$ as $h, k \to 0$. Hence $w_{i,j} \to 0$ and therefore $|u(x_i, t_j) - U_{i,j}| \to 0$ as $\Delta x$ and $\Delta t \to 0$ if $0 < r = (k/h^2) \le \frac{1}{2}$.

We could analyze the other finite difference methods the same way, but this is a tedious and long procedure. Since there is a connection between stability and convergence, the results obtained through an analysis of convergence correspond to those obtained for stability. Therefore, let us discuss the stability of finite difference equations.

We can rarely obtain an exact solution to a system of difference equations. In numerical solutions, round-off errors are introduced. At the grid point $(x_2, t_0)$, we use $U(x_2, t_0) + \epsilon$ rather than $U(x_2, t_0)$. Thus at the grid $(x_2, t_0)$, we introduce a round-off error $\epsilon$. If the solution procedure is continued exactly with $U_{2,0} + \epsilon$ [in other words, no new errors are introduced (merely a mathematical assumption)], then $\bar{U}(x, t)$ is obtained as an exact solution. Then $\bar{U}(x, t) - U(x, t)$ denotes the departure of the solution due to an error $\epsilon$ at $(x_2, t_0)$. When errors are introduced at more than one point, cumulative departures result. Since the heat equation is linear, the effect of the succession of errors is just the sum of the effects of the individual errors. We follow just one error and if this single error does not grow in magnitude, we call the method **stable**. The cumulative effect of all errors will not grow because of linearity.

As an illustration, consider the forward difference method for the problem

$$\frac{\partial^2 u}{\partial x^2} = \frac{\partial u}{\partial t} \quad 0 < x < 1 \quad t > 0 \qquad u(0, t) = u(1, t) = 0 \quad t > 0$$
$$u(x, 0) = 0 \quad 0 \le x \le 1 \tag{15.4.7}$$

The exact solution is $u(x, t) = 0$.

Let $h = \frac{1}{4} = \Delta x$ and $r = 1$. Then $k = 1/16$ and Equation (15.3.9) becomes

$$U_{i,j+1} = U_{i-1,j} - U_{i,j} + U_{i+1,j} \tag{15.4.8}$$

From Equation (15.4.7), we have $U_{0,0} = U_{1,0} = U_{2,0} = U_{3,0} = U_{4,0} = 0$. Also $U_{0,j} = U_{4,j} = 0$ for $j = 1, 2, \ldots, M$.

Consider an error $\epsilon$ at $(x_2, 0)$. Then $U_{2,0} = \epsilon$. For $j = 0$, Equation (15.4.8) becomes $U_{i,1} = U_{i-1,0} - U_{i,0} + U_{i+1,0}$. For $i = 1$, $U_{1,1} = \epsilon$; $i = 2$, $U_{2,1} = -\epsilon$; $i = 3$, $U_{3,1} = \epsilon$. Similarly compute $U_{1,2}, U_{2,2}$, and so forth. The computed values are shown in parentheses in Table 15.4.1 along with the exact value 0.

**Table 15.4.1**

| $i/j$ | (END) $x_0$ | $x_1$ | $x_2$ | $x_3$ | $x_4$ (END) |
|---|---|---|---|---|---|
| 0 | 0 | 0 | $0(\epsilon)$ | 0 | 0 |
| 1/16 | 0 | $0(\epsilon)$ | $0(-\epsilon)$ | $0(\epsilon)$ | 0 |
| 2/16 | 0 | $0(-2\epsilon)$ | $0(3\epsilon)$ | $0(-2\epsilon)$ | 0 |
| 3/16 | 0 | $0(5\epsilon)$ | $0(-7\epsilon)$ | $0(5\epsilon)$ | 0 |
| 4/16 | 0 | $0(-12\epsilon)$ | $0(17\epsilon)$ | $0(-12\epsilon)$ | 0 |
| 5/16 | 0 | $0(29\epsilon)$ | $0(-41\epsilon)$ | $0(29\epsilon)$ | 0 |
| 6/16 | 0 | $0(-70\epsilon)$ | $0(99\epsilon)$ | $0(-70\epsilon)$ | 0 |

Errors are growing and will continue to grow. The method is unstable, because an error $\epsilon$ at $x = 0.5$ and $t = 0$ has been magnified by a factor of 99 after six time steps. Since $\epsilon$ depends on the number of decimals carried in calculations, there may be a delay before we see the error build-up. Sooner or later, the error build-up will be observed.

Consider $h = \frac{1}{4}$ and $r = \frac{1}{2}$. Then $k = 1/32$ and Equation (15.3.9) becomes

$$U_{i,j+1} = \frac{1}{2}(U_{i-1,j} + U_{i+1,j}) \tag{15.4.9}$$

Introduce an error $\epsilon$ at $(x_2, 0)$. Using $U_{0,0} = U_{1,0} = U_{3,0} = U_{4,0} = 0$, $U_{2,0} = \epsilon$, and $U_{0,j} = U_{4,0} = 0$ for $j = 1, 2, \ldots, M$ in Equation (15.4.9), we compute $U_{i,j}$. The computed values are shown in parentheses in Table 15.4.2 along with the exact value 0.

**Table 15.4.2**

| $i/j$ | (END) $x_0$ | $x_1$ | $x_2$ | $x_3$ | $x_4$ (END) |
|-------|-------------|-------|-------|-------|-------------|
| 0 | 0 | 0 | $0(\epsilon)$ | 0 | 0 |
| 1/32 | 0 | $0(\epsilon/2)$ | 0 | $0(\epsilon/2)$ | 0 |
| 2/32 | 0 | 0 | $0(\epsilon/2)$ | 0 | 0 |
| 3/32 | 0 | $0(\epsilon/4)$ | 0 | $0(\epsilon/4)$ | 0 |
| 4/32 | 0 | 0 | $0(\epsilon/4)$ | 0 | 0 |
| 5/32 | 0 | $0(\epsilon/8)$ | 0 | $0(\epsilon/8)$ | 0 |
| 6/32 | 0 | 0 | $0(\epsilon/8)$ | 0 | 0 |

The quantities in parentheses show the effect of an error $\epsilon$ at $(x_2, 0)$; it has decreased by $\frac{1}{8}$ after six time steps. Thus the effect of an error $\epsilon$ at $(x_2, 0)$ decreases as $t$ increases.

If the method is stable, then, in principle, the computational errors can be made arbitrarily small. Thus stability is the most pressing problem for any method. If we are assured that a meaningful computation can be carried out, then we think about accuracy and computational efficiency. For the forward difference method, for $0 < r \leq \frac{1}{2}$, the method is convergent and from our example it seems to be stable. In fact, the formal statement of the relationship between stability and convergence is known as the Lax Equivalence Theorem. Before we state it, we need to understand consistency.

The term **consistency** means that the difference equation approximates the required partial differential equation and not some other partial differential equation. Alternatively, we can say that the difference equation is compatible with the required differential equation if the local truncation error goes to zero as $h, k \rightarrow 0$. For example, the local truncation error for the forward difference method is given by Equation (15.3.12)

$$-\frac{k}{2}\frac{\partial^2 u}{\partial t^2}(x_i, \eta_j) + \frac{h^2}{12}\frac{\partial^4 u}{\partial t^4}(\xi_i, t_j) \tag{15.4.10}$$

The limits of these terms go to zero as $h, k \rightarrow 0$ no matter how this limit is taken. Thus the forward difference equation [Equation (15.3.9)] is consistent with the original partial differential equation. Most of the time, consistency is taken for granted but the following example shows that we need to be cautious. The local truncation error for the DuFort–Frankel method (Section 15.5) is

$$-\frac{k^2}{6}\frac{\partial^3 u}{\partial t^4} + \frac{h^2}{12}\frac{\partial^4 u}{\partial x^4} - \left(\frac{k}{h}\right)^2\frac{\partial^2 u}{\partial t^2} \tag{15.4.11}$$

If $k$ goes to zero faster than $h$ goes to zero, then Equation (15.4.11) goes to zero. If $h$ and $k$ go to zero at the same rate, then Equation (15.4.11) does not go to zero. Thus the DuFort–Frankel method [Equation (15.5.7)] is not consistent. The following theorem gives the condition for convergence.

**Theorem 15.4.1**    (**Lax Equivalence Theorem**)    Given a linear well-posed problem and a finite difference approximation to it that satisfies the consistency condition, stability is the necessary and sufficient condition for convergence.    ∎

For the proof, see Richtmeyer and Morton (1967). For nonlinear cases, it is a conjecture. Thus we need to discuss stability in a more analytical way. We consider the **Fourier** or **Von Neumann method** developed by J. Von Neumann and the matrix method.

Consider the propagating effect of a single row of errors along the line at $t = 0$. Assume that the error function $E(x, 0)$ can be represented by a finite Fourier series as

$$E(x, 0) = \sum_{i=0}^{N} A_i\, e^{\sqrt{-1}\lambda_i x} \tag{15.4.12}$$

where $A_i$ are the Fourier coefficients, $\lambda_i$ are the frequencies, and $(N + 1)$ is the number of mesh points on the line. At each point on the line, the right-hand side of Equation (15.4.12) reduces to the exact error. Because of the linearity of the partial differential equation, we need to consider only the effect of one term $e^{\sqrt{-1}\lambda x}$ where $\lambda$ is a constant. The complete effect is then obtained by linear superposition of all such errors. Further assume that a separation of time and space variables can be made and that at time $t$ the error is given by

$$e(x, t) = e^{\sqrt{-1}\lambda x} \cdot e^{\gamma t} \tag{15.4.13}$$

where $\gamma$ and $\lambda$ are constants. Note that $e(x, 0) = e^{\sqrt{-1}\lambda x}$ and

$$e(x_i, t_j) = e^{\sqrt{-1}\lambda x_i} \cdot e^{\gamma t_j} = e^{\sqrt{-1}\lambda i h} \cdot e^{\gamma j k} = e_{i,j} \tag{15.4.14}$$

Now we need to investigate how $e^{\gamma t_j}$ will behave as we proceed to $t_j + k$ or $j + 1$. We need to determine whether the amplitude of the error will grow or die as we proceed to the next time level $j + 1$.

Consider the Richardson formula [Equation (15.3.16)]

$$U_{i,j+1} = U_{i,j-1} + 2r(U_{i+1,j} - 2U_{i,j} + U_{i-1,j}) \tag{15.4.15}$$

Let $U_{i,j+1}$ be the exact solution of Equation (15.4.15) and $\bar{U}_{i,j+1}$ be the computed solution (so it contains rounding errors). Then

$$e_{i,j+1} = U_{i,j+1} - \bar{U}_{i,j+1} \tag{15.4.16}$$

is the error due to round-off. Since $\bar{U}_{i,j+1}$ also satisfies Equation (15.4.15), we can verify that $e_{i,j+1}$ satisfies

$$e_{i,j+1} = e_{i,j-1} + 2r(e_{i+1,j} - 2\,e_{i,j} + e_{i-1,j}) \tag{15.4.17}$$

Substituting Equation (15.4.14) in Equation (15.4.17) and simplifying, Equation (15.4.17) reduces to

$$e^{\gamma k} = e^{-\gamma k} + 2r(e^{\sqrt{-1}\lambda h} - 2 + e^{-\sqrt{-1}\lambda h}) \tag{15.4.18}$$

Using $e^{\sqrt{-1}\lambda h} = \cos \lambda h + \sqrt{-1}\,\sin \lambda h$ and $\cos \lambda h = 1 - 2\sin^2(\lambda h /2)$ in Equation (15.4.18) and simplifying, we get

$$e^{2\gamma k} + 8re^{\gamma k}\sin^2(\lambda h /2) - 1 = 0 \tag{15.4.19}$$

Solving this quadratic equation and selecting the negative sign, we get

$$e^{\gamma k} = -4r\sin^2\frac{\lambda h}{2} - \sqrt{16r^2\sin^4\frac{\lambda h}{2} + 1} \tag{15.4.20}$$

Thus $e^{\gamma k} < -1$, since $r = \Delta t/(\Delta x)^2 = h/k^2 > 0$. Consequently

$$|e^{\gamma k}| > 1 \quad \text{for all } r \tag{15.4.21}$$

The amplification factor is

$$\left|\frac{e_{i,j+1}}{e_{i,j}}\right| = |e^{\gamma k}| > 1 \tag{15.4.22}$$

therefore $(e^{\gamma k})^j$ increases as $j$ increases. Hence, the Richardson method is unstable for all $r$.

By using a similar procedure, we can show that the forward difference method [Equation (15.3.9)] is stable if $0 < r \leq \frac{1}{2}$ (Exercise 3). By accident, we arrive at the same restriction imposed on us by the convergence requirements. Stability consideration forces us to select $k \leq h^2/2$ for the forward difference method; $h$ must be small in order to represent the derivative accurately by the finite difference formula. Thus the corresponding $k$ must be significantly small. If $h = 0.1$, then $k$ must be smaller than 0.005. Since the accuracy of the forward difference method is of the order $(h^2 + k)$, it is quite common to select $h = 0.01$. Hence $k$ must be smaller than 0.00005. With such small values of $k$, we need an inordinate amount of computation time. Fortunately, the backward difference method is stable for all $r$ (Exercise 4); however, it is an implicit method.

Next we consider the matrix method. The backward difference formula [Equation (15.3.21)] is

$$rU_{i-1,j} - (1 + 2r)\,U_{i,j} + rU_{i+1,j} = -U_{i,j-1} \tag{15.4.23}$$

for $i = 1, 2, \ldots, N - 1$ and $j = 1, 2, \ldots, M$. In the matrix form

$$\begin{bmatrix} -(1+2r) & r & 0 & \cdots & 0 & 0 \\ r & -(1+2r) & r & \cdots & 0 & 0 \\ \vdots & \vdots & \vdots & \ddots & \vdots & \vdots \\ 0 & 0 & 0 & \cdots & -(1+2r) & r \\ 0 & 0 & 0 & \cdots & r & -(1+2r) \end{bmatrix}$$

$$\begin{bmatrix} U_{1,j} \\ U_{2,j} \\ \vdots \\ U_{N-2,j} \\ U_{N-1,j} \end{bmatrix} = \begin{bmatrix} -rU_{0,j} - U_{1,j-1} \\ -U_{2,j-1} \\ \vdots \\ -U_{N-2,j-1} \\ -U_{N-1,j-1} - rU_{N,j} \end{bmatrix} \tag{15.4.24}$$

for $j = 1, 2, \ldots, M$. Using $U_{0,j} = c_1$ and $U_{N,j} = c_2$ for all $j$, $U_{i,0} = f(x_i)$ for $i = 0, 1, \ldots, N$ and $\mathbf{U}^{(j)} = [U_{1,j}, U_{2,j}, \ldots, U_{N-1,j}]'$, Equation (15.4.24) becomes

$$A\mathbf{U}^{(j)} = -\mathbf{U}^{(j-1)} - \mathbf{b} \tag{15.4.25}$$

The computed values are given by $\tilde{\mathbf{U}}^{(j)} = [\tilde{U}_{1,j}\,\tilde{U}_{2,j}, \ldots, \tilde{U}_{N-1,j}]'$ and satisfy Equation (15.4.25). Thus

$$A\tilde{\mathbf{U}}^{(j)} = -\tilde{\mathbf{U}}^{(j-1)} - \mathbf{b} \tag{15.4.26}$$

Subtracting Equation (15.4.26) from Equation (15.4.25), we get

$$A(\mathbf{U}^{(j)} - \tilde{\mathbf{U}}^{(j)}) = -(\mathbf{U}^{(j-1)} - \tilde{\mathbf{U}}^{(j-1)}) \tag{15.4.27}$$

Define the error vector $\mathbf{e}^{(j)} = \mathbf{U}^{(j)} - \tilde{\mathbf{U}}^{(j)}$. Then Equation (15.4.27) reduces to

$$A\mathbf{e}^{(j)} = -\mathbf{e}^{(j-1)} \qquad \text{(15.4.28)}$$

Multiplying both sides of Equation (15.4.28) by $A^{-1}$, we get

$$\mathbf{e}^{(j)} = -A^{-1}\mathbf{e}^{(j-1)} \qquad \text{(15.4.29)}$$

If $\rho(A^{-1}) < 1$, then $\mathbf{e}^{(j)} \to 0$ as $j \to \infty$. It can be verified that the eigenvalues of $A$ are given by

$$\lambda_i = 1 + 4r\left[\sin\left(\frac{i\pi}{2N}\right)\right]^2 \quad \text{for } i = 1, 2, \ldots, N - 1 \qquad \text{(15.4.30)}$$

Since $r > 0$, $\lambda_i > 1$ for $i = 1, 2, \ldots, N - 1$. The eigenvalues of $A^{-1}$ are the reciprocals of the eigenvalues of $A$, and therefore $\rho(A^{-1}) < 1$. Thus the backward difference method is stable for all $r$.

There is a third method called the **energy method** for analyzing stability. For its discussion, see Richtmeyer and Morton (1967).

## 15.5   THE CRANK–NICOLSON AND DuFORT–FRANKEL METHODS

The forward difference method converges for $r \leq \frac{1}{2}$, the Richardson method (central difference method) is unstable for all $r$, and the backward difference method is stable for all $r$; in other words, for the backward difference method, $\Delta t$ can be selected independently of $\Delta x$. However, the accuracy of the backward difference method is of the order $((\Delta x)^2 + \Delta t)$. This prompted people to improve the accuracy of the backward difference method. The best known method is the Crank–Nicolson method. From Equation (15.3.20), we have the backward difference form

$$\frac{U_{i-1,j} - 2U_{i,j} + U_{i+1,j}}{h^2} = \frac{U_{i,j} - U_{i,j-1}}{k} \qquad \text{(15.5.1)}$$

Since the space derivative in Equation (15.5.1) is centered at the grid location $(i, j)$, we attempt to center the approximation for $\partial u/\partial t$ by using

$$\left(\frac{\partial u}{\partial t}\right)_{i,j} = \frac{U_{i,j+1} - U_{i,j-1}}{2k} - \frac{k^2}{6}\frac{\partial^3 u}{\partial t^3}(x_i, \eta_j) \qquad \text{(15.5.2)}$$

which is of the second order. We know that this leads to a disastrous Richardson method. Alternatively, we think that the time derivative in Equation (15.5.1) is centered at $(i, j - \frac{1}{2})$. Thus

$$\left(\frac{\partial u}{\partial t}\right)_{i,j-1/2} = \frac{U_{i,j} - U_{i,j-1}}{2(k/2)} - \frac{k^2}{24}\frac{\partial^3 u}{\partial t^3}(x_i, \eta_{j-1/2}) \qquad \text{(15.5.3)}$$

which is of the second order. We approximate $\partial^2 u/\partial x^2$ at $(i, j - \frac{1}{2})$ by the average of the second order central differences at $(i, j - 1)$ and $(i, j)$. Substituting these two approximations in

$$\left(\frac{\partial^2 u}{\partial x^2}\right)\left(x_i, t_j - \frac{k}{2}\right) = \frac{\partial u}{\partial t}\left(x_i, t_j - \frac{k}{2}\right)$$

we get

$$\frac{[U_{i-1,j-1} - 2U_{i,j-1} + U_{i+1,j-1} + U_{i-1,j} - 2U_{i,j} + U_{i+1,j}]}{2h^2} = \frac{U_{i,j} - U_{i,j-1}}{k} \quad \textbf{(15.5.4)}$$

Equation (15.5.4) can be written as

$$rU_{i-1,j} - (2r + 2) U_{i,j} + rU_{i+1,j} = -rU_{i-1,j-1} + (2r - 2) U_{i,j-1} - rU_{i+1,j-1}$$
$$\textbf{(15.5.5)}$$

where $r = k/h^2 = \Delta t/(\Delta x)^2$.

The Crank–Nicolson formula [Equation (15.5.5)] can also be obtained by combining the forward difference formula at $t_{j-1}$

$$\frac{U_{i-1,j-1} - 2U_{i,j-1} + U_{i+1,j-1}}{h^2} = \frac{U_{i,j} - U_{i,j-1}}{k}$$

and the backward difference formula at $t_j$

$$\frac{U_{i-1,j} - 2U_{i,j} + U_{i+1,j}}{h^2} = \frac{U_{i,j} - U_{i,j-1}}{k}$$

It can be verified (Exercise 5) that Equation (15.5.5) is stable for all $r$. The local truncation error for Equation (15.5.5) is

$$-\frac{k^2}{12} \frac{\partial^3 u}{\partial t^3} (x_i, \eta_{j-1/2}) + \frac{h^2}{12} \frac{\partial^4 u}{\partial x^4} (\xi_i, t_j)$$

where $\eta_{j-1/2} \in (t_j, t_{j+1})$ and $\xi_i \in (x_{i-1}, x_{i+1})$. Error-wise, this is a distinct improvement over previous methods. However, it requires more computation per time step than the backward difference since it requires more computation time to evaluate the right-hand side of Equation (15.5.5).

The net points in space and time are shown in Figure 15.5.1 at a typical step.

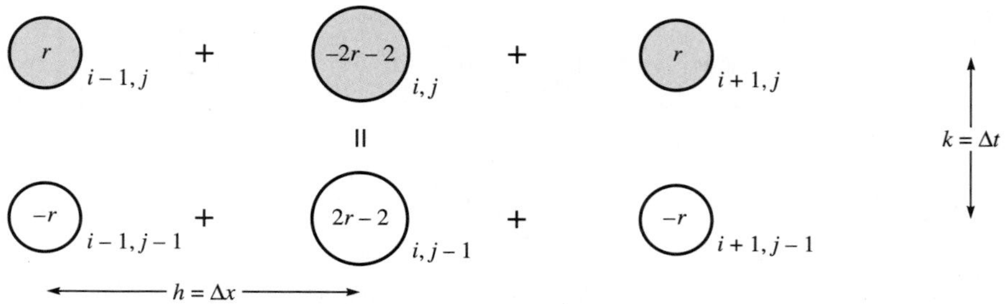

Fig. 15.5.1

**Algorithm**

**Subroutine** Crank($ba$, $bb$, $N$, $f$, $bc_0$, $bc_1$, $M$, $T$, **z**)
*Comment: This subroutine approximates the solution of*

$$\frac{\partial^2 u}{\partial x^2} = \frac{\partial u}{\partial t} \qquad u(ba, t) = bc_0 \quad u(bb, t) = bc_1 \qquad u(x, 0) = f(x) \qquad ba \le x \le bb$$

*using the Crank–Nicolson formula [Equation (15.5.5)]. We use ba for a, bb for b, $bc_0$ for $c_0$, $bc_1$ for $c_1$ to avoid confusion with the parameters of the Subroutine Tridi. We solve the tridiagonal system of Equation (15.5.5) using the LU decomposition method given by Equations (7.7.4), (7.7.6), and (7.7.7) by calling the Subroutine Tridi(**a**, **b**, **c**, **s**, N, **x**) from Section 7.7. We have $\Delta x = h = (bb - ba)/N$, $k = ak = \Delta t = T/M$, and $z_i = U_{i,j}$ for $i = 0, 1, \ldots, N$ and $j = 0, 1, \ldots, M$.*

1. **Print** $ba$, $bb$, $N$, $c_0$, $c_1$, $M$, $T$
2. $h = (bb - ba)/N$
3. $ak = T/M$
4. $r = ak/h^2$
5. time $= 0$
6. **Repeat** steps 7–9 **with** $i = 0, 1, \ldots, N$
   7. $ai = i$
   8. $x = ba + (ai) h$
   9. $z_i = f(x)$
10. **Print** time and $z$
11. $z_0 = bc_0$
12. $z_N = bc_1$
13. **Repeat** steps 14–16 **with** $i = 1, 2, \ldots, N - 1$
   14. $a_i = r$
   15. $b_i = -(2r + 2)$
   16. $c_i = r$
17. time $=$ time $+ ak$
18. **Repeat** step 19 **with** $i = 1, 2, \ldots, N - 1$
   19. $s_i = -rz_{i-1} + (2r - 2) z_i - rz_{i+1}$
20. $s_1 = s_1 - a_1 z_0$
21. $s_{N-1} = s_{N-1} - c_{N-1} z_N$
22. **Call** Tridi(**a**, **b**, **c**, **s**, $N - 1$, **z**)
23. **If** time $< T$, **then** return to 17
24. **Print** time and **z** and exit.    ◀

## The DuFort–Frankel Method

The Richardson formula [Equation (15.3.15)] has two advantages:

1. It is an explicit method. Explicit methods are easy to program while implicit methods require involved programming.
2. It has a truncation error of order $(h^2 + k^2)$.

Unfortunately, it is unstable for all $r > 0$. DuFort and Frankel replaced $2U_{i,j}$ in the Richardson formula

$$\frac{U_{i-1,j} - 2U_{i,j} + U_{i+1,j}}{h^2} = \frac{U_{i,j+1} - U_{i,j-1}}{2k} \tag{15.5.6}$$

by $U_{i,j-1} + U_{i,j+1}$. This yields

$$\frac{U_{i-1,j} - (U_{i,j-1} + U_{i,j+1}) + U_{i+1,j}}{h^2} = \frac{U_{i,j+1} - U_{i,j-1}}{2k} \tag{15.5.7}$$

Solving Equation (15.5.7) for $U_{i,j+1}$, we get

$$(1 + 2r)\,U_{i,j+1} = 2r(U_{i-1,j} + U_{i+1,j}) + (1 - 2r)\,U_{i,j-1} \tag{15.5.8}$$

where as usual $r = k/h^2 = \Delta t/(\Delta x)^2$.

In order to examine consistency, consider (Exercise 9)

$$\frac{u(x_i - h, t_j) - [u(x_i, t_j - k) + u(x_i, t_j + k)] + u(x_i + h, t_j)}{h^2}$$

$$- \frac{u(x_i, t_j + k) - u(x_i, t_j - k)}{2k}$$

$$= \left( \frac{\partial^2 u}{\partial x^2}(x_i, t_j) - \frac{\partial u}{\partial t}(x_i, t_j) \right) - \frac{k^2}{6}\frac{\partial^3 u}{\partial t^3}(x_i, t_j) + \frac{h^2}{12}\frac{\partial^4 u}{\partial x^4}(x_i, t_j)$$

$$- \frac{k^2}{h^2}\frac{\partial^2 u}{\partial t^2}(x_i, t_j) + \cdots \tag{15.5.9}$$

If $k$ goes to zero faster than $h$, then Equation (15.5.8) is consistent with $\partial^2 u/\partial x^2 = \partial u/\partial t$. But if $h$ and $k \to 0$ at the same rate, that is, $k/h = c = $ fixed, then the truncation error in Equation (15.5.9) tends to $-c^2\,(\partial^2 u/\partial t^2)$. Thus Equation (15.5.8) would be consistent with

$$\frac{\partial^2 u}{\partial x^2} = c^2\frac{\partial^2 u}{\partial t^2} + \frac{\partial u}{\partial t} \tag{15.5.10}$$

which is hyperbolic.

It can be shown (Exercise 6) that Equation (15.5.8), although explicit, is stable for all $r$. Though stable, it is not very accurate when a large time step is used.

In Figure 15.5.2, the mesh points involved at a typical step are shown.

A number of other methods are available for the treatment of parabolic equations (Richtmeyer and Morton 1967).

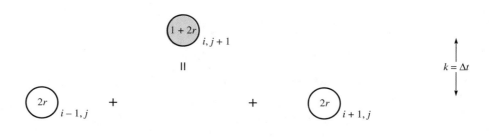

**Fig. 15.5.2**

**EXAMPLE 15.5.1**    Compute an approximate solution of

$$\frac{\partial^2 u}{\partial x^2} = \frac{\partial u}{\partial t} \quad 0 < x < 1 \quad t > 0$$
$$u(0, t) = u(1, t) = 0 \quad t > 0$$
$$u(x, 0) = \sin \pi x \quad 0 \le x \le 1$$

using (1) the Crank–Nicolson method and (2) the DuFort–Frankel method.

**1.** Consider $h = \frac{1}{4}$ and $r = 1$. Then $k = rh^2 = 1(0.25)^2 = 0.0625$. The Crank–Nicolson formula [Equation (15.5.5)] reduces to

$$U_{i-1,j} - 4U_{i,j} + U_{i+1,j} = -U_{i-1,j-1} - U_{i+1,j-1} \qquad \textbf{(15.5.11)}$$

for $i = 1, 2$, and 3 and $j = 1, 2, \ldots$. The nodal points and initial values $U_{i,0}$ are given in Example 15.3.1.

For $j = 1$ and $i = 1, 2$, and 3, Equation (15.5.11) reduces to

$$U_{0,1} - 4U_{1,1} + U_{2,1} = -U_{0,0} - U_{2,0}$$
$$U_{1,1} - 4U_{2,1} + U_{3,1} = -U_{1,0} - U_{3,0}$$
$$U_{2,1} - 4U_{3,1} + U_{4,1} = -U_{2,0} - U_{4,0}$$

This can be written as

$$\begin{bmatrix} -4 & 1 & 0 \\ 1 & -4 & 1 \\ 0 & 1 & -4 \end{bmatrix} \begin{bmatrix} U_{1,1} \\ U_{2,1} \\ U_{3,1} \end{bmatrix} = -\begin{bmatrix} U_{0,0} + U_{2,0} + U_{0,1} \\ U_{1,0} + U_{3,0} \\ U_{2,0} + U_{4,0} + U_{4,1} \end{bmatrix} \qquad \textbf{(15.5.12)}$$

$$= -\begin{bmatrix} 1 \\ \sin \dfrac{\pi}{4} + \sin \dfrac{3\pi}{4} \\ 1 \end{bmatrix} = -\begin{bmatrix} 1 \\ \sqrt{2} \\ 1 \end{bmatrix}$$

Solving Equation (15.5.12), we get $U_{1,1} = 0.38673$, $U_{2,1} = 0.54692$, and $U_{3,1} = 0.38673$.

For $j = 2$ and $i = 1, 2$, and 3, Equation (15.5.11) reduces to

$$U_{0,2} - 4U_{1,2} + U_{2,2} = -U_{0,1} - U_{2,1}$$
$$U_{1,2} - 4U_{2,2} + U_{3,2} = -U_{1,1} - U_{3,1}$$
$$U_{2,2} - 4U_{3,2} + U_{4,2} = -U_{2,1} - U_{4,1}$$

This can be written as

$$\begin{bmatrix} -4 & 1 & 0 \\ 1 & -4 & 1 \\ 0 & 1 & -4 \end{bmatrix} \begin{bmatrix} U_{1,2} \\ U_{2,2} \\ U_{3,2} \end{bmatrix} = -\begin{bmatrix} U_{0,1} + U_{2,1} + U_{0,2} \\ U_{1,1} + U_{3,1} \\ U_{2,1} + U_{4,1} + U_{4,2} \end{bmatrix} = -\begin{bmatrix} 0.54692 \\ 0.77346 \\ 0.54692 \end{bmatrix}$$

$$\textbf{(15.5.13)}$$

Solving Equation (15.5.13), we get $U_{1,2} = 0.21151$, $U_{2,2} = 0.29912$, and $U_{3,2} = 0.21151$. Continuing we compute the values up to $t = 0.5$. The computed values are given in Table 15.5.1.

**Table 15.5.1**

| $x_i$ | Exact Solution | Crank–Nicolson Method | \|Actual Error\| | DuFort–Frankel Method |
|-------|------|------|------|------|
| 0.00 | 0.00000 | 0.00000 | 0.00000 | 0.00000 |
| 0.25 | 0.00509 | 0.00566 | 0.00057 | −0.00010 |
| 0.50 | 0.00719 | 0.00801 | 0.00082 | −0.00014 |
| 0.75 | 0.00509 | 0.00566 | 0.00057 | −0.00010 |
| 1.00 | 0.00000 | 0.00000 | 0.00000 | 0.00000 |

2.  For $h = \frac{1}{4}$ and $r = 1$, the DuFort–Frankel formula [Equation (15.5.8)] reduces to

$$3U_{i,j+1} = 2(U_{i-1,j} + U_{i+1,j}) - U_{i,j-1} \qquad (15.5.14)$$

for $i = 1, 2,$ and 3 and $j = 1, 2, \ldots$.

   For $j = 0$, we cannot use Equation (15.5.14) because $U_{i,-1}$ for $i = 1, 2,$ and 3 are not known. Just as in the Richardson method, we need to supply the values $U_{1,1}, U_{2,1},$ and $U_{3,1}$. The values of $U_{1,1}, U_{2,1},$ and $U_{3,1}$ can be generated using the Crank–Nicolson or any other method. We use the values $U_{1,1} = 0.38673, U_{2,1} = 0.54692,$ and $U_{3,1} = 0.38673$ computed by using the Crank–Nicolson method. For $j = 1$ and $i = 1, 2,$ and 3, Equation (15.5.14) reduces to

$$U_{1,2} = \frac{2}{3}(U_{0,1} + U_{2,1}) - \frac{1}{3}U_{1,0} = \frac{2}{3}(0 + 0.54692) - \frac{1}{3}\sin\frac{\pi}{4} = 0.12891$$

$$U_{2,2} = \frac{2}{3}(U_{1,1} + U_{3,1}) - \frac{1}{3}U_{2,0} = \frac{2}{3}(0.38673 + 0.38673) - \frac{1}{3}\sin\frac{\pi}{2} = 0.18231$$

$$U_{3,2} = \frac{2}{3}(U_{2,1} + U_{4,1}) - \frac{1}{3}U_{3,0} = \frac{2}{3}(0.54692 + 0) - \frac{1}{3}\sin\frac{3\pi}{4} = 0.12891$$

The computed values at $t = 0.5$ are given in Table 15.5.1

   The values of \|actual error\| for the Crank–Nicolson method given in Table 15.5.1 are very small although $h = \Delta x = \frac{1}{4} = 0.25$ and $k = \Delta t = 1/16 = .0625$. This explains the Crank–Nicolson method's popularity. The computed results given by the DuFort–Frankel method in Table 15.5.1 are absurd because of the large time step.

   In Table 15.5.2, the computed results for $h = \Delta x = \frac{1}{4} = 0.25$ and $k = \Delta t = 0.01$ are given.

**Table 15.5.2**

| $x_i$ | Exact Solution | Crank–Nicolson Method | \|Actual Error\| | DuFort–Frankel Method | \|Actual Error\| |
|-------|------|------|------|------|------|
| 0.00 | 0.00000 | 0.00000 | 0.00000 | 0.00000 | 0.00000 |
| 0.25 | 0.00509 | 0.00650 | 0.00141 | 0.00461 | 0.00048 |
| 0.50 | 0.00719 | 0.00919 | 0.00200 | 0.00652 | 0.00067 |
| 0.75 | 0.00509 | 0.00650 | 0.00141 | 0.00461 | 0.00048 |
| 1.00 | 0.00000 | 0.00000 | 0.00000 | 0.00000 | 0.00000 |

The values of |actual error| given by both methods are reasonably small since $h = \Delta x = 0.25$. ■ ■ ■

## EXERCISES

1. Use $h = \frac{1}{4}$ and $r = \frac{1}{2}$ to compute an approximate solution of

$$\frac{\partial^2 u}{\partial x^2} = \frac{\partial u}{\partial t} \quad 0 < x < 1 \quad t > 0$$

$$u(0, t) = u(1, t) = 0 \quad t > 0$$

$$u(x, 0) = \begin{cases} 0 & \text{if } x \neq 1/2 \\ \epsilon & \text{if } x = 1/2 \end{cases} \quad 0 \leq x \leq 1$$

using the Richardson formula [Equation (15.3.16)] up to $j = 6$. Use $U_{i,1} = 0$ for $i = 1, 2,$ and 3. Compare this solution with the exact solution $u(0, t) = 0$ of

$$\frac{\partial^2 u}{\partial x^2} = \frac{\partial u}{\partial t} \quad 0 < x < 1 \quad t > 0$$

$$u(0, t) = u(1, t) = 0 \quad t > 0 \qquad u(x, 0) = 0 \quad 0 \leq x \leq 1$$

2. Use $h = \frac{1}{3}$ and $r = 1$ to compute an approximate solution of

$$\frac{\partial^2 u}{\partial x^2} = \frac{\partial u}{\partial t} \quad 0 < x < 1 \quad t > 0$$

$$u(0, t) = u(1, t) = 0 \quad t > 0$$

$$u(x, 0) = \begin{cases} 0 & \text{if } x \neq 1/3 \\ \epsilon & \text{if } x = 1/3 \end{cases} \quad 0 \leq x \leq 1$$

using the backward difference formula [Equation (15.3.21)] up to $j = 6$. Compare this solution with the exact solution $u(0, t) = 0$ of

$$\frac{\partial^2 u}{\partial x^2} = \frac{\partial u}{\partial t} \quad < x < 1 \quad t > 0 \qquad u(0, t) = u(1, t) = 0 \quad t > 0$$

$$u(x, 0) = 0 \quad 0 \leq x \leq 1$$

3. Carry out the Fourier stability analysis for the forward difference formula [Equation (15.3.9)].
4. Carry out the Fourier stability analysis for the backward difference formula [Equation (15.3.21)].
5. Carry out the Fourier stability analysis for the Crank–Nicolson formula [Equation (15.5.5)].
6. Carry out the Fourier stability analysis for the DuFort–Frankel formula [Equation (15.5.8)].
7. Compute an approximate solution of

$$\frac{\partial^2 u}{\partial x^2} = \frac{\partial u}{\partial t} \quad 0 < x < 1 \quad t > 0$$

$$u(0, t) = u(1, t) = 0 \quad t > 0$$

$$u(x, 0) = \frac{1}{2} \sin \pi x \quad 0 \leq x \leq 1$$

using the Crank–Nicolson method. Use $h = \frac{1}{3}$, $r = \frac{1}{2}$, and $M = 3$.

8. Compute an approximate solution of

$$\frac{\partial^2 u}{\partial x^2} = \frac{\partial u}{\partial t} \quad 0 < x < 1 \quad t > 0$$
$$u(0, t) = u(1, t) = 0 \quad t > 0$$
$$u(x, 0) = \frac{1}{2} \sin \pi x \quad 0 \le x \le 1$$

using the DuFort–Frankel method. Use $h = \frac{1}{3}$, $r = \frac{1}{2}$, and $M = 3$. (*Hint*: Use $U_{i,1}$ from Exercise 7).

9. Verify Equation (15.5.9) by expanding each term of the left side of Equation (15.5.9) in terms of a Taylor's series around $(x_i, t_j)$ and collecting like terms.

## COMPUTER EXERCISES

**C.1.** Write a subroutine to generate difference equations given by Equation (15.5.5) for use of the Crank–Nicolson method.

**C.2.** Write a subroutine to solve Equation (15.5.5) by *LU* decomposition using Equations (7.7.4), (7.7.6), and (7.7.7).

**C.3.** Write a computer program using Exercises C.1 and C.2 to approximate a solution of

$$\frac{\partial^2 u}{\partial x^2} = \frac{\partial u}{\partial t} \quad a < x < b \quad t > 0$$
$$u(a, t) = c_1 \quad u(b, t) = c_2 \quad t > 0$$
$$u(x, 0) = f(x) \quad a \le x \le b$$

Using your computer program of Exercise C.3, compute the solutions of the parabolic partial differential equations in Exercises C.4–C.6. Use $h = 0.1$ and $0.01$ and $r = 0.1, 0.5,$ and $2.0$. Compare your solutions to exact solutions at $t = 0.5$ and $1.0$.

**C.4.**

$$\frac{\partial^2 u}{\partial x^2} = \frac{\partial u}{\partial t} \quad 0 < x < 1 \quad t > 0$$
$$u(0, t) = u(1, t) = 0 \quad t > 0$$
$$u(x, 0) = \frac{1}{2} \sin \pi x \quad 0 \le x \le 1$$

The exact solution is given by $u(x, t) = (e^{-\pi^2 t} \sin \pi x)/2$.

**C.5.**

$$\frac{\partial^2 u}{\partial x^2} = \frac{\partial u}{\partial t} \quad -1 < x < 1 \quad t > 0$$
$$u(-1, t) = u(1, t) = 0 \quad t > 0$$
$$u(x, 0) = \cos \frac{\pi x}{2} \quad -1 \le x \le 1$$

The exact solution is given by $u(x, t) = e^{-(\pi^2/4)t} \cos(\pi x/2)$.

**C.6.**

$$\frac{\partial^2 u}{\partial x^2} = \frac{\partial u}{\partial t} \quad 0 < x < \pi \quad t > 0$$

$$u(0, t) = u(\pi, t) = 0 \quad t > 0$$

$$u(x, 0) = \cos\left(x - \frac{\pi}{2}\right) \quad 0 < x < \pi$$

The exact solution is given by $u(x, t) = e^{-t} \cos\left(x - \left(\frac{\pi}{2}\right)\right)$.

**C.7.** Write a subroutine to generate difference equations given by Equation (15.5.8) for use of the DuFort–Frankel method.

**C.8.** Write a computer program using Exercise C.7 to approximate the solution of

$$\frac{\partial^2 u}{\partial x^2} = \frac{\partial u}{\partial t} \quad a < x < b \quad t > 0$$

$$u(a, t) = c_1 \quad u(b, t) = c_2 \quad t > 0$$
$$u(x, 0) = f(x) \quad a \le x \le b$$

**C.9.** Using your computer program of Exercise C.7, approximate the solution of the parabolic partial differential equation in Exercise C.5. Use $h = 0.1$ and 0.01 and $r = 0.1$, 0.5, and 2.0. Compare your solutions to the exact solutions and to the solutions given by the Crank–Nicolson method at $t = 0.5$ and 1.0.

**C.10.** The temperature $u(r, t)$ in a hollow sphere is given by

$$\frac{\partial^2 u}{\partial r^2} + \frac{2}{r} \frac{\partial u}{\partial r} = \frac{\partial u}{\partial t} \quad a < r < b \quad t > 0$$

$$u(a, t) = c \quad u(b, t) = d \quad t > 0$$
$$u(r, 0) = f(r) \quad a \le r \le b$$

Approximate the temperature at $t = 0.5$ and 1.0 when $a = 1$, $b = 3/2$, $c = 1$, $d = 0$, and $f(r) = 4(3/2 - r)^2$. (*Hint*: Use $u(r) = v(r)/r$.)

**C.11.** A homogeneous rod of length five initially has a linear rise of temperature from 0°C to 100°C. The end at 0°C is kept at 0°C while the hot end is cooled rapidly. Approximate the temperature at $t = 0.5$ and 1.0 by solving

$$\frac{\partial^2 u}{\partial x^2} = \frac{\partial u}{\partial t} \quad 0 < x < 5 \quad t > 0$$

$$u(0, t) = 0 \quad u(5, t) = 100 \, e^{-0.5t} \quad t > 0$$
$$u(x, 0) = 20 \, x \quad 0 \le x \le 5$$

## 15.6  PARABOLIC EQUATIONS IN TWO DIMENSIONS

We consider various techniques for solving the two dimensional heat flow equation

$$\frac{\partial^2 u}{\partial x^2} + \frac{\partial^2 u}{\partial y^2} = \frac{\partial u}{\partial t} \quad 0 < x, \quad y < 1 \quad t > 0 \qquad \textbf{(15.6.1a)}$$

subject to the boundary conditions

$$u(x, 0, t) = 0 \quad u(x, 1, t) = 0 \quad 0 \le x \le 1 \quad t > 0$$
$$u(0, y, t) = 0 \quad u(1, y, t) = 0 \quad 0 \le y \le 1 \quad t > 0 \qquad \textbf{(15.6.1b)}$$

and initial condition

$$u(x, y, 0) = f(x, y) \quad 0 \leq x, \quad y \leq 1 \tag{15.6.1c}$$

Partition the interval [0, 1] along both the $x$-axis and $y$-axis into $N$ equal parts of width $h$. Further, partition the interval [0, $T$] along the $t$-axis into $M$ equal parts of width $k$. Thus $\Delta x = \Delta y = h = 1/N$ and $\Delta t = k = T/M$. The region in $(x, y, t)$ space is covered by the rectangular grids parallel to the axes. The lines $x = x_i$, $y = y_j$, and $t = t_l$ are called the **grid lines** and their intersections are called the **mesh points** of the grid. At each interior mesh point $(x_i, y_j, t_l)$ for $i, j = 1, 2, \ldots, N - 1$ and $l = 1, 2, \ldots, M$, Equation (15.6.1a) is satisfied and we have

$$\frac{\partial^2 u}{\partial x^2}(x_i, y_j, t_l) + \frac{\partial^2 u}{\partial y^2}(x_i, y_j, t_l) = \frac{\partial u}{\partial t}(x_i, y_j, t_l) \tag{15.6.2}$$

The simplest replacement of Equation (15.6.2) consists of approximating the space derivatives by the centered second order differences and the time derivative by the forward difference. Using the notations $U(x_i, y_j, t_l) = U_{i,j,l}$, $U(x_i + h, y_j, t_l) = U_{i+1,j,l}$, $U(x_i - h, y_j, t_l) = U_{i-1,j,l}$ and so forth, we obtain from Equation (15.6.2)

$$\frac{U_{i-1,j,l} - 2U_{i,j,l} + U_{i+1,j,l}}{h^2} + \frac{U_{i,j-1,l} - 2U_{i,j,l} + U_{i,j+1,l}}{h^2} = \frac{U_{i,j,l+1} - U_{i,j,l}}{k} \tag{15.6.3}$$

Solving Equation (15.6.3) for $U_{i,j,l+1}$, we get

$$U_{i,j,l+1} = r[U_{i-1,j,l} + U_{i+1,j,l} + U_{i,j-1,l} + U_{i,j+1,l}] + (1 - 4r) U_{i,j,l} \tag{15.6.4}$$

for $i, j = 1, 2, \ldots, N - 1$ and $l = 0, 1, \ldots, M - 1$ where

$$r = \frac{k}{h^2} = \frac{\Delta t}{(\Delta x)^2} \tag{15.6.5}$$

As before, the forward difference formula (15.6.4) may be unstable and therefore we perform a stability analysis. Let $e_{i,j,l+1} = U_{i,j,l+1} - \tilde{U}_{i,j,l+1}$ where $U_{i,j,l+1}$ is the exact solution of Equation (15.6.4) and $\tilde{U}_{i,j,l+1}$ is the computed solution of Equation (15.6.4). Since $U_{i,j,l+1}$ and $\tilde{U}_{i,j,l+1}$ satisfy Equation (15.6.4), we get

$$e_{i,j,l+1} = r[e_{i-1,j,l} + e_{i+1,j,l} + e_{i,j-1,l} + e_{i,j+1,l}] + (1 - 4r) e_{i,j,l} \tag{15.6.6}$$

By extending the arguments of the one-dimensional case, the error $e_{i,j,l}$ is given by

$$e_{i,j,l} = e^{\sqrt{-1}\alpha x_i} \cdot e^{\sqrt{-1}\beta y_j} \cdot e^{\gamma t_l} \tag{15.6.7}$$

Substituting Equation (15.6.7) in Equation (15.6.6) and simplifying, we get

$$\begin{aligned}
e^{\gamma k} &= r[e^{-\sqrt{-1}\alpha h} + e^{\sqrt{-1}\alpha h} + e^{-\sqrt{-1}\beta h} + e^{-\sqrt{-1}\beta h}] + (1 - 4r) \\
&= 2r[\cos \alpha h + \cos \beta h] + (1 - 4r) \\
&= 2r\left(1 - 2\sin^2\frac{\alpha h}{2} + 1 - 2\sin^2\frac{\beta h}{2}\right) + (1 - 4r) \\
&= 1 - 4r\left(\sin^2\frac{\alpha h}{2} + \sin^2\frac{\beta h}{2}\right)
\end{aligned} \tag{15.6.8}$$

For stability, we need $|e^{\gamma k}| < 1$, therefore

$$-1 \leq 1 - 4r\left(\sin^2\frac{\alpha h}{2} + \sin^2\frac{\beta h}{2}\right) \leq 1 \tag{15.6.9}$$

If $r \leq 1/\left[2\left(\sin^2 \dfrac{\alpha h}{2} + \sin^2 \dfrac{\beta h}{2}\right)\right]$, then Equation (15.6.9) is satisfied. Since $\alpha$ and $\beta$ are arbitrary, $r \leq \frac{1}{4}$. The stability condition $r \leq \frac{1}{4}$ differs from the stability condition for the one-dimensional explicit method by a factor of $\frac{1}{2}$. If $\Delta x \neq \Delta y$, then the stability condition (Exercise 1) is given by

$$\Delta t \leq \frac{1}{2\left[(\Delta x)^{-2} + (\Delta y)^{-2}\right]} \tag{15.6.10}$$

Since this is too restrictive, implicit methods are preferable.

In one dimension, the Crank–Nicolson method was used as an implicit method. We extend it in two dimensions by taking the averages of $\partial^2 u/\partial x^2$ and $\partial^2 u/\partial y^2$ at $t = (l - 1)\, \Delta t$ and $t = l\, \Delta t$ in Equation (15.6.2) to get

$$\frac{1}{2h^2}\left[U_{i-1,j,l-1} + U_{i+1,j,l-1} + U_{i,j-1,l-1} + U_{i,j+1,l-1} - 4U_{i,j,l-1} + U_{i-1,j,l}\right.$$

$$\left. + U_{i+1,j,l} + U_{i,j-1,l} + U_{i,j+1,l} - 4U_{i,j,l}\right] = \frac{1}{k}\left[U_{i,j,l} - U_{i,j,l-1}\right] \tag{15.6.11}$$

It can be shown that Equation (15.6.11) is stable for all $r$ by writing it in the form

$$-rU_{i-1,j,l} - rU_{i+1,j,l} - rU_{i,j-1,l} - rU_{i,j+1,l} + (2 + 4r)\, U_{i,j,l}$$
$$= (2 - 4r)\, U_{i,j,l-1} + r\, (U_{i-1,j,l-1} + U_{i+1,j,l-1} + U_{i,j-1,l-1} + U_{i,j+1,l-1}) \tag{15.6.12}$$

where $r = k/h^2$. Since there are five unknowns per equation in Equation (15.6.12), attempts were made to get an implicit method which has a tridiagonal matrix. Peaceman and Rachford (1955) and Douglas (1955) developed an alternating direction implicit (A.D.I.) method which is widely used. In this method we approximate $\partial^2 u/\partial x^2$ by the centered second difference at time $t = (l - 1)\, \Delta t = (l - 1)\, k$ and $\partial^2 u/\partial y^2$ by the centered second difference at time $t = l\, \Delta t = l\, k$. Thus Equation (15.6.2) is written as

$$r[U_{i-1,j,l-1} - 2U_{i,j,l-1} + U_{i+1,j,l-1} + U_{i,j-1,l} - 2U_{i,j,l} + U_{i,j+1,l}] = U_{i,j,l} - U_{i,j,l-1}$$
$$\tag{15.6.13}$$

for $i, j = 1, 2, \ldots, N - 1$. This is a tridiagonal system. At the next time step ($t = (l + 1)\, k = t_{l+1}$), we reverse the order. We approximate $\partial^2 u/\partial x^2$ by the centered second difference at time $t = (l + 1)\, \Delta t = t_{l+1}$ and $\partial^2 u/\partial y^2$ by the centered second difference at time $t = l\, \Delta t = t_l$. Then Equation (15.6.2) becomes

$$r[U_{i-1,j,l+1} - 2U_{i,j,l+1} + U_{i+1,j,l+1} + U_{i,j-1,l} - 2U_{i,j,l} + U_{i,j+1,l}] = U_{i,j,l+1} - U_{i,j,l}$$
$$\tag{15.6.14}$$

for $i, j = 1, 2, \ldots, N - 1$. This procedure is repeated until $t$ becomes $T$.

---

**E X A M P L E   15.6.1**

Demonstrate the alternating direction implicit (A.D.I.) method using two steps to compute an approximate solution of

$$\frac{\partial^2 u}{\partial x^2} + \frac{\partial^2 u}{\partial y^2} = \frac{\partial u}{\partial t} \quad 0 < x, y < 1 \quad t > 0$$

subject to the boundary conditions

$$u(0, y, t) = u(1, y, t) = 0 \quad 0 \leq y \leq 1 \quad t > 0$$
$$u(x, 0, t) = u(x, 1, t) = 0 \quad 0 \leq x \leq 1 \quad t > 0$$

and initial condition

$$u(x, y, 0) = \sin \pi x \sin \pi y \quad 0 \le x, y \le 1$$

Consider $\Delta x = \Delta y = \frac{1}{4}$ and $r = 1$. Therefore $k = (\Delta x)^2 = 0.0625$. Since $u(x, y, 0) = \sin \pi x \, \sin \pi y$, $u(0, 0, 0) = U_{0,0,0} = 0$, $u(0, \frac{1}{4}, 0) = U_{0,1,0} = 0$, $u(0, \frac{1}{2}, 0) = U_{0,2,0} = 0$, $u(0, \frac{3}{4}, 0) = U_{0,3,0} = 0$, $u(0, 1, 0) = U_{0,4,0} = 0$. Similarly we find

$$U_{1,0,0} = 0 \qquad U_{1,1,0} = \frac{1}{2} \qquad U_{1,2,0} = \frac{1}{\sqrt{2}} \qquad U_{1,3,0} = \frac{1}{2} \qquad U_{1,4,0} = 0$$

$$U_{2,0,0} = 0 \qquad U_{2,1,0} = \frac{1}{\sqrt{2}} \qquad U_{2,2,0} = 1 \qquad U_{2,3,0} = \frac{1}{\sqrt{2}} \qquad U_{2,4,0} = 0$$

$$U_{3,0,0} = 0 \qquad U_{3,1,0} = \frac{1}{2} \qquad U_{3,2,0} = \frac{1}{\sqrt{2}} \qquad U_{3,3,0} = \frac{1}{2} \qquad U_{3,4,0} = 0$$

$$U_{4,0,0} = U_{4,1,0} = U_{4,2,0} = U_{4,3,0} = U_{4,4,0} = 0$$

For $l = 1$, Equation (15.6.13) becomes

$$U_{i-1,j,0} - 2U_{i,j,0} + U_{i+1,j,0} + U_{i,j-1,1} - 2U_{i,j,1} + U_{i,j+1,1} = U_{i,j,1} - U_{i,j,0}$$

This can be written as

$$-U_{i,j-1,1} + 3U_{i,j,1} - U_{i,j+1,1} = U_{i-1,j,0} - U_{i,j,0} + U_{i+1,j,0} \qquad \textbf{(15.6.15)}$$

For $i = 1$ and $j = 1, 2$ and $3$, Equation (15.6.15) gives

$$\begin{aligned} -U_{1,0,1} + 3U_{1,1,1} - U_{1,2,1} &= U_{0,1,0} - U_{1,1,0} + U_{2,1,0} \\ -U_{1,1,1} + 3U_{1,2,1} - U_{1,3,1} &= U_{0,2,0} - U_{1,2,0} + U_{2,2,0} \\ -U_{1,2,1} + 3U_{1,3,1} - U_{1,4,1} &= U_{0,3,0} - U_{1,3,0} + U_{2,3,0} \end{aligned} \qquad \textbf{(15.6.16)}$$

Substituting the initial values, Equation (15.6.16) reduces to

$$\begin{bmatrix} 3 & -1 & 0 \\ -1 & 3 & -1 \\ 0 & -1 & 3 \end{bmatrix} \begin{bmatrix} U_{1,1,1} \\ U_{1,2,1} \\ U_{1,3,1} \end{bmatrix} = \begin{bmatrix} -\dfrac{1}{2} + \dfrac{1}{\sqrt{2}} \\ -\dfrac{1}{\sqrt{2}} + 1 \\ -\dfrac{1}{2} + \dfrac{1}{\sqrt{2}} \end{bmatrix} = \begin{bmatrix} 0.20711 \\ 0.29289 \\ 0.20711 \end{bmatrix}$$

Solving this system, we get

$$U_{1,1,1} = 0.13060 \qquad U_{1,2,1} = 0.18470 \qquad U_{1,3,1} = 0.13060$$

For $i = 2$ and $j = 1, 2$, and $3$, Equation (15.6.15) leads to

$$\begin{bmatrix} 3 & -1 & 0 \\ -1 & 3 & -1 \\ 0 & -1 & 3 \end{bmatrix} \begin{bmatrix} U_{2,1,1} \\ U_{2,2,1} \\ U_{2,3,1} \end{bmatrix} = \begin{bmatrix} 1 - \dfrac{1}{\sqrt{2}} \\ \sqrt{2} - 1 \\ 1 - \dfrac{1}{\sqrt{2}} \end{bmatrix} = \begin{bmatrix} 0.29289 \\ 0.41421 \\ 0.29289 \end{bmatrix}$$

This leads to $U_{2,1,1} = 0.18470$, $U_{2,2,1} = 0.26210$, and $U_{2,3,1} = 0.18470$.

For $i = 3$ and $j = 1, 2$, and $3$, we get from Equation (15.6.16)

$$U_{3,1,1} = 0.13060 \qquad U_{3,2,1} = 0.18470 \qquad U_{3,3,1} = 0.13060$$

For $l = 1$ and $r = 1$, Equation (15.6.14) leads to

$$-U_{i-1,j,2} + 3U_{i,j,2} - U_{i+1,j,2} = -U_{i,j,1} + U_{i,j-1,1} + U_{i,j+1,1} \qquad \textbf{(15.6.17)}$$

We get from Equation (15.6.17), for $j = 1$ and $i = 1, 2$, and 3,

$$U_{1,1,2} = 0.03424 \qquad U_{2,1,2} = 0.04863 \qquad U_{3,1,2} = 0.03424$$

for $j = 2$ and $i = 1, 2$, and 3,

$$U_{1,2,2} = 0.04811 \qquad U_{2,2,2} = 0.06784 \qquad U_{3,2,2} = 0.04811$$

and, for $j = 3$ and $i = 1, 2$, and 3,

$$U_{1,3,2} = 0.03424 \qquad U_{2,3,2} = 0.04863 \qquad U_{3,3,2} = 0.03424$$

Similarly we can continue.

■ ■ ■

**Algorithm**

**Subroutine** ADI($N, f, M, T, U$)
Comment: This program approximates the solution of

$$\frac{\partial^2 u}{\partial x^2} + \frac{\partial^2 u}{\partial y^2} = \frac{\partial u}{\partial t} \qquad u(x, 0, t) = u(x, 1, t) = 0 \quad u(0, y, t) = u(1, y, t) = 0$$
$$u(x, y, 0) = f(x, y) \quad 0 \le x, y \le 1$$

using the A.D.I. method given by Equations (15.6.13) and (15.6.14). We solve the tridiagonal system using $LU$ decomposition method given by Equations (7.7.4), (7.7.6), and (7.7.7) by calling the subroutine Tridi (**a**, **b**, **c**, **s**, $N$, **x**) from Section 7.7.
  1 **Print** $N, M, T$
  2 $h = 1/N$
  3 $ak = T/M$
  4 $r = ak/h^2$
  5 time $= 0$
  6 **Repeat** steps 7–12 **with** $j = 0, 1, \dots, N$
      7 $aj = j$
      8 $y = aj * h$
       9 **Repeat** steps 10–12 **with** $i = 0, 1, \dots, N$
          10 $ai = i$
          11 $x = ai * h$
          12 $U_{i,j} = f(x, y)$
  13 **Print** time and $U$
  14 time $=$ time $+ ak$
  15 **Repeat** steps 16–25 **with** $i = 1, 2, \dots, N - 1$
      16 **Repeat** steps 17–20 **with** $j = 1, 2, \dots, N - 1$
          17 $a_j = r$
          18 $b_j = -(1 + 2r)$
          19 $c_j = r$
          20 $s_j = -rU_{i-1,j} - (1 - 2r) U_{i,j} - rU_{i+1,j}$
      21 $s_1 = s_1 - a_1 U_{i,0}$
      22 $s_{N-1} = s_{N-1} - c_{N-1} U_{i,N}$
      23 **Call** Tridi(**a**, **b**, **c**, **s**, $N - 1$, **z**)

24 **Repeat** step 25 **with** $j = 1, 2, \ldots, N - 1$
   25 $U_{i,j} = z_j$
26 **Repeat** steps 27–28 **with** $i = 0, 1, \ldots, N$
   27 $U_{i,0} = 0$
   28 $U_{i,N} = 0$
29 **Repeat** steps 30–31 **with** $j = 0, 1, \ldots, N$
   30 $U_{0,j} = 0$
   31 $U_{N,j} = 0$
32 time = time + $ak$
33 **Repeat** steps 34–40 **with** $j = 1, 2, \ldots, N - 1$
   34 **Repeat** step 35 **with** $i = 1, 2, \ldots, N - 1$
      35 $s_i = -rU_{i,j-1} - (1 - 2r)\,U_{i,j} - rU_{i,j+1}$
   36 $s_1 = s_1 - a_1\,U_{0,j}$
   37 $s_{N-1} = s_{N-1} - c_{N-1}U_{N,j}$
   38 **Call** Tridi($\mathbf{a}, \mathbf{b}, \mathbf{c}, \mathbf{s}, N - 1, \mathbf{z}$)
   39 **Repeat** step 40 **with** $i = 1, 2, \ldots, N - 1$
      40 $U_{i,j} = z_i$
41 **If** time $> T$, **then print** time and $U$ and exit
42 time = time + $ak$
43 **Repeat** steps 44–50 **with** $i = 1, 2, \ldots, N - 1$
   44 **Repeat** step 45 **with** $j = 1, 2, \ldots, N - 1$
      45 $s_j = -rU_{i-1,j} - (1 - 2r)\,U_{i,j} - rU_{i+1,j}$
   46 $s_1 = s_1 - a_1\,U_{0,j}$
   47 $s_{N-1} = s_{N-1} - c_{N-1}\,U_{i,N}$
   48 **Call** Tridi($\mathbf{a}, \mathbf{b}, \mathbf{c}, \mathbf{s}, N - 1, \mathbf{z}$)
   49 **Repeat** step 50 **with** $j = 1, 2, \ldots, N - 1$
      50 $U_{i,j} = z_j$
51 **If** time $< T$, **then** return to 32
52 **Print** time and $U$ and exit. ◀

---

**EXERCISES**

1. Derive the stability condition [Equation (15.6.10)] for the forward difference formula of the two-dimensional heat equation [Equation (15.6.1a)] if $\Delta x \neq \Delta y$.
2. Prove that Equation (15.6.12) is stable for all values of $r$.
3. Use Equation (15.6.12) with $h = \frac{1}{3}$ and $r = 1$ to compute an approximate solution at $t = \frac{1}{9}$ of

$$\frac{\partial^2 u}{\partial x^2} + \frac{\partial^2 u}{\partial y^2} = \frac{\partial u}{\partial t} \quad 0 < x, y < 1 \quad t > 0$$

subject to the boundary conditions

$$u(x, 0, t) = u(x, 1, t) = 0 \quad 0 \le x \le 1 \quad t > 0$$
$$u(0, y, t) = u(1, y, t) = 0 \quad 0 \le y \le 1 \quad t > 0$$

and initial condition

$$u(x, y, 0) = \sin \pi x \sin \pi y \quad 0 \le x, y \le 1$$

4. Use Equation (15.6.12) with $h = \frac{1}{3}$ and $r = 1$ to compute an approximate solution at $t = \frac{1}{9}$ of

$$\frac{\partial^2 u}{\partial x^2} + \frac{\partial^2 u}{\partial y^2} = \frac{\partial u}{\partial t} \quad 0 < x, y < 1 \quad t > 0$$

subject to the boundary conditions

$$u(x, 0, t) = u(x, 1, t) = 0 \quad 0 \le x \le 1 \quad t > 0$$
$$u(0, y, t) = u(1, y, t) = 0 \quad 0 \le y \le 1 \quad t > 0$$

and initial condition

$$u(x, y, 0) = \cos \frac{\pi x}{2} \cos \frac{\pi y}{2} \quad 0 \le x, y \le 1$$

5. Use Equation (15.6.12) with $h = \frac{1}{3}$ and $r = \frac{1}{2}$ to compute an approximate solution at $t = \frac{1}{9}$ of

$$\frac{\partial^2 u}{\partial x^2} + \frac{\partial^2 u}{\partial y^2} = \frac{\partial u}{\partial t} \quad 0 < x, y < 1 \quad t > 0$$

subject to the boundary conditions

$$u(x, 0, t) = u(x, 1, t) = 0 \quad 0 \le x \le 1 \quad t > 0$$
$$u(0, y, t) = u(1, y, t) = 0 \quad 0 \le y \le 1 \quad t > 0$$

and initial condition

$$u(x, y, 0) = \sin \pi x \sin \pi y \quad 0 \le x, y \le 1$$

6. Compute an approximate solution at $t = \frac{1}{9}$ of

$$\frac{\partial^2 u}{\partial x^2} + \frac{\partial^2 u}{\partial y^2} = \frac{\partial u}{\partial t} \quad 0 < x, y < 1 \quad t > 0$$

subject to the boundary conditions

$$u(x, 0, t) = u(x, 1, t) = 0 \quad 0 \le x \le 1 \quad t > 0$$
$$u(0, y, t) = u(1, y, t) = 0 \quad 0 \le y \le 1 \quad t > 0$$

and initial condition

$$u(x, y, 0) = \cos \frac{\pi x}{2} \cos \frac{\pi y}{2} \quad 0 \le x, y \le 1$$

using the alternating direction implicit (A.D.I.) method with $h = \frac{1}{3}$ and $r = \frac{1}{2}$.

7. Compute an approximate solution at $t = \frac{1}{9}$ of

$$\frac{\partial^2 u}{\partial x^2} + \frac{\partial^2 u}{\partial y^2} = \frac{\partial u}{\partial t} \quad 0 < x, y < 1 \quad t > 0$$

subject to the boundary conditions

$$u(x, 0, t) = u(x, 1, t) = 0 \quad 0 \le x \le 1 \quad t > 0$$
$$u(0, y, t) = u(1, y, t) = 0 \quad 0 \le y \le 1 \quad t > 0$$

and initial condition

$$u(x, y, 0) = \sin \pi x \sin \pi y \quad 0 \le x, y \le 1$$

using the alternating direction implicit (A.D.I.) method with $h = \frac{1}{3}$ and $r = \frac{1}{2}$.

## COMPUTER EXERCISES

**C.1.**  Write a subroutine to generate difference equations given by Equations (15.6.13) and (15.6.14).

**C.2.**  Write a subroutine to solve a system of equations given by Equations (15.6.13) and (15.6.14) using $LU$ decomposition method given by Equations (7.7.4), (7.7.6), and (7.7.7).

**C.3.**  Using Exercises C.1 and C.2, approximate the solution at $t = 0.5$ and 1.0 of

$$\frac{\partial^2 u}{\partial x^2} + \frac{\partial^2 u}{\partial y^2} = \frac{\partial u}{\partial t} \quad 0 < x, y < 1 \quad t > 0$$

subject to the boundary conditions

$$u(x, 0, t) = u(x, 1, t) = 0 \quad 0 \le x \le 1 \quad t > 0$$
$$u(0, y, t) = u(1, y, t) = 0 \quad 0 \le y \le 1 \quad t > 0$$

and initial condition

$$u(x, y, 0) = \cos \frac{\pi x}{2} \cos \frac{\pi y}{2} \quad 0 \le x, y \le 1$$

Use $h = 0.1$ and 0.01 and $r = 0.1$ and 0.5. Compare your solutions to the exact solutions at $t = 0.5$ and 1. The exact solution is given by $u(x, y, t) = e^{-(\pi^2 t)/2} \cos(\pi x/2) \cos(\pi y/2)$.

**C.4.**  Using Exercises C.1 and C.2, approximate the solution at $t = 0.5$ and 1.0 of

$$\frac{\partial^2 u}{\partial x^2} + \frac{\partial^2 u}{\partial y^2} = \frac{\partial u}{\partial t} \quad 0 < x, y < 1 \quad t > 0$$

subject to the boundary conditions

$$u(x, 0, t) = u(x, 1, t) = 0 \quad 0 \le x \le 1 \quad t > 0$$
$$u(0, y, t) = u(1, y, t) = 0 \quad 0 \le y \le 1 \quad t > 0$$

and initial condition

$$u(x, y, 0) = \sin \pi x \sin \pi y \quad 0 \le x, y \le 1$$

Use $h = 0.1$ and 0.01 and $r = 0.1$ and 0.5. Compare your solutions to the exact solutions at $t = 0.5$ and 1.0. The exact solution is given by $u(x, y, t) = e^{-\pi^2 t} \sin \pi x \sin \pi y$.

## 15.7   ELLIPTIC EQUATIONS

The temperature distribution $u$ on a plate satisfies the two-dimensional heat equation

$$\frac{\partial^2 u}{\partial x^2} + \frac{\partial^2 u}{\partial y^2} = \frac{\partial u}{\partial t} \qquad (15.7.1)$$

Suppose the temperature $u$ on the plate is steady; that is, $u$ is independent of $t$. Then Equation (15.7.1) reduces to

$$\frac{\partial^2 u}{\partial x^2} + \frac{\partial^2 u}{\partial y^2} = 0 \qquad (15.7.2)$$

This equation is called the **two-dimensional Laplace equation**. If there is a heat source on the plate, then the temperature $u$ satisfies

$$\frac{\partial^2 u}{\partial x^2} + \frac{\partial^2 u}{\partial y^2} + f(x, y, t) = \frac{\partial u}{\partial t} \qquad (15.7.3)$$

If $u$ and $f$ are independent of $t$, then Equation (15.7.3) reduces to

$$\frac{\partial^2 u}{\partial x^2} + \frac{\partial^2 u}{\partial y^2} = -f(x, y) = q(x, y) \qquad (15.7.4)$$

This equation is called **Poisson's equation**.

Equations (15.7.2) and (15.7.4) are elliptic partial differential equations that are of great importance. In this section, we are concerned with the numerical solution of Equation (15.7.4) over a unit square $R = \{(x, y) \mid 0 \le x, y \le 1\}$ subject to a boundary condition

$$u(x, y) = g(x, y) \qquad (15.7.5)$$

on the boundary of the unit square.

In order to approximate the solution of Equations (15.7.4) and (15.7.5), a network of grid points is first established as shown in Figure 15.7.1 with grid spacing $\Delta x = h = 1/N$ and $\Delta y = k = 1/M$ where $M$ and $N$ are positive integers.

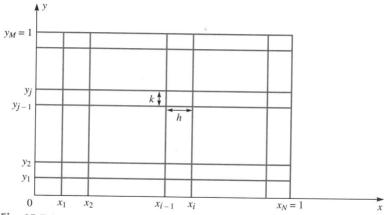

**Fig. 15.7.1**

At each interior grid point $(x_i, y_j)$, Equation (15.7.4) is satisfied and we have

$$\frac{\partial^2 u}{\partial x^2}(x_i, y_j) + \frac{\partial^2 u}{\partial y^2}(x_i, y_j) = q(x_i, y_j) \qquad \textbf{(15.7.6)}$$

Approximating the partial derivatives $\partial^2 u/\partial x^2$ and $\partial^2 u/\partial y^2$ at $(x_i, y_j)$ in Equation (15.7.6) by the centered second order differences, we get

$$\frac{1}{h^2}[U_{i-1,j} - 2U_{i,j} + U_{i+1,j}] + \frac{1}{k^2}[U_{i,j-1} - 2U_{i,j} + U_{i,j+1}] = q(x_i, y_j) = q_{i,j} \quad \textbf{(15.7.7)}$$

for $i = 1, 2, \ldots, N - 1$ and $j = 1, 2, \ldots, M - 1$. The central difference form of Equation (15.7.7) has local truncation error

$$-\left(\frac{h^2}{12}\frac{\partial^4 u}{\partial x^4}(\xi_i, y_j) + \frac{k^2}{12}\frac{\partial^4 u}{\partial x^4}(x_i, \eta_j)\right) \qquad \textbf{(15.7.8)}$$

where $0 < \xi_i < 1$ and $0 < \eta_j < 1$.

The mesh points involved in a typical step of Equation (15.7.7) for $h = k$ are shown in Figure 15.7.2.

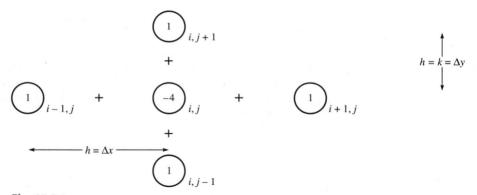

**Fig. 15.7.2**

Equation (15.7.7) involves five points as shown in Figure 15.7.2 and therefore Equation (15.7.7) is called the **five-point formula**. If smaller grids are used, then the truncation error [Equation (15.7.8)] becomes small. More accurate representations of the derivatives in Equation (15.7.6) can be used and often a nine-point formula (Exercise 7) is used in place of Equation (15.7.7).

**EXAMPLE 15.7.1**   Compute an approximate solution of

$$\frac{\partial^2 u}{\partial x^2} + \frac{\partial^2 u}{\partial y^2} = 0 \quad 0 < x, y < 1$$

with the boundary conditions

$$u(x, 0) = \sin 2\pi x \quad u(x, 1) = 0 \quad 0 \le x \le 1$$
$$u(0, y) = 0 \quad u(1, y) = \sin 2\pi y \quad 0 \le y \le 1$$

Let $N = M = 4$. Then $h = k = \frac{1}{4}$ and the boundary conditions give

$$U_{0,0} = 0 \qquad U_{1,0} = \sin\frac{\pi}{2} = 1 \qquad U_{2,0} = \sin\pi = 0 \qquad U_{3,0} = \sin\frac{3\pi}{2} = -1$$

$$U_{4,0} = 0$$

$$U_{0,4} = U_{1,4} = U_{2,4} = U_{3,4} = U_{4,4} = 0 \qquad U_{0,0} = U_{0,1} = U_{0,2} = U_{0,3} = U_{0,4} = 0$$

$$U_{4,0} = 0 \qquad U_{4,1} = \sin\frac{\pi}{2} = 1 \qquad U_{4,2} = \sin\pi = 0 \qquad U_{4,3} = \sin\frac{3\pi}{2} = -1$$

$$U_{4,4} = 0$$

The five-point formula [Equation (15.7.7)] reduces to

$$U_{i-1,j} - 4U_{i,j} + U_{i+1,j} + U_{i,j-1} + U_{i,j+1} = 0 \qquad \textbf{(15.7.9)}$$

for $i, j = 1, 2, 3$. There are nine interior points as shown in Figure 15.7.3.

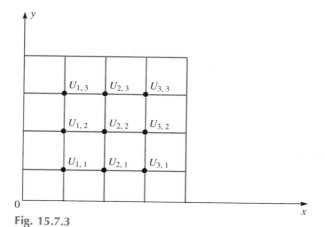

**Fig. 15.7.3**

We can order these unknowns in many ways. We choose to follow the **natural ordering** which is defined as the order of unknowns from left to right and up; for example,

$$\mathbf{u} = [U_{1,1}, U_{2,1}, U_{3,1}, U_{1,2}, U_{2,2}, U_{3,2}, U_{1,3}, U_{2,3}, U_{3,3}]'$$

For $j = 1$ and $i = 1, 2, 3$, Equation (15.7.9) gives

$$U_{0,1} - 4U_{1,1} + U_{2,1} + U_{1,0} + U_{1,2} = 0$$
$$U_{1,1} - 4U_{2,1} + U_{3,1} + U_{2,0} + U_{2,2} = 0$$
$$U_{2,1} - 4U_{3,1} + U_{4,1} + U_{3,0} + U_{3,2} = 0$$

For $j = 2$ and $i = 1, 2, 3$, Equation (15.7.9) reduces to

$$U_{0,2} - 4U_{1,2} + U_{2,2} + U_{1,1} + U_{1,3} = 0$$
$$U_{1,2} - 4U_{2,2} + U_{3,2} + U_{2,1} + U_{2,3} = 0$$
$$U_{2,2} - 4U_{3,2} + U_{4,2} + U_{3,1} + U_{3,3} = 0$$

For $j = 3$ and $i = 1, 2, 3$, Equation (15.7.9) gives

$$U_{0,3} - 4U_{1,3} + U_{2,3} + U_{1,2} + U_{1,4} = 0$$
$$U_{1,3} - 4U_{2,3} + U_{3,3} + U_{2,2} + U_{2,4} = 0$$
$$U_{2,3} - 4U_{3,3} + U_{4,3} + U_{3,2} + U_{3,4} = 0 \qquad \textbf{(15.7.10)}$$

The system of Equation (15.7.10) has the form

$$\begin{bmatrix} -4 & 1 & 0 & 1 & 0 & 0 & 0 & 0 & 0 \\ 1 & -4 & 1 & 0 & 1 & 0 & 0 & 0 & 0 \\ 0 & 1 & -4 & 0 & 0 & 1 & 0 & 0 & 0 \\ 1 & 0 & 0 & -4 & 1 & 0 & 1 & 0 & 0 \\ 0 & 1 & 0 & 1 & -4 & 1 & 0 & 1 & 0 \\ 0 & 0 & 1 & 0 & 1 & -4 & 0 & 0 & 1 \\ 0 & 0 & 0 & 1 & 0 & 0 & -4 & 1 & 0 \\ 0 & 0 & 0 & 0 & 1 & 0 & 1 & -4 & 1 \\ 0 & 0 & 0 & 0 & 0 & 1 & 0 & 1 & -4 \end{bmatrix} \begin{bmatrix} U_{1,1} \\ U_{2,1} \\ U_{3,1} \\ U_{1,2} \\ U_{2,2} \\ U_{3,2} \\ U_{1,3} \\ U_{2,3} \\ U_{3,3} \end{bmatrix} = \begin{bmatrix} 1 \\ 0 \\ 0 \\ 0 \\ 0 \\ 0 \\ 0 \\ 0 \\ -1 \end{bmatrix}$$
$$\textbf{(15.7.11)}$$

Thus Equation (15.7.10) can be written as

$$A\mathbf{u} = \mathbf{b} \qquad \textbf{(15.7.12)}$$

where $A$ is partitioned into blocks in a natural way so that Equation (15.7.11) may be written as

$$\begin{bmatrix} B & I & O \\ I & B & I \\ O & I & B \end{bmatrix} \begin{bmatrix} \mathbf{u}_1 \\ \mathbf{u}_2 \\ \mathbf{u}_3 \end{bmatrix} = \begin{bmatrix} \mathbf{b}_1 \\ \mathbf{b}_2 \\ \mathbf{b}_3 \end{bmatrix} \qquad \textbf{(15.7.13)}$$

where $I$ is the identity matrix,

$$B = \begin{bmatrix} -4 & 1 & 0 \\ 1 & -4 & 1 \\ 0 & 1 & -4 \end{bmatrix}$$

and $\mathbf{u}_j$ and $\mathbf{b}_j$ are the vectors with the $j$th row of grid points as shown in Figure 15.7.1. We can use the Gaussian elimination method to solve Equation (15.7.11), but the element that starts out as zero can become nonzero. Thus the Gaussian elimination method requires more storage space than the iterative methods. Since $A$ in Equation (15.7.11) is diagonally dominant, the Jacobi and Gauss–Seidel methods converge for any choice of $\mathbf{u}^{(0)}$. Solving Equation (15.7.11) by the Gauss–Seidel method gives

$$\begin{array}{lll} U_{1,1} = 0.28572 & U_{2,1} = 0.07143 & U_{3,1} = 0.00000 \\ U_{1,2} = 0.07143 & U_{2,2} = 0.00000 & U_{3,2} = -0.07143 \\ U_{1,3} = 0.00000 & U_{2,3} = -0.07143 & U_{3,3} = -0.28572 \end{array}$$

The errors at grid points are given by

$$\begin{array}{lll} e_{1,1} = 0.069261 & e_{2,1} = 0.02830 & e_{3,1} = 0.00000 \\ e_{1,2} = 0.02830 & e_{2,2} = 0.00000 & e_{3,2} = -0.02829 \\ e_{1,3} = 0.00000 & e_{2,3} = -0.02829 & e_{3,3} = -0.69254 \end{array}$$

The errors are quite large because the mesh size we used is coarse.  ■ ■ ■

In most applications, we need a fine mesh size to get a good approximation of the exact solution. It is not uncommon to divide the intervals $0 \leq x \leq 1$ and $0 \leq y \leq 1$ into 100 subintervals each. This gives in turn $10^4$ interior grid points and therefore a linear system with $10^4$ equations. We can use the SOR method, where the optimal $\omega_o$ is given by Equation (11.6.11)

$$\omega_o = \frac{2}{1 + \sqrt{1 - (\rho(B_J))^2}} = \frac{2}{1 + \sqrt{1 - \cos^2 \dfrac{\pi}{N+1}}} \qquad \textbf{(15.7.14)}$$

## EXERCISES

1. Compute an approximate solution of

$$\frac{\partial^2 u}{\partial x^2} + \frac{\partial^2 u}{\partial y^2} = 0 \quad 0 < x, y < 1$$

with the boundary conditions

$$u(x, 0) = \sin \pi x \quad u(x, 1) = 0 \quad 0 \leq x \leq 1$$
$$u(0, y) = 0 \quad u(1, y) = 0 \quad 0 \leq y \leq 1$$

Use $h = k = \frac{1}{3}$ and compare the approximate solution with the exact solution

$$u(x, y) = \sin \pi x \left[ \cosh \pi y - \coth \pi \sinh \pi y \right]$$

2. Compute an approximate solution of

$$\frac{\partial^2 u}{\partial x^2} + \frac{\partial^2 u}{\partial y^2} = 2e^{x+y} \quad 0 < x, y < 1$$

with the boundary conditions

$$u(x, 0) = e^x \quad u(x, 1) = e^{x+1} \quad 0 \leq x \leq 1$$
$$u(0, y) = e^y \quad u(1, y) = e^{y+1} \quad 0 \leq y \leq 1$$

Use $h = k = \frac{1}{3}$ and compare the approximate solution with the exact solution $u(x, y) = e^{x+y}$.

3. Compute an approximate solution of

$$\frac{\partial^2 u}{\partial x^2} + \frac{\partial^2 u}{\partial y^2} = 0 \quad 0 < x, y < 1$$

with the boundary conditions

$$u(x, 0) = u(x, 1) = x(1 - x) \quad 0 \leq x \leq 1$$
$$u(0, y) = u(1, y) = y(y - 1) \quad 0 \leq y \leq 1$$

Use $h = k = \frac{1}{3}$ and compare the approximate solution with the exact solution $u(x, y) = x(1 - x) + y(y - 1)$.

4. Compute an approximate solution of

$$\frac{\partial^2 u}{\partial x^2} + \frac{\partial^2 u}{\partial y^2} = 0 \quad 0 < x, y < 1$$

with the boundary conditions

$$\frac{\partial u}{\partial y}(x, 0) = \frac{\partial u}{\partial y}(x, 1) = 0 \quad 0 \le x \le 1 \qquad u(0,y) = u(1, y) = 1 \quad 0 \le y \le 1$$

Use $h = k = \frac{1}{3}$ and compare the approximate solution with the exact solution $u(x, y) = 1$.

5. Compute an approximate solution of

$$\frac{\partial^2 u}{\partial x^2} + \frac{\partial^2 u}{\partial y^2} = 0 \quad 0 < x, y < 1$$

with the boundary conditions

$$u(x, 0) = \cos \pi x \quad u(x, 1) = \cdot 0 \quad 0 \le x \le 1$$
$$\frac{\partial u}{\partial x}(0, y) = \frac{\partial u}{\partial x}(1, y) = 0 \quad 0 \le y \le 1$$

Use $h = k = \frac{1}{3}$ and compare the approximate solution with the exact solution

$$u(x, y) = \cos \pi x \frac{\sinh \pi (y - 1)}{\sinh(-\pi)}$$

6. Verify that the higher-order formula for Equation (15.7.6) is given by

$$\frac{1}{h^2}[-U_{i-2,j} + 16U_{i-1,j} - 30U_{i,j} + 16U_{i+1,j} - U_{i+2,j}]$$

$$+ \frac{1}{k^2}[-U_{i,j-2} + 16U_{i,j-1} - 30U_{i,j} + 16U_{i,j+1} - U_{i,j+2}] = q_{i,j} \qquad \textbf{(15.7.15)}$$

Sketch the mesh points involved in a typical step of Equation (15.7.15). What is a serious drawback of using Equation (15.7.15)?

7. If $h = k$, then verify the so-called **nine-point formula** for Equation (15.7.6)

$$\frac{1}{6h^2}[U_{i+1,j+1} + U_{i-1,j+1} + U_{i+1,j-1} + U_{i-1,j-1}$$

$$+ 4(U_{i+1,j} + U_{i-1,j} + U_{i,j+1} + U_{i,j-1}) - 20U_{i,j}] = q_{i,j} \qquad \textbf{(15.7.16)}$$

for $i = j = 1, 2, \ldots, N - 1$. Sketch the mesh points involved in a typical step of Equation (15.7.16).

8. Compute an approximate solution of

$$\frac{\partial^2 u}{\partial x^2} + \frac{\partial^2 u}{\partial y^2} = 0 \quad 0 < x, y < 1$$

with the boundary conditions

$$u(x,0) = \sin \pi x \quad u(x, 1) = 0 \quad 0 \le x \le 1$$
$$u(0, y) = 0 \quad u(1, y) = 0 \quad 0 \le y \le 1$$

using Equation (15.7.16) with $h = k = \frac{1}{3}$. Compare the approximate solution with the exact solution $u(x, y) = \sin \pi x [\cosh \pi y - \coth \pi \sinh \pi y]$.

**COMPUTER EXERCISES**

**C.1.**  Write a subroutine to solve Equation (15.7.7) using the Gauss–Seidel method or the SOR method with $\omega$ given by Equation (15.7.14).

**C.2.**  Compute an approximate solution of

$$\frac{\partial^2 u}{\partial x^2} + \frac{\partial^2 u}{\partial y^2} = 0 \quad 0 < x, y < 1$$

with the boundary conditions

$$u(x, 0) = \sin \pi x \quad u(x, 1) = 0 \quad 0 \le x \le 1$$
$$u(0, y) = 0 \quad u(1, y) = 0 \quad 0 \le y \le 1$$

using various values of $N = M$. Compare the approximate solution with the exact solution $u(x, y) = \sin \pi x \, [\cosh \pi y - \coth \pi \sinh \pi y]$.

**C.3.**  A rectangular, thin, homogeneous, and thermally conducting steel plate has length 20 cm and height 10 cm. Find the steady-state temperature at the interior points if one of the 10 cm edges is held at 100°C and the other three edges are held at 0°C.

**C.4.**  Compute an approximate solution $u$ of the Laplace equation $(\partial^2 u/\partial x^2) + (\partial^2 u/\partial y^2) = 0$ in the annular region between two concentric squares shown in the following figure.

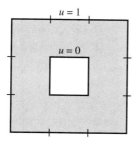

**C.5.**  When an arbitrary cross sectional shaped cylinder bar is twisted about its axis, the twisting produces shear stresses on the cross sectional planes which, in turn, cause the planes to wrap. The Prandtl stress function $\psi$ satisfies

$$\Delta^2 \psi = -2 \quad \text{inside the region } (\Omega) \text{ of cross section}$$

and

$$\psi = 0 \quad \text{on the perimeter of cross section}$$

The torsional rigidity $c$ is defined by $c = \iint_\Omega \psi \, dx \, dy$. Compute the torsional rigidity $c$ of a square (two unit length) shaped cylinder bar. (*Hint*: Use Simpson's rule twice.)

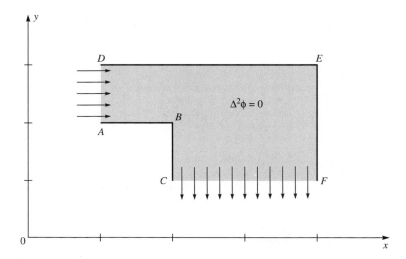

**C.6.**   The flow of an incompressible ideal fluid through a two-dimensional channel with a right angle bend as shown in the figure above is given by

$$\frac{\partial^2 \phi}{\partial x^2} + \frac{\partial^2 \phi}{\partial y^2} = 0 \quad \text{inside the channel}$$

$$\frac{\partial \phi}{\partial y} = 0 \quad \text{along DE and AB} \qquad \frac{\partial \phi}{\partial x} = 0 \quad \text{along BC and EF}$$

$$u = \frac{\partial \phi}{\partial x} = 1 \quad \text{along AD} \qquad v = \frac{\partial \phi}{\partial y} = -1 \quad \text{along CF}$$

Find $u$ and $v$ inside the channel for at least ten points.

## 15.8   HYPERBOLIC EQUATIONS

We are concerned with a typical hyperbolic equation

$$\frac{\partial^2 u}{\partial x^2} - \frac{\partial^2 u}{\partial t^2} = F(x, t)$$

in two independent variables $x$ and $t$. This equation, known as the **one-dimensional wave equation**, arises in the analysis of the vertical displacement of vibrating strings.

Consider a uniform string of length one and constant density $\rho = 1$ stretched between two supports. The string rests on the interval $[0, 1]$ with its ends fixed at $x = 0$ and 1. If the string is released from its resting position, then it vibrates. We assume that the string vibrates in a vertical plane. The vertical displacement $u(x, t)$ of a point at $x$ of the string at time $t$ is found by solving

$$\frac{\partial^2 u}{\partial x^2} = \frac{\partial^2 u}{\partial t^2} \quad 0 < x < 1 \quad t > 0 \tag{15.8.1}$$

subject to the boundary conditions

$$u(0, t) = u(1, t) = 0 \quad t > 0 \tag{15.8.2a}$$

and initial conditions

$$u(x, 0) = f(x) \quad \frac{\partial u}{\partial t}(x, 0) = g(x) \quad 0 \le x \le 1 \tag{15.8.2b}$$

The d'Alembert solution of Equations (15.8.1), (15.8.2a), and (15.8.2b) is given by

$$u(x, t) = \frac{1}{2}\left[ \tilde{f}_o(x + t) + \tilde{f}_o(x - t) + \tilde{G}_e(x + t) + \tilde{G}_e(x - t) \right] \tag{15.8.3}$$

where $\tilde{f}_o$ is an odd periodic extension of $f$ with period two and $\tilde{G}_e$ is an even periodic extension of $G$ with period two where

$$G(x) = \int_0^x g(s)\,ds \tag{15.8.4}$$

Let $(x_0, t_0)$ be any point in the $xt$ plane. Then

$$u(x_0, t_0) = \frac{1}{2}\left[ \tilde{f}_o(x_0 + t_0) + \tilde{f}_o(x_0 - t_0) + \tilde{G}_e(x_0 + t_0) + \tilde{G}_e(x_0 - t_0) \right] \tag{15.8.5}$$

Thus the value of $u$ at $(x_0, t_0)$ depends only on the values of the initial displacement $f$ at $x_0 - t_0$ and $x_0 + t_0$, and on the initial velocity $g$ over $[x_0 - t_0, x_0 + t_0]$. This means that a change in the initial conditions at points outside $[x_0 - t_0, x_0 + t_0]$ has no effect on the solution at $(x_0, t_0)$.

In order to solve Equations (15.8.1), (15.8.2a), and (15.8.2b), we rarely need numerical methods since the d'Alembert solution [Equation (15.8.3)] provides $u(x, t)$ for arbitrary $x$ and $t$. If a partial differential equation contains nonlinear terms or boundary and initial conditions are complex, then we need numerical techniques. In order to reduce Equation (15.8.1) to a difference equation, we introduce a network of grid points as shown in Figure 15.3.2 with grid spacing $\Delta x = h = 1/N$ and $\Delta t = k = T/M$ where $M$ and $N$ are positive integers. At each of the interior mesh points $(x_i, t_j)$, Equation (15.8.1) is satisfied and we have

$$\frac{\partial^2 u}{\partial x^2}(x_i, t_j) = \frac{\partial^2 u}{\partial t^2}(x_i, t_j) \tag{15.8.6}$$

Replacing $\partial^2 u/\partial x^2$ and $\partial^2 u/\partial t^2$ at $(x_i, t_j)$ in Equation (15.8.6) by centered second order differences, we get

$$\frac{U_{i-1,j} - 2U_{i,j} + U_{i+1,j}}{h^2} = \frac{U_{i,j-1} - 2U_{i,j} + U_{i,j+1}}{k^2} \tag{15.8.7}$$

for $i = 1, 2, \ldots, N - 1$ and $j = 1, 2, \ldots, M - 1$. Solving Equation (15.8.7) for $U_{i,j+1}$, we get

$$U_{i,j+1} = p^2(U_{i-1,j} + U_{i+1,j}) + 2(1 - p^2)U_{i,j} - U_{i,j-1} \tag{15.8.8}$$

for $i = 1, 2, \ldots, N - 1$ and $j = 1, 2, \ldots, M - 1$, where

$$p = \frac{k}{h} = \frac{\Delta t}{\Delta x} \tag{15.8.9}$$

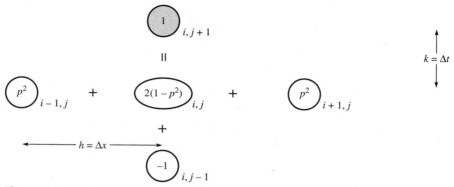

**Fig. 15.8.1**

The mesh points in a typical step of Equation (15.8.8) are shown in Figure 15.8.1.

It is clear from Figure 15.8.1 and Equation (15.8.8) that $U_{i,j+1}$ can be calculated explicitly from the values of $U$ at $t_j$ and $t_{j-1}$ levels. In order to get $U_{i,1}$, we need $U_{i,-1}$. We can use Equation (15.8.2b) to find $U_{i,-1}$. Using the second-order central difference formula

$$\frac{\partial u}{\partial t}(x_i, 0) = \frac{u(x_i, t_1) - u(x_i, t_{-1})}{2k} - \frac{k^2}{6}\frac{\partial^3 u}{\partial t^3}(x_i, \eta_i)$$

in Equation (15.8.2b), we get

$$U_{i,-1} = U_{i,1} - 2kg_i \quad \text{for } i = 1, 2, \ldots, N-1 \tag{15.8.10}$$

where $g_i = g(x_i)$. For $j = 0$, Equation (15.8.8) reduces to

$$U_{i,1} = p^2(U_{i-1,0} + U_{i+1,0}) + 2(1 - p^2)U_{i,0} - U_{i,-1} \tag{15.8.11}$$

Substituting $U_{i,-1}$ from Equation (15.8.10) in Equation (15.8.11), we get

$$U_{i,1} = \frac{p^2}{2}(U_{i-1,0} + U_{i+1,0}) + (1 - p^2)U_{i,0} + kg_i \tag{15.8.12}$$

Using $U_{i,0}$ and $U_{i,1}$, we calculate $U_{i,2}$ using Equation (15.8.8). Using $U_{i,1}$ and $U_{i,2}$, $U_{i,3}$ is calculated. We continue to compute the rest.

Using the Fourier stability method, it can be verified (Exercise 6) that Equation (15.8.8) is stable if $p \leq 1$. Further, we can prove (Exercise 7) that the truncation error for Equation (15.8.7) vanishes completely when $p = 1$. Thus the difference formula [Equation (15.8.8)] for $p = 1$ gives

$$U_{i,j+1} = U_{i-1,j} + U_{i+1,j} - U_{i,j-1} \tag{15.8.13}$$

an exact representation of Equation (15.8.1).

**EXAMPLE 15.8.1**   Compute an approximate solution of

$$\frac{\partial^2 u}{\partial x^2} = \frac{\partial^2 u}{\partial t^2} \quad 0 < x < 1 \quad t > 0$$

subject to the boundary conditions

$$u(0, t) = u(1, t) = 0 \quad t > 0$$

and initial conditions

$$u(x, 0) = \sin \pi x \quad \frac{\partial u}{\partial t}(x, 0) = 0 \quad 0 \le x \le 1$$

using Equation (15.8.13). The exact solution is given by $u(x, t) = \sin \pi x \cos \pi t$. Let $h = \frac{1}{4}$. Then $k = \frac{1}{4}$ since $p = 1$. Since $u(x, 0) = \sin \pi x$, we have

$$U_{1,0} = \sin \frac{\pi}{4} = \frac{1}{\sqrt{2}} = 0.70711 \qquad U_{2,0} = \sin \frac{\pi}{2} = 1.0$$

$$U_{3,0} = \sin \frac{3\pi}{4} = \frac{1}{\sqrt{2}} = 0.70711$$

For $p = 1$, Equation (15.8.12) reduces to

$$U_{i,1} = \frac{1}{2}(U_{i-1,0} + U_{i+1,0}) \tag{15.8.14}$$

For $i = 1, 2$, and 3, Equation (15.8.14) gives

$$U_{1,1} = \frac{1}{2}(U_{0,0} + U_{2,0}) = \frac{1}{2} = 0.5 \qquad U_{2,1} = \frac{1}{2}(U_{1,0} + U_{3,0}) = \frac{1}{\sqrt{2}} = 0.70711$$

$$U_{3,1} = \frac{1}{2}(U_{2,0} + U_{4,0}) = \frac{1}{2} = 0.5$$

For $j = 1$, Equation (15.8.13) reduces to

$$U_{i,2} = U_{i-1,1} + U_{i+1,1} - U_{i,0} \tag{15.8.15}$$

For $i = 1, 2$, and 3, Equation (15.8.15) gives

$$U_{1,2} = U_{0,1} + U_{2,1} - U_{1,0} = 0 + \frac{1}{\sqrt{2}} - \frac{1}{\sqrt{2}} = 0.0$$

$$U_{2,2} = U_{1,1} + U_{3,1} - U_{2,0} = 0.5 + 0.5 - 1.0 = 0.0$$

$$U_{3,2} = U_{2,1} + U_{4,1} - U_{3,0} = \frac{1}{\sqrt{2}} + 0 - \frac{1}{\sqrt{2}} = 0.0$$

In Table 15.8.1, the computed values of $U_{i,j}$ are given.

**Table 15.8.1**

| $j$ | $t = kj$ | $U(0,j)$ | $U(1,j)$ | $U(2,j)$ | $U(3,j)$ | $U(4,j)$ |
|---|---|---|---|---|---|---|
| 0 | 0.00 | 0.00000 | 0.70711 | 1.00000 | 0.70711 | 0.00000 |
| 1 | 0.25 | 0.00000 | 0.50000 | 0.70711 | 0.50000 | 0.00000 |
| 2 | 0.50 | 0.00000 | 0.00000 | 0.00000 | 0.00000 | 0.00000 |
| 3 | 0.75 | 0.00000 | −0.50000 | −0.70711 | −0.50000 | 0.00000 |
| 4 | 1.00 | 0.00000 | −0.70711 | −1.00000 | −0.70711 | 0.00000 |

The values given in Table 15.8.1 are also the values of the analytical solution $u(x, t) = \sin \pi x \cos \pi t$. This is not a surprise since the $p = 1$ case has zero local truncation error (Exercise 7). ∎

We can develop implicit methods in many ways to solve Equations (15.8.1)–(15.8.2b). The simplest method is to approximate $\partial^2 u/\partial t^2$ by centered second order differences while $\partial^2 u/\partial x^2$ is approximated by the average of the second order central differences at $(i, j + 1)$ and $(i, j - 1)$ (a Crank–Nicolson approximation). Thus

$$\frac{1}{2h^2} [U_{i-1,j+1} - 2U_{i,j+1} + U_{i+1,j+1} + U_{i-1,j-1} - 2U_{i,j-1} + U_{i+1,j-1}]$$

$$= \frac{1}{k^2} [U_{i,j-1} - 2U_{i,j} + U_{i,j+1}] \qquad \textbf{(15.8.16)}$$

for $i = 1, 2, \ldots, N - 1$ and $j = 1, 2, \ldots, M - 1$. Equation (15.8.16) can be written as

$$p^2 U_{i-1,j+1} - 2(1 + p^2) U_{i,j+1} + p^2 U_{i+1,j+1}$$
$$= 2(1 + p^2) U_{i,j-1} - p^2 U_{i-1,j-1} - p^2 U_{i+1,j-1} - 4U_{i,j} \qquad \textbf{(15.8.17)}$$

for $i = 1, 2, \ldots, N - 1$ and $j = 1, 2, \ldots, M - 1$.

The system (15.8.17) is tridiagonal and can be solved by *LU* decomposition method discussed in Section 7.7. Further, Equation (15.8.17) is stable (Exercise 11) for any positive value of $p$.

We can fairly easily use various techniques to solve a two-dimensional wave problem.

## EXERCISES

1. The tranverse displacement $\hat{u}$ of a point at a distance $\hat{x}$ from one end of a vibrating string of length $l$ at time $\hat{t}$ satisfies $\partial^2 \hat{u}/\partial \hat{x}^2 = c^2 (\partial^2 \hat{u}/\partial \hat{t}^2)$ where $c$ is a constant. Show that $x = \hat{x}/l$, $u = \hat{u}/l$, and $t = c\hat{t}/l$ reduces $\partial^2 \hat{u}/\partial \hat{x}^2 = c^2 (\partial^2 \hat{u}/\partial \hat{t}^2)$ to the nondimensional form $\partial^2 u/\partial x^2 = \partial^2 u/\partial t^2$.

2. Show that the independent variables defined by $\xi = x + t$ and $\eta = x - t$ transform $\partial^2 u/\partial x^2 = \partial^2 u/\partial t^2$ to $\partial^2 u/\partial \xi \partial \eta = 0$. Hence, deduce that the solution of $\partial^2 u/\partial x^2 = \partial^2 u/\partial t^2$ with initial conditions $u(x, 0) = f(x)$, $\partial u/\partial t (x, 0) = g(x)$ is given by

$$u(x, t) = \frac{1}{2} \left[ f(x + t) + f(x - t) + \int_{x-t}^{x+t} g(s)\, ds \right]$$

3. Compute an approximate solution of

$$\frac{\partial^2 u}{\partial x^2} = \frac{\partial^2 u}{\partial t^2} \quad 0 < x < 1 \quad t > 0$$

subject to the boundary conditions

$$u(0, t) = u(1, t) = 0 \quad t > 0$$

and initial conditions

$$u(x, 0) = 0 \quad \frac{\partial u}{\partial t}(x, 0) = \sin \pi x \quad 0 \le x \le 1$$

using Equation (15.8.13) with $h = \frac{1}{4}$. Compare your results at $t = 0.5$ and $1.0$ with the exact solution $u(x, t) = (\sin \pi x \sin \pi t)/\pi$.

4.  Compute an approximate solution of

$$\frac{\partial^2 u}{\partial x^2} = \frac{\partial^2 u}{\partial t^2} \quad 0 < x < 1 \quad t > 0$$

subject to the boundary conditions

$$u(0, t) = u(1, t) = 0 \quad t > 0$$

and initial conditions

$$u(x, 0) = 0 \quad \frac{\partial u}{\partial t}(x, 0) = 1 \quad 0 \le x \le 1$$

using Equation (15.8.13) with $h = \frac{1}{4}$.

5.  Compute an approximate solution of

$$\frac{\partial^2 u}{\partial x^2} = \frac{\partial^2 u}{\partial t^2} \quad 0 < x < 1 \quad t > 0$$

subject to the boundary conditions

$$u(0, t) = u(1, t) = 0 \quad t > 0$$

and initial conditions

$$u(x, 0) = \sin \pi x \quad \frac{\partial u}{\partial t}(x, 0) = 0 \quad 0 \le x \le 1$$

using Equation (15.8.17) with $h = \frac{1}{3}$. Compare your results at $t = 1.0$ with the exact solution $u(x, t) = \sin \pi x \cos \pi t$.

6.  Using the Fourier stability method, verify that Equation (15.8.8) is stable if $p \le 1$.

7.  Expand

$$\frac{u(x_i - h, t_j) - 2u(x_i, t_j) + u(x_i + h, t_j)}{h^2} - \frac{u(x_i, t_j - k) - 2u(x_i, t_j) + u(x_i, t_j + k)}{k^2}$$

using Taylor's Theorem A.9 around $(x_i, t_j)$ and then verify that the truncation error is zero when $p = 1$.

8.  Use Equation (15.8.8) with $h = \frac{1}{4}$ and $p = 2$ to compute an approximate solution of

$$\frac{\partial^2 u}{\partial x^2} = \frac{\partial^2 u}{\partial t^2} \quad 0 < x < 1 \quad t > 0$$

subject to the boundary conditions

$$u(0, t) = u(1, t) = 0 \quad t > 0$$

and initial conditions

$$u(x, 0) = \begin{cases} 0 & \text{if } x \ne 1/2 \\ \epsilon & \text{if } x = 1/2 \end{cases} \quad \frac{\partial u}{\partial t}(x, 0) = 0 \quad 0 \le x \le 1$$

Compare your results with the exact solution $u(x, t) = 0$ of

$$\frac{\partial^2 u}{\partial x^2} = \frac{\partial^2 u}{\partial t^2} \quad 0 < x < 1 \quad t > 0 \qquad u(0, t) = u(1, t) = 0 \quad t > 0$$

$$u(x, 0) = \frac{\partial u}{\partial t}(x, 0) = 0 \quad 0 \le x \le 1$$

9. Use Equation (15.8.8) with $h = \frac{1}{4}$ and $p = \frac{1}{2}$ to compute an approximate solution at $t = \frac{1}{2}$ of

$$\frac{\partial^2 u}{\partial x^2} = \frac{\partial^2 u}{\partial t^2} \quad 0 < x < 1 \quad t > 0$$

subject to the boundary conditions

$$u(0, t) = u(1, t) = 0 \quad t > 0$$

and initial conditions

$$u(x, 0) = \begin{cases} 0 & \text{if } x \ne 1/2 \\ \epsilon & \text{if } x = 1/2 \end{cases} \qquad \frac{\partial u}{\partial t}(x, 0) = 0 \quad 0 \le x \le 1$$

Compare your results with the exact solution $u(x, t) = 0$ at $t = \frac{1}{2}$ of

$$\frac{\partial^2 u}{\partial x^2} = \frac{\partial^2 u}{\partial t^2} \quad 0 < x < 1 \quad t > 0 \qquad u(0, t) = u(1, t) = 0 \quad t > 0$$

$$u(x, 0) = \frac{\partial u}{\partial t}(x, 0) = 0 \quad 0 \le x \le 1$$

10. Compute an approximate solution of

$$\frac{\partial^2 u}{\partial x^2} = \frac{\partial^2 u}{\partial t^2} \quad 0 < x < 1 \quad t > 0$$

subject to the boundary conditions

$$u(0, t) = u(1, t) = 0 \quad t > 0$$

and initial conditions

$$u(x, 0) = 0 \quad \frac{\partial u}{\partial t}(x, 0) = \sin \pi x \quad 0 \le x \le 1$$

using Equation (15.8.17) with $h = \frac{1}{4}$. Compare your results at $t = 0.5$ and $1.0$ with the exact solution $u(x, t) = (\sin \pi x \sin \pi t)/\pi$.

11. Using the Fourier stability method, verify that Equation (15.8.17) is stable for any positive value of $p$.

## COMPUTER EXERCISES

**C.1.** Write a subroutine to generate difference equations given by Equation (15.8.8).

**C.2.** Write a computer program using Exercise C.1 to approximate the solution of

$$\frac{\partial^2 u}{\partial x^2} = \frac{\partial^2 u}{\partial t^2} \quad 0 < x < 1 \quad t > 0$$

subject to the boundary conditions

$$u(0, t) = u(1, t) = 0 \quad t > 0$$

and initial conditions

$$u(x, 0) = \begin{cases} x & \text{if } 0 < x \le 1/2 \\ 1 - x & \text{if } 1/2 < x \le 1 \end{cases} \qquad \frac{\partial u}{\partial t}(x, 0) = 0 \quad 0 \le x \le 1$$

This corresponds to the vibration of the string lifted in the middle and then released. Use $h = 0.1$ and $0.01$ and $p = 0.5$, $1$, and $2$. Sketch the graph of $U$ at various time $t$.

**C.3.** A uniform circular membrane of radius $a$ is fixed rigidly around its circumference and is displaced. When the membrane is released, vibration starts. The displacement $\hat{u}(\hat{r}, \hat{t})$ satisfies

$$\frac{\partial^2 \hat{u}}{\partial \hat{r}^2} + \frac{1}{\hat{r}} \frac{\partial \hat{u}}{\partial \hat{r}} = \frac{1}{c^2} \frac{\partial^2 \hat{u}}{\partial \hat{t}^2}$$

$$\hat{u}(a, t) = 0 \quad t > 0$$

$$\hat{u}(\hat{r}, 0) = \left(1 - \frac{\hat{r}^2}{a^2}\right) \quad \frac{\partial \hat{u}}{\partial \hat{t}}(\hat{r}, 0) = 0 \quad 0 < \hat{r} < a$$

Reduce it to the nondimensional form using $u = \hat{u}$, $r = \hat{r}/a$, and $t = c\hat{t}/a$. Then approximate the displacement $u$ at various points at $t = 0.5$ and $1.0$.

## SUGGESTED READINGS

W. F. Ames, *Numerical Methods for Partial Differential Equations*, 3rd ed., Academic Press, New York, 1992.

J. Douglas Jr., *On the numerical integration of $\partial^2 u/\partial x^2 + \partial^2 u/\partial y^2 = \partial u/\partial t$ by implicit methods*, J. Soc. Indust. Appl. Math. **3** (1955), 42–65.

E. C. DuFort and S. P. Frankel, *Stability conditions in the numerical treatment of parabolic differential equations*, Math. Tables and other Aids to Computation **7**, (1953), 135–152.

P. R. Garabedian, *Partial Differential Equations*, John Wiley & Sons, Inc., New York, 1964.

M. K. Jain, *Numerical Solutions of Differential Equations*, 2nd ed., Wiley Eastern, New Delhi, 1984.

L. Lapidus and G. F. Pinder, *Numerical Solution of Partial Differential Equations in Science and Engineering*, Wiley–Interscience, New York, 1982.

V. A. Patel, *Time-dependent solutions of the viscous incompressible flow past a circular cylinder by the method of series truncation*, Comput. Fluids **4** (1976), 13–27.

V. A. Patel, *Kármán vortex street behind a circular cylinder by the series truncation method*, J. Comput. Phys. **28** (1978), 14–42.

D. W. Peaceman and H. H. Rachford, *The numerical solution of parabolic and elliptic differential equations*, J. Soc. Indust. App. Math. **3** (1955), 28–41.

R. D. Richtmeyer and K. W. Morton, *Difference Methods for Initial-Value Problems*, 2nd ed., Interscience Publishers, New York, 1967.

M. Van Dyke, *An Album Of Fluid Motion*, The Parabolic Press, Stanford, California, 1981.

# APPENDIX A

## Review of Calculus

In this appendix we list some basic results from calculus that are used in this text. The exact statement of a given theorem frequently varies from one book to another; therefore we give the exact form for our use. Also, we present other results used in the text that the reader may not be familiar with. In some cases the proof is given. The proofs of the results can be found in any calculus or advanced calculus text.

Let $R$ denote the set of all real numbers and $f$ be defined on an open interval containing $x_0 \in R$. Then $f$ is said to have the **limit** $L$ at $x_0$, written as $\lim_{x \to x_0} f(x) = L$, if, for any given real number $\epsilon > 0$ there exists a real number $\delta > 0$ such that if $0 < |x - x_0| < \delta$, then $|f(x) - L| < \epsilon$. Further $f$ is said to be **continuous** at $x_0$ if $\lim_{x \to x_0} f(x) = f(x_0)$.

**Theorem A.0**

(**Sandwich Theorem**)   Let $f(x) \le h(x) \le g(x)$ for every $x$ in an open interval containing $x_0$, except possibly at $x_0$ and $\lim_{x \to x_0} f(x) = L = \lim_{x \to x_0} g(x)$. Then $\lim_{x \to x_0} h(x) = L$.  ∎

$f$ is said to be **differentiable** at $x_0$ if $\lim_{x \to x_0} [f(x) - f(x_0)]/(x - x_0)$ exists. If this limit exists, then it is denoted by $f'(x_0)$ and is called the derivative of $f$ at $x_0$.

We use the following notations:

$$[a, b] = \{x \mid a \le x \le b\} \qquad (a, b) = \{x \mid a < x < b\}$$
$$(a, b] = \{x \mid a < x \le b\} \qquad [a, b) = \{x \mid a \le x < b\}$$

**Theorem A.1**

(**Extreme Value Theorem**)   Let $f$ be continuous on a closed interval $[a, b]$. Then $f$ assumes a minimum value $m$ and a maximum value $M$ at least once in $[a, b]$; that is, there exists at least two numbers $x_m$ and $x_M$ in $[a, b]$ such that $m = f(x_m) \le f(x) \le f(x_M) = M$ for any $x$ in $[a, b]$.  ∎

**Theorem A.2**

(**Intermediate Value Theorem**)   Let $f$ be continuous on a closed interval $[a, b]$, $m = \min_{a \le x \le b} f(x)$, and $M = \max_{a \le x \le b} f(x)$. Then for any number $k$ in $[m, M]$, there is at least one number $c$ in $[a, b]$ such that $f(c) = k$.  ∎

**Corollary A.2.1**    If $f$ is continuous on a closed interval $[a, b]$ and $f(a)f(b) \leq 0$, then there is at least one number $c$ in $[a, b]$ such that $f(c) = 0$. ∎

This corollary is used extensively to find the roots of $f(x) = 0$.

**Corollary A.2.2**    Let $f$ be continuous on a closed interval $[a, b]$ and $\alpha_1, \alpha_2, \ldots, \alpha_N$ be any nonnegative numbers such that $\Sigma_{i=1}^{N} \alpha_i = A$. Then for each set of $N$ numbers $x_i \in [a, b]$ for $i = 1, 2, \ldots, N$, there is at least one number $\xi$ in $[a, b]$ such that $\Sigma_{i=1}^{N} \alpha_i f(x_i) = Af(\xi)$. ∎

**Theorem A.3**    (**Mean Value Theorem**)    Let $f$ be continuous on a closed interval $[a, b]$ and differentiable on the open interval $(a, b)$. Then there is at least one number $c$ in $(a, b)$ such that $f'(c) = [f(b) - f(a)]/(b - a)$. ∎

**Corollary A.3.1**    (**Rolle's Theorem**)    Let $f$ be continuous on a closed interval $[a, b]$ and differentiable on the open interval $(a, b)$ and $f(a) = f(b)$. Then there is at least one number $c$ in $(a, b)$ such that $f'(c) = 0$. ∎

**Corollary A.3.2**    (**Generalized Rolle's Theorem**)    Let $f$ be $N$ times differentiable on a closed interval $[a, b]$ and $f(x_0) = f(x_1) = \cdots = f(x_N)$ where $x_0, x_1, \ldots, x_N$ are distinct numbers in $[a, b]$. Then there is at least one number $c$ in $(a, b)$ such that $f^{(N)}(c) = 0$. ∎

**Theorem A.4**    (**Fundamental Theorem of Calculus**)    Let $f$ be continuous on a closed interval $[a, b]$ and $g$ be defined by $g(x) = \int_a^x f(t)dt$. Then $g'(x) = f(x)$ for all $x$ in $[a, b]$. Further, if $f'$ is continuous on $[a, b]$, then $\int_a^b f'(x)dx = f(b) - f(a)$. ∎

**Theorem A.5**    (**Second Mean Value Theorem for Definite Integrals**)    Let $f$ be continuous on a closed interval $[a, b]$ and $w$ be an integrable function which does not change sign on $[a, b]$. Then there is at least one number $c$ in $[a, b]$ such that

$$\int_a^b f(x)w(x)dx = f(c) \int_a^b w(x)dx$$

***Proof***    For convenience, let $w(x) \geq 0$ for all $x$ in $[a, b]$. Let

$$h(u) = f(u) \int_a^b w(x)dx - \int_a^b f(x)w(x)dx = \int_a^b [f(u) - f(x)]w(x)dx$$

Since $\int_a^b w(x)dx$ and $\int_a^b f(x)w(x)dx$ are constants and $f$ is continuous on $[a, b]$, $h$ is continuous on $[a, b]$. Further $f$ is continuous, therefore by Theorem A.1 (Extreme Value Theorem), $f$ attains its minimum and maximum values on $[a, b]$. Let $m = f(x_m) = \min_{a \leq x \leq b} f(x)$ and $M = f(x_M) = \max_{a \leq x \leq b} f(x)$. Then $h(x_m) = \int_a^b [f(x_m) - f(x)]w(x)dx$.

Since $f(x_m) - f(x) \leq 0$ for all $x$ in $[a, b]$, then $h(x_m) \leq 0$. Similarly we can show that $h(x_M) \geq 0$. Thus $h(x_m)h(x_M) \leq 0$ and therefore by Corollary A.2.1, there is at least one number $c$ in $[a, b]$ such that $h(c) = 0$. This proves the theorem. ∎

**Corollary A.5.1**

(**First Mean Value Theorem for Definite Integrals**)  Let $f$ be continuous on a closed interval $[a, b]$. Then there is at least one number $c$ in $[a, b]$ such that $\int_a^b f(x)dx = f(c)(b - a)$. ∎

## Sequences and Series

Let $\{x_n\}$ be an infinite sequence of real numbers. This sequence is said to **converge** to a real number $x$, called the limit, if, for any given real number $\epsilon > 0$, there exists a positive integer $N$ such that for all $n > N$, we have $|x_n - x| < \epsilon$. This is written as $\lim_{n \to \infty} x_n = x$.

**Theorem A.6**

If $\{x_n\}$ is a bounded monotonic sequence, then the sequence $\{x_n\}$ converges. ∎

An expression of the form $\sum_{n=1}^{\infty} a_n = a_1 + a_2 + \cdots + a_n + \cdots$ is called an **infinite series** and $a_n$ is called the $n$th term of the series.

Let $S_1 = a_1, S_2 = a_1 + a_2, S_3 = a_1 + a_2 + a_3, \ldots, S_n = a_1 + a_2 + a_3 + \cdots + a_n$. Then the sequence $\{S_n\}$ is called the **sequence of partial sums** of the series $\sum_{i=1}^{\infty} a_i$. If the sequence $\{S_n\}$ converges to $L$, then $\sum_{n=1}^{\infty} a_n$ is said to converge to $L$ and we write $\sum_{n=1}^{\infty} a_n = L$. $L$ is called the **sum of the series**. If the sequence $\{S_n\}$ diverges, then the infinite series $\sum_{n=1}^{\infty} a_n$ is said to diverge.

The **geometric series** $a + ar + ar^2 + \cdots + ar^{n-1} + \cdots$ arises frequently in applications.

**Theorem A.7**

The geometric series $\sum_{n=1}^{\infty} ar^{n-1}$ where $a \neq 0$, (1) converges to $a/(1 - r)$ if $|r| < 1$ and (2) diverges if $|r| \geq 1$. ∎

A **power series** in $x$ is a series of the form $\sum_{n=1}^{\infty} a_n x^n$.

**Theorem A.8**

(**Taylor's Theorem**)  Let $f$ be $(N + 1)$ times differentiable and its $(N + 1)$th derivative be continuous on a closed interval $[a, b]$ and let $x_0 \in [a, b]$. Then for every $x \in [a, b]$, there exists $\xi(x)$ between $x_0$ and $x$ with $f(x) = P_N(x) + R_N(x)$ where

$$P_N(x) = f(x_0) + \frac{f'(x_0)}{1!}(x - x_0) + \frac{f''(x_0)}{2!}(x - x_0)^2 + \cdots + \frac{f^{(N)}(x_0)}{N!}(x - x_0)^N$$

and

$$R_N(x) = \frac{f^{(N+1)}(\xi(x))}{(N + 1)!}(x - x_0)^{N+1} = \frac{1}{N!}\int_{x_0}^{x}(x - t)^N f^{(N+1)}(t)dt$$

$P_N(x)$ is called the **Nth degree Taylor polynomial** for $f(x)$ about $x_0$ and $R_N(x)$ is called the **remainder term** (or **truncation error**) associated with $P_N(x)$. If $x_0 = 0$, then the Taylor polynomial is called the **Maclaurin polynomial**. The infinite series obtained by taking the limit of $P_N(x)$ as $N \to \infty$ is called the **Taylor series**.

One of the most important theorems used in numerical analysis is Taylor's Theorem and we will use it extensively. This theorem gives a relatively simple method for the approximation of $f$ by polynomials.

*Proof*   By the Fundamental Theorem of Calculus (Theorem A.4),

$$\int_{x_0}^{x} f'(t)dt = f(x) - f(x_0) \tag{A.1}$$

Using integration by parts and Equation (A.1), we get

$$\int_{x_0}^{x} (x - t)f''(t)dt = -(x - x_0)f'(x_0) + \int_{x_0}^{x} f'(t)dt$$
$$= -(x - x_0)f'(x_0) + f(x) - f(x_0) \tag{A.2}$$

Again, using integration by parts and Equation (A.2), we get

$$\frac{1}{2!}\int_{x_0}^{x} (x - t)^2 f^{(3)}(t)dt = -\frac{(x - x_0)^2}{2!}f''(x_0) + \int_{x_0}^{x} (x - t)f''(t)dt$$
$$= -\frac{(x - x_0)^2}{2!}f''(x_0) - (x - x_0)f'(x_0) + f(x) - f(x_0)$$

Repeating this $N$ times yields

$$\frac{1}{N!}\int_{x_0}^{x} (x - t)^N f^{(N+1)}(t)dt = -\frac{(x - x_0)^N}{N!}f^{(N)}(x_0)$$
$$- \frac{(x - x_0)^{N-1}}{(N - 1)!}f^{(N-1)}(x_0) \cdots$$
$$- (x - x_0)f'(x_0) - f(x_0) + f(x) \tag{A.3}$$

Equation (A.3) can be written as

$$f(x) = f(x_0) + f'(x_0)(x - x_0) + \cdots + \frac{f^{(N)}(x_0)}{N!}(x - x_0)^N + R_N$$

where

$$R_N = \frac{1}{N!}\int_{x_0}^{x} (x - t)^N f^{(N+1)}(t)dt \tag{A.4}$$

Since $(x - t)$ does not change its sign on $[x_0, x]$, $(x - t)^N$ remains of the same sign on $[x_0, x]$. Hence, by Theorem A.5, there exists $\xi$ between $x_0$ and $x$ such that

$$R_N(x) = \frac{1}{N!}\int_{x_0}^{x} (x - t)^N f^{(N+1)}(t)dt = \frac{f^{(N+1)}(\xi)}{N!}\int_{x_0}^{x} (x - t)^N dt = \frac{f^{(N+1)}(\xi)}{(N + 1)!}(x - x_0)^{N+1}$$

This proves the theorem.   ∎

**EXAMPLE A.1**    Find the third-degree Taylor polynomial for $f(x) = e^x$ about $x_0 = 1$. Use this polynomial to approximate $f(0.9)$, and find an upper bound for the truncation error in this approximation.

We can verify that the third-degree Taylor polynomial $P_3(x)$ about $x_0 = 1$ is given by

$$P_3(x) = e + \frac{e}{1!}(x - 1) + \frac{e}{2!}(x - 1)^2 + \frac{e}{3!}(x - 1)^3$$

and the remainder $R_3(x)$ is given by

$$R_3(x) = \frac{e^{\xi(x)}}{4!}(x - 1)^4$$

for some $\xi(x)$ between $x_0$ and $x$. With $x = 0.9$, the Taylor polynomial is

$$P_3(0.9) = e\left[1 + \frac{(0.9 - 1)}{1!} + \frac{(0.9 - 1)^2}{2!} + \frac{(0.9 - 1)^3}{3!}\right] \approx 2.45959$$

This compares well with the exact value, $e^{0.9} = 2.45960$ to six significant digits. The remainder is

$$|R_3(x)| = |R_3(0.9)| = \frac{e^\xi}{4!}(0.9 - 1)^4 \quad \text{where } 0.9 < \xi < 1$$

Since $\xi$ is unknown, we pick the maximum value of $e^\xi$ in the closed interval $[0.9, 1]$. Thus

$$|R_3(0.9)| = \left|\frac{e^\xi}{4!}(-0.1)^4\right| \leq \frac{e^1}{4!}(-0.1)^4 \approx 0.00001. \quad \blacksquare\ \blacksquare\ \blacksquare$$

Let $f(x, y)$ be a function of two independent variables $x$ and $y$. Then the earlier Taylor's Theorem can be extended for the expansion of $f(x, y)$ about a given point $(x_0, y_0)$.

**Theorem A.9**    (**Taylor's Theorem in two dimensions**)    Let $f(x, y)$ be $(N + 1)$ times continuously differentiable for all $(x, y)$ in a neighborhood $D$ of the given point $(x_0, y_0)$ in the $xy$-plane. Then, for every $(x, y) \in D$ there exists a point $(r, s)$ somewhere on the line segment joining the points $(x, y)$ and $(x_0, y_0)$ with

$$f(x, y) = P_N(x, y) + R_N(x, y)$$

where

$$P_N(x, y) = f(x_0, y_0) + \left[(x - x_0)\frac{\partial f}{\partial x}(x_0, y_0) + (y - y_0)\frac{\partial f}{\partial y}(x_0, y_0)\right]$$

$$+ \frac{1}{2!}\left[(x - x_0)^2\frac{\partial^2 f}{\partial x^2}(x_0, y_0) + 2(x - x_0)(y - y_0)\frac{\partial^2 f}{\partial x \partial y}(x_0, y_0)\right.$$

$$\left. + (y - y_0)^2\frac{\partial^2 f}{\partial y^2}(x_0, y_0)\right]$$

$$+ \cdots + \frac{1}{N!}\left[\sum_{i=0}^{N}\binom{N}{i}(x - x_0)^{N-i}(y - y_0)^i\frac{\partial^N f}{\partial x^{N-i}\partial y^i}(x_0, y_0)\right]$$

and

$R_N(x, y)$

$$= \frac{1}{(N+1)!} \left[ \sum_{i=0}^{N+1} \binom{N+1}{i} (x - x_0)^{N+1-i} (y - y_0)^i \frac{\partial^{N+1} f}{\partial x^{N+1-i} \partial y^i} (r, s) \right] \Bigg|_{\substack{r = x_0 + \tau(x - x_0) \\ s = y_0 + \tau(y - y_0)}}$$

for some $0 \leq \tau(x, y) \leq 1$.

$P_N(x, y)$ is called the **Taylor polynomial of degree $N$ in two variables** and $R_N(x, y)$ is called the **remainder term** associated with $P_N(x, y)$.

Similar results can be extended for the functions of more than two variables.

**Proof**  Let $h = (x - x_0)$, $k = (y - y_0)$, and define $F(t) = f(x + ht, y + kt)$ where $x$, $y$, $h$, and $k$ are independent of $t$. Then by Taylor's Theorem

$$F(t) = F(0 + t) = F(0) + \frac{F'(0)}{1} t + \frac{F''(0)}{2!} t^2$$
$$+ \cdots + \frac{F^{(N)}(0)}{N!} t^N + \frac{F^{(N+1)}(\tau)}{(N+1)!} t^{N+1} \qquad \text{(A.5)}$$

where $\tau$ is between $0$ and $t$.

Let $r = x + ht$ and $s = y + kt$. Then $F(t) = f(r(t), s(t))$, so

$$F'(t) = \frac{d}{dt} (F(t)) = \frac{\partial f}{\partial r} \frac{dr}{dt} + \frac{\partial f}{\partial s} \frac{ds}{dt} = h \frac{\partial f}{\partial r} + k \frac{\partial f}{\partial s}$$

$$= \left( h \frac{\partial}{\partial r} + k \frac{\partial}{\partial s} \right) f(r, s)$$

$$F''(t) = \frac{d}{dt} (F'(t)) = h \frac{\partial}{\partial r} \left( h \frac{\partial f}{\partial r} + k \frac{\partial f}{\partial s} \right) + k \frac{\partial}{\partial s} \left( h \frac{\partial f}{\partial r} + k \frac{\partial f}{\partial s} \right)$$

$$= h^2 \frac{\partial^2 f}{\partial r^2} + 2hk \frac{\partial^2 f}{\partial r \partial s} + k^2 \frac{\partial^2 f}{\partial s^2} \quad \text{by assuming} \quad \frac{\partial^2 f}{\partial r \partial s} = \frac{\partial^2 f}{\partial s \partial r}$$

$$= \left( h \frac{\partial}{\partial r} + k \frac{\partial}{\partial s} \right)^2 f(r, s)$$

The analogy with the binomial expansion is evident, and so

$$F^{(i)}(t) = \left( h \frac{\partial}{\partial r} + k \frac{\partial}{\partial s} \right)^i f(r, s) \quad \text{for } i = 0, 1, \ldots, N + 1$$

Hence

$$F^{(i)}(0) = \left( h \frac{\partial}{\partial x} + k \frac{\partial}{\partial y} \right)^i f(x, y) \quad \text{for } i = 0, 1, \ldots, N + 1$$

Let $t = 1$. Then Equation (A.5) gives

$$F(1) = F(0) + \frac{F'(0)}{1} + \frac{F''(0)}{2!} + \cdots + \frac{F^{(N)}(0)}{N!} + \frac{F^{(N+1)}(\tau)}{(N+1)!} \qquad \text{(A.6)}$$

Substituting $F(1)$, $F(0)$, $F'(0)$, ..., $F^{(N)}(0)$, and $F^{(N+1)}(\tau)$ in Equation (A.6), we get

$$f(x + h, y + k) = f(x, y) + \left( h\frac{\partial f}{\partial x}(x, y) + k\frac{\partial f}{\partial y}(x, y) \right)$$

$$+ \frac{1}{2!}\left( h^2\frac{\partial^2 f}{\partial x^2}(x, y) + 2hk\frac{\partial^2 f}{\partial x\partial y}(x, y) + k^2\frac{\partial^2 f}{\partial y^2}(x, y) \right) + \cdots$$

$$+ \frac{1}{N!}\left( h\frac{\partial}{\partial x} + k\frac{\partial}{\partial y} \right)^N f(x, y)$$

$$+ \frac{1}{(N+1)!}\left( h\frac{\partial}{\partial r} + k\frac{\partial}{\partial s} \right)^{N+1} f(r, s)\Bigg|_{\substack{r=x+h\tau \\ s=y+k\tau}} \tag{A.7}$$

This is perhaps the most compact form, but in order to write it in a form which resembles the one-dimensional form, replace $(x, y)$ by $(x_0, y_0)$, and $h$ by $(x - x_0)$, and $k$ by $(y - y_0)$ respectively in Equation (A.7). Then we get

$$f(x, y) = f(x_0 + (x - x_0), y_0 + (y - y_0)) = f(x_0, y_0)$$

$$+ \left[ (x - x_0)\frac{\partial f}{\partial x}(x_0, y_0) + (y - y_0)\frac{\partial f}{\partial y}(x_0, y_0) \right]$$

$$+ \frac{1}{2!}\left[ (x - x_0)^2\frac{\partial^2 f}{\partial x^2}(x_0, y_0) + 2(x - x_0)(y - y_0)\frac{\partial^2 f}{\partial x\partial y}(x_0, y_0) \right.$$

$$\left. + (y - y_0)^2\frac{\partial^2 f}{\partial y^2}(x_0, y_0) \right]$$

$$+ \cdots + \frac{1}{N!}\left[ \sum_{i=0}^{N}\binom{N}{i}(x - x_0)^{N-i}(y - y_0)^i\frac{\partial^N f}{\partial x^{N-i}\partial y^i}(x_0, y_0) \right]$$

$$+ \frac{1}{(N+1)!}\left[ (x - x_0)\frac{\partial}{\partial r} + (y - y_0)\frac{\partial}{\partial s} \right]^{N+1} f(r, s)\Bigg|_{\substack{r=x_0+\tau(x-x_0) \\ s=y_0+\tau(y-y_0)}} \quad\blacksquare$$

**EXAMPLE A.2**    Find the Taylor polynomial of second degree for $f(x, y) = e^{x+y}$ about $(0, 1)$.

Since $f(x, y) = e^{x+y}$, $f_x(x, y) = \dfrac{\partial f}{\partial x}(x, y) = e^{x+y}$, $f_y(x, y) = \dfrac{\partial f}{\partial y}(x, y) = e^{x+y}$, and

$$f_{xx}(x, y) = \frac{\partial^2 f}{\partial x^2}(x, y) = f_{yy}(x, y) = \frac{\partial^2 f}{\partial^2 y}(x, y) = f_{xy}(x, y) = \frac{\partial^2 f}{\partial x\partial y}(x, y) = e^{x+y}$$

Hence

$$P_2(x, y) = f(0, 1) + [(x - 0)f_x(0, 1) + (y - 1)f_y(0, 1)]$$

$$+ \frac{1}{2!}[(x - 0)^2 f_{xx}(0, 1) + 2(x - 0)(y - 1)f_{xy}(0, 1) + (y - 1)^2 f_{yy}(0, 1)]$$

$$= e + [xe + (y - 1)e] + \frac{1}{2!}[x^2 e + 2x(y - 1)e + (y - 1)^2 e]$$

$$= e\left\{ x + y + \frac{x^2}{2} + x(y - 1) + \frac{(y - 1)^2}{2} \right\}$$   ■ ■ ■

## EXERCISES

1. Prove Corollary A.2.1.
2. Prove that there is a number $c$ between 0 and 2 such that $f(c) = 0$ where $f(x) = x^3 - 1$ using Corollary A.2.1.
3. Prove that $x^3 - 1 = 0$ has a root between 0 and 2.
4. Prove Corollary A.2.2.
5. Prove that $|\sin u - \sin v| \le |u - v|$ for any real numbers $u$ and $v$ by using the Mean Value Theorem.
6. Prove by Rolle's Theorem that the equation $x^3 + x^2 + x + 1 = 0$ cannot have more than one real root.
7. Prove Corollary A.3.2 for $N = 4$.
8. Prove that there exists at least one number $c$ in $[a, b]$ such that $\int_a^b (x - a)(x - b)F(x)dx = F(c)\int_a^b (x - a)(x - b)dx = -[(b - a)^3/6]F(c)$ where $F$ is continuous on $[a, b]$.
9. Prove Corollary A.5.1.
10. Let $f$ be continuous on $[a, b]$ and $\int_a^b f(x)dx = 0$. Then prove that there is at least one number $c$ in $[a, b]$ such that $f(c) = 0$.
11. Determine whether the following infinite series converges or diverges. If it converges, find its sum.

    (a) $\sum_{n=1}^{\infty} e^{-n}$   (b) $0.3 + 0.03 + 0.003 + \cdots + 3 \times 10^{-n} + \cdots$

12. Let $f(x) = \cos x$. Determine the third-degree Taylor polynomial of $f$ at $x_0 = \pi/3$ and its remainder. Use this polynomial to approximate $\cos 62°$ and an upper bound for the error in this approximation. (*Hint:* $62° = (\pi/180)62$ radian.)
13. Use the Taylor polynomial to express the polynomial $P(x) = 2x^3 + 5x^2 + 6x + 1$ as a polynomial in powers of $(x + 1)$.
14. Show that the formula $1/\sqrt{1 + x} = 1 - (1/2)x$ is accurate to two decimal places if $|x| \le 0.1$.

**15.** Let $f(x, y) = 1/\sqrt{1 + x + y}$. Determine the second-degree Taylor polynomial of $f$ at $x_0 = 0$, $y_0 = 0$.

**16.** Let $f(x, y) = 1/(xy)$. Determine the second-degree Taylor polynomial of $f$ at $x_0 = 1$, $y_0 = -1$.

**17.** (Binomial series) If $k$ is any real number, then verify that the Maclaurin series for $f(x) = (1 + x)^k = 1 + kx + \dfrac{k(k - 1)}{2!} x^2 + \dfrac{k(k - 1)(k - 2)}{3!} x^3 + \cdots$ for all $x$ such that $|x| < 1$.

# APPENDIX B

## Review of Linear Algebra

In this appendix we review some material on matrices, vector spaces, eigenvalues and eigenvectors that is taught in most undergraduate linear algebra courses. No derivations are included and only some results are summarized.

An $N \times M$ (read $N$ by $M$) matrix $A$ is a rectangular array of real or complex numbers with $N$ rows and $M$ columns and is denoted by

$$
A = \begin{bmatrix}
a_{11} & a_{12} & \dots & a_{1j} & \dots & a_{1M} \\
a_{21} & a_{22} & \dots & a_{2j} & \dots & a_{2M} \\
\dots & \dots & \dots & \dots & \dots & \dots \\
a_{i1} & a_{i2} & \dots & a_{ij} & \dots & a_{iM} \\
\dots & \dots & \dots & \dots & \dots & \dots \\
a_{N1} & a_{N2} & \dots & a_{Nj} & \dots & a_{NM}
\end{bmatrix}
$$

The numbers in the array are called the **elements** of the matrix. The entry $a_{ij}$ is at the intersection of the $i$th row and $j$th column.

Two matrices $A$ and $B$ are said to be **equal** if both are of the same size, say $N \times M$, and if $a_{ij} = b_{ij}$ for each $i = 1, 2, \dots, N$ and $j = 1, 2, \dots, M$.

Addition and subtraction are defined only for matrices of the same size. Let $A$ and $B$ be $N \times M$ matrices. Then the sum of $A$ and $B$ is the matrix $C = A + B$ of order $N \times M$, obtained by adding together corresponding elements, whose $(i, j)$th element $c_{ij} = a_{ij} + b_{ij}$. Similarly, subtraction can be defined.

Let $A$ be an $N \times M$ matrix and $\alpha$ be any scalar. Then the scalar multiple of $A$ by $\alpha$ is the matrix $C = \alpha A$, of order $N \times M$, obtained by multiplying each element of $A$ by $\alpha$, whose $(i, j)$th element $c_{ij} = \alpha a_{ij}$.

Let $A$ be an $N \times L$ matrix and $B$ be $L \times M$ matrix. Then the matrix product of $A$ and $B$ is the matrix $C = AB$, of order $N \times M$ whose $(i, j)$th element $c_{ij} = \Sigma_{k=1}^{L} a_{ik}b_{kj}$. Matrix multiplication is not always commutative; in other words, $AB = BA$ is not always true. However, associativity holds true; that is, $A(BC) = (AB)C$.

**EXAMPLE B.1**

Let $A = \begin{bmatrix} 1 & 2 \\ -1 & 3 \end{bmatrix}$, $B = \begin{bmatrix} 3 & 1 & 0 \\ 4 & 0 & 1 \end{bmatrix}$, and $C = \begin{bmatrix} 1 & 3 \\ -1 & 0 \end{bmatrix}$. Then

$$
A + C = \begin{bmatrix} 1 + 1 & 2 + 3 \\ -1 - 1 & 3 + 0 \end{bmatrix} = \begin{bmatrix} 2 & 5 \\ -2 & 3 \end{bmatrix}
$$

$$
\alpha A = \begin{bmatrix} \alpha \cdot 1 & \alpha \cdot 2 \\ \alpha \cdot (-1) & \alpha \cdot (3) \end{bmatrix} = \begin{bmatrix} \alpha & 2\alpha \\ -\alpha & 3\alpha \end{bmatrix}
$$

and

$$AB = \begin{bmatrix} 1 & 2 \\ -1 & 3 \end{bmatrix} \begin{bmatrix} 3 & 1 & 0 \\ 4 & 0 & 1 \end{bmatrix}$$

$$= \begin{bmatrix} 1\cdot3 + 2\cdot4 & 1\cdot1 + 2\cdot0 & 1\cdot0 + 2\cdot1 \\ -1\cdot3 + 3\cdot4 & -1\cdot1 + 3\cdot0 & -1\cdot0 + 3\cdot1 \end{bmatrix}$$

$$= \begin{bmatrix} 11 & 1 & 2 \\ 9 & -1 & 3 \end{bmatrix}$$

We cannot compute $A + B$ because $A$ and $B$ are of different sizes. ∎ ∎ ∎

Let $A$ be any $N \times M$ matrix. Then exchanging the rows for the columns produces its **transpose**, denoted by $A'$, of order $M \times N$, whose $(i, j)$th element $a'_{ij} = a_{ji}$.

**EXAMPLE B.2**    If $A = \begin{bmatrix} 1 & 2 \\ -1 & 3 \end{bmatrix}$, then $A' = \begin{bmatrix} 1 & -1 \\ 2 & 3 \end{bmatrix}$. ∎ ∎ ∎

We can prove: (1) $(A')' = A$    (2) $(A + B)' = A' + B'$    (3) $(AB)' = B' A'$.

An $N \times N$ matrix is called a **square matrix** of order $N$. A square matrix $A$ is called **symmetric** if $A' = A$ and **skew-symmetric** if $A' = -A$.

An **identity** matrix $I_N$ is a square matrix of order $N$ whose $(i, j)$th element is given by the Kronecker delta

$$\delta_{ij} = \begin{cases} 1 & \text{if } i = j \\ 0 & \text{if } i \neq j \end{cases}$$

**EXAMPLE B.3**

$$I_2 = \begin{bmatrix} 1 & 0 \\ 0 & 1 \end{bmatrix} \qquad I_3 = \begin{bmatrix} 1 & 0 & 0 \\ 0 & 1 & 0 \\ 0 & 0 & 1 \end{bmatrix} \qquad I_4 = \begin{bmatrix} 1 & 0 & 0 & 0 \\ 0 & 1 & 0 & 0 \\ 0 & 0 & 1 & 0 \\ 0 & 0 & 0 & 1 \end{bmatrix}$$

are identity matrices. ∎ ∎ ∎

For convenience, we may denote the identity matrix $I_N$ simply by $I$.

A **diagonal** matrix $D$ is a square matrix in which all the elements not on the main diagonal are zero; that is, whose $(i, j)$th element $d_{ij} = 0$ if $i \neq j$.

An **upper triangular** matrix is a square matrix having only zeros below the main diagonal; that is, whose $(i, j)$th element $u_{ij} = 0$ if $i > j$.

A **lower triangular** matrix is a square matrix having only zeros above the main diagonal; that is, whose $(i, j)$th element $l_{ij} = 0$ if $i < j$.

## EXAMPLE B.4

$$\begin{bmatrix} 5 & 0 & 0 \\ 0 & 0 & 0 \\ 0 & 0 & 2 \end{bmatrix}$$

is a diagonal matrix. Also, it is a symmetric, lower, and upper triangular matrix.

$$\begin{bmatrix} 3 & 2 & 4 \\ 0 & 4 & -3 \\ 0 & 0 & 6 \end{bmatrix}$$

is an upper triangular matrix.

$$\begin{bmatrix} 5 & 0 & 0 \\ 6 & 4 & 0 \\ 8 & 9 & 10 \end{bmatrix}$$

is a lower triangular matrix.   ■ ■ ■

Let $A$ be a square matrix of order $N$. If there is a square matrix $B$ of order $N$ such that $AB = BA = I$, then $B$ is called an **inverse** of $A$ and is denoted by $A^{-1}$. If $A$ has an inverse, then $A$ is said to be **invertible**.

We can prove: (1) $(A^{-1})^{-1} = A$   (2) $(AB)^{-1} = B^{-1}A^{-1}$   (3) $(A^{-1})^t = (A^t)^{-1}$. A square matrix $A$ is called an **orthogonal** matrix if $A^t A = A A^t = I$.

## EXAMPLE B.5

$\begin{bmatrix} \cos \theta & -\sin \theta \\ \sin \theta & \cos \theta \end{bmatrix}$ is an orthogonal matrix.   ■ ■ ■

Let $A = \begin{bmatrix} a_{11} & a_{12} \\ a_{21} & a_{22} \end{bmatrix}$. Then the **determinant** of $A$, written as det $A$ or $|A|$, is defined by det $A = |A| = a_{11}a_{22} - a_{12}a_{21}$. In order to define a $3 \times 3$ determinant, we use $2 \times 2$ determinants and for a $4 \times 4$ determinant, we use $3 \times 3$ determinants, and so on. Let us define the minor and cofactor of a given matrix $A$ in order to define det $A$ in general.

Let $A$ be a $3 \times 3$ matrix and let $M_{ij}$ be the matrix obtained from $A$ by deleting the $i$th row and $j$th column of $A$. Then $M_{ij}$ is called the $ij$th **minor** of $A$.

Let

$$A = \begin{bmatrix} a_{11} & a_{12} & a_{13} \\ a_{21} & a_{22} & a_{23} \\ a_{31} & a_{32} & a_{33} \end{bmatrix}$$

Then the minor $M_{23}$ of $A$ is $\begin{bmatrix} a_{11} & a_{12} \\ a_{31} & a_{32} \end{bmatrix}$. The $ij$th **cofactor** of $A$, denoted by $A_{ij}$, is given by $A_{ij} = (-1)^{i+j} \det M_{ij}$. Thus $A_{23} = (-1)^{2+3} \det M_{23} = -\begin{vmatrix} a_{11} & a_{12} \\ a_{31} & a_{32} \end{vmatrix} = -(a_{11}a_{32} - a_{12}a_{31})$. The determinant of $A$ is given by

$$\det A = |A| = a_{11}A_{11} + a_{12}A_{12} + a_{13}A_{13}$$

$$= a_{11}\begin{vmatrix} a_{22} & a_{23} \\ a_{32} & a_{33} \end{vmatrix} - a_{12}\begin{vmatrix} a_{21} & a_{23} \\ a_{31} & a_{33} \end{vmatrix} + a_{13}\begin{vmatrix} a_{21} & a_{22} \\ a_{31} & a_{32} \end{vmatrix}$$

$$= a_{11}(a_{22}a_{33} - a_{23}a_{32}) - a_{12}(a_{21}a_{33} - a_{23}a_{31}) + a_{13}(a_{21}a_{32} - a_{22}a_{31})$$

Let $A$ be an $N \times N$ matrix given by

$$A = \begin{bmatrix} a_{11} & a_{12} & \dots & a_{1N} \\ a_{21} & a_{22} & \dots & a_{2N} \\ \vdots & \vdots & \ddots & \vdots \\ a_{N1} & a_{N2} & \dots & a_{NN} \end{bmatrix} \tag{B.1}$$

The determinant of $A$ is given by

$$\det A = |A| = a_{11}A_{11} + a_{12}A_{12} + \cdots + a_{1N}A_{1N} = \sum_{i=1}^{N} a_{1i}A_{1i}$$

The determinant can also be evaluated by cofactor expansion along any row or any column of $A$ and is given by

$$\det A = \sum_{i=1}^{N} a_{ji}A_{ji} \qquad j = 1, 2, \dots, N \tag{B.2}$$

or

$$\det A = \sum_{i=1}^{N} a_{ij}A_{ij} \qquad j = 1, 2, \dots, N \tag{B.3}$$

Let $A$ and $B$ be $N \times N$ matrices. Then we can prove:

$$\begin{aligned} &\text{(1)} \ \det(AB) = \det A \cdot \det B \\ &\text{(2)} \ \det A^t = \det A \\ &\text{(3)} \ \det A^{-1} = \frac{1}{\det A} \quad \text{if } \det A \neq 0 \end{aligned} \tag{B.4}$$

The linear system of $N$ equations in $N$ unknowns given by

$$\begin{aligned} a_{11}x_1 + a_{12}x_2 + \cdots + a_{1N}x_N &= b_1 \\ a_{21}x_1 + a_{22}x_2 + \cdots + a_{2N}x_N &= b_2 \\ \vdots \qquad \vdots \qquad\qquad \vdots \qquad &\ \vdots \\ a_{N1}x_1 + a_{N2}x_2 + \cdots + a_{NN}x_N &= b_N \end{aligned} \tag{B.5}$$

can be written as

$$A\mathbf{x} = \mathbf{b} \tag{B.6}$$

where $A$ is given by Equation (B.1), $\mathbf{x} = [x_1, x_2, \dots, x_N]^t$, and $\mathbf{b} = [b_1, b_2, \dots, b_N]^t$.

**Theorem B.1**    Let a system of $N$ equations in $N$ unknowns be given by Equation (B.6). Then the following statements are equivalent:

1. $A\mathbf{x} = \mathbf{b}$ has a unique solution for any $\mathbf{b}$.
2. $A^{-1}$ exists.
3. $A\mathbf{x} = \mathbf{0}$ has only the trivial solution $\mathbf{x} = \mathbf{0}$, where $\mathbf{0} = [0, 0, \ldots, 0]^t$.
4. $\det A \neq 0$. ∎

**Theorem B.2**    (**Cramer's Rule**)   If $A\mathbf{x} = \mathbf{b}$ is a system of $N$ linear equations in $N$ unknowns such that $\det A \neq 0$. Then the unique solution is given by

$$x_i = \frac{\det (A_i)}{\det A} \quad \text{for } i = 1, 2, \ldots, N \tag{B.7}$$

where $A_i$ is the matrix obtained by replacing the entries in the $i$th column of $A$ by $\mathbf{b} = [b_1, b_2, \ldots, b_N]^t$. ∎

**E X A M P L E   B.6**    Solve the following system of equations using Cramer's rule:

$$\begin{aligned} x_1 + x_2 - x_3 &= 1 \\ x_1 \phantom{{}+ x_2} + x_3 &= 2 \\ x_2 - x_3 &= 0 \end{aligned}$$

We have

$$A = \begin{bmatrix} 1 & 1 & -1 \\ 1 & 0 & 1 \\ 0 & 1 & -1 \end{bmatrix}$$

From Equation (B.7), we get

$$A_1 = \begin{bmatrix} 1 & 1 & -1 \\ 2 & 0 & 1 \\ 0 & 1 & -1 \end{bmatrix} \quad A_2 = \begin{bmatrix} 1 & 1 & -1 \\ 1 & 2 & 1 \\ 0 & 0 & -1 \end{bmatrix} \quad A_3 = \begin{bmatrix} 1 & 1 & 1 \\ 1 & 0 & 2 \\ 0 & 1 & 0 \end{bmatrix}$$

Therefore $x_1 = \det(A_1)/\det A = -1/-1 = 1$, $x_2 = \det(A_2)/\det A = -1/-1 = 1$, and $x_3 = \det(A_3)/\det A = -1/-1 = 1$. ∎∎∎

For systems of equations with more than three unknowns, Cramer's rule is hardly used, though it is a useful formula to study properties of the solution without solving the system.

## Vector Spaces

Let $V$ be a set of objects on which two operations (called addition and scalar multiplication) are defined. The addition operation associates with each pair of elements $\mathbf{u}$ and $\mathbf{v}$ in $V$ an element $\mathbf{u} + \mathbf{v}$ called the sum of $\mathbf{u}$ and $\mathbf{v}$. The multiplication operation associates with each scalar $\alpha$ ($\alpha$ is a member of the set of real numbers $R$ or the set of complex numbers $C$ or any field $F$) and each element $\mathbf{u}$ in $V$ an element $\alpha\mathbf{u}$ called the scalar multiple of **u** by $\alpha$. If the following conditions are satisfied by all elements **u**, **v**, and **w** in $V$ and all scalars $\alpha$ and $\beta$ in $F$, then we call $V$ a **vector space** over the field $F$ and the elements in $V$ **vectors**.

Addition conditions:

1. If **u** and **v** are in $V$, then **u** + **v** is also in $V$ (closure property)
2. **u** + **v** = **v** + **u** (commutative)
3. **u** + (**v** + **w**) = (**u** + **v**) + **w** (associative)
4. There is an element **0** in $V$ such that **u** + **0** = **0** + **u** = **u** for all **u** in $V$
5. For every **u** in $V$, there exists an element $-$**u** in $V$ such that **u** + ($-$**u**) = ($-$**u**) + **u** = **0**

Scalar multiplication conditions:

1. If $\alpha$ is any element in $F$ and **u** is any element in $V$, then $\alpha$**u** is in $V$ (closure property)
2. $\alpha($**u** + **v**$)$ = $\alpha$**u** + $\alpha$**v**
3. $(\alpha + \beta)$**u** = $\alpha$**u** + $\beta$**u**
4. $(\alpha\beta)$**u** = $\alpha(\beta$**u**$)$
5. $1$**u** = **u** where $1$ is the identity of field $F$

**EXAMPLE B.7**

1. $V = R^N = \{(x_1, x_2, \ldots, x_N)|x_1, x_2, \ldots, x_N \text{ are real numbers}\}$ and $F = R$. Then with the usual vector addition and scalar multiplication, $V$ is a vector space over $R$.
2. Let $V$ be the set of $N \times M$ matrices with real elements and $F$ be the set of real numbers. Then with the matrix addition and scalar multiplication of matrices, $V$ forms a vector space over $R$.
3. Let $V = P_N$ be the set of polynomials with real coefficients of degree less than or equal to $N$. Then with the polynomial addition and scalar multiplication of polynomials, $V$ forms a vector space over $R$.
4. Let $V = \{f \mid f \text{ is a continuous function on } [0, 1]\}$. Then define $(f + g)(x) = f(x) + g(x)$ and $(\alpha f)(x) = \alpha f(x)$ where $\alpha \in R$. Then $V$ forms a vector space over $R$. ■ ■ ■

Let $V$ be a vector space over a field $F$ and let $\{$**u**$_1$, **u**$_2$, $\ldots$, **u**$_N\} \subset V$. Then $\{$**u**$_1$, **u**$_2$, $\ldots$, **u**$_N\}$ is called **linearly dependent** if there are scalars $\alpha_1, \alpha_2, \ldots, \alpha_N$ in $F$ with at least one nonzero scalar such that

$$\alpha_1\mathbf{u}_1 + \alpha_2\mathbf{u}_2 + \cdots + \alpha_N\mathbf{u}_N = \mathbf{0} \qquad \text{(B.8)}$$

Let $\alpha_i \neq 0$ in Equation (B.8). Then

$$\mathbf{u}_i = -\frac{\alpha_1}{\alpha_i}\mathbf{u}_1 - \frac{\alpha_2}{\alpha_i}\mathbf{u}_2 - \cdots - \frac{\alpha_{i-1}}{\alpha_i}\mathbf{u}_{i-1} - \frac{\alpha_{i-2}}{\alpha_i}\mathbf{u}_{i+1} - \cdots - \frac{\alpha_N}{\alpha_i}\mathbf{u}_N$$

In this case, we say that **u**$_i$ is a linear combination of **u**$_1$, **u**$_2$, $\ldots$, **u**$_{i-1}$, **u**$_{i+1}$, $\ldots$, **u**$_N$.

If the vectors **u**$_1$, **u**$_2$, $\ldots$, **u**$_N$ are not linearly dependent, then they are called **linearly independent**. That is, $\alpha_1\mathbf{u}_1 + \alpha_2\mathbf{u}_2 + \cdots + \alpha_N\mathbf{u}_N = \mathbf{0}$ implies $\alpha_1 = \alpha_2 = \cdots = \alpha_N = 0$.

**EXAMPLE B.8**

Let $P_3$ be the set of polynomials with real coefficients of degree less than or equal to three. Determine whether $\{1 - x, 1 + x, x^2 - x, x^3 - x\}$ is linearly independent or dependent in $P_3$.

To determine this, we examine

$$\alpha_1(1 - x) + \alpha_2(1 + x) + \alpha_3(x^2 - x) + \alpha_4(x^3 - x) = \mathbf{0}$$
$$= 0 \cdot 1 + 0 \cdot x + 0 \cdot x^2 + 0 \cdot x^3$$

This simplifies to

$$(\alpha_1 + \alpha_2) \cdot 1 + (-\alpha_1 + \alpha_2 - \alpha_3 - \alpha_4)x + \alpha_3 x^2 + \alpha_4 x^3 = 0 \cdot 1 + 0 \cdot x + 0 \cdot x^2 + 0 \cdot x^3$$

This holds true for every real $x$, hence the coefficients of $1$, $x$, $x^2$, and $x^3$ must be zero. This yields the system of four linear equations

$$
\begin{aligned}
\alpha_1 + \alpha_2 \qquad\qquad &= 0 \\
-\alpha_1 + \alpha_2 - \alpha_3 - \alpha_4 &= 0 \\
\alpha_3 \qquad\quad &= 0 \\
\alpha_4 &= 0
\end{aligned}
$$

We verify that $\alpha_1 = \alpha_2 = \alpha_3 = \alpha_4 = 0$ is the only solution, so the set $\{1 - x, 1 + x, x^2 - x, x^3 - x\}$ is linearly independent in $P_3$. ■ ■ ■

A finite set of vectors $\{\mathbf{u}_1, \mathbf{u}_2, \ldots, \mathbf{u}_N\} \subset V$ is a **basis** for $V$ if it spans $V$ and is linearly independent.

**EXAMPLE B.9**

Show that the set $\{1 - x, 1 + x, x^2 - x, x^3 - x\}$ is a basis of $P_3$.

Let $ax^3 + bx^2 + cx + d$ be an arbitrary element in $P_3$. Then find $\alpha_1, \alpha_2, \alpha_3$, and $\alpha_4$ such that

$$
\begin{aligned}
ax^3 + bx^2 + cx + d &= \alpha_1(1 - x) + \alpha_2(1 + x) + \alpha_3(x^2 - x) + \alpha_4(x^3 - x) \\
&= \alpha_4 x^3 + \alpha_3 x^2 + (-\alpha_1 + \alpha_2 - \alpha_3 - \alpha_4)x + (\alpha_1 + \alpha_2)
\end{aligned}
$$

This gives

$$
\begin{aligned}
\alpha_4 &= a \\
\alpha_3 \qquad\quad &= b \\
-\alpha_1 + \alpha_2 - \alpha_3 - \alpha_4 &= c \\
\alpha_1 + \alpha_2 \qquad\qquad &= d
\end{aligned}
$$

We can verify that $\alpha_1 = (d - a - c - b)/2$, $\alpha_2 = (a + b + c + d)/2$, $\alpha_3 = b$, and $\alpha_4 = a$. Hence the vectors $1 - x$, $1 + x$, $x^2 - x$, and $x^3 - x$ span $P_3$. Further, let $a = b = c = d = 0$. Then $\alpha_1 = \alpha_2 = \alpha_3 = \alpha_4 = 0$. Hence $\{1 - x, 1 + x, x^2 - x, x^3 - x\}$ is linearly independent (Example B.8) and thus forms a basis of $P_3$. ■ ■ ■

The **dimension** of a vector space $V$ is the number of vectors that make up a basis. Since $\{1 - x, 1 + x, x^2 - x, x^3 - x\}$ is a basis for $P_3$, the dimension of $P_3$ is four.

## Eigenvalues and Eigenvectors

Let $A$ be a given $N \times N$ matrix. If there is a vector $\mathbf{x} = [x_1, x_2, \ldots, x_N]' \neq \mathbf{0}$ and a scalar $\lambda$ such that $A\mathbf{x} = \lambda\mathbf{x}$, then $\lambda$ is called an **eigenvalue** of $A$ and $\mathbf{x}$ is said to be an **eigenvector** of $A$ corresponding to $\lambda$. The set of all eigenvalues of $A$ is called the **spectrum** of $A$.

The equation

$$A\mathbf{x} = \lambda\mathbf{x} \tag{B.9}$$

can be written as

$$(A - \lambda I)\mathbf{x} = \mathbf{0}$$

This represents the set of homogeneous linear equations given by

$$\begin{bmatrix} a_{11} - \lambda & a_{12} & \cdots & a_{1N} \\ a_{21} & a_{22} - \lambda & \cdots & a_{2N} \\ \vdots & \vdots & \ddots & \vdots \\ a_{N1} & a_{N2} & \cdots & a_{NN} - \lambda \end{bmatrix} \begin{bmatrix} x_1 \\ x_2 \\ \vdots \\ x_N \end{bmatrix} = \begin{bmatrix} 0 \\ 0 \\ \vdots \\ 0 \end{bmatrix}$$

The system has nonzero solutions if and only if $\det(A - \lambda I) = 0$. Thus $\det(A - \lambda I) = 0$ is an equation to determine $\lambda$. The polynomial $|A - \lambda I|$ is called the **characteristic polynomial** of $A$ and the equation

$$|A - \lambda I| = 0 \tag{B.10}$$

is called the **characteristic equation** of $A$. Once the eigenvalues are determined by solving Equation (B.10), the eigenvalues are substituted into equation $(A - \lambda I)\mathbf{x} = \mathbf{0}$ to find the corresponding eigenvectors $\mathbf{x}$.

**EXAMPLE B.10**

Find the eigenvalues and eigenvectors of the matrix $A = \begin{bmatrix} 1 & 2 \\ 2 & 1 \end{bmatrix}$.

$$\text{Since } A - \lambda I = \begin{bmatrix} 1 & 2 \\ 2 & 1 \end{bmatrix} - \lambda \begin{bmatrix} 1 & 0 \\ 0 & 1 \end{bmatrix} = \begin{bmatrix} 1 - \lambda & 2 \\ 2 & 1 - \lambda \end{bmatrix}$$

the characteristic polynomial of $A$ is given by

$$0 = |A - \lambda I| = \begin{vmatrix} 1 - \lambda & 2 \\ 2 & 1 - \lambda \end{vmatrix} = (1 - \lambda)^2 - 4 = \lambda^2 - 2\lambda - 3$$

or

$$\lambda^2 - 2\lambda - 3 = (\lambda + 1)(\lambda - 3) = 0 \tag{B.11}$$

The solutions of Equation (B.11) are $-1$ and 3. Thus the eigenvalues of $A$ are $-1$ and 3.

For $\lambda = -1$, $(A - \lambda I)\mathbf{x} = \mathbf{0}$ becomes

$$\begin{bmatrix} 2 & 2 \\ 2 & 2 \end{bmatrix} \begin{bmatrix} x_1 \\ x_2 \end{bmatrix} = \begin{bmatrix} 0 \\ 0 \end{bmatrix}$$

Solving this system, we get $x_1 = -x_2 = t$ where $t$ is any scalar. Hence any nonzero scalar multiple of $\begin{bmatrix} 1/\sqrt{2} \\ -1/\sqrt{2} \end{bmatrix}$ is an eigenvector of $A$ corresponding to $\lambda = -1$. For $\lambda = 3$, $(A - \lambda I)\mathbf{x} = \mathbf{0}$ becomes

$$\begin{bmatrix} -2 & 2 \\ 2 & -2 \end{bmatrix} \begin{bmatrix} x_1 \\ x_2 \end{bmatrix} = \begin{bmatrix} 0 \\ 0 \end{bmatrix}$$

Solving this system, we get $x_1 = x_2 = s$ where $s$ is any scalar. Hence any nonzero scalar multiple of $\begin{bmatrix} 1/\sqrt{2} \\ 1/\sqrt{2} \end{bmatrix}$ is an eigenvector of $A$ corresponding to $\lambda = 3$.

■ ■ ■

## EXERCISES

1.  Given $A = \begin{bmatrix} 1 & 2 & 0 \\ 3 & 0 & 5 \end{bmatrix}$ and $B = \begin{bmatrix} -1 & 0 & -2 \\ 0 & 4 & 1 \end{bmatrix}$. Find (a) $3A$
    (b) $A + 3B$    (c) $D$ if $A + D = B$.
2.  Multiply the following, if possible:

    (a) $[1 \quad 2 \quad 3] \begin{bmatrix} 3 \\ 2 \\ 1 \end{bmatrix}$    (b) $\begin{bmatrix} 3 \\ 2 \\ 1 \end{bmatrix} [1 \quad 2 \quad 3]$    (c) $\begin{bmatrix} 1 & 0 & 1 \\ 2 & 3 & 1 \end{bmatrix} \begin{bmatrix} 3 & 0 \\ 1 & 1 \\ 0 & 2 \end{bmatrix}$

3.  If $A$ is an $N \times N$ matrix, then prove that
    (a) $A + A^t$ is symmetric.
    (b) $A - A^t$ is skew-symmetric.
    (c) $AA^t$ is symmetric.
4.  Let $A$ be an invertible matrix and $c$ a nonzero scalar. Then prove that
    (a) $A^{-1}$ is invertible and $(A^{-1})^{-1} = A$.
    (b) $(cA)^{-1} = (1/c)A^{-1}$.
    (c) $A^n$ is invertible and $(A^n)^{-1} = (A^{-1})^n$ for $n = 1, 2, \ldots$.
5.  Find the determinant and inverse (if it exists) of the following matrices:

    (a) $\begin{bmatrix} 2 & 1 & 3 \\ 1 & 1 & 1 \\ 0 & 0 & 1 \end{bmatrix}$    (b) $\begin{bmatrix} 1 & 0 & 2 & 1 \\ 2 & -1 & 1 & 0 \\ 1 & 0 & 0 & 3 \\ -1 & 0 & 2 & 1 \end{bmatrix}$

6.  Prove that $\det A^{-1} = 1/\det A$ if $\det A \neq 0$.
7.  Solve the following systems of linear equations using Cramer's rule:

    (a) $\begin{aligned} x_1 + 2x_2 &= 5 \\ 2x_1 + x_2 &= 4 \end{aligned}$    (b) $\begin{aligned} x_1 + x_2 + x_3 &= 1 \\ x_1 - x_2 + x_3 &= 1 \\ x_1 + x_2 - x_3 &= 1 \end{aligned}$

**8.** Determine whether the set $M_{22} = \left\{ \begin{bmatrix} a & b \\ c & d \end{bmatrix} \middle| a, b, c, d \in R \right\}$ with matrix addition and scalar multiplication forms a vector space over $R$.

**9.** Prove that $(-1, 2, 1)$, $(3, -1, 0)$, $(1, 1, -1)$ is a basis for $R^3$.

**10.** Determine the eigenvalues and eigenvectors of the following matrices:

(a) $\begin{bmatrix} 2 & -2 \\ -2 & 5 \end{bmatrix}$   (b) $\begin{bmatrix} 4 & 0 & 1 \\ -2 & 1 & 0 \\ -2 & 0 & 1 \end{bmatrix}$

**11.** Prove that $\lambda = 0$ is an eigenvalue of a matrix $A$ if and only if $A$ is not invertible.

**12.** Let $\lambda$ be an eigenvalue of the $N \times N$ matrix $A$ and $\mathbf{x} \neq \mathbf{0}$ be an associated eigenvector.

(a) Prove that $\lambda^2$ is an eigenvalue of $A^2$ with eigenvector $\mathbf{x}$.

(b) Prove that $1/\lambda$ is an eigenvalue of $A^{-1}$ (if it exists) with eigenvector $\mathbf{x}$.

(c) Generalize parts (a) and (b) to $A^N$ and $(A^{-1})^N$ for integers $N \geq 2$.

# APPENDIX C

## Review of Ordinary Differential Equations

In this appendix we review some material on ordinary differential equations that is taught in most undergraduate ordinary differential equation courses. No derivations are included and only some results are summarized.

An equation containing derivatives of one or more dependent variables with respect to one or more independent variables is called a differential equation. Some examples of differential equations are:

$$(1) \quad \frac{d^2y}{dx^2} + xy = x \sin x$$

$$(2) \quad \frac{d^3y}{dx^3} + 3\left(\frac{dy}{dx}\right)^2 + 6y = 0$$

$$(3) \quad \frac{\partial u}{\partial x} - \frac{\partial u}{\partial y} - u^2 = y - x$$

$$(4) \quad \frac{\partial^2 u}{\partial x^2} - \frac{\partial^2 u}{\partial y^2} = 0$$

An ordinary differential equation is a differential equation containing ordinary derivatives with respect to a single independent variable. Notice that differential equations (1) and (2) are ordinary differential equations. A partial differential equation is a differential equation containing partial derivatives with respect to more than one independent variable. Differential equations (3) and (4) are partial differential equations.

The highest order derivative present in a differential equation is called the order of the differential equation. Differential equations (1) and (4) are of the second order while (2) and (3) are of the third and first orders, respectively.

A **linear differential equation** is a differential equation in which (a) every derivative and every dependent variable is of the first degree only, and (b) no product of derivatives or dependent variables occurs. If a differential equation is not linear, then it is called a **nonlinear differential equation**. Differential equations (1) and (4) are linear, while (2) and (3) are nonlinear.

Consider a linear first-order differential equation

$$\frac{dy}{dx} + P(x)y = Q(x) \tag{C.1}$$

Multiplying Equation (C.1) by an integrating factor

$$\mu(x) = e^{\int P(x)dx} \tag{C.2}$$

we get

$$\mu(x)\frac{dy}{dx} + \mu(x)P(x)y = \frac{d}{dx}[\mu(x)y] = \mu(x)Q(x) \tag{C.3}$$

Integrating Equation (C.3) with respect to $x$, we get

$$\mu(x)y = \int \mu(x)Q(x)dx + c$$

where $c$ is an arbitrary constant. Hence the general solution of Equation (C.1) is given by

$$y(x) = \frac{1}{\mu(x)}\left(\int \mu(x)Q(x)dx + c\right) \tag{C.4}$$

---

**EXAMPLE C.1**

Find the general solution of

$$\frac{dy}{dx} - y = e^{3x} \tag{C.5}$$

An integrating factor is given by

$$\mu(x) = e^{\int P(x)dx} = e^{\int(-1)dx}$$

Multiplying Equation (C.5) by $e^{-x}$ leads to

$$e^{-x}\frac{dy}{dx} - e^{-x}y = \frac{d}{dx}(e^{-x}y) = e^{2x}$$

Integrating both sides with respect to $x$ leads to

$$e^{-x}y = \int e^{2x}\,dx = \frac{e^{2x}}{2} + c \tag{C.6}$$

where $c$ is an arbitrary constant. Solving Equation (C.6) for $y$ gives the general solution

$$y(x) = \frac{e^{3x}}{2} + ce^{x} \qquad \blacksquare\blacksquare\blacksquare$$

Consider a second-order linear ordinary differential equation

$$a\frac{d^2y}{dx^2} + b\frac{dy}{dx} + cy = g(x) \tag{C.7}$$

where $a \neq 0$, $b$, and $c$ are real numbers. The general solution of Equation (C.7) may be written as

$$y(x) = y_c(x) + y_p(x)$$

where $y_c(x)$ is the general solution of the **corresponding homogeneous equation** (replace $g(x)$ by 0 in Equation (C.7))

$$a\frac{d^2y}{dx^2} + b\frac{dy}{dx} + cy = 0 \qquad \text{(C.8)}$$

of (C.7) and $y_p(x)$ is a particular solution of (C.7) containing no arbitrary constants.

Let $y = e^{rx}$ be a solution of Equation (C.8). Then substituting $y = e^{rx}$, $dy/dx = re^{rx}$, and $d^2y/dx^2 = r^2e^{rx}$ in Equation (C.8) and simplifying, we get

$$ar^2 + br + c = 0 \qquad \text{(C.9)}$$

which is called the **characteristic** or **auxiliary equation** of (C.8). In Table C.1 we summarize the general solution of Equation (C.8) corresponding to three different cases of the roots of Equation (C.9).

**Table C.1** The General Solution of Equation (C.8)

| Case | Roots of Eq. (C.9) | Solution of Eq. (C.8) | General Solution of Eq. (C.8) |
|------|--------------------|-----------------------|-------------------------------|
| I | Distinct real: $r_1$, $r_2$ | $e^{r_1x}$, $e^{r_2x}$ | $y = c_1e^{r_1x} + c_2e^{r_2x}$ |
| II | Repeated real root: $r_1$ | $e^{r_1x}$, $xe^{r_1x}$ | $y = e^{r_1x}(c_1 + c_2x)$ |
| III | Complex roots: $\alpha + I\beta$, $\alpha - I\beta$ | $e^{\alpha} \cos \beta x$, $e^{\alpha} \sin \beta x$ | $y = e^{\alpha x}(c_1 \cos \beta x + c_2 \sin \beta x)$ |

where $I = \sqrt{-1}$, $c_1$, and $c_2$ are arbitrary constants.

**EXAMPLE C.2**    Find the general solution of

$$\frac{d^2y}{dx^2} - 3\frac{dy}{dx} + 6y = 0 \qquad \text{(C.10)}$$

The characteristic equation of (C.10) is $r^2 - 3r + 6 = 0$ which has roots

$$r = \frac{3 \pm \sqrt{9 - 24}}{2} = \frac{3}{2} \pm I\frac{\sqrt{15}}{2}$$

Hence the general solution of Equation (C.10) is given by

$$y = e^{(3/2)x}\left(c_1 \cos \frac{\sqrt{15}}{2}x + c_2 \sin \frac{\sqrt{15}}{2}x\right)$$

where $c_1$ and $c_2$ are arbitrary constants.    ■ ■ ■

If $g(x)$ in Equation (C.7) is a polynomial $P_n(x) = a_nx^n + a_{n-1}x^{n-1} + \cdots + a_1x + a_0$ or exponential function $e^{\alpha x}$ or $\sin(px + q)$ or $\cos(px + q)$ or any combination of these functions, then the particular solution $y_p(x)$ of Equation (C.7) is usually determined by the method of **undetermined coefficients**.

**EXAMPLE C.3**

Find a particular solution of

$$\frac{d^2y}{dx^2} - 3\frac{dy}{dx} + 6y = x^2 \tag{C.11}$$

Applying the method of undetermined coefficients, we assume $y_p = Ax^2 + Bx + C$ where $A$, $B$, and $C$ are constants to be determined. Substitution of $y_p = Ax^2 + Bx + C$, $dy_p/dx = 2Ax + B$, and $d^2y_p/dx^2 = 2A$ in Equation (C.11) yields $A = B = \frac{1}{6}$ and $C = \frac{1}{36}$; hence, a particular solution of Equation (C.11) is

$$y_P(x) = \frac{1}{6}\left(x^2 + x + \frac{1}{6}\right)$$

■ ■ ■

**EXAMPLE C.4**

Find a particular solution of

$$\frac{d^2y}{dx^2} - 4\frac{dy}{dx} = 3e^{4x} \tag{C.12}$$

Assume that $y_p = Ae^{4x}$. Substituting $y_p = Ae^{4x}$, $dy_p/dx = 4Ae^{4x}$, and $d^2y_p/dx^2 = 16Ae^{4x}$ in Equation (C.12) yields

$$16Ae^{4x} - 16Ae^{4x} = 0 \cdot e^{4x} = 3e^{4x}$$

We cannot determine $A$ and therefore $y_p$ is not of the form $Ae^{4x}$. This difficulty arises because $e^{4x}$ is a solution of the corresponding homogeneous equation $d^2y/dx^2 - 4(dy/dx) = 0$ of Equation (C.12). The method of undetermined coefficients has to be modified when the nonhomogeneous term $g(x)$ is also a solution of the corresponding homogeneous equation. Let us multiply the form of $y_p$ by $x$; that is, $y_p = Axe^{4x}$. Then substituting $y_p = Axe^{4x}$, $dy_p/dx = Ae^{4x}(4x + 1)$, and $d^2y_p/dx^2 = 4Ae^{4x}(4x + 2)$ in Equation (C.12) yields

$$4Ae^{4x}(4x + 2) - 4Ae^{4x}(4x + 1) = 4Ae^{4x} = 3e^{4x}$$

Hence $A = \frac{3}{4}$ and therefore a particular solution is given by

$$y_p = \frac{3}{4}xe^{4x} \tag{C.13}$$

■ ■ ■

Thus if any term of the nonhomogeneous term $g(x)$ is a solution of the corresponding homogeneous equation, then replace $y_p$ by $x^s y_p$ where $s$ is the smallest nonnegative integer such that no term in $x^s y_p$ is a solution of the corresponding homogeneous equation. In Table C.2, the second column gives the form of $y_p$ for a given $g(x)$ in the first column.

**Table C.2**   Method of Undetermined Coefficients

| $g(x)$ | Choice of $y_p(x)$ |
|---|---|
| $a$ | |
| $x^n$ | $x^s (A_n x^n + A_{n-1} x^{n-1} + \cdots + A_1 x + A_0)$ |
| $a_n x^n + a_{n-1} x^{n-1} + \cdots + a_1 x + a_0$ | |
| $a e^{px}$ | $A x^s e^{px}$ |
| $(a_n x^n + a_{n-1} x^{n-1} + \cdots + a_1 x + a_0) e^{px}$ | $x^s (A_n x^n + A_{n-1} x^{n-1} + \cdots + A_1 x + A_0) e^{px}$ |
| $a \sin qx$ | |
| $a \cos qx$ | $x^s (A \cos qx + B \sin qx)$ |
| $a \cos qx + b \sin qx$ | |
| $(a \cos qx + b \sin qx) e^{px}$ | $x^s (A \cos qx + B \sin qx) e^{px}$ |
| $(a \cos qx + b \sin qx)(a_n x^n + a_{n-1} x^{n-1} + \cdots + a_1 x + a_0)$ | $x^s (A \cos qx + B \sin qx)(A_n x^n + A_{n-1} x^{n-1} + \cdots + A_1 x + A_0)$ |

where $s$ is the smallest nonnegative integer ($s = 0$, $1$, or $2$) such that $y_p(x)$ contains no term which is a solution of the corresponding homogeneous equation.

**EXAMPLE C.5**

Find the general solution of

$$\frac{d^2y}{dx^2} - 4\frac{dy}{dx} = 3e^{4x} \tag{C.14}$$

First find the general solution of the corresponding homogeneous equation

$$\frac{d^2y}{dx^2} - 4\frac{dy}{dx} = 0 \tag{C.15}$$

of (C.14). The characteristic equation of (C.15) is given by $r^2 - 4r = 0$ which has roots $r = 0$ and $r = 4$. Hence the general solution of Equation (C.15) is given by

$$y_c(x) = c_1 + c_2 e^{4x}$$

where $c_1$ and $c_2$ are arbitrary constants. From Example C.4, a particular solution is $y_p(x) = \frac{3}{4} x e^{4x}$. Hence, the general solution of Equation (C.14) is given by

$$y(x) = y_c(x) + y_p(x) = c_1 + c_2 e^{4x} + \frac{3}{4} x e^{4x} \qquad \blacksquare\ \blacksquare\ \blacksquare$$

Consider an $n$th order linear ordinary differential equation

$$a_0 \frac{d^ny}{dx^n} + a_1 \frac{d^{n-1}y}{dx^{n-1}} + \cdots + a_{n-1}\frac{dy}{dx} + a_n y = g(x) \tag{C.16}$$

where $a_0, a_1, \ldots, a_n$ are constants. The general solution of Equation (C.16) may be written as

$$y(x) = y_c(x) + y_p(x)$$

where $y_c$ is the general solution of the corresponding homogeneous equation (replace $g(x)$ by 0 in Equation (C.16)) of (C.16)

$$a_0 \frac{d^n y}{dx^n} + a_1 \frac{d^{n-1} y}{dx^{n-1}} + \cdots + a_{n-1} \frac{dy}{dx} + a_n y = 0 \tag{C.17}$$

and $y_p(x)$ is a particular solution of Equation (C.16) containing no arbitrary constants.

It is natural to extend techniques of the second-order linear ordinary differential equations with constant coefficients and assume that $y = e^{rx}$ is a solution of Equation (C.17). Then substituting $y = e^{rx}$, $dy/dx = re^{rx}$, $d^2 y/dx^2 = r^2 e^{rx}$, ..., $d^n y/dx^n = r^n e^{rx}$ in Equation (C.17) and simplifying, we get

$$a_0 r^n + a_1 r^{n-1} + \cdots + a_{n-1} r + a_n = 0 \tag{C.18}$$

which is called the **characteristic** or **auxiliary equation** of (C.17). Denote the roots of Equation (C.18) by $r_1, r_2, \ldots, r_n$.

If $r_1, r_2, \ldots, r_n$ are real and no two of them are equal, then the general solution of Equation (C.17) is given by

$$y(x) = c_1 e^{r_1 x} + c_2 e^{r_2 x} + \cdots + c_n e^{r_n x} \tag{C.19}$$

where $c_1, c_2, \ldots, c_n$ are arbitrary constants.

If $r_1, r_2, \ldots, r_n$ are real and $r_1$ is repeated $m$ times, then the general solution of Equation (C.17) is

$$y(x) = (c_1 + c_2 x + \cdots + c_m x^m) e^{r_1 x} + c_{m+1} e^{r_{m+1} x} + c_{m+2} e^{r_{m+2} x} + \cdots + c_n e^{r_n x} \tag{C.20}$$

where $c_1, c_2, \ldots, c_n$ are arbitrary constants.

Since $a_0, a_1, \ldots, a_n$ are real numbers, if the characteristic equation [Equation (C.18)] has complex roots, then they must occur in pairs $\alpha + I\beta$ and $\alpha - I\beta$. If Equation (C.18) has $\alpha + I\beta$, $\alpha - I\beta$ and $r_3, \ldots, r_n$ real and unequal, then the general solution of Equation (C.17) is

$$y(x) = e^{\alpha x} (c_1 \cos \beta x + c_2 \sin \beta x) + c_3 e^{r_3 x} + c_4 e^{r_4 x} + \cdots + c_n e^{r_n x} \tag{C.21}$$

where $c_1, c_2, \ldots, c_n$ are arbitrary constants.

If $\alpha + I\beta$ is repeated $s$ times, the complex conjugate $\alpha - I\beta$ is also repeated $s$ times, $r_{2s+1}, r_{2s+2}, \ldots, r_n$ are real and unequal roots, then the general solution of Equation (C.17) is given by

$$y(x) = e^{\alpha x} [(c_1 + c_2 x + c_3 x^2 + \cdots + c_s x^s) \cos \beta x \\ + (c_{s+1} + c_{s+2} x + \cdots + c_{2s} x^s) \sin \beta x] + c_{2s+1} e^{r_{2s+1} x} + \cdots + c_n e^{r_n x} \tag{C.22}$$

where $c_1, c_2, \ldots, c_n$ are arbitrary constants.

If $g(x)$ is of an appropriate form, then the method of undetermined coefficients is used to determine a particular solution $y_p(x)$ of Equation (C.16). In fact, we use Table C.2 to determine the form of $y_p(x)$ where $s$ is the smallest integer such that $y_p(x)$ contains no term which is a solution of the corresponding homogeneous equation.

**EXAMPLE C.6**    Find the general solution of

$$\frac{d^3y}{dx^3} - 2\frac{d^2y}{dx^2} + \frac{dy}{dx} = e^x + \sin x \qquad \text{(C.23)}$$

First find the general solution of the corresponding homogeneous equation

$$\frac{d^3y}{dx^3} - 2\frac{d^2y}{dx^2} + \frac{dy}{dx} = 0 \qquad \text{(C.24)}$$

The characteristic equation of Equation (C.24) is $r^3 - 2r^2 + r = r(r^2 - 2r + 1) = r(r - 1)^2 = 0$ which has roots $r = 0$ and $r = 1$. The root $r = 1$ is a repeated root. Hence the general solution of Equation (C.24) is

$$y_c(x) = c_1 + (c_2 x + c_3)e^x \qquad \text{(C.25)}$$

Using the superposition principle, we find a particular solution of Equation (C.23) as the sum of particular solutions of

$$\frac{d^3y}{dx^3} - 2\frac{d^2y}{dx^2} + \frac{dy}{dx} = e^x \quad \text{and} \quad \frac{d^3y}{dx^3} - 2\frac{d^2y}{dx^2} + \frac{dy}{dx} = \sin x \qquad \text{(C.26)}$$

For the first equation, our initial guess for $y_{p_1}(x)$ is $Ae^x$; since $e^x$ and $xe^x$ are solutions of the corresponding homogeneous equation [Equation (C.24)], we multiply our guess by $x^2$. Thus

$$y_{p_1}(x) = Ax^2 e^x$$

For the second equation, our initial guess for $y_{p_2}(x)$ is $B \cos x + C \sin x$; since $\cos x$ and $\sin x$ are not solutions of Equation (C.24), there is no need to modify them. Hence

$$y_{p_2}(x) = B \cos x + C \sin x$$

We determine the constants by substituting $y_{p_1}(x)$ and $y_{p_2}(x)$ in the corresponding equations [Equation (C.26)], obtaining $A = \frac{1}{2}$, $B = 0$, and $C = \frac{1}{2}$. The general solution of Equation (C.23) is therefore

$$y(x) = y_c(x) + y_{p_1}(x) + y_{p_2}(x) = c_1 + (c_2 x + c_3)e^x + \frac{1}{2}x^2 e^x + \frac{1}{2}\sin x \quad \blacksquare\blacksquare\blacksquare$$

## EXERCISES

Find the general solution of the ordinary differential equations in Exercises 1–3.

1.  $\dfrac{dy}{dx} + 4y = x$

2.  $x\dfrac{dy}{dx} + y = \dfrac{1}{x^3}$

3.  $(x^2 + 1)\dfrac{dy}{dx} = (x^2 + 1) - 2xy$

**4.** A first-order ordinary differential equation which can be written in the form

$$\frac{dy}{dx} + P(x)y = Q(x)y^n$$

where $n \neq 1$ is any real number is called a Bernoulli equation. Show that this equation reduces to a linear differential equation

$$\frac{1}{1-n}\frac{dv}{dx} + P(x)v = Q(x)$$

first by dividing it with $y^n$ and then substituting $v = y^{1-n}$.

**5.** Solve $\dfrac{dy}{dx} + 5y = xy^2$ using Exercise 4.

Find the general solution of the ordinary differential equations in Exercises 6–8.

**6.** $\dfrac{d^2y}{dx^2} - 3\dfrac{dy}{dx} + 2y = 0$

**7.** $\dfrac{d^2y}{dx^2} + 4\dfrac{dy}{dx} + 4y = 0$

**8.** $\dfrac{d^2y}{dx^2} + 4\dfrac{dy}{dx} + 8y = 0$

Find a particular solution of the ordinary differential equations in Exercises 9–11.

**9.** $\dfrac{d^2y}{dx^2} - 2\dfrac{dy}{dx} - 3y = 5e^x + 2\cos x$

**10.** $\dfrac{d^2y}{dx^2} - 3\dfrac{dy}{dx} + 2y = x + e^x$

**11.** $\dfrac{d^2y}{dx^2} + y = \cos x$

Find the general solution of the ordinary differential equations in Exercises 12–17.

**12.** $\dfrac{d^2y}{dx^2} + 2\dfrac{dy}{dx} - 3y = 5e^x + 2\cos x$

**13.** $\dfrac{d^2y}{dx^2} + 3\dfrac{dy}{dx} + 2y = x^2 + e^{-x}$

**14.** $\dfrac{d^2y}{dx^2} + y = \sin x$

**15.** $\dfrac{d^3y}{dx^3} - 2\dfrac{d^2y}{dx^2} - \dfrac{dy}{dx} + 2y = 0$

**16.** $\dfrac{d^6y}{dx^6} = x$

**17.** $\dfrac{d^4y}{dx^4} + 5\dfrac{d^2y}{dx^2} + 4y = \sin x \cos x$

Find the general solution of the Cauchy–Euler equations in Exercises 18–19 using the substitution $y = x^r$.

**18.** $x^2\dfrac{d^2y}{dx^2} - 2x\dfrac{dy}{dx} + 2y = 0$

**19.**  $3x^2 \dfrac{d^2y}{dx^2} - 2x \dfrac{dy}{dx} - 12y = 0$

Find the general solution of the Cauchy–Euler equations in Exercises 20–21 using the substitution $x = e^t$.

**20.**  $x^2 \dfrac{d^2y}{dx^2} - 5x \dfrac{dy}{dx} + 8y = x^2$

**21.**  $3x^2 \dfrac{d^2y}{dx^2} - 2x \dfrac{dy}{dx} - 12y = 0$

# APPENDIX D

## Review of Difference Equations

We briefly discuss difference equations. Difference equations occur in mathematical models when either the independent variable is discrete or it is convenient to treat the independent variable as a discrete variable.

The Malthusian law of population growth states that the rate of change of population is proportional to the population present; in other words, if $y(t)$ denotes the population at time $t$, then $dy/dt = ky$ where $k$ is a constant of proportionality. However, in the case of some insect populations where one generation dies out before the next generation hatches, it is convenient to assume that the increase in size from one generation to the next is proportional to the size of the former generation. Let $y_i$ be the size of the population of the $i$th generation. Then

$$y_{i+1} - y_i = \alpha y_i$$

where $\alpha$ is a constant of proportionality. This can be written as

$$y_{i+1} - (1 + \alpha)y_i = 0$$

which is a difference equation of order one. In some cases, the realistic model may be

$$y_{i+2} + \alpha y_{i+1} + \beta y_i = 0$$

which assumes that the population of the $(i + 2)$nd generation depends linearly on the previous two generations. This is a difference equation of order two.

An **nth-order difference equation** is a sequence of equations of the form

$$F(y_{i+n}, y_{i+n-1}, \ldots, y_{i+1}, y_i, i) = 0 \quad \text{for } i = 0, 1, \ldots \tag{D.1}$$

where $F$ is a given function and $n$ is a positive integer.

Some examples of difference equations are:

1.  $3y_{i+1} - 2y_i = 0$    for $i = 0, 1, \ldots$
2.  $y_{i+1} - (i + 1)y_i = -2i + 1$    for $i = 0, 1, \ldots$
3.  $y_{i+3}^2 - y_{i+1}y_i = i^2$    for $i = 0, 1, \ldots$
4.  $y_{i+2} - (\cos \gamma_i)y_{i+1} - y_i = 0$    for $i = 0, 1, \ldots$
5.  $y_{i+5} - y_{i+1} + y_i = 1$    for $i = 0, 1, \ldots$

Equations 1 and 2 are of order one, while 3 is of order three, 4 is of order two, and 5 is of order five.

A solution of the difference equation [Equation (D.1)] is a sequence $\{y_i\}$ which satisfies Equation (D.1). For example, $y_i = c(\frac{2}{3})^i$ for $i = 0, 1, \ldots$, where $c$ is an arbitrary constant, is a solution of Equation 1. The general solution of an $n$th-order difference equation contains $n$ arbitrary constants. This situation is similar to an $n$th-order ordinary differential equation. There are many similarities between difference equations and ordinary differential equations.

A difference equation is said to be **linear** if it can be written in the form

$$a_n(i)y_{n+i} + a_{n-1}(i)y_{n+i-1} + \cdots + a_1(i)y_{n+1} + a_0(i)y_n = f(i) \quad \text{for } i = 0, 1, \ldots$$

$$\text{(D.2)}$$

where the coefficients $a_n, a_{n-1}, \ldots, a_0$, and $f$ are functions of $i$. Equations 1, 2, 4, and 5 are linear. If $f(i) = 0$ in Equation (D.2), then Equation (D.2) is called **homogeneous**. This is similar to ordinary differential equations. If $f(i) \neq 0$, then Equation (D.2) is called **nonhomogeneous**. Equations 1 and 4 are homogeneous, while 2, 3, and 5 are nonhomogeneous.

If $a_n, a_{n-1}, \ldots, a_0$ in Equation (D.2) are constants, then we say that Equation (D.2) is an **$n$th-order linear difference equation with constant coefficients**. Equations 1 and 5 are linear difference equations with constant coefficients.

Let us find the general solution of a linear second-order difference equation with constant coefficients

$$ay_{i+2} + by_{i+1} + cy_i = q_i \quad \text{for } i = 0, 1, \ldots \quad \text{(D.3)}$$

where $a \neq 0$, $b$, and $c$ are real numbers. As in the case of differential equations, we first find the general solution of the corresponding homogeneous equation

$$ay_{i+2} + by_{i+1} + cy_i = 0 \quad \text{for } i = 0, 1, \ldots \quad \text{(D.4)}$$

For ordinary homogeneous differential equations with constant coefficients, we looked for a solution of the form $y = e^{rx}$; for homogeneous difference equations with constant coefficients, we look for a solution of the form

$$y_i = r^i \qquad r \neq 0 \quad \text{(D.5)}$$

Substitution of $y_i = r^i$, $y_{i+1} = r^{i+1}$, and $y_{i+2} = r^{i+2}$ in Equation (D.4) leads to

$$ar^2 + br + c = 0 \quad \text{(D.6)}$$

which is called the **characteristic equation** of (D.4). In Table D.1, we summarize the general solution of Equation (D.4) corresponding to three cases of the roots of Equation (D.6).

**Table D.1**   The General Solution of Equation (D.4)

| Case | Roots of Eq. (D.6) | Solution of Eq. (D.4) | General Solution of Eq. (D.4) |
|------|--------------------|------------------------|-------------------------------|
| I | Distinct real: $r_1, r_2$ | $r_1^i, r_2^i$ | $y = c_1 r_1^i + c_2 r_2^i$ |
| II | Repeated root: $r_1$ | $r_1^i, ir_1^i$ | $y = c_1 r_1^i + c_2 ir_1^i$ |
| III | Complex roots: $\alpha + I\beta, \alpha - I\beta$ $(I = \sqrt{-1})$ | $r_1^i \cos i\theta, r_1^i \sin i\theta$ where $r_1 = \sqrt{\alpha^2 + \beta^2}$, $\cos \theta = \alpha/r, \sin \theta = \beta/r$ | $y = r_1^i (c_1 \cos i\theta + c_2 \sin i\theta)$ |

where $c_1$ and $c_2$ are arbitrary constants.

We want to verify Case III in Table D.1. Let $B_1 = \alpha + I\beta$ and $B_2 = \alpha - I\beta$ where $I = \sqrt{-1}$. We express $B_1$ in polar coordinates by expressing $B_1 = r_1 e^{I\theta} = r(\cos\theta + I\sin\theta) = \alpha + I\beta$. Comparing real and imaginary parts, we get $\alpha = r_1\cos\theta$, $\beta = r_1\sin\theta$. Hence $r_1 = \sqrt{\alpha^2 + \beta^2}$ and $\tan\theta = \beta/\alpha$. Similarly $B_2 = r_1 e^{-I\theta}$. The general solution of Equation (D.4) is given by

$$
\begin{aligned}
y_i &= C_1 B_1^i + C_2 B_2^i \\
&= C_1(r_1 e^{I\theta})^i + C_2(r_1 e^{-I\theta})^i \\
&= r_1^i [C_1 e^{Ii\theta} + C_2 e^{-Ii\theta}] \\
&= r_1^i [C_1(\cos i\theta + I\sin i\theta) + C_2(\cos i\theta - I\sin i\theta)] \\
&= r_1^i [(C_1 + C_2)\cos i\theta + I(C_1 - C_2)\sin i\theta] \\
&= r_1^i [c_1 \cos i\theta + c_2 \sin i\theta]
\end{aligned}
$$

where $c_1 = C_1 + C_2$ and $c_2 = I(C_1 - C_2)$. Hence the result in Case III of Table D.1 is verified.

**EXAMPLE D.1**

Find the general solution of

$$y_{i+2} + 4y_{i+1} + 4y_i = 0 \quad \text{for } i = 0, 1, \ldots \tag{D.7}$$

The characteristic equation of (D.7) is $r^2 + 4r + 4 = (r + 2)^2 = 0$. Since $-2$ is the repeated root, the general solution of Equation (D.7) is

$$y_i = c_1(-2)^i + c_2 i(-2)^i \quad \text{for } i = 0, 1, \ldots \qquad ∎∎∎$$

**EXAMPLE D.2**

Find the general solution of

$$y_{i+2} - 2y_{i+1} + 2y_i = 0 \quad \text{for } i = 0, 1, \ldots \tag{D.8}$$

The characteristic equation of (D.8) is given by $r^2 - 2r + 2 = 0$ which has roots $1 \pm I$. Hence $r_1 = \sqrt{\alpha^2 + \beta^2} = \sqrt{1^2 + 1^2} = \sqrt{2}$ and $\cos\theta = \alpha/r_1 = 1/\sqrt{2}$, $\sin\theta = \beta/r_1 = 1/\sqrt{2}$. Therefore $\theta = \pi/4$. Hence the general solution of Equation (D.8) is

$$y_i = (\sqrt{2})^i \left[ c_1 \cos\frac{\pi i}{4} + c_2 \sin\frac{\pi i}{4} \right] \quad \text{for } i = 0, 1, \ldots \qquad ∎∎∎$$

The method of undetermined coefficients is used to find a particular solution of Equation (D.3) when $q_i$ in Equation (D.3) is a linear combination of sequences of the following types:

1. $i^m$, where $m$ is zero or a positive integer,
2. $p^i$, where $p$ is a constant, and
3. $\cos\alpha i$ or $\sin\alpha i$, where $\alpha$ is a constant.

**EXAMPLE D.3**    Find a particular solution of

$$y_{i+2} - 2y_{i+1} + y_i = 3^i + \sin\frac{\pi i}{4} \quad \text{for } i = 0, 1, \ldots \qquad \textbf{(D.9)}$$

Here $q_i$ is a linear combination of $3^i$ and $\sin((\pi i)/4)$. Using a superposition principle (Exercise 15), we find a particular solution of Equation (D.9) as the sum of the particular solutions of

$$y_{i+2} - 2y_{i+1} + y_i = 3^i \quad \text{and} \quad y_{i+2} - 2y_{i+1} + y_i = \sin\frac{\pi i}{4} \quad \text{for } i = 0, 1, \ldots \quad \textbf{(D.10)}$$

For the first equation of (D.10), we treat $3^i$ as we treated $e^{px}$ in Section C; our initial guess $y_i^{p_1}$ is $A \cdot 3^i$. Hence

$$y_i^{p_1} = A \cdot 3^i$$

Substituting $y_i^{p_1} = A \cdot 3^i$, $y_{i+1}^{p_1} = A \cdot 3^{i+1}$, and $y_{i+2}^{p_1} = A \cdot 3^{i+2}$ in the first equation of (D.10) , we get

$$A \cdot 3^{i+2} - 2A \cdot 3^{i+1} + A \cdot 3^i = 3^i$$

Dividing by $3^i$ leads to $9A - 6A + A = 4A = 1$. Thus $A = \frac{1}{4}$. Hence $y_i^{p_1} = \frac{1}{4} \cdot 3^i$.

For the second equation of (D.10), our guess for $y_i^{p_2}$ is $y_i^{p_2} = B\cos(\pi i/4) + C\sin(\pi i/4)$. Therefore

$$
\begin{aligned}
y_{i+1}^{p_2} &= B\cos\frac{\pi(i+1)}{4} + C\sin\frac{\pi(i+1)}{4} \\[2mm]
&= B\left(\cos\frac{\pi i}{4}\cos\frac{\pi}{4} - \sin\frac{\pi i}{4}\sin\frac{\pi}{4}\right) + C\left(\sin\frac{\pi i}{4}\cos\frac{\pi}{4} + \cos\frac{\pi i}{4}\sin\frac{\pi}{4}\right) \\[2mm]
&= \left(\cos\frac{\pi i}{4}\right)\left(\frac{B+C}{\sqrt{2}}\right) + \left(\sin\frac{\pi i}{4}\right)\left(\frac{-B+C}{\sqrt{2}}\right)
\end{aligned}
$$

and

$$
\begin{aligned}
y_{i+2}^{p_2} &= B\cos\frac{\pi(i+2)}{4} + C\sin\frac{\pi(i+2)}{4} \\[2mm]
&= B\left(\cos\frac{\pi i}{4}\cos\frac{\pi}{2} - \sin\frac{\pi i}{4}\sin\frac{\pi}{2}\right) + C\left(\sin\frac{\pi i}{4}\cos\frac{\pi}{2} + \cos\frac{\pi i}{4}\sin\frac{\pi}{2}\right) \\[2mm]
&= -B\sin\frac{\pi i}{4} + C\cos\frac{\pi i}{4} \qquad\qquad\qquad\qquad\qquad \textbf{(D.11)}
\end{aligned}
$$

Substitution of Equation (D.11) in the second equation of Equation (D.10) leads to

$$-B\sin\frac{\pi i}{4} + C\cos\frac{\pi i}{4} - \sqrt{2}\left[(B+C)\cos\frac{\pi i}{4} + (-B+C)\sin\frac{\pi i}{4}\right]$$

$$+ B\cos\frac{\pi i}{4} + C\sin\frac{\pi i}{4} = \sin\frac{\pi i}{4}$$

or

$$[-B - \sqrt{2}(-B + C) + C] \sin \frac{\pi i}{4}$$

$$+ [C - \sqrt{2}(B + C) + B] \cos \frac{\pi i}{4} = \sin \frac{\pi i}{4} \qquad \textbf{(D.12)}$$

Comparing the coefficients of $\sin((\pi i)/4)$ and $\cos((\pi i)/4)$ in Equation (D.12), we get

$$(-B + C) - \sqrt{2}(-B + C) = 1$$
$$(B + C) - \sqrt{2}(B + C) = 0$$

It can be verified that $B = -1/[2(1 - \sqrt{2})]$ and $C = 1/[2(1 - \sqrt{2})]$. Hence a particular solution of Equation (D.9) is

$$y_i^p = \frac{1}{4} \cdot 3^i + \frac{1}{2(1 - \sqrt{2})} \left( -\cos \frac{\pi i}{4} + \sin \frac{\pi i}{4} \right) \quad \text{for } i = 0, 1, \ldots \quad \blacksquare \blacksquare \blacksquare$$

**EXAMPLE D.4**

Find a particular solution of

$$y_{i+2} - 4y_{i+1} = 3 \cdot 4^i \quad \text{for } i = 0, 1, \ldots \qquad \textbf{(D.13)}$$

Assume that $y_i^p = A \cdot 4^i$. Substituting $y_i^p = A \cdot 4^i$, $y_{i+1}^p = A \cdot 4^{i+1}$, and $y_{i+2}^p = A \cdot 4^{i+2}$ in Equation (D.13) yields

$$A \cdot 4^{i+2} - A \cdot 4^{i+2} = 0 \cdot 4^{i+2} = 3 \cdot 4^i$$

Hence we cannot determine $A$ and therefore $y_i^p$ is not of the form $A \cdot 4^i$. This difficulty arises because $4^i$ is a solution of the corresponding homogeneous difference equation $y_{i+2} - 4y_{i+1} = 0$ of Equation (D.13). The method of undetermined coefficients has to be modified when the nonhomogeneous term $q_i$ is also a solution of the corresponding homogeneous equation. Let us multiply the form of $y_i^p$ by $i^s$ where $s$ is the smallest positive integer such that $y_i^p = Ai^s \cdot 4^i$ is not a solution of $y_{i+2} - 4y_{i+1} = 0$. This is similar to $x^s$ in Section C. Thus $y_i^p = Ai4^i$. Then substituting $y_i^p = Ai4^i$, $y_{i+1}^p = A(i + 1)4^{i+1}$, and $y_{i+2}^p = A(i + 2)4^{i+2}$ in Equation (D.13) yields

$$A(i + 2)4^{i+2} - 4A(i + 1)4^{i+1} = 3 \cdot 4^i \qquad \textbf{(D.14)}$$

Dividing Equation (D.14) by $4^i$ we get

$$16A(i + 2) - 16A(i + 1) = 3$$

Hence $A = \frac{3}{16}$ and therefore a particular solution of Equation (D.13) is

$$y_i^p = \frac{3}{16} i4^i \quad \text{for } i = 0, 1, \ldots \qquad \textbf{(D.15)}$$

$\blacksquare \blacksquare \blacksquare$

Thus if any term of the nonhomogeneous term $q_i$ is a solution of the corresponding homogeneous equation, then replace $y_i^p$ by $i^s y_i^p$, where $s$ is the smallest positive integer such that no term in $i^s y_{i+1}^p$ is a solution of the corresponding homogeneous equation. In Table D.2, the second column gives the form of $y_i^p$ for a given $q_i$ in the first column.

**Table D.2** Method of Undetermined Coefficients

| $r_i$ | Choice of $y_i^p$ |
|---|---|
| $a$ | |
| $i^n$ | $i^s(A_n i^n + A_{n-1} i^{n-1} + \cdots + A_1 i + A_0)$ |
| $a_n i^n + a_{n-1} i^{n-1} + \cdots + a_1 i + a_0$ | |
| $b^i$ | $i^s \cdot Bb^i$ |
| $a \cos \alpha i$ | |
| $a \sin \alpha i$ | $i^s(A \cos \alpha i + B \sin \alpha i)$ |
| $a \cos \alpha i + b \sin \alpha i$ | |

where $s$ is the smallest positive integer such that $y_i^p$ contains no term which is a solution of the corresponding homogeneous equation.

Now, we determine the general solution of a linear second-order difference equation with constant coefficients

$$ay_{i+2} + by_{i+1} + cy_i = q_i \quad \text{for } i = 0, 1, \ldots \qquad \textbf{(D.3)}$$

where $a \neq 0$, $b$, and $c$ are constants.

The general solution of Equation (D.3) may be written as

$$y_i = y_i^c + y_i^p$$

where $y_i^c$ is the general solution of the corresponding homogeneous equation (replace $q_i$ by 0 in Equation (D.3))

$$ay_{i+2} + by_{i+1} + cy_i = 0 \quad \text{for } i = 0, 1, \ldots \qquad \textbf{(D.4)}$$

of (D.3) and $y_i^p$ is a particular solution of Equation (D.3) containing no arbitrary constant.

**EXAMPLE D.5**

Find the general solution of

$$y_{i+2} - 4y_{i+1} = 3 \cdot 4^i \quad \text{for } i = 0, 1, \ldots \qquad \textbf{(D.16)}$$

First find the general solution of the corresponding homogeneous equation

$$y_{i+2} - 4y_{i+1} = 0 \quad \text{for } i = 0, 1, \ldots \qquad \textbf{(D.17)}$$

of (D.16). The characteristic equation of (D.17) is given by $r^2 - 4r = 0$ which has roots $r = 0$ and $r = 4$. Hence the general solution of Equation (D.17) is given by

$$y_i^c = c_1 + c_2 4^i$$

where $c_1$ and $c_2$ are arbitrary constants. From Example D.4, a particular solution is $y_i^p = (3/16)i4^i$. Combining both solutions, the general solution of Equation (D.16) is given by

$$y_i = y_i^c + y_i^p = c_1 + c_2 4^i + \frac{3}{16} i4^i \quad \text{for } i = 0, 1, \ldots \quad \blacksquare \blacksquare \blacksquare$$

In a similar fashion, we can solve higher- and lower-order linear difference equations with constant coefficients.

**EXAMPLE D.6**

Find the general solution of a first-order linear difference equation

$$y_{i+1} + ay_i = b \quad \text{for } i = 0, 1, \ldots \tag{D.18}$$

where $a \neq 0$ and $b$ are given constants.

First find the general solution of the corresponding homogeneous difference equation

$$y_{i+1} + ay_i = 0 \tag{D.19}$$

of (D.18). The characteristic equation of (D.19) is given by $r + a = 0$ which gives $r = -a$. Hence the general solution of Equation (D.19) is given by

$$y_i^c = c(-a)^i \quad \text{for } i = 0, 1, \ldots$$

where $c$ is an arbitrary constant. From Table D.2, a particular solution of Equation (D.18) is $y_i^p = B$. Substituting $y_i^p = B$ and $y_{i+1}^p = B$ in Equation (D.18), we get $B + aB = B(1 + a) = b$. Thus $B = b/(1 + a)$ (where $1 + a \neq 0$). The general solution of Equation (D.18) is

$$y_i = y_i^c + y_i^p = c(-a)^i + \frac{b}{1 + a} \quad 1 + a \neq 0 \quad \text{for } i = 0, 1, \ldots$$

If $a = -1$, then we verify (Exercise 11) that the general solution of Equation (D.18) is

$$y_i = c + bi \quad \text{for } i = 0, 1, \ldots$$

where $c$ is an arbitrary constant.    ■ ■ ■

**EXERCISES**

Find the general solution of the difference equations in Exercises 1–3 for $i = 0, 1, \ldots$.

1. $y_{i+2} + 3y_{i+1} + 2y_i = 0$
2. $y_{i+2} - 4y_{i+1} + 4y_i = 0$
3. $y_{i+2} - y_{i+1} + y_i = 0$

Find a particular solution of the difference equations in Exercises 4–6 for $i = 0, 1, \ldots$.

4. $y_{i+2} + y_i = i^2$
5. $y_{i+2} - y_i = \sin 2i$
6. $y_{i+2} - 6y_{i+1} + 9y_i = 3^i$

Find the general solution of the difference equations in Exercises 7–13 for $i = 0, 1, \ldots$.

**7.** $y_{i+2} - 2y_{i+1} + y_i = \cos \dfrac{\pi i}{4}$

**8.** $y_{i+2} + 6y_{i+1} + 5y_i = 4^i + 2$

**9.** $y_{i+2} + y_i = 2$

**10.** $y_{i+2} - y_i = \cos \dfrac{i\pi}{2}$

**11.** $y_{i+1} - y_i = b$    where $b$ is a constant

**12.** $y_{i+3} + y_{i+2} + y_{i+1} + y_i = i$

**13.** $y_{i+2} - 2y_{i+1} + y_i = 0$     $y_0 = 1, y_1 = 2$

**14.** If $\{y_i\}$ and $\{z_i\}$ are solutions of

$$a_{n+i}(i)y_{n+i} + a_{n+i-1}(i)y_{n+i-1} + \cdots + a_1(i)y_{n+1} + a_0(i)y_n$$
$$= 0 \quad \text{for } i = 0, 1, \ldots \qquad \textbf{(D.20)}$$

then show that $\{c_1 y_i + c_2 z_i\}$ is also a solution of Equation (D.20), where $c_1$ and $c_2$ are arbitrary constants.

**15.** Superposition Principle. If $\{y_i\}$ is a solution of $y_{i+2} + p_i y_{i+1} + q_i y_i = r_i$ and $\{z_i\}$ is a solution of $y_{i+2} + p_i y_{i+1} + q_i y_i = s_i$, then show that $\{c_1 y_i + c_2 z_i\}$ is a solution of

$$y_{i+2} + p_i y_{i+1} + q_i y_i = c_1 r_i + c_2 s_i \quad \text{for } i = 0, 1, \ldots$$

where $c_1$ and $c_2$ are constants.

**16.** The sequence $\{1, 1, 2, 3, 5, 8, 11, \ldots\}$ is called the Fibonacci sequence. Note that each term after the second is the sum of the two preceding terms. If $y_i$ denotes the Fibonacci sequence, then the Fibonacci sequence is the unique solution of

$$y_{i+2} = y_{i+1} + y_i \quad \text{for } i = 0, 1, \ldots \qquad y_0 = 1 \quad y_1 = 1$$

Find $y_i$.

**17.** Let $r_1$ and $r_2$ be the roots of the characteristic equation of a difference equation

$$ay_{i+2} + by_{i+1} + cy_i = 0 \quad \text{for } i = 0, 1, \ldots \qquad \textbf{(D.4)}$$

and $L = \max(|r_1|, |r_2|)$. Then show that $L < 1$ is a necessary and sufficient condition for all solutions of Equation (D.4) to approach zero as $i$ approaches infinity.

# Appendix E

## Orthogonal Polynomials*

In this appendix we develop orthogonal vectors and orthogonal polynomials. We need Chebyshev and Legendre polynomials for interpolation and integration respectively. Both families of orthogonal polynomials have interesting properties.

The dot product $\mathbf{u} \cdot \mathbf{v} = u_1 v_1 + u_2 v_2 + u_3 v_3$ of two vectors $\mathbf{u} = (u_1, u_2, u_3)$ and $\mathbf{v} = (v_1, v_2, v_3)$ in $R^3$ can be generalized. An inner product on a real vector space $V$ is an operation on $V$ that assigns to every pair of vectors $\mathbf{u}$ and $\mathbf{v}$ in $V$ a real number $\langle \mathbf{u}, \mathbf{v} \rangle$, called the **inner product** of $\mathbf{u}$ and $\mathbf{v}$, satisfying the following conditions:

1. $\langle \mathbf{u}, \mathbf{u} \rangle \geq 0$ and $\langle \mathbf{u}, \mathbf{u} \rangle = 0$ if and only if $\mathbf{u} = \mathbf{0}$
2. $\langle \mathbf{u}, \mathbf{v} \rangle = \langle \mathbf{v}, \mathbf{u} \rangle$ for all $\mathbf{u}$ and $\mathbf{v} \in V$
3. $\langle \alpha \mathbf{u} + \beta \mathbf{v}, \mathbf{w} \rangle = \alpha \langle \mathbf{u}, \mathbf{w} \rangle + \beta \langle \mathbf{v}, \mathbf{w} \rangle$ for all $\mathbf{u}, \mathbf{v}, \mathbf{w} \in V$ and all $\alpha, \beta \in R$.

A vector space $V$ with inner product is known as an **inner product** space $V$.

**EXAMPLE E.1**

1. Let $R^3 = V = \{\mathbf{u} = (u_1, u_2, u_3) | u_1, u_2, u_3 \in R\}$ and define

$$\langle \mathbf{u}, \mathbf{v} \rangle = \mathbf{u} \cdot \mathbf{v} = u_1 v_1 + u_2 v_2 + u_3 v_3$$

Then we can verify that $R^3$ becomes an inner product space.

2. Let $V = C[a, b] = \{f \,|\, f \text{ is a continuous function on } [a, b]\}$ and define

$$\langle f, g \rangle = \int_a^b f(x)g(x)dx$$

Then we verify that $V = C[a, b]$ is an inner product space.

3. Let $w$ be a positive continuous function on $[a, b]$, $V = C[a, b]$, and define

$$\langle f, g \rangle = \int_a^b w(x)f(x)g(x)dx \tag{E.1}$$

Then we can show that (Exercise 2) Equation (E.1) defines an inner product on $C[a, b]$. The function $w$ is called the weight function for this inner product. When $w(x) = 1$, Equation (E.1) reduces to the inner product defined in (2). ∎ ∎ ∎

Consider $\mathbf{e}_1 = (1, 0)$ and $\mathbf{e}_2 = (0, 1)$ in $R^2$. Then $\langle \mathbf{e}_1, \mathbf{e}_2 \rangle = \mathbf{e}_1 \cdot \mathbf{e}_2 = 0$ and $\mathbf{e}_1$ and $\mathbf{e}_2$ are orthogonal in $R^2$. We are interested in generalizing orthogonal basis, a generalization of orthogonal basis in $R^n$. Two vectors $\mathbf{u}$ and $\mathbf{v}$ in an inner product space $V$ are said to be **orthogonal** if $\langle \mathbf{u}, \mathbf{v} \rangle = 0$.

**EXAMPLE E.2**    Let $w(x) = 1, f(x) = \sin x$, and $g(x) = \cos x$. Then

$$\langle f, g \rangle = \int_{-\pi}^{\pi} 1 \sin x \cos x \, dx = \frac{\sin^2 x}{2} \Bigg|_{-\pi}^{\pi} = 0$$

Thus $\sin x$ and $\cos x$ are orthogonal in $C[-\pi, \pi]$ with respect to the inner product defined in Equation (E.1) with $w(x) = 1$.    ■ ■ ■

A set of vectors $\{\mathbf{u}_1, \mathbf{u}_2, \ldots, \mathbf{u}_N\}$ in an inner product space $V$ is said to be an **orthogonal set** if $\mathbf{u}_i \neq \mathbf{0}$ for all $i$ and

$$\langle \mathbf{u}_i, \mathbf{u}_j \rangle = 0 \quad \text{whenever } i \neq j \tag{E.2}$$

If in addition to Equation (E.2),

$$\langle \mathbf{u}_i, \mathbf{u}_i \rangle = 1 \quad \text{for } i = 1, 2, \ldots, N \tag{E.3}$$

then the set is said to be an **orthonormal set**.

**EXAMPLE E.3**    Prove that the set

$$\left\{ \frac{\sin x}{\sqrt{\pi}}, \frac{\sin 2x}{\sqrt{\pi}} \cdots, \frac{\sin Nx}{\sqrt{\pi}} \right\}$$

is orthonormal in $C[-\pi, \pi]$ with respect to the inner product defined in Equation (E.1) with $w(x) = 1$.

$$\langle \mathbf{u}_i, \mathbf{u}_j \rangle = \frac{1}{\pi} \int_{-\pi}^{\pi} \sin ix \sin jx \, dx$$

$$= \frac{1}{2\pi} \int_{-\pi}^{\pi} [\cos(i - j)x - \cos(i + j)x] dx$$

$$= \frac{1}{2\pi} \left[ \frac{\sin(i - j)x}{i - j} - \frac{\sin(i + j)x}{i + j} \right] \Bigg|_{-\pi}^{\pi} = 0 \quad \text{for } j \neq i$$

For $j = i$,

$$\langle \mathbf{u}_i, \mathbf{u}_j \rangle = \langle \mathbf{u}_i, \mathbf{u}_i \rangle$$

$$= \frac{1}{\pi} \int_{-\pi}^{\pi} \sin^2 ix \, dx = \frac{1}{2\pi} \int_{-\pi}^{\pi} (1 - \cos 2ix) dx$$

$$= \frac{1}{2\pi} \left[ x - \frac{\sin 2ix}{2i} \right] \Bigg|_{-\pi}^{\pi} = 1$$

Therefore $\langle \mathbf{u}_i, \mathbf{u}_j \rangle = \delta_{ij}$. Hence $\{\sin x/\sqrt{\pi}, \sin 2x/\sqrt{\pi} \cdots, \sin Nx/\sqrt{\pi}\}$ is orthonormal in $C[-\pi, \pi]$.    ■ ■ ■

**Theorem E.1**    If $\{\mathbf{u}_1, \mathbf{u}_2, \ldots, \mathbf{u}_N\}$ is an orthogonal set in an inner product space $V$, then $\{\mathbf{u}_1, \mathbf{u}_2, \ldots, \mathbf{u}_N\}$ is also linearly independent in $V$.

***Proof***    Assume that the set $\{\mathbf{u}_1, \mathbf{u}_2, \ldots, \mathbf{u}_N\}$ is linearly dependent. Therefore there exist scalars $\alpha_1, \alpha_2, \ldots, \alpha_N$ not all zero such that

$$\alpha_1 \mathbf{u}_1 + \alpha_2 \mathbf{u}_2 + \cdots + \alpha_i \mathbf{u}_i + \cdots + \alpha_N \mathbf{u}_N = \mathbf{0} \tag{E.4}$$

Then taking the inner product with $\mathbf{u}_i$ on both sides of Equation (E.4), we get

$$\begin{aligned}
\langle \alpha_1 \mathbf{u}_1 &+ \alpha_2 \mathbf{u}_2 + \cdots + \alpha_i \mathbf{u}_i + \cdots + \alpha_N \mathbf{u}_N, \mathbf{u}_i \rangle \\
&= \alpha_1 \langle \mathbf{u}_1, \mathbf{u}_i \rangle + \alpha_2 \langle \mathbf{u}_2, \mathbf{u}_i \rangle + \cdots + \alpha_i \langle \mathbf{u}_i, \mathbf{u}_i \rangle + \cdots + \alpha_N \langle \mathbf{u}_N, \mathbf{u}_i \rangle \\
&= \langle \mathbf{0}, \mathbf{u}_i \rangle = 0
\end{aligned}$$

Since $\langle \mathbf{u}_j, \mathbf{u}_i \rangle = 0$ if $j \neq i$, this reduces to $\alpha_i \langle \mathbf{u}_i, \mathbf{u}_i \rangle = 0$. But $\langle \mathbf{u}_i, \mathbf{u}_i \rangle \neq 0$, therefore it follows that $\alpha_i = 0$. Since $\mathbf{u}_i$ was any vector from the orthogonal set $\{\mathbf{u}_1, \mathbf{u}_2, \ldots, \mathbf{u}_N\}$, $\alpha_i = 0$ for all $i$ in Equation (E.4). Hence $\mathbf{u}_1, \mathbf{u}_2, \ldots, \mathbf{u}_N$ are linearly independent in $V$.    ∎

The next theorem follows from this theorem.

**Theorem E.2**    An orthogonal set is a basis for an $N$-dimensional inner product space if and only if it contains $N$ vectors.    ∎

Let $\{\mathbf{u}_1, \mathbf{u}_2, \ldots, \mathbf{u}_N\}$ be an orthogonal basis for an $N$-dimensional inner product space $V$. Then any vector $\mathbf{u} \in V$ can be expressed as

$$\mathbf{u} = c_1 \mathbf{u}_1 + c_2 \mathbf{u}_2 + \cdots + c_i \mathbf{u}_i + \cdots + c_N \mathbf{u}_N$$

Taking inner product with $\mathbf{u}_i$ on both sides yields

$$\begin{aligned}
\langle \mathbf{u}, \mathbf{u}_i \rangle &= \langle c_1 \mathbf{u}_1 + c_2 \mathbf{u}_2 + \cdots + c_i \mathbf{u}_i + \cdots + c_N \mathbf{u}_N, \mathbf{u}_i \rangle \\
&= c_1 \langle \mathbf{u}_1, \mathbf{u}_i \rangle + c_2 \langle \mathbf{u}_2, \mathbf{u}_i \rangle + \cdots + c_i \langle \mathbf{u}_i, \mathbf{u}_i \rangle + \cdots + c_N \langle \mathbf{u}_N, \mathbf{u}_i \rangle \\
&= c_i \langle \mathbf{u}_i, \mathbf{u}_i \rangle \quad \text{since } \langle \mathbf{u}_j, \mathbf{u}_i \rangle = 0 \text{ if } j \neq i
\end{aligned}$$

Thus

$$c_i = \frac{\langle \mathbf{u}, \mathbf{u}_i \rangle}{\langle \mathbf{u}_i, \mathbf{u}_i \rangle} \quad \text{for } i = 1, 2, \ldots, N$$

**Theorem E.3**    If $\{\mathbf{u}_1, \mathbf{u}_2, \ldots, \mathbf{u}_N\}$ is an orthogonal basis for an $N$-dimensional inner product space $V$, then any $\mathbf{u} \in V$ can be expressed as

$$\mathbf{u} = c_1 \mathbf{u}_1 + c_2 \mathbf{u}_2 + \cdots + c_i \mathbf{u}_i + \cdots + c_N \mathbf{u}_N \tag{E.5}$$

where

$$c_i = \frac{\langle \mathbf{u}, \mathbf{u}_i \rangle}{\langle \mathbf{u}_i, \mathbf{u}_i \rangle} \quad \text{for } i = 1, 2, \ldots, N \qquad ∎$$

Let $\{p_0(x), p_1(x), \ldots,\}$ be a sequence of polynomials with degree of $p_i(x) = i$ for each $i$ and $\langle p_i, p_j \rangle = 0$ if $i \neq j$. Then $\{p_0(x), p_1(x), \ldots,\}$ is said to be a sequence of orthogonal polynomials.

---

**EXAMPLE E.4**

The Chebyshev polynomial of degree $n$ is given by

$$T_n(x) = \cos(n \cos^{-1}x) \quad \text{for } n = 0, 1, \ldots \tag{E.6}$$

and the set of Chebyshev polynomials is orthogonal in $C[-1, 1]$ with respect to the inner product defined by

$$\langle T_i, T_j \rangle = \int_{-1}^{1} \frac{1}{\sqrt{1 - x^2}} T_i(x)T_j(x)dx \tag{E.7}$$

The properties of Chebyshev polynomials are described in some detail in Chapter 4 (Section 4.7). The first few Chebyshev polynomials are

$$\begin{array}{ll} T_0(x) = 1 & T_2(x) = 2x^2 - 1 \\ T_1(x) = x & T_3(x) = 4x^3 - 3x \end{array}$$

■ ■ ■

---

**EXAMPLE E.5**

The Legendre polynomial of degree $n$ is given by

$$P_n(x) = \frac{(-1)^n}{2^n\, n!} \frac{d^n}{dx^n} [(1 - x^2)^n] \quad \text{for } n = 1, 2, \ldots \tag{E.8}$$

with $P_0(x) = 1$ and the set of Legendre polynomials is orthogonal in $C[-1, 1]$ with respect to the inner product defined by

$$\langle P_i, P_j \rangle = \int_{-1}^{1} P_i(x)P_j(x)dx \tag{E.9}$$

This is described in some detail in Chapter 6 (Section 6.6). The first few Legendre polynomials are

$$\begin{array}{ll} P_0(x) = 1 & P_2(x) = \frac{1}{2}(3x^2 - 1) \\[2mm] P_1(x) = x & P_3(x) = \frac{1}{2}(5x^3 - 3x) \end{array}$$

■ ■ ■

---

**EXAMPLE E.6**

Express $x^2 + x + 1$ in terms of Chebyshev polynomials $T_0(x)$, $T_1(x)$, and $T_2(x)$ given in Example E.4 using Equation (E.5).

We have

$$x^2 + x + 1 = c_0 T_0(x) + c_1 T_1(x) + c_2 T_2(x)$$

where $c_0$, $c_1$, and $c_2$ are given by Equation (E.5). Using Equation (E.7), we get

$$c_0 = \frac{\langle x^2 + x + 1, T_0 \rangle}{\langle T_0, T_0 \rangle} = \frac{\displaystyle\int_{-1}^{1} \frac{(x^2 + x + 1) \cdot 1}{\sqrt{1 - x^2}}\, dx}{\displaystyle\int_{-1}^{1} \frac{1}{\sqrt{1 - x^2}}\, dx}$$

Let $x = \cos\theta$. Then $dx = -\sin\theta\, d\theta$ and $\theta = \pi$ when $x = -1$ and $\theta = 0$ when $x = 1$. Hence

$$\int_{-1}^{1} \frac{x^2 + x + 1}{\sqrt{1 - x^2}}\, dx = -\int_{\pi}^{0} \frac{(\cos^2\theta + \cos\theta + 1) \sin\theta}{\sin\theta}\, d\theta$$

$$= \int_{0}^{\pi} (\cos^2\theta + \cos\theta + 1)\, d\theta = \frac{3}{2}\pi$$

and

$$\int_{-1}^{1} \frac{1}{\sqrt{1 - x^2}}\, dx = -\int_{\pi}^{0} \frac{1}{\sin\theta} \sin\theta\, d\theta = \pi$$

Therefore

$$c_0 = \frac{\displaystyle\int_{-1}^{1} \frac{(x^2 + x + 1)}{\sqrt{1 - x^2}}\, dx}{\displaystyle\int_{-1}^{1} \frac{dx}{\sqrt{1 - x^2}}} = \frac{\frac{3\pi}{2}}{\pi} = \frac{3}{2}$$

Similarly we show that $c_1 = 1$ and $c_2 = \frac{1}{2}$. Hence $x^2 + x + 1 = (\frac{1}{2})(2x^2 - 1) + 1 \cdot x + (\frac{3}{2})$. ■ ■ ■

We have seen orthogonal and orthonormal bases. The Gram–Schmidt process constructs an orthogonal basis of an inner product space $V$ from its given basis.

## The Gram–Schmidt Process

For simplicity, consider a vector space $V$ of dim 3. Let $\{\mathbf{v}_1, \mathbf{v}_2, \mathbf{v}_3\}$ be a basis of $V$. We want to construct an orthogonal basis of an inner product space $V$. We need to have an inner product for the vector space $V$.

First we select $\mathbf{u}_1 = \mathbf{v}_1$.

We want to construct $\mathbf{u}_2$ orthogonal to $\mathbf{u}_1$. It is convenient to use only $\mathbf{u}_1$ and $\mathbf{v}_2$ to determine $\mathbf{u}_2$. Since $\mathbf{u}_2 \in V$, $\mathbf{u}_2 = \mathbf{v}_2 - c\mathbf{u}_1$ where $c$ is to be determined. If $\mathbf{u}_2$ is orthogonal to $\mathbf{u}_1$, then taking the inner product with $\mathbf{u}_1$ on both sides of

$$\mathbf{u}_2 = \mathbf{v}_2 - c\mathbf{u}_1$$

we get

$$0 = \langle \mathbf{u}_2, \mathbf{u}_1 \rangle = \langle \mathbf{v}_2 - c\mathbf{u}_1, \mathbf{u}_1 \rangle = \langle \mathbf{v}_2, \mathbf{u}_1 \rangle - c\langle \mathbf{u}_1, \mathbf{u}_1 \rangle$$

Hence $c = \langle \mathbf{v}_2, \mathbf{u}_1 \rangle / \langle \mathbf{u}_1, \mathbf{u}_1 \rangle$. Thus $\mathbf{u}_2 = \mathbf{v}_2 - [\langle \mathbf{v}_2, \mathbf{u}_1 \rangle / \langle \mathbf{u}_1, \mathbf{u}_1 \rangle] \mathbf{u}_1$ is orthogonal to $\mathbf{u}_1$.

We want to find $c_1$ and $c_2$ such that $\mathbf{u}_3 = \mathbf{v}_3 - c_1 \mathbf{u}_1 - c_2 \mathbf{u}_2$ is orthogonal to $\mathbf{u}_1$ and $\mathbf{u}_2$. If $\mathbf{u}_3$ is orthogonal to $\mathbf{u}_1$, then

$$\begin{aligned} 0 = \langle \mathbf{u}_3, \mathbf{u}_1 \rangle &= \langle \mathbf{v}_3, \mathbf{u}_1 \rangle - c_1 \langle \mathbf{u}_1, \mathbf{u}_1 \rangle - c_2 \langle \mathbf{u}_2, \mathbf{u}_1 \rangle \\ &= \langle \mathbf{v}_3, \mathbf{u}_1 \rangle - c_1 \langle \mathbf{u}_1, \mathbf{u}_1 \rangle \end{aligned}$$

Hence $c_1 = \langle \mathbf{v}_3, \mathbf{u}_1 \rangle / \langle \mathbf{u}_1, \mathbf{u}_1 \rangle$. Similarly we find $c_2 = \langle \mathbf{v}_3, \mathbf{u}_2 \rangle / \langle \mathbf{u}_2, \mathbf{u}_2 \rangle$. Hence

$$\mathbf{u}_3 = \mathbf{v}_3 - \frac{\langle \mathbf{v}_3, \mathbf{u}_1 \rangle}{\langle \mathbf{u}_1, \mathbf{u}_1 \rangle} \mathbf{u}_1 - \frac{\langle \mathbf{v}_3, \mathbf{u}_2 \rangle}{\langle \mathbf{u}_2, \mathbf{u}_2 \rangle} \mathbf{u}_2$$

This process can be generalized to a vector space of dim $N$. Let $S = \{\mathbf{v}_1, \mathbf{v}_2, \ldots, \mathbf{v}_N\}$ be a basis of a vector space $V$. The following vectors form an orthogonal basis of an inner product space $V$:

$$\begin{aligned} \mathbf{u}_1 &= \mathbf{v}_1 \\ \mathbf{u}_2 &= \mathbf{v}_2 - \frac{\langle \mathbf{v}_2, \mathbf{u}_1 \rangle}{\langle \mathbf{u}_1, \mathbf{u}_1 \rangle} \mathbf{u}_1 \\ \mathbf{u}_3 &= \mathbf{v}_3 - \frac{\langle \mathbf{v}_3, \mathbf{u}_1 \rangle}{\langle \mathbf{u}_1, \mathbf{u}_1 \rangle} \mathbf{u}_1 - \frac{\langle \mathbf{v}_3, \mathbf{u}_2 \rangle}{\langle \mathbf{u}_2, \mathbf{u}_2 \rangle} \mathbf{u}_2 \\ &\vdots \\ \mathbf{u}_N &= \mathbf{v}_N - \frac{\langle \mathbf{v}_N, \mathbf{u}_1 \rangle}{\langle \mathbf{u}_1, \mathbf{u}_1 \rangle} \mathbf{u}_1 - \cdots - \frac{\langle \mathbf{v}_N, \mathbf{u}_{N-1} \rangle}{\langle \mathbf{u}_{N-1}, \mathbf{u}_{N-1} \rangle} \mathbf{u}_{N-1} \end{aligned} \qquad \textbf{(E.10)}$$

It is easy (Exercise 11) to construct an orthonormal set $\{\mathbf{u}_1 / \|\mathbf{u}_1\|, \mathbf{u}_2 / \|\mathbf{u}_2\|, \ldots, \mathbf{u}_N / \|\mathbf{u}_N\|\}$ where $\|\mathbf{u}_i\| = \sqrt{\langle \mathbf{u}_i, \mathbf{u}_i \rangle}$ for $i = 1, 2, \ldots, N$ from a given orthogonal set $\{\mathbf{u}_1, \mathbf{u}_2, \ldots, \mathbf{u}_N\}$.

**EXAMPLE E.7**

Construct an orthonormal basis of $P_3[-1, 1]$, a subspace of $C[-1, 1]$, using the Gram–Schmidt process if the inner product is defined by Equation (E.1) with $w(x) = 1$, $b = -1$, and $a = 1$. Start with the standard basis $\{1, x, x^2, x^3\}$ of $P_3$.

We have $\mathbf{v}_1 = 1$, $\mathbf{v}_2 = x$, $\mathbf{v}_3 = x^2$, and $\mathbf{v}_4 = x^3$.

Since $\mathbf{u}_1 = \mathbf{v}_1$, we have $\mathbf{u}_1 = 1$.

We want to find $\mathbf{u}_2 = \mathbf{v}_2 - (\langle \mathbf{v}_2, \mathbf{u}_1 \rangle / \langle \mathbf{u}_1, \mathbf{u}_1 \rangle)\mathbf{u}_1$. We have

$$\langle \mathbf{u}_1, \mathbf{u}_1 \rangle = \int_{-1}^{1} 1 \cdot 1 \, dx = x\Big|_{-1}^{1} = 2$$

$$\langle \mathbf{v}_2, \mathbf{u}_1 \rangle = \int_{-1}^{1} x \cdot 1 \, dx = \frac{x^2}{2}\Big|_{-1}^{1} = 0$$

Hence

$$\mathbf{u}_2 = \mathbf{v}_2 - \frac{\langle \mathbf{v}_2, \mathbf{u}_1 \rangle}{\langle \mathbf{u}_1, \mathbf{u}_1 \rangle} \mathbf{u}_1 = x - 0 = x$$

To find $\mathbf{u}_3$, we need

$$\langle \mathbf{v}_3, \mathbf{u}_1 \rangle = \int_{-1}^{1} x^2 \, dx = \frac{x^3}{3} \bigg|_{-1}^{1} = \frac{2}{3}$$

$$\langle \mathbf{v}_3, \mathbf{u}_2 \rangle = \int_{-1}^{1} x^3 \, dx = \frac{x^4}{4} \bigg|_{-1}^{1} = 0$$

$$\langle \mathbf{u}_2, \mathbf{u}_2 \rangle = \int_{-1}^{1} x^2 \, dx = \frac{x^3}{3} \bigg|_{-1}^{1} = \frac{2}{3}$$

This gives

$$\mathbf{u}_3 = \mathbf{v}_3 - \frac{\langle \mathbf{v}_3, \mathbf{u}_1 \rangle}{\langle \mathbf{u}_1, \mathbf{u}_1 \rangle} \mathbf{u}_1 - \frac{\langle \mathbf{v}_3, \mathbf{u}_2 \rangle}{\langle \mathbf{u}_2, \mathbf{u}_2 \rangle} \mathbf{u}_2$$

$$= x^2 - \frac{2}{3} - 0 = x^2 - \frac{2}{3}$$

Similarly we obtain $\mathbf{u}_4 = x^3 - (6/5)x$.

Thus $\{1, x, x^2 - (\frac{2}{3}), x^3 - (6/5)x\}$ is an orthogonal basis of $P_3$.

To construct an orthonormal set, we compute $\|\mathbf{u}_1\|^2$, $\|\mathbf{u}_2\|^2$, $\|\mathbf{u}_3\|^2$, and $\|\mathbf{u}_4\|^2$. We have

$$\|\mathbf{u}_1\|^2 = \langle \mathbf{u}_1, \mathbf{u}_1 \rangle = \langle 1, 1 \rangle = 2$$

$$\|\mathbf{u}_2\|^2 = \langle \mathbf{u}_2, \mathbf{u}_2 \rangle = \langle x, x \rangle = \frac{2}{3}$$

$$\|\mathbf{u}_3\|^2 = \langle \mathbf{u}_3, \mathbf{u}_3 \rangle = \left\langle x^2 - \frac{2}{3}, x^2 - \frac{2}{3} \right\rangle = \frac{2}{5}$$

$$\|\mathbf{u}_4\|^2 = \langle \mathbf{u}_4, \mathbf{u}_4 \rangle = \left\langle x^3 - \frac{6}{5}x, x^3 - \frac{6}{5}x \right\rangle = \frac{2}{7}$$

Finally, the orthonormal basis of $P_3$ is $\{\frac{1}{2}, (3x)/2, 5(x^2 - \frac{2}{3})/2, 7(x^3 - (6/5)x)/2\}$. These are called the first four normalized Legendre polynomials.  ■ ■ ■

**EXAMPLE E.8**    Find an orthonormal basis for the column space of

$$A = \begin{bmatrix} 1 & 0 & 1 \\ 1 & 1 & 0 \\ 0 & 1 & 1 \end{bmatrix}$$

We have column vectors

$$\mathbf{v}_1 = \begin{bmatrix} 1 \\ 1 \\ 0 \end{bmatrix} \qquad \mathbf{v}_2 = \begin{bmatrix} 0 \\ 1 \\ 1 \end{bmatrix} \qquad \mathbf{v}_3 = \begin{bmatrix} 1 \\ 0 \\ 1 \end{bmatrix}$$

Since $\mathbf{u}_1 = \mathbf{v}_1$, we have

$$\mathbf{u}_1 = \begin{bmatrix} 1 \\ 1 \\ 0 \end{bmatrix}$$

We want $\mathbf{u}_2 = \mathbf{v}_2 - [\langle \mathbf{v}_2, \mathbf{u}_1 \rangle / \langle \mathbf{u}_1, \mathbf{u}_1 \rangle] \, \mathbf{u}_1$. We have

$$\langle \mathbf{v}_2, \mathbf{u}_1 \rangle = \mathbf{v}_2' \mathbf{u}_1 = \begin{bmatrix} 0 & 1 & 1 \end{bmatrix} \begin{bmatrix} 1 \\ 1 \\ 0 \end{bmatrix} = 1$$

$$\langle \mathbf{u}_1, \mathbf{u}_1 \rangle = \mathbf{u}_1' \mathbf{u}_1 = \begin{bmatrix} 1 & 1 & 0 \end{bmatrix} \begin{bmatrix} 1 \\ 1 \\ 0 \end{bmatrix} = 2$$

Hence

$$\mathbf{u}_2 = \mathbf{v}_2 - \frac{\langle \mathbf{v}_2, \mathbf{u}_1 \rangle}{\langle \mathbf{u}_1, \mathbf{u}_1 \rangle} \mathbf{u}_1 = \begin{bmatrix} 0 \\ 1 \\ 1 \end{bmatrix} - \frac{1}{2} \begin{bmatrix} 1 \\ 1 \\ 0 \end{bmatrix} = \begin{bmatrix} -1/2 \\ 1/2 \\ 1 \end{bmatrix}$$

Continuing, we have

$$\mathbf{u}_3 = \mathbf{v}_3 - \frac{\langle \mathbf{v}_3, \mathbf{u}_1 \rangle}{\langle \mathbf{u}_1, \mathbf{u}_1 \rangle} \mathbf{u}_1 - \frac{\langle \mathbf{v}_3, \mathbf{u}_2 \rangle}{\langle \mathbf{u}_2, \mathbf{u}_2 \rangle} \mathbf{u}_2$$

$$= \begin{bmatrix} 1 \\ 0 \\ 1 \end{bmatrix} - \frac{1}{2} \begin{bmatrix} 1 \\ 1 \\ 0 \end{bmatrix} - \frac{1}{3} \begin{bmatrix} -1/2 \\ 1/2 \\ 1 \end{bmatrix} = \begin{bmatrix} 2/3 \\ -2/3 \\ 2/3 \end{bmatrix}$$

To construct an orthonormal set, we compute

$$\|\mathbf{u}_1\|^2 = \langle \mathbf{u}_1, \mathbf{u}_1 \rangle = \mathbf{u}_1' \mathbf{u}_1 = 2$$

$$\|\mathbf{u}_2\|^2 = \langle \mathbf{u}_2, \mathbf{u}_2 \rangle = \mathbf{u}_2' \mathbf{u}_2 = \frac{3}{2}$$

$$\|\mathbf{u}_3\|^2 = \langle \mathbf{u}_3, \mathbf{u}_3 \rangle = \mathbf{u}_3' \mathbf{u}_3 = \frac{4}{3}$$

Hence the orthonormal basis for the column space of $A$ is

$$\left\{ \begin{bmatrix} 1/\sqrt{2} \\ 1/\sqrt{2} \\ 0 \end{bmatrix}, \begin{bmatrix} -1/\sqrt{6} \\ 1/\sqrt{6} \\ 1/\sqrt{6} \end{bmatrix}, \begin{bmatrix} 1/\sqrt{3} \\ -1/\sqrt{3} \\ 1/\sqrt{3} \end{bmatrix} \right\}$$

∎

**Theorem E.4**    If $\{ p_0(x), p_1(x), \ldots \}$ is a sequence of orthogonal polynomials, then

(a)   $\{ p_0(x), p_1(x), \ldots, p_N(x) \}$ is a basis of $P_N$ where $P_N$ is a vector space of polynomials of degree less than or equal to $N$ and

(b)   $\langle q, p_N \rangle = 0$ where $q(x)$ is a polynomial of degree less than $N$.

*Proof*

(a)  Since dim $P_N = N + 1$ and $p_0, p_1, \ldots, p_N$ are $N + 1$ linearly independent in $P_N$ by Theorem E.1, these $N + 1$ vectors form a basis of $P_N$.

(b)  Since $q(x)$ is a polynomial of degree less than $N$,

$$q(x) = c_0 p_0(x) + c_1 p_1(x) + \cdots + c_{N-1} p_{N-1}(x)$$

Taking the inner product with $p_N(x)$ on both sides, we obtain

$$\begin{aligned}\langle q, p_N \rangle &= \langle c_0 p_0 + c_1 p_1 + \cdots + c_{N-1} p_{N-1}, p_N \rangle \\ &= c_0 \langle p_0, p_N \rangle + c_1 \langle p_1, p_N \rangle + \cdots + c_{N-1} \langle p_{N-1}, p_N \rangle \\ &= 0 \quad \text{since } \langle p_i, p_N \rangle = 0 \quad \text{for } i = 0, 1, \ldots, N - 1\end{aligned}$$

Hence $p_N$ is orthogonal to every polynomial of degree less than $N$.    ■

**Theorem E.5**    If $\{p_0, p_1, \ldots\}$ is a sequence of orthogonal polynomials with respect to the inner product defined by Equation (E.1), then $p_N$ has $N$ real and distinct zeros in the interval $(a, b)$.

This is an extremely useful theorem and we use it in Chapters 4 and 6.

*Proof*  Assume that $p_N$ changes its sign for the real zeros $x_1, x_2, \ldots, x_m$ in the interval $(a, b)$. Then

$$p_N(x) = (x - x_1)^{l_1}(x - x_2)^{l_2} \cdots (x - x_m)^{l_m} q(x)$$

where $l_1, l_2, \ldots, l_m$ are odd positive integers and $q(x)$ does not change sign in $(a, b)$ and $q(x_i) \neq 0$ for $i = 1, 2, \ldots, m$. We clearly have $m \leq N$. We want to show that $m = N$.

Let

$$r(x) = (x - x_1)(x - x_2) \cdots (x - x_m)$$

Then

$$p_N(x)r(x) = (x - x_1)^{l_1+1} (x - x_2)^{l_2+1} \cdots (x - x_m)^{l_m+1} q(x)$$

does not change sign in $(a, b)$ since $l_1 + 1, l_2 + 1, \ldots, l_m + 1$ are even powers and $q(x)$ does not change sign in $(a, b)$. Hence

$$\langle p_N, r \rangle = \int_a^b w(x) p_N(x) r(x) dx \neq 0$$

But by Theorem E.4(b), $p_N(x)$ is orthogonal to all polynomials of degree less than $N$. Hence $\deg(r(x)) = m \geq N$. Hence $m = N$.    ■

**EXAMPLE E.9**    The zeros of the third-degree Chebyshev polynomial $T_3(x) = 4x^3 - 3x = x(4x^2 - 3)$ are $-\sqrt{3}/2$, 0, and $\sqrt{3}/2$ and are in the interval $(-1, 1)$.    ■ ■ ■

For more details about orthogonal vectors and polynomials, refer to Leon (1986).

**EXERCISES**

1. Show that $\langle \mathbf{u}, \mathbf{v} \rangle = \mathbf{u} \cdot \mathbf{v} = \mathbf{u}\mathbf{v}^t = u_1v_1 + u_2v_2 + u_3v_3$, where $\mathbf{u} = (u_1, u_2, u_3)$ and $\mathbf{v} = (v_1, v_2, v_3)$, defines an inner product on $R^3$.

2. Show that $\langle f, g \rangle = \int_a^b w(x)f(x)g(x)dx$ where $w$ is a positive continuous function on $[a, b]$ defines an inner product on $C[a, b]$.

3. In $C[-\pi, \pi]$ with the inner product defined by Equation (E.1) with $w(x) = 1$, find
   (a) $\langle \sin x, \cos x \rangle$      (b) $\langle \sin ix, \sin jx \rangle$
   (c) $\langle \sin ix, \cos jx \rangle$   where $i$ and $j$ are nonnegative integers.

4. Let

$$D_2 = \left\{ \begin{bmatrix} d_{11} & 0 \\ 0 & d_{22} \end{bmatrix} \middle| d_{11}, d_{22} \in R \right\}$$

   denote the vector space of $2 \times 2$ diagonal matrices and define $\langle A, B \rangle = a_{11}b_{11} + a_{22}b_{22}$ where

$$A = \begin{bmatrix} a_{11} & 0 \\ 0 & a_{22} \end{bmatrix} \in D_2 \text{ and } B = \begin{bmatrix} b_{11} & 0 \\ 0 & b_{22} \end{bmatrix} \in D_2.$$

   Prove that $D_2$ is an inner product space.

5. In $C[-\pi, \pi]$ with the inner product defined by Equation (E.1) with $w(x) = 1/\sqrt{1 - x^2}$, find
   (a) $\langle 2x^2 - 1, x \rangle$      (b) $\langle 2x^2 - 1, 2x^2 - 1 \rangle$      (c) $\langle x, 4x^3 - 3x \rangle$

6. Show that the set $\{ 1/2\pi, \cos x/\sqrt{\pi}, \ldots, \cos Nx/\sqrt{\pi} \}$ is orthonormal in $C[-\pi, \pi]$ with respect to the inner product defined by $\langle f, g \rangle = \int_{-\pi}^{\pi} f(x)g(x)dx$.

7. Show that the set $\{ 1/2\pi, \sin x/\sqrt{\pi}, \cos x/\sqrt{\pi} \ldots \sin Nx/\sqrt{\pi}, \cos Nx/\sqrt{\pi} \}$ is orthonormal in $C[-\pi, \pi]$ with respect to the inner product defined by $\langle f, g \rangle = \int_{-\pi}^{\pi} f(x)g(x)dx$.

8. Show that the set of polynomials $\{1, x, 2x^2 - 1, 4x^3 - 3x\}$ is orthogonal in $C[-1, 1]$ with respect to the inner product defined by $\langle p_i, p_j \rangle = \int_{-1}^{1} 1/\sqrt{1 - x^2}\, p_i(x)p_j(x)dx$.

9. The Chebyshev polynomials of the second kind are defined by

$$U_{n-1} = \frac{1}{n} T'_n(x) \quad \text{for } n = 1, 2, \ldots$$

   where $T_n$ is the $n$th degree Chebyshev polynomial.
   (a) Find $U_0(x)$, $U_1(x)$, and $U_2(x)$.
   (b) Show that the set $\{U_0, U_1, U_2\}$ is orthogonal with respect to the inner product defined by

$$\langle U_i, U_j \rangle = \int_{-1}^{1} \sqrt{1 - x^2}\, U_i(x)U_j(x)dx$$

10. If $\mathbf{u} - a\mathbf{v}$ where $\mathbf{v} \neq \mathbf{0}$ is orthogonal to $\mathbf{v}$ for any choice of $\mathbf{u}$, then prove that $a = (\langle \mathbf{u}, \mathbf{v} \rangle)/\langle \mathbf{v}, \mathbf{v} \rangle$.

11. Prove that if $\{\mathbf{u}_1, \mathbf{u}_2, \ldots, \mathbf{u}_N\}$ is an orthogonal basis of an inner product space $V$, then $\{\mathbf{u}_1/\|\mathbf{u}_1\|, \mathbf{u}_2/\|\mathbf{u}_2\|, \ldots, \mathbf{u}_N/\|\mathbf{u}_N\|\}$ where $\|u_i\| = \sqrt{\langle \mathbf{u}_i, \mathbf{u}_i \rangle}$ for $i = 1, 2, \ldots, N$ is an orthonormal basis of the inner product space $V$.

12. Given the basis $\{(1, 2), (2, 3)\}$ for $R^2$, use the Gram–Schmidt process to obtain an orthonormal basis of $R^2$. (*Hint:* Use $\langle \mathbf{u}, \mathbf{v} \rangle = \mathbf{u} \cdot \mathbf{v}$)

13. Given the basis $\{(1, 0, 1), (3, 1, -1), (1, 2, 3)\}$ for $R^3$, use the Gram–Schmidt process to obtain an orthonormal basis for $R^3$. (*Hint:* Use $\langle \mathbf{u}, \mathbf{v} \rangle = \mathbf{u} \cdot \mathbf{v}$)

14. Find an orthonormal basis for the column space of

$$A = \begin{bmatrix} 1 & 3 \\ 2 & 4 \end{bmatrix}$$

15. Find an orthonormal basis for the column space of

$$A = \begin{bmatrix} 1 & 1 & 2 \\ 0 & 0 & 1 \\ 0 & 1 & 0 \end{bmatrix}$$

16. Find the zeros of the fourth-degree Legendre polynomial $P_4(x) = (35/8)[x^4 - (6/7)x^2 + (3/35)]$ (Table 6.6.1) and verify Theorem E.5.

## SUGGESTED READINGS

**Calculus**

R. Ellis and D. Gulick, *Calculus with Analytic Geometry,* 4th ed., Harcourt Brace Jovanovich, Publishers, San Diego, 1990.

R. E. Larson, R. P. Hostetler, and B. H. Edwards, *Calculus,* 4th ed., D. C. Heath and Company, Lexington, Massachusetts, 1990.

E. Swokowski, *Calculus with Analytic Geometry,* 2nd Alt. ed., PWS–KENT Publishing Company, Boston, 1988.

**Linear Algebra**

S. J. Leon, *Linear Algebra with Applications,* 2nd ed., Macmillan Publishing Company, New York, 1990.

G. Strang, *Linear Algebra and its Applications,* 3rd ed., Harcourt Brace Jovanovich, Publishers, San Diego, 1988.

**Differential Equations and Difference Equations**

M. M. Guterman and Z. H. Nitecki, *Differential Equations: A First Course,* 3rd ed., Saunders College Publishing, Philadelphia, 1991.

R. K. Nagle and E. B. Saff, *Fundamentals of Differential Equations,* 2nd ed., The Benjamin/Cummings Publishing Company, Inc., Redwood City, California, 1989.

S. L. Ross, *Differential Equations,* 2nd ed., Xerox College Publishing, Lexington, Massachusetts, 1974.

# ANSWERS TO SELECTED ODD-NUMBERED EXERCISES

# CHAPTER 1

## Sections 1.1, 1.2 *(p. 10)*

1. (a) $0.123456 \times 10^3$    (b) $0.27653 \times 8^2$
   (c) $0.101011101 \times 2^5$    (d) $0.AB168 \times 16^2$

3. The elements in the decimal system are

   $0, \pm\frac{1}{4}, \pm\frac{5}{16}, \pm\frac{3}{8}, \pm\frac{1}{2}, \pm\frac{9}{16}, \pm\frac{5}{8}, \pm\frac{3}{4}, \pm\frac{7}{8}, \pm1, \pm\frac{5}{4}, \pm\frac{5}{2},$
   $\pm\frac{7}{4}, \pm2, \pm\frac{5}{2}, \pm3, \pm\frac{7}{2}$

7. (a) $852.045 = 0.\underline{852045} \times 10^3$
   $852.01 = 0.\underline{85201} \times 10^3$    four
   (b) $0.000452 = 0.\underline{452} \times 10^{-3}$
   $0.00041 = 0.\underline{41} \times 10^{-3}$    one
   (c) $1.2345 = 0.\underline{12345} \times 10^1$
   $1.234 = 0.\underline{1234} \times 10^1$    four

13. (a) $\ln \dfrac{x}{y}$   (b) $\dfrac{x^2}{2!} + \dfrac{x^3}{3!}$   (c) $\dfrac{\sin^2 x}{1 + \cos x}$   (d) $\dfrac{e^x}{e^y}$

(e) $-\dfrac{2c}{b + \sqrt{b^2 - 4ac}}$ if $b > 0$ and
   $-\dfrac{2c}{b - \sqrt{b^2 - 4\,ac}}$ if $b < 0$

## Section 1.3 *(p. 15)*

3. The absolute propagated error $= 0.0056$. The absolute round-off error $= 0.005$. The exact absolute error $= 0.0106$. The upper error bound $= 0.029396$.

5. $x_1 + x_2 + x_3 + x_4 = 3.093707;$
   $((x_1 \oplus x_2) \oplus x_3) \oplus x_4 = 0.3091 \times 10^1 = 3.091;$
   $((x_4 \oplus x_3) \oplus x_1) \oplus x_2 = 0.3092 \times 10^1 = 3.092;$

# CHAPTER 2

## Sections 2.1, 2.2, 2.3, and 2.4 *(p. 24)*

1. (a) $[1/2, 1]$; $[3/4, 1]$; $[3/4, 7/8]$; $[13/16, 7/8]$
   (b) $[\pi/4, \pi/2]$; $[\pi/4, 3\pi/8]$; $[5\pi/16, 3\pi/8]$; $[5\pi/16, 11\pi/32]$
   (c) $[0, \pi/4]$; $[0, \pi/8]$; $[\pi/16, \pi/8]$; $[\pi/16, 3\pi/32]$
   (d) $[3/2, 2]$; $[7/4, 2]$; $[7/4, 15/8]$; $[29/16, 15/8]$

3. (a) $0.73576$; $0.83952$; $0.85383$; $0.85259$
   (b) $0.87980$; $1.01859$; $1.03023$; $1.02986$
   (c) $1.32053$; $1.62296$; $1.60367$; $1.60865$
   (d) $1.66667$; $1.81632$; $1.84300$; $1.83922$

9. $n \geq 17$

## Sections 2.5, 2.6, and 2.7 *(p. 34)*

1. Converges linearly with $c = 1/3$

3. No, since $c \nmid 1$

11.

| $n$ | $r_n$ | $\hat{r}_n$ |
|---|---|---|
| 16 | 7.33228 | 9.00000 |
| 17 | 7.49905 | 9.00000 |
| 18 | 7.64915 | 9.00000 |
| 19 | 7.78423 | 9.00000 |
| 20 | 7.90581 | |
| 21 | 8.01523 | |

## Section 2.8 *(p. 45)*

1. $r_1 = 3/2 = 1.5$; $r_2 = 17/12 \approx 1.41667$;
   $r_3 = 577/408 \approx 1.41422$; $r_4 = 665857/470832 \approx 1.41421$

3. $r_1 = 5/4$; $r_2 = 355/256 \approx 1.38672$; $r_3 = 1.41342$;
   $r_4 = 1.41421$

7. (a) $\dfrac{(m-1)}{2|r|}$    (b) $\dfrac{(m+1)}{2|r|}$

9. (a) $r_1 = 2.27489$; $r_2 = 1.64303$; $r_3 = 1.16773$;
   $r_4 = 0.91604$
   (b) $r_1 = -0.88382$; $r_2 = -4.82321$;
   $r_3 = -3.13484$; $r_4 = -1.98442$
   (c) $r_1 = 2.22590$; $r_2 = 1.79554$; $r_3 = 1.63462$;
   $r_4 = 1.60933$
   (d) $r_1 = 2.30000$; $r_2 = 1.95170$; $r_3 = 1.84847$;
   $r_4 = 1.83936$

15. $r_1 = 11/8 = 1.375$; $r_2 = 1.41420$; $r_3 = 1.41421$;
   $r_4 = 1.41421$

**Section 2.9** *(p. 53)*

1. (a) 1. $x` = \cos x$
   2. $x = \cos^{-1} x$

3. $x = x + A(x - \cos x)$ where $A$ is any
   nonzero number

4. $x = x - \dfrac{x - \cos x}{1 + \sin x}$

(b) 1. $x = x^3 - 1$
   2. $x = \sqrt[3]{x + 1}$
   3. $x = x + A(x^3 - x - 1)$ where $A$ is any
      nonzero number
   4. $x = x - \dfrac{x^3 - x - 1}{3x^2 - 1}$

(c) 1. $x = \dfrac{e^{x/2}}{2}$
   2. $x = 2 \ln 2x$
   3. $x = x + A(e^{x/2} - 2x)$ where $A$ is any
      nonzero number
   4. $x = x - \dfrac{e^{x/2} - 2x}{(e^{x/2}/2) - 2}$

9. (a) $[0, 3]$    (b) none    (c) $[0.1, 1]$

---

**CHAPTER 3**

---

**Sections 3.1, 3.2, and 3.3** *(p. 68)*

5. $\begin{bmatrix} 0 & 1 & 0 \\ 0 & 0 & 1 \\ -2 & 0 & 3 \end{bmatrix}$; $-(\lambda^3 - 3\lambda^2 + 2) = 0$

7. $P(1) = 6$; $P'(1) = 15$    9. $-0.73$; $1.00$; $2.73$

11. $-0.86$; $-0.34$; $0.34$; $0.86$

13.

| $i$ | $r_i$ | $s_i$ |
|-----|---------|-----------|
| 0 | 1.00000 | 1.00000 |
| 1 | 2.80000 | 4.80000 |
| 2 | 5.80488 | -13.76488 |
| 3 | 4.02354 | -3.00396 |

**Section 3.4** *(p. 75)*

1. $P(x) = x^2 + 3x + 2$; $a_2 = 1$, $a_1 = 3$, and $a_0 = 2$

| $k$ | $e_0^{(k)}$ | $q_1^{(k)}$ | $e_1^{(k)}$ | $q_2^{(k)}$ | $e_2^{(k)}$ |
|-----|-------------|-------------|-------------|-------------|-------------|
| 0 | | -3.00000 | | 0.00000 | |
| | 0.00000 | | 0.66667 | | 0.00000 |
| 1 | | -2.33333 | | -0.66667 | |
| | 0.00000 | | 0.19048 | | 0.00000 |
| 2 | | -2.14286 | | -0.85714 | |
| | 0.00000 | | 0.07619 | | 0.00000 |
| 3 | | -2.06667 | | -0.93333 | |
| | 0.00000 | | 0.03441 | | 0.00000 |

**3.** $P(x) = x^2 - 2x + 1$; $a_2 = 1$, $a_1 = -2$, $a_0 = 1$

| k | $e_0^{(k)}$ | $q_1^{(k)}$ | $e_1^{(k)}$ | $q_2^{(k)}$ | $e_2^{(k)}$ |
|---|---|---|---|---|---|
| 0 | | 2.00000 | | 0.00000 | |
| | 0.00000 | | -0.50000 | | 0.00000 |
| 1 | | 1.50000 | | 0.50000 | |
| | 0.00000 | | -0.16667 | | 0.00000 |
| 2 | | 1.33333 | | 0.66667 | |
| | 0.00000 | | -0.08333 | | 0.00000 |
| 3 | | 1.25000 | | 0.75000 | |
| | 0.00000 | | -0.05000 | | 0.00000 |

**5.** $x = y + 2$; $P*(y + 2) = (y + 2)^2 + 1 = y^2 + 4y + 5$;
$a_2 = 1$, $a_1 = 4$, $a_0 = 5$

| k | $e_0^{(k)}$ | $q_1^{(k)}$ | $e_1^{(k)}$ | $q_2^{(k)}$ | $e_2^{(k)}$ |
|---|---|---|---|---|---|
| 0 | | -4.00000 | | 0.00000 | |
| | 0.00000 | | 1.25000 | | 0.00000 |
| 1 | | -2.75000 | | -1.25000 | |
| | 0.00000 | | 0.56818 | | 0.00000 |
| 2 | | -2.18182 | | -1.81818 | |
| | 0.00000 | | 0.47348 | | 0.00000 |
| 3 | | -1.70834 | | -2.29166 | |
| | 0.00000 | | 0.63515 | | 0.00000 |

---

### C H A P T E R  4

### Sections 4.2 and 4.3 *(p. 86)*

**1.** $P_2(0.23) = 0.97365$

**5.** $P_3(x) = -\dfrac{1}{4} x^3 + \dfrac{5}{4} x^2 - x$; $P_3(3) = \dfrac{3}{2}$

**7.** $Q_2(x) = 4x^2 - 8x + 5$

### Section 4.4 *(p. 94)*

**1.** (a)

| $x_i$ | $f(x_i)$ | $f[x_{i+1}, x_i]$ | $f[x_{i+2}, x_{i+1}, x_i]$ | $f[x_{i+3}, x_{i+2}, x_{i+1}, x_i]$ |
|---|---|---|---|---|
| 1.00 | 0.84147 | | | |
| | | 0.53200 | | |
| 1.02 | 0.85211 | | -0.43340 | |
| | | 0.51033 | | 0.41917 |
| 1.05 | 0.86742 | | -0.40825 | |
| | | 0.49400 | | |
| 1.06 | 0.87236 | | | |

(b) $P_2(x) = 0.84147 + 0.532(x - 1) -$
0.4334$(x - 1)(x - 1.02)$
$P_3(x) = P_2(x) + 0.41917(x - 1)(x - 1.02)$
$(x - 1.05)$

(c) $P_2(1.01) = 0.84683; P_3(1.01) = 0.84683$

(d) For $N = 2$, |truncation error| =
$|-\sin \xi(x - 1.0)(x - 1.02)/2!| \leq 0.00005$;
for $N = 3$, |truncation error| =
$|-\cos \xi(x - 1.0)(x - 1.02)(x - 1.05)/3!| \leq$
$6.6 \times 10^{-7}$; for $N = 2$, |actual error| =
0.00000; for $N = 3$, |actual error| = 0.00000.

11. $P(x) = 1.0 + 0.79719(x - 1.54308) -$
$0.35003(x - 1.54308)(x - 1.66852) +$
$0.20356(x - 1.54308)(x - 1.66852)(x - 1.81066) -$
$0.12405(x - 1.54308)(x - 1.66852)$
$(x - 1.81066)(x - 1.97091); P(1.75) = 1.15882$

## Section 4.5 *(p. 102)*

1. (a)

| $x_i$ | $f(x_i)$ | $\Delta$ | $\Delta^2$ | $\Delta^3$ |
|---|---|---|---|---|
| 1.00 | 0.84147 | | | |
| | | 0.01064 | | |
| 1.02 | 0.85211 | | $-0.00035$ | |
| | | 0.01029 | | 0.00002 |
| 1.04 | 0.86240 | | $-0.00033$ | |
| | | 0.00996 | | |
| 1.06 | 0.87236 | | | |

(b) For $N = 2, f(1.01) \approx P_2(0.5) = 0.84683$;
$f(1.05) \approx P_2(2.5) = 0.86741$; for $N = 3$,
$f(1.01) \approx P_3(0.5) = 0.84683; f(1.05) \approx$
$P_3(2.5) = 0.86742$

(c) For $N = 2, f(1.01) \approx P_2(-2.5) = 0.84684$;
$f(1.05) \approx P_2(-0.5) = 0.86742$; for $N = 3$,
$f(1.01) \approx P_3(-2.5) = 0.84683; f(1.05) \approx$
$P_3(-0.5) = 0.86742$

(d) Almost equal. This is due to round-off error.

## Section 4.6 *(p. 108)*

1. $P_3(0.23) = 0.28948$; |actual error| = 0.00000;
|truncation error| $\leq \dfrac{(5.80439)(0.03)^2(0.07)^2}{24} =$
$1.06656 \times 10^{-6}$

7. $P_3(x) = x^2$

**9.** $P_5(x) = L_0^3(x)\Big\{f(x_0)\Big[(6L_0'^2(x_0) - \dfrac{3}{2}L_0''(x_0))(x - x_0)^2 - 3L_0'(x_0)(x - x_0) + 1\Big]$

$\qquad + f'(x_0)(x - x_0)[1 - 3(x - x_0)L_0'(x_0)] + \dfrac{1}{2}f''(x_0)(x - x_0)^2$

$\qquad + L_1^3(x)\Big\{f(x_1)\Big[(6L_1'^2(x_1) - \dfrac{3}{2}L_1''(x_1))(x - x_1)^2 - 3L_1'(x_1)(x - x_1) + 1\Big]$

$\qquad + f'(x_1)(x - x_1)[1 - 3(x - x_1)L_1'(x_1)] + \dfrac{1}{2}f''(x_1)(x - x_1)^2$

where $L_0(x) = \dfrac{x - x_1}{x_0 - x_1}$ and $L_1(x) = \dfrac{x - x_0}{x_1 - x_0}$.

## Section 4.7 *(p. 117)*

**1.** $P_1(x) = \dfrac{(e + e^{-1})}{2} + \dfrac{(e - e^{-1})}{2}x = 1.54308 +$
$1.1752x$

$\tilde{P}_1(x) = \dfrac{e^{1/\sqrt{2}} + e^{-1/\sqrt{2}}}{2} + \dfrac{e^{1/\sqrt{2}} - e^{-1/\sqrt{2}}}{\sqrt{2}}x =$
$1.26059 + 1.08544x$

| $x$ | $e^x$ | $P_1(x)$ | $|P_1(x) - e^x|$ | $\tilde{P}_1(x)$ | $|e^x - \tilde{P}_1(x)|$ |
|---|---|---|---|---|---|
| $-1.0$ | 0.36788 | 0.36788 | 0.00000 | 0.17515 | 0.19273 |
| $-0.5$ | 0.60653 | 0.95548 | 0.34895 | 0.71787 | 0.11134 |
| 0.0 | 1.00000 | 1.54308 | 0.54308 | 1.26059 | 0.26059 |
| 0.5 | 1.64872 | 2.13068 | 0.48196 | 1.80331 | 0.15459 |
| 1.0 | 2.71828 | 2.71828 | 0.00000 | 2.34603 | 0.37225 |

**3.** $T_4(x) = 8x^4 - 8x^2 + 1$, $T_5(x) = 16x^5 - 20x^3 +$
$5x$, $T_6(x) = 32x^6 - 48x^4 + 18x^2 - 1$, $T_7(x) =$
$64x^7 - 112x^5 + 56x^3 - 7x$, $T_8(x) = 128x^8 -$
$256x^5 + 160x^4 - 32x^3 + 1$

**5.** $x^2 = \dfrac{1}{2}\cdot T_0 + 0 \cdot T_1 + \dfrac{1}{2}\cdot T_2 = \dfrac{1}{2} + 0 +$
$\dfrac{1}{2}(2x^2 - 1)$

**9.** Local maximum at $x = -1/2$: $T_3(-1/2) = 1$; local
minimum at $x = 1/2$: $T_3(1/2) = -1$

**15.** $e^x \approx \dfrac{23041}{23040} + \dfrac{383}{384}x + \dfrac{1917}{3840}x^2 + \dfrac{17}{96}x^3 + \dfrac{7}{160}x^4$

# CHAPTER 5

## Section 5.2 (p. 127)

**1.** (a) and (b): $f'(x) = \cos x$; exact value: $f'(1) = \cos 1$
$= 0.54030$

| $h$ | Eq. (5.2.14) | Actual Error | Eq. (5.2.31) | Actual Error |
|---|---|---|---|---|
| 0.10000 | 0.49740 | 0.04290 | 0.53940 | 0.00090 |
| 0.01000 | 0.53600 | 0.00430 | 0.54000 | 0.00030 |
| 0.00100 | 0.54000 | 0.00030 | 0.54000 | 0.00030 |
| 0.00010 | 0.60000 | 0.05970 | 0.55000 | 0.00970 |
| 0.00001 | 1.00000 | 0.45970 | 0.50000 | 0.04030 |

**3.** (a) and (b): $f'(x) = e^x$; exact value: $f'(1) = e^1 = 2.71828$

| $h$ | Eq. (5.2.16) | Actual Error | Eq. (5.2.31) | Actual Error |
|---|---|---|---|---|
| 0.10000 | 2.58680 | 0.13148 | 2.72285 | 0.00457 |
| 0.01000 | 2.70500 | 0.01328 | 2.71850 | 0.00022 |
| 0.00100 | 2.72000 | 0.00172 | 2.72000 | 0.00172 |
| 0.00010 | 2.70000 | 0.01828 | 2.70000 | 0.01828 |
| 0.00001 | 3.00000 | 0.28172 | 3.00000 | 0.28172 |

**5.** (a) and (b): $f'(x) = 3x^2$; exact value: $f'(1) = 3.0$

| $h$ | Eq. (5.2.14) | Actual Error | Eq. (5.2.31) | Actual Error |
|---|---|---|---|---|
| 0.10000 | 3.31000 | 0.31000 | 3.01000 | 0.01000 |
| 0.01000 | 3.03010 | 0.03010 | 3.00000 | 0.00000 |
| 0.00100 | 3.00000 | 0.00000 | 3.00000 | 0.00000 |
| 0.00010 | 3.00000 | 0.00000 | 3.00000 | 0.00000 |
| 0.00001 | 3.00000 | 0.00000 | 3.00000 | 0.00000 |

**7.** $J_0'(5.3) \approx 0.347$; $J_0'(5.5) \approx 0.3416$; $J_0'(5.8) \approx 0.31595$

**15.** (a) $f(x, y) = x^2 e^y$; $f_x(x, y) = 2xe^y$; $f_y(x, y) = x^2 e^y$; exact values: $f_x(1, 2) = 2e^2 = 14.77811$; $f_y(1, 2) = e^2 = 7.38906$

| $h$ | $f_x(1, 2)$ | Actual Error | $f_y(1, 2)$ | Actual Error |
|---|---|---|---|---|
| 0.10000 | 14.77810 | 0.00001 | 7.40140 | 0.01234 |
| 0.01000 | 14.77800 | 0.00011 | 7.38950 | 0.00044 |
| 0.00100 | 14.78000 | 0.00189 | 7.39000 | 0.00094 |
| 0.00001 | 15.00000 | 0.22189 | 7.50000 | 0.11094 |

(b) $f(x, y) = \cos \dfrac{x}{y}$; $f_x(x, y) = -\dfrac{1}{y}\sin\dfrac{x}{y}$; $f_y(x, y) = \dfrac{x}{y^2}\sin\dfrac{x}{y}$; exact values: $f_x(1, 2) = -0.23971$; $f_y(1, 2) = 0.11986$

| $h$ | $f_x(1, 2)$ | Actual Error | $f_y(1, 2)$ | Actual Error |
|---|---|---|---|---|
| 0.10000 | −0.23965 | 0.00006 | 0.12048 | 0.00059 |
| 0.01000 | −0.24000 | 0.00029 | 0.11950 | 0.00036 |
| 0.00100 | −0.24000 | 0.00029 | 0.12000 | 0.00014 |
| 0.00001 | 0.00000 | 0.23971 | 0.00000 | 0.11986 |

## Sections 5.3 and 5.4 *(p. 135)*

1. (a) and (b): $f''(x) = -\sin x$; exact value: $f''(1) = -\sin 1 = -0.84147$

| $h$ | $f''(1)$ | Actual Error |
|---|---|---|
| 0.10000 | −0.84000 | 0.00147 |
| 0.01000 | −0.80000 | 0.04147 |
| 0.00500 | −0.66667 | 0.17480 |

3. (a) and (b): $f''(x) = e^x$; exact value: $f''(0) = e^0 = 1$

| $h$ | $f''(0)$ | Actual Error |
|---|---|---|
| 0.10000 | 1.00100 | 0.00100 |
| 0.01000 | 1.0000 | 0.00000 |
| 0.00500 | 0.66667 | 0.33333 |

5. (a) and (b): $f''(x) = 6x$; exact value: $f''(0) = 0$

| $h$ | $f''(0)$ | Actual Error |
|---|---|---|
| 0.10000 | 0.00000 | 0.00000 |
| 0.01000 | 0.00000 | 0.00000 |
| 0.00500 | 0.00000 | 0.00000 |

7. $J_0''(5.3) = 0.012$; $J_0''(5.5) = -0.05467$; $J_0''(5.8) = -0.12700$ using

$$f''(x_0) = \frac{f(x_0) - f(x_0 - h) - 2f(x_0 - 2h) + 3f(x_0 - 3h) - f(x_0 - 4h)}{h^2}$$
$$+ \frac{46h^2}{4!}f^{(4)}(\xi)$$

where $\xi$ is in the interval $[x_0 - 4h, x_0]$.

**9.**

| $h$ | $x$ | $\ln x$ | $D(h)$ | $D_1(h)$ | Actual Error |
|---|---|---|---|---|---|
| — | 3.00000 | 1.09861 | — | | — |
| 0.4000 | 3.4000 | 1.22378 | 0.31293 | — | 0.00086 |
| 0.2000 | 3.2000 | 1.16315 | 0.32270 | 0.33247 | 0.00023 |
| 0.1000 | 3.1000 | 1.13140 | 0.32790 | 0.33310 | 0.00003 |
| 0.0500 | 3.0500 | 1.11514 | 0.33060 | 0.33330 | 0.00007 |
| 0.0250 | 3.0250 | 1.10691 | 0.33200 | 0.33340 | 0.00027 |
| 0.0125 | 3.0125 | 1.10277 | 0.33280 | 0.33360 | |

**11.** $h_o = \sqrt[6]{\dfrac{240\epsilon}{M_6}}$ where $|f^{(6)}(x)| \le M_6$ for all $x$ in the interval $[x_0 - 2h, x_0 + 2h]$

**13.** $\dfrac{3h}{2} f^{(4)}(\xi)$ where $\xi$ is in $[x_0, x_0 + 3h]$

# CHAPTER 6

## Section 6.2 *(p. 144)*

**1.** (a) 1/2; exact value: 1/2
(b) 1.75393; exact value: $e^1 - 1 = 1.71828$
(c) 0.94806; exact value: 1

**5.** (a) 3/8; exact value: 1/3
(b) 5/16; exact value: 1/4

## Section 6.3 *(p. 151)*

**1.** (a) 1/2; exact value: 1/2
(b) 1.71886; exact value: $e^1 - 1 = 1.71828$
(c) 1.00228; exact value: 1

**5.** $\int_a^b f(x)\,dx \approx ((b-a)/2)(f(a) + f(b)) + ((b-a)^2/12)(f'(a) - f'(b))$

## Section 6.4 *(p. 159)*

**1.** (a) $N = 3, 1.71854; N = 4, 1.71828; N = 5, 1.71828$
(b) $N = 3, 0.45977; N = 4, 0.45970; N = 5, 0.45970$

**3.** (a) $N = 3$, $|\text{truncation error}| \le \dfrac{e}{(3)^4(80)}$
$= 0.00042$

$N = 4$, $|\text{truncation error}| \le \dfrac{2e}{(4)^6(975)}$
$= 1.36132 \times 10^{-6}$

$N = 6$, $|\text{truncation error}| \le \dfrac{11e}{(5)^5(12096)}$
$= 7.91034 \times 10^{-7}$

(b) $N = 3$, $|\text{truncation error}| \le \dfrac{\sin 1}{(3)^4(80)}$
$= 0.00013$

$N = 4$, $|\text{truncation error}| \le \dfrac{2\sin 1}{(4)^6(975)}$
$= 4.21410 \times 10^{-7}$

$N = 6$, $|\text{truncation error}| \le \dfrac{11\sin 1}{(5)^5(12096)}$
$= 2.44873 \times 10^{-7}$

**9.** (a) $N = 2, 1.64872; N = 3, 1.67167$
(b) $N = 2, 0.47943; N = 3, 0.47278$

**11.** (a) $N = 2$, $|\text{truncation error}| \le 0.11326$, $|\text{actual error}| = 0.06956$
$N = 3$, $|\text{truncation error}| \le 0.07551$, $|\text{actual error}| = 0.04661$

(b) $N = 2$, $|\text{truncation error}| \le 0.0356$, $|\text{actual error}| = 0.01973$
$N = 3$, $|\text{truncation error}| \le 0.02337$, $|\text{actual error}| = 0.01308$

## Section 6.5 *(p. 164)*

**1.** $T_{3,0} = 1.72052$

**3.**

| $k$ | $T_{k,0}$ | $T_{k,1}$ | $T_{k,2}$ | $T_{k,3}$ |
|---|---|---|---|---|
| 0 | 0.75000 | | | |
| 1 | 0.70833 | 0.69444 | | |
| 2 | 0.69702 | 0.69325 | 0.69317 | |
| 3 | 0.69412 | 0.69315 | 0.69314 | 0.69314 |

**5.** $T_{2,1} = 0.74682$

**7.** (a)

| k | $T_{k,0}$ | $T_{k,1}$ | $T_{k,2}$ |
|---|-----------|-----------|-----------|
| 0 | 1/2 | | |
| 1 | 1/2 | 1/2 | |
| 2 | 1/2 | 1/2 | 1/2 |

(b)

| k | $T_{k,0}$ | $T_{k,1}$ | $T_{k,2}$ |
|---|-----------|-----------|-----------|
| 0 | 1.85914 | | |
| 1 | 1.75393 | 1.71886 | |
| 2 | 1.72722 | 1.71832 | 1.71828 |

## Section 6.6 *(p. 170)*

**1.** (a)  two-point: 2.34270; three-point: 2.35034; exact value: 2.35040
(b) two-point: 1.93582; three-point: 2.00139; exact value: 2
(c) two-point: 0.40541; three-point: 0.40546; exact value: 0.40547

**3.**  0.40546

**7.** $\int_{-1}^{1} f(x)\, dx \approx \frac{2}{3}\left[f\left(-\frac{1}{\sqrt{2}}\right) + f(0) + f\left(\frac{1}{\sqrt{2}}\right)\right]$    $\int_{-1}^{1} e^x\, dx$

$\approx 2.34746$

**11.**  $x^2 + 8x + 1 = \frac{2}{3}P_2 + 8P_1 + \frac{4}{3}P_0$

## Section 6.7 *(p. 173)*

**1.**  exact value: 1.79     **3.**  exact value: 0.78540

---

■ **C H A P T E R  7** ■

## Section 7.2 *(p. 184)*

**1.** (a) $x_1 = 1, x_2 = 1, x_3 = 1$ with and without partial pivoting
(b) $x_1 = 53/45, x_2 = 17/45, x_3 = 11/15$ with and without partial pivoting

**3.** (a) $x_1 = 1, x_2 = 1, x_3 = 1$     (b) $x_1 = 53/45, x_2 = 17/45, x_3 = 11/15$

**5.**  Partial pivoting: no solution     without partial pivoting: no solution     exact solution: $x_1 = 10{,}001, x_2 = -10{,}000$

**7.** (a) $x_1 = 0, x_2 = 1$     (b) $x_1 = 1, x_2 = 1$

**11.**  Sixth step without partial pivoting

## Sections 7.3 and 7.4 *(p. 193)*

**1.** (a) $x_1 = 1, x_2 = 1, x_3 = 1$     (b) $x_1 = 53/45, x_2 = 17/45, x_3 = 11/15$

**3.** (a) multiplications: $N^2$     additions/subtractions: $N(N - 1)$
(b) multiplications: $N^3$     additions/subtractions: $N^2(N - 1)$

**7.**

| | Gaussian Elimination | | | Gauss–Jordan | | |
|---|---|---|---|---|---|---|
| $N$ | Divisions | Multiplications | Additions/ Subtractions | Divisions | Multiplications | Additions/ Subtractions |
| 20 | 210 | 2850 | 2850 | 400 | 3990 | 3990 |
| 40 | 820 | 22100 | 22100 | 1600 | 31980 | 31980 |
| 1000 | $5.005 \times 10^5$ | $3.33833 \times 10^8$ | $3.33833 \times 10^8$ | $10^6$ | $5.00000 \times 10^8$ | $5.00000 \times 10^8$ |

## Section 7.5 *(p. 204)*

**1.** (a) $LU = \begin{bmatrix} 2 & 0 & 0 \\ -4 & -3 & 0 \\ 6 & -6 & 1 \end{bmatrix} \begin{bmatrix} 1 & 1/2 & 1/2 \\ 0 & 1 & -1/3 \\ 0 & 0 & 2 \end{bmatrix}$     (b) $LU = \begin{bmatrix} 2 & 0 & 0 & 0 \\ -1 & -2 & 0 & 0 \\ 2 & 1 & 1 & 0 \\ 1 & 2 & 1 & 2 \end{bmatrix} \begin{bmatrix} 1 & 2 & 1 & 2 \\ 0 & 1 & -1 & -2 \\ 0 & 0 & 1 & 2 \\ 0 & 0 & 0 & 1 \end{bmatrix}$

**3.** (a) $LU = \begin{bmatrix} 1 & 0 & 0 \\ 2 & 1 & 0 \\ 3/2 & 13/14 & 1 \end{bmatrix} \begin{bmatrix} 2 & 8 & 6 \\ 0 & -14 & -14 \\ 0 & 0 & 5 \end{bmatrix}$     (b) $LU = \begin{bmatrix} 1 & 0 & 0 \\ 2 & 1 & 0 \\ 1 & -2 & 1 \end{bmatrix} \begin{bmatrix} 1 & 1 & 1 \\ 0 & 1 & -1 \\ 0 & 0 & -5 \end{bmatrix}$

**5.** (a) 3     (b) 1

## Section 7.6 *(p. 210)*

**1.** $\begin{bmatrix} -1 & 0 \\ 0 & 2 \end{bmatrix}\begin{bmatrix} -1 & 0 \\ 0 & 2 \end{bmatrix}, \begin{bmatrix} 1 & 0 \\ 0 & -2 \end{bmatrix}\begin{bmatrix} 1 & 0 \\ 0 & -2 \end{bmatrix}, \begin{bmatrix} -1 & 0 \\ 0 & -2 \end{bmatrix}\begin{bmatrix} -1 & 0 \\ 0 & -2 \end{bmatrix}, \begin{bmatrix} 1 & 0 \\ 0 & 2 \end{bmatrix}\begin{bmatrix} 1 & 0 \\ 0 & 2 \end{bmatrix}$

**3.** No

**9.** $R = \dfrac{1}{\sqrt{a_{33}}} \begin{bmatrix} \sqrt{(a_{11}a_{33} - a_{13}^2) - \dfrac{(a_{21}a_{33} - a_{13}a_{23})^2}{a_{22}a_{33} - a_{23}^2}} & \dfrac{a_{21}a_{33} - a_{13}a_{23}}{\sqrt{a_{22}a_{33} - a_{23}^2}} & a_{13} \\ 0 & \sqrt{a_{22}a_{33} - a_{23}^2} & a_{23} \\ 0 & 0 & a_{33} \end{bmatrix}$

## Sections 7.7 and 7.8 *(p. 218)*

**1.** (a) $z_1 = \frac{1}{4}$, $z_2 = \frac{1}{15}$, $z_3 = \frac{1}{56}$; $x_1 = \frac{15}{56}$, $x_2 = \frac{1}{14}$, $x_3 = \frac{1}{56}$

(b) $z_1 = 1$, $z_2 = \frac{2}{3}$, $z_3 = \frac{1}{2}$; $x_1 = \frac{3}{2}$, $x_2 = 1$, $x_3 = \frac{1}{2}$

**3.** divisions: $2N - 1$     multiplications: $3N - 3$     additions/subtractions: $3N - 3$

**5.** $L = \begin{bmatrix} 2 & 0 & 0 & 0 & 0 & 0 \\ -1 & 3/2 & 0 & 0 & 0 & 0 \\ 0 & -1 & 4/3 & 0 & 0 & 0 \\ 0 & 0 & -1 & 5/4 & 0 & 0 \\ 0 & 0 & 0 & -1 & 6/5 & 0 \\ 0 & 0 & 0 & 0 & -1 & 7/6 \end{bmatrix}$     $U = \begin{bmatrix} 1 & -1/2 & 0 & 0 & 0 & 0 \\ 0 & 1 & -2/3 & 0 & 0 & 0 \\ 0 & 0 & 1 & -3/4 & 0 & 0 \\ 0 & 0 & 0 & 1 & -4/5 & 0 \\ 0 & 0 & 0 & 0 & 1 & -5/6 \\ 0 & 0 & 0 & 0 & 0 & 1 \end{bmatrix}$

Generalizations for $N \times N$ matrix $A$:

$L = \begin{bmatrix} 2 & 0 & 0 & \cdots & 0 & 0 \\ -1 & 3/2 & 0 & \cdots & 0 & 0 \\ 0 & -1 & 4/3 & \cdots & 0 & 0 \\ \vdots & \vdots & \vdots & \ddots & \vdots & \vdots \\ 0 & 0 & 0 & \cdots & N/(N-1) & 0 \\ 0 & 0 & 0 & \cdots & -1 & (N+1)/N \end{bmatrix}$     $U = \begin{bmatrix} 1 & -1/2 & 0 & \cdots & 0 & 0 \\ 0 & 1 & -2/3 & \cdots & 0 & 0 \\ 0 & 0 & 1 & \cdots & 0 & 0 \\ \vdots & \vdots & \vdots & \ddots & \vdots & \vdots \\ 0 & 0 & 0 & \cdots & 1 & -(N-1)/N \\ 0 & 0 & 0 & \cdots & 0 & 1 \end{bmatrix}$

**7.**  (a) $\begin{bmatrix} 1 & 0 \\ -1 & 1/3 \end{bmatrix}$   (b) $\begin{bmatrix} 8 & -13/3 & -7/3 \\ -2 & 5/3 & 2/3 \\ -2 & 2/3 & 2/3 \end{bmatrix}$

---

# CHAPTER 8

## Section 8.1 (p. 227)

**1.**  (a) $\dfrac{dy_1}{dx} = y_2$

$\dfrac{dy_2}{dx} = y_2^2 - y_1 + x^2$

$y_1(0) = 1 \qquad y_2(0) = 2$

(b) $\dfrac{dy_1}{dx} = y_2$

$\dfrac{dy_2}{dx} = y_3$

$\dfrac{dy_3}{dx} = 3y_3 + y_2 - y_1^2 + e^x$

$y_1(0) = 1 \qquad y_2(0) = 1 \qquad y_3(0) = 0$

(c) $\dfrac{dy_1}{dx} = y_2$

$\dfrac{dy_2}{dx} = -p(x)\, y_2 - q(x)\, y_1 + g(x)$

$y_1(0) = \alpha_1 \qquad y_2(0) = \alpha_2$

**3.**  (a) $|x + 1| \le 1/4$   (b) $|x| \le \sqrt{2}/2$

**5.**  $\dfrac{d^2 y_1}{dx^2} - (a_{11} + a_{22})\dfrac{dy_1}{dx}$

$+ (a_{11}\, a_{22} - a_{12}\, a_{21})\, y_1$

$+ a_{22}\, f_1(x) - f_1'(x) - a_{12}\, f_2(x) = 0$

or

$\dfrac{d^2 y_2}{dx^2} - (a_{11} + a_{22})\dfrac{dy_2}{dx}$

$+ (a_{11}\, a_{22} - a_{12}\, a_{21})\, y_2 + a_{21}\, f_2(x)$

$- f_2'(x) - a_{21}\, f_1(x) = 0$

## Section 8.2 (p. 232)

**1.**

| $x_i$ | $Y_i$ | $y(x_i)$ | Actual Error |
|---|---|---|---|
| 0.0 | 0.00000 | 0.00000 | 0.00000 |
| 0.1 | 0.00000 | 0.00033 | 0.00033 |
| 0.2 | 0.00100 | 0.00267 | 0.00167 |
| 0.3 | 0.00500 | 0.00900 | 0.00400 |
| 0.4 | 0.01400 | 0.02133 | 0.00733 |

**3.**

| $x_i$ | $Y_i$ | $y(x_i)$ | Actual Error |
|---|---|---|---|
| 0.0 | 1.00000 | 1.00000 | 0.00000 |
| 0.1 | 1.00000 | 0.99504 | 0.00496 |
| 0.2 | 0.99010 | 0.98058 | 0.00952 |
| 0.3 | 0.97106 | 0.95783 | 0.01323 |
| 0.4 | 0.94433 | 0.92848 | 0.01585 |

**5.**

| $x_i$ | $Y_i$ | $y(x_i)$ | Actual Error |
|---|---|---|---|
| 0.0 | 0.00000 | 0.00000 | 0.00000 |
| 0.1 | 0.01111 | 0.00517 | 0.00594 |
| 0.2 | 0.03457 | 0.02140 | 0.01317 |
| 0.3 | 0.07174 | 0.04986 | 0.02188 |
| 0.4 | 0.12416 | 0.09182 | 0.03234 |

**7.**

| $x_i$ | $Y_i$ | $y(x_i)$ | Actual Error |
|---|---|---|---|
| 0.0 | 0.00000 | 0.00000 | 0.00000 |
| 0.1 | 0.00033 | 0.00033 | 0.00000 |
| 0.2 | 0.00200 | 0.00267 | 0.00067 |
| 0.3 | 0.00833 | 0.00900 | 0.00067 |
| 0.4 | 0.02000 | 0.02133 | 0.00133 |

**9.**  $Y_i = \dfrac{Y_0}{(1 + 5h)^i}$ for $i = 0, 1, \ldots$

## Section 8.3 (p. 239)

**3.**

| $x_i$ | $n = 1$ $Y_i$ | $n = 2$ $Y_i$ | $n = 3$ $Y_i$ |
|---|---|---|---|
| 0.0 | 0.00000 | 0.00000 | 0.00000 |
| 0.1 | 0.00000 | 0.00500 | 0.00517 |
| 0.2 | 0.01000 | 0.02103 | 0.02140 |
| 0.3 | 0.03100 | 0.04923 | 0.04986 |
| 0.4 | 0.06410 | 0.09090 | 0.09182 |

## Section 8.4 *(p. 244)*

**1.**  $L_T \leq h^2/2$

**3.**  $|\text{global truncation error}| \leq h\, e^{1/2} (e^{1/2} - 1)/2$

**7.**  $|\text{global errors}| \leq \dfrac{e^{1/5}\, 10^{-5}}{2} + \dfrac{5}{2}\left(\dfrac{h}{2} + \dfrac{10^{-5}}{2h}\right)(e^{1/5} - 1)$

**9.**  $Y_i = \dfrac{c_1}{(1 + 5h)^i}\quad i = 0, 1, \dots$

## Section 8.5 *(p. 256)*

**1.**

| $x_i$ | Eq. (8.5.16) | Actual Error | Eq. (8.5.17) | Actual Error | Eq. (8.5.18) | Actual Error |
|-------|--------------|--------------|--------------|--------------|--------------|--------------|
| 0.0 | 1.00000 | 0.00000 | 1.00000 | 0.00000 | 1.00000 | 0.00000 |
| 0.1 | 0.90950 | 0.00041 | 0.90975 | 0.00066 | 0.90967 | 0.00058 |
| 0.2 | 0.83396 | 0.00063 | 0.83434 | 0.00101 | 0.83422 | 0.00089 |
| 0.3 | 0.76997 | 0.00074 | 0.77042 | 0.00119 | 0.77027 | 0.00104 |

**3.**

| $x_i$ | Eq. (8.5.16) | Actual Error | Eq. (8.5.17) | Actual Error | Eq. (8.5.18) | Actual Error |
|-------|--------------|--------------|--------------|--------------|--------------|--------------|
| 0.0 | 1.00000 | 0.00000 | 1.00000 | 0.00000 | 1.00000 | 0.00000 |
| 0.1 | 0.90550 | 0.00034 | 0.90525 | 0.00009 | 0.90533 | 0.00017 |
| 0.2 | 0.82193 | 0.00066 | 0.82145 | 0.00018 | 0.82161 | 0.00034 |
| 0.3 | 0.75014 | 0.00096 | 0.74946 | 0.00028 | 0.74969 | 0.00051 |

**5.**

| $x_i$ | $Y_i$ | Actual Error |
|-------|-------|--------------|
| 0.0 | 1.00000 | 0.00000 |
| 0.1 | 0.90908 | 0.00001 |
| 0.2 | 0.83331 | 0.00002 |
| 0.3 | 0.76921 | 0.00002 |

**7.**

| $x_i$ | $Y_i$ | Actual Error |
|-------|-------|--------------|
| 0.0 | 1.00000 | 0.00000 |
| 0.1 | 1.01005 | 0.00000 |
| 0.2 | 1.04081 | 0.00000 |
| 0.3 | 1.17351 | 0.00000 |

## Section 8.6 *(p. 264)*

**5.**

| $i$ | $x_i$ | $Y_i$ | $\hat{Y}_i$ | $|\hat{Y}_i - y(x_i)|$ | $h_i$ |
|-----|-------|-------|-------------|------------------------|-------|
| 0 | 0.00000 | 0.00000 | — | — | — |
| 1 | 0.10000 | $-0.00500$ | $-0.00517$ | 0.00000 | 0.10000 |
| 2 | 0.32045 | $-0.05517$ | $-0.05714$ | 0.00016 | 0.22045 |

## Section 8.7 *(p. 273)*

**1.**

| $x_i$ | $Y_{1,i}$ | $Y_{2,i}$ |
|-------|-----------|-----------|
| 0.00 | 2.00000 | 0.00000 |
| 0.25 | 2.00000 | 0.50000 |
| 0.50 | 2.50000 | 1.00000 |
| 0.75 | 3.50000 | 1.62500 |
| 1.00 | 5.12500 | 2.50000 |

**3.**

| $x_i$ | $Y_{1,i}$ | $Y_{2,i}$ |
|-------|-----------|-----------|
| 0.00 | 2.00000 | 0.00000 |
| 0.25 | 2.25521 | 0.52083 |
| 0.50 | 3.08551 | 1.17458 |

**5.**

| $x_i$ | $Y_{1,i}$ | $Y_{2,i}$ |
|-------|-----------|-----------|
| 0.00 | 0.00000 | 0.50000 |
| 0.25 | 0.17188 | 1.01563 |
| 0.50 | 0.51807 | 1.99854 |

## CHAPTER 9

## Section 9.2 *(p. 284)*

**1.**

| $x_i$ | $Y_i$ | $y(x_i)$ | Actual Error |
|-------|-------|----------|--------------|
| 0.0 | — | 0.00000 | — |
| 0.1 | — | 0.00517 | — |
| 0.2 | 0.02095 | 0.02140 | 0.00045 |
| 0.3 | 0.04883 | 0.04986 | 0.00103 |
| 0.4 | 0.09011 | 0.09182 | 0.00171 |

**3.**

| $x_i$ | $Y_i$ | $y(x_i)$ | Actual Error |
|-------|-------|----------|--------------|
| 1.0 | — | 2.50000 | — |
| 1.1 | — | 2.36818 | — |
| 1.2 | 2.27025 | 2.26667 | 0.00358 |
| 1.3 | 2.19411 | 2.18846 | 0.00565 |
| 1.4 | 2.13554 | 2.12857 | 0.00697 |

**5.**

| $x_i$ | $Y_i$ | $y(x_i)$ | Actual Error |
|-------|-------|----------|--------------|
| 0.0 | — | 0.00000 | — |
| 0.1 | 0.00526 | 0.00517 | 0.00009 |
| 0.2 | 0.02160 | 0.02140 | 0.00020 |
| 0.3 | 0.05019 | 0.04986 | 0.00033 |
| 0.4 | 0.09232 | 0.09182 | 0.00050 |

## Section 9.3 (p. 293)

**1.**

| $x_i$ | $Y_i^{(p)}$ | $Y_i$ | $y(x_i)$ | Actual Error |
|-------|-------------|-------|----------|--------------|
| 0.0 | — | — | 0.00000 | — |
| 0.1 | — | — | 0.00517 | — |
| 0.2 | 0.02095 | 0.02148 | 0.02140 | 0.00008 |
| 0.3 | 0.04944 | 0.05003 | 0.04986 | 0.00017 |
| 0.4 | 0.09146 | 0.09210 | 0.09182 | 0.00028 |

**3.**

| $x_i$ | $Y_i^{(p)}$ | $Y_i$ | $y(x_i)$ | Actual Error |
|-------|-------------|-------|----------|--------------|
| 1.0 | — | — | 2.50000 | — |
| 1.1 | — | — | 2.36818 | — |
| 1.2 | 2.27025 | 2.26594 | 2.26667 | 0.00073 |
| 1.3 | 2.19034 | 2.18728 | 2.18846 | 0.00118 |
| 1.4 | 2.12932 | 2.12711 | 2.12857 | 0.00146 |

**5.**

| $x_i$ | $Y_i^{(p)}$ | $Y_i$ | $y(x_i)$ | Actual Error |
|-------|-------------|-------|----------|--------------|
| 1.0 | — | — | −1.00000 | — |
| 1.1 | — | — | −0.90909 | — |
| 1.2 | — | — | −0.83333 | — |
| 1.3 | — | — | −0.76923 | — |
| 1.4 | −0.71444 | −0.71427 | −0.71429 | 0.00002 |

**7.**

| $x_i$ | $Y_1$ | Actual Error | $Y_2$ | Actual Error |
|-------|-------|--------------|-------|--------------|
| 0.0 | 2.00000 | — | 0.00000 | — |
| 0.1 | 2.04013 | — | 0.20134 | — |
| 0.2 | 2.16214 | — | 0.41075 | — |
| 0.3 | 2.37093 | — | 0.63665 | — |
| 0.4 | 2.67486 | 0.00001 | 0.88811 | 0.00000 |

Exact solution:

$$y_1 = e^{2x} + e^{-2x}; \quad y_2 = \frac{1}{2}\left(e^{2x} - e^{-2x}\right)$$

## Section 9.4 (p. 303)

**5.**  (a) $Y_5 = 0$; $e^{-5} - Y_5 = 0.00674$
   (b) $Y_5 = 1/2^5 = 0.03125$; $e^{-5} - Y_5 = -0.0245$
   (c) $Y_5 = 1/2^5 = 0.03125$; $e^{-5} - Y_5 = -0.0245$

**7.**  Exact solution: $y(x) = -(2/99)e^{-100x} + (101/99)e^{-x}$
   (a) $Y_{1,4} = -131.87610$;
       $Y_{2,4} = 13253.87610$, $|Y_{1,4} - y(0.4)| = 132.55996$
   (b) $Y_{1,4} = 0.69681$, $Y_{2,4} = -0.69668$,
       $|Y_{1,4} - y(0.4)| = 0.01295$

**9.**  Exact solution: $y(x) = \dfrac{e^{-50x}}{2501} - \dfrac{\cos x}{2501} + \dfrac{50 \sin x}{2501}$
   (a) $Y_4 = 0.10982$, $|Y_4 - y(0.4)| = 0.10240$
   (b) $Y_4 = 0.00741$, $|Y_4 - y(0.4)| = 0.00001$

# CHAPTER 10

## Section 10.2 (p. 314)

**1.**  $y = -\dfrac{2}{3} + \dfrac{5}{2}x$    **3.**  $y = \dfrac{5}{2} - \dfrac{12}{5}x + x^2$

**5.**  $k \approx 9.98$    **7.**  $x = 6/5$, $y = 8/15$

**11.**  $x_1 = 290/819 \approx 0.35409$, $x_2 = 391/819 \approx 0.47741$,
   $x_3 = -23/819 \approx -0.02808$

**Section 10.3** *(p. 319)*

1.  $a_0 = 511.136/466.416 = 1.096$.

3.  (a)  $a_0 \sum\limits_{i=1}^{N} w_i x_i + a_1 \sum\limits_{i=1}^{N} w_i x_i^{3/2} = \sum\limits_{i=1}^{N} w_i \sqrt{x_i} \, y_i$

    $a_0 \sum\limits_{i=1}^{N} w_i x_i^{3/2} + a_1 \sum\limits_{i=1}^{N} w_i x_i^2 = \sum\limits_{i=1}^{N} w_i x_i y_i$

    (b)  $a_0 \sum\limits_{i=1}^{N} w_i (\sin a_1 x_i)^2 - \sum\limits_{i=1}^{N} w_i y_i \sin a_1 x_i = 0$

    $a_0 \sum\limits_{i=1}^{N} w_i x_i (\sin a_1 x_i)(\cos a_1 x_i) - \sum\limits_{i=1}^{N} w_i x_i y_i \cos a_1 x_i = 0$

    (c)  $a_0 \sum\limits_{i=1}^{N} w_i + a_1 \sum\limits_{i=1}^{N} \dfrac{w_i}{x_i^2} = \sum\limits_{i=1}^{N} w_i y_i$

    $a_0 \sum\limits_{i=1}^{N} \dfrac{w_i}{x_i^2} + a_1 \sum\limits_{i=1}^{N} \dfrac{w_i}{x_i^4} = \sum\limits_{i=1}^{N} \dfrac{w_i y_i}{x_i^2}$

5.  $\gamma = 2.35, \; c = 725437.35$

**Section 10.4** *(p. 328)*

1.  $[1] \begin{bmatrix} \sqrt{5} & 0 \end{bmatrix} \begin{bmatrix} 1/\sqrt{5} & 2/\sqrt{5} \\ -2/\sqrt{5} & 1/\sqrt{5} \end{bmatrix}$

3.  $\begin{bmatrix} 1/\sqrt{2} & 1/\sqrt{2} & 0 \\ 0 & 0 & 1 \\ 1/\sqrt{2} & -1/\sqrt{2} & 0 \end{bmatrix} \begin{bmatrix} 2 & 0 \\ 0 & 0 \\ 0 & 0 \end{bmatrix} \begin{bmatrix} 1/\sqrt{2} & 1/\sqrt{2} \\ 1/\sqrt{2} & -1/\sqrt{2} \end{bmatrix}$

5.  $Q \Sigma Q^t$ where $\Sigma$ is the eigenvalue matrix

7.  $\begin{bmatrix} 1/\sqrt{5} & -2/\sqrt{5} \\ 2/\sqrt{5} & 1/\sqrt{5} \end{bmatrix} \begin{bmatrix} \sqrt{5} \\ 0 \end{bmatrix} [1]$

13.  $\mathbf{x} = \begin{bmatrix} x_1 \\ x_2 \end{bmatrix} = \begin{bmatrix} 1.525 \\ 1.525 \end{bmatrix}$

**Section 10.5** *(p. 338)*

3.  $P(x) = \dfrac{107}{2} - \dfrac{19}{3} \cos \dfrac{\pi}{12} x - 4\sqrt{3} \sin \dfrac{\pi}{12} x - \cos \dfrac{\pi}{6} x;$
    $P(11) = 57°C, \; P(17) = 62.7°C.$

5.  $f(x) = \dfrac{\pi}{2} + \sum\limits_{k=1}^{\infty} \dfrac{2((-1)^k - 1)}{\pi k^2} \cos kx$

7.  $P(x) = \dfrac{\pi}{2} - \dfrac{\pi}{2} \cos x$

9.  $f(x) = \dfrac{4\pi^2}{3} + \sum\limits_{k=1}^{\infty} \left( \dfrac{4}{k^2} \cos kx - \dfrac{4\pi}{k} \sin kx \right)$

11.  $P(x) = \dfrac{7\pi^2}{8} - \dfrac{\pi^2}{2} \cos x - \pi^2 \sin x$

13.  $P(x) = \dfrac{1}{2}$

## Section 10.6 *(p. 344)*

**3.** $c_0 = 0.38825$, $c_1 = 0.21617 - 0.07952\,I$, $c_2 = 0.17942$, $c_3 = 0.21617 + 0.07952\,I$ where $I = \sqrt{-1}$

## Section 10.7 *(p. 353)*

**1.** $S(x) = \begin{cases} x & 0 \le x \le 1 \\ 1 + 7(x - 1) & 1 \le x \le 2 \end{cases}$

**3.** $S(x) = \begin{cases} -(x/2) + (3x^3/2) & 0 \le x \le 1 \\ 1 + 4(x - 1) + (9(x - 1)^2/2) - (3(x - 1)^3/2) & 1 \le x \le 2 \end{cases}$

**5.** $S(x) = \begin{cases} 1 - 0.10990x - 0.54657x^3 & 0 \le x \le 0.4 \\ 0.92106 - 0.37225(x - 0.4) - 0.65588(x - 0.4)^2 + 0.36438(x - 0.4)^3 & 0.4 \le x \le 1 \end{cases}$

$S(0.5) = 0.87764$

**7.** $S(x) = \begin{cases} -9 + 12(x + 2) - 6(x + 2)^2 + (x + 2)^3 & -2 \le x \le 0 \\ -1 + x^3 & 0 \le x \le 1 \end{cases}$

$S(-1) = -2$

## Section 10.8 *(p. 359)*

**1.** $e^x = \dfrac{(1 + (1/2)x)}{1 - (1/2)x}$

**3.**

| $x$ | $e^x$ | $\dfrac{1 + (1/2)x}{1 - (1/2)x}$ | Actual Error | $1 + x + \dfrac{x^2}{2}$ | Actual Error |
|-----|-------|----------------------------------|--------------|--------------------------|--------------|
| 0.1 | 1.10517 | 1.10526 | 0.00009 | 1.10500 | 0.00017 |
| 0.5 | 1.64872 | 1.66667 | 0.01795 | 1.62500 | 0.02372 |
| 1.0 | 2.71828 | 3.00000 | 0.28172 | 2.50000 | 0.21828 |

**5.** $\dfrac{x}{1 + (1/3)x^2}$

**7.** (a) $x + \dfrac{1}{x + 1}$   (b) $1 + \dfrac{2}{x - 2 + (1/(x + 1))}$

**9.** $\dfrac{x + (1/2)x^2}{1 + x + (1/6)x^2} = 3 - \dfrac{2}{(1/6)x + (11/12) + ((13/24)/(x + (1/2)))}$

# CHAPTER 11

## Sections 11.1 and 11.2 *(p. 370)*

**1.** (a)

| $k$ | $x_1^{(k)}$ | $x_2^{(k)}$ |
|-----|-------------|-------------|
| 0 | 0 | 0 |
| 1 | 1.25 | 0.8 |
| 2 | 1.05 | 1.05 |
| 3 | 0.9875 | 1.01 |
| 4 | 0.9975 | 0.9975 |

(b)

| $k$ | $x_1^{(k)}$ | $x_2^{(k)}$ |
|-----|-------------|-------------|
| 0 | 0 | 0 |
| 1 | 1.25 | 1.05 |
| 2 | 0.9875 | 0.9975 |
| 3 | 1.00063 | 1.00013 |
| 4 | 0.99997 | 0.99999 |

**3.** (a)

| $k$ | $x_1^{(k)}$ | $x_2^{(k)}$ |
|---|---|---|
| 0 | 0 | 0 |
| 1 | $-4.0$ | 5.0 |
| 2 | 21.0 | 21.0 |
| 3 | 101.0 | $-79.0$ |
| 4 | $-399.0$ | $-399.0$ |

(b)

| $k$ | $x_1^{(k)}$ | $x_2^{(k)}$ |
|---|---|---|
| 0 | 0 | 0 |
| 1 | $-4.0$ | 21.0 |
| 2 | 101.0 | $-399.0$ |
| 3 | $-1999.0$ | 8001.0 |
| 4 | 40001.0 | $-159999.0$ |

**5.** (a) $-D^{-1}(L + U) = \begin{bmatrix} 0 & -1/3 & 1/3 \\ -1/5 & 0 & -1/5 \\ -1/8 & 1/8 & 0 \end{bmatrix}$    eigenvalues:

$-1/\sqrt[3]{60}$, $(1 \pm I\sqrt{3})/(2\sqrt[3]{60})$ where $I = \sqrt{-1}$

(b) $-(D + L)^{-1} U = \begin{bmatrix} 0 & -1/3 & 1/3 \\ 0 & 1/15 & -4/15 \\ 0 & 1/20 & -3/40 \end{bmatrix}$

eigenvalues: 0, $(-1 \pm I\sqrt{479})/240$ where $I = \sqrt{-1}$

## Section 11.3 (p. 376)

**1.** (a) $\|\mathbf{x}\|_1 = 10$,    $\|\mathbf{x}\|_2 = \sqrt{30}$,    $\|\mathbf{x}\|_\infty = 4$

(b) $\|\mathbf{x}\|_1 = |\cos i| + |\sin i| + 10^{-i}$,

$\|\mathbf{x}\|_2 = \sqrt{1 + 10^{-2i}}$, $\|\mathbf{x}\|_\infty = \begin{cases} 10^{-i} & \text{if } i \leq 0 \\ |\cos i| \text{ or } |\sin i| & \text{if } i \geq 1 \end{cases}$

(c) $\|\mathbf{x}\|_1 = 15/4$,    $\|\mathbf{x}\|_2 = \sqrt{85}/4$,    $\|\mathbf{x}\|_\infty = 2$

**5.** (a) converges to $[0, 0, 0]^t$

(b) converges to $[0, 1, 1]^t$

(c) does not converge

## Section 11.4 (p. 383)

**1.** (a) $\|A\|_1 = 1, \|A\|_2 = 1, \|A\|_\infty = 1$

(b) $\|A\|_1 = 5, \|A\|_2 = 5.92921, \|A\|_\infty = 8$

(c) $\|A\|_1 = 4, \|A\|_2 = \sqrt{(15 + \sqrt{29})/2} = 3.19258, \|A\|_\infty = 4$

**3.** (a) yes    (b) yes    (c) no

**9.** (a) no convergent    (b) convergent

**11.** $\begin{bmatrix} \frac{3}{4} & 0 & 0 \\ 0 & \frac{3}{4} & 0 \\ 0 & 0 & \frac{3}{4} \end{bmatrix}^{-1} = (I - A)^{-1}$

$= \begin{bmatrix} 1 + (1/4) + (1/4^2) + \cdots & 0 & 0 \\ 0 & 1 + (1/4) + (1/4^2) + \cdots & 0 \\ 0 & 0 & 1 + (1/4) + (1/4^2) + \cdots \end{bmatrix}$

$= \begin{bmatrix} \frac{4}{3} & 0 & 0 \\ 0 & \frac{4}{3} & 0 \\ 0 & 0 & \frac{4}{3} \end{bmatrix}$

### Section 11.5 *(p. 388)*

1. both methods converge. The Jacobi method:
   $R = \log \sqrt{20} = 0.65051$. The Gauss–Seidel method:
   $R = \log 20 = 1.30103$

3. The Jacobi method: $R = \log (1/0.94443) = 0.02483$

5. (a) no value exists
   (b) no value exists
   (c) $|\alpha| < 1/|\beta|$

### Section 11.6 *(p. 394)*

1. (a) $B_J = \begin{bmatrix} 0 & \frac{1}{2} & 0 \\ \frac{1}{2} & 0 & \frac{1}{2} \\ 0 & \frac{1}{2} & 0 \end{bmatrix}$   (b) $-1/\sqrt{2},\ 0,\ 1/\sqrt{2}$

   (c) $\omega_o = 4\,(1 - \sqrt{1/2}) = 1.17157$

3. (a) $B_J = \begin{bmatrix} 0 & \frac{1}{4} & 0 \\ \frac{1}{4} & 0 & \frac{1}{4} \\ 0 & \frac{1}{4} & 0 \end{bmatrix}$   (b) $-1/(2\sqrt{2}),\ 0,\ 1/(2\sqrt{2})$

   (c) $\omega_o = 16\,(1 - \sqrt{7/8}) = 1.03337$

5.

| $k$ | $\hat{x}_1^{(k)}$ | $\hat{x}_2^{(k)}$ | $\hat{x}_3^{(k)}$ | $x_1^{(k)}$ | $x_2^{(k)}$ | $x_3^{(k)}$ |
|---|---|---|---|---|---|---|
| 0 | 0.00000 | 0.00000 | 0.00000 | — | — | — |
| 1 | 0.75000 | 0.68750 | 0.92188 | 0.77503 | 0.71044 | 0.95264 |
| 2 | 0.92761 | 0.97006 | 0.99252 | 0.93270 | 0.97872 | 0.99385 |
| 3 | 0.99468 | 0.99713 | 0.99928 | 0.99675 | 0.99774 | 0.99946 |

## CHAPTER 12

### Section 12.2 and 12.3 *(p. 408)*

3. $g_1(x_1, x_2) = -\sqrt{1 - x_2^2}$, $g_2(x_1, x_2) = e^{x_1}$;
   $S = \{(x_1, x_2) \mid -1 \le x_1 \le 0,\ -1/\sqrt{2} < x_2 < 1/\sqrt{2}\}$

5. (a)

| $k$ | $x_1^{(k)}$ | $x_2^{(k)}$ |
|---|---|---|
| 0 | 0.2 | 0.2 |
| 1 | 0.08 | 0.0 |
| 2 | 0.0064 | 0.0064 |
| 3 | 0.00008 | 0.0 |

(b)

| $k$ | $x_1^{*(k)}$ | $x_2^{*(k)}$ |
|---|---|---|
| 0 | 0.2 | 0.2 |
| 1 | 0.08 | $-0.03360$ |
| 2 | 0.00753 | $-0.00107$ |
| 3 | 0.00006 | 0.00000 |

(c)

| $k$ | $x_1^{(k)}$ | $x_2^{(k)}$ |
|---|---|---|
| 0 | 1.0 | 1.0 |
| 1 | 2.0 | 0.0 |
| 2 | 4.0 | 4.0 |
| 3 | 32.0 | 0.0 |

**7.**

| $k$ | $x_1^{(k)}$ | $x_2^{(k)}$ |
|---|---|---|
| 0 | 1.00000 | 1.00000 |
| 1 | 1.25000 | 1.75000 |
| 2 | 1.23611 | 1.58135 |
| 3 | 1.23607 | 1.57233 |

**11.** $x_1^{(1)} = -1.0$, $x_2^{(1)} = 1.0$, $x_3^{(1)} = -2.0$

**13.**

| $k$ | $x^{(k)}$ | $y^{(k)}$ |
|---|---|---|
| 0 | 0.00000 | 0.00000 |
| 1 | 1.00000 | 0.50000 |
| 2 | 1.30000 | 0.65000 |
| 3 | 1.33293 | 0.66646 |
| 4 | 1.33333 | 0.66667 |

exact values: $x = 4/3$, $y = 2/3$

## Section 12.4 *(p. 415)*

**1.**

| $k$ | $t^{(k)}$ | $x_1^{(k)}$ | $x_2^{(k)}$ |
|---|---|---|---|
| 0 | — | 0.00000 | 0.00000 |
| 1 | 0.5 | 1.00000 | 1.00000 |
| 2 | 0.0 | 1.00000 | 1.00000 |

**3.**

| $k$ | $t^{(k)}$ | $x_1^{(k)}$ | $x_2^{(k)}$ | $x_3^{(k)}$ |
|---|---|---|---|---|
| 0 | — | 0.00000 | 0.00000 | 0.00000 |
| 1 | 0.25146 | 1.25731 | 0.75439 | 0.75439 |
| 2 | 0.97725 | 0.98299 | 0.98299 | 0.98299 |

**5.**

| $k$ | $t^{(k)}$ | $x_1^{(k)}$ | $x_2^{(k)}$ |
|---|---|---|---|
| 0 | — | 0.00000 | 0.00000 |
| 1 | 0.05882 | 0.94118 | 0.00000 |
| 2 | 0.25000 | 0.94118 | 0.47059 |
| 3 | 0.03590 | 1.07037 | 0.37939 |

---

# CHAPTER 13

## Section 13.1 *(p. 426)*

**1.** $P = \begin{bmatrix} -1 & 0 \\ 5 & 1 \end{bmatrix}$,  $D = \begin{bmatrix} 1 & 0 \\ 0 & 2 \end{bmatrix}$

**3.** $P = \begin{bmatrix} 1 & 0 & 0 \\ 0 & (1+\sqrt{5})/2 & (1-\sqrt{5})/2 \\ 0 & 1 & 1 \end{bmatrix}$,

$D = \begin{bmatrix} 3 & 0 & 0 \\ 0 & (3+\sqrt{5})/2 & 0 \\ 0 & 0 & (3-\sqrt{5})/2 \end{bmatrix}$

**5.** Not possible

**11.** $\begin{bmatrix} \frac{1}{2} & 1 \\ 0 & \frac{1}{2} \end{bmatrix}$

**13.** $\begin{bmatrix} 0 & 1 & 0 \\ 0 & 0 & 1 \\ 0 & 0 & 0 \end{bmatrix}$

**19.** Positive definite: (a), (b)

## Sections 13.2, 13.3, and 13.4 *(p. 437)*

**1.** Eigenvalues: $3 + I$ and $3 - I$

**3.** Eigenvalues: $-2$, 1, and 3

**5.**

| $k$ | $\lambda_1^{(k+1)}$ | $z^{(k)}$ |
|---|---|---|
| 0 | $-4.0$ | $(1.0, 1.0)$ |
| 1 | $-25.0/17.0$ | $(-1.0, -3.0/5.0)$ |
| 2 | $-174.0/74.0$ | $(1.0, 5.0/7.0)$ |

Eigenvalues: $-2$ and 1

**7.**

| $k$ | $\lambda_1^{(k+1)}$ | $z^{(k)}$ |
|---|---|---|
| 0 | 2.0 | $(1.0, 1.0, 1.0)$ |
| 1 | 2.0 | $(1.0, 1.0, 1.0)$ |
| 2 | 2.0 | $(1.0, 1.0, 1.0)$ |

Eigenvalues: $-1$, $-1$, and 2

**11.**

| $k$ | $\lambda_1^{(k+1)}$ | $z^{(k)}$ |
|---|---|---|
| 0 | $-3.0/2.0$ | $(1.0,\ 1.0)$ |
| 1 | $1.0/5.0$ | $(-1.0,\ -1.0/2.0)$ |
| 2 | $5.0/2.0$ | $(-1.0,\ 0)$ |

Eigenvalues: $-2$ and $1$

## Sections 13.5 and 13.6 *(p. 447)*

**1.** $Q = \dfrac{1}{2}\begin{bmatrix} \sqrt{2} & -\sqrt{2} \\ \sqrt{2} & \sqrt{2} \end{bmatrix}$,

$D = \begin{bmatrix} 3 & 0 \\ 0 & 1 \end{bmatrix}$

**3.** $Q = \dfrac{1}{2}\begin{bmatrix} \sqrt{2 + \sqrt{2}} & -\sqrt{2 - \sqrt{2}} \\ \sqrt{2 - \sqrt{2}} & \sqrt{2 + \sqrt{2}} \end{bmatrix}$,

$D = \begin{bmatrix} 4 + \sqrt{2} & 0 \\ 0 & 4 - \sqrt{2} \end{bmatrix}$

**5.** $A^{(1)} = \begin{bmatrix} 0 & 0 & \sqrt{2} \\ 0 & 2 & -\sqrt{2} \\ \sqrt{2} & -\sqrt{2} & 2 \end{bmatrix}$

$A^{(2)} = \begin{bmatrix} -0.73205 & 0.65012 & 0 \\ 0.65012 & 2 & -1.25593 \\ 0 & -1.25593 & 2.73205 \end{bmatrix}$

$A^{(3)} = \begin{bmatrix} -0.73205 & 0.52005 & -0.39013 \\ 0.52005 & 1.05785 & 0.0 \\ -0.39013 & 0.0 & 3.67420 \end{bmatrix}$

**9.** $\begin{bmatrix} 0 & 3.60555 & 0 \\ 3.60555 & 0.92308 & 0.38462 \\ 0 & 0.38462 & -0.92308 \end{bmatrix}$

**11.** $\begin{bmatrix} 1 & 3.16228 & 0 \\ 3.16228 & 3.10000 & 0.70000 \\ 0 & 0.70000 & -2.10000 \end{bmatrix}$

## Sections 13.7 and 13.8 *(p. 459)*

**1.** $P = \begin{bmatrix} -1/\sqrt{3} & -1/\sqrt{3} & -1/\sqrt{3} \\ -1/\sqrt{3} & (2 + \sqrt{3})/(3 + \sqrt{3}) & -1/(3 + \sqrt{3}) \\ -1/\sqrt{3} & -1/(3 + \sqrt{3}) & (2 + \sqrt{3})/(3 + \sqrt{3}) \end{bmatrix}$

**3.** $P = \begin{bmatrix} -\dfrac{3 + \sqrt{15}}{(5 + \sqrt{15})} & -\dfrac{2\,(3 + \sqrt{15})}{3\,(5 + \sqrt{15})} & -\dfrac{3 + \sqrt{15}}{3\,(5 + \sqrt{15})} & \dfrac{3 + \sqrt{15}}{3\,(5 + \sqrt{15})} \\ -\dfrac{2\,(3 + \sqrt{15})}{3\,(5 + \sqrt{15})} & \dfrac{11 + 3\sqrt{15}}{3\,(5 + \sqrt{15})} & -\dfrac{2}{3\,(5 + \sqrt{15})} & -\dfrac{2}{3\,(5 + \sqrt{15})} \\ -\dfrac{3 + \sqrt{15}}{3\,(5 + \sqrt{15})} & -\dfrac{2}{3\,(5 + \sqrt{15})} & \dfrac{14 + 3\sqrt{15}}{3\,(5 + \sqrt{15})} & -\dfrac{1}{3\,(5 + \sqrt{15})} \\ -\dfrac{3 + \sqrt{15}}{3\,(5 + \sqrt{15})} & -\dfrac{2}{3\,(5 + \sqrt{15})} & -\dfrac{1}{3\,(5 + \sqrt{15})} & \dfrac{14 + 3\sqrt{15}}{3\,(5 + \sqrt{15})} \end{bmatrix}$

**5.** $A^{(1)} = \begin{bmatrix} 0 & -\sqrt{13} & 0 \\ -\sqrt{13} & 12/13 & -5/13 \\ 0 & -5/13 & -12/13 \end{bmatrix}$

**7.** $A^{(1)} = \begin{bmatrix} 1 & -\sqrt{10} & 0 \\ -\sqrt{10} & 31/10 & -7/10 \\ 0 & -7/10 & -21/10 \end{bmatrix}$

## Sections 13.9 and 13.10 *(p. 470)*

**3.** (a) $\begin{bmatrix} 0 & -1 \\ -1 & 0 \end{bmatrix}\begin{bmatrix} -1 & -1 \\ 0 & -2 \end{bmatrix}$     (b) $\begin{bmatrix} \cos\theta & \sin\theta \\ \sin\theta & -\cos\theta \end{bmatrix}\begin{bmatrix} 1 & 0 \\ 0 & -1 \end{bmatrix}$

**5.** (a) Not possible     (b) $A_4 = \begin{bmatrix} \dfrac{(\cos^2 2\theta - \sin^2\theta)^2 - \sin^2\theta\cos^2\theta}{\cos\theta\cos 2\theta(\cos^2\theta - \sin^2\theta)} & -\sin\theta \\ \dfrac{\sin\theta\cos^2\theta}{(\cos^2 2\theta - \sin^2\theta)^2} & \dfrac{\cos\theta\cos 2\theta}{(\cos^2 2\theta - \sin^2\theta)} \end{bmatrix}$

# CHAPTER 14

## Sections 14.1 and 14.2 *(p. 481)*

**3.** (a) There is no $\alpha$     (b) $\alpha \neq k^2\pi^2$ where $k = 1, 2, \ldots$
(c) $\alpha = k^2\pi^2$ where $k = 1, 2, \ldots$

**5.** (a) $y(x) = y(x, t) = ((1 + t)(e^x - e^{-x})/2) - x$
(b) $y(1) = y(1, 0) = ((e^1 - e^{-1})/2) - 1$
(c) $y(1) = y(1, -2) = -((e^1 - e^{-1})/2) - 1$     (d) $t = -1$

## Section 14.3 *(p. 490)*

**1.** $Y_0 = 0, Y_1 = 0.50542, Y_2 = 0.87228, Y_3 = 1$

**3.** $Y_0 = 0, Y_1 = 0.33333, Y_2 = 0.66667, Y_3 = 1$

**5.**

| $i$ | $x_i$ | Exact | $Y_i$ using (a) | $Y_i$ using (b) |
|---|---|---|---|---|
| 0 | 0 | 0.00000 | 0.17803 | 0.00822 |
| 1 | 1/3 | 0.31831 | 0.51136 | 0.30038 |
| 2 | 2/3 | 0.55133 | 0.70450 | 0.51019 |
| 3 | 1 | 0.63662 | 0.70450 | 0.58013 |

Exact solution: $y(x) = (2/\pi) \sin (\pi/2)x$

**7.**

| $i$ | $x_i$ | Exact | $Y_i$ using (a) | $Y_i$ using (b) |
|---|---|---|---|---|
| 0 | 0 | 0.00000 | 0.00000 | 0.00000 |
| 1 | 1/3 | $-0.04212$ | 0.00055 | $-0.02937$ |
| 2 | 2/3 | $-0.01419$ | 0.07523 | 0.01206 |
| 3 | 1 | 0.16166 | 0.30643 | 0.20298 |

**9.** $Y_0 = 0.65031, Y_1 = 0.68644, Y_2 = 0.79884, Y_3 = 1$

**11.** $Y_0 = 0.00000, Y_1 = 0.00187, Y_2 = 0.00224, Y_3 = 0.00000$     exact solution: $y(x) = x^2(2 - 3x + x^2)/5!$

## Sections 14.4 and 14.5 *(p. 498)*

**3.** $a_{-1} = a_1, a_0 = -2a_1, B_{-1} = -(h^2 a_1)/12,$
$B_0 = -(5h^2 a_1)/6, B_1 = -(h^2 a_1)/12$

**5.** $[u''/(1 + x^2)^{1/2}] - [u(2 + x^2)/(1 + x^2)^{5/2}] = 1;$
$u(-1) = \sqrt{2}; u(1) = 0$

**7.** $Y_0 = 0, Y_1 = 0.33333, Y_2 = 0.66667, Y_3 = 1$

**9.** $Y_0 = 0.64324, Y_1 = 0.68145, Y_2 = 0.79608, Y_3 = 1$

**11.** $Y_0 = 0.00000, Y_1 = 0.33333, Y_2 = 0.66667,$
$Y_3 = 1.00001$

**13.** $Y_0 = 4, Y_1 = 2.29596, Y_2 = 1.46845, Y_3 = 1$

**Section 14.6** *(p. 503)*

1.  eigenvalues: $\lambda_n = (2n - 1)^2/4$
    eigenvectors: $y_n = \cos\sqrt{\lambda_n}\, x$ for $n = 1, 2, \ldots$

3.  eigenvalues: $\lambda_n = -(n\pi)^2$
    eigenvectors: $y_n = e^{-x} \sin\sqrt{-\lambda_n}\, x$ for $n = 1, 2, \ldots$

5.  $\lambda_1 = -8.51472$, $\lambda_2 = -25.48528$
    exact eigenvalues: $\lambda_1 = -\pi^2 = -9.86960$, $\lambda_2 = -4\pi^2 = -39.47842$

7.  $\lambda_1 = -9.81818$, $\lambda_2 = -36.00000$

**Section 14.7** *(p. 508)*

1.  (a) $Y_2(x) = -(81/416)(x - x^2) - (9/52)(x^2 - x^3)$
    (b) $Y_2(x) = -(97/517)x(1 - x) - (24/141)x^2(1 - x)$

3.  (a) $Y_2(x) = -(46161/246137)(x - x^2) - (413/2437)(x^2 - x^3)$
    (b) $Y_2(x) = -(71/369)(x - x^2) - (7/41)(x^2 - x^3)$

5.  (a) $Y_1(x) = x + (1 - x^2)$
    (b) $Y_1(x) = x + (5/6)(1 - x^2)$

**Section 14.8** *(p. 515)*

1.  $S(x) = \begin{cases} -(3/44)(2x) & 0 \le x \le \frac{1}{2} \\ -(3/44)(2 - 2x) & \frac{1}{2} \le x \le 1 \end{cases}$

3.  $S(x) = \begin{cases} -(257/686)(2x) & 0 \le x \le \frac{1}{2} \\ -(257/686)(2 - 2x) - (370/686)(2x - 1) & \frac{1}{2} \le x \le 1 \end{cases}$

5.  $S(x) = \begin{cases} (3 - 2x) + 1.36712(2x - 2) & 1 \le x \le \frac{3}{2} \\ 1.36712(4 - 2x) + 2(2x - 3) & \frac{3}{2} \le x \le 2 \end{cases}$

**CHAPTER 15**

**Section 15.2** *(p. 524)*

1.  Hyperbolic; $\xi = -8x + y$, $\eta = 2x + y$; $\dfrac{\partial^2 u}{\partial\xi\partial\eta} = 0$

3.  Parabolic; $\xi = x$, $\eta = 3x + y$; $\dfrac{\partial^2 u}{\partial\xi^2} + \dfrac{\partial u}{\partial\eta} = \xi(\eta - 3\xi)$

5.  Elliptic; $\xi = -2x + y$, $\eta = -x$; $\dfrac{\partial^2 u}{\partial\xi^2} + \dfrac{\partial^2 u}{\partial\eta^2} - e^{-\eta} u = \sin(\xi - 2\eta)$

## Section 15.3 (p. 533)

**1.** $r = 1$

| $t/x_i$ | 0 | 1/3 | 2/3 | 1 |
|---|---|---|---|---|
| 0 | 0 | $\sqrt{3}/4$ | $\sqrt{3}/4$ | 0 |
| 1/9 | 0 | 0 | 0 | 0 |
| 2/9 | 0 | 0 | 0 | 0 |
| 1/3 | 0 | 0 | 0 | 0 |

$r = 1/2$

| $t/x_i$ | 0 | 1/3 | 2/3 | 1 |
|---|---|---|---|---|
| 0 | 0 | $\sqrt{3}/4$ | $\sqrt{3}/4$ | 0 |
| 1/18 | 0 | $\sqrt{3}/8$ | $\sqrt{3}/8$ | 0 |
| 2/18 | 0 | $\sqrt{3}/16$ | $\sqrt{3}/16$ | 0 |
| 3/18 | 0 | $\sqrt{3}/32$ | $\sqrt{3}/32$ | 0 |

**3.** $r = 1$

| $t/x_i$ | 0 | 1/3 | 2/3 | 1 |
|---|---|---|---|---|
| 0 | 0 | $\sqrt{3}/4$ | $\sqrt{3}/4$ | 0 |
| 1/9 | 0 | $\sqrt{3}/8$ | $\sqrt{3}/8$ | 0 |
| 2/9 | 0 | $\sqrt{3}/16$ | $\sqrt{3}/16$ | 0 |
| 1/3 | 0 | $\sqrt{3}/32$ | $\sqrt{3}/32$ | 0 |

$r = 1/2$

| $t/x_i$ | 0 | 1/3 | 2/3 | 1 |
|---|---|---|---|---|
| 0 | 0 | $\sqrt{3}/4$ | $\sqrt{3}/4$ | 0 |
| 1/18 | 0 | $\sqrt{3}/6$ | $\sqrt{3}/6$ | 0 |
| 2/18 | 0 | $\sqrt{3}/9$ | $\sqrt{3}/9$ | 0 |
| 3/18 | 0 | $2\sqrt{3}/27$ | $2\sqrt{3}/27$ | 0 |

**5.** Using $U_{0,j} = (4U_{1,j} - U_{2,j} - h)/3$, the backward difference in time, and the central difference for space, we get

| $t/x_i$ | 0 | 1/3 | 2/3 | 1 |
|---|---|---|---|---|
| 0 | 1.00000 | 1.00000 | 1.00000 | 1.00000 |
| 1/18 | 0.83583 | 0.95763 | 0.98971 | 1.00000 |
| 2/18 | 0.79518 | 0.92388 | 0.97664 | 1.00000 |
| 3/18 | 0.76273 | 0.89630 | 0.96369 | 1.00000 |

## Sections 15.4 and 15.5 (p. 547)

**1.**

| $t/x_i$ | 0 | 1/4 | 1/2 | 2/3 | 1 |
|---|---|---|---|---|---|
| 0 | 0 | 0 | $\epsilon$ | 0 | 0 |
| 1/32 | 0 | 0 | 0 | 0 | 0 |
| 2/32 | 0 | 0 | $\epsilon$ | 0 | 0 |
| 3/32 | 0 | $\epsilon$ | $-2\,\epsilon$ | $\epsilon$ | 0 |
| 4/32 | 0 | $-4\,\epsilon$ | $7\,\epsilon$ | $-4\,\epsilon$ | 0 |
| 5/32 | 0 | $16\,\epsilon$ | $-24\,\epsilon$ | $16\,\epsilon$ | 0 |
| 6/32 | 0 | $-60\,\epsilon$ | $87\,\epsilon$ | $-60\,\epsilon$ | 0 |

**7.**

| $t/x_i$ | 0 | 1/3 | 2/3 | 1 |
|---|---|---|---|---|
| 0 | 0 | 0.43301 | 0.43301 | 0 |
| 1/18 | 0 | 0.25981 | 0.25981 | 0 |
| 2/18 | 0 | 0.15588 | 0.15588 | 0 |
| 3/18 | 0 | 0.09353 | 0.09353 | 0 |

## Section 15.6 (p. 554)

**3.** $U_{1,1,1} = U_{2,1,1} = U_{1,2,1} = U_{2,2,1} = 0$

**5.** $U_{1,1,1} = U_{2,1,1} = U_{1,2,1} = U_{2,2,1} = \frac{1}{4}$,
$U_{1,1,2} = U_{2,1,2} = U_{1,2,2} = U_{2,2,2} = 1/12$

**7.** $U_{1,1,1} = U_{2,1,1} = U_{1,2,1} = U_{2,2,1} = \frac{1}{4}$,
$U_{1,1,2} = U_{2,1,2} = U_{1,2,2} = U_{2,2,2} = 1/12$

## Section 15.7 (p. 561)

**1.** $U_{1,1} = 3\sqrt{3}/16 = 0.32476$    $U_{2,1} = 3\sqrt{3}/16 = 0.32476$
$U_{1,2} = \sqrt{3}/16 = 0.10825$    $U_{2,2} = \sqrt{3}/16 = 0.10825$
exact values: $u(\frac{1}{3}, \frac{1}{3}) = u_{1,1} = 0.29986$
$u(\frac{2}{3}, \frac{1}{3}) = u_{2,1} = 0.29986$    $u(\frac{1}{3}, \frac{2}{3}) = u_{1,2} = 0.09369$
$u(\frac{2}{3}, \frac{2}{3}) = u_{2,2} = 0.09369$

**3.** $U_{1,1} = 0$    $U_{2,1} = 0$    $U_{1,2} = 0$    $U_{2,2} = 0$
exact values: $u(\frac{1}{3}, \frac{1}{3}) = u_{1,1} = 0$    $u(\frac{2}{3}, \frac{1}{3}) =$
$u_{2,1} = 0$    $u(\frac{1}{3}, \frac{2}{3}) = u_{1,2} = 0$
$u(\frac{2}{3}, \frac{2}{3}) = u_{2,2} = 0$

**5.** $U_{1,1} = 2/15 = 0.13333$    $U_{2,1} = -2/15 = -0.13333$
$U_{1,2} = 1/30 = 0.03333$    $U_{2,2} = -1/30 = -0.03333$
exact values: $u(\frac{1}{3}, \frac{1}{3}) = u_{1,1} = 0.17312$
$u(\frac{2}{3}, \frac{1}{3}) = u_{2,1} = -0.17312$    $u(\frac{1}{3}, \frac{2}{3}) = u_{1,2} = 0.05409$    $u(\frac{2}{3}, \frac{2}{3}) = u_{2,2} = -0.05409$

## Section 15.8 (p. 568)

**3.**

| $t/x_i$ | 0 | 1/4 | 1/2 | 3/4 | 1 |
|---|---|---|---|---|---|
| 0.5 | 0 | $\frac{1}{4} = 0.25$ | $\frac{\sqrt{2}}{4} \approx 0.35355$ | $\frac{1}{4} = 0.25$ | 0 |
| Exact ($t = 0.5$) | 0 | 0.22508 | 0.31831 | 0.22508 | 0 |
| 1.0 | 0 | 0 | 0 | 0 | 0 |
| Exact ($t = 1.0$) | 0 | 0 | 0 | 0 | 0 |

**5.**

| $t/x_i$ | 0 | 1/3 | 2/3 | 1 |
|---|---|---|---|---|
| 0 | 0 | $\sqrt{3}/2$ | $\sqrt{3}/2$ | 0 |
| 1/3 | 0 | $\sqrt{3}/4$ | $\sqrt{3}/4$ | 0 |
| Exact ($t = 1/3$) | 0 | $\sqrt{3}/4$ | $\sqrt{3}/4$ | 0 |
| 2/3 | 0 | $-\sqrt{3}/6$ | $-\sqrt{3}/6$ | 0 |
| Exact ($t = 2/3$) | 0 | $-\sqrt{3}/4$ | $-\sqrt{3}/4$ | 0 |
| 1 | 0 | $-17\sqrt{3}/36$ | $-17\sqrt{3}/36$ | 0 |
| Exact ($t = 1$) | 0 | $-\sqrt{3}/2$ | $-\sqrt{3}/2$ | 0 |

**9.**

| $t/x_i$ | 0 | 1/4 | 1/2 | 3/4 | 1 |
|---|---|---|---|---|---|
| 0 | 0 | 0 | $\epsilon$ | 0 | 0 |
| 1/8 | 0 | $\epsilon/8$ | $3\epsilon/4$ | $\epsilon/8$ | 0 |
| 2/8 | 0 | $3\epsilon/8$ | $3\epsilon/16$ | $3\epsilon/8$ | 0 |
| 3/8 | 0 | $31\epsilon/64$ | $-9\epsilon/32$ | $31\epsilon/64$ | 0 |
| 4/8 | 0 | $9\epsilon/32$ | $-47\epsilon/128$ | $9\epsilon/32$ | 0 |

# APPENDICES

## Appendix A (p. 580)

**11.** (a) $\dfrac{1}{e-1}$    (b) $\dfrac{1}{3}$

**13.** $P(x) = -2 + 2(x+1) - (x+1)^2 + 2(x+1)^3$

**15.** $P_2(x, y) = 1 - \dfrac{1}{2}(x+y) + \dfrac{3}{8}(x+y)^2$

## Appendix B (p. 590)

**1.** (a) $\begin{bmatrix} 3 & 6 & 0 \\ 9 & 0 & 15 \end{bmatrix}$    (b) $\begin{bmatrix} -2 & 2 & -6 \\ 3 & 12 & 8 \end{bmatrix}$

(c) $D = \begin{bmatrix} -2 & -2 & -2 \\ -3 & 4 & -4 \end{bmatrix}$

**5.** (a) $1;$ $\begin{bmatrix} 1 & -1 & -2 \\ -1 & 2 & 1 \\ 0 & 0 & 1 \end{bmatrix}$

(b) $-12;$ $\begin{bmatrix} 1/2 & 0 & 0 & -1/2 \\ 4/3 & -1 & -1/6 & -5/6 \\ 1/3 & 0 & -1/6 & 1/6 \\ -1/6 & 0 & 1/3 & 1/6 \end{bmatrix}$

**7.** (a) $x_1 = 1, x_2 = 2$    (b) $x_1 = 1, x_2 = 0, x_3 = 0$

## Appendix C (p. 598)

**1.** $y = \dfrac{1}{16}(4x - 1) + c_1 e^{-4x}$

**3.** $y = \dfrac{1}{(x^2 + 1)}\left[\dfrac{x^3}{3} + x + c_1\right]$

**5.** $y = \dfrac{25}{1 + 5x + 25c_1e^{5x}}$

**7.** $y = e^{-2x}(c_1 + c_2x)$

**9.** $y_p = -\dfrac{5}{4}e^x - \dfrac{2}{5}\cos x - \dfrac{1}{5}\sin x$

**11.** $y_p = \dfrac{1}{2}x \sin x$

**13.** $y = c_1e^{-x} + c_2e^{-2x} + \dfrac{1}{2}x^2 - \dfrac{3}{2}x + \dfrac{7}{4} + xe^{-x}$

**15.** $y = c_1e^{-x} + c_2e^x + c_3e^{2x}$

**17.** $y = (c_1 \cos x + c_2 \sin x)$
$\qquad + (c_3 \cos 2x + c_4 \sin 2x) + \dfrac{x}{24}\cos 2x$

**19.** $y = c_1x^3 + c_2x^{-4/3}$

**21.** $y = c_1x^3 + c_2x^{-4/3}$

## Appendix D (p. 607)

**1.** $y_i = c_1(-1)^i + c_2(-2)^i$

**3.** $y_i = c_1 \cos\dfrac{\pi i}{3} + c_2 \sin\dfrac{\pi i}{3}$

**5.** $y_i^p = -\dfrac{\cos(2i - 2)}{2 \sin 2}$

**7.** $y_i = c_1 + c_2 i + \dfrac{1}{2(1 - \sqrt{2})}\left[\cos\dfrac{\pi i}{4} + \sin\dfrac{\pi i}{4}\right]$

**9.** $y_i = c_1 \cos\dfrac{\pi i}{2} + c_2 \sin\dfrac{\pi i}{2} + 1$

**11.** $y_i = c_1 + bi$

**13.** $y_i = 1 + i$

## Appendix E (p. 618)

**3.** (a) $\langle \sin x, \cos x \rangle = 0$     (b) $\langle \sin ix, \sin jx \rangle = 0$ if $i \neq j$ and $\langle \sin ix, \sin jx \rangle = \pi$ if $i = j$
(c) $\langle \sin ix, \cos jx \rangle = 0$

**5.** (a) $\langle 2x^2 - 1, x \rangle = 0$     (b) $\langle 2x^2 - 1, 2x^2 - 1 \rangle = \pi/2$     (c) $\langle x, 4x^3 - 3x \rangle = 0$

**9.** (a) $U_0 = 1$, $U_1 = 2x$, and $U_2 = 4x^2 - 1$

**13.** $\{(1/\sqrt{2}, 0, 1/\sqrt{2}), (2/3, 1/3, -2/3), (-1/(3\sqrt{2}), 4/(3\sqrt{2}), 1/(3\sqrt{2}))\}$

**15.** $\left\{\begin{bmatrix} 1 \\ 0 \\ 0 \end{bmatrix}, \begin{bmatrix} 0 \\ 0 \\ 1 \end{bmatrix}, \begin{bmatrix} 0 \\ 1 \\ 0 \end{bmatrix}\right\}$

# INDEX